HANDBOOK OF INDUSTRIAL DRYING

SECOND EDITION
REVISED AND EXPANDED

VOLUME
1

HANDBOOK OF INDUSTRIAL DRYING

SECOND EDITION
REVISED AND EXPANDED

VOLUME
1

ARUN S. MUJUMDAR

Department of Chemical Engineering
McGill University
Montreal, Quebec, Canada

CRC Press
Taylor & Francis Group
Boca Raton London New York

CRC Press is an imprint of the
Taylor & Francis Group, an **informa** business

First published 1995 by Marcel Dekker, Inc.

CRC Press
Taylor & Francis Group
6000 Broken Sound Parkway NW, Suite 300
Boca Raton, FL 33487-2742

Reissued 2019 by CRC Press

© 1995 by Taylor & Francis Group, LLC
CRC Press is an imprint of Taylor & Francis Group, an Informa business

No claim to original U.S. Government works

A Library of Congress record exists under LC control number:

Publisher's Note
The publisher has gone to great lengths to ensure the quality of this reprint but points out that some imperfections in the original copies may be apparent.

Disclaimer
The publisher has made every effort to trace copyright holders and welcomes correspondence from those they have been unable to contact.

ISBN 13: 978-0-367-25770-5 (hbk)
ISBN 13: 978-0-367-25773-6 (pbk)
ISBN 13: 978-0-429-28977-4 (ebk)

Visit the Taylor & Francis Web site at http://www.taylorandfrancis.com and the
CRC Press Web site at http://www.crcpress.com

Foreword to the Second Edition

The second edition of the *Handbook of Industrial Drying* continues the tradition of the editor and the publisher as international leaders in providing information in the field of industrial drying. The authors are knowledgable of the subjects and have been drawn from among the world's authorities in industry, academia, government, and consulting. Some fifty authors from fifteen countries have written forty-three chapters plus three appendices. There are twenty-one new chapters, plus two new appendices. All chapters have been updated or revised. There is over sixty percent new material, making this edition practically a new volume.

The mark of an outstanding handbook is that it provides current information on a subject—in this case multidisciplinary in nature—understandable to a broad audience. A balanced approach of covering principles and practices provides a sound basis for the presentations. Students, academics, consultants, and industry people can find information to meet their needs. Researchers, designers, manufacturers, and sales people can benefit from the book as they consider elements or components related to drying as well as the system itself.

New material has been added to provide the latest information on minimizing environmental impacts, increasing energy efficiency, maintaining quality control, improving safety of operation, and improving the control of drying systems. New sections or chapters have been added to cover in detail microwave drying; infrared drying; impinging stream dryers; use of superheated steam and osmotic dehydration; and drying of biotechnological materials, tissue and towels, peat, coal, and fibrous materials.

The information in this book can be categorized as product related, equipment related, and the relationship between the two—the system of drying. For products not specifically covered, or for the design of dryers not detailed, users can select closely related applicable information to meet many needs. The user may want to pursue a subject in considerably more detail. Pertinent references, but not voluminous overwhelming bibliographies, are included at the end of each chapter. An appendix devoted to an annotated bibliography is also included.

Carl W. Hall

Foreword to the First Edition

The *Handbook of Industrial Drying* fills an important need and is of immeasurable value in the field of drying. Academics, students, and industry people — from sales to research — can learn much from the combination of principles and practices used throughout. The presentation of principles does not overwhelm the coverage of equipment and systems. More appropriate theories will develop as a result of the description of equipment and systems. For example, a description of dryers, particularly industrial dyers, is lacking in many research articles; this handbook provides such information.

The authors have distilled much information from extensive literature to provide generic information as contrasted with details of a specific drying system of a particular manufacturer. The users can extrapolate the use of drying systems, by design and management, to a variety of products. As a special feature, a complete listing of books written on the subject of drying is included.

The authors, a blend of students, faculty, and those in industry, represent experience with different kinds of drying systems, different applications of principles, and different products. The book provides excellent coverage of the cross-disciplinary nature of drying by utilizing well-known authors from many countries of the world. Dr. Mujumdar and these associates have assembled an excellent up-to-date handbook.

The common thread throughout the book is the movement of heat and moisture as well as the movement and handling of products. Also included are instrumentation, sensors, and controls that are important for quality control of products and efficiency of operation. The emphasis on the design of equipment to expedite these processes in an economical manner is appropriate and useful.

The word *handbook* is sometimes used disparagingly to describe a reference for quick answers to limited questions or problems. In that sense this book is more than a handbook — the knowledge base provided permits the user to build different systems for products other than those covered.

Carl W. Hall

Preface to the Second Edition

The second edition of the *Handbook of Industrial Drying* is a testimonial to the success of the first edition published in 1987. Interest in the drying operation has continued to increase on a truly global scale over the past decade. For example, over 1500 papers have been presented at the biennial International Drying Symposia (IDS) since its inception in 1978. *Drying Technology—An International Journal* published some 2000 pages in seven issues in 1993 compared with just over three hundred only a decade earlier. The growth in drying R&D is stimulated by the need to design and operate dryers more efficiently and to produce products of higher quality.

A handbook is expected to provide the reader with critical information and advice on appropriate use of such information compiled in a readily accessible form. It is intended to bring together widely scattered information and knowhow in a coherent format. Since drying of solids is a multidisciplinary field—indeed, a discipline by itself—it is necessary to call on the expertise of individuals from different disciplines, different industrial sectors, and several countries. A quick perusal of the list of contributors will indicate a balanced blend of authorship from industry as well as academia. An attempt has been made to provide the key elements of fundamentals along with details of industrial dryers and special aspects of drying in specific industries, e.g., foods, pulp and paper, and pharmaceuticals.

The first edition contained twenty-nine chapters and two appendixes; this one contains forty-three chapters and three appendices. Aside from the addition of new chapters to cover topics missing from the first one, a majority of earlier chapters have been updated—some fully rewritten with new authorship. This edition contains over sixty percent new updated material. Thus, this book will be a valuable addition even to the bookshelves that already hold the first edition.

This revised and expanded edition follows the same general organization as the first with additions made to each of the four parts to eliminate some of the weaknesses of the first edition. For example, an extensive chapter is added in Part I on transport properties needed for dryer calculations. Chapters on infrared drying and the novel impinging stream dryers are added to Part II. Part III contains the largest enhancement with ten new chapters while Part IV is completely new except for the chapter on humidity measurements.

A two-volume set of this magnitude must depend on the direct and indirect contributions of a large number of individuals and organizations. Clearly it is impossible to name them all. I am grateful to all the contributors for the valuable time and effort they devoted to this project. The companies and publishers who have permitted us to reproduce some of their copyrighted artwork are acknowledged for their support. Appropriate credits are given in the text where applicable. Exergex Corporation, Brossard, Quebec, Canada provided all the secretarial and related assistance over a three-year period. Without it this revision would have been nearly impossible.

Over the past two years most of my graduate students and postdoctoral fellows of McGill University have provided me with very enthusiastic assistance in various forms in connection with this project. In particular, I wish to express my thanks to Dr. T. Kudra for his continued help in various ways. Purnima, Anita, and Amit Mujumdar kindly word-processed numerous chapters and letters and helped me keep track of the incredible paperwork involved. The encouragement I received from Dr. Carl W. Hall was singularly valuable in keeping me going on this project while handling concurrently the editorial responsibilities for *Drying Technology—An International Journal* and a host of other books. Finally, the staff at Marcel Dekker, Inc., have been marvellous; I sincerely appreciate their patience and faith in this project.

Arun S. Mujumdar

Preface to the First Edition

Drying of solids is one of the oldest and most common unit operations found in diverse processes such as those used in the agricultural, ceramic, chemical, food, pharmaceutical, pulp and paper, mineral, polymer, and textile industries. It is also one of the most complex and least understood operations because of the difficulties and deficiencies in mathematical descriptions of the phenomena of simultaneous — and often coupled and multiphase — transport of heat, mass, and momentum in solid media. Drying is therefore an amalgam of science, technology, and art (or know-how based on extensive experimental observations and operating experience) and is likely to remain so, at least for the foreseeable future.

Industrial as well as academic interest in solids drying has been on the rise for over a decade, as evidenced by the continuing success of the Biennial Industrial Drying Symposia (IDS) series. The emergence of several book series and an international journal devoted exclusively to drying and related areas also demonstrates the growing interest in this field. The significant growth in research and development activity in the western world related to drying and dewatering was no doubt triggered by the energy crunch of the early 1970s, which increased the cost of drying several-fold within only a few years. However, it is worth noting that continued efforts in this area will be driven not only by the need to conserve energy, but also by needs related to increased productivity, better product quality, quality control, new products and new processes, safer and environmentally superior operation, etc.

This book is intended to serve both the practicing engineer involved in the selection or design of drying systems and the researcher as a reference work that covers the wide field of drying principles, various commonly used drying equipment, and aspects of drying in important industries. Since industrial dryers can be finely categorized into over two hundred variants and, furthermore, since they are found in practically all major industrial sectors, it is impossible within limited space to cover all aspects of drying and dryers. We have had to make choices. In view of the availability of such publications as *Advances in Drying* and the *Proceedings of the International Drying Symposia*, which emphasize research and development in solids drying, we decided to concentrate on various practical aspects of commonly used industrial dryers following a brief introduction to the basic principles, classification and selection of dryers, process calculation schemes, and basic experimental techniques in drying. For detailed information on the fundamentals of drying, the reader is referred to various textbooks in this area.

The volume is divided into four major parts. Part I covers the basic principles, definitions, and process calculation methods in a general but concise fashion. The second part is devoted to a series of chapters that describe and discuss the more commonly used industrial dryers. Novel and less prevalent dryers have been excluded from coverage; the reader will find the necessary references in Appendix B, which lists books devoted to drying and related areas in English as well as other languages. Part III is devoted to the discussion of current drying practices in key industrial sectors in which drying is a significant if not necessarily dominant operation. Some degree of repetition was unavoidable since various dryers are discussed under two possible categories. Most readers will, however, find such information complementary as it is derived from different sources and generally presented in different contexts.

Because of the importance of gas humidity measurement techniques, which can be used to monitor and control the convective drying operation, Part IV includes a chapter that discusses such techniques. Energy savings in drying via the application of energy recovery techniques, and process and design modifications, optimization and control, and new drying techniques and nonconventional energy sources are also covered in some depth in the final part of the book.

Finally, it is my pleasant duty to express my sincerest gratitude to the contributors from industry and academia, from various parts of the world, for their continued enthusiasm and interest in completing this major project. The comments and criticisms received from over twenty-five reviewers were very valuable in improving the contents within the limitations of space. Many dryer manufacturers assisted me and the contributors directly or indirectly, by providing nonproprietary information about their equipment. Dr. Maurits Dekker, Chairman of the Board, Marcel Dekker, Inc., was instrumental in elevating the level of my interest in drying so that I was able to undertake the major task of compiling and editing a handbook in a truly multidisciplinary area whose advancement depends on closer industry-academia interaction and cooperation. My heartfelt thanks go to Chairman Mau for his kindness, continuous encouragement, and contagious enthusiasm throughout this project.

Over the past four years, many of my graduate students provided me with enthusiastic assistance in connection with this project. In particular, I wish to thank Mainul Hasan and Victor Jariwala for their help and support. In addition, Purnima and Anita Mujumdar kindly word-processed countless drafts of numerous chapters. Without the assistance of my coauthors, it would have been impossible to achieve the degree of coverage attained in this book. I wish to record my appreciation of their efforts. Indeed, this book is a result of the combined and sustained efforts of everyone involved.

Arun S. Mujumdar

Contents

VOLUME 2

Contributors

Janusz Adamiec   Faculty of Process and Environmental Engineering, Technical University of Łódź, Łódź, Poland

Roberto Bruttini*   Department of Chemical Engineering and Biochemical Processing Institute, University of Missouri—Rolla, Rolla, Missouri

D. K. Das Gupta   Defence Food Research Laboratory, Siddarthanagar, Mysore, India

Iva Filková   IFP Process Engineering, Prague, Czech Republic

Carl W. Hall   Engineering Information Services, Arlington, Virginia

Mainul Hasan   Department of Mining and Metallurgical Engineering, McGill University, Montreal, Quebec, Canada

Svend Hovmand   Niro Atomizer, Inc., Columbia, Maryland

Bing Huang   Department of Chemical Engineering, McGill University, Montreal, Quebec, Canada

James Y. Hung   Hung International, Appleton, Wisconsin

László Imre   Department for Energy, Technical University of Budapest, Budapest, Hungary

K. S. Jayaraman   Defence Food Research Laboratory, Siddarthanagar, Mysore, India

Digvir S. Jayas   Department of Agricultural Engineering, University of Manitoba, Winnipeg, Manitoba, Canada

Peter L. Jones   EA Technology Ltd., Capenhurst, Chester, United Kingdom

Rami Y. Jumah   Department of Chemical Engineering, McGill University, Montreal, Quebec, Canada

---

Current affiliation: Criofarma-Freeze Drying Equipment, Turin, Italy.

Władysław Kamiński   Faculty of Process and Environmental Engineering, Technical University of Łódź, Łódź, Poland

Roger B. Keey   Department of Chemical and Process Engineering, University of Canterbury, Christchurch, New Zealand

John J. Kelly   University College Dublin, Dublin, Ireland

Bilgin Kisakürek   Department of Chemical Engineering, Ankara University, Ankara, Turkey

Tadeusz Kudra*   Department of Chemical Engineering, McGill University, Montreal, Quebec, Canada

Markku J. Lampinen   Department of Energy Engineering, Helsinki University of Technology, Espoo, Finland

Andrzej Lenart   Department of Food Engineering, Warsaw Agricultural University, Warsaw, Poland

Piotr P. Lewicki   Department of Food Engineering, Warsaw Agricultural University, Warsaw, Poland

Athanasios I. Liapis   Department of Chemical Engineering and Biochemical Processing Institute, University of Missouri–Rolla, Rolla, Missouri

D. Marinos-Kouris   Department of Chemical Engineering, National Technical University, Zografos, Athens, Greece

Adam S. Markowski   Faculty of Process and Environmental Engineering, Technical University of Łódź, Łódź, Poland

Z. B. Maroulis   Department of Chemical Engineering, National Technical University, Zografos, Athens, Greece

Valentin Meltser   Belorussian Academy of Sciences, Minsk, Belorussia

Anilkumar S. Menon   Mount Sinai Hospital, Toronto, Ontario, Canada

Károly Molnár   Department of Chemical and Food Engineering, Technical University of Budapest, Budapest, Hungary

James G. Moore†   Blaw-Knox Food and Chemical Equipment Company, Buffalo, New York

Arun S. Mujumdar   Department of Chemical Engineering, McGill University, Montreal, Quebec, Canada

Kari T. Ojala   Department of Energy Engineering, Helsinki University of Technology, Espoo, Finland

Zdzisław Pakowski   Faculty of Process and Environmental Engineering, Technical University of Łódź, Łódź, Poland

Elizabeth Pallai   Research Institute of Chemical Engineering, Hungarian Academy of Sciences, Veszprém, Hungary

---

Current affiliation: CANMET, Energy Diversification Research Laboratory, Varennes, Quebec, Canada.
†Deceased.

Jerzy Pikoń   Silesian Technical University, Gliwice, Poland

Jan Písecký   Niro A/S, Soeborg, Denmark

Osman Polat   Procter & Gamble Co., Cincinnati, Ohio

Vijaya G. S. Raghavan   Department of Agricultural Engineering, Macdonald Campus of McGill University, Ste.-Anne-de-Bellevue, Quebec, Canada

Cristina Ratti   Planta Piloto de Ingeniería Química, Bahía Blanca, Argentina

Howard N. Rosen*   U.S. Department of Agriculture – Forest Service, Carbondale, Illinois

Robert F. Schiffmann   R. F. Schiffmann Associates, Inc., New York, New York

Shahab Sokhansanj   Department of Agricultural and Biosource Engineering, University of Saskatchewan, Saskatoon, Saskatchewan, Canada

Czesław Strumiłło   Faculty of Process and Environmental Engineering, Technical University of Łódź, Łódź, Poland

Lloyd F. Sturgeon   S.T. Hudson Engineers, Inc., Camden, New Jersey

Tibor Szentmarjay   Research Institute of Chemical Engineering, Hungarian Academy of Sciences, Veszprém, Hungary

Zbigniew T. Sztabert   Industrial Chemistry Research Group, Warsaw, Poland

Peter Wiederhold   General Eastern Instruments Corporation, Woburn, Massachusetts

Richard J. Wimberger   Spooner Industries, Inc., Green Bay, Wisconsin

Roland Wimmerstedt   Chemical Engineering Department, Lund University, Lund, Sweden

Romuald Żyłła   Department of Process and Environmental Engineering, Technical University of Łódź, Łódź, Poland

---

*_Current affiliation_: Forest Products and Harvesting Research Staff, U.S. Department of Agriculture – Forest Service, Washington, D.C.

HANDBOOK
OF
INDUSTRIAL
DRYING

SECOND EDITION
REVISED AND EXPANDED

Drying of Solids: Principles, Classification, and Selection of Dryers

Arun S. Mujumdar
McGill University
Montreal, Quebec, Canada

Anilkumar S. Menon
Mount Sinai Hospital
Toronto, Ontario, Canada

1. INTRODUCTION

Drying commonly describes the process of thermally removing volatile substances (moisture) to yield a solid product. Moisture held in loose chemical combination, present as a liquid solution within the solid or even trapped in the microstructure of the solid, which exerts a vapor pressure less than that of pure liquid, is called *bound* moisture. Moisture in excess of bound moisture is called *unbound* moisture

When a wet solid is subjected to thermal drying, two processes occur simultaneously:

1. Transfer of energy (mostly as heat) from the surrounding environment to evaporate the surface moisture
2. Transfer of internal moisture to the surface of the solid and its subsequent evaporation due to process 1

The rate at which drying is accomplished is governed by the rate at which the two processes proceed. Energy transfer as heat from the surrounding environment to the wet solid can occur as a result of convection, conduction, or radiation and in some cases as a result of a combination of these effects. Industrial dryers differ in type and design, depending on the principal method of heat transfer employed. In most cases heat is transferred to the surface of the wet solid and thence to the interior. However, in dielectric, radiofrequency, or microwave freeze drying, energy is supplied to generate heat internally within the solid and flows to the exterior surfaces.

Process 1, the removal of water as vapor from the material surface, depends on the external conditions of temperature, air humidity and flow, area of exposed surface, and pressure.

Process 2, the movement of moisture internally within the solid, is a function of the physical nature of the solid, the temperature, and its moisture content. In a drying operation any one of these processes may be the limiting factor governing the rate of drying, although they both proceed simultaneously throughout the drying cycle. In the following sections we shall discuss the terminology and some of the basic concepts behind the two processes involved in drying.

2.   EXTERNAL CONDITIONS (PROCESS 1)

Here the essential external variables are temperature, humidity, rate and direction of air flow, the physical form of the solid, the desirability of agitation, and the method of supporting the solid during the drying operation (1). External drying conditions are especially important during the initial stages of drying when unbound surface moisture is being removed. In certain cases, for example, in materials like ceramics and timber in which considerable shrinkage occurs, excessive surface evaporation after the initial free moisture has been removed sets up high moisture gradients from the interior to the surface. This is liable to cause overdrying and excessive shrinkage and consequently high tension within the material, resulting in cracking and warping. In these cases surface evaporation should be retarded through the employment of high air relative humidities while maintaining the highest safe rate of internal moisture movement by heat transfer.

Surface evaporation is controlled by the diffusion of vapor from the surface of the solid to the surrounding atmosphere through a thin film of air in contact with the surface. Since drying involves the interphase transfer of mass when a gas is brought in contact with a liquid in which it is essentially insoluble, it is necessary to be familiar with the equilibrium characteristics of the wet solid. Also, since the mass transfer is usually accompanied with the simultaneous transfer of heat, due consideration must be given to the enthalpy characteristics.

2.1.   Vapor-Liquid Equilibrium and Enthalpy for a Pure Substance Vapor-Pressure Curve

When a liquid is exposed to a dry gas, the liquid evaporates, that is, forms vapor and passes into the gaseous phase. If m_w is the mass of vapor in the gaseous phase, then this vapor exerts a pressure over the liquid, the *partial pressure*, which, assuming ideal gas behavior for the vapor, is given by

$$P_W V = \frac{m_W}{M_W} RT \quad \text{or} \quad P_W V_W = RT \tag{1}$$

The maximum value of P_W that can be reached at any temperature is the saturated vapor pressure P_W^0. If the vapor pressure of a substance is plotted against temperature, a curve such as TC of Figure 1 is obtained. Also plotted in the figure are the solid-liquid equilibrium curve (melting curve) and the solid-vapor (sublimation) curve. The point T in the graph at which all three phases can coexist is called the *triple point*. For all conditions along the curve TC, liquid and vapor may coexist, and these points correspond with the *saturated liquid* and the *saturated vapor state*. Point C is the critical point at which distinction between the liquid and vapor phases disappears, and all properties of the liquid, such as density, viscosity, and refractive index, are identical with those of the vapor. The substance above the critical temperature is called a gas, the temperature corresponding to a pressure at each point on the curve TC is the *boiling point*, and that corresponding to a pressure of 101.3 kPa is the *normal boiling point*.

The Clausius–Clapeyron Equation

Comprehensive tables of vapor pressure data of common liquids, such as water, common refrigerants, and others, may be found in References 2 and 3. For most liquids, the vapor pressure data are obtained at a few discrete temperatures, and it might frequently be necessary to interpolate between or extrapolate beyond these measurement points. At

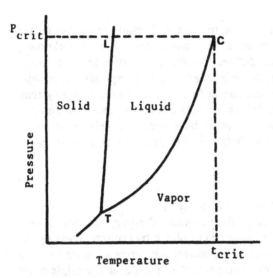

Figure 1 Vapor pressure of a pure liquid.

a constant pressure, the Clausius–Clapeyron equation relates the slope of the vapor pressure-temperature curve to the latent heat of vaporization through the relation

$$\frac{dP^0_W}{dT} = \frac{\Delta H_W}{T(V_W - V_L)} \tag{2}$$

where V_W and V_L are the specific molar volumes of saturated vapor and saturated liquid, respectively, and ΔH_W is the molar latent heat of vaporization. Since the molar volume of the liquid is very small compared with that of the vapor, we neglect V_L and substitute for V_W from Eq. [1] to obtain

$$d \ln P^0_W = \frac{\Delta H_W}{RT^2} dT \tag{3}$$

Since ΔH_W could be assumed to be a constant over short temperature ranges, Eq. [3] can be integrated to

$$\ln P^0_W = -\frac{\Delta H_W}{RT} + \text{constant} \tag{4}$$

and this equation can be used for interpolation. Alternatively, reference-substance plots (6) may be constructed. For the reference substance,

$$d \ln P^0_R = \frac{\Delta H_R}{RT^2} dT \tag{5}$$

Dividing Eq. [3] by Eq. [5] and integrating provides

$$\ln P^0_W = \frac{M_W \Delta H_W}{M_R \Delta H_R} \ln P^0_R + \text{constant} \tag{6}$$

The reference substance chosen is one whose vapor-pressure data are known.

Enthalpy

All substances have an internal energy due to the motion and relative position of the constituent atoms and molecules. Absolute values of the internal energy, u, are unknown, but numerical values relative to an arbitrarily defined baseline at a particular temperature can be computed. In any steady flow system there is an additional energy associated with forcing streams into a system against a pressure and in forcing streams out of the system. This flow work per unit mass is PV, where P is the pressure and V the specific volume. The internal energy and the flow work per unit mass have been conveniently grouped together into a composite energy called the enthalpy H. The enthalpy is defined by the expression

$$H = u + PV \qquad\qquad\qquad\qquad [7]$$

and has the units of energy per unit mass (J kg^{-1} or N·m kg^{-1}).

Absolute values of enthalpy of a substance like the internal energy are not known. Relative values of enthalpy at other conditions may be calculated by arbitrarily setting the enthalpy to zero at a convenient reference state. One convenient reference state for zero enthalpy is liquid water under its own vapor pressure of 611.2 Pa at the triple-point temperature of 273.16 K (0.01 °C).

The isobaric variation of enthalpy with temperature is shown in Figure 2. At low pressures in the gaseous state, when the gas behavior is essentially ideal, the enthalpy is almost independent of the pressure, so the isobars nearly superimpose on each other. The curves marked "saturated liquid" and "saturated vapor," however, cut across the constant pressure lines and show the enthalpies for these conditions at temperatures and pressures

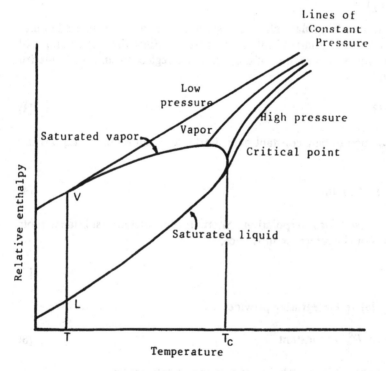

Figure 2   Typical enthalpy-temperature diagram for a pure substance.

corresponding to the equilibrium vapor pressure relationship for the substance. The distance between the saturated vapor and saturated liquid curves, such as the distance $V-L$ corresponds to the latent heat of vaporization at a temperature T. Both T and $V-L$ are dependent on pressure, the distance $V-L$ decreases and becomes zero at the critical temperature T_C. Except near the critical temperature, the enthalpy of the liquid is almost independent of pressure until exceedingly high pressures are reached.

Heat Capacity

The *heat capacity* is defined as the heat required to raise the temperature of a unit mass of substance by a unit temperature. For a constant pressure process, the heat capacity C_P is given by

$$C_P = \left(\frac{\partial Q}{\partial T} \right)_P \qquad [8]$$

where the heat flow Q is the sum of the internal energy change ∂u and the work done against pressure $P \, \partial V$. Equation [8] may be expanded as follows:

$$C_P = \left(\frac{\partial u}{\partial T} \right)_P + P \left(\frac{\partial V}{\partial T} \right)_P = \left(\frac{\partial H}{\partial T} \right)_P \qquad [9]$$

The slope of the isobars of Figure 2 yields the heat capacities.

In drying calculation, it is more convenient to use the mean values of heat capacity over a finite temperature step.

$$\overline{C}_P = \left(\frac{\Delta Q}{\Delta T} \right)_P = \frac{1}{(T_2 - T_1)} \int_{T_1}^{T_2} C_P \, dT \qquad [10]$$

Second-order polynomials in temperature have been found to adequately describe the variation of C_P with temperature in the temperature ranges 300–1500 K (4), but for the temperature changes normally occurring in drying the quadratic term can be neglected.

Thus if

$$C_P = a + bT \qquad [11]$$

Then from Eq. [10],

$$\overline{C}_P = a + \frac{1}{2} b \, (T_1 + T_2) = C_P (T_{av}) \qquad [12]$$

The mean heat capacity is the heat capacity evaluated at the arithmetic mean temperature T_{av}.

From Eqs. [9] and [10], the enthalpy of the pure substance can be estimated from its heat capacity by

$$H = \overline{C}_P \theta \qquad [13]$$

where θ denotes the temperature difference or excess over the zero enthalpy reference state. Heat capacity data for a large number of liquids and vapors are found in Reference 5.

2.2. Vapor-Gas Mixtures

When a gas or gaseous mixture remains in contact with a liquid surface, it will acquire vapor from the liquid until the partial pressure of the vapor in the gas mixture equals the vapor pressure of the liquid at the existing temperature. In drying applications, the gas frequently is air and the liquid is water. Although common concentration units (partial pressure, mole fraction, and others) based on total quantity of gas and vapor are useful, for operations that involve changes in vapor content of a vapor-gas mixture without changes in the amount of gas, it is more convenient to use a unit based on the unchanging amount of gas.

Humid air is a mixture of water vapor and gas, composed of a mass m_W of water vapor and a mass m_G of gas (air). The *moisture content* or *absolute humidity* can be expressed as

$$Y = \frac{m_W}{m_G} \qquad\qquad [14]$$

The total mass can be written in terms of Y and m_G as

$$m_G + m_W = m_G(1 + Y) \qquad\qquad [15]$$

Using the gas law for vapor and air fractions at constant total volume V and temperature T,

$$m_G = \frac{P_G V}{RT} M_G \qquad \text{and} \qquad m_W = \frac{P_W V}{RT} M_W \qquad\qquad [16]$$

Thus,

$$Y = \frac{P_W}{P_G} \frac{M_W}{M_G} \qquad\qquad [17]$$

Using Dalton's law of partial pressures,

$$P = P_W + P_G \qquad\qquad [18]$$

and

$$Y = \frac{P_W}{P - P_W} \frac{M_W}{M_G} \qquad\qquad [19]$$

When the partial pressure of the vapor in the gas equals the vapor pressure of the liquid, an equilibrium is reached and the gas is said to be *saturated* with vapor. The ideal saturated absolute humidity is then

$$Y_s = \frac{P_W^0}{P - P_W^0} \frac{M_W}{M_G} \qquad\qquad [20]$$

The *relative humidity* ψ of a vapor-gas mixture is a measure of its fractional saturation with moisture and is defined as the ratio of the partial pressure of the vapor P_W to the saturated pressure P_W^0 at the same temperature. Thus ψ is given by

$$\psi = \frac{P_W}{P_W^0} \qquad\qquad [21]$$

Equation [19] may now be written as

$$Y = \frac{M_W}{M_G} \frac{\psi P_W^0}{P - \psi P_W^0} \qquad\qquad [22]$$

For water vapor and air when $M_W = 18.01$ kg/kmol and $M_G = 28.96$ kg/kmol, respectively, Eq. [22] becomes

$$Y = 0.622 \frac{\psi P_W^0}{P - \psi P_W^0}$$ [23]

2.3. Unsaturated Vapor-Gas Mixtures: Psychrometry in Relation to Drying

If the partial pressure of the vapor in the vapor-gas mixture is for any reason less than the vapor pressure of the liquid at the same temperature, the vapor-gas mixture is said to be unsaturated. As mentioned earlier, two processes occur simultaneously during the thermal process of drying a wet solid, namely, heat transfer to change the temperature of the wet solid and to evaporate its surface moisture and the mass transfer of moisture to the surface of the solid and its subsequent evaporation from the surface to the surrounding atmosphere. Frequently, the surrounding medium is the drying medium, usually heated air or combustion gases. Consideration of the actual quantities of air required to remove the moisture liberated by evaporation is based on psychometry and the use of humidity charts. The following are definitions of expressions used in psychrometry (6).

Dry Bulb Temperature

This is the temperature of a vapor-gas mixture as ordinarily determined by the immersion of a thermometer in the mixture.

Dew Point

This is the temperature at which a vapor gas mixture becomes saturated when cooled at a constant total pressure out of contact with a liquid (i.e., at constant absolute humidity). The concept of the dew point is best illustrated by referring to Figure 3, a plot of the absolute humidity versus temperature for a fixed pressure and the same gas. If an unsaturated mixture initially at point F is cooled at constant pressure out of contact of liquid, the gas saturation increases until the point G is reached, when the gas is fully saturated. The temperature at which the gas is fully saturated is called the dew point T_D. If the temperature is reduced an infinitesimal amount below T_D, the vapor will condense and the process follows the saturation curve.

While condensation occurs the gas always remains saturated. Except under specially

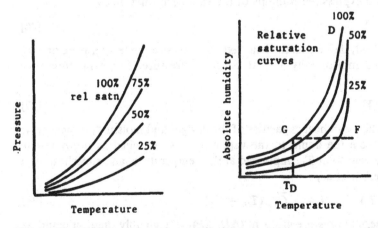

Figure 3 Two forms of psychrometric charts.

controlled circumstances, supersaturation will not occur and no vapor-gas mixture whose coordinates lie to the left of the saturation curve will result.

Humid Volume

The humid volume V_H of a vapor-gas mixture is the volume in cubic meters of 1 kg of dry gas and its accompanying vapor at the prevailing temperature and pressure. The volume of an ideal gas or vapor at 273 K and 1 atmosphere (101.3 kPa) is 22.4 m^3/kg·mol. For a mixture with an absolute humidity Y at T_G (K) and P (atm), the ideal gas law gives the humid volume as

$$V_H = \left(\frac{1}{M_G} + \frac{Y}{M_W} \right) 22.4 \frac{T}{273.14} \frac{1}{P}$$

$$V_H = 0.082 \left(\frac{1}{M_G} + \frac{Y}{M_W} \right) \frac{T}{P} \qquad [24]$$

When the mass of dry gas in the vapor-gas mixture is multiplied by the humid volume, the volume of the vapor-gas mixture is obtained. The humid volume at saturation is computed with $Y = Y_S$, and the specific volume of the dry gas can be obtained by substituting $Y = 0$. For partially saturated mixtures, V_H may be interpolated between values for 0% and 100% saturation at the same temperature and pressure.

Enthalpy

Since the enthalpy is an extensive property, it could be expected that the enthalpy of a humid gas is the sum of the partial enthalpies of the constituents and a term to take into account the heat of mixing and other effects. The humid enthalpy I_G is defined as the enthalpy of a unit mass of dry gas and its associated moisture. With this definition of enthalpy,

$$I_G = H_{GG} + YH_{GW} + \Delta H_{GM} \qquad [25]$$

where H_{GG} is the enthalpy of dry gas, H_{GW} is the enthalpy of moisture, and ΔH_{GM} is the residual enthalpy of mixing and other effects. In air saturated with water vapor, this residual enthalpy is only -0.63 kJ kg^{-1} at 60°C (333.14 K) (3) and is only 1% of H_{GG}; thus it is customary to neglect the influences of this residual enthalpy.

It is sometimes convenient to express the enthalpy in terms of specific heat. Analogous to Eq. [13], we could express the enthalpy of the vapor-gas mixture by

$$I_G = \overline{C}_{PY}\theta + \Delta H_{V0} Y \qquad [26]$$

$\overline{C}_{PY}$ is called the *humid heat*, defined as the heat required to raise the temperature of 1 kg of gas and its associated moisture by 1°K at constant pressure. For a mixture with absolute humidity Y,

$$\overline{C}_{PY} = \overline{C}_{PG} + \overline{C}_{PW} Y \qquad [27]$$

where $\overline{C}_{PG}$ and $\overline{C}_{PW}$ are the mean heat capacities of the dry gas and moisture, respectively.

The path followed from the liquid to the vapor state is described as follows. The liquid is heated up to the dew point T_D, vaporized at this temperature, and superheated to the dry bulb temperature T_G. Thus,

$$H_{GW} = \overline{C}_{LW}(T_D - T_0) + \Delta H_{VD} + \overline{C}_{PW}(T_G - T_D) \qquad [28]$$

However, since the isothermal pressure gradient $(\Delta H/\Delta P)_T$ is negligibly small, it could be assumed that the final enthalpy is independent of the vaporization path followed. For the sake of convenience it could be assumed that vaporization occurs at 0°C (273.14 K), at which the enthalpy is zero, and then directly superheated to the final temperature T_G. The enthalpy of the vapor can now be written as

$$H_{GW} = \overline{C}_{PW}(T_G - T_0) + \Delta H_{V0}$$ [29]

and the humid enthalpy given by

$$I_G = \overline{C}_{PG}(T_G - T_0) + Y(\overline{C}_{PW}(T_G - T_0) + \Delta H_{V0})$$ [30]

Using the definition for the humid heat capacity, Eq. [30] reduces to

$$I_G = \overline{C}_{PY}(T_G - T_0) + \Delta H_{V0} Y$$ [31]

In Eq. [31] the humid heat is evaluated at $(T_G + T_0)/2$ and ΔH_{V0}, the latent heat of vaporization at 0°C (273.14 K). Despite its handiness, the use of Eq. [31] is not recommended above a humidity of 0.05 kg kg^{-1}. For more accurate work, it is necessary to resort to the use of Eq. [28] in conjunction with Eq. [25]. In Eq. [28] it should be noted that $\overline{C}_{LW}$ is the mean capacity of liquid moisture between T_0 and T_D, $\overline{C}_{PW}$ is the mean capacity of the moisture vapor evaluated between T_D and T_G, and ΔH_{VD} is the latent heat of vaporization at the dew point T_D. The value of ΔH_{VD} can be approximately calculated from a known latent heat value at temperature T_0 by

$$\frac{\Delta H_{VD}}{\Delta H_{V0}} - \left(\frac{T_D - T_C}{T_0 - T_C}\right)^{1/3}$$ [32]

where T_C is the critical temperature. Better and more accurate methods of estimating ΔH_{VD} are available in References 5 and 7.

2.4. Enthalpy-Humidity Charts

Using Eqs. [23], [25], and [28], the enthalpy humidity diagram for unsaturated air ($\psi < 1$) can be constructed using the parameters ψ and θ. In order to be able to follow the drying process we need access to enthalpy-humidity values. There seems to be no better, convenient, and cheaper way to store this data than in graphic form. The first of these enthalpy-humidity charts is attributed to Mollier. Mollier's original enthalpy-humidity chart was drawn with standard rectangular coordinates (Fig. 4), but in order to extend the area over which it can be read, an oblique-angle system of coordinates is chosen for $I_G = f(Y)$.

In the unsaturated region, it can be seen from Eq. [30] that I_G varies linearly with the humidity Y and the temperature T_G. If zero temperature (0°C) is taken as the datum for zero enthalpy, then

$$I_G = \overline{C}_{PG}\theta + Y(\overline{C}_{PW}\theta + \Delta H_{V0})$$ [33]

where θ is the temperature in °C.

The isotherms (θ = constant) cut the ordinate ($Y = 0$) at a value $\overline{C}_{PG}\theta$ (the dry gas enthalpy). If the isenthalpic lines (I_G = constant) are so inclined that they fall with a slope $-\Delta H_{V0}$, and if only $\Delta H_{V0}Y$ were taken into account in the contribution of vapor to the vapor-gas enthalpy, then the isotherms would run horizontally, but because of the contribution of $\overline{C}_{PW}\theta Y$, they increase with Y for $\theta > 0$°C and decrease with Y for $\theta < 0$°C . Contours of relative humidity ψ are also plotted. The region above the curve $\psi = 1$ at which air is saturated corresponds to an unsaturated moist gas; the region below the curve corresponds to fogging conditions. At a fixed temperature air cannot take up more

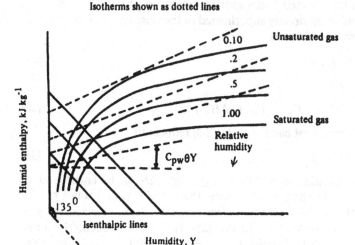

Figure 4   An enthalpy-humidity diagram for a moist gas.

than a certain amount of vapor. Liquid droplets then precipitate due to oversaturation, and this is called the *cloud* or *fog state*.

Detailed enthalpy-humidity diagrams are available elsewhere in this handbook and in Reference 10.

A humidity chart is not only limited to a specific system of gas and vapor but is also limited to a particular total pressure. The thermophysical properties of air may be generally used with reasonable accuracy for diatomic gases (3), so that charts developed for mixtures in air can be used to describe the properties of the same moisture vapor in a gas such as nitrogen.

Charts other than those of moist air are often required in the drying of fine chemicals and pharmaceutical products. These are available in Refs. 3, 8, and 9.

Adiabatic Saturation Curves

Also plotted on the psychrometric chart are a family of adiabatic saturation curves. The operation of adiabatic saturation is indicated schematically in Figure 5. The entering gas is contacted with a liquid and as a result of mass and heat transfer between the gas and liquid the gas leaves at conditions of humidity and temperature different from those at the entrance. The operation is adiabatic since no heat is gained or lost by the surroundings. Doing a mass balance on the vapor results in

$$G_V = G_G(Y_{out} - Y_{in}) \qquad [34]$$

The enthalpy balance yields

$$I_{G_{in}} + (Y_{out} - Y_{in})I_{LW} = I_{G_{out}} \qquad [35]$$

Substituting for I_G from Eq. [31], we have

$$\overline{C}_{PY_{in}}(T_{in} - T_0) + \Delta H_{V0}Y_{in} + (Y_{out} - Y_{in})\overline{C}_{LW}(T_L - T_0)$$
$$= \overline{C}_{PY_{out}}(T_{out} - T_0) + \Delta H_{V0}Y_{out} \qquad [36]$$

Now, if a further restriction is made that the gas and the liquid phases reach equilibrium

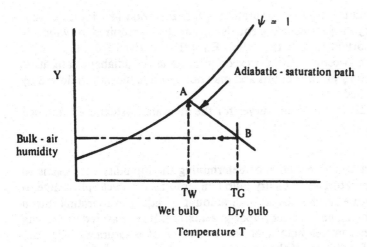

Figure 5 A temperature-humidity diagram for moist air.

when they leave the system (i.e., the gas-vapor mixture leaving the system is saturated with liquid), then $T_{out} = T_{GS}$, $I_{G_{out}} = I_{GS}$, and $Y_{out} = Y_{GS}$ where T_{GS} is the adiabatic saturation temperature and Y_{GS} is the absolute humidity saturated at T_{GS}. Still further, if the liquid enters at the adiabatic saturation temperature T_{GS}, that is, $T_L = T_{GS}$, Eq. [36] becomes

$$\overline{C}_{PY_{in}} (T_{in} - T_0) + \Delta H_{V0} Y_{in} + (Y_{GS} - Y_{in})\overline{C}_{LW}(T_{GS} - T_0)$$
$$= \overline{C}_{PY_{GS}} (T_{GS} - T_0) + \Delta H_{V0}Y_{GS} \qquad [37]$$

or substituting for $\overline{C}_{PG}$ from Eq. [27]

$$\overline{C}_{PY_{in}} (T_{in} - T_0) + Y_{in}\overline{C}_{PW_{in}} (T_{in} - T_0) + \Delta H_{V0}Y_{in} + (Y_{GS} - Y_{in})\overline{C}_{LW}(T_{GS} - T_0)$$
$$= \overline{C}_{PG_{GS}} (T_{GS} - T_0) + \overline{C}_{PW_{GS}} Y_{GS} (T_{GS} - T_0) + \Delta H_{V0} Y_{GS} \qquad [38]$$

Assuming that the heat capacities are essentially constant over the temperature range involved, $\overline{C}_{PG_{in}} = \overline{C}_{PG_{GS}} = \overline{C}_{PG}$ and $\overline{C}_{PW_{in}} = \overline{C}_{PW_{GS}} = \overline{C}_{PW}$. Further subtracting $Y_{in} \overline{C}_{PW}T_{GS}$ from both sides of Eq. [38] and simplifying, we have

$$\overline{C}_{PY}(T_{in} - T_{GS}) = (Y_{GS} - Y_{in})[(\overline{C}_{PW}(T_{GS} - T_0) + \Delta H_{V0} - \overline{C}_{LW}(T_{GS} - T_0)] \qquad [39]$$

From Figure 2 the quantity in square brackets is equal to ΔH_{VS}, and thus,

$$\overline{C}_{PY} (T_{in} - T_{GS}) = (Y_{GS} - Y_{in}) \Delta H_{VS} \qquad [40]$$

or

$$T_{in} - T_{GS} = (Y_{GS} - Y_{in}) \frac{\Delta H_{VS}}{\overline{C}_{PY}} \qquad [41]$$

Equation [41] represents the "adiabatic saturation curve" on the psychrometric chart, which passes through the points $A(Y_{GS}, T_{GS})$ on the 100% saturation curve, ($\psi = 1$) and $B(Y_{in}, T_{in})$, the initial condition. Since the humid heat contains the term Y_{in}, the curve is not straight but is curved slightly concave upward. Knowing the adiabatic saturation temperature and the actual gas temperature, the actual gas humidity can be easily obtained as the absolute humidity from the saturation locus. Equation [40] indicates that the sensible heat given up by the gas in cooling equals the latent heat required to evaporate the added vapor. It is important to note that, since Eq. [41] is derived from the overall mass and energy balances between the initial gas conditions and the adiabatic saturation conditions, it is applicable only at these points and may not describe the path followed by the gas as it becomes saturated.

A family of these adiabatic saturation curves for the air-water system are contained in the psychrometric charts (10).

Wet Bulb Temperature

One of the oldest and best-known methods of determining the humidity of a gas is to measure its "wet bulb temperature" and its dry bulb temperature. The wet bulb temperature is the steady temperature reached by a small amount of liquid evaporating into a large amount of rapidly moving unsaturated vapor-gas mixture. It is measured by passing the gas rapidly past a thermometer bulb kept wet by means of a saturated wick and shielded from the effects of radiation. If the gas is unsaturated, some liquid is evaporated from the wick into the gas stream, carrying with it the associated latent heat. This latent heat is taken from within the liquid in the wick, and the wick is cooled. As the temperature of the wick is lowered, sensible heat is transferred by convection from the gas stream and by radiation from the surroundings. At steady state, the net heat flow to the wick is zero and the temperature is constant.

The heat transfer to the wick can be written as

$$q = (h_C + h_R)A(T_G - T_W) \tag{42}$$

where h_C and h_R are the convective and radiative heat transfer coefficients, respectively, T_G is the gas temperature, T_W is the temperature indicated by thermometer. By using h_R, it assumed that radiant heat transfer can be approximated

$$q_R = h_R A(T_G - T_W) \tag{43}$$

The rate of mass transfer from the wick is

$$N_G = KA(Y_W - Y_G) \tag{44}$$

An amount of heat given by

$$q = N_G \Delta H_{VW} \tag{45}$$

is associated with this mass transfer. Since under steady conditions all the heat transferred to the wick is utilized in mass transfer, from Eqs. [42], [44] and [45] we have

$$T_G - T_W = \frac{K \, \Delta H V_W}{h_C + h_R} (Y_W - Y_G) \tag{46}$$

The quantity $T_G - T_W$ is called the *wet bulb depression*. In order to determine the humidity Y_G from Eq. [46], predictable values of $K \, \Delta H_{VW}/(h_C + h_R)$ must be obtained. This ratio of coefficients depends upon the flow, boundary, and temperature conditions encountered. In measuring the wet bulb temperature, several precautions are taken to ensure reproducible values of $K \, \Delta H_V/(h_C + h_R)$. The contribution by radiation is minimized by shielding the wick. The convective heat transfer can be enhanced by making the gas movement past the bulb rapid, often by swinging the thermometer through the gas, as in the sling psychrometer, or by inserting the wet bulb thermometer in a constriction in the gas flow path. Under these conditions. Eq. [46] reduces to

$$T_G - T_W = \frac{K \, \Delta H_{VW}}{h_C} (Y_W - Y_G) \qquad [47]$$

For turbulent flow past a wet cylinder, such as a wet bulb thermometer, the accumulated experimental data give

$$\frac{h_C}{K} = 35.53 \left(-\frac{\mu}{\rho D} \right)^{0.56} \quad \text{J/mol} \cdot {}^\circ\text{C} \qquad [48]$$

when air is the noncondensable gas and

$$\frac{h_C}{K} = \overline{C}_{PY} \left(\frac{\text{Sc}}{\text{Pr}} \right)^{0.56} \qquad [49]$$

for other gases. Equation [49] is based on heat and mass transfer experiments with various gases flowing normal to cylinders. For pure air, $\text{Sc} \cong \text{PR} \cong 0.70$ and $h_C/K = 29.08$ J/mol $\cdot$ $^\circ$C from Eqs. [48] and [49]. Experimental data for the air-water system yield values of h_C/K ranging between 32.68 and 28.54 J/mol $\cdot$ $^\circ$C. The latter figure is recommended (11). For the air-water system, the h_C/K value can be replaced by $\overline{C}_{PY}$ within moderate ranges of temperature and humidity, provided flow is turbulent. Under these conditions, Eq. [47] becomes identical to the adiabatic saturation curve Eq. [41] and thus the adiabatic saturation temperature is the same as the wet bulb temperature for the air-water system. For systems other than air-water, they are not the same, as can be seen from the psychrometric charts given by Perry (7).

It is worthwhile pointing out here that, although the adiabatic saturation curve equation does not reveal anything of the enthalpy-humidity path of either the liquid phase or gas phase at various points in the contacting device (except for the air-water vapor system), each point within the system must conform with the wet bulb relation, which requires that the heat transferred be exactly consumed as latent heat of vaporization of the mass of liquid evaporated. The identity of h_C/K with $\overline{C}_{PY}$ was first found empirically by Lewis and hence is called the Lewis relation. The treatment given here on the wet bulb temperature applies only in the limit of very mild drying conditions when the vapor flux becomes directly proportional to the humidity potential ΔY. This is the case in most drying operations.

A more detailed treatment using a logarithmic driving force for vapor flux and the concept of the humidity potential coefficient ϕ while accounting for the influence of the moisture vapor flux on the transfer of heat to the surface, namely, the Ackermann correction ϕ_E, has been given in Ref. 3. The concept of Luikov number Lu, which is essentially the ratio of the Prandtl number Pr to the Schmidt number Sc, has also been introduced.

2.5. Types of Psychrometric Representation

As stated previously, two processes occur simultaneously during the thermal process of drying a wet solid: heat transfer, to change the temperature of the wet solid, and mass transfer of moisture to the surface of a solid accompanied by its evaporation from the surface to the surrounding atmosphere, which in convection or direct dryers is the drying medium. Consideration of the actual quantities of air required to remove the moisture liberated by evaporation is based on psychrometry and the use of humidity charts. This procedure is extremely important in the design of forced convection, pneumatic, and rotary dryers. The definitions of terms and expressions involved in psychrometry have been discussed in Section 2.3.

There are different ways of plotting humidity charts. One procedure involves plotting the absolute humidity against the dry bulb temperature. A series of curves is obtained for different percentage humidity values from saturation downward (Fig. 3). On this chart, the saturation humidities are plotted from vapor pressure data with the help of Eq. [23] to give curve GD. The curve for humidities at 50% saturation is plotted at half the ordinate of curve GD. All curves at constant percentage saturation reach infinity at the boiling point of the liquid at the prevailing pressure.

Another alternative is the graphic representation of conditions of constant relative saturation on a vapor pressure temperature chart (Fig. 3). The curve for 50% relative saturation shows a partial pressure equal to one-half the equilibrium vapor pressure at any temperature.

A common method of portraying humidity charts is by using the enthalpy-humidity chart indicated earlier (10).

3. INTERNAL CONDITIONS (PROCESS 2)

After having discussed the factors and definitions related to the external conditions of air temperature and humidity, attention will now be paid to the solid characteristics.

As a result of heat transfer to a wet solid, a temperature gradient develops within the solid while moisture evaporation occurs from the surface. This produces a migration of moisture from within the solid to the surface, which occurs through one or more mechanisms, namely, diffusion, capillary flow, internal pressures set up by shrinkage during drying, and, in the case of indirect (conduction) dryers, through a repeated and progressive occurring vaporization and recondensation of moisture to the exposed surface. An appreciation of this internal movement of moisture is important when it is the controlling factor, as occurs after the critical moisture content, in a drying operation carried to low final moisture contents. Variables such as air quality, which normally enhance the rate of surface evaporation, become of decreasing importance except to promote higher heat transfer rates. Longer residence times, and, where permissible, higher temperatures become necessary. In the case of such materials as ceramics and timber, in which considerable shrinkage occurs, excessive surface evaporation sets up high moisture gradients from the interior toward the surface, which is liable to cause overdrying, excessive shrinkage, and, consequently, high tension, resulting in cracking or warping. In such cases, it is essential not to incur too high moisture gradients by retarding surface evaporation through the employment of high air relative humidities while maintaining the highest safe rate of internal moisture movement by virtue of heat transfer. The temperature gradient set up in the solid will also create a vapor pressure gradient, which will in turn result in moisture vapor diffusion to the surface; this will occur simultaneously with liquid moisture movement.

3.1. Moisture Content of Solids

The moisture contained in a wet solid or liquid solution exerts a vapor pressure to an extent depending upon the nature of moisture, the nature of solid, and the temperature. A wet solid exposed to a continuous supply of fresh gas continues to lose moisture until the vapor pressure of the moisture in the solid is equal to the partial pressure of the vapor in the gas. The solid and gas are then said to be in *equilibrium*, and the moisture content of the solid is called the *equilibrium moisture content* under the prevailing conditions. Further exposure to this air for indefinitely long periods will not bring about any additional loss of moisture. The moisture content in the solid could be reduced further by exposing it to air of lower relative humidity. Solids can best be classified as follows (12):

Nonhygroscopic capillary-porous media, such as sand, crushed minerals, nonhygroscopic crystals, polymer particles, and some ceramics. The defining criteria are as follows. (1) There is a clearly recognizable pore space; the pore space is filled with liquid if the capillary-porous medium is completely saturated and is filled with air when the medium is completely dry. (2) The amount of physically bound moisture is negligible; that is, the material is nonhygroscopic. (3) The medium does not shrink during drying.

Hygroscopic-porous media, such as clay, molecular sieves, wood, and textiles. The defining criteria are as follows. (1) There is a clearly recognizable pore space. (2) There is a large amount of physically bound liquid. (3) Shrinkage often occurs in the initial stages of drying. This category was further classified into (a) hygroscopic capilliary-porous media (micropores and macropores, including bidisperse media, such as wood, clays, and textiles), and (b) strictly hygroscopic media (only micropores, such as silica gel, alumina, and zeolites).

Colloidal (nonporous) media, such as soap, glue, some polymers (e.g., nylons), and various food products. The defining criteria are as follows: (1) There is no pore space (evaporation can take place only at the surface). (2) All liquid is physically bound.

It should be noted that such classifications are applicable only to homogeneous media that could be considered as continua for transport.

Since a wet solid is usually swollen compared with its condition when free of moisture and its volume changes during the drying process, it is not convenient to express moisture content in terms of volume. The moisture content of a solid is usually expressed as the moisture content by weight of bone-dry material in the solid, X. Sometimes a wet basis moisture content W, which is the moisture-solid ratio based on the total mass of wet material, is used. The two moisture contents are related by the expression.

$$X = \frac{W}{1 - W} \qquad [50]$$

Water may become *bound* in a solid by retention in capillaries, solution in cellular structures, solution with the solid, or chemical or physical adsorption on the surface of the solid. *Unbound moisture* in a hygroscopic material is the moisture in excess of the equilibrium moisture content corresponding to saturation humidity. All the moisture content of a nonhygroscopic material is unbound moisture. *Free moisture content* is the moisture content removable at a given temperature and may include both bound and unbound moisture.

In the immediate vicinity of the interface between free water and vapor, the vapor pressure at equilibrium is the saturated vapor pressure. Very moist products have a vapor pressure at the interface almost equal to the saturation vapor pressure. If the concentration of solids is increased by the removal of water, then the dissolved hygroscopic solids produce a fall in the vapor pressure due to osmotic forces. Further removal of water finally results in the surface of the product being dried. Water now exists only in the interior in very small capillaries, between small particles, between large molecules, and bound to the molecules themselves. This binding produces a considerable lowering of vapor pressure. Such a product can therefore be in equilibrium only with an external atmosphere in which the vapor pressure is considerably decreased.

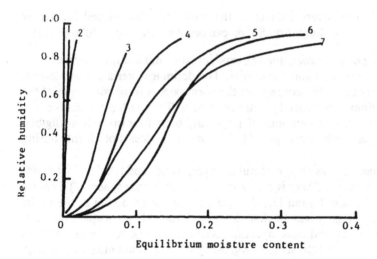

Figure 6   Typical equilibrium moisture isotherms at room temperature for selected substances: (1) asbestos fiber, (2) PVC (50°C), (3) wood charcoal, (4) Kraft paper, (5) jute, (6) wheat, (7) potatoes.

3.2.   Moisture Isotherms (10)

A dry product is called *hygroscopic* if it is able to bind water with a simultaneous lowering of vapor pressure. Different products vary widely in their hygroscopic properties. The reason for this is their molecular structure, their solubility, and the extent of reactive surface.

Sorption isotherms measured experimentally under isothermal conditions are used to describe the hygroscopic properties of a product. A graph is constructed in which the moisture bound by sorption per unit weight is plotted against relative humidity, and vice versa. Such isotherms are shown in Figures 6 and 7. From Figure 7 it is seen that molecular sieves are highly hygroscopic but polyvinyl chloride (PVC) powder is mildly hygroscopic. Potatoes and milk exhibit intermediate hygroscopicity.

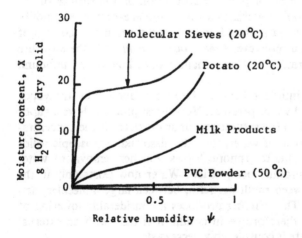

Figure 7   Shapes of sorption isotherms for materials of varying hygroscopicity.

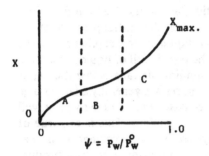

Figure 8 A typical isotherm (see text for explanation of areas within dashed lines).

Figure 8 shows the shape of the sorption isotherm characteristic of many dry food products. If the partial pressure of the external atmosphere P_W is nearly zero, then the equilibrium moisture inside the dry product will also be almost zero. Section A of the curve represents a region in which the monomolecular layers are formed, although there may be multimolecular layers in some places toward the end of A. Section B is a transitional region in which double and multiple layers are mainly formed. Capillary condensation could also have taken place. In section C the slope of the curve increases again, which is attributed mainly to increasing capillary condensation and swelling. The maximum hygroscopicity X_{max} is achieved when the solid is in equilibrium with air saturated with moisture ($\psi = 1$).

Sorption-Desorption Hysteresis

The equilibrium moisture content of a product may be different depending on whether the product is being wetted (sorption or absorption) or dried (desorption) (Fig. 9). These differences are observed to varying degrees in almost all hygroscopic products.

One of the hypotheses used to explain hysteresis is to consider a pore connected to its

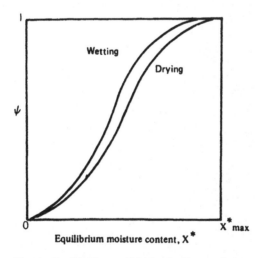

Figure 9 Wetting and drying isotherms for a typical hygroscopic solid.

surroundings by a small capillary (10). During absorption, as the relative humidity rises, the capillary begins to fill while the pore is empty. Only when the partial pressure of the vapor in air is greater than the vapor pressure of the liquid in the capillary will the moisture move into the pore. Starting from saturation the pore is full of liquid. This fluid can only escape when the partial pressure of the surrounding air falls below the vapor pressure of the liquid in the capillary. Since the system of pores has generally a large range of capillary diameters, it follows that differences between adsorption and desorption will be observed. This theory assumes that the pore is a rigid structure. This is not true of foods or synthetic materials, although these show hysteresis. The explanation is that contraction and swelling are superimposed on the drying and wetting processes, producing states of tension in the interior of the products and leading to varying equilibrium moisture contents depending on whether desorption or absorption is in progress.

Temperature Variations and Enthalpy of Binding

Moisture isotherms pertain to a particular temperature. However, the variation in equilibrium moisture content for small changes of temperature ($< 10°C$) is neglected (3). To a first approximation, the temperature coefficient of the equilibrium moisture content is proportional to the moisture content at a given relative humidity.

$$\left(\frac{\partial X^*}{\partial T}\right)_\psi = -AX^* \tag{51}$$

The coefficient A lies between 0.005 and 0.01 K^{-1} for relative humidities between 0.1 and 0.9 for such materials as natural and synthetic fibers, wood, and potatoes. A could be taken to increase linearly with ψ. So for $\psi = 0.5$ there is a 0.75% fall in moisture content for each degree Kelvin rise in temperature. The extent of absorption-desorption hysteresis becomes smaller with increasing temperature.

Figure 10 shows moisture isotherms at various temperatures. The binding forces decrease with increasing temperature; that is, less moisture is absorbed at higher temperatures at the same relative humidity. Kessler (10) has shown that the slope of a plot of

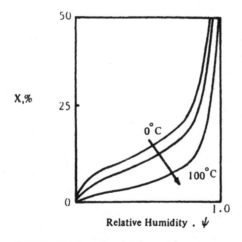

Figure 10 Sorption isotherms for potatoes.

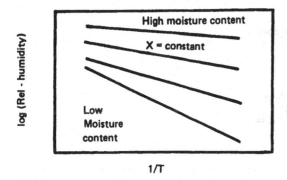

Figure 11 Determination of the heat of sorption from sorption isotherms.

$\ln (P_W/P_W^0)$ versus $1/T$ at constant X (Fig. 11) gives the enthalpy of binding. The variation of enthalpy of binding versus moisture content is shown in Figure 12. From the figure it is seen that in the region where monomolecular layers are formed, enthalpies of binding are very high.

3.3. Determination of Sorption Isotherms (10)

The sorption isotherms are established experimentally starting mostly with dry products. The initial humidity of the air with which the product is in equilibrium should be brought to extremely low values using either concentrated sulfuric acid or phosphorus pentoxide, so that the moisture content of the product is close to zero at the beginning. The product is then exposed to successively greater humidities in a thermostatically controlled atmosphere. Sufficient time must be allowed for equilibrium between the air and solid to be attained. Using thin slices of the product, moving air and especially vacuum help to establish equilibrium quickly. This is especially important for foodstuffs: there is always the danger of spoilage. There are severe problems associated with the maintenance of

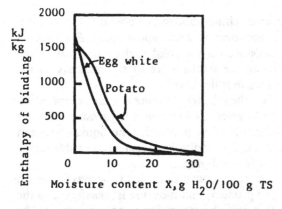

Figure 12 Enthalpy of sorption as a function of the hygroscopic moisture content. (Egg white data by Nemitz; potato data by Krischer.)

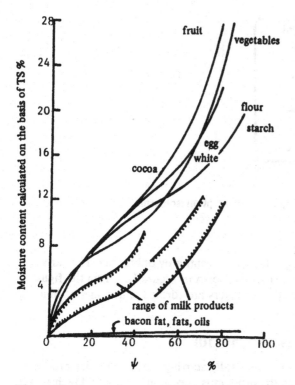

Figure 13   Range of sorption isotherms of various foods at room temperature.

constant humidity and temperature. These problems could be alleviated by using sulfuric acid-water mixtures and saturated salt solutions to obtain different relative humidities (10,13). Figure 13 depicts the absorption isotherms of a range of food products. Further information on solid moisture characteristics, enthalpy of wetting, and sorption isotherms are available in References 3 and 10.

4. MECHANISM OF DRYING

As mentioned above, moisture in a solid may be either unbound or bound. There are two methods of removing unbound moisture: evaporation and vaporization. *Evaporation* occurs when the vapor pressure of the moisture on the solid surface is equal to the atmospheric pressure. This is done by raising the temperature of the moisture to the boiling point. This kind of phenomenon occurs in roller dryers.

If the material being dried is heat sensitive, then the temperature at which evaporation occurs, that is, the boiling point, could be lowered by lowering the pressure (vacuum evaporation). If the pressure is lowered below the triple point, then no liquid phase can exist and the moisture in the product is frozen. The addition of heat causes sublimation of ice directly to water vapor as in the case of freeze drying.

Second, in *vaporization*, drying is carried out by convection, that is, by passing warm air over the product. The air is cooled by the product, and moisture is transferred to the air by the product and carried away. In this case the saturation vapor pressure of the moisture over the solid is less than the atmospheric pressure.

A preliminary necessity to the selection of a suitable type of dryer and design and sizing thereof is the determination of the drying characteristics. Information also required are the solid handling characteristics, solid moisture equilibrium, and material sensitivity to temperature, together with the limits of temperature attainable with the particular heat source. These will be considered later and in other sections of this book.

The drying behavior of solids can be characterized by measuring the moisture content loss as a function of time. The methods used are humidity difference, continuous weighing, and intermittent weighing. Descriptions of these methods are available in References 3 and 13.

Figure 14 qualitatively depicts a typical drying rate curve of a hygroscopic product. Products that contain water behave differently on drying according to their moisture content. During the first stage of drying the drying rate is constant. The surface contains free moisture. Vaporization takes place from there, and some shrinkage might occur as the moisture surface is drawn back toward the solid surface. In this stage of drying the rate-controlling step is the diffusion of the water vapor across the air-moisture interface and the rate at which the surface for diffusion is removed. Toward the end of the constant rate period, moisture has to be transported from the inside of the solid to the surface by capillary forces and the drying rate may still be constant. When the average moisture content has reached the *critical moisture content* X_c, the surface film of moisture has been so reduced by evaporation that further drying causes dry spots to appear upon the surface. Since, however, the rate is computed with respect to the overall solid surface area, the drying rate falls even though the rate per unit wet solid surface area remains constant. This gives rise to the second drying stage or the first part of the falling rate period, the period of unsaturated surface drying. This stage proceeds until the surface film of liquid is entirely evaporated. This part of the curve may be missing entirely, or it may constitute the whole falling rate period.

On further drying (the second falling rate period or the third drying stage), the rate at which moisture may move through the solid as a result of concentration gradients between the deeper parts and the surface is the controlling step. The heat transmission now

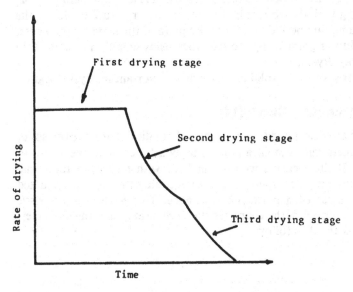

Figure 14 Typical rate-of-drying curve, constant drying conditions.

consists of heat transfer to the surface and heat conduction in the product. Since the average depth of the moisture level increases progressively and the heat conductivity of the dry external zones is very small, the drying rate is increasingly influenced by the heat conduction. However, if the dry product has a relatively high bulk density and a small cavity volume with very small pores, drying is determined not so much by heat conduction but by a rather high resistance to diffusion within the product. The drying rate is controlled by diffusion of moisture from the inside to the surface and then mass transfer from the surface. During this stage some of the moisture bound by sorption is being removed. As the moisture concentration is lowered by the drying, the rate of internal movement of moisture decreases. The rate of drying falls even more rapidly than before and continues until the moisture content falls down to the equilibrium value X^* for the prevailing air humidity and then drying stops. The transition from one drying stage to another is not sharp, as indicated in Figure 14.

In actual practice, the original feedstock may have a high moisture content and the product may be required to have a high residual moisture content so that all the drying may occur in the constant rate period. In most cases however both phenomena exist, and for slow-drying materials most of the drying may occur in the falling rate period. As mentioned earlier, in the constant rate period the rate of drying is determined by the rate of evaporation. When all the exposed surface of the solid ceases to be wetted, vapor movement by diffusion and capillarity from within the solid to the surface are the rate-controlling steps. Whenever considerable shrinkage occurs, as in the drying of timber, pressure gradients are set up within the solid and these may assume importance. In this case, as in the case of materials that "caseharden," that is, form a hard impermeable skin, it is essential to retard evaporation and bring it in step with the rate of moisture movement from the interior. This could be achieved by increasing the relative humidity of the drying air. With solids, in which the initial moisture content is relatively low and the final moisture content required is extremely low, the falling rate period becomes important. Dryness times are long. Air velocities will be important only to the extent to which they enhance heat transfer rates. Air temperature, humidity, material thickness, and bed depth all become important. When the rate of diffusion is the controlling factor, particularly when long drying periods are required to attain low moisture contents, the rate of drying during the falling rate period varies as the square of the material thickness, which indicates the desirability of granulating the feedstock using agitation or using thin layers in case of crossflow tray dryers.

Thus the drying characteristics of the solid are extremely important in dryer design.

4.1. Characteristic Drying Rate Curve (14)

When the drying rate curves are determined over a range of conditions for a given solid, the curves appear to be geometrically similar and are simply a function of the extent to which drying has occurred. If these curves were normalized with respect to the initial drying rate and average moisture content, then all the curves could often be approximated to a single curve, "characteristic" of a particular substance. This is the *characteristic drying curve*. The normalized variables, the characteristic drying rate f and the characteristic moisture content ϕ, are defined as follows:

$$f = \frac{N_v}{N_w}$$

and

$$\phi = \frac{\overline{X} - X^*}{X_{cr} - X^*}$$

where N_v is the rate of drying for a unit surface, N_W is the rate when the body is fully saturated or the initial drying rate, $\overline{X}$ is the average moisture content in the body, and $\overline{X}_{cr}$ is the corresponding critical point value. X^* is the equilibrium moisture content.

If a solid's drying behavior is to be described by the characteristic curve, then its properties must satisfy the following two criteria:

1. The critical moisture content $\overline{X}_{cr}$ is invariant and independent of initial moisture content and external conditions.
2. All drying curves for a specific substance are geometrically similar so that the shape of the curve is unique and independent of external conditions.

These criteria are restrictive, and it is quite unlikely that any solid will satisfy them over an exhaustive range of conditions; nevertheless, the concept is widely used and often utilized for interpolation and prediction of dryer performance (3,17). The use of the mean moisture content as an index of the degree of drying contains the implicit assumption that the extent of drying at a mean moisture content will also depend on the relative extensiveness of the exposed surface per unit volume of material. Thus, similar drying behavior may be expected only in the case of materials that are unchanged in form. A typical characteristic drying curve is shown in Figure 15.

Further information on the characteristic drying curve, extrapolation procedures used, and the theoretical developments in examining the range of validity of the characteristic drying rate model are available (3,13,25). The various types of characteristic drying curves have been depicted schematically in Figure 16.

5. CLASSIFICATION AND SELECTION OF DRYERS

With a very few exceptions, most products from today's industry undergo drying at some stage or another. A product must be suitable for either subsequent processing or sale. Materials need to have a particular moisture content for processing, molding, or pelleting. Powders must be dried to suitable low moisture contents for satisfactory packaging. Whenever products must be heated to high temperatures, as in ceramic and metallurgical processes, predrying at lower temperatures ahead of firing kilns is advantageous for energy savings. Cost of transport (as in the case of coal) depends on the moisture content of the product, and a balance must be struck between the cost of conveying and the cost of drying. Excessive drying is wasteful; not only is more heat, that is, expense, involved than is necessary, but often overdrying results in a degraded product, as in the case of paper and timber. Consideration must be given to methods involved in energy savings in dryers (15). Examples of products and the types of precautions that need to be taken during drying are highlighted. Thermal drying is an essential stage in the manufacture of colors and dyes. Many inorganic colors and most organic dyes are heat sensitive; the time-temperature effect in drying may be critical in arriving at a correct shade of color or in the elimination of thermal degradation. Drying is normally accomplished at low temperatures and in the absence of air. The most widely used dryer is the recirculation type truck and tray compartment dryer. Most pharmaceuticals and fine chemicals require drying before packaging. Large turbo-tray and through circulation dryers are employed. Excessively heat-sensitive products, such as antibiotics and blood plasma, call for special treatment, like freeze drying or high vacuum tray drying. Continuous rotary dryers are usually used to handle large tonnages of natural ores, minerals, and heavy chemicals.

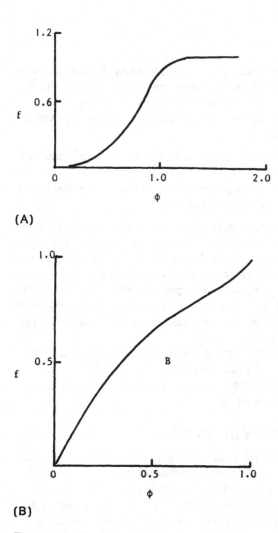

(A)

(B)

Figure 15 Experimental drying rates for (A) molecular sieves 13-X particles with $X_0 = 0.27$, diameter = 2.2 mm, air velocity = 4.4 m·s^{-1}, $T_G \cong 36.5$–97°C; (B) silica-gel particles with diameter = 3.0 mm, air velocity = 1 m·s^{-1}, T_G = 54–68°C, T_W = 25–29°C.

The largest demand for drying equipment is for the continuous drying of paper, which is done on cylinder or "can" dryers. The temperature and humidity conditions are important to the consistency of the paper. Thermal drying is essential in the foodstuffs and agricultural fields. Spray drying and freeze drying are also widely used. In the ceramic industry, drying is a vital operation. Great care must be exercised because of the considerable shrinkage that occurs in drying. Thus control of humidity is important. Drying is also widely used in the textile industry. The need for product quality puts grave constraints on the dryer chosen and dryer operation (13). Quality depends on the end use of the product. For many bulk chemicals, handling considerations determine moisture content requirements. For foodstuffs, flavor retention, palatability, and rehydration properties are important. Timber must retain its strength and decorative properties after drying.

Drying of Solids

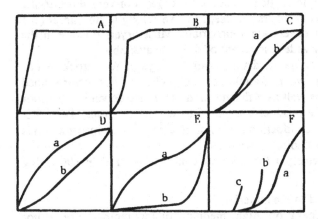

Figure 16 Examples of normalized drying rate curves for different types of media: (A) sand, clay, silica gel, paper pulp, leather; (B) sand, plastic-clay mix, silica-brick mix, ceramic plate, leather; (C) [a] fir wood and [b] cyprus wood; (D) [a] paper, wool, and [b] potatoes, tapioca tuber, rice flour; (E) [a] rye bread, yeast, and [b] butter and margarine; (F) [a] wheat corns, [b and c] represent curves at lower values of initial moisture content.

The choice of end moisture content is largely dictated by storage and stability requirements. The end moisture content determines the drying time and conditions required for drying. Overdrying should be avoided. Internal moisture gradients within particles and interparticle moisture content variation are important. Temperature restrictions may arise because of degradation, phase changes, discoloration and staining, flammability of dust, and other factors. Thermal sensitivity fixes the maximum temperature to which the substance can be exposed for the drying time. In spray and pneumatic dryers, the retention time of a few seconds permits drying heat-sensitive materials at higher temperatures. Many hygroscopic materials shrink on drying. The extent of shrinkage is linearly related to the moisture content change below the hygroscopic limit. Case hardening and tensile cracking pose problems. Details on how some of these problems are overcome by a compromise between energy efficiency and drying time are available (13).

The first subdivision is based on methods of heat transfer, namely, (a) conduction heating, (b) convection heating, (c) radiant heating, and (d) dielectric heating. Freeze drying is classified as a special case of conduction heating. The next subdivision is the type of drying vessel: tray, rotating drum, fluidized bed, pneumatic, or spray. Dryer classification based on physical form of the feed has also been done. These are indicated in Reference 16.

Sloan (18) has identified some 20 types of dryers and classified them according to whether they are batch or continuous, conduct heat exchange through direct contact with gases or by heat exchange through vessel walls, and according to the motion of the equipment. Such a classification, though helpful in distinguishing and describing discrete systems, does not go far in relating the discrete systems to the process problems they are supposed to handle. McCormick (19) has tried to tackle the problem from the user's point of view. A total of 19 types of dryers were classified according to how well they handle different materials.

Schlünder (23) has given a classification of dryers that encompasses the physical state of the product as well as the dwell time of the product in the dryer. For very short drying times (<1 min), flash, spray, or drum dryers are used. For very long drying times (>1 h), only tunnel, truck, or conveyor dryers are appropriate. Most dryers operate in the intermediate range, for which a very wide assortment of dryers is available.

Table 1 gives a summary of the type of dryer versus the type of feed stock, which may be a slurry, paste, filter cake, powder, granules, crystal, pellet, or fibrous or shaped material. Since thermal sensitivity as well as efficiency and dryer size depend to a major extent on the thermal conditions the product is exposed to within the dryer, Table 2 is presented to classify convection and conduction dryers on this basis. Such information is often helpful in narrowing down the choice of dryers.

Keey (3,25) has noted three principal factors that could be utilized in classifying dryers:

1. Manner in which heat is supplied to the material
2. Temperature and pressure of operation (high, medium, or low temperature; atmospheric or vacuum drying)
3. Manner in which the material is handled within the dryer

Further subclassification is of course possible but generally unnecessary. For example, a given dryer may be operated in batchwise or continuous mode.

5.1. Heating Methods

Convection

Convection is possibly the most common mode of drying particulate or sheet-form or pasty solids. Heat is supplied by heated air or gas flowing over the surface of the solid. Heat for evaporation is supplied by convection to the exposed surface of the material and the evaporated moisture carried away by the drying medium. Air (most common), inert gas (such as N_2 for drying solids wet with organic solvent), direct combustion gases, or superheated steam (or solvent vapor) can be used in convective drying systems.

Such dryers are also called *direct* dryers. In the initial constant rate drying period (drying in which surface moisture is removed), the solid surface takes on the wet bulb temperature corresponding to the air temperature and humidity conditions at the same location. In the falling rate period the solids' temperature approaches the dry bulb temperature of the medium. These factors must be considered when drying heat-sensitive solids.

When drying with superheated vapors, the solids' temperature corresponds to the saturation temperature at the operating pressure, for example, 100°C for steam at 1 atm. For solids susceptible to oxidation or denaturation, for example, in the presence of oxygen, the heat sensitivity can be quite different in a steam environment. Product quality may differ as well. This can only be ascertained via laboratory tests.

Examples of convective (direct) dryers are air suspension dryers, such as fluid bed, flash, rotary, or spray dryers; air impingement dryers for paper or pulp; packed bed or through dryers; and conveyor-truck-tunnel dryers.

Conduction

Conduction or indirect dryers are more appropriate for thin products or for very wet solids. Heat for evaporation is supplied through heated surfaces (stationary or moving) placed within the dryer to support, convey, or confine the solids. The evaporated moisture is carried away by vacuum operation or by a stream of gas that is mainly a carrier of

Table 1 Dryer Selection versus Feedstock Form

Nature of feed	Liquids			Cakes			Free-flowing solids				
	Solution	Slurry	Pastes	Centrifuge	Filter	Powder	Granule	Fragile crystal	Pellet	Fiber	Formed solids
Convection Dryers											
Belt conveyer dryer								X	X	X	X
Flash dryer	X			X	X	X	X			X	
Fluid bed dryer		X		X	X	X	X		X		
Rotary dryer				X	X	X	X		X	X	
Spray dryer	X	X	X								X
Tray dryer (batch)				X	X	X	X	X	X	X	
Tray dryer (continuous)				X	X	X	X	X	X	X	
Conduction Dryers											
Drum dryer	X	X	X								
Steam jacket rotary dryer				X	X	X	X		X	X	
Steam tube rotary dryer				X	X	X	X	X	X	X	
Tray dryer (batch)				X	X	X	X	X	X	X	X
Tray dryer (continuous)				X	X	X	X		X	X	

Table 2 Solids' Exposures to Heat Conditions

Dryers	Typical residence time within dryer				
	0–10 (sec)	10–30 (sec)	5–10 (min)	10–60 (min)	1–6 (h)
Convection					
Belt conveyor dryer				X	
Flash dryer	X				
Fluid bed dryer				X	
Rotary dryer				X	
Spray dryer		X			
Tray dryer (batch)					X
Tray dryer (continuous)				X	
Conduction					
Drum dryer		X			
Steam jacket rotary dryer				X	
Steam tube rotary dryer				X	
Tray dryer (batch)					X
Tray dryer (continuous)				X	

moisture. Vacuum operation is recommended for heat-sensitive solids. Because the enthalpy lost with the drying air in convective dryers is large, their thermal efficiency tends to be low. For conduction dryers the thermal efficiency is higher. Paddle dryers for drying of pastes, rotary dryers with internal steam tubes, and drum dryers for drying thin slurries are examples of indirect dryers.

A more efficient dryer can be designed for some operations that combines advantages of both direct and indirect heating, for example, a fluid bed dryer with immersed heating tubes or coils for drying of heat-sensitive polymer or resin pellets. Such a dryer can be only one-third the size of a purely convective fluid bed dryer for the same duty.

It is noteworthy that sometimes one can operate the same apparatus in direct, indirect, or combined modes. For example, a vibrated fluid bed dryer can be purely convective (e.g., drying of tea), purely conductive (e.g., vacuum drying of pharmaceutical granules), or combined direct-indirect (e.g., drying of pulverized coal with immersed heating tubes). The drying medium could be steam for such products as coal.

Radiation

Various sources of electromagnetic radiation with wavelengths ranging from the solar spectrum to microwave (0.2 m to 0.2 μm). Solar radiation barely penetrates beyond the skin of the material, which absorbs only a part of the incident radiation depending on its wavelength. Infrared radiation is often used in drying coatings, thin sheets, and films, for example (4–8 μm band). Although most moist materials are poor conductors of 50–60 Hz current, the impedance falls dramatically at radiofrequencies (RF); such radiation can be used to heat the solid volumetrically, thus reducing internal resistance to heat transfer. Energy is absorbed selectively by the water molecules: as the product gets drier less energy is used. The capital as well as operating costs are high, and so these techniques are useful for drying high unit value products or for final correction of moisture profile wherein only small quantities of hard-to-get moisture are removed, as in moisture profiling of paper using RF heating.

Combined mode drying with convection (e.g., infrared plus air jets or microwave with impingement for drying of sheet-form foodstuffs) is also commercially feasible.

5.2. Temperature and Pressure of Operation

Most dryers are operated at near atmospheric pressures. A slight positive pressure avoids in-leakage from outside, which may not be permissible in some cases. If no leakage is permitted to the outside, then a slight negative pressure is used.

Vacuum operation is expensive and is recommended only if the product must be dried at low temperatures or in the absence of oxygen or has flavors that are generated at medium- or high-temperature operation. High-temperature operation tends to be more efficient since lower gas flow rates and smaller equipment may be used for a given evaporation duty. Availability of low-temperature waste heat or energy from solar collectors may dictate the choice of a lower temperature operation. These dryers will then be large in size.

Freeze drying is a special case of drying under vacuum at a temperature below the triple point of water; here water (ice) sublimes directly into water vapor. Although the heat required for sublimation is severalfold lower than for evaporation, vacuum operation is expensive. Freeze drying of coffee, for example, costs two to three times more than spray drying. On the other hand, the product quality and flavor retention are better.

5.3. Conveying of Material in Dryer

Handling the material to be dried is of course one of the key considerations in dryer selection. This is best illustrated in Table 3. In some cases the material may be treated or preformed to make it suitable for handling in a particular dryer, for example, reslurrying of filter cake to make it pumpable for atomization and spray drying or pelletizing pasty materials. This, of course, costs extra and must be considered in an overall evaluation. The final product characteristics and quality requirements also govern the choice of dryer. In some cases a combination of two or more different types of dryers may be the optimal

Table 3 Capacity and Energy Consumption for Selected Dryers

Method	Typical dryer	Typical materials
Material not conveyed	Tray dryer	Wide range of pastes, granules
Material falls by gravity	Rotary dryer	Free-flowing granules
Material conveyed mechanically	Screw-conveyor, paddle	Wet sludges, pastes
Transported on trucks	Tunnel dryer	Wide range of materials
Sheet-form materials, supported on rolls	Cylinder dryers	Paper, textiles, pulp
Conveyed on bands	Band, conveyor dryer	Wide range of solids (pellets, grains)
Material suspended in air	Fluid bed, flash	Free-flowing granules
Slurries or solutions atomized in air	Spray dryer	Milk, coffee, etc.

Note: Most dryers may operate continuously, semicontinuously, or batchwise. Labor costs are high for tray and tunnel dryers.

strategy if the product-handling properties change significantly as it dries or if its heat sensitivity changes during the process of drying.

It should be pointed out that many new dryers cannot conveniently fit the classification suggested earlier. For example, a pulsed combustion dryer for pasty solids or waste sludge, the Remaflam process for drying textiles by controlled combustion of solvent (alcohol) on the wet fabric itself, and vibrated bed drying of pastes cannot be placed under any one single category of dryers. Such a "coarse" classification is still of interest in that it allows one to "home in" on a limited number of possible dryers, which can then be evaluated in depth. One must look very carefully into some of the newer and novel dryers although they are not even mentioned in most textbooks or handbooks (22). Many of them have the potential to supplant some of the age-old drying technologies, at least in some industrial applications.

Dittman (20) has proposed a structured classification of dryers according to two general classes and five subclasses. The general classes are adiabatic or nonadiabatic dryers. Adiabatic dryers are further subclassified according to whether drying gases pass through the material (for permeable solids or beds of solids) or across the surface. Nonadiabatic dryers are categorized according to the mode of heat supply, such as heat applied through a heat exchange surface or direct radiation, and according to the mode of moisture carryover, for example, moisture removal by vacuum or by a carrier gas.

Several difficulties are encountered in the selection of dryers. These mainly arise because there is no standard set of systematized laboratory tests using standardized apparatus to provide key data on the drying characteristics of materials. The real mechanics of liquid removal from the solid is not really understood, nor is the operation of many dryers. A systematic comprehensive classification of existing dryers has yet to be agreed upon. There is also a lack of a reliable procedure for scaling up laboratory data and even pilot plant data for some types of dryers.

In spite of the above-mentioned lacunae, dryers have still to be selected and some prior information is required to facilitate this job. This includes (a) flowsheet quantities, such as dry solid quantity, total liquid to be removed, and the source of the wet material; (b) batch or continuous feed physical characteristics, such as source of feed, presence of any previous dewatering stage, like filtration, mechanical pressing, or centrifuging, method of supplying material to the dryer, particle size distribution in the wet feed, physical characteristics and handleability, and abrasive properties of wet and dry materials; (c) feed chemical properties, such as toxicity, odor problems, whether the material can be dried with hot combustion gases containing carbon dioxide, sulfur dioxide, some nitrogen oxides, and traces of partially burnt hydrocarbon, fire and explosion hazards, temperature limitations, temperatures of relevant phase changes, and corrosive properties; (d) dry product specification and properties, such as moisture content, removal of solvent odor, particle size distribution, bulk density, maximum percentage of impurities, desired granular or crystalline form, flow properties, and temperature to which the dried product must be cooled before storage; and (e) drying data obtained from a pilot plant or laboratory as well as previous experience of the drying performance of similar materials in a full-scale plant. Information on solvent recovery, product loss, and site conditions would be an added bonus.

The best method of selection involves using past experience. One of the preliminary ways of selecting dryers is based on the nature of the feed (16). There is little difficulty in handling liquid feeds, and the choice of the equipment is normally limited to (a) spray dryer, (b) drum dryer, atmospheric or vacuum, and (c) agitated batch vacuum dryer. Other considerations that might influence the final choice are the need for small product losses and a clean plant, solvent recovery or the need to use an inert atmosphere in which an agitated vacuum dryer is preferred, and the temperature sensitivity of the material. The agitated vacuum dryer has a long residence time, the through circulation dryer, a moderate temperature and moderate residence time. The drum dryer can have a high mean temperature with a short contact time; the spray dryer has a short contact time with a wide range of operating temperatures. The above-mentioned selection is applicable to pumpable suspensions of fine solids, excluding pastes.

For the continuous drying of pastes and sludges, in which the solids are in a finely divided state, dust problems are a major consideration. However, the choice between batch and continuous operation is difficult. The batch dryers normally used are tray atmospheric or vacuum, agitated batch atmospheric or vacuum, and rotary atmospheric or vacuum. The vacuum operation is preferable in cases of solvent recovery, fire, or toxic hazards or when temperature limitations are necessary. Dryers used in continuous operation are (a) spray, where atomization itself poses a considerable problem; (b) fluidized bed, where dispersion of the feed in a deep bed is difficult; (c) continuous band circulation, suitable if dust-free product is required; (d) pneumatic, requires mixing of feed with dry product to facilitate dispersion of wet solid in the gas entering the dryer; and (e) continuous rotary, direct or indirect; here too blending of wet feed with dry product is necessary to facilitate handling. If the feed contains fine particles, the indirect mode of heat transfer is normally preferred. In case of free-blowing wet powders (particle size <300 μm) all the dryers used for pastes and sludges could be used with the inclusion of the vertical rotating shelf dryer. For granular crystalline solids with particle sizes >300 μm, the direct rotary dryer is commonly used. With this type of dryer crystal breakage is a problem that can be overcome by proper flight design. For particles larger than 25 mesh, a through circulation dryer using a moving band or a vibrating screen could be used. Fibrous solids hold a considerable amount of water but dry quite easily. These materials are often temperature sensitive due to their high specific surface, and care should be taken to keep the air temperature down. Through circulation tests at various temperatures should establish the need, if any, to avoid overheating. Apart from this, fibrous materials could be treated in a similar way to any of the solids mentioned above.

The final selection of the dryer will usually represent a compromise between the total cost, operating cost, quality of the product, safety consideration, and convenience of installation. It is always wise, in case of uncertainty, to run preliminary tests to ascertain both design and operating data and also the suitability of the dryer for the particular operation. On certain types of dryers full-scale tests are the only way of establishing reliable design and operating data, but in some cases, such as through circulation dryers, there are reliable techniques for establishing reliable data from laboratory tests. Details on test techniques and as to when tests are needed are available in References 13 and 16. It should be noted that specialized drying techniques, such as vacuum freeze drying, microwave freeze drying, and dielectric or infrared heating or drying, have not been considered here. Details of some of these specialized techniques are given elsewhere in this handbook.

6. EFFECT OF ENERGY COSTS, SAFETY, AND ENVIRONMENTAL FACTORS ON DRYER SELECTION

Escalating energy costs and increasingly stringent legislation on pollution, working conditions, and safety have a direct bearing on the design as well as selection of industrial dryers. Lang (27) has discussed the effects of these factors on design of suspension dryers for particulates, such as spray, flash, and fluidized bed dryers. These factors must be considered during the phase of selection of dryers. In some cases a choice between competing drying systems exists; in others one must incorporate these factors at the design stage.

For a given dryer system (including preprocessing, such as mechanical dewatering, centrifugation, evaporation, and pressing, and postprocessing, such as product collection, cooling, agglomeration, granulation, and scrubbing), in general several energy-saving flowsheets may be devised, including gas recycle, closed cycle operation, self-inertization, multistage drying, and exhaust incineration. Areas of conflict may exist between legal requirements, hygienic operation, and energy efficiency. Lang gives the following possible scenarios of conflict:

1. Explosion vents could be a hygiene problem.
2. Dust in recycling streams fouls heat exchanger surfaces or causes difficulties in direct combustion systems.
3. Thermal expansion joints or fire-extinguishing equipment can cause product buildup and hence a fire hazard.
4. High product collection efficiency for particulate dryers means high pressure drop and increased fan noise.

Note that unnecessarily stringent product specifications can cause significant increase in dryer costs, both capital and operating.

In selecting energy-saving drying systems, it is important to note the following (mainly for particulate drying, but some factors are of general applicability):

1. When handling a thermally sensitive product, recycled exhaust must be totally free of product if the stream is to pass through or near a burner.
2. Recycling increases humidity level in drying, which may increase the equilibrium moisture content to unacceptable levels in some cases.
3. To avoid passing dust in recycled gas through air heaters, if fresh makeup air is heated and mixed with recycled gas, to obtain high mixture temperature (say, 400°C), the fresh gas must be heated to a temperature too high for simple materials of construction. Exotic metals or refractories are needed, which can cause product contamination or a source of ignition if it reaches high enough temperatures.
4. In multiple-stage drying, heat economy requires that the first-stage drying give a partially dried product, which is sometimes too sticky to handle.

A drying installation may cause air pollution by emission of dust and gases. Even plumes of clean water vapor are unacceptable in some areas. Particulates below the range 20–50 mg/nm^3 of exhaust air are a common requirement. High-efficiency dust collection is essential. It is important to operate the dryer under conditions conducive to production of coarse product. On the other hand, larger products take longer to dry. Cyclones, bag filters, scrubbers, and electrostatic precipitators are commonly used for particle collection and gas cleaning on particulate material drying of materials in other forms, such as pulp sheets.

For removal of noxious gaseous pollutants, one may resort to absorption, adsorption, or incineration; the last operation is becoming more common.

Although rare, care must be taken in drying of airborne materials that can catch fire. Reduction in oxygen content (by recycle) can suppress explosion hazard. Should explosion occur, suitable explosion vents must be included to avoid buildup of excessive pressure in the system. Elimination of sources of ignition is not acceptable as adequate assurance against fire or explosion hazard. When an explosion risk exists, buildup of product in the dryer or collector must be avoided. For example, venting doors on the roof of large-volume spray dryers must be flush with the interior surface to prevent any buildup of product. It is often cheaper to design a partially or fully inertized drying system than to design larger dryer chambers to withstand high internal pressures (5–6 psig).

Finally, local legislation about noise levels must be considered at the selection and design state. Depending on the stringency of noise requirements, the cost of acoustic treatment can be as much as 20% of the total system cost. For air suspension dryers, the fan is the main noise generator. Other sources, such as pumps, gearboxes, compressors, atomization equipment, burners, and mixers, also contribute to noise. Low fan noise requires a low pressure drop in the system, which is in conflict with the high pressure drop required for higher collection efficiency.

A simple direct-fired dryer with once through air flow heating, especially with gas, can be achieved with little noise generation. On the other hand, a more efficient air recirculation unit often requires high-noise burners and ancillaries.

The aforementioned discussion is intended to provide the specifier of drying equipment some practical factors that should be considered at the stage of selection of the dryer system. Rarely, if ever, is it possible to select a dryer that meets all criteria. In most cases, however, one can modify the dryer system design or operation to meet all the essential specifications of the user.

Menon and Mujumdar (15) have indicated some energy-saving measures involved in conditioning of the feed, dryer design, and heat recovery from the exhaust stream, including the use of heat pumps. Some novel techniques, like displacement drying, steam drying, drying using superheated steam, radiofrequency drying, press drying, and combined impingement and through drying, have been proposed.

Although prior experience is a guide commonly used in specifying dryers, it is important to recognize that earlier dryers were often specified in times when energy costs were minimal and requirements of product quality and/or production rates were different. There is also a variation in energy costs from one geographic location to another and certainly from one country to another. It is strongly recommended that the process engineer make a selection of dryer on the basis of current conditions and geographic location while taking into account future expected trends. In many instances the most widely used dryers for a specific product have been found to be poor choices under prevailing conditions.

Table 4 gives a summary of typical drying capacities and energy consumption in existing common industrial dryers.

7. DESIGN OF DRYERS

The engineer concerned with process design has to choose for a given dryer those conditions that enable the specified properties of the product to be obtained. The performance characteristics of alternative systems should also be assessed before the final choice of a dryer type. Almost always some small-scale tests are needed to determine the material's drying characteristics required to predict the way in which the raw material would behave in the actual unit. A flowchart illustrating the various steps needed to design a dryer are shown in Reference 3.

For a preliminary estimate of dryer size, Toei (24) gives a simple method based on

Table 4 Capacity and Energy Consumption for Selected Dryers

Dryer type	Typical evaporation capacity (kg H_2O/h·m^2 or kg H_2O/h·m^3)	Typical energy consumption (kJ/kg of H_2O evaporated)
Tunnel dryer	–	5500–6000
Band dryer	–	4000–6000
Impingement dryer	50 m^{-2}	5000–7000
Rotary dryer	30–80 m^{-3}	4600–9200
Fluid bed dryer		4000–6000
Flash dryer	5–100 m^{-3}	4500–9000
	(depends on particle size)	
Spray dryer	1–30 m^{-3}	4500–11,500
Drum dryer (for pastes)	6–20 m^{-2}	3200–6500

Note: Figures are only approximate and are based on current practice. Better results can often be obtained by optimizing operating conditions and using advanced technology to modify the earlier designs.

data obtained from operating industrial dryers. For convection dryers, the rate of heat transfer (kcal/h) is given by

$$q = (ha)(V)(t - t_m) \qquad \text{for batch type}$$

and

$$q = (ha)(V)(t - t_m)_{l_m} \qquad \text{for continuous type}$$

For conduction dryers,

$$q = UA(t_k - t_m)$$

Here, t_m is the product temperature (k); t is the inlet temperature (k), $(t - t_m)_{l_m}$ is the logarithmic mean (k) of the temperature differences between the hot air and the product at the inlet and outlet, respectively, ha is the volumetric heat transfer coefficient (kcal/sec·K·m^3), U is the overall heat transfer coefficient (kcal/sec·K·m^2), A is the heating area in contact with the product (m^2), and t_k is the temperature of the heat source (K).

Table 4 is excerpted from Toei (24). Here the volume used to define ha includes void volumes, for example, above and below a fluid bed dryer. Thus it is the overall volume and is subject to significant variation. It is also dependent on the critical moisture of the solids. Note the units when using this table. An estimate must be made a priori about the drop in air temperature across the dryer and rise of product temperature in order to use Table 5 for a very rough sizing of a dryer.

Extensive work has been done in developing theories of drying, such as the theory of simultaneous transport, theories involving flow through porous media, and simplified models, like the wetted surface model and the receding-plane model. The characteristic drying rate curve has also found use in design. Details of these theories are available in References 3, 13, 17, and 26.

In designing a dryer, basically one or more of the following sources of information are used: (a) information obtained from customers, (b) previous experience in the form of files on dryers sold and tended or (c) pilot-plant tests and/or bench-scale tests. For

Table 5 Approximate Values of *ha* (kcal/h·°C·m³) for Various Dryer Types

Type	ha	$(t - t_m)_{l_m}$ (°C)	Inlet hot air temperature (°C)
Convection			
Rotary	100 ~ 200	Countercurrent: 80 150	200 ~ 600
		Cocurrent: 100 180	300 ~ 600
Flash	2000 ~ 6000	Parallel flow only: 100 ~ 180	400 ~ 600
Fluid bed	2000 ~ 6000	50 ~ 150	100 ~ 600
Spray	20–80	Counterflow: 80 ~ 90	200 ~ 300
	(large five)	Cocurrent: 70 ~ 170	200 ~ 450
Tunnel	200 ~ 300	Counterflow: 30 ~ 60	100 ~ 200
		Cocurrent: 50 ~ 70	100 ~ 200
Jet flow	$h = 100 \sim 150$	30 ~ 80	60 ~ 150
Conduction	U (kcal/h·°C·m²)	$t_k - t_m$ (°C)	
Drum	100–200	50–80	
Agitated through	60–130	50–100	
rotary with steam	(smaller for sticky solids)		
tubes, etc.			

design a computer is usually used; the input parameters are based on hard data and partly on experience. Pilot-plant tests ensure that the material can be processed in the desired manner. However, the scaleup procedures are by no means straightforward. The theory in the computer program is restricted to heat and mass balances. The design is based on drying times guessed by the designer, who may be guided by drying data obtained in bench-scale and pilot-plant tests. Design procedures for various dryers have been outlined in References 7 and 16. Factors affecting product quality should be borne in mind before deciding on a dryer. (For discussion of some new drying technologies, see Reference 28.)

8. CLOSING REMARKS

Recent studies have resulted in significant advances in the understanding of the thermodynamics of drying hygroscopic materials, kinetics of drying, evaporation of multicomponent mixtures from porous bodies, behavior of particulate motion in various dryers, and so on. In general the empirical knowledge gained in the past two decades has been of considerable value in the design of industrial dryers, including modeling of dryers and control. On the other hand the understanding at the microscopic level of the drying mechanisms remain at a rudimentary level in the sense that modeling "drying" remains a complex and challenging task. Numerous textbooks (29–33) have appeared in recent years that focus on one or more aspects of drying. The interested reader is referred to the series *Advances in Drying* (26,34) and the journal *Drying Technology* (35) for more recent developments in the field of drying. Appendix B of this handbook consists of an annotated, selective bibliography of books on drying and closely related topics. The reader will find this compilation especially helpful in locating information or obtaining details that could not be included in this handbook. In addition, the series *Drying of Solids* (36–38) contains valuable recent information on drying technology. The technical catalogs published by various manufacturers of drying and ancillary equipment are also very valuable. The proceedings of the biennial international drying symposia (IDS), published in the *Drying* series provide a useful guide to current research and development (R&D) in drying (39–41).

It should be stressed that one must think in terms of the drying *system* and not just the dryer when examining an industrial dehydration problem. The preprocessing steps (feeding, dewatering, etc.) of the feed as well as the postprocessing steps (e.g., cooling, granulation, blending, etc.) and cleaning of the dryer emissions are often just as important as the dryer itself. In view of the increasingly stringent environmental regulations coming in force around the world it is not unusual for the dryer itself to cost only a small fraction of the total drying system cost. This edition of the handbook contains details concerning feeders as well as treatment of dryer emissions.

Selection of dryers and drying systems is another important area of practical significance. The cost of a poorly selected drying system is often underestimated since the user must pay for it over the entire life span of the system. If past experience is used as the sole guide, we automatically eliminate the potential benefits of specifying some of the newer dryers marketed recently around the world. Indeed, to cover the broad spectrum of new and special drying technologies that have appeared in the marketplace in the past decade, a new chapter (Chap. 36) is devoted exclusively to this subject. The reader is urged to go through that chapter in conjunction with any specific drying application to become familiar with some of the newly developed technologies that may afford some advantages over the conventional drying technologies. Use of superheated steam rather than hot air as the drying medium for direct dryers has attracted considerable attention in recent years. The topic is covered in some detail in a separate chapter (Chap. 35) from the applications viewpoint.

Scaleup of dryers is perhaps the central issue of most significance in the design of dryers. Information on this subject is rather limited and widely scattered. Genskow (42) provides perhaps the only compilation of papers dealing with the scaleup of several dryer types. For design, analysis or scaleup of dryers, Houska, Valchar, and Viktorin (43) have presented the general equations and methodology as applied to a number of industrial dryers. Unfortunately, the extreme diversity of products and dryer types precludes development of a single design package.

ACKNOWLEDGMENT

The authors are grateful to Purnima Mujumdar for her prompt and efficient typing of this article.

NOMENCLATURE

A	heat transfer area, m^2
C_{LW}	heat capacity of liquid moisture, $J \cdot kg^{-1} K^{-1}$
C_P	heat capacity at constant pressure, $J \cdot kg^{-1} K^{-1}$
C_{PG}	heat capacity of dry gas, $J \cdot kg^{-1} K^{-1}$
C_{PW}	heat capacity of moisture vapor, $J \cdot kg^{-1} K^{-1}$
C_{PY}	humid heat, $J \cdot kg^{-1} K^{-1}$
D	molecular diffusivity, $m^2 s^{-1}$

Note: 1 denotes dimensionless entity.

E	efficiency of the dryer, 1
f	relative drying rate, 1
G_G	dry gas flow rate, $kg \cdot s^{-1}$
G_V	evaporation rate, $kg \cdot s^{-1}$
ha	volumetric heat transfer coefficient, $kcal \cdot m^{-3} K^{-1}$
h_C	convective heat transfer coefficient, $W \cdot m^{-2} K^{-1}$
h_R	radiative heat transfer coefficient, $W \cdot m^{-2} K^{-1}$
H	enthalpy, $J \cdot kg^{-1}$
H_{GG}	dry gas enthalpy, $J \cdot kg^{-1}$
H_{GW}	moisture vapor enthalpy, $J \cdot kg^{-1}$
I_G	humid enthalpy, $J \cdot kg^{-1}$
I_{GS}	enthalpy of dry gas, $J \cdot kg^{-1}$
I_{LW}	enthalpy of added moisture, $J \cdot kg^{-1}$
K	mass transfer coefficient $kg \cdot m^{-2} s^{-1}$
m_G	mass of dry air, kg
m_W	moisture of mass vapor, kg
M_G	molar mass of dry gas, $kg \cdot mol^{-1}$
M_W	molar mass of moisture $kg \cdot mol^{-1}$
N	rate of drying per unit surface area, $kg \cdot m^{-2} sec^{-1}$
N_G	molar gas flow per unit area, $mol \cdot m^2 sec^{-1}$
P_G	partial pressure of dry gas, Pa
P_W	partial pressure of moisture vapor, Pa
P_W^0	vapor pressure, Pa
P	total pressure, Pa
q	heat flux, $W \cdot m^{-2}$
Q	heat quality, J
R	gas constant, $J \cdot mol^{-1} K^{-1}$
T	temperature, K
T_{av}	average temperature, K
T_C	critical temperature, K
T_D	dew point temperature, K
T_G	gas temperature (dry bulb), K
T_{GS}	adiabatic saturation temperature, K
T_0	initial or reference temperature, K
T_W	wet bulb temperature, K
u	internal energy, $J \cdot kg^{-1}$
U	overall heat transfer coefficient $kcal \cdot sec^{-1} m^{-2}$
V_H	humid volume, $m^3 kg^{-1}$
V_W	specific molar volume, $m^3 mol^{-1}$
V_L	specific molar volume of liquid moisture, $m^3 mol^{-1}$
W	moisture content (wet basis), 1
X	moisture content (dry basis), 1
X^*	equilibrium moisture content, 1
X_{cr}	critical moisture content, 1
X_0	initial moisture content, 1
Y	humidity (mass ratio vapor/dry gas), 1
Y_{GS}	humidity at adiabatic saturation temperature, 1
Y_S	saturation humidity, 1
Y_W	wet bulb humidity, 1

Greek Symbols

ΔH_{GM}	residual gas-mixing enthalpy, $J \cdot kg^{-1}$
ΔH_R	molar latent heat of reference substance, $J \cdot mol^{-1}$
ΔH_{VD}	latent heat of vaporization at T_D, $J \cdot kg^{-1}$
ΔH_{V0}	latent heat of vaporization at T_0, $J \cdot kg^{-1}$
ΔH_{VS}	latent heat of vaporization at T_S, $J \cdot kg^{-1}$
ΔH_{VW}	latent heat of vaporization at T_W, $J \cdot kg^{-1}$
ΔH_W	molar latent heat of vaporization, $J \cdot mol^{-1}$
θ	temperature difference, K
θ^*	temperature difference on saturation, K
ρ	gas density, $kg \cdot m^{-3}$
ϕ	humidity-potential coefficient, l
ϕ	characteristic moisture content, l
ϕ_E	Ackermann correction, l
ψ	relative humidity, l

Dimensionless Groups

Lu	Luikov number
Nu	Nusselt number
Pr	Prandtl number
Sc	Schmidt number

REFERENCES

1. Williams-Gardner, A., *Industrial Drying*, Leonard Hill, London, 1971, Chaps, 2, 3, and 4.
2. Weast, R. C., *CRC Handbook of Physics and Chemistry*, Section D, 53 Ed., The Chemical Rubber Co., Cleveland, 1973.
3. Keey, R. B., *Introduction to Industrial Drying Operations*, 1st Ed. Pergamon, New York, 1978, Chap. 2.
4. Hougen, O. A., K. M. Watson, and R. A. Ragatz, *Chemical Process Principles*, 2nd Ed., Wiley, New York, 1954, Vol. 1, p. 257.
5. Reid, R. C., J. M. Prausnitz, and T. K. Sherwood, *The Properties of Gases and Liquids*, 3rd Ed., McGraw-Hill, New York, 1977, Appendix A.
6. Treybal, R. E., *Mass Transfer Operations*, 2nd Ed., McGraw-Hill, New York, 1968, Chaps. 7 and 12.
7. Perry, J. H., *Chemical Engineering Handbook*, 5th Ed., McGraw-Hill, New York, pp. 20.7–20.8.
8. Wilke, C. R., and D. T. Wasan, A new correlation for the psychrometric ratio, *AIChE-IChE Symp. Series*, *6*, 21–26, 1965.
9. Moller, J. T., and O. Hansen, Computer-drawn H-X diagram and design of closed-cycle dryers, *Proc. Eng.*, *53*, 84–86, 1972.
10. Kessler, H. G., *Food Engineering and Dairy Technology*, Verlag, A. Kessler, Freising, Germany, 1981, Chaps. 8, 9, and 10.
11. Foust, A. S., L. S. Wenzel, L. W. Clump, L. Maus, and L. B. Andersen, *Principles of Unit Operations*, 2nd Ed., Wiley, New York, 1980, Chap. 17, p. 431.
12. van Brackel, J., Mass transfer in convective drying, In *Advances in Drying*, Vol. 1, A. S. Mujumdar, ed., Hemisphere, New York, 1980, pp. 217–268.
13. Ashworth, J. C., Use of Bench Scale Tests for Dryer Design, Industrial Drying Short Course, Department of Chemical Engineering, McGill University, 1978.

14. Keey, R. B., and M. Suzuki, On the characteristic drying curve, *Int. J. Heat Mass Transfer*, *17*, 1455–1464, 1974.

15. Menon, A. S., and A. S. Mujumdar, Energy saving in the drying of solids, *Indian Chem. Eng.*, *14*(2), 8–13, 1982.

16. Nonhebel, G., and A. A. H. Moss, *Drying of Solids in the Chemical Industry*, Butterworths, London, 1971, Chap. 3.

17. Keey, R. B., Theoretical foundations in drying technology, In *Advances in Drying*, A. S. Mujumdar, ed., Hemisphere, New York, 1980, Vol. 1, pp. 1–22.

18. Sloan, C. E., Drying systems and equipment, *Chem. Eng.*, June 19, 1967, p. 167.

19. McCormick, P. Y., *Chemical Engineering Handbook*, J. H. Perry, 5th Ed., McGraw-Hill, New York, 1973.

20. Dittman, F. W., How to classify a drying process, *Chem. Eng.*, Jan. 17, 106–108, 1977.

21. Mujumdar, A. S., ed., *Advances in Drying*, Vol. 1 (1980), Vol. 2 (1982), Vol. 3 (1984), Hemisphere, New York.

22. Mujumdar, A. S., ed., *Drying '80*, Vols. 1 and 2 (1980), *Drying '82* (1982), *Drying '84* (1984), *Drying '85* (1985), and *Drying '86* (1986), Hemisphere/Springer-Verlag, New York.

23. Schlünder, E. U., *Handbook of Heat Exchange Design*, E. U. Schlünder et al., eds. Hemisphere, New York, 1982.

24. Toei, R., Course Notes on Drying Technology, Asian Institute of Technology, Bangkok, Thailand, 1980.

25. Keey, R. B., *Drying: Principles and Practice*, Pergamon, Oxford, 1972.

26. Kroll, K., *Trockner und Trockungsverfahren*, Springer-Verlag, Berlin, 1959.

27. Lang, R. W., in Proc. 1st Int. Drying Symp., Montreal, A. S. Mujumdar, ed., Science Press, Princeton, New Jersey, 1978.

28. Mujumdar, A. S., ed., Drying of Solids, Sarita, New Delhi, 1990, pp. 17–71.

29. Strumillo, C., and Kudra, T., *Drying: Principles, Applications and Design*, Gordon and Breach, New York, 1987.

30. Cook, E. M., and Dumont, D., *Process Drying Practice*, McGraw-Hill, New York, 1991.

31. Keey, R. B., *Drying of Loose and Particulate Materials*, Hemisphere, New York, 1992.

32. Van't Land, C. M., *Industrial Drying Equipment*, Marcel Dekker, New York, 1991.

33. Vergnaud, J. M., *Drying of Polymeric and Solid Materials*, Springer-Verlag, London, 1992.

34. Mujumdar, A. S., ed., *Advances in Drying*, Vol. 5 (1992), Hemisphere/Taylor Francis, New York.

35. Mujumdar, A. S., ed., *Drying Technology—An International Journal*, Marcel Dekker, New York (1982–).

36. Mujumdar, A. S., ed., *Drying of Solids—Recent International Developments*, Wiley-Eastern, New York, 1987.

37. Mujumdar, A. S., ed., *Drying of Solids*, Sarita Prakashan, Nauchandi Grounds, Meerut, U.p., India, 1990.

38. Mujumdar, A. S., ed., *Drying of Solids*, Oxford/IBH, New Delhi, India, and International Publishers, New York, 1992.

39. Mujumdar, A. S., and Roques, M., eds., *Drying 89*, Hemisphere/Taylor Francis, New York, 1989.

40. Mujumdar, A. S., and Filkova, I., eds., *Drying '91*, Elsevier, Amsterdam, The Netherlands, 1991.

41. Mujumdar, A. S., ed., *Drying '92*, Elsevier, Amsterdam, The Netherlands, 1992.

42. Genskow, L. R., Guest ed., *Scale-Up of Dryers*, special issue of *Drying Technology*, *12*(1–2), 1994.

43. Houska, K., Valchar, J., and Viktorin, Z., Computer-aided design of dryers, In *Advances in Drying*, A. S. Mujumdar, ed., Hemisphere, New York, 1987, Vol. 4, pp. 1–98.

2
Experimental Techniques in Drying

Károly Molnár
Technical University of Budapest
Budapest, Hungary

1. INTRODUCTION

The calculation of drying processes requires knowledge of a number of characteristics of drying techniques, such as the characteristics of the material, the coefficients of conductivity and transfer, and the characteristics of shrinkage. In most cases these characteristics cannot be calculated by analysis, and it is emphasized in the description of mathematical models of the physical process that the so-called global conductivity and transfer coefficients, which reflect the total effect on the partial processes, must frequently be interpreted as experimental characteristics. Consequently, these characteristics can be determined only by adequate experiments. With experimental data it is possible to apply analytical or numerical solutions of simultaneous heat and mass transfer to practical calculations.

The general aim of drying experiments is as follows (1):

1. Choice of adequate drying equipment
2. Establishing the data required for planning
3. Investigation of the efficiency and capacity of existing drying equipment
4. Investigation of the effect of operational conditions on the shape and quality of the product
5. Study of the mechanism of drying

When the aim of experiments is the choice of adequate drying equipment and the determination of the data required for planning, the effect of the different variables must be examined. It is practical to carry out such a series of experiments in small or pilot-plant equipment, which operates from both thermal and materials-handling aspects as does actual equipment, and by means of which the effect of the parameters to be examined can be studied over a wide range of conditions.

Experiments are frequently carried out in equipment of plant scale, for example, in order to establish data for certain materials, required for planning, to select the type of adequate drying equipment, or to test the efficiency of existing equipment or its suitability for drying other materials. In general, the efficiency and, further, the heat and mass balances, can be determined from data obtained with equipment of plant size.

Owing to the diversity of the aims of experimental investigations and the large number of drying characteristics, the techniques of experimental determination are extremely diverse and in many cases very specific. Therefore, without attempting to be complete, only the methods generally applied are presented here; some special cases are mentioned with reference to the sources of the literature.

2. DETERMINATION OF MOISTURE CONTENT

2.1. Determination of the Moisture Content of Solid Materials

Although determination of the moisture content of wet materials appears simple, the results obtained are often not sufficiently accurate, namely, many materials may suffer, on heating, not only moisture loss but also chemical changes (oxidation, decomposition, destructive distillation, and others), and this may change the material as well. At the same time the adsorbed water must be distinguished from the so-called water of crystallization, which is frequently a very complex problem.

On selecting the techniques of moisture determination one must take into account the desired accuracy, the case of the procedure, the length of the investigation, and the complexity of the required instruments and equipment. Possible methods of measuring the distribution of moisture content during drying are presented separately.

Direct Methods

The direct methods consist essentially of determination of the moisture content of a sample by drying carried out in a drying oven with or without blow through of air, or by drying in a vacuum chamber or in a vacuum desiccator. The sample material must be prepared in every case in the following way. The material is disintegrated into pieces of 1–2 mm³, and a sample of known mass (4–5 g) is placed into a previously dried and weighed glass container, which is put into the drying chamber and dried at 102–105 °C. The measurement of mass is carried out at ambient temperature, previously allowing the sample to be cooled in a desiccator. The drying process may be considered complete when the difference between the values obtained for the moisture content of the material by two consecutive measurements does not exceed ±0.05%. The literature indicates that this process is faster when drying is carried out at 130–150 °C. However, our investigations proved that results obtained in this way may deviate by 0.5%–1.0%. Thus, the quick method appears to be suitable only for an approximate determination of the moisture content of a material.

The drying period may be significantly shortened by blowing air through the drying chamber, provided this air has been previously heated to 102–105 °C, generally by means of electricity (2).

Drying of foods and other materials sensitive to heat is carried out at 60 °C in a vacuum desiccator, in order to prevent their decomposition. In this case drying requires several days.

Indirect Methods

Under industrial conditions the moisture present in material must be determined by faster methods, such as by electrical methods of which three main varieties have become widespread: moisture determination based on the change of the ohmic dc resistance, measurement of the electrostatic capacitance (dielectric constant of the material), and measurement of the loss in an ac field. Other quick methods are the chemical methods developed mainly for the most frequently occurring case, when the moisture is water, such as the Karl–Fischer analysis based on the chemical reaction of iodine in the presence of water (3), the distillation method, in which moisture is determined by distillation with toluene, and the extraction method, which is carried out with absolute ethanol.

For the determination of factors required for the description of the moisture transport mechanism of wet materials, the distribution of the moisture content of the dried material and even its change during the drying process must be known. These measurements are usually carried out under laboratory conditions. Two main varieties of these measurements are widespread:

1. Mechanical disintegration of the material for the rapid determination of the moisture content of the individual elements and, on compressing the specimen, its further drying (4,5)
2. Special adaptation of one of the above-mentioned electrical methods (indirect) (6)

2.2. Determination of Moisture Content of Gases

The following methods have become widespread for the determination of the moisture content of gases: determination of the absolute value of moisture content by gravimetry or barometry (direct method), measurement of wet gas particles in air by determination of the dry and wet bulb temperatures and the dew point temperature (indirect method), and measurement of a property of the wet gas that depends on the moisture content, such as the absorptivity of the wet gas to electromagnetic waves.

Most frequently applied is the indirect method—determining the dry and wet bulb temperatures or measuring the dew point temperature, which requires a slightly more expensive instrument. However, on using these methods, the determination of the moisture content of the wet gas cannot be carried out in every case with sufficient accuracy. When high accuracy is required, one of the absolute methods must be applied. (In case of air of 50°C temperature and 50% relative moisture content, a change of 1°C in the temperature of the wet bulb results in a 9% change of the absolute moisture content, or a change of 1°C in the dew point temperature results in a 7% change of the absolute moisture content, respectively.)

The absolute determination of the moisture content of gases can be carried out by the absorption method (allowing gas to flow through a silica-gel bed or allowing it to be absorbed by methanol and then determining the water content by Karl–Fischer titration, for example). A detailed description of these and other special methods can be found in the literature (3,7).

3. EXPERIMENTAL DETERMINATION OF THE SORPTION EQUILIBRIUM CHARACTERISTICS OF MATERIALS

3.1. Interpretation of the Equilibrium Moisture Content of Materials

A significant part of materials contains a multitude of capillaries, micropores and macropores, and cells and micelles of various dimensions and shapes. In these materials the potential sites of moisture are determined by structural buildup.

The moisture-binding properties of materials are affected by their interior structural buildup. Thus, the equilibrium characteristics determined by measurement may be applied only to structural materials strictly identical with the material used for the measurement.

Although the materials may approximate the equilibrium moisture content either by moisture uptake (adsorption) or by drying (desorption), the value of equilibrium vapor pressure generally depends, for capillary-porous materials, upon the direction of the approximation of the equilibrium moisture content. This phenomenon is known as *sorption hysteresis*. For this reason it is necessary to distinguish adsorption equilibrium moisture from desorption equilibrium moisture.

The equilibrium moisture content is developed as a result of an interaction between the material and the environment: $\overline{X}^* = \overline{X}^*(p_v, T)$. Changes in the moisture content of the material are due to conditions (p_v, T) prevailing on the surface of the material. After a sufficiently long time—with steady-state limit conditions—an internal moisture diffusion balance takes place until the equilibrium moisture content is attained. In an equilibrium state a steady internal moisture distribution exists. Although theoretically an infinite time is needed for its formation, a practically acceptable accurate approximation may be attained by a number of procedures and methods with a finite time.

3.2. Interpretation of the Equilibrium Vapor Pressure

The vapor pressure at which a material of a given $\overline{X}$ moisture content is at a given temperature T is a sorption equilibrium (i.e., it does not lose and does not take up any moisture) is the equilibrium vapor pressure $p_v^* = p_v^*(T)_{\overline{X}}$.

Generally, in drying knowledge of the equilibrium vapor pressure at a constant temperature is needed. Thus, since when T is constant p_v^* is constant, the equilibrium relative vapor content was applied in drying as a characteristic of the vapor pressure.

$$\psi = \frac{p_v}{p_{ov}^*} \tag{1}$$

3.3. Characteristic Functions of Sorption Equilibrium

Since the function $\overline{X}^* = \overline{X}^*(T, p_v)$ expresses sorption equilibrium, sorption isotherms $\overline{X}^* = \overline{X}^*(p_v)_T$, sorption isobars $\overline{X}^* = \overline{X}^*(T)_{p_v}$, and sorption isosteres $p_v = p_v(T)_{X^*}$ are applied to drying. Since sorption isotherms are most frequently applied, the method of their determination is dealt with here. Sorption isotherms are determined from point to point:

$$\overline{X}_1^*(\Psi_1)_T, \overline{X}_2^*(\Psi_2)_T \cdots \overline{X}_n^*(\Psi_n)_T$$

Each pair of values determining a "point" is in general the result of a measurement.
The elements of this measurement are as follows:

1. Presentation of the pair of values to be measured on the condition that T is constant
2. Measurement of the value of p_v and Ψ during the measurement or at the end of the measurement
3. At the end of the measurement the determination of equilibrium moisture content $\overline{X}^*$ of the sample of material

Widespread methods of measurement in practice differ from each other decisively in the presentation of p_v and Ψ with respect to $\overline{X}^*$. Techniques of determining the moisture content of wet gases are described in detail in Section 2.2 and those of the materials are described in Section 2.1. In general, two main methods of developing the pair of values to be measured are applied.

Measuring Techniques Carried Out at a Constant Vapor Pressure

During measurement p_v and Ψ are kept at a constant value until the moisture $\overline{X}$ of the material sample attains the constant equilibrium value of $\overline{X}{}^*$. Practically, the sample is dried to equilibrium by means of a gas held at a constant state.

Measuring Techniques Based on Developing an Equilibrium Vapor Pressure

Equilibrium vapor pressure is developed by measurement of the material sample itself, and measurement is continued until the values of p_v and Ψ become constant in the measuring space. In preparing a drying (desorption) isotherm, the source of moisture serves as the sample.

3.4. Measuring Techniques at Constant Vapor Pressure

The gravimetric, the bithermal, and the method based on the interpolation of weight changes are the best known, but of these the gravimetric method is the most widespread. This method is described here in detail; references are provided for the others (8–10,29). Constant vapor pressure is maintained by means of an acid or salt solution of specified concentration. It is known that the moisture content of wet air, in equilibrium with an acid or salt solution of known concentration, is constant at a given temperature. Partial pressures of water vapor developed with the use of solutions of sulfuric acid can be determined, as shown in Figure 1. The relative vapor pressure above an aqueous solution of sulfuric acid as a function of concentration and temperature can be calculated by the correlation (2)

$$\log \frac{p_v}{p_{ov}^*} = \left(a_1 - \frac{a_2}{T}\right) + \frac{1}{\log p_{ov}^*} \tag{2}$$

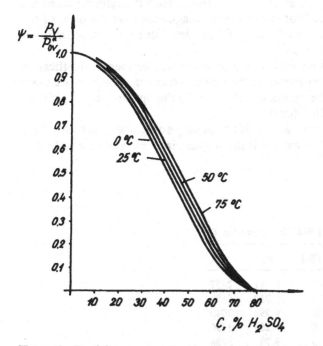

Figure 1   Partial pressures of water vapor developed with the use of solutions of sulfuric acid.

where values for constants a_1 and a_2 are given in Table 1. The term p_{ov}^* in correlation [2] is the saturation vapor pressure of water referred to the given temperature (Table 2).

The relative equilibrium vapor pressures developed above saturated salt solutions are summarized in Table 3 (2,11,12).

With the gravimetric method a closed measuring space must exist in the thermostat so that a constant vapor pressure is formed and maintained in the manner already described. The equilibrium moisture content of the sample placed in the equilibrium air space serves as a source for one point of the isotherm.

The static and the dynamic methods have become widespread; their names indicate a stagnant volume of air in the measuring space and air flow, respectively. The static measuring arrangement is shown in Figure 2, and that of a dynamic system is shown in Figure 3.

A variant of the dynamic method may be, for example, an air conditioner of high accuracy that maintains the constant state of a sample placed in the air flow, which is kept at given T and Ψ values (13). Development of the equilibrium moisture content of the material is attained when the weight of the sample becomes constant. This may be checked by gravimetric measurements.

The sorption equilibrium of the system material-air-solution in principle is attained only after an infinite time. However, the constant weight is observed after a finite time.

The measurement period of the static method is, particularly with high Ψ values, very long (several hundred hours). Thus this method cannot be applied to perishable materials even when the thickness δ of the material sample is small and its specific surface area high. The mass of the sample cannot be significantly decreased because of the hazard of increasing the relative error of measurement originating in the inaccuracy of the weight measurement.

With change in the moisture content of the sample during investigation the concentration of the solution is also altered, as are p_v and Ψ. This change is negligible when the mass of the solution is higher by at least two orders of magnitude than the mass of the sample. But even in this case it is advisable to repeatedly determine the concentration of the solution when weight constancy is achieved.

With the dynamic method of measurement the period of measurement is significantly shorter because over the course of evaporation the material transfer resistance decreases to a greater extent. According to the suggestion of Likov (2) the optimum rinsing rate of the measuring space is $0.2\text{--}0.3 \times 10^{-4}$ (m^3/s).

The gravimetric method is simple and yields reliable results. There are also fully automated measuring systems (14) provided with computerized evaluation possibilities (15).

Table 1 Values of a_1 and a_2 against H_2SO_4 Concentration C

C (%)	a_1	a_2	C (%)	a_1	a_2
10	8.925	2259	60	8.841	2457
20	8.922	2268	70	9.032	2688
30	8.864	2271	80	9.293	3040
40	8.84	2299	90	9.265	3390
50	8.832	2357	95	9.79	3888

Table 2 Saturation Vapor Pressure of Water Referred to the Given Temperature

Temperature T		Pressure $P \cdot 10^{-4}$	Temperature T		Pressure $P \cdot 10^{-4}$
°C	K	(Pa)	°C	K	(Pa)
0	273.15	0.061076	51	324.15	1.296047
1	274.15	0.065656	52	325.15	1.361163
3	276.15	0.075747	53	326.15	1.429221
4	277.15	0.081287	54	327.15	1.500123
5	278.15	0.008891	55	328.15	1.573967
6	279.15	0.093477	56	329.15	1.650950
7	280.15	0.100126	57	330.15	1.731168
8	281.15	0.107206	58	331.15	1.814623
9	282.15	0.114728	59	332.15	1.901509
10	283.15	0.122711	60	333.15	1.991731
11	284.15	0.131174	61	334.15	2.085874
12	285.15	0.140157	62	335.15	2.183941
13	286.15	0.149669	63	336.16	2.284949
14	287.15	0.159741	64	337.15	2.390861
15	288.15	0.170410	65	338.15	2.500696
16	289.15	0.181698	66	339.15	2.614453
17	290.15	0.193642	67	340.15	2.733113
18	291.15	0.206234	68	341.15	2.855696
19	292.15	0.219571	69	342.15	2.984164
20	293.15	0.233692	70	343.15	3.116553
21	294.15	0.248599	71	344.15	3.252846
22	295.15	0.264289	72	345.15	3.396043
23	296.15	0.280764	73	346.15	3.543134
24	297.15	0.298220	74	347.15	3.696126
25	298.15	0.316657	75	348.15	3.854994
26	299.15	0.335976	76	349.15	4.018765
27	300.15	0.356374	77	350.15	4.189401
28	301.15	0.377850	78	351.15	4.364940
29	302.15	0.400406	79	352.15	4.547344
30	303.15	0.424138	80	353.15	4.735631
31	304.15	0.449145	81	354.15	4.930784
32	305.15	0.375328	82	355.15	5.132801
33	306.15	0.502885	83	356.15	5.341682
34	307.15	0.531815	84	357.15	5.557429
35	308.15	0.562215	85	358.15	5.780040
36	309.15	0.593989	86	359.15	6.010496
37	310.15	0.627429	87	360.15	6.248797
38	311.15	0.662439	88	361.15	6.494944
39	312.15	0.669116	89	362.15	6.748937
40	313.15	0.737460	90	363.15	7.010774
41	314.15	0.777765	91	364.15	7.280457
42	315.15	0.819836	92	365.15	7.560927
43	316.15	0.863868	93	366.15	7.849243
44	317.15	0.909959	94	367.15	8.146384
45	318.15	0.958208	95	368.15	8.452352
46	319.15	1.008516	96	369.15	8.769106
47	320.15	1.061178	97	370.15	9.094687
48	321.15	1.116193	98	371.15	9.430075
49	322.15	1.173562	99	372.15	9.778211
50	323.15	1.233480	100	373.15	10.13223

Table 3   Relative Equilibrium Vapor Pressures Developed Above Saturated Salt Solutions

Salt	T (°C)	$\Psi = \dfrac{p_v}{p^*_{ov}}$	Salt	T (°C)	$\Psi = \dfrac{p_v}{p^*_{ov}}$
BaCl$_2$·2H$_2$O	24.5	0.88	KI	100	0.562
CaCl$_2$·6H$_2$O	2	0.398	K$_2$S	100	0.562
	10	0.38	KNO$_2$	20	0.45
	18.5	0.35	LiCl·H$_2$O	20	0.15
	20.0	0.323		73.06	0.112
	24.5	0.31		88.96	0.0994
CaHPO$_4$·H$_2$O	22.8	0.952	LiSO$_4$	24.7	0.865
	30.1	0.94	MgCl$_2$	41.7	0.33
	39.0	0.954		50.8	0.311
Ca(NO$_3$)$_2$·4H$_2$O	18.5	0.56		62.3	0.301
	24.5	0.51		79.5	0.287
CaSO$_4$·5H$_2$O	20	0.98	Mg(C$_2$H$_3$O$_2$)$_2$·4H$_2$O	20	0.65
CdBR$_2$	24	0.896	Mg(NO$_3$)$_2$·6H$_2$O	18.5	0.56
	31.6	0.857		24.5	0.52
	41.5	0.835		33.9	0.505
CdSO$_4$	23.9	0.906		42.6	0.472
	31.4	0.865		54.8	0.428
CuCl$_2$·2H$_2$O	28.9	0.67		76.3	0.3275
	36.1	0.669	MgSO$_4$	30	0.942
CuSO$_4$	22.6	0.974		40.8	0.865
	29.5	0.97		55.1	0.843
	38.7	0.964		95.4	0.776
	91.74	0.888	NH$_4$Cl	20.0	0.792
CH$_4$N$_2$O	26.6	0.77		25.0	0.793
	34.8	0.719		30.0	0.795
	46.1	0.658	NH$_4$Cl + KNO$_3$	20	0.726
C$_4$H$_6$O$_6$	24.25	0.877		25	0.716
	32.43	0.818		30	0.686
	43.27	0.760	NH$_4$H$_2$PO$_4$	20	0.931
CrO$_3$	20	0.35		25	0.93
H$_2$C$_2$O$_4$·2H$_2$O	20	0.76		30	0.929
H$_3$PO$_5$·5H$_2$O	24.5	0.09		40	0.9048
KCl	24.5	0.866	(NH$_4$)$_2$SO$_4$	25	0.811
	32.3	0.825		30	0.811
	41.8	0.833		108.2	0.75
	55.4	0.832	NH$_4$Br	95.6	0.772
	95.5	0.773	NH$_4$NO$_3$	28	0.706
KBr	20	0.84		42	0.488
	10	0.692		52.9	0.469
KC$_2$H$_3$O$_2$	168.0	0.13		76	0.332
	20	0.20			0.229
K$_2$CO$_3$·2H$_2$O	18.5	0.44	NaBr	100	
	24.5	0.43	NaBr·2H$_2$O	20	0.58
KCNS	20	0.47	NaBrO$_3$	20	0.92
K$_2$CrO$_4$	20	0.88	NaC$_2$H$_3$O$_2$·H$_2$O	20	0.76
KF	100	0.229	NaCl	26.82	0.758
K$_2$HPO$_4$	20	0.92		34.2	0.743
KHSO$_4$	20	0.86		96.8	0.738
			NaCl + KClO$_3$	16.4	0.366

Table 3   Continued

Salt	T (°C)	$\Psi = \dfrac{p_v}{p_{ov}^*}$	Salt	T (°C)	$\Psi = \dfrac{p_v}{p_{ov}^*}$
$NaCl + KNO_3$	16.4	0.326	$Na_4P_2O_7$	22.6	0.973
$NaCl + KNO_3 +$	16.4	0.305	$Na_2SO_3 \cdot 7H_2O$	20	0.95
$NaNO_3$				23.5	0.9215
$NaC_2H_3O_2 \cdot 3H_2O$	20	0.76		30.9	0.894
$Na_2CO_3 \cdot 10H_2O$	18.5	0.92		40.8	0.867
	24.5	0.87		55.4	0.83
$NaClO_3$	20	0.75		94.9	0.785
	100	0.54	$Na_2S_2O_3 \cdot 5H_2O$	20	0.93
$Na_2Cr_2O_7 \cdot 2H_2O$	20	0.52	$Na_2SO_4 \cdot 10H_2O$	20	0.93
NaF	100	0.966		31.3	0.8746
$Na_2HPO_4 \cdot 2H_2O$	20	0.95	$Pb(NO_3)_2$	20	0.98
$NaHSO_4 \cdot 2H_2O$	20	0.52		103.5	0.884
NaI	100	0.504	$TICI$	100.1	0.997
$NaNO_2$	20	0.66	Tl_2SO_4	104.7	0.848
$NaNO_3$	27.4	0.7295	$ZnCl_2 \cdot \frac{1}{2}H_2O$	20	0.10
	35.1	0.708	$Zn(NO_3)_2 \cdot 6H_2O$	20	0.42
	59.0	0.702	$ZnSO_4 \cdot$	5	0.947
	102.0	0.654	$ZnSO_4 \cdot 7H_2O$	20	0.90

3.5.   Measuring Techniques Based on Developing an Equilibrium Vapor Pressure

The sample of a wet material placed in a closed measuring space adjusts, and after a sufficiently long period, the vapor pressure of the space reaches equilibrium with the moist solid. If the change of moisture content of the solid is to be negligible, the moisture in the sample must be higher by at least three orders of magnitude than the amount of water required for saturation of the space. The saturation of the measuring space practically takes place at a constant moist solid vapor pressure, and the time required to attain equilibrium can be fairly long, with knowledge of the air mass present in the measuring

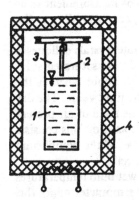

Figure 2   The measuring arrangement of a static method: (*1*) sulfuric acid or salt solutions, (*2*) sample, (*3*) measuring place, (*4*) thermostat.

Experimental Techniques

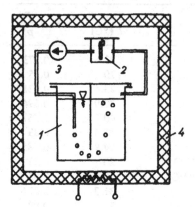

Figure 3 Schematic diagram of gravimetric dynamic system: (*1*) bubbling recipient with sulfuric acid or salt solution, (*2*) sample, (*3*) air pump, (*4*) thermostat.

space, the evaporation surface of the material sample, and the average evaporation coefficient (8). For this measurement determination of the partial pressure and of the relative vapor content of the space are of fundamental importance.

Vapor pressure can be measured by manometer on freezing out the water vapor. After insertion of the material sample, the measuring space is evacuated at constant temperature. The pressure value can be read on an accurate micromanometer. The measuring space is removed from the thermostat, water vapor is frozen out, and the manometer is read again. The difference between the two values read on the manometer is equal to the equilibrium vapor pressure of the material.

If the medium in the measuring space is allowed to flow, the measurement will be lowered since the coefficient of evaporation increases. However, care must be taken during measurement to maintain the mass flow of the flowing air at a constant value.

The essence of the isotonic method (16) is that solutions of acids or salts of various concentration are placed into the closed space—the mass of these solutions being negligible in comparison with the mass of the material sample—together with the material sample itself. These solutions act as sources of absorbents of moisture, respectively, from the material sample. At a given temperature over the course of measurement their composition approaches equilibrium concentration of the solution, corresponding to the equilibrium pressure.

In a method utilizing the deliquescence of crystalline salts, the salt crystals deliquesce in a space with a vapor pressure that exceeds the vapor pressure of their saturated solution. In this way it is possible to draw from their state conclusions about whether the vapor pressure of the measuring space is actually below or above the limit value. The dry bulb temperature and the dew point temperature of air determine the state of air present in the measuring space and also the partial pressures. The method based on this consists essentially of measuring the dew point of the air. The measurement of the equilibrium vapor tension of the air can be carried out by means of hygroscopic cell.

The state of the air is unequivocally determined by the dry and wet bulb temperatures of the air. Methods based on the measurement of the wet bulb thermometer apply this principle. Figure 4 presents the principle of the measuring equipment of a method of this type.

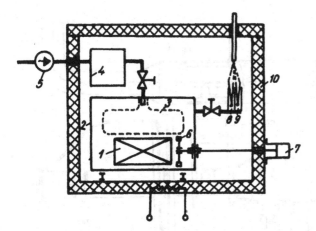

Figure 4 Schematic diagram of the arrangement based on measuring the wet bulb temperature: (*1*) sample, (*2*) measuring chamber, (*3*) foil ballon, (*4*) dashpot, (*5*) air pump, (*6*) paddle wheel with magnetic clutch, (*7*) motor, (*8*) thermometer, (*9*) wet bulb thermometer, (*10*) thermostat.

4. TECHNIQUES AND EQUIPMENT OF THE INVESTIGATION OF DRYING KINETICS

In drying, the heat transfer coefficient h and the mass transfer coefficient K between the drying gas and the wet material, the heat diffusivity κ_h and moisture (mass) diffusivity κ_m coefficients, and the thermal conductivity λ_h and moisture conductivity λ_m coefficients are the parameters of interest. Although several methods are employed for the determination of these parameters, here only those techniques and equipments will be described that allow their determination via a drying experiment.

From the simpler drying experiments quite a few data may be obtained on the basis of which—in fractional drying—conclusions can be drawn concerning the time required for the prescribed change in moisture content and the transport characteristics listed above.

When the heat and mass diffusivity properties of the drying material are also needed, a technique of investigation and measurement is applied such that the distribution of temperature and moisture content can also be measured during drying, along with the thickness of the dried material. When, however, during investigation, the above-mentioned distributions are not measured and only the integral-average moisture content of the dried material and eventually the temperature of the drying surface (or the average temperature) are measured, the model used to analyze the data is the so-called "lumped parameter" type as opposed to the "distributed parameter" model applied when local values are measured.

4.1. Description of the Measuring Equipment

Experimental measuring equipment developed for drying by convection is shown in Figure 5. The equipment is in fact a controlled climate air duct. The wet material is placed on a balance in the measuring space. The temperature and moisture content of the drying air is measured by wet and dry bulb thermometers. Other on-line humidity measuring devices may be used for convenience.

The flow rate can be determined by means of an orifice meter. Prior to beginning the

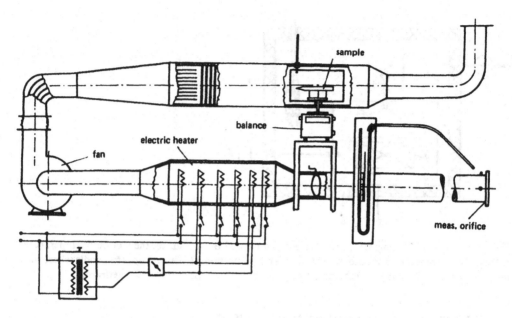

Figure 5 Drying apparatus.

measurement, the fan is started and the desired air flow adjusted by a throttle valve; the temperature and moisture content of the air are controlled by means of an electric heater and a steam valve. In the next section the development of an adequate value of flow rate may be ensured by means of screens. The time-varying mass of the sample to be dried is measured continuously and recorded.

4.2. Drying Experiments (Lumped Approach)

To determine the drying rate, the mass of a sample placed in the air (constant temperature, humidity, and velocity) flow must be measured as a function of time. In order to obtain results that can be applied for scaleup the following aspects must be taken into account: the sample must not be too small and the conditions of drying must if possible be identical to the conditions anticipated in the industrial unit (1,17).

1. The sample must be placed in a similar way in the laboratory unit.
2. The ratios of the drying surface to the nondrying surface must be identical.
3. The conditions of heat transfer by radiation must be similar.
4. The temperature and the moisture content of air (drying gas) and its velocity and direction must be identical with respect to the sample.

It is practical to carry out several experiments with material samples of various thicknesses and to determine in every case the dry matter content of the sample in the way described in Section 2.

When the temperature, moisture content, and velocity of the drying air are constant, drying takes place under constant drying conditions. This way of drying is ensured by the equipment sketched in Figure 5.

Since the wet mass is

$$m_{SW} = m_W + m_S \tag{3}$$

and the moisture content is

$$X = \frac{m_W}{m_S} \qquad [4]$$

thus

$$\frac{m_{SW}}{m_S} = 1 + X$$

That is,

$$X = \frac{m_{SW} - m_S}{m_S} \qquad [5]$$

These correlations indicate that, with knowledge of the drying wet mass as a function of time $[m_{SW} = m_{SW}(\tau)]$ and of the bone-dry mass of the sample, it is possible to plot the moisture content of the sample as a function of time (Fig. 6).

This curve can be directly applied for the determination of time of drying greater masses to a prescribed lower moisture content, provided the drying is carried out under identical conditions. However, better information is obtained when, on the basis of Figure 6, the drying rate is plotted against the moisture content of the material. The drying rate is defined by

$$N_W = -\frac{1}{A_s} \frac{dm_{SW}}{d\tau} = \frac{m_S}{A_s} \frac{d\overline{X}}{d\tau} \qquad [6]$$

In Eq. [6], A_s denotes the contact surface of the drying gas and the dried material, and a cross-section perpendicular to the air flow in through circulation drying.

According to Eq. [6], with the knowledge of the weight loss curve we can plot Figure

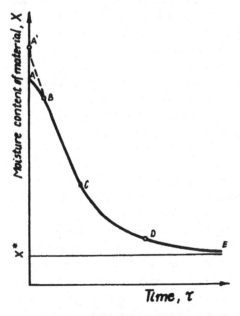

Figure 6   Batch, convection drying curve.

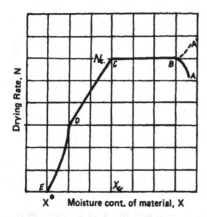

Figure 7   Drying rate curve.

7, which represents the drying rate. If the drying rate is known under one set of constant drying conditions, it can be extrapolated within limits to other conditions (17). Figure 8 shows the variation of drying surface temperature with time while Figure 9 summarizes the characteristic parameters of the drying process.

Determination of transport coefficients by a drying experiment is achieved as follows. During constant rate drying, the temperature of the drying surface is, in a purely convective drying, the so-called wet bulb temperature. With this value, the steady-state heat flux is

$$j_q = h(T_G - T_W) \tag{7}$$

where j_q can also be calculated using the constant drying rate

$$j_q = N_{Wc}\,\Delta H \tag{8}$$

Thus,

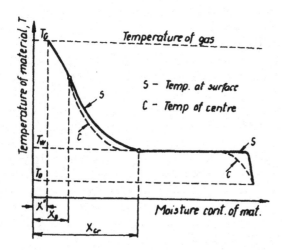

Figure 8   Temperature curves of a wet material at the surface and the middle of the sample.

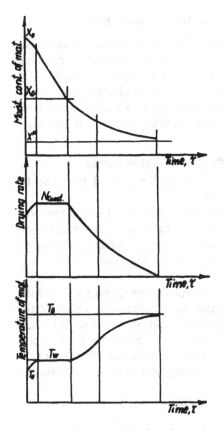

Figure 9 Characteristic functions of a convective drying.

$$h = N_{Wc}\left(\frac{\Delta H}{T_G - T_W}\right) \tag{9}$$

In terms of the mass transfer coefficient,

$$N_{Wc} = K(Y_W - Y_G) \tag{10}$$

where Y_W is the moisture content of the saturated air at temperature T_W, the coefficient of evaporation (mass transfer) is therefore

$$K = \frac{N_{Wc}}{Y_W - Y_G} \tag{11}$$

The heat and mass transfer coefficients on the gas side, characterizing the steady state, may be considered constant during the entire drying process, under constant drying conditions. During the falling rate period, the process is significantly affected by the internal heat and moisture diffusivities. Therefore, determination of these coefficients is also necessary. It may be noted here that the transport coefficients on the gas side also change during this period owing to the continuous decrease in vapor diffusivity of the material in the surface layer (18).

4.3.   Techniques of Investigation with Distributed Parameters

In the initial constant-rate drying period knowledge of the heat diffusivity and heat conductivity coefficients of the wet material is necessary because they are the controlling factors for heat transport within the material. In this section of drying it is presumed that the pores of the wet material are saturated by moisture. Consequently, the above-mentioned characteristics of a heterogeneous material consisting of a solid skeleton and, within this, a capillary system (filled with moisture) of diversified sizes and shapes must be determined.

Determination of the Thermal Conductivity and Diffusivity of Wet Materials

The method for drying a plane slab is described, but a similar method may also be developed for bodies of other geometric shapes (2,8,30).

During drying of wet capillary-porous slabs, when only surface moisture is removed and the value of the so-called phase-change criterion ϵ is zero—that is, no evaporation takes place within the material itself—the development of the temperature of the wet material at the surface and in the plane of symmetry can be determined analytically (8,19).

Using an apparatus of the type shown in Figure 5, the surface temperature T_s and the slab midplane temperature T_z are monitored in addition to the change in weight of the sample. In the initial transient period, surface temperature variation is plotted as a function of the corresponding temperatures at the midplane (Fig. 10). Figure 11 shows a plot of the logarithm of the absolute difference between T_s and T_z versus time. Except at very short times, the curve is linear as expected from the analytical solution of transient one-dimensional diffusion of mass or heat.

The surface temperature is expressed by the function (20)

$$\frac{T_s - T_W}{T_O - T_W} = \sum_{k=1}^{k=\infty} 2 \left(\frac{\sin \delta_k \cos \delta_k}{\delta_k + \sin \delta_k \cos \delta_k} \right) \exp \left(-\delta_k^2 \, Fo \right) \qquad [12]$$

and the temperature of the center of the slab by the function

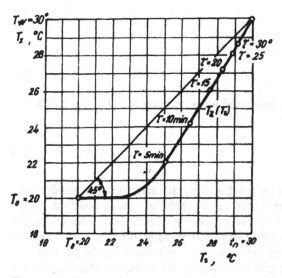

Figure 10   Temperature on the surface versus the temperature on the symmetry plane.

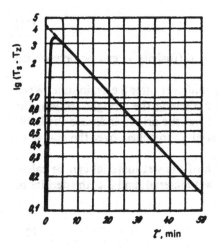

Figure 11 Plot of log $(T_s - T_z)$ versus time.

$$\frac{T_Z - T_W}{T_O - T_W} = \sum_{k=1}^{k=\infty} 2\left(\frac{\sin \delta_k}{\delta_k + \sin \delta_k \cos \delta_k}\right) \exp\left(-\delta_k^2 \text{Fo}\right) \qquad [13]$$

where the eigenvalues δ_k are roots of the transcendent equation

$$\text{ctg } \delta_k = \frac{k}{\text{Bi}_{hm}} \qquad [14]$$

In the above equations the Biot and Fourier numbers are

$$\text{Bi}_{hm} = \frac{h}{\lambda} Z\left(1 + \frac{h_m}{h}\right) \qquad [15]$$

and

$$\text{Fo} = \frac{\kappa_r}{Z^2} \qquad [16]$$

The coefficient h_m in Eq. [15] is given by

$$h_m = \Delta H K S^* \qquad [17]$$

where

$$S^* = \frac{Y_W - Y_G}{T_W - T_d} \cong \frac{Y_s - Y_G}{T_s - T_d} = \text{constant} \qquad [18]$$

is the equation of the equilibrium curve, considered a straight line. It appears from Eqs. [12] and [13] that after certain time (Fourier number) only the first term of the series is predominant. Neglecting further terms we obtain

$$\frac{T_Z - T_W}{T_s - T_W} = \frac{dT_Z}{dT_s} = \frac{1}{\cos \delta_1} \qquad [19]$$

and

$$\frac{d[\ln(T_s - T_z)]}{d\tau} = -\frac{\delta_1^2 \kappa_h}{Z^2} \qquad\qquad [20]$$

In Eq. [19] δ_1 is an eigenvalue, with which the thermal diffusivity κ_h can be determined from Eq. [20].

The function $T_z(T_s)$ shown in Figure 10 becomes nonlinear after a certain time, when the superficial moisture disappears. However, on lengthening the straight section, the point of intersection of unit steepness (45°), starting from the origin, indicates the limiting temperature T_W that the temperature of the material approaches, over the course of drying of the surface moisture.

Knowing κ, the thermal conductivity coefficient λ of the wet material can be calculated from Eq. [21] when c_{SW}, the specific heat of the wet material, and $\overline{\rho}_{Os}$, the density of the dry material, are known.

$$\kappa = \frac{\lambda}{\overline{\rho}_{Os} c_{SW}} \qquad\qquad [21]$$

In an example of the determination of the thermal diffusivity and thermal conductivity, the wet material is a suspension of water and powdered chalk. Its characteristics are given as $T_O = 14.3°C$, $Z = 40$ mm $= 4 \times 10^{-2}$ m, $c_{SW} = 2.51$ kJ/kg°C, and $\rho_{Os} = 1500$ kg/m³. The drying gas is air. Its characteristics are $T_G = 52.5°C$, $T_d = 10.8°C$, $P = 1$ bar, and $c_{PG} = 1.05$ kJ/kg°C. Values of temperatures observed during drying are given in Figure 12.

T_W extrapolated from Figure 12 is $T_W = 25.2°C$. According to Eq. [19],

$$\frac{dT_Z}{dT_s} = \frac{1}{\cos\delta_1} = \frac{T_{Z81'} - T_{Z24'}}{T_{s81'} - T_{s24'}} = \frac{20.73 - 16.13}{24.18 - 23.17} = 4.55$$

where

$$\cos\delta_1 = 0.22$$

that is, the first eigenvalue is $\delta_1 = 1.35$.

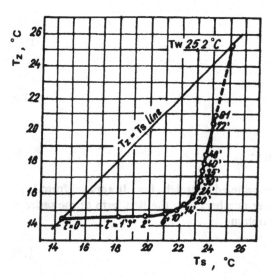

Figure 12 Experimental $T_Z(T_s)$ curve for suspension of water and powdered chalk.

Figure 13 shows the logarithm of the difference between the temperatures at the surface and in the center plane plotted against time. Using the data of Figure 13, it follows according to Eq. [20] that

$$-\frac{d[\ln\,(T_s - T_Z)]}{d\tau} = \frac{\delta_1^2}{Z^2} = \frac{\ln\,(T_{s24'} - T_{Z24'}) - \ln\,(T_{s81'} - T_{Z81'})}{81 - 24} = 0.755\,\,1/h$$

Thus the heat diffusivity of the given wet material is

$$= \frac{\lambda}{c_{SW}\bar{\rho}_{Os}} = 0.755\,\frac{Z^2}{\delta_1^2} = 0.755\,\frac{16 \times 10^{-4}}{1.35^2} = 0.655 \times 10^{-3}\,\,m^2/h$$

and the thermal conductivity is

$$\lambda = c_{SW}\bar{\rho}_{Os} = (0.655 \times 10^{-3})(2.51)(1500) = 2.47\,\,kJ/m \cdot h\,°C$$

Determination of the Mass Diffusivity and Moisture Conductivity Coefficient of Wet Materials

The mass diffusivity coefficient can be determined from the drying curve already discussed. It has been shown for drying of a plane slab (20), that in the falling rate period of drying as the equilibrium moisture content is approached, the time rate of change to the weight of the slab as a function of its weight is a linear function, given by

$$\frac{-(dm/d\tau)}{m - m^*} = \nu_1^2\,\frac{m}{Z^2} = \nu_1^2\,\frac{\lambda_m}{c_m\bar{\rho}_{Os}Z^2} \qquad [22]$$

The eigenvalue ν_1 in Eq. [22] is defined by

$$ctg\,\,\nu_1 = \frac{\nu_1}{Bi_m} = \frac{\nu_1\lambda_m}{S^{**}KZc_m} \qquad [23]$$

where S^{**} in Eq. [23] is given by

$$S^{**} = \frac{P_{Vs} - P_{VW}}{X_s - X_W} \qquad [24]$$

therefore the slope of the sorption isotherm (assuming that the isotherms are not strongly dependent on temperature).

On the basis of Eqs. [22] and [23], we obtain

$$\nu_1 ctg\nu_1 = \frac{(-dm/d\tau)\bar{\rho}_{Os}Z}{(m - m^*)KS^{**}} \qquad [25]$$

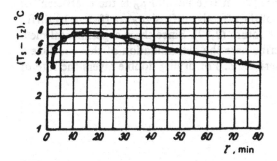

Figure 13 Experimental log $(T_s - T_z) - \tau$ functions for suspension of water and powdered chalk.

The right side of Eq. [25] contains known data when the weight of the wet material is measured as a function of time and when we have already determined, according to the previous section, the mass transfer coefficient K from data obtained in the constant rate drying period. The first eigenvalue can be determined numerically (by trial and error or by iteration) from the transcendental equation. (The value of the ν ranges between 0 and π [Ref. 2].) The mass diffusivity coefficient κ_m can be calculated directly from Eq. [22] if ν_1 is known. When the moisture capacity coefficient c_m is determined separately (2,19), then the moisture conductivity coefficient λ_m can also be determined from Eq. [23].

Determination of the Thermal Conductivity and Effective Diffusivity Coefficient of the Dry Material

In the course of drying of macroporous and capillary-porous materials—in the falling rate period of drying—the moisture from the pores of the material is removed, and an evaporation front penetrates from the surface of the material into the interior. At a given time the primarily wet material can be divided into two parts separated by an evaporation front: a part already dried and a part that is still wet. The moisture evaporated at the plane of phase change must pass through the pores of the already dried layer in order to leave the material. At the same time the heat required for drying has to pass through the already dried layer to carry out the evaporation. This model of drying is known as the *penetrating* or *receding front model* and is described in the literature (2,4,21,22).

Besides drying with a penetrating front, other drying phenomena also occur in which, in the course of the drying, development of a layer inhibits the drying process (in drying two-layer materials, such as salami varieties, seeds coated with an earth layer, and so on). The mathematical model of the drying process of such two-layer materials of constant thickness is described in detail in the literature (23,24). For a description of drying by convection of double-layer materials, in the drying section with a penetrating front, and also of the macroporous and capillary-porous bodies—besides the material characteristics and transport coefficients—knowledge of the thermal diffusivity of the dry layer and the effective diffusivity coefficient value is also necessary. When the mean free path of the molecules being removed is shorter than the characteristic dimension of the capillary, the vapor diffusion process through the pores of the dry layer is the so-called molecular diffusion process. However, the length of the diffusion path and the cross-section available for vapor flow may change the direction of the diffusion. For this case, Krischer and Kröll (25) defined an effective diffusivity coefficient

$$D_a = \frac{D_{VG}}{\mu_D} \tag{26}$$

D_{VG} is the conventional diffusivity of the vapor in free air and μ_D is the coefficient of resistance due to diffusion. Determination of the effective diffusivity coefficient by diffusion experiments is widely known in the literature (2,19). Since these experimental measurements are relatively complex, we describe here the determination of this coefficient by means of a simple series of drying experiments. For drying double layers the rate of drying can be defined as (22)

$$j_m = K_G(p_{v\xi} - p_{VG}) \tag{27}$$

where

$$K_G = \frac{1}{1/k_G M_V + \xi(RT/D_a M_V)} \tag{28}$$

is the overall mass transfer coefficient, p_{VG} the partial pressure of vapor of the bulk gas phase, and $p_{V\xi}$ the vapor pressure at the evaporation plane (see Fig. 14).

It can be shown that a steady state exists in which the rate of drying is

$$j_{me} = \frac{1}{1/k_G M_V + \xi(RT/D_a M_V)} (p_{V\xi e} - p_{VG})$$ [29]

where $p_{V\xi e}$ is the equilibrium partial vapor pressure corresponding to the equilibrium temperature of the site evaporation front. The equilibrium temperature is given by

$$T_{\xi e} = \frac{T_d + T_G(K_H/\Delta H \, SK_G)}{1 + (K_H/\Delta H \, SK_G)}$$ [30]

In Eq. [30],

$$S = \frac{p_{V\xi e} - p_{VG}}{T_{\xi e} - T_d}$$ [31]

and the overall heat transfer coefficient is

$$K_H = \frac{1}{1/h + \xi/\Lambda_I}$$ [32]

On rearranging Eq. [29], we obtain

$$\frac{P_{V\xi e} - P_{VG}}{j_{me}} = \frac{1}{k_G M_V} + \xi \frac{RT}{D_a M_V}$$ [33]

Equation [33] can be used as the basis for determining D_a if the mass transfer coefficient k_G is known. Other necessary parameters are measurable from the weight-loss curve.

For the determination of the thermal conductivity of the dry layer, a steady state yields

$$j_{qe} = j_{me} \Delta H$$ [34]

or

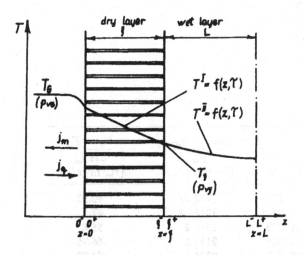

Figure 14 The double-layer model investigated.

Table 4 Data Measured on the Stabilized State

No. of Experiments	$\frac{\xi}{Z}$	ξ (mm)	$j_{me}10^5$ (kg/m^2·s)	$T_{\xi e}$ (°C)	T_d (°C)	T_G (°C)
1	0.2	4	8.14	40.0	10.0	48.0
2	0.3	6	6.44	40.9	6.5	48.6
3	0.5	10	4.53	41.9	11.5	48.0
4	0.7	14	3.25	42.0	12.0	47.2
5	0.8	16	3.11	42.2	11.5	47.7

$$K_H(T_G - T_{\xi e}) = j_{me}\,\Delta H \tag{35}$$

Substituting K_H given by Eq. [32] into Eq. [35] and rearranging we obtain

$$\frac{T_G - T_{\xi e}}{\Delta H\,j_{me}} = \frac{1}{h} + \xi\frac{1}{\lambda_I} \tag{36}$$

This equation can be used to determine λ_I if all other parameter values are known.

In an experimental apparatus utilizing constant (and known) drying conditions, change in weight of the two-layer slab as well as the temperature of the evaporation front must be measured as functions of time. Appropriate plots of Eqs. [33] and [36] using this data yield the effective and thermal conductivity of the dry layer. An example of the determination of the thermal conductivity and effective diffusivity coefficients of dry layers follows.

A two-layer slab consisting of a sand layer and a layer of gypsum is used in this illustration. Five different thicknesses of the sand layer were employed. The flow rate of the drying air was the same in all experiments. The measured data at steady state are summarized in Table 4. Table 5 gives the calculated results using Eqs. [33] and [36]. Plots corresponding to Eqs. [33] and [36] are presented in Figures 15 and 16. The transfer coefficients are determined from the slopes of the linear regions of the curves. The results are

$$k_G = 5.56 \times 10^{-4} \quad \text{kmol/m}^2\text{bar·s}$$
$$D_a = 9.1 \times 10^{-6} \quad \text{m}^2/\text{s}$$
$$h = 0.031 \quad \text{kJ/m}^2\text{·s·K}$$
$$\lambda_I = 0.4 \quad \text{J/m·s·K}$$

Table 5 Calculated Results (from Table 4)

No. of Experiments	ξ (mm)	$p_{v\xi e}$ (bar)	p_{VG} (bar)	$\dfrac{p_{v\xi e} - p_{VG}}{j_{me}}$ (bar·m^2·s/kg)	$\dfrac{T_G - T_{\xi e}}{\Delta H\,j_{me}}$ (°C·m^2·s/kJ)
1	4	0.07374	0.01251	752.21	40.91
2	6	0.07889	0.009871	1071.72	47.2
3	10	0.08317	0.01383	1519.2	55.8
4	14	0.0836	0.01429	2131.2	66.75
5	16	0.08449	0.01383	2271.6	73.72

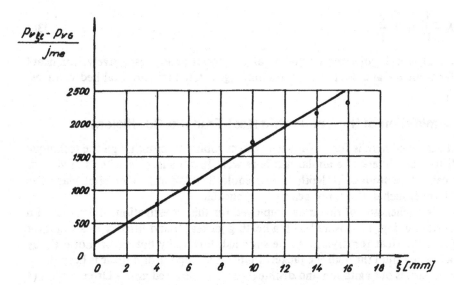

Figure 15 Determination of the mass transfer coefficient and the effective diffusivity.

5. DRYING OF FIXED AND MOVING BEDS

In through circulation fixed bed dryers and in moving or suspended bed dryers (spouted beds, fluidized beds, rotary dryers, and so on), the actual interfacial area participating in heat and mass transfer is unknown. For such cases it is appropriate to define volumetric transfer coefficients for heat and mass transfer, h_v and K_v, respectively, as follows:

$$h_V = h \frac{dA_s}{dV} \approx h \frac{A_s}{V} \qquad [37]$$

and

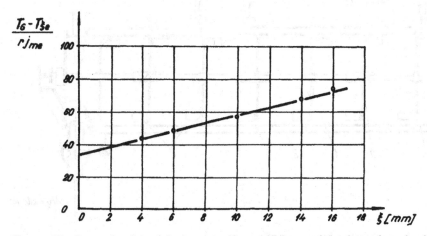

Figure 16 Determination of the heat transfer coefficient and the thermal conductivity.

$$K_V = K \frac{dA_s}{dV} \approx K \frac{A_s}{V} \qquad\qquad\qquad [38]$$

Here dA_s and A_s are the differential and total interfacial areas, respectively, for heat/mass transfer while dV and V are the corresponding differential and total bed volumes, respectively.

5.1. Determination of the Volumetric Heat Transfer Coefficient

The method described here is for a plant-size, direct rotary dryer although the technique is applicable to other dryers of granular media as well. In a plant scale, effects of various parameters cannot be studied in depth as one could in a laboratory or pilot-plant size equipment. Hence, such data are not generally applicable.

Figure 17 is a schematic of the dryer employed for this investigation. It consists of a cylindrical shell (1 in Fig. 17) covered with a heating jacket (2) and fitted with an agitator (3) within the shell. Heat for drying can be supplied directly by hot air or indirectly by circulating a heating fluid through the jacket. In direct drying the drying air is heated.

The stirrer carries out agitation and disintegration of the dried grains Driving rings (4 in Fig. 17) and plates (5) are attached in a staggered fashion on the agitator in order to increase the turbulence of the drying medium and the residence time of the material to be dried. The scraper-stirrer plates (8), which stir the material and feed it into the flow of the drying gas, are fastened to these rings and plates, respectively. The material to be dried is fed through the hopper (6) to the feeder (7). Exit air leaves through the dust separator (9).

The parameters measured include flow rate of the solids feed, mass flow rate of air, solids temperature at entry and exit, and the humidity of the drying air entering and

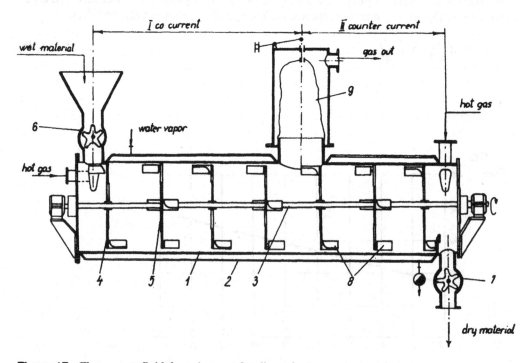

Figure 17   The contact-fluid dryer (see text for discussion).

leaving the dryer. The volumetric heat transfer coefficient can then be calculated from the definition

$$Q_H = hA_s \Delta T_m \tag{39}$$

where the average temperature difference for cocurrent flow is the logarithmic average value

$$\Delta T_m = \Delta T_{\log} = \frac{(T_{G\,\text{in}} - T_{SW\,\text{in}}) - (T_{G\,\text{out}} - T_{SW\,\text{out}})}{\ln \dfrac{T_{G\,\text{in}} - T_{SW\,\text{in}}}{T_{G\,\text{out}} - T_{SW\,\text{out}}}} \tag{40}$$

Dividing both sides of Eq. [39] by the empty volume of the equipment V results in

$$\frac{Q_H}{V} = h \frac{A_s}{V} \Delta T_m \tag{41}$$

that is,

$$\frac{Q_H}{V} = h_V \Delta T_m \tag{42}$$

The transfer rate is given by

$$Q_H = G_T c_{pG}(T_{G\,\text{in}} - T_{G\,\text{out}}) = L_{SW} c_{SW}(T_{SW\,\text{out}} - T_{SW\,\text{in}}) \tag{43}$$

On the basis of Eqs. [42] and [43] the volumetric heat transfer coefficient can also be determined as

$$h_V = \frac{G_T c_{pG}(T_{G\,\text{in}} - T_{G\,\text{out}})}{V \, \Delta T_{\log}} = \frac{L_{SW} c_{SW}(T_{SW\,\text{out}} - T_{SW\,\text{in}})}{V \, \Delta T_{\log}} \tag{44}$$

In these experiments already dried granular material must be used in order to produce a process of pure sensible heat transfer.

For a direct-heated rotary dryer, the volumetric heat transfer coefficient can be estimated from the following empirical correlation:

$$h_V = 2950 \, n\phi^{0.173} G_0^{0.9} \tag{45}$$

where ϕ is the fractional hold-up (volume of granular material as a fraction of the total volume of the dryer), n is the rpm of the agitator, and G_0 is the mass flow rate of the drying air.

5.2. Determination of the Heat and Mass Transfer Coefficients in Through Circulation Drying

In general one needs only the volumetric heat/mass transfer coefficients for design purposes. In some special cases (e.g., drying alfalfa plants), the material to be dried may consist of a mixture of materials with distinctly different characteristics (the stems and the leaves in the case of alfalfa). The stems have a well-defined smaller surface area while the leaves have large areas which may, however, be covered, therefore not all areas are available for heat/mass transfer. The drying rates are quite dissimilar for the two components, therefore care must be exercised in drying whole alfalfa plants. References 26 and 27 give the necessary details for the determination of the volumetric as well as real heat/mass transfer coefficients. These references give a methodology for evaluation of the interfacial area for heat/mass transfer on the basis of certain assumptions.

The volumetric mass and heat transfer coefficients can be determined simply from the definitions (Eqs. [46], [47]) for an adiabatic operation. (See Ref. 28 for details.)

$$K_V = \frac{G_T}{AH} \frac{Y_{G\,out} - Y_{G\,in}}{\Delta Y_{log}}$$ [46]

and

$$h_V = \frac{G_T c_{WG}}{AH} \frac{T_{G\,in} - T_{G\,out}}{\Delta T_{log}}$$ [47]

where

$$\Delta Y_{log} = \frac{(Y_W - Y_{G\,in}) - (Y_W - Y_{G\,out})}{\ln \dfrac{Y_W - Y_{G\,in}}{Y_W - Y_{G\,out}}}$$ [48]

and

$$\Delta T_{log} = \frac{(T_{G\,in} - T_W) - (T_{G\,out} - T_W)}{\ln \dfrac{T_{G\,in} - T_W}{T_{G\,out} - T_W}}$$ [49]

Further parameters can be seen in Figures 18 and 19. When the saturation is nonadiabatic and the bed height is sufficiently small, it is evident according to Reference 28 that

$$\overline{\frac{1}{T_s - T_G}} \cong \frac{1}{\overline{T}_s - \overline{T}_G}$$ [50]

and

$$\left(\overline{\frac{1}{Y_s - Y_G}}\right) \cong \frac{1}{\overline{Y}_s - \overline{Y}_G}$$ [51]

The volumetric transport coefficients are in this case

$$K_V(\tau) = \frac{G_T}{A\,\Delta H} \frac{\Delta Y_G}{\overline{\overline{Y}}_s - \overline{Y}_G}$$ [52]

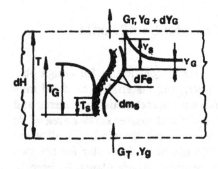

Figure 18 Elementary part of lucerne bed.

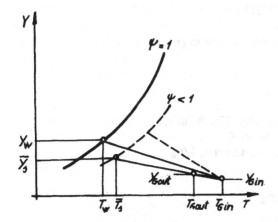

Figure 19 Saturation of air during circulation through bed.

and

$$h_V(\tau) = \frac{G_T c_{wG}}{A \,\Delta H} \frac{\Delta T_G}{\overline{T}_s - \overline{T}_G}$$ [53]

In the above correlations,

$$\Delta Y_G = Y_{G\,\text{out}} - Y_{G\,\text{in}}$$ [54]

and

$$\Delta T_G = T_{G\,\text{out}} - T_{G\,\text{in}}$$ [55]

$\overline{T}_s$ can be calculated by iteration from the correlation

$$\frac{\Delta Y_G}{\Delta T_G} = \frac{\dfrac{M_V}{M_G} \dfrac{p_{OV}^*}{P} \, \Psi(\overline{T}_s, \overline{X}) - \overline{Y}_G}{\overline{T}_s - \overline{T}_G}$$ [56]

and with $\overline{T}_s$ known, $\overline{Y}_s$ can also be determined:

$$\overline{Y}_s = \frac{M_V}{M_G} \frac{p_{OV}^*(\overline{T}_s)}{P} \, \Psi(\overline{T}_s, \overline{X})$$ [57]

6. CONCLUSION

Simple experimental methods are discussed in this chapter for determination of the key equilibrium and heat/mass transfer parameters need in engineering calculations for common dryers. The coverage is by no means all-inclusive. The reader is referred to the literature cited for details.

NOMENCLATURE

A surface, m^2
Bi   Biot number
c_m moisture capacity, kg/kg

c_{SW}	specific heat capacity of wet material, J/kg·K
D_a	effective diffusivity coefficient, m²/s
D_{VG}	molecular diffusivity coefficient between vapor and gas, m²/s
Fo	Fourier number
G_T	total gas flow rate, kg/s
G_0	total gas flow, kg/m²s
h	heat transfer coefficient, W/m²K
h_m	modified material transfer coefficient, Eq. [17], W/m²K
h_v	volumetric heat transfer coefficient, W/m³K
ΔH	sum of heat of evaporation plus heat of moisture, J/kg
j_m	mass flow density, kg/m²s
j_q	heat flow density, W/m²
k_G	mass transfer coefficient, kmol/m²sPa
K	mass transfer coefficient, kg/m²sPa
K_G	overall mass transfer coefficient, kg/m²sPa
K_H	overall heat transfer coefficient, W/m²K
K_V	volumetric mass transfer coefficient, kg/m²s
L_{SW}	mass flow of wet material, kg/s
m	mass, kg
n	revolutions per minute, 1/s
N_W	drying rate, kg/m²s
N_{We}	constant drying rate, kg/m²s
p	partial pressure, Pa
p_O	partial vapor pressure of pure component, Pa
Q	heat rate, W
R	gas constant, J/kgK
S	slope of equilibrium curve defined by Eq. [31], Pa/K
$S*$	slope of equilibrium curve, K⁻¹
$S**$	slope of equilibrium curve defined by Eq. [24], Pa
T	temperature, K
T_m	average temperature, K
X	moisture content of material, dry basis
Y	moisture content of air, dry basis
Z	half thickness of the plane, m
V	volume, m³

Subscripts

d	belonging to the dew point
e	equilibrium value
G	gas
h	heat transfer
in	entering
log	logarithmic mean value
m	mass transfer
O	initial value
out	leaving
s	surface
S	solid

V vapor
W wet, wet bulb
Z in the symmetry plane of a plate
ξ at the border of the dry and wet layer

Superscripts

$^{-}$ average value
$*$ equilibrium

Greek Symbols

δ thickness, m
δ_k eigenvalue, defined by Eq. [14]
ϕ fractional holdup of solids
ϵ phase change criterion
κ_h heat diffusivity, m^2/s
κ_m moisture diffusivity coefficient, m^2/s
λ_h thermal conductivity, W/mK
λ_m moisture conductivity, kg/ms K
ν eigenvalue, defined by Eq. [23]
Ψ relative moisture content
μ_D diffusion-resistance coefficient
$\bar{\rho}_{Os}$ bulk density of solids, kg/m^3
τ time, s, h
ξ thickness of dry zone, m
Δ symbol of difference

REFERENCES

1. J. H. Perry, *Chemical Engineers' Handbook*, McGraw-Hill, New York, 1967.
2. A. V. Likov, *Theory of Drying* (in Hungarian), Nehézipari Könyvkiadó, Budapest, 1952.
3. R. B. Keey, *Drying Principles and Practice*, Pergamon Press, Oxford, 1972.
4. R. Toei and S. Hayashi, *Memoirs Fac. Eng. Kyoto Univ.*, *25*: 457 (1963).
5. R. Toei, S. Hayashi, S. Sawada, and T. Fujitani, *Kagakukogaku*, *29*: 525 (1965).
6. P. Toei and M. Okasaki, *Chemical Physical Journal* (in Russian), *19*: 464 (1970).
7. A. Wexler, *Humidity and Moisture*, Reinhold, New York, 1965.
8. L. Imre, *Handbook of Drying* (in Hungarian), Müszaki Könyvkiadó, Budapest, 1974.
9. R. H. Stokes, *J. Amer. Chem. Soc.*, *67*: 1689 (1945).
10. A. H. Landrock and B. E. Proctor, *Food Technol.*, *S. 15*: 332 (1961).
11. C. D. Hogdman, *Handbook of Chemistry and Physik*, Chem. Rubber Publ. Co., Cleveland, 1960.
12. J. D'Ans and E. Lax, *Taschenbuc für Chemiker und Physiker*, Bd. I, Springer, Berlin and New York, 1967.
13. S. Kamei, *Dechema-Monogr.*, *32*: 305 (1959).
14. E. Robens and G. Sandstede, *Chem. Ing. Technik*, *40*: 957 (1968).
15. M. Büchner and E. Robens, *Kolloid-Z. Z. Polym.*, *248*, 1020 (1971).
16. W. R. Bousfield, *Trans. Faraday Soc.*, *13*: 401 (1918).
17. R. E. Treybal, *Mass-Transfer Operations*, McGraw-Hill, New York, 1968.
18. S. Szentgyörgyi, *Period. Polytech.*, *Mech. Eng.*, *24*: 137 (1980).
19. A. V. Likov, *Heat and Mass Transfer in Capillary-porous Bodies*, Pergamon Press, Oxford, 1966.

20. S. Szentgyörgyi, *Handbook of Drying* (L. Imre, ed.) (in Hungarian), Müszaki Könyvkiadó, Budapest, 1974.
21. S. Szentgyörgyi and K. Molnár, *Proceedings of the First International Symposium on Drying*, Montreal, 1978, p. 92.
22. K. Molnár, *Acta. Tech.*, *Tech. Sci.* (in Hungarian), *52*: 93 (1976).
23. K. Molnár, Investigation of Batch Convective Drying of Capillary Pored Materials Bared on Double-Layer Model, Thesis of Cand. Sc., Budapest, 1978, pp. 113–130.
24. K. Molnár, Fifth International Heat Pipe Conference, Tsukuba, Japan, 1984.
25. O. Krischer and K. Kröll, *Trocknungstechnik*, Springer, Berlin, 1956.
26. L. Imre, I. L. Kiss, I. Környey, and K. Molnár, *Drying '80* (A. S. Mujumdar, ed.), Hemisphere Publ. Co., New York, 1980, p. 446.
27. L. Imre and K. Molnár, *Proc. 3rd Int. Drying Symp.*, (I. C. Ashwort, ed.), Drying Res. Ltd., Wolverhampton, 1982, Vol. 2, p. 73.
28. K. Molnár and S. Szentgyörgyi, *Period. Polytech.*, *Mech. Eng.*, *28*: 261 (1983).
29. L. Imre, *Measuring and Automation* (in Hungarian), *8, 9, 10*: 302 (1963).
30. S. Szentgyörgyi, *Advances in Drying* (A. S. Mujumdar, ed.), Hemisphere Publ. Co., Washington, 1987.

3
Basic Process Calculations in Drying

Zdzisław Pakowski
Technical University of Łódź
Łódź, Poland

Arun S. Mujumdar
McGill University
Montreal, Quebec, Canada

1. OBJECTIVES

In industrial practice three situations may arise that call for process calculations of dryers:

1. Selection of suitable dryer type and size for a given product to optimize the capital and operating costs
2. Finding operating conditions and data for the ancillary equipment for a selected dryer or a dryer already in use for each new product to be dried
3. Determination of the optimal operating conditions for a dryer already in operation

The objectives of each of these tasks must be specified before meaningful process calculations are undertaken.

The first problem requires previous knowledge of what dryer types are applicable to the drying of a given material. Generally, this information comes from prior industrial experience. However, drying of new products requires experimental work. It is also important to examine new drying technology before the final selection is made. Subsequently, simplified calculations of dryer volume, drying time, heat and electricity consumption, volumetric flow rate of gas required, and so on, can be made and on this basis an optimum selection made.

The second type of problem is a further development of the first one limited to one dryer. Here more accurate calculations of all process parameters and data for auxiliary equipment design or selection are to be prepared. The basic parameters that need to be calculated are material holdup in the dryer (or batch size in batch dryers) and material residence time necessary to complete the specified drying duty. The ratio of these two parameters gives the average material throughput for which the dryer is designed.

Let us consider a continuously operated adiabatic dryer. The set of parameters that describes the operating conditions of such a dryer is given in Figure 1. The set of known parameters consists of dry solid throughput, inlet and outlet solid moisture content and temperature, and, if gas entering the dryer is not recycled, inlet gas humidity. In the design process we should be able to determine all unknown values. Usually the situation

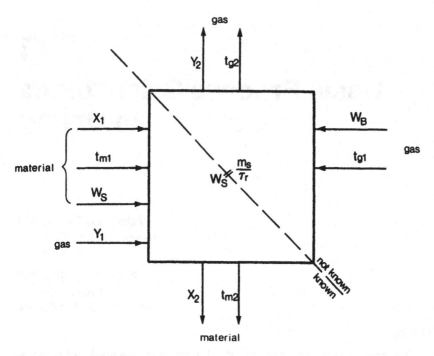

Figure 1 Structure of information preceding process calculations of a dryer.

is more complicated as, in the case of many dryers, the holdup (or drying time) is affected by velocities of solids and gas. Thus additional unknowns are added to the set.

The third type of problem involves mathematical modeling of an existing dryer; that is, for a given dryer and material, all relations between the parameters shown in Figure 1 are known. The optimum values of these parameters are to be found after the appropriate "target function" is formulated. Usually the drying cost per unit product is sought to be minimized. Other factors may also be specified. In case of multiple targets multiobjective optimization is employed.

All these problems are solved using a large experimental database and experience; it is scarcely possible to carry out a thorough process design of a dryer on a theoretical basis alone.

In this chapter water is considered the liquid to be removed by thermal means using air as the drying medium. Similar calculation procedures are also applicable to other liquids (e.g., organic solvents) and gases (e.g., nitrogen), which are insoluble and non-reacting with these liquids.

An approach to drying of solids containing mixtures of solvents differs to a large extent from drying of a single component moisture. This subject is presented in detail in Pakowski (1).

2. TECHNICAL PERFORMANCE FACTORS OF DRYERS

For a realistic technoeconomic evaluation of a dryer installation, certain technical performance factors must be defined. The following performance factors are often used in practice:

Energy efficiency (EE) = $\dfrac{\text{energy used for moisture evaporation}}{\text{total energy supplied to the dryer}}$ %

Thermal efficiency (TE) = $\dfrac{\text{mass of water evaporated}}{\text{amount of heat supplied}}$ kg/kJ

Adiabatic thermal efficiency (ATE)

$\quad$ = $\dfrac{\text{mass of water evaporated at adiabatic saturated temperature}}{\text{amount of heat supplied}}$ kg/kJ

Specific heat consumption (SHC) = $\dfrac{\text{amount of heat supplied}}{\text{mass of water evaporated}}$ = $\dfrac{1}{TE}$ kJ/kg

Specific power consumption (SPC) = $\dfrac{\text{amount of electrical power consumed}}{\text{material throughput}}$ kJ/kg

Specific volume (SV) of dryer = $\dfrac{\text{volume of the drying chamber}}{\text{material throughput}}$ $\text{m}^3 \cdot \text{s/kg}$

Specific gas consumption (SGC) = $\dfrac{\text{dry air flow rate}}{\text{dry material throughput}}$ kg/kg

3. HEAT AND MASS BALANCES OVER DRYERS

Figure 1 is a schematic representation of an adiabatic dryer with six variables on the unknown side. To determine these unknown values, six equations are required. Unfortunately, only five equations are at our disposal. These are

Mass balance equation
Enthalpy balance equation
Mass transfer kinetic equation (rate equation)
Heat transfer kinetic equation (rate equation)
Residence time equation

For an ideal adiabatic dryer (i.e., dryer with no heat or mass losses, with direct heat supply from incoming gas) in steady state these equations are as follows:

$$W_S(X_2 - X_1) = W_B(Y_1 - Y_2) \tag{1}$$

$$W_S(i_{m2} - i_{m1}) = W_B(i_{g1} - i_{g2}) \tag{2}$$

$$W_S(X_2 - X_1) = -(w_D)_m A \tag{3}$$

$$W_S(i_{m2} - i_{m1}) = \alpha A(\Delta t)_m - (h_{Av}w_D)_m A \tag{4}$$

$$W_S = \dfrac{m_S}{\tau_r} \tag{5}$$

where

$$A = m_S a_S/\rho_S = \text{interfacial area} \tag{6}$$

$$i_g = (c_A + c_B)t + \Delta h_{v0}\, Y \tag{7}$$

$$i_m = (c_{AL} + c_B)t - \Delta h_s X \tag{8}$$

$$(w_D)_m = k_Y(Y^* - Y)_m \tag{9}$$

$$(\Delta t)_m = \alpha\,(t_g - t_m)_m \tag{10}$$

The drying rate definition in Eq. [9] is shown in a form suitable mainly for the constant drying rate period. In the falling rate period it may be defined in a form of the so-called characteristic drying curve (see Ref. 2) or calculated on the basis of suitable models of drying kinetics. Please note the sign convention used in Eqs. [3], [4], [9], and [10]: heat flux is positive when heat is transferred from gas to solid and drying rate is positive when mass is transferred from solid to gas phase.

In this situation one variable must be assumed to allow one to solve the equation set. Generally the drying gas flow rate W_B is assumed. In most cases the inlet gas temperature t_{g1} may also be assumed, which allows calculation of the solids exit temperature t_{m2} (placed previously on the known side). The set of Eqs. [1] through [5] can be extended to fit any continuous dryer. In particular, heat transfer Eqs. [2] and [4] must be modified to allow the inevitable heat losses or input of additional heat in case of indirect dryers, for example. A more general form of Eq. [2] can be written as

$$W_S(i_{m2} - i_{m1}) = W_B(i_{g1} - i_{g2}) + q_c - q_t + \Delta q_t + q_m \tag{11}$$

where

q_c = indirectly supplied heat
q_t = heat losses
Δq_t = heat carried in minus heat carried away by conveyors or other transport devices
q_m = all mechanical work input into the drying chamber

For graphic representation of the balance equations on the enthalpy-humidity charts (see Sec. 4), Eq. [11] is rearranged into the following more convenient form:

$$\frac{i_{g2} - i_{g1}}{Y_2 - Y_1} = \left(\frac{W_S}{W_B} (i_{m1} - i_{m2}) + \frac{\Sigma q}{W_B} \right) \frac{1}{Y_2 - Y_1} = \Delta \tag{12}$$

Here, Σq = algebraic sum of heat gains and losses over the dryer (gains have positive signs, losses negative). Additional enthalpy terms also must be introduced into Eq. [2], which can be written in the form

$$W_S(i_{m2} - i_{m1}) - \Sigma q_S = W_B(i_{g1} - i_{g2}) + \Sigma q_g = \alpha A \, (\Delta t_m) - (h_{Av} w_D)_m A \tag{13}$$

where

Σq_S = sum of heat gains (+) and losses (−) for the solid phase
Σq_g = sum of heat gains and losses for the gas phase

The above set of equations is written in terms of input-output values. This makes the heat transfer driving force $(\Delta t)_m$ as well as the drying rate $(w_D)_m$ difficult to estimate since these vary continuously with distance within the dryer.

In many cases local driving forces can be used. Unless we model an ideally mixed dryer, differential forms of equations equivalent to Eqs. [1] through [5] can be written for a differential dryer element. In these equations local heat transfer driving force and the local drying rate must be used. For a cocurrent or countercurrent dryer element illustrated in Figure 2 (cut perpendicularly to the direction of solid flow, which is the positive direction of the ℓ axis), the following set of equations can be written.

$$\frac{d(W_S X)}{d\ell} \pm \frac{d(W_B Y)}{d\ell} + \frac{d(m'_s X)}{d\tau} = 0 \tag{14}$$

$$\frac{d(W_S i_m)}{d\ell} \pm \frac{d(W_B i_g)}{d\ell} + \frac{d(m'_s i_m)}{d\tau} = 0 \tag{15}$$

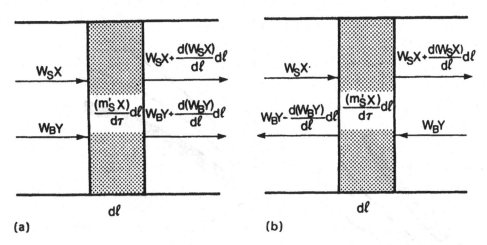

Figure 2 Scheme of a differential dryer length element: (a) cocurrent; (b) countercurrent.

$$\frac{1}{S}\frac{d(W_s X)}{d\ell} + \frac{1}{S}\frac{d(m_s' X)}{d\tau} = -w_D a_V \tag{16}$$

$$\frac{1}{S}\frac{d(W_s i_m)}{d\ell} + \frac{1}{S}\frac{d(m_s' i_m)}{d\tau} = (\alpha \, \Delta t - h_{Av} w_D) a_V \tag{17}$$

where m_s' = mass loading per length of the dryer (kg/m). The positive sign for the second term of both Eqs. [14] and [15] is used for cocurrent; negative is used for countercurrent. Change of gas phase concentration is neglected here as small compared with that of the solid phase. At steady-state operation all accumulation terms become zero and W_B and W_S become constants.

In effect, the following simple set of equations is obtained:

$$W_s dX \pm W_B dY = 0 \tag{18}$$

$$W_s di_m \pm W_B di_g = 0 \tag{19}$$

$$W_s dX = -w_D a_V dV \tag{20}$$

$$W_s di_m = (\alpha \, \Delta t - h_{Av} w_D) a_V dV \tag{21}$$

This set of equations holds for parallel plug flow of both media. It must be modified for the case of crossflow or when axial dispersion in one or both phases must be considered. It must not be used when perfect mixing of phases can be assumed to take place in the drying chamber. In the last case both moisture content and temperature change stepwise and simple input-output analysis are applicable.

In the case of a nonadiabatic dryer, differential dryer balances must include local heat losses and gains in Eqs. [19] and [21].

4. HEAT AND MASS BALANCE CALCULATIONS
ON HUMIDITY CHARTS

Although almost all drying problems can be handled using a computer, a traditional engineering tool, humidity (or psychrometric) charts for graphic solution of many problems, still offers a simple, fast, and visually appealing method of approach. Two types of humidity charts for solution of drying problems are in use: psychrometric and enthalpy-humidity charts.

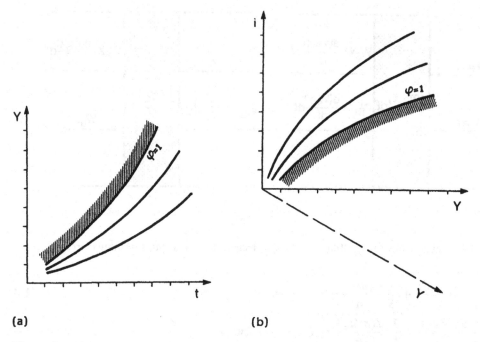

Figure 3 Scheme of basic humidity charts used in drying: (a) psychrometric chart; (b) enthalpy-humidity chart.

Psychrometric charts, most commonly used to determine air humidity on the basis of wet and dry bulb temperatures, are prepared in orthogonal coordinates with temperature as the abscissa and humidity as the ordinate. Most psychrometric charts do not generally provide an enthalpy scale. If this scale is provided, the isotherms are curvilinear because of the choice of the coordinate system. To facilitate readout in the low-humidity range, enthalpy charts are commonly drawn in oblique coordinates with enthalpy as the ordinate and gas humidity as the abscissa, which is inclined to the ordinate at an angle greater than 90°, as shown in Figure 3. Using the definition of enthalpy, an additional grid of temperature can be placed on this chart, thus allowing for simultaneous readouts of enthalpy, humidity, and temperature.

Both types of charts are suitable for heat and mass balance calculations of dryers. Kinetic calculations cannot be done on these charts as they require the fourth variable—time.

4.1. Construction of Psychrometric and Enthalpy-Humidity Charts

Psychrometric charts have a linear temperature scale as the abscissa and a linear humidity scale as the ordinate. Lines of constant relative humidity varying from 100% down to any required value are then drawn according to vapor pressure data using the formula

$$Y = \frac{M_A}{M_B} \frac{\varphi p_s(t)}{P - \varphi p_s(t)} \qquad [22]$$

Furthermore, constant adiabatic saturation lines are drawn according to the equations

$$\frac{t - t_{as}}{Y - Y_{as}} = -\frac{\Delta h_{v,as}}{c_H} \quad \text{or} \quad \frac{i_g - i_{g,as}}{Y - Y_{as}} = +c_{AL}t_{as} \tag{23}$$

and constant wet bulb temperature lines according to the approximate equation

$$\frac{t - t_w}{Y - Y_w} = -\left[\text{Le}^{-2/3} \frac{M_A/M_B}{M_A/M_B + Y_w} (1 + Y_w) \right] \frac{\Delta h_{vw}}{c_H} \tag{24}$$

A psychrometric chart for air and toluene is shown in Figure 4 as an illustration.

Most of the dryer balance calculations can be performed on enthalpy charts. Almost all dryer manufacturers routinely generate their own enthalpy-humidity charts (generally by means of a computer graphics facility) for each new system they encounter. To facilitate generating enthalpy-humidity charts, the principles of their construction will be outlined here.

The basis of the chart (Fig. 5) is an oblique coordinate system with linear scales on

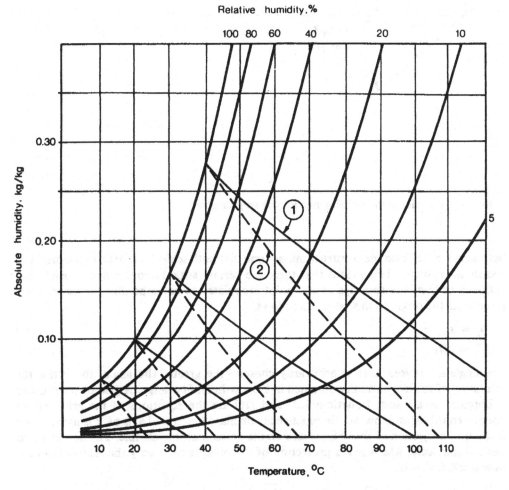

Figure 4 Psychrometric charts for toluene-air mixture: (1) adiabatic saturation lines; (2) wet bulb lines.

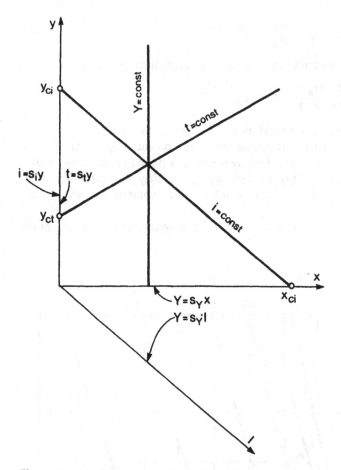

Figure 5   Principles of the enthalpy-humidity chart geometry.

both axes (to facilitate the construction, assume that both scales have zero in the origin of coordinate system). This means that there are certain scale factors s_i and s'_Y such that enthalpy or humidity readings at any point are related to corresponding distances measured from the origin to this point, as follows:

$$i_g = s_i l_i \tag{25}$$

$$Y = s'_Y l_Y \tag{26}$$

The angle α between axes is arbitrarily chosen—the greater the angle, the better the resolution of relative humidity obtained in the low-humidity range. A grid of coordinates is obtained as a result. Equations [25] and [26] are the equations of the grid lines in oblique coordinates. However, in practice an auxiliary horizontal axis of Y is used, which is simply a projection of the oblique axis on the horizontal straight line. The new axis also has a linear scale, which is the projection of the oblique axis, only the scaling factor is now smaller; that is,

$$s_Y = s'_Y \cos\left(\alpha - \frac{\pi}{4}\right) \tag{27}$$

In x,y coordinates of the plotting paper,

$$i_g = s_i y \qquad [28]$$

$$Y = s_Y x \qquad [29]$$

In these coordinates the equation of constant Y lines is

$$x = \frac{Y}{s_Y} \qquad [30]$$

and of constant enthalpy lines (isoenthalps),

$$y = A_i x + \frac{i_g}{s_i} \qquad [31]$$

As seen from Fig. 5, the slope of isoenthalps is negative. Isoenthalps intercept the vertical axis at intercept $y_{ci} = i_g/s_i$ and the horizontal axis at the point x_{ci}. The enthalpy scale factor is selected in such a way that

$$y_{ci} s_i = \Delta h_{vo} x_{ci} s_Y \qquad [32]$$

Using Eq. [31] we find the relationship

$$A_i = -\Delta h_{vo} \frac{s_Y}{s_i} \qquad [33]$$

which can be used to find s_i if the slope of isoenthalps is already selected or to find the slope if scaling factors s_i and s_Y are determined.

The equation for isotherms can be easily found. The local slope of any isotherm can be calculated as $\Delta y/\Delta x$. The y can be calculated using the equation of the isoenthalp, and finally we get

$$\frac{\Delta y}{\Delta x} = \frac{A_i \Delta x + \Delta i_g/s_i}{\Delta x} \qquad [34]$$

At constant temperature, the enthalpy increment is*

$$\Delta i_g = c_A t \Delta Y + \Delta h_{vo} \Delta Y \qquad [35]$$

Using Eq. [29] to express Y, we obtain

$$\frac{\Delta y}{\Delta x} = \frac{A_i \Delta x + (c_A t s_Y \Delta x + \Delta h_{vo} s_Y \Delta x)}{\Delta x} = A_i + (c_A t + \Delta h_{vo}) \frac{s_Y}{s_i} \qquad [36]$$

Substituting Eq. [33] for A_i, we finally get

$$\frac{\Delta y}{\Delta x} = c_A t \frac{s_Y}{s_i} = A_t \qquad [37]$$

The slope of isotherm will then depend only on temperature, that is, isotherms of higher temperatures have greater slope than those for lower ones. However, for constant temperature this slope is constant. This proves that isotherms are straight lines and as such can be described by the equation

$$y = A_t x + y_{ct} \qquad [38]$$

---

*Please note that the specific heats are taken as averaged in the given temperature range.

The value of the intercept y_{ct} can be easily found from the fact that at $y = 0$ enthalpy is given by

$$i_g = c_B t \qquad\qquad [39]$$

Because the enthalpy scale is linear, if c_B is assumed constant one can also draw a linear scale for temperature such that

$$t = s_t y \qquad\qquad [40]$$

Letting $y_{ct} = t/s_t$ in Eq. [38], the equation for the isotherm becomes

$$y = A_t x + \frac{t}{s_t} \qquad\qquad [41]$$

where, on the basis of Eq. [39],

$$s_t = \frac{s_i}{c_B} \qquad\qquad [42]$$

Substituting i, Y, and t given by transformed Eqs. [29], [31], and [41] into Eqs. [22], [23], and [26], equations for constant relative humidity, constant adiabatic saturation temperature, and constant wet bulb temperature lines can be derived. As an illustration, a chart for an air-water system is given in Figure 6.

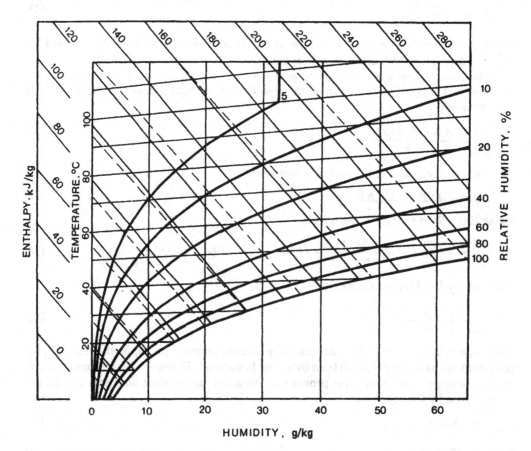

Figure 6   Enthalpy-humidity chart for water-air mixture.

More details on computerized enthalpy-humidity chart construction can be found in Ref. 3. This was the method used to produce such charts in Appendix A of this book.

4.2. Input-Output Balances on Humidity Charts

Any change of enthalpy or humidity or both of a gas-vapor mixture described by point A in the enthalpy-humidity chart (Fig. 7-a) moves the point A to a new position B. The direction of this change is given by the ratio ϵ:

$$\epsilon = \frac{\Delta i_\ell}{\Delta Y} \tag{43}$$

A scale for various values of ϵ is provided in enthalpy-humidity charts for rapid calculations. This scale can be used in describing quantitatively some of the basic operations, such as heating, adding moisture at constant temperature, or adiabatic saturation, which are shown in Figures 7-b, c, and d. Graphic determination of the properties of a mixture obtained as a result of the mixing of two humid gases can be performed using calculated ϵ values. Heat and mass balance (Fig. 8) for the adiabatic case yield

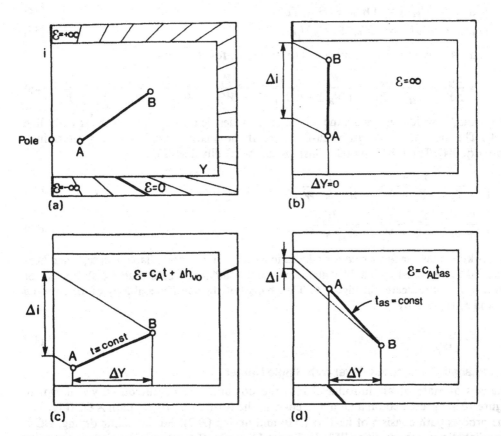

Figure 7 Basic process paths on enthalpy-humidity charts: (a) heating and humidification of gas; (b) heating at constant humidity; (c) humidification at constant temperature; (d) adiabatic saturation process.

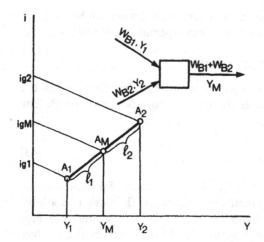

Figure 8   Graphic representation of adiabatic mixing of two humid gases.

$$W_{B1}Y_1 + W_{B2}Y_2 = (W_{B1} + W_{B2})Y_M \qquad [44]$$

$$W_{B1}i_{g1} + W_{B2}i_{g2} = (W_{B1} + W_{B2})i_{gM} \qquad [45]$$

If A_1 and A_2 represent the humid gases to be mixed, for A_1 we have

$$\epsilon = \frac{\Delta i_g}{\Delta Y} = \frac{i_{gM} - i_{g1}}{Y_M - Y_1} = \frac{(W_{B1}i_{g1} + W_{B2}i_{g2})/(W_{B1} + W_{B2}) - i_{g1}}{(W_{B1}Y_1 + W_{B2}Y_2)/(W_{B1} + W_{B2}) - Y_1} = \frac{i_{g2} - i_{g1}}{Y_2 - Y_i} \qquad [46]$$

The point M, which represents the mixture properties, is thus located on the straight line A_1A_2. The location of M can be easily found if the mixture humidity Y_M is determined from Eq. [44]. The following relationships can be readily derived:

$$\frac{\ell_1}{\ell_1 + \ell_2} = \frac{Y_M - Y_1}{Y_2 - Y_1} = \frac{W_{B2}}{W_{B1} + W_{B2}} \qquad [47]$$

$$\frac{\ell_2}{\ell_1 + \ell_2} = \frac{Y_2 - Y_M}{Y_2 - Y_1} = \frac{W_{B1}}{W_{B1} + W_{B2}} \qquad [48]$$

This is known as the *lever rule* for adiabatic mixing. In the nonadiabatic case, when heat is added or removed, point M_o calculated for the adiabatic case must be shifted up or down to M, as indicated in Figure 9. The change of the specific enthalpy of air between M_o and M is given by

$$\Delta i_g = \frac{q}{W_{B1} + W_{B2}} = q_M \qquad [49]$$

Nonadiabatic Continuous Dryer with Single Heater

Changes of state of air introduced into the nonadiabatic continuous dryer shown in Figure 10 are presented on a humidity chart in the form of a process path ABC (Fig. 11). The process path consists of heating (AB) and drying (BC). For adiabatic drying, BC is the adiabatic saturation line (BD). In Figure 11, point E represents the surface tempera- ture for constant rate drying; for the falling drying rate period, point E corresponds to conditions of air at equilibrium with the solids leaving the dryer. In principle, true equilib-

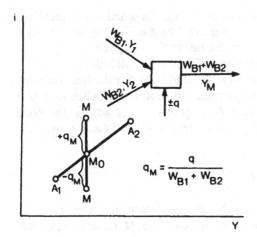

Figure 9   Graphic representation of nonadiabatic mixing of two humid gases.

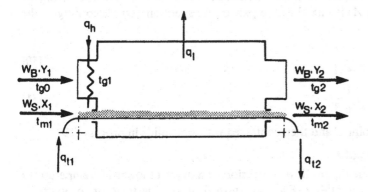

Figure 10   Scheme for nonadiabatic dryer.

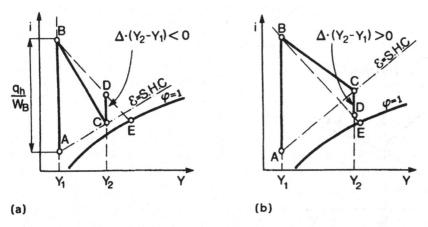

(a)   (b)

Figure 11   Graphic representation of nonadiabatic drying air paths: (a) dryer with heat losses; (b) dryer with heat addition.

rium of air and solid requires an infinitely long time of contact. In practice the actual exit air conditions for an adiabatic dryer are given by point D. The distance between D and E corresponds to the degree of equilibrium reached; the ratio $(Y_D - Y_A)/(Y_E - Y_A)$ is the dryer efficiency corresponding to the Murphree stage efficiency in distillation.

In nonadiabatic conditions, working in the same range of humidity the final enthalpy of air is given in Eq. [12], which graphically corresponds to subtracting (when $\Delta < 0$) a portion $\Delta(Y_2 - Y_1)$ from the enthalpy of point D (Figure 11-a) or adding the same portion (when $\Delta > 0$, Fig. 11-b) to get point C, which represents actual conditions of the exiting air. Here Δ is equal to

$$\Delta = \frac{1}{W_B(Y_2 - Y_1)} [W_S(i_{m1} - i_{m2}) + \Sigma q]$$

A similar way of finding the final air enthalpy graphically is presented in Figure 12.

Note that specific heat consumption is simply the slope of line AC connecting incoming and exiting air conditions.

Continuous Dryer with Partial Recycle of Gas

A dryer with partial recycle of air is shown schematically in Figure 13. Here, entering air with a flow rate W_B (point A) is mixed with a part of the spent air (D). According to the lever rule, we have

$$\frac{AB}{AD} = \frac{rW_B}{W_B + rW_B} = \frac{r}{r + 1} \qquad [50]$$

$$\frac{BD}{AD} = \frac{W_B}{W_B + rW_B} = \frac{1}{r + 1} \qquad [51]$$

The mixture is heated to point C and then contacted with wet solids in process CD.

Dryer with Internal Heating of Air

Placing several heaters in the dryer allows operation of a dryer at lower air temperatures to reach the same final gas humidity (Fig. 14). Instead of heating inlet air to point B_o, three heaters may be used in sections to achieve the same final humidity of the dryer exhaust gas (point C).

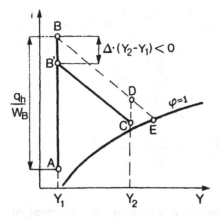

Figure 12 An alternative method of construction of the air path in a nonadiabatic dryer.

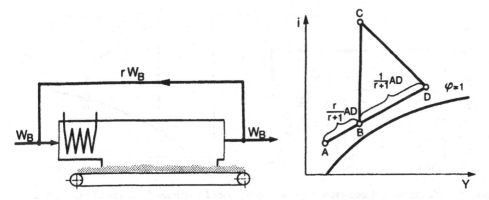

Figure 13 Scheme for a continuous dryer with partial recycle of spent air and graphic representation of this process on an enthalpy-humidity chart.

Closed-Cycle Dryer

Closed gas circulation is appropriate when moisture is to be recovered or an expensive inert gas is used for drying. Referring to Figure 15, the gas initially at A is heated up to point B and then contacted with the wet solid in a dryer, where it is humidified and cooled to point C. The exhaust gas is cooled in a condenser to its dew point temperature D. As condensation proceeds along the saturation line, the gas humidity decreases down to A. This cycle is then repeated. A heat pump may also be used for heating and condensing the vapor. Sometimes the vapor may be dehumidified in a wet scrubber using the solvent in a liquid form. This type of drying cycle is shown in Figure 16.

Dryer Heated by Combustion Gases

Direct combustion of liquid or gaseous fuels in a dryer increases not only the temperature of gas but also its humidity as all fuels contain hydrogen molecules. This is reflected in a finite positive slope of heating portion AB of the process path ABC (Fig. 17). The slope can be calculated using stoichiometric equations for combustion of the fuel used. In this

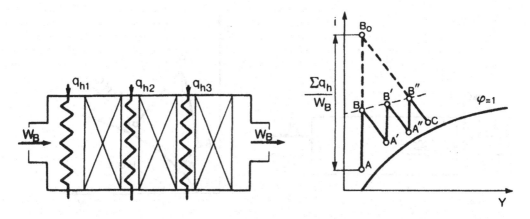

Figure 14 Scheme for a dryer with internal heaters and corresponding air path on an enthalpy-humidity chart.

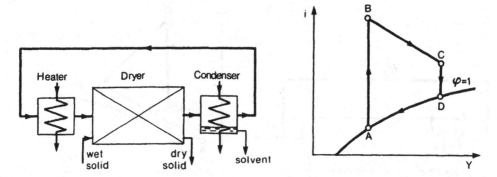

Figure 15 Scheme for a closed-cycle dryer with condenser for dehumidification of exit gases and corresponding air path on humidity chart.

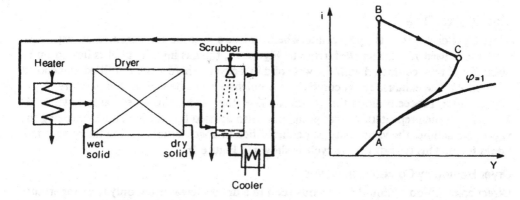

Figure 16 Scheme for a closed-cycle dryer with scrubber for dehumidification of exit gases and corresponding air path on a humidity chart.

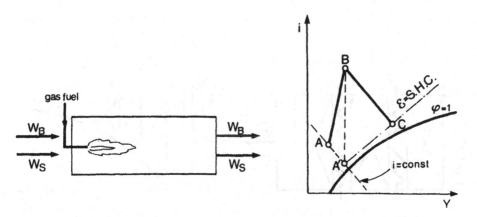

Figure 17 Scheme for a dryer directly heated by burning a hydrocarbon fuel and corresponding air path on humidity chart.

case, the *SHC* can be calculated as a slope of the line AA' in Figure 17, where A' is the point of intersection of isoenthalp A and constant humidity line Y_B.

4.3. Differential Balances Using Humidity Charts

Analytic integration of the set of differential Eqs. [14] through [17] yields the variation of solid moisture content, gas humidity, and temperatures of both media within the dryer with time. The variation of humidity and temperature of the drying gas in the dryer can be represented on a humidity chart in the form of a process path (2,4–6) for the cases of externally as well as internally controlled drying rate periods. However, only in the former case can the process calculations be entirely based on humidity charts. This will be presented in Section 5. Presentation of the process path for an internally controlled dryer requires the knowledge of actual drying rate, which can be calculated on the basis of the formulas in Section 6.1.

5. PROCESS CALCULATIONS IN EXTERNALLY CONTROLLED DRYING

Drying is externally controlled if heat and mass transfer resistance is located on the gas side only. This is, for example, the case in the constant drying rate period. Internal resistance to heat and mass transfer is negligible if the Biot number is less than 0.1 for both heat and mass transfer. Usually the Biot number for mass transfer for drying should be defined as

$$Bi_D = \frac{k_Y(d/2)}{\rho_S D_{AS} A^*} \qquad [52]$$

which is much larger than the Biot number for heat transfer, defined as

$$Bi_H = \frac{\alpha(d/2)}{\lambda_S} \qquad [53]$$

However, in the case of porous materials with large pores and for small, submillimeter particles, the condition $Bi_D < 0.1$ is often fulfilled.

For constant rate drying we can write

$$w_D = k_Y(\Delta Y)_m \qquad [54]$$

and because of lack of temperature and moisture content gradients inside the solid (small Biot numbers), the average moisture content and temperature will be the same as local values of these two parameters at the solid surface.

The mass transfer driving force $(\Delta Y)_m$ is the difference between the humidity of gas at the solid-gas interface and that in the bulk of the gas phase. Humidity at the solid interface can be easily found from the experimentally determined sorption isotherms as $\phi^* = f(X, t_m)$ if thermodynamic equilibrium between solid surface and gas in the laminar sublayers is assumed. This gas, in equilibrium with the moist solid surface, will be termed *equilibrium gas*. Its parameters can be shown on a humidity chart. Since the bulk gas and equilibrium gas paths can be plotted on humidity charts, one can then find the local and averaged driving forces for heat and mass transfer on the gas side. This enables one to separate balance calculations from the kinetic calculations. This facilitates selection of the dryer volume after all the balance calculations are made. This method, when computerized, is also suitable for calculations of a dryer with partial recycle of gas or material.

5.1. Principles of Moist Solid-Humid Gas Contact

When bulk gas A contacts with equilibrium gas E (Fig. 18), the parameters of gas A approach those of gas E along the straight line AE; the slope of this line is given by

$$m = \frac{i_g^* - i_g}{Y^* - Y} \qquad [55]$$

In fact, with a limited amount of air A and large amount of water held at a constant temperature t_m, the gas would eventually become saturated, reaching point E along AE. In typical drying operations the amount of gas is comparable to the amount of solid being dried. Any change in the location of point A to A_1 will usually effect a movement of point E to E_1. The new contact path will be along the line A_1E_1. If, however, the integration step $\Delta Y = Y_1 - Y$ is small, Eq. [55] will allow determination of the bulk gas enthalpy change as follows:

$$m = \frac{i_g^* - i_g}{Y^* - Y} = \frac{i_{g1} - i_g}{Y_1 - Y} = \frac{\Delta i_g}{\Delta Y}$$

$$\Delta i_g = m \, \Delta Y \qquad [56]$$

All four possible situations in moist solid/humid gas contact (locations of points A and E) are sketched in Figure 19.

5.2. Construction of Process Paths for Cocurrent Flow

As input data for such a construction, initial temperature t_{m1} and moisture content X_1 of the solid plus temperature t_{g1} and humidity Y_1 of the entering gas must be known or specified. The equilibrium data in the form of sorption isotherms should also be given. Sorption isotherms are usually given in a graphic form but in many cases can be approximated by a suitable empirical equation. Both methods yield a dependence of equilibrium relative humidity on the solid moisture content and temperature $\varphi^* = f(X, t_m)$. To start

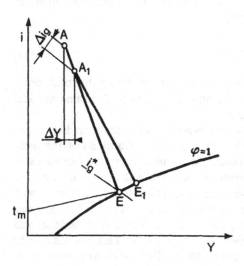

Figure 18 Graphic representation of contact of humid bulk gas A and gas in equilibrium with wet solid E.

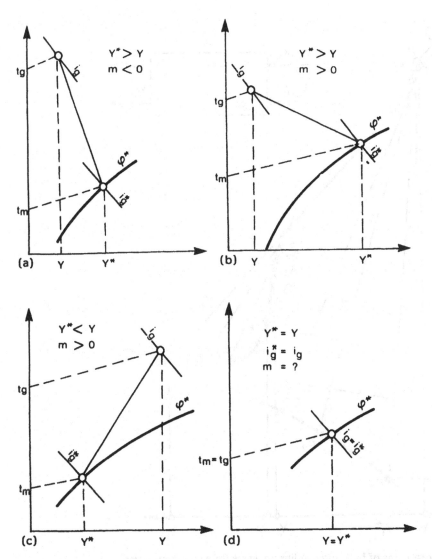

Figure 19 Possible situations in moist solid/humid gas contact operations and their schematic representation on humidity charts: (a, b) drying of solid; (c) humidification of solid; (d) thermodynamic equilibrium.

the construction, unit air consumption W_B/W_S must be assumed. The initial parameters of the incoming gas are represented by the point A_1 in Figure 20. This point is a starting point of the bulk gas properties path. The initial parameters of the gas in equilibrium with incoming solid are represented by a point E_1. Relative humidity of this point $\varphi_1^\dagger$ is read from sorption isotherms for t_{m1} and X_1, as shown in Figure 21. (Here, the process starts in the unbound moisture range; i.e., $\varphi_1^\dagger = 1$.) During contact, the bulk gas properties A_1 tend to equilibrium gas properties E_1. However, as the bulk gas reaches A_2, transferring heat to the solid and gaining in humidity, the equilibrium gas will shift to E_2. If the change of humidity between A_1 and A_2 is ΔY (step of integration), the new material moisture content will be

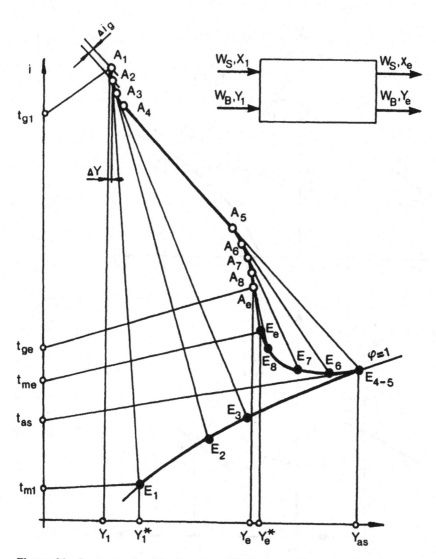

Figure 20 Process paths of bulk and equilibrium gases for a cocurrent dryer.

$$X_2 = X_1 - \Delta Y \frac{W_B}{W_S} \tag{57}$$

where

$$\Delta Y = Y_2 - Y_1 \tag{58}$$

The enthalpy change of bulk gas $\Delta i_g = i_{g2} - i_{g1}$ can be calculated on the basis of Eq. [52] as

$$\Delta i_g = i_{g2} - i_{g1} = m(Y_2 - Y_1) \tag{59}$$

so that the new solid temperature will be

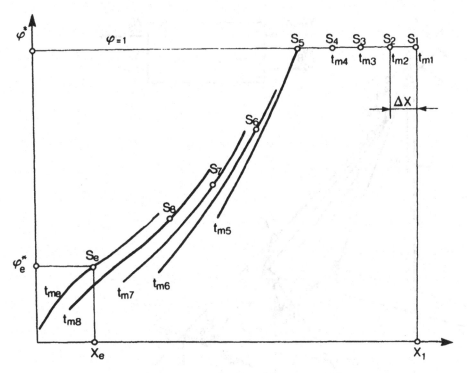

Figure 21   Representation of the drying process on sorption isotherms.

$$t_{m2} = \frac{(c_S + c_{AL}X_1)t_{m1} - \Delta h_s \, \Delta X - \Delta i_g(W_B/W_S)}{c_S + c_{AL}X_2}$$ [60]

where

$$\Delta X = X_2 - X_1$$

Knowing t_{m2} and X_2, point S_2 on the sorption isotherms can be found, which gives Y_2, so we now can plot point E_2 on the humidity chart. This sequence of steps can be repeated. The assumed step ΔY will give the point A_3 on line A_2E_2, and so on, until the point denoted A_4 is reached. Points A_4 and E_4 lie on an adiabatic saturation line. Beyond this, drying is in the constant rate period and point E_4 does not move any longer but point A goes down to A_5. At this point the constant drying rate period ends, which on the sorption isotherms corresponds to point S_5 as the last point for which $\phi = 1$. Further changes take place in the falling drying rate period until the final moisture content X_e and/or solid temperature t_{me} is reached.

5.3.   Construction of the Process Paths for Countercurrent Flow

In the case of the construction of the process paths for countercurrent flow, the incoming gas, whose equilibrium gas parameters are represented by point E_1, contacts the exiting bulk gas A_e (Fig. 22). To perform the aforementioned procedure, parameters of point A_e are calculated from an overall heat and mass balance, assuming a certain reasonable

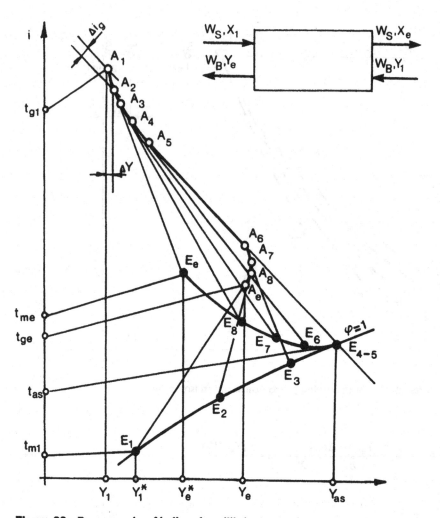

Figure 22 Process paths of bulk and equilibrium gases for a countercurrent dryer.

temperature difference between the gas and the material at the other end of the dryer. Then calculations analogous to these for countercurrent flows are performed. The resulting process path is given in Figure 22. In practice, even if the input-output moisture mass and enthalpy balances are satisfied, the solution does not necessarily converge to the inlet gas parameters. A trial-and-error procedure is needed to obtain convergence within desired accuracy. The method shown above is presented in a form suitable for numerical computations. A fully graphical method earlier developed by Mushtayev may be found in References 5 and 6.

Care must be exercised when computing drying of solids with large heat of sorption when drying starts in the unbound moisture region. Most temperature-dependent sorption isotherm equations used for calculation of the heat of sorption give discontinuity at critical moisture content. This causes numerical algorithms to fail. Smoothing the heat of the sorption curve so it gently approaches zero at critical point is recommended.

5.4.   Calculation of the Dryer Volume

Dryer volume can be calculated on the basis of known transfer coefficient and averaged driving forces over the entire dryer. For input-output balances, mean heat or mass transfer driving forces can be calculated as the mean logarithmic values

$$(\Delta t)_m = \frac{\Delta t_1 - \Delta t_2}{\ln \frac{\Delta t_1}{\Delta t_2}} \tag{61}$$

$$(\Delta Y)_m = \frac{\Delta Y_1 - \Delta Y_2}{\ln \frac{\Delta Y_1}{\Delta Y_2}} \tag{62}$$

where $\Delta t = t_g - t_m$ and $\Delta Y = Y^* - Y$. Equations [61] and [62] hold both for cocurrent and countercurrent drying. However, they are suitable only for ideal plug flow of both phases. In a dryer with perfect mixing, the mean driving forces are simply

$$(\Delta t)_m = \Delta t_2 \tag{63}$$

and

$$(\Delta Y)_m = \Delta Y_2 \tag{64}$$

In dryers with other types of flow, the driving forces lie between the maximum value obtained for plug flow and the minimum value applicable for ideal mixing. These can be expressed in terms of an empirical coefficient π as

$$(\Delta t)_m = \pi (\Delta t)_{m,\text{plug flow}} \tag{65}$$

and

$$(\Delta Y)_m = \pi (\Delta Y)_{m,\text{plug flow}} \tag{66}$$

When differential balances were used, the plug flow driving forces can be easily calculated as

$$(\Delta t)_m = \frac{t_{g2} - t_{g1}}{N_{tH}} \tag{67}$$

$$(\Delta Y)_m = \frac{Y_2 - Y_1}{N_{tM}} \tag{68}$$

where N_t represents the number of transfer units for heat and mass transfer, defined as

$$N_{tH} = \int_{t_{g1}}^{t_{g2}} \frac{dt}{t_g - t_m} \tag{69}$$

$$N_{tM} = \int_{Y_1}^{Y_2} \frac{dY}{Y^* - Y} \tag{70}$$

respectively. On the basis of mean driving forces, the gas-solid interfacial area can be found using heat and mass transfer coefficients, which can be easily translated into material volume using specific area ratio $a_V (m^2/m^3)$. Dryer volume can be calculated if the void fraction or concentration of solids is known. For particulate systems this yields the following formula for dryer bed volume:

$$V = A \frac{d\phi_s}{6(1 - \epsilon)} \tag{71}$$

5.5. Crossflow Dryer Calculations

Crossflow of gas and solids is encountered in many industrial dryers, such as band dryers, through dryers for textiles and paper, and trough fluid bed dryers. The difference between the crossflow dryer and cocurrent and countercurrent dryers is basically in the impossibility of finding an equivalent driving force for the whole dryer. Instead, the total gas flow rate must be calculated. The scheme of a crossflow dryer is shown in Figure 23. For a differentially thin element of bed length, the gas flow rate is dW_B such that

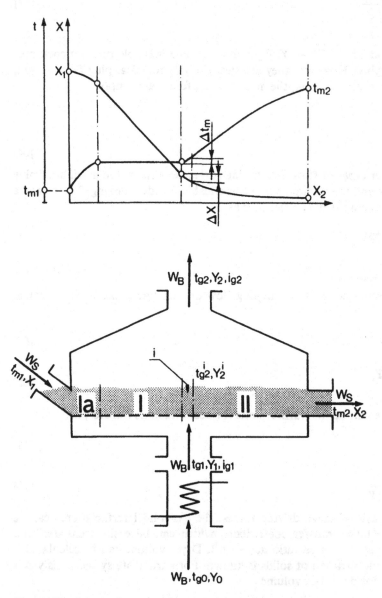

Figure 23 Scheme for a crossflow dryer: (Ia) warm-up zone; (I) zone of constant drying rate; (II) zone of falling drying rate.

$$W_B = \int_0^L dW_B \tag{72}$$

When writing the differential balances and kinetic equations for such an element of an adiabatic dryer, the following equations are readily obtained:

$$W_S dX + dW_B \Delta Y = 0 \tag{73}$$

$$W_S di_m + dW_B \Delta i_g = 0 \tag{74}$$

$$W_S dX = -dW_B \Delta Y = w_D a_V dV \tag{75}$$

$$W_S di_m = -dW_B \Delta i_g = (q - h_{Av}) a_V dV \tag{76}$$

In finite difference form these equations can be expressed as

$$W_S(X^{i-1} - X^i) = -\Delta W_B(Y_1 - Y_2^i) \tag{77}$$

$$W_S(i_m^{i-1} - i_m^i) = -\Delta W_B(i_{g1} - i_{g2}^i) \tag{78}$$

$$W_S(X^{i-1} - X^i) = k_Y a_V (\Delta Y)_m \Delta V \tag{79}$$

$$W_S(i_m^{i-1} - i_m^i) = [\alpha(\Delta t)_m - h_{Av}k_Y(\Delta Y)_m]a_V \Delta V \tag{80}$$

A differential ith bed element is shown in Figure 24.

The mean driving forces $(\Delta t)_m$ and $(\Delta Y)_m$ can be found if an appropriate type of flow of both phases is assumed. For example, in thin bed fluid bed dryers, ideal mixing of solids and plug flow of gas can be assumed to yield the following expressions:

$$(\Delta t)_m = \frac{t_{g1} - t_{g2}^i}{\ln \dfrac{t_{g1} - t_m^i}{t_{g2} - t_m^i}} \tag{81}$$

and

$$(\Delta Y)_m = \frac{Y_2^i - Y_1}{\ln \dfrac{Y^{*i} - Y_1}{Y^{*i} - Y_2^i}} \tag{82}$$

However, in most crossflow drying applications, thermodynamic equilibrium between the moist material and the exit gas is reached (except for very thin beds). In the case of perfect mixing of the material, Eqs. [79] and [80] are simplified to

$$Y_2^i = Y^*(t_m^i, X^i) \tag{83}$$

and

$$t_{g2}^i = t_m^i \tag{84}$$

The following integration procedure may be used to calculate the flow rate of a crossflow dryer. First, the step of integration ΔW_B is assumed. Then, having the temperatures and humidities of all incoming streams to the first bed element, Eqs. [77] through [80] are solved to find the parameters of the exit streams from this element. These are used in the next step. In each step ΔW_B is added to the previous value; when the material reaches the final required moisture content (or temperature), the value of $\Sigma \Delta W_B$ is the total gas flow rate required for drying.

If material flow has certain axial dispersion this can be taken into account in the balance equations, which can be now written to allow for the dispersion.

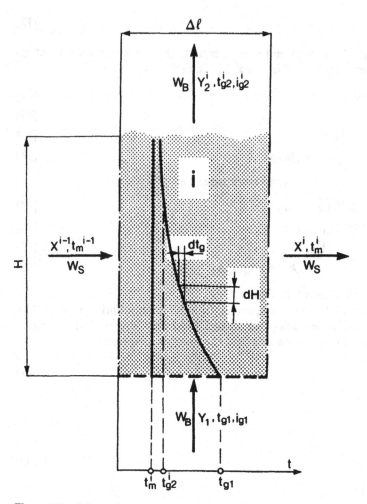

Figure 24 Scheme for an element of a crossflow dryer.

$$u_m \frac{dX}{d\ell} = E_\ell \frac{d^2X}{d\ell^2} - \frac{w_D a_V}{\rho_S(1 - \epsilon)} \qquad [85]$$

$$u_m \frac{di_m}{d\ell} = E_\ell \frac{d^2i_m}{d\ell^2} + \frac{(q - h_{A_V} w_D)a_V}{\rho_S(1 - \epsilon)} \qquad [86]$$

In these equations,

$$u_m = \frac{W_S}{\rho_S(1 - \epsilon)S} \qquad [87]$$

In addition, the following two kinetic equations for the gas phase are needed:

$$W_B \, \Delta Y = w_D a_V S \, d\ell \qquad [88]$$

$$W_B \, \Delta i_g = (q - h_{A_V} w_D)a_V S \, d\ell \qquad [89]$$

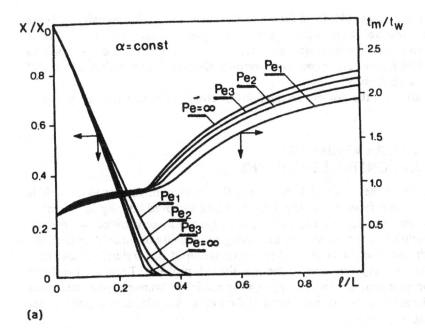

(a)

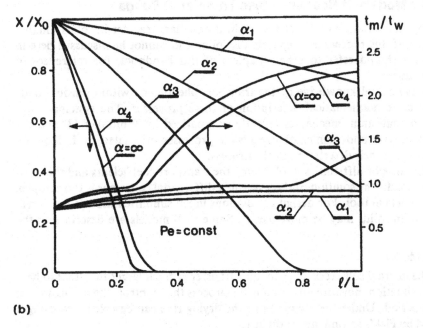

(b)

Figure 25 Solutions for crossflow dryer model: (a) effect of axial dispersion at constant heat transfer coefficient; (b) effect of heat transfer coefficient at constant axial dispersion.

Solution of Eqs. [85] through [89] for a fluid bed dryer are shown schematically in Figure 25 for various heat transfer coefficients and Peclet numbers. (Mass transfer coefficients were calculated from the Chilton–Colburn analogy.) Note that $Pe = \infty$ corresponds to the case of ideal plug flow and $\alpha = \infty$ corresponds to the case of equilibrium of gas and solid at the exit of each element.

More information on the process calculations of crossflow dryers may be found in Reference 7.

6. PROCESS CALCULATIONS IN INTERNALLY CONTROLLED DRYING

When the dimensions of the moist solid are large or the moisture diffusivity in the solid is low and when moisture from inside the body is being removed, drying is internally controlled. This occurs for Bi_D and $Bi_H > 1$. In this case any intensification of heat and mass transfer condition on the gas side by increasing gas velocity or turbulence does not help reduce the drying time. In these conditions evaluation of the drying time and thus the material holdup and dryer volume becomes a difficult problem. There are two basic ways to solve the problem: solving an appropriate model of transient heat and mass transfer within the solid, or use of experimental data on drying kinetics obtained under appropriate drying conditions.

6.1. Analytic Models of Heat and Mass Transfer in Solids

Due to the complicated nature of simultaneous heat and mass transfer in solids, mathematical modeling of this process is a separate discipline and cannot be discussed here in detail. The reader should refer to other chapters of this handbook for references to appropriate literature.

In general, the actual knowledge of the transport phenomena within porous solids allows quite accurate mathematical description of these processes. The systematic approach based on continuum mechanics recently summarized by Whittaker (8) and the approach based on thermodynamics of irreversible processes originated by Luikov are very comprehensive in their description of the process.

However, because of difficulties in obtaining the transport coefficients and the complex structure of solids, the solution of the governing differential equations is too complex to be used as a practical tool. For design calculations much simplified models for movement of moisture in solids may be more useful. Some such models are described in the following.

Pure Diffusion Model

In most cases the internal solid resistance to mass transfer is much larger than that to heat transfer. In this situation moisture diffusion is the process that controls the drying rate in the falling rate period. Under these conditions the drying rate can be calculated using a model described by Fick's second law of diffusion,

$$\frac{\partial X}{\partial \tau} = D_{AS} \frac{\partial^2 X}{\partial z^2} \tag{90}$$

For given boundary conditions this equation can be solved for any given geometry. Table 1 provides solutions of the Fick's diffusion equation for a few basic geometries. They are presented in terms of the average moisture content X_m, which can be used to calculate actual drying rate since

Table 1 Solutions of Fick's Law for Some Simple Geometries

Geometry	BC	Dimensionless average moisture content
Flat plate $2b$ (thick drying at both sides)	$\tau = 0;\ -b < z < b;\ X = X_o$ $\tau > 0;\ z = \pm b;\ X = X^*$	$\Phi_m = \dfrac{8}{\pi^2}$ $\displaystyle\sum_{n=1}^{\infty} \frac{1}{(2n-1)^2} \exp\left[-(2n-1)^2 \frac{\pi^2}{4b}(D_{AS}\tau/b)\right]$
Infinitely long cylinder of radius R	$\tau = 0;\ 0 < r < R;\ X = X_o$ $\tau > 0;\ r = R;\ X = X^*$	$\Phi_m = 4 \displaystyle\sum_{n=1}^{\infty} \frac{1}{R^2\alpha_n^2} \exp(-D_{AS}\alpha_n^2\tau)$ where α_n — positive roots of the equation $J_o(R\alpha_n) = 0$
Sphere of radius R	$\tau = 0;\ 0 < r < R;\ X = X_o$ $\tau > 0;\ r = R;\ X = X^*$	$\Phi_m = \dfrac{6}{\pi^2} \displaystyle\sum_{n=1}^{\infty} \frac{1}{n^2} \exp\left[\frac{-n^2\pi^2}{R}(D_{AS}\tau/R)\right]$

Source: From Reference 9. In these equations, $\Phi_m = (X_m - X^*)/(X_o - X^*)$.

$$w_D = -\frac{m_S}{A}\frac{dX_m}{d\tau} \qquad [91]$$

For more complicated geometries, usually numerical calculations are required in which drying rate is calculated over the whole solid surface. The diffusion coefficient D_{AS} usually must be determined experimentally as it varies with moisture content and temperature generally in an unpredictable manner. Considerable data on measured moisture diffusivities for various moist solids are available in the literature (10,11).

Receding Plane (or Front) Model

Experimental observations indicate that during drying of some porous bodies a distinct evaporation zone may be observed. It divides the solid into dry and wet parts, and as drying progresses this zone recedes, as shown in Figure 26, increasing the ratio of dry to

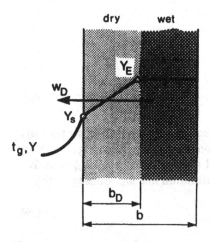

Figure 26 Scheme of the receding front model.

wet parts of the solid. Assuming that moisture within the solids is not bound to the solid lattice—that is, the solid is nonhygroscopic—the following analysis (2) is applicable. Using modified mass transfer coefficients, the drying rate at the solid surface and the evaporation front are, respectively,

$$w_D = K_o \phi_s (Y_s - Y)$$ [92]

where ϕ_s is approximately equal to $(M_A/M_B)/(M_A/M_B + Y_s)$ and

$$w_D = K_s \phi_E (Y_E - Y_s)$$ [93]

The mass transfer coefficient K_s within the solid can be written using Krischer's diffusion resistance coefficient μ_D as

$$K_s = \frac{cD_{AS}}{b_D} \mu_D M_B$$ [94]

where c is a total molar concentration of gas inside pores of the solid. In this equation, μ_D is the ratio of the solid porosity (voidage) to the tortuosity of the pores and thus takes into account (at least partially) the solid structural properties. In practice, μ_D varies between 3 and 30.

The evaporation rate can be expressed using a newly defined overall mass transfer coefficient K_T as

$$w_D = K_T \phi_E (Y_E - Y)$$ [95]

Adding resistances and taking advantage of the fact that usually $\phi_E \simeq \phi_s$, we find

$$K_T = \frac{K_0}{1 + \text{Bi}_M}$$ [96]

where $\text{Bi}_M = K_0/K_s$. Similarly, for heat transfer, the overall heat transfer coefficient is given by

$$U = \frac{\alpha}{1 + \text{Bi}_H}$$ [97]

where $\text{Bi}_H = \alpha b_D/\lambda_S$. Assuming that all heat input is used for evaporation of moisture, we have

$$\frac{Y_E - Y}{t_E - t_s} = \frac{U}{K_T \phi_E \, \Delta h_{vE}}$$

or, using Eqs. [96] and [97],

$$\frac{Y_E - Y}{t_E - t_s} = \frac{\alpha}{K_0} \frac{1 + \text{Bi}_M}{1 + \text{Bi}_H} \frac{1}{\phi_E \, \Delta h_{vE}} = m_E$$ [98]

This expression represents the slope of a line connecting points representing the bulk gas and gas at equilibrium with the evaporation front on a humidity chart (Fig. 27). Note that, for solids without internal resistance to heat and mass transfer (i.e., $\text{Bi}_M = 0$, $\text{Bi}_H = 0$), this slope is equal to the slope of the adiabatic saturation line m_w. Consequently,

$$\frac{m_E}{m_w} = \frac{K_0}{\alpha} \frac{\lambda}{cD_{AS}M_B} \frac{1}{\mu_D}$$ [99]

Using Chilton–Colburn analogy to express the mass transfer coefficient,

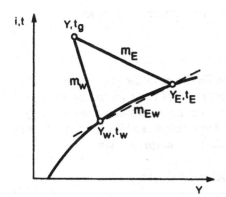

Figure 27   Schematic representation of the receding front model on a humidity chart.

$$K_0 = \frac{\beta \alpha}{C_p} \, \text{Le}^{-2/3} \qquad\qquad\qquad [100]$$

and, since for an air-water system, $\beta \text{Le}^{-1/3} \cong 1$, we obtain

$$\frac{m_E}{m_w} = \frac{1}{\mu_D} \qquad\qquad\qquad [101]$$

Now using Figure 27 we find that

$$Y_E - Y = (Y_E - Y_w) + (Y_w - Y)$$

or

$$m_E(t_E - t_g) = m_E[(t_E - t_g) - (t_H - t_g)] + m_w(t_H - t_g) \qquad [102]$$

which with Eq. [97] can be rewritten as

$$\frac{t_g - t_E}{t_w - t_g} = \frac{1 - m_w/m_{Ew}}{1 - \mu_D m_w/m_{Ew}} \qquad\qquad [103]$$

The ratio of drying rates on falling and constant drying rate periods can now be calculated as

$$\frac{w_{DII}}{w_{DI}} = \frac{U(t_g - t_E)}{\alpha(t_g - t_w)} \qquad\qquad [104]$$

Using Eqs. [97] and [103], this leads to

$$\frac{w_{DII}}{w_{DI}} = \frac{1}{1 + \text{Bi}_M \dfrac{\text{Bi}_H/\text{Bi}_M - m_w/m_{Ew}}{1 - m_w/m_{Ew}}} \qquad [105]$$

Note that the drying rate depends on moisture content as Bi_M depends on the thickness of the dry zone within the solid. Using the Biot number expressed in terms of the total body thickness Bi_{Mt}, we have

$$\text{Bi}_M = \text{Bi}_{Mt} \frac{b_D}{b} \qquad\qquad\qquad [106]$$

The actual thickness of the dry zone may be related to the moisture content as

$$\frac{X - X^*}{X_{cr} - X^*} = 1 - \frac{b_D}{b}$$ [107]

Introducing Eqs. [106] and [107] into Eq. [105], we finally have the expression for the drying rate in the falling drying rate period. The limitation of the receding evaporation front model presented here is that it applies to porous nonhygroscopic bodies only. Hallström (12) has presented a similar model for hygroscopic solids. Receding front models have also been applied successfully to freeze drying and spray drying.

6.2. Use of Experimentally Measured Kinetics of Drying

In many cases performing drying kinetic experiments is much easier than measuring the coefficients for analytic methods. This is especially true in the case of particulate solids. In a typical kinetic experiment, a sample of wet solid is dried under specified drying conditions and the moisture content is monitored as it varies with time. Unfortunately, the data from such a batch experiment can be used only for modeling of other batch dryers or a plug flow dryer. In both these dryers materials having different drying times do not mix. However, in most dryers for particulate solids, mixing is inevitable. As a result, the exiting solids are composed of particles with a residence time distribution. In addition, heat and mass is also transferred by contact between particles. In effect, the exit material moisture content in a continuous dryer will not necessarily be the same as that measured in a batch dryer having the same residence (drying) time.

To describe the average moisture content of solids leaving an arbitrary dryer, the concept of residence time distribution (RTD) functions is used. According to this concept, the average moisture content at the dryer exit is given by

$$X_m = \int_0^\infty E(\tau) X(X_0, \tau) d\tau$$ [108]

where $E(\tau)$ is the external RTD function. The drying kinetic curves $X(X_0, \tau)$ are usually described by the following equations:

$$-\frac{dX}{d\tau} = A \qquad \text{in constant rate period}$$ [109]

$$-\frac{dX}{d\tau} = A \frac{X - X_e^*}{X_{cr} - X_e^*}$$ [110]

Equations [109] and [110] can be integrated using the nomenclature of Figure 28. The following equations are obtained upon integration:

$$X - X_e^* = (X_0 - X_e^*)\left(1 - \frac{\tau}{\tau_c}\right) \qquad \text{for } X \le X_{cr}$$ [111]

and

$$X - X_e^* = (X_{cr} - X_e^*) \exp[-B(\tau - \tau_{cr})] \qquad \text{for } X > X_{cr}$$ [112]

where

$$B = \frac{X_0 - X_e^*}{X_{cr} - X_e^*} \frac{1}{\tau_c}$$

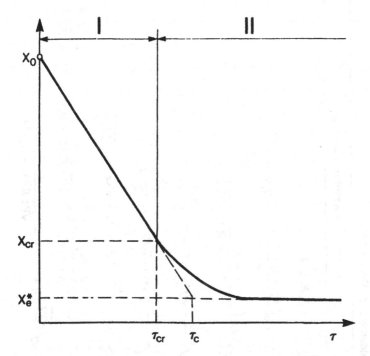

Figure 28 Typical experimentally measured drying kinetics: (I) constant drying rate period; (II) falling drying rate period.

The symbol τ_c used here is a constant and corresponds to the hypothetical time in which drying would be completed if the material were dried in the constant rate period only.

Expressions for calculation of the average solids moisture content for various types of flow are given in Table 2, together with the appropriate E functions. The formulas of Table 2 are presented in terms of the following dimensionless variables:

$$X_s = \frac{X_m - X_e^*}{X_0 - X_e^*} \qquad R_x = \frac{X_0 - X_{cr}}{X_0 - X_e^*} \qquad R_t = \frac{\tau_c}{\tau}$$

In practice, parameters τ_c, X_{cr}, and X_e^* are generally found from experimentally measured kinetic curves and the function E is evaluated for a specific dryer. Using the general formula of Eq. [108] or one of the equations of Table 2, the mean residence time, which gives the required average exit moisture content, may be calculated using a trial-and-error procedure. Note that the mean residence time is given by the formula

$$\tau_r = \frac{m_s}{W_S(1 + X_{1-2})} \tag{113}$$

where m_S is the material holdup in the dryer and X_{1-2} is the material moisture content averaged between the input and output. Residence time corresponds here to τ in the formula for R_x. The procedure described here does not take into consideration the contact heat and mass transfer between solids having different residence times. This, however, can be avoided if the kinetic experiments are performed in the following way. A continuously operated dryer is at a certain time fed with a small batch of traced wet solids, which are added to the main stream of the feed. At this moment the time measurement is started

Table 2 Dimensionless Outlet Average Moisture Content Predicted by Models of Residence Time Distribution

Model	E function of RTD	Dimensionless moisture content
Plug-flow bed	$E(\tau) = \delta(\tau - \tau_r)$	$X_s = 1 - \dfrac{1}{R_t}$ for $R_x R_t \geq 1$ $X_s = (1 - R_x)\exp\left(\dfrac{R_x - 1/R_t}{1 - R_x}\right)$ for $R_x R_t \leq 1$
Perfectly mixed bed	$E(\tau) = \dfrac{1}{\tau_r}e^{-\tau/\tau_r}$	$X_s = 1 - \dfrac{1}{R_t}e^{-R_x R_t}\left[\dfrac{1 - R_x}{1 + R_t(1 - R_x)} - \dfrac{1}{R_t}\right]$
Plug flow with axial dispersion	$E(\tau) = \dfrac{1}{\sigma\sqrt{2\pi}}\exp\left(\dfrac{(\tau - \tau_r)^2}{2\sigma^2}\right)$ where Pe ≥ 10 $\dfrac{\sigma^2}{\tau_r^2} = \dfrac{2}{Pe}$	$X_s = 0.5\left[erf\dfrac{\sqrt{Pe}}{2} + erf\left((R_x R_t - 1)\dfrac{\sqrt{Pe}}{2}\right)\right]$ $- \dfrac{0.5}{R_t}\left[erf\dfrac{\sqrt{Pe}}{2} + erf\left((R_x R_t - 1)\dfrac{\sqrt{Pe}}{2}\right)\right]$ $- \dfrac{2}{\sqrt{\pi Pe}}\left[\exp\left(-\dfrac{Pe}{4}(R_x R_t - 1)^2\right) - \exp\left(-\dfrac{Pe}{4}\right)\right]$ $+ 0.5(1 - R_x)\exp\left(\dfrac{R_x}{1 - R_x}\right)\exp\left[\dfrac{1}{Pe \cdot R_t^2(1 - R_x)^2} - \dfrac{1}{R_t(1 - R_x)}\right]$ $\cdot\left\{1 - erf\left[\dfrac{\sqrt{Pe}}{2}(R_x R_t - 1) + \dfrac{1}{\sqrt{Pe}\,R_t(1 - R_x)}\right]\right\}$

n-Perfectly mixed uniform beds

$$E(\tau) = \frac{n}{\tau_r} \cdot \frac{\left(n\dfrac{\tau}{\tau_r}\right)^{n-1}}{(n-1)!} \exp\left(-n\frac{\tau}{\tau_r}\right)$$

$$X_s = 1 - \frac{1}{R_t} - e^{-nR_xR_t}\left\{\sum_{i=0}^{n-1}\frac{(nR_xR_t)^i}{i!} - \frac{1}{R_t}\sum_{i=0}^{n}\frac{(nR_xR_t)^i}{i!}\right.$$

$$\left. - \frac{(1-R_x)}{\left[1+\dfrac{1}{n(1-R_x)R_t}\right]^n}\sum_{i=0}^{n-1}\frac{\left(nR_xR_t+\dfrac{R_x}{1-R_x}\right)^i}{i!}\right\}$$

Cholette and Cloutier model[a]

$$E(\tau) = \frac{1}{\tau_r}\frac{(1-x)^2}{y}\exp\left(-\frac{1-x}{y}\frac{\tau}{\tau_r}\right) + x\delta(\tau)$$

$$X_s = 1 - \frac{y}{R_t} - \exp\left(-\frac{1-x}{y}R_xR_t\right)\left[\frac{(1-x)y(1-R_x)}{y+(1-x)R_t(1-R_x)} - \frac{y}{R_t}\right]$$

Source: From Reference 13.

[a] A model of bed with dead zones and bypassing; x, fraction of gas stream bypassing; y, fraction of bed volume occupied by dead zones; (1 − y), fraction of bed volume perfectly mixed; δ, Dirac's delta function.

and as the traced particles arrive at the dryer exit their residence times are recorded and their moisture contents are analyzed. In this manner a drying curve under real drying conditions can be measured and used in Eq. [108]. A drawback of this method is that it requires the experiments to be performed in a dryer with material flow corresponding to that of the dryer being designed.

7. BATCH DRYER CALCULATIONS

In batch drying, conditions inside the dryer vary both in space and in time, which makes the calculations more complicated than in the case of continuous drying. In general, the following three most important cases can be distinguished:

Drying of ideally mixed beds of solids having low internal resistance to heat and mass transfer, such as batch fluid or spouted beds
Drying of large solids, such as bricks and timber
Through drying, for example, drying of fixed beds of solids

The first two cases are relatively easily handled. In the first case, because of the close-to-ideal mixing, the spatial gradients of moisture content and temperature are negligible so the problem is one dimensional. To solve the problem, that is, to find how moisture content and temperature of the product vary in time, the relations of Section 3 (Eqs. [1] through [4]) can be used. Substituting mass of batch m_S instead of material flow rate W_S and using m_B instead of W_B ($m_B = W_B \, \Delta\tau$) allows one to calculate the moisture content and temperature of the material after a certain increment of time $\Delta\tau$. If the increments of time used are small, a reasonably smooth drying curve can be generated. Drying rate in this case is calculated from the gas side (resistance) as negligible internal particle resistance to mass transfer is typically assumed.

   In the second case, especially when the external drying conditions are fairly uniform (this is true when frequent reversals of gas flow direction are made, as in drying of timber), the drying rate can be calculated according to the diffusion theories of Section 6.1. In the same manner, a fixed bed dried by conduction, radiation, or convection from the surface can be handled, however, only if the drying rate from the surface is uniform, which is often not true in practice. For example, in a typical industrial shelf dryer, air distribution is not ideal and the material close to the air inlet dries out while that close to exit is still wet. A dryer of this type is presented in Figure 29. Typical shelf dryers operate with partial recycle of the spent air. For short times the graphic representation of the process on the humidity chart is exactly the same as for a continuous dryer. However, at longer times there is a continuous shift of point D due to the continuously changing average moisture content and temperature of the material. This implies shift of points C and also B if the recycle ratio is held constant. Finally, accurate calculations of the drying time in such a dryer are very complex and rather unlikely to succeed without strong experimental backup.

   On the contrary, the problem of batch through drying of deep beds has been studied by many investigators, especially in connection with drying of grains, and theoretical and experimental results are widely available (4).

   The basic theoretical approach to this problem will be presented here for the simple case of a material with sorption properties that do not depend on temperature. This is true for constant rate drying or for materials with negligible heats of sorption. Under these conditions, the temperature front traveling across the solids will not have any

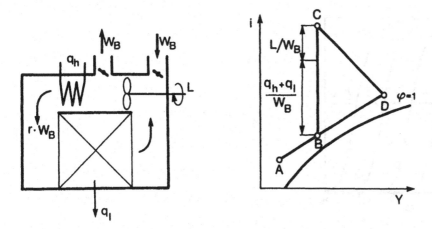

Figure 29 Batch dryer with partial recycle of air and corresponding air path on humidity chart.

substantial influence on the drying rate. The process of drying can then be analyzed as a process of pure mass transfer.

Taking a differential element of bed, thickness dz, the following mass balance can be written:

$$(W_B S \, d\tau)dY = -[\rho_S(1 - \epsilon)S \, dz] \, dX - (\rho_q \epsilon \, S \, dz) \, dY \qquad [114]$$

Because of the much steeper solids moisture gradient than the gradient of gas humidity, the third term of this equation is negligible. Thus,

$$-\rho_S(1 - \epsilon) \frac{\partial X}{\partial t} = W_B \frac{\partial Y}{\partial z} \qquad [115]$$

Adding the rate equation for moisture transport,

$$-dX[\rho_S(1 - \epsilon)S \, dz] = (w_D a_V S \, dz)d\tau \qquad [116]$$

which can be simplified to

$$-\rho_S(1 - \epsilon) \frac{\partial X}{\partial \tau} = w_D a_V \qquad [117]$$

upon the introduction of the following dimensionless variables:

dimensionless moisture content: $\Phi = \dfrac{X - X^*}{X_{cr} - X^*}$ $\qquad [118]$

dimensionless drying rate: $f = \dfrac{w_D}{w_{DI}}$ $\qquad [119]$

dimensionless distance: $\zeta = \dfrac{k_Y a_V}{W_B} z$ $\qquad [120]$

dimensionless time: $\Theta = \dfrac{W_B}{\rho_S H(1 - \epsilon)(X_{cr} - X^*)} \tau$ $\qquad [121]$

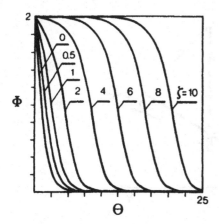

Figure 30 Solutions of the Eq. [122] for $f = \Phi$.

Equations [115] and [117] can be combined into the following dimensionless equation:

$$\frac{\partial \Phi}{\partial \Theta} = \frac{\partial}{\partial \Theta}\left(-\frac{1}{f}\frac{\partial \Phi}{\partial \zeta}\right) \tag{122}$$

This equation can be solved by numerical methods for any $f(\Phi)$. However, for the linear form of this relationship (e.g., $f = \Phi$) Eq. [122] can be solved analytically. The solution of Eq. [122] for this case is presented in Figure 30. It represents the front of decreasing moisture content of the solid as it travels along the bed.

In the conditions when the drying curve is given by $f = 1, \Phi \geq 1; f = \Phi, \Phi < 1$; the total drying time can be calculated from the simple relationship

$$\Theta = \frac{1}{(Y_w - Y)_1}(\Phi_1 - 1 - \ln \Phi_2) \tag{123}$$

where Φ_1 is the initial moisture content and Φ_2 the final moisture content. X^* corresponds to equilibrium with the incoming air.

This relatively simple analysis becomes very complex in the case of solids with large heats of sorption, complicated drying kinetics, or when the drying air is recycled. Again, experimental verification of the calculation procedure is desirable.

CONCLUSION

Drying is a procedure of unit operations; however, while other operations for years have grown their own unified mathematical description, drying was left with basically only an experimental drying kinetics curve and drying time calculated on the basis of this curve. Only in the last two decades researchers and engineers realized that drying may be described by the same unified approach as any other heat and mass transfer operation. Since then, driving forces in both phases, numbers of transfer units, process paths, phase equilibria, RTD analysis, and so on entered into drying calculations. While a book representing a fully unified approach to drying calculations remains yet to be seen, numerous books covering various areas of drying calculations appeared in the meantime. A general approach is presented by Keey (2), Strumillo and Kudra (14), and Sazhin (15). Calculation methods of drying of granular products, apparently easier to describe mathematically, are presented by Planovskii et al. (5), Frolov (16), Mushtayev and Ulyanov (6), Houška et al. (17), and Keey (4). Drying calculations for processes limited by solid-side diffusion are well described by Rudobashta (10). Calculations of drying of food products are presented by Ginzburg (11). Finally, van't Land presents an industrial approach to drying and dryer calculations (18).

ACKNOWLEDGMENT

The authors gratefully acknowledge the assistance of Purnima Mujumdar in the preparation of this typescript.

NOMENCLATURE

A	total interfacial area, m^2
A^*	local slope of equilibrium isotherm $Y^* = f(X)$
a_S	interfacial area per unit volume of solid, m^2/m^3
a_V	interfacial area per unit volume of the system, m^2/m^3
b	thickness, m
b_D	thickness of the dry zone, m
c	specific heat capacity, $kJ/kg \cdot K$
c_H	humid heat, $kJ/kg \cdot K$
D_{AS}	diffusivity of moisture in solid, m^2/s
d	diameter of particle, m
E_t	axial dispersion coefficient, m^2/s
f	relative drying rate defined by Eq. [119]
h_{Av}	specific enthalpy of moisture vapor, kJ/kg of A
Δh_s	heat of sorption, kJ/kg of S
Δh_v	latent heat of vaporization, kJ/kg of A
i	specific enthalpy of humid medium, kJ/kg of B or S
K	mass transfer coefficient, $kg/m^2 \cdot s$
k_Y	modified mass transfer coefficient, $kg/m^2 \cdot s$
L	length of the dryer, m
ℓ	distance from the dryer inlet, m
M	molar mass, kg/kmol
m	slope of the line on enthalpy-humidity chart, kJ/(kg/kg)
m	mass, kg
m'	mass loading per unit length of the dryer, kg/m
N_{tH}	number of heat transfer units
N_{tM}	number of mass transfer units
P	ambient pressure, kPa
p_s	saturated vapor pressure, kPa
q	heat transfer rate, kW
q_h	heat input from heaters, kW
q_l	heat losses, kW
q_t	heat carried with transporters, kW
r	recycle ratio

S	cross-sectional area normal to flow direction, m^2
s_i	scaling factor of enthalpy, $(kJ/kg)/m$
s_Y	scaling factor of humidity, $1/m$
t	temperature, K, °C
U	overall heat transfer coefficient, $W/m^2 \cdot K$
u	velocity, m/sec
W	mass flow rate, kg/s
w_D	drying rate, $kg/m^2 \cdot s$
X	moisture content, kg of A/kg of S
Y	humidity, kg of A/kg of B
z	distance coordinate, m

Greek Symbols

α	heat transfer coefficient, $W/m^2 \cdot K$
β	correcting coefficient of psychrometry
Δ	parameter defined by Eq. [12], kJ/kg
ϵ	directional coefficient at humidity chart, kJ/kg
ϵ	voidage
ζ	relative distance defined by Eq. [120]
Θ	relative time defined by Eq. [121]
λ	heat conductivity, $W/m \cdot K$
μ_D	diffusion-resistance coefficient
π	correction coefficient of driving force
ρ	density, kg/m^3
τ	time, s
τ_r	residence time, s
Φ	relative moisture content defined by Eq. [118]
ϕ_s	shape factor
φ	relative humidity

Superscript

| * | at equilibrium |

Subscripts

A	moisture (active diffusing component)
as	adiabatic saturation conditions
B	dry gas (gaseous inert)
e	end of drying
g	humid gas
L	liquid
M	mixture
m	wet material
m	mean value
o	triple point of water
p	at constant pressure
S	bone-dry solid
w	wet bulb conditions

0   ambient conditions
1   inlet
2   outlet

Dimensionless Numbers

Bi_H Biot number for heat transfer, Eq. [53]
Bi_D Biot number for mass transfer, Eq. [52]
Le   Lewis number $= \lambda_g/D_{AB}c_p\rho_g$
Pe   Peclet number $= u_m L/E_t$

REFERENCES

1. Pakowski, Z., *Advances in Drying*, Vol. 5, A. S. Mujumdar, ed., Hemisphere, New York, 1992.
2. Keey, R., *Introduction to Industrial Drying Operations*, Pergamon Press, Oxford, 1978.
3. Pakowski, Z., *Hung. Journ. Ind. Chem.*, 14, 1986, pp. 225-236.
4. Keey, R., *Drying of Loose and Particulate Materials*, Hemisphere, New York, 1992.
5. Planovskii, A. N., Mushtayev, V. I., Ulyanov, V. M., *Drying of Disperse Materials in Chemical Industry*, Khimiya, Moscow, 1979 (in Russian).
6. Mushtayev, V. I., Ulyanov, V. M., *Drying of Disperse Solids*, Khimiya, Moscow, 1988 (in Russian).
7. Strumillo, C., Pakowski, Z., Zylla, R., *Zhurn. Prikl. Khimii*, 9, 1986, pp. 2108-2115.
8. Whittaker, S., *Advances in Drying*, Vol. 2, A. S. Mujumdar, ed., Hemisphere, New York, 1980.
9. Crank, J., *The Mathematics of Diffusion*, Oxford University Press, London, 1956.
10. Rudobashta, S. P., *Mass Transfer in Systems with Solid Phase*, Khimiya, Moscow, 1980 (in Russian).
11. Ginzburg, A. S., *Calculations and Design of Food Industry Dryers*, Agropromizdat, Moscow, 1985 (in Russian).
12. Hallström, A., in *Drying '82*, A. S. Mujumdar, ed., Hemisphere, New York, 1982.
13. Vanderschuren, J., Delvosalle, C., in *Drying '82*, A. S. Mujumdar, ed., Hemisphere, New York, 1982.
14. Strumillo, C., Kudra, T., *Drying: Principles, Applications and Design*, Gordon and Breach, London, 1986.
15. Sazhin, B. S., *Principles of Drying Technology*, Khimiya, Moscow, 1984 (in Russian).
16. Frolov, V. F., *Modeling of Drying of Disperse Solids*, Khimiya, Leningrad, 1987 (in Russian).
17. Houška, K., Valchar, J., Viktorin, Z., *Advances in Drying*, Vol. 4, A. S. Mujumdar, ed., Hemisphere, New York, 1987.
18. van't Land, C. M., *Industrial Drying Equipment*, Marcel Dekker, New York, 1991.

4

Transport Properties in the Drying of Solids

D. Marinos-Kouris and Z. B. Maroulis
National Technical University
Athens, Greece

1. INTRODUCTION

Drying is a complicated process involving simultaneous heat, mass, and momentum transfer phenomena, and effective models are necessary for process design, optimization, energy integration, and control. The development of mathematical models to describe drying processes has been a topic of many research studies for several decades. Undoubtedly, the observed progress has limited empiricism to a large extent. However, the design of dryers is still a mixture of science and practical experience. Thus the prediction of Luikov that by 1985 "would obviate the need for empiricism in selecting optimum drying conditions," represented an optimistic perspective, which, however, shows that the efforts must be increased (1). Presently, more and more sophisticated drying models are becoming available, but a major question that still remains is the measurement or determination of the parameters used in the models. The measurement or estimation of the necessary parameters should be feasible and practical for general applicability of a drying model.

In the early 1970s, Nonhebel and Moss stated that "the choice of drying plant, or design of special plant to meet unprecedented conditions" would require use of 34 parameters (2). Regardless of the truth of such a statement, that is, of the actual number of parameters necessary for the design of a dryer, there is an obvious need for a large amount of data. Nowadays, the completeness and accuracy of such data reflect to a large extent our ability to perform effective process design. It should be noted that in spite of the intense activities in the drying literature (*Drying Technology Journal, Advances in Drying, Drying, International Drying Symposium* etc.), the problem of property data still remains an important one in view of the fact that such data are widely scattered and not systematically evaluated. Moreover, while the need "for accurate design data is increasing, the rate of accumulation of new data is not increasing fast enough" (3). The lack of data is expected to continue and, as noted by Keey, "it is probably unrealistic to expect complete hygrothermal data for materials of commercial interest" (4).

Out of the full set of thermophysical properties necessary for the analysis of drying of a material, this chapter examines only those that are critical. As such, we consider the thermodynamic and transport properties, which are usually incorporated in a drying model as model parameters, and which are:

Effective moisture diffusivity
Effective thermal conductivity
Air boundary heat and mass transfer coefficients
Drying constant
Equilibrium material moisture content

Effective thermal conductivity and effective moisture diffusivity are related to internal heat and mass transfer, respectively, while air boundary heat and mass transfer coefficients are related to external heat and mass transfer, respectively. The above transport properties are usually coefficients in the corresponding flow rate/driving force relationship. The equilibrium material moisture content, on the other hand, is usually related to the mass transfer driving force.

The above transport properties in conjunction with a transport phenomena mechanistic model can adequately describe the drying kinetics, but sometimes an additional property, the drying constant, is also used. The drying constant is essentially a combination of the above transport properties and it must be used in conjunction with the so-called thin-layer model.

Effective moisture diffusivity and effective thermal conductivity are in general functions of material moisture content and temperature, as well as of the material structure. Air boundary coefficients are functions of the conditions of the drying air, that is humidity, temperature, and velocity, as well as system geometry. Equilibrium moisture content of a given material is a function of air humidity and temperature. The drying constant is a function of material moisture content, temperature, and thickness, as well as air humidity, temperature, and velocity.

The required accuracy of the above properties depends on the controlling resistance to heat and mass transfer. If, for example, drying is controlled by the internal moisture diffusion, then the effective moisture diffusivity must be known with high accuracy. This situation is valid when large particles are drying with air of high velocity. Drying of small particles with low velocity of air is controlled by the external mass transfer, and the corresponding coefficient should be known with high accuracy. But there are situations in which heat transfer is the controlling resistance. This happens, for example, in drying of solids with high porosity, in which high mass and low heat transfer rates are obtained.

The purpose of this chapter is to examine the above properties related to drying processes, particularly drying kinetics. Most of the following topics are discussed for each property:

Definition
Methods of experimental measurement
Data compilation
Effect of various factors
Theoretical estimation

The statement of Poersch (quoted in Ref. 4) that it is possible for someone to dry a product based on experience and without theoretical knowledge but not the reverse is worth repeating here. To this we may add the comment that it is impossible to efficiently dry a product without complete and precise thermophysical data.

2. MOISTURE DIFFUSIVITY

2.1. Definition

Diffusion in solids during drying is a complex process that may involve molecular diffusion, capillary flow, Knudsen flow, hydrodynamic flow, or surface diffusion. If we combine all these phenomena into one, the effective diffusivity can be defined from Fick's second law:

$$\partial X/\partial t = D \nabla^2 X \qquad\qquad [1]$$

where D (m^2/s) is the effective diffusivity, X (kg/kg db) is the material moisture content, and t (s) is the time.

The moisture transfer in heterogeneous media can be conveniently analyzed by using Fick's law for homogeneous materials, in which the heterogeneity of the material is accounted for by the use of an effective diffusivity.

Equation [1] shows the time change of the material moisture distribution, that is, it describes the movement of moisture within the solid. The previous equation can be used for design purposes in cases in which the controlling mechanism of drying is the diffusion of moisture.

Pakowski and Mujumdar (5) describe the use of Eq. [1] for the calculation of the drying rate, while Strumillo and Kudra (6) describe its use in calculating the drying time. Solutions of the Fickian equation for a variety of initial and boundary conditions are exhaustively described by Crank (7).

2.2. Methods of Experimental Measurement

There is no standard method for the experimental determination of diffusivity. The diffusivity in solids can be determined using the methods presented in Table 1. These methods have been developed primarily for polymeric materials (7–9). Table 1 also includes the relevant entries in the "References" section for the application of the methods in food systems.

Sorption Kinetics

The sorption (adsorption or desorption) rate is measured with a sorption balance (spring or electrical) while the solid sample is kept in a controlled environment. Assuming negligible surface resistance to mass transfer, the method is based on Fick's diffusion equation.

Table 1 Methods for the Experimental Measurement of Moisture Diffusivity

Method	Reference nos.
Sorption kinetics	8
Permeation methods	8
Concentration-distance curves	10–12
Other methods: Radiotracer methods	8
Nuclear magnetic resonance (NMR)	8, 13, 14
Electron spin resonance (ESR)	8, 15
Drying technique: Simplified methods	16
Regular regime method	17–19
Numerical solution — regression analysis	See Section 7

Permeation Method

The permeation method is a steady-state method applied to a film of material. According to this method, the permeation rate of a diffusant through a material of known thickness is measured under constant, well-defined, surface concentrations. The analysis is also based on Fick's diffusion equation.

Concentration-Distance Curves

The concentration-distance curves method is based on the measurement of the distribution of the diffusant concentration as a function of time. Light interference methods, as well as radiation adsorption or simply gravimetric methods, can be used for concentration measurements. Various sample geometries can be used, for example semiinfinite solid, two joint cylinders with the same or different material and so on. The analysis is based on the solution of Fick's equation.

Other Methods

Modern methods for the measurement of moisture profiles lead to diffusivity measurement methods. Such methods discussed in the literature are radiotracer methods, nuclear magnetic resonance (NMR), electron spin resonance (ESR), and the like.

Drying Methods

The simplified, regular regime, and regression analysis methods are particularly relevant for drying processes. In them, the samples are placed in a dryer and moisture diffusivity is estimated from drying data. All the drying methods are based on Fick's equation of diffusion, and they differ with respect to the solution methodology. The following analysis is considered.

Simplified Methods. Fick's equation is solved analytically for certain sample geometries under the following assumptions:

Surface mass transfer coefficient is high enough so that the material moisture content at the surface is in equilibrium with the air conditions.

Air drying conditions are constant.

Moisture diffusivity is constant, independent of material moisture content and temperature.

The analytical solution for slab, spherical, or cylindrical samples is used in the analysis. Several alternatives exist concerning the methodology of estimation of diffusivity using the above equations. They are discussed in the COST 90bis project of European Economic Community (EEC) (16). These alternatives differ essentially on the variable on which a regression analysis is applied.

Regular Regime Method. The regular regime method is based on the experimental measurement of the *regular regime curve*, which is the drying curve when it becomes independent of the initial concentration profile. Using this method, the concentration-dependent diffusivity can be calculated from one experiment.

Numerical Solution–Regression Analysis Method. The regression analysis method can be considered as a generalization of the other two types of methods. It can estimate simultaneously some additional transport properties; it is analyzed in detail in Section 7.

2.3. Data Compilation

Effective diffusivities, reported in the literature, have been usually estimated from drying or sorption rate data. Experimental data are scarce because of the effect of the experimental method, the method of analysis, the variations in composition and structure of the

Table 2 Effective Moisture Diffusivity in Some Materials

Classification*	Material	Water content (kg/kg db)	Temperature (°C)	Diffusivity (m²/s)	Reference nos.
	Food				
1	Alfalfa stems	<3.70	26	2.6E-12–2.6E-09	23
2	Apple	0.12	60	6.5E-12–1.2E-10	24
		0.15–7.00	30–76	1.2E-10–2.6E-10	25
3	Avocado		31–56	1.1E-10–3.3E-10	26
4	Beet		65	1.5E-09	26
5	Biscuit	0.10–0.65	20–100	9.4E-10–9.7E-08	27
6	Bread	0.10–0.70	20–100	2.5E-09–5.5e-07	27
7	Carrot	0.03–11.6	42–80	9.0E-10–3.3E-09	28
8	Corn	0.05–0.23	40	1.0E-12–1.0E-10	29
		0.19–0.27	36–62	7.2E-11–3.3E-10	30
9	Fish muscle	0.05–0.30	30	8.1E-11–3.4E-10	31
10	Garlic	0.20–1.60	22–58	1.1E-11–2.0E-10	32
11	Milk foam	0.20	40	1.1E-09	33
	skim	0.25–0.80	30–70	1.5E-11–2.5E-10	34
12	Muffin	0.10–0.65	20–100	8.4E-10–1.5E-07	27
13	Onion	0.05–18.7	47–81	7.0E-10–4.9E-09	35
14	Pasta, semolina	0.01–0.25	40–125	3.0E-13–1.5E-10	36
	corn based	0.10–0.40	40–80	5.0E-11–1.3E-10	37
	durum wheat	0.16–0.35	50–90	2.5E-12–5.6E-11	38
15	Pepper, green	0.04–16.2	47–81	5.0E-10–9.2E-09	35
16	Pepperoni	0.19	12	4.7E-11–5.7E-11	39
17	Potato	0.60	54	2.6E-10	40
		<4.00	65	4.0E-10	41
		0.15–3.50	65	1.7E-09	42
		0.01–7.20	39–82	5.0E-11–2.7E-09	43
18	Rice	0.18–0.36	60	1.3E-11–2.3E-11	44
		0.28–0.64	40–56	1.0E-11–6.9E-11	45
19	Soybeans, defatted	0.05	30	2.0E-12–5.4E-12	46
20	Starch, gel	0.10–0.30	25	1.0E-12–2.3E-11	47
		0.20–3.00	30–50	1.0E-10–1.2E-09	48
		0.75	25–140	1.0E-10–1.5E-09	49
	granular	0.10–0.50	25–140	5.0E-10–3.0E-09	49
21	Sugar beet	2.50–3.60	40–80	4.0E-10–1.3E-09	50,51
22	Tapioca root	0.16–1.95	97	9.0E-10	52,53
23	Turkey	0.04	22	8.0E-15	54
24	Wheat	0.12–0.30	21–80	6.9E-12–2.8E-10	55
		0.13–0.20	20	3.3E-10–3.7E-09	56
	Other materials				
1	Asbestos cement	0.10–0.60	20	2.0E-09–5.0E-09	20
2	Avicel (FMC Corp)		37	5.0E-09–5.0E-08	57
3	Brick powder	0.08–0.16	60	2.5E-08–2.5E-06	58
4	Carbon, activated		25	1.6E-05	59
5	Cellulose acetate	0.05–0.12	25	2.0E-12–3.2E-12	60
6	Clay brick	0.20	25	1.3E-08–1.4E-08	61
7	Concrete	0.10–0.40	20	5.0E-10–1.2E-08	20
	Concrete, pumice	0.20	25	1.8E-08	61
8	Diatomite	0.05–0.50	20	3.0E-09–5.0E-09	20

(*continued*)

Table 2   Continued

Classification*	Material	Water content (kg/kg db)	Temperature (°C)	Diffusivity (m²/s)	Reference nos.
9	Glass wool	0.10–1.80	20	2.0E-09–1.5E-08	20
	Glass spheres, 10 μm	0.01–0.22	60	1.84E-8 ± 0.94E-8	16
10	Hyde clay	0.10–0.40		5.0E-09–1.0E-08	62
11	Kaolin clay	<0.50	45	1.5E-08–1.5E-07	20
12	Model system		68	3.1E-09	63
13	Peat	0.30–2.50	45	4.0E-08–5.0E-08	20
14	Sand	<0.15	45	8.0E-08–1.5E-07	20
	Sand, sea	0.07–0.13	60	2.5E-08–2.5E-06	58
	Sand	0.05–0.10		1.0E-07–1.0E-06	64
15	Silica alumina	0.59–1.18	60	2.5E-08–2.5E-06	58
16	Silica gel		25	3.0E-06–5.6E-06	59
17	Tobacco leaf		30–50	3.2E-11–8.1E-11	65
18	Wood, soft		40–90	5.0E-10–2.5E-09	66
	Wood, yellow poplar	1.00	100–150	1.0E-08–2.5E-08	67

*Classification number for each material used in Fig. 1.

examined materials, and so on. Data of effective diffusion coefficients are available for inorganic materials in Reference 20, for polymers in Reference 8, and for foods in References 21 and 22.

Table 2 gives some literature values of the effective diffusivity of moisture in various materials. A number of data from the forementioned bibliographic entries are also included in Table 2. New data up to 1992 are also incorporated. Foods are the most investigated materials in the literature, and they are presented separately. Table 2 was prepared for the needs of this chapter, that is, in order to show the range of variation of diffusivity for various materials and not to present some experimental values. That is why most of the data are presented as ranges.

The data of Table 2 are further displayed in Figures 1–4. The moisture diffusivity is plotted versus number of material for food and other materials in Figure 1. Diffusivities in foods have values in the range 10^{-13} m²/s to 10^{-6}, and most of them (82%) are accumulated in the region 10^{-11} to 10^{-8}. Diffusivities of other materials have values in the range 10^{-12} to 10^{-5}, while most of them (58%) are accumulated in the region 10^{-9} to 10^{-7}. These results also are clarified in the histograms of Figure 2. Diffusivities in foods are less than those in other materials. This is because of the complicated biopolymer structure of food and, probably, the stronger binding of water in them.

The influence of material moisture content and temperature from the statistical point of view is shown in Figures 3 and 4. Figure 3 shows the diffusivities versus the material moisture content for all the materials. The positive effect of material moisture content on diffusivity is evident. The same trend is noted in Figure 4 with regard to the temperature. It should be noted that the observed trends in the previous figures are the result of examining different materials at various temperatures and moistures and from various sources. The influence of material moisture content and temperature for each material is discussed in the next section.

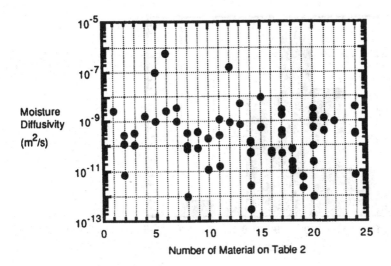

Food Materials

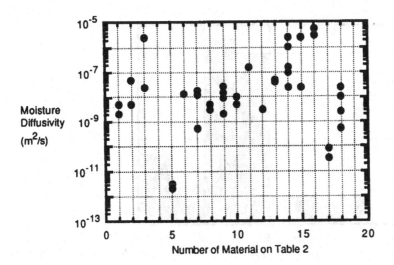

Other Materials

Figure 1   Moisture diffusivity in various materials (data from Table 2).

In general, comparison among diffusivities reported in the literature is difficult because of the different methods of estimation and the variation of composition, especially for foods. However, on the basis of Figures 3 and 4, it is concluded that the differences in diffusivity among materials are less than that between temperature or material moisture content of the same material. Diffusivities of other solutes in various materials are also presented in the literature (e.g., see Ref. 68).

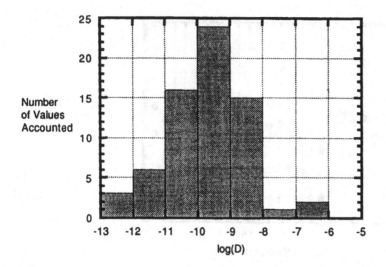

Food Materials

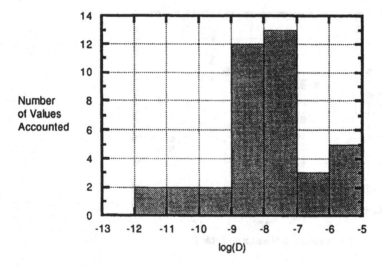

Other Materials

Figure 2 Histograms of diffusivities in various materials (data from Table 2).

2.4. Factors Affecting Diffusivity

Moisture diffusivity depends strongly on temperature and, often, very strongly on the moisture content, but there are few reliable figures. In porous materials the void fraction affects diffusivity significantly, and the pore structure and distribution do so even more.

The temperature dependence of the diffusivity can generally be described by the Arrhenius equation, which takes the form

$$D = D_o \exp(-E/RT)$$ [2]

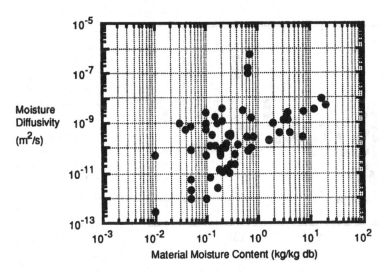

Food Materials

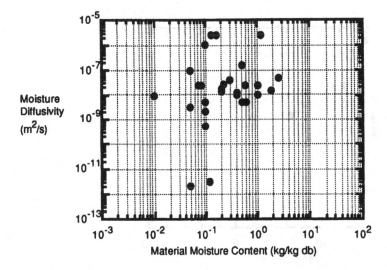

Other Materials

Figure 3   Moisture diffusivity versus material moisture content (data from Table 2).

where D_O (m²/s) is the Arrhenius factor, E (kJ/kmol) is the activation energy for diffusion, R (kJ/kmol/K) the gas constant, and T(K) the temperature.

The moisture content dependence of the diffusivity can be introduced in the Arrhenius equation by considering either the activation energy or the Arrhenius factor as an empirical function of moisture. Both modifications can be considered simultaneously. Other empirical equations not based on the Arrhenius equation can be used.

The moisture diffusivity is an increasing function of the temperature and moisture of the material. Yet, in certain categories of polymers, deviation from this kind of behavior

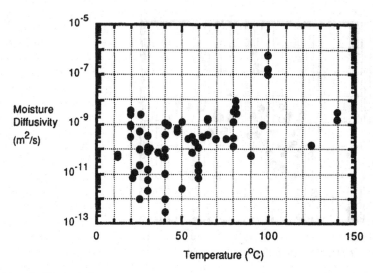

Food Materials

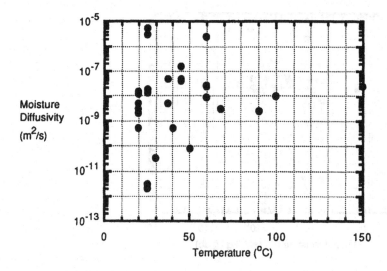

Other Materials

Figure 4   Moisture diffusivity versus material temperature (data from Table 2).

have been observed. For instance, for several of the less hydrophilic polymers (e.g., polymethacrylates and polycrylates) the moisture diffusivity decreases with increasing water content. On the other hand, the moisture diffusivity appears to be independent of the concentration — and hence constant — for some hydrophobic polyolefins.

Table 3 gives some relationships that describe simultaneous dependence of the diffusivity upon temperature and moisture. Some rearrangement of the equations proposed has been done in order to present them in a uniform format. Table 4 lists parameter values for typical equations of Table 3.

Table 3 Effect of Material Moisture Content and Temperature on Diffusivity

Equation no.	Materials of application	Equation	No. of parameters	Reference nos.
1	Apple, carrot, starch	$D(X,T) = a_0 \exp(a_1 X) \exp(-a_2/T)$	3	49, 69, 70
2	Bread, biscuit, muffin	$D(X,T) = a_0 \exp\left(\sum_{i=1}^{3} a_i X^i\right)$ $\exp(-a_2/T)$	5	27
3	Polyvinylalcohol	$D(X,T) = a_0 \exp\left(\sum_{i=1}^{10} a_i X^i\right)$ $\exp(-a_2/T)$	12	71
4	Vegetables	$D(X,T) = a_0 \exp(-a_1/X)$ $\exp(-a_2/T)$	3	72
5	Glucose, coffee extract, skim milk, apple, potato, animal feed	$D(X,T) = a_0 \exp[-a_1(1/T - 1/a_2)]$ $a_1 = a_{10} + a_{11}\exp(-a_{12}X)$	5	18
6	Silica gel	$D(X,T) = a_0 \exp(-a_1/T)$ $a_1 = a_{10} + a_{11}X$	3	73
7	Clay brick, burned clay, pumice concrete	$D(X,T) = a_0 X^{a_1} T^{a_2}$	3	61
8	Corn	$D(X,T) = a_0 \exp(a_1 X) \exp(-a_2/T)$ $a_1 = a_{11}T + a_{10}$	4	30
9	Rough rice	$D(X,T) = a_1 \exp(a_2 X)$ $a_1 = a_{10} \exp(a_{11}T),$ $a_2 = a_{20} \exp(a_{21}T + a_{22}T^2)$	5	74, 75
10	Wheat	$D(X,T) = a_0 + a_1 X + a_2 X^2$ $a_0 = a_{01} \exp(a_{02}T),$ $a_1 = a_{11} \exp(a_{12}T),$ $a_2 = a_{21} \exp(a_{22}T)$	6	76
11	Semolina, extruded	$D(X,T) = a_0 \exp(-a_1/T)$ $\dfrac{a_2\exp(-a_3/T)}{1 + a_2\exp(-a_3/T)}$ $a_3 = F(a_{3i}, X)$	>4	77
12	Porous starch	$D(X,T) = (a_0 + a_1 X^{a_2}) \exp(-a_3/T)$ $a_0 = F(\epsilon),$	>5	78

D, moisture diffusivity; X, material moisture content; T, temperature; a_i, constants; ϵ = porosity

Equations 1 to 4 in Table 3 suggest that the material moisture content can be taken into account by considering the preexponential factor of the Arrhenius equation as a function of material moisture content. Polynomial functions of first order can be considered (Eq. 1), as well as of higher order (Eqs. 2 or 3). The exponential function can also be used (Eq. 4).

Equations 5 and 6 in Table 3 are obtained by considering the activation energy for diffusion as a function of material moisture content. Equations 7 to 10 are not based on the Arrhenius form. They are empirical and they use complicated functions concerning the discrimination of the moisture and temperature effects (except, of course, Eq. 7). Equation 11 is more sophisticated as it considers different diffusivities of bound and free water and introduces the functional dependence of material moisture content on the

Table 4 Application Examples

Material	Equation	Constants	Reference nos.
Clay brick, burned clay	$D = D_0(T/T_0)^{a_T}(X/X_0)^{a_X}$	$D_0 = 7.36 \times 10^{-9} \text{ m}^2/\text{s}, T_0 = 273\text{K}, a_T = 9.5, X_0 = 0.35 \text{ kg/kg db}, a_X = 0.5$ for clay brick; $D_0 = 1.11 \times 10^{-9} \text{ m}^2/\text{s}, T_0 = 273\text{K}, a_T = 6.5, X_0 = 0.40 \text{ kg/kg db}, a_X = 0.5$ for burned clay	61
Polyvinyl-alcohol	$D = D_0 \exp[-E/R(1/T - 1/T_0)]$, $D_0 = \Sigma a_i X^i$	$T_0 = 298\text{K}, E = 3.05 \times 10^4 \text{ J/mol}, R = 8.314 \text{ J/molK}, a_0 = -0.104015 \times 10^2, a_1 = 0.363457 \times 10^2, a_2 = -0.469291 \times 10^3, a_3 = 0.634869 \times 10^4, a_4 = -0.517559 \times 10^5, a_5 = 0.250188 \times 10^6, a_6 = -0.747613 \times 10^6, a_7 = 0.139929 \times 10^7, a_8 = -0.159715 \times 10^7, a_9 = 0.101503 \times 10^7, a_{10} = -0.274672 \times 10^6$	71
Potato, carrot	$D = D_0 \exp(-X_0/X) \exp(-T_0/T)$	$D_0 = 2.41 \times 10^{-7} \text{ m}^2/\text{s}, X_0 = 7.62 \times 10^{-2} \text{ kg/kg db}, T_0 = 1.49 \times 10^{+3} {}^\circ\text{C for potato}; D_0 = 2.68 \times 10^{-4} \text{ m}^2/\text{s}, X_0 = 8.92 \times 10^{-2} \text{ kg/kg db}, T_0 = 3.68 \times 10^{+3} {}^\circ\text{C for carrot}$	72
Silica gel	$D = D_0 \exp((E_0 - E_1 X)/T)$	$D_0 = 5.71 \times 10^{-7} \text{ m}^2/\text{s}, E_0 = 2450\text{K}, E_1 = 1400\text{K}/(\text{kg/kg db})$	73

binding energy of desorption. Equation 12 introduces the effect of porosity on moisture diffusivity.

With regard to the number of the parameters involved (a significant measure concerning the regression analysis), it is concluded that at least 3 parameters are needed (Eqs. 1, 5 and 7).

Equations 5 and 7 in Table 3 were applied to potato and clay brick, respectively, and the results are presented in Figure 5. Both materials exhibit typical behavior. Diffusivity at low moisture content shows a steep descent when the moisture content decreases.

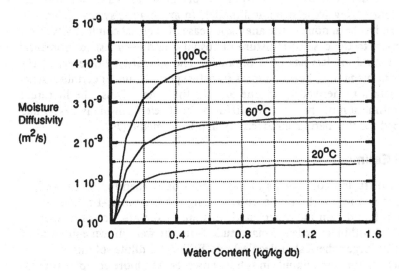

Potato [Data from (72)]

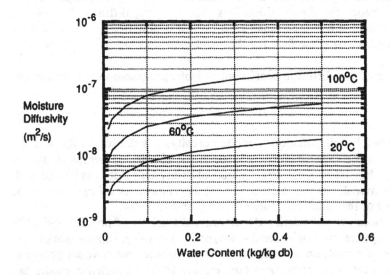

Clay Brick [Data from (61)]

Figure 5   Effect of material moisture content and temperature on moisture diffusivity.

The equations listed in Table 3 resulted from fitting to experimental data. The reason for the success of this procedure is the apparent simple dependence of diffusivity upon the material moisture content and temperature, which, as stated above, can be described even by three parameters only. The equations of Table 3 have been chosen by the respective researchers as the most appropriate for the material listed.

A single relation for the dependence of diffusivity upon the material moisture content and temperature general enough so as to apply to all the materials would be especially useful. It is expected that such a relation will be proposed in the near future.

The effect of pore structure and distribution on moisture diffusion can be examined by considering the material as a two- (or multi-) phase (dry material, water, air in voids, etc.) system and by considering some structural models to express the system geometry. Although a lot of work has been done in the analogous case of thermal conductivity, little attention has been given to the case of moisture diffusivity, and even less experimental validation of the structural models has been obtained. The similarity, however, of the relevant transport phenomena (i.e., heat and mass transfer) permits, under certain restrictions, the use of conclusions derived from one area in the other. Thus, the literature correlations for the estimation of the effective diffusion coefficient, in many cases, had been initially developed for the thermal conductivity in porous media (79).

2.5. Theoretical Estimation

The prediction of the diffusion coefficients of gases from basic thermophysical and/or molecular properties is possible with great accuracy using the Chapman–Enskog kinetic theory. Diffusivities in liquids, on the other hand, in spite of the absence of a rigorous theory, can be estimated within an order of magnitude from the well-known equations of Stokes and Einstein (for large spherical molecules) and Wilke (for dilute solutions).

Diffusion of gases, vapors, and liquids in solids, however, is a more complex process than the diffusion in fluids because of the heterogeneous structure of the solid and its interactions with the diffusing components. As a result, it has not yet been possible to develop an effective theory for the diffusion in solids. Usually, diffusion in solids is handled by the researchers in a manner analogous to heat conduction. In the following paragraphs typical methods are described for the development of semiempirical correlations for diffusivity.

For the estimation of the diffusion coefficient in isotropic macroporous media, the relation

$$D = (\delta\epsilon/\tau^2)D_A \qquad [4]$$

has been proposed (79). In this equation, ϵ is the porosity, τ is the tortuosity, δ is the constrictivity, and D_A is the vapor diffusivity in air in the absence of porous media. In spite of its simplicity, Eq. [4] will not attain practical utility unless it is validated with additional pore space models, its parameters (ϵ, τ, δ) determined for a large number of systems, and the effect of the solid's moisture properly accounted for.

An equation has been derived relating the effective diffusivity of porous foodstuffs to various physical properties such as molecular weight, bulk density, vapor space permeability, water activity as a function of material moisture content, water vapor pressure, thermal conductivity, heat of sorption, and temperature (80). A predictive model has been proposed to obtain effective diffusivities in cellular foods. The method requires data for composition, binary molecular diffusivities, densities, membrane and cell wall permeabilities, molecular weights, and water viscosity and molar volume (81). The effect of moisture upon the effective diffusivity is taken into account via the binding energy of sorption in an equation suggested in Reference 77.

3. THERMAL CONDUCTIVITY

3.1. Definition

The *thermal conductivity* of a material is a measure of its ability to conduct heat. It can be defined using Fourier's law for homogeneous materials:

$$\partial T/\partial t = (k/\varrho c_p)\nabla^2 T \tag{3}$$

where k (kW/m·K) is the thermal conductivity, ϱ (kg/m^3) the density, c_p (kJ/kg·K) the specific heat of the material, T (K) the temperature, and t (s) the time. The quantity $(k/\varrho c_p)$ is the thermal diffusivity. For heterogeneous materials, the effective thermal conductivity is used in conjunction with Fourier's law.

Equation [3] is used in cases in which heat transfer during drying takes place via conduction (internally controlled drying). This, for example, is the situation when drying large particles, relatively immobile, that are immersed in the heat transfer medium.

As far as heat and mass transfer is concerned, the drying process is internally controlled whenever the respective Biot number (Bi$_H$, Bi$_M$) is greater than 1 (5).

3.2. Methods of Experimental Measurement

The effective thermal conductivity can be determined using the methods presented on Table 5, which includes the relevant references. Measurement techniques for thermal conductivity can be grouped into steady-state and transient-state methods. Transient methods are more popular because they can be run for as short as 10 seconds, during which time the moisture migration and other property changes are kept minimal.

Steady-State Methods

In steady-state methods, the temperature distribution of the sample is measured at steady state, with the sample placed between a heat source and a heat sink. Different geometries can be used, those for longitudinal heat flow and radial heat flow.

Longitudinal Heat Flow (Guarded Hot Plate)

The longitudinal heat flow (guarded hot plate) method is regarded as the most accurate and most widely used apparatus for the measurement of thermal conductivity of poor

Table 5 Methods for the Experimental Measurement of Thermal Conductivity

Method	Reference nos.
Steady-state method	
Longitudinal heat flow (guarded hot plate)	82
Radial heat flow	83
Unsteady-state method	
Fitch	84, 85
Plane heat source	86
Probe method	87, 88

conductors of heat. The method is most suitable for dry homogeneous specimens in slab forms. The details of the technique are given by the American Society for Testing and Materials (ASTM) Standard C-177 (82).

Radial Heat Flow

While the longitudinal heat flow methods are most suitable for slab specimens, the radial heat flow techniques are used for loose, unconsolidated, powder or granular materials. The methods can be classified as follows:

Cylinder with/without end guards
Sphere with central heating source
Concentric cylinder comparative method

Unsteady-State Methods

Transient-state or unsteady-state methods make use of either a line source of heat or plane sources of heat. In both cases, the usual procedure is to apply a steady heat flux to the specimen, which must be initially in thermal equilibrium, and to measure the temperature rise at some point in the specimen, resulting from this applied flux (83). The Fitch method is one of the most common transient methods for measuring the thermal conductivity of poor conductors. The method was developed in 1935 and was described in the National Bureau of Standards Research Report No. 561. Experimental apparatus is commercially available.

Probe Method

The probe method is one of the most common transient methods using a line heat source. The method is simple and quick. The probe is a needle of good thermal conductivity that is provided with a heater wire over its length and some means of measuring the temperature at the center of its length. Having the probe embedded in the sample, the temperature response of the probe is measured in a step change of heat source and the thermal conductivity is estimated using the transient solution of Fourier's law. Detailed descriptions as well as the necessary modifications for the application of the forementioned methods in food systems are given in References 83, 89 and 90.

3.3. Data Compilation

Despite the limited data of effective moisture diffusivity, a lot of data are reported in the literature for thermal conductivity. Data for mainly homogeneous materials are available in handbooks such as the *Handbook of Chemistry and Physics* (91), the *Chemical Engineers' Handbook* (92), *ASHRAE Handbook of Fundamentals* (93), Rohsenow and Choi (94), and many others. For foods and agricultural products, data are available in References 83, 88, and 95–97. For selected pharmaceutical materials, data are presented by Pakowski and Mujumdar (98).

Some data for thermal conductivity are presented for example in Table 6. These values are distributed as shown in Figure 6. The distribution is different from that of moisture diffusivity (Fig. 2), which is normal. For thermal conductivity, the values are uniformly distributed in the range 0.25 to 2.25 W/m/K, while a lot of data are accumulated below 0.25 W/m/K.

3.4. Factors Affecting Thermal Conductivity

The thermal conductivity of homogeneous materials depends on temperature and composition, and empirical equations are used for its estimation. For each material, polynomial

Table 6 Effective Thermal Conductivity in Some Materials

Material	Temperature (°C)	Thermal conductivity (W/mK)	Reference nos.
Aerogel, silica	38	0.022	94
Asbestos	427	0.225	94
Bakelite	20	0.232	94
Beef, 69.5% water	−18	0.622	99
Beef fat, 9% water	−10	0.311	100
Brick, common	20	0.173–0.346	94
Brick, fire clay	800	1.37	94
Carrots	−15−−19	0.622	101
Concrete	20	0.813–1.40	94
Corkboard	38	0.043	94
Diatomaceous earth	38	0.052	94
Fiber insulating board	38	0.042	94
Fish	−20	1.50	100
Fish, cod and haddock	−20	1.83	102
Fish muscle	−23	1.82	103
Glass, window	20	0.882	94
Glass wool, fine	38	0.054	94
Glass wool, packed	38	0.038	94
Ice	0	2.21	94
Magnesia	38	0.067	94
Marble	20	2.77	94
Paper		0.130	94
Peach	18–27	1.12	104
Peas	18–27	1.05	104
Peas	−12−−20	0.501	101
Plums	−13−−17	0.294	101
Potato	−10−−15	1.09	101
Potato flesh	18–27	1.05	104
Rock wool	38	0.040	94
Rubber, hard	0	0.150	94
Strawberries	18–27	1.35	104
Turkey breast	−25	0.167	100
Turkey leg	−25	1.51	100
Wood, oak	21	0.207	94

functions of first or higher order are used to express the temperature effect. A large number of empirical equations for the calculations of thermal conductivity as a function of temperature and/or humidity are available in the literature (83,92).

For heterogeneous materials, the effect of geometry must be considered using structural models. Utilizing Maxwell's and Eucken's work in the field of electricity, Luikov et al. (105) initially used the idea of an elementary cell, as representative of the model structure of materials, in order to calculate the effective thermal conductivity of powdered systems and solid porous materials. In the same paper, a method is proposed for the estimation of the effective thermal conductivity of mixtures of powdered and solid porous materials.

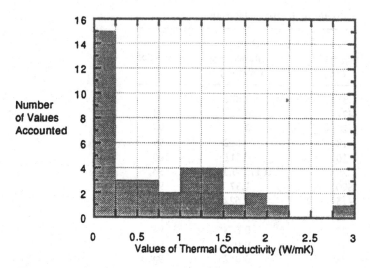

Figure 6 Distribution of thermal conductivity values (data from Table 5).

Since then, a number of structural models have been proposed, some of which are given in Table 7. The perpendicular model assumes that heat conduction is perpendicular to alternate layers of the two phases, while the parallel model assumes that the two phases are parallel to heat conduction. In the mixed model, heat conduction is assumed to take place by a combination of parallel and perpendicular heat flow. In the random model, the two phases are assumed to be randomly mixed. The Maxwell model assumes that one phase is continuous, while the other phase is dispersed as uniform spheres. Several other models have been reviewed in References 107, 110, and 111, among others.

The use of some of these structural models to calculate the thermal conductivity of a hypothetical porous material is presented in Figure 7. The parallel model gives the larger value for the effective thermal conductivity, while the perpendicular model gives the

Table 7 Structural Models for Thermal Conductivity in Heterogeneous Materials

Model	Equation	Reference nos.
Perpendicular (series)	$1/k = (1 - \epsilon)/k_1 + \epsilon/k_2$	106, 107
Parallel	$k = (1 - \epsilon)k_1 + \epsilon k_2$	106, 107
Mixed	$1/k = \dfrac{1 - F}{(1 - \epsilon)k_1 + \epsilon k_2} + F\left(\dfrac{1 - \epsilon}{k_1} + \dfrac{\epsilon}{k_2}\right)$	106, 107
Random	$k = k_1^{(1-\epsilon)}k_2^{\epsilon}$	106, 107
Effective medium theory	$k = k_1[b + (b^2 + 2(k_1/k_2)/(Z - 2))^{1/2}]$ $b = [Z(1 - \epsilon)/2 - 1 + (k_2/k_1)(\epsilon Z/2 - 1)]/(Z - 2)$	108
Maxwell	$k = \dfrac{k_2[k_1 + 2k_2 - 2(1 - \epsilon)(k_2 - k_1)]}{k_1 + 2k_2 + (1 - \epsilon)(k_2 - k_1)}$	109

k, effective thermal conductivity; k_i, thermal conductivities of phase i; ϵ, void fraction of phase 2; F, Z, parameters

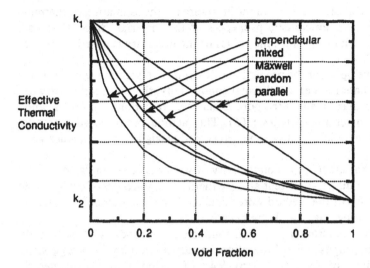

Figure 7 Effect of geometry on the thermal conductivity of heterogeneous materials using structural models.

lower. All other models predict values in-between. The use of structural models has been successfully extended to foods (108,112), which exhibit a more complex structure than that of other materials, while this structure often changes during the heat conduction.

A systematic general procedure for selecting suitable structural models, even in multiphase systems, has been proposed in Reference 113. The method is based on a model discrimination procedure. If a component has unknown thermal conductivity, the method estimates the dependence of the temperature on the unknown thermal conductivity, and the suitable structural models simultaneously.

An excellent example of applicability of the above is in the case of starch, a useful material in extrusion. The granular starch consists of two phases, the wet granules and the air/vapor mixture in the intergranular space. The starch granule also consists of two phases, the dry starch and the water. Consequently, the thermal conductivity of the granular starch depends on the thermal conductivities of pure materials (that is dry pure starch, water, air, and vapor, all functions of temperature) and the structures of granular starch and the starch granule. It has been shown that the parallel model is the best model for both the granular starch and the starch granule (113). These results led to simultaneous experimental determination of the thermal conductivity of dry pure starch versus temperature. Dry pure starch is a material that cannot be isolated for direct measurement.

3.5. Theoretical Estimation

As in the case of the diffusion coefficient, the thermal conductivity in fluids can be predicted with satisfactory accuracy using theoretical expressions, such as the formulas of Chapman and Enskog for monoatomic gases, of Eucken for polyatomic ones, or of Bridgman for pure liquids. The thermal conductivity of solids, however, has not yet been predicted using basic thermophysical and/or molecular properties, just like the analogous diffusion coefficient. Usually, the thermal conductivities of solids must be established experimentally since they depend upon a large number of factors that cannot easily be measured or predicted.

A large number of correlations are listed in the literature for the estimation of thermal conductivity as a function of characteristic properties of the material. Such relations, however, have limited practical utility since the values of the necessary properties are not readily available.

A method has been developed for the prediction of thermal conductivity as a function of temperature, porosity, material skeleton thermal conductivity, thermal conductivity of the gas in the porous, mechanical load on the porous material, radiation, and optical and surface properties of the material's particles (105). The method produced satisfactory results for a wide range of materials (quartz sand, powdered plexiglass, perlite, silica gel, etc.).

It has been proposed that the thermal conductivity of wet beads of granular material be estimated as a function of material content and the thermal conductivity of each of the three phases (114). The results of the method were validated in a small number of materials such as crushed marble, slate, glass, and quartz sand.

Empirical equations for estimating the thermal conductivity of foods as a function of their composition have been proposed in the literature. In particular, it has been suggested that the thermal conductivity of foods is a first-degree function of the constituents' concentrations (water, protein, fat, carbohydrate, etc.) (97).

4. INTERPHASE HEAT AND MASS TRANSFER COEFFICIENTS

4.1. Definition

The interphase heat transfer coefficient is related to heat transfer through a relative stagnant layer of the flowing air, which is assumed to adhere to the surface of the solid during drying (generally heating or cooling). It may be defined as the proportionality factor in the equation (Newton's law)

$$Q = h_H A \, (T_A - T) \tag{5}$$

where h_H (kW/m^2/K) is the surface heat transfer coefficient at the material-air interface, Q (kW) is the rate of heat transfer, A (m^2) is the effective surface area, T (K) is the solid temperature at the interface, and T_A (K) is the bulk air temperature.

By analogy, a surface mass transfer coefficient can be defined using the following equation:

$$J = h_M A (X_A - X_{AS}) \tag{6}$$

where h_M (kg/m^{2}s) is the surface mass transfer coefficient at the material-air interface, J (kg/s) is the rate of mass transfer, A (m^2) is the effective surface area, X_{AS} (kg/kg), and X_A (kg/kg) are the air humidities at the solid interface and the bulk air.

Equations [5] and [6] are used in cases in which the drying is externally controlled. This occurs when the Biot number (Bi$_H$, Bi$_M$) for heat and mass transfer is less than 0.1 (5).

Volumetric heat and mass transfer coefficients are often used instead of surface heat and mass transfer coefficients. They can be defined using the equations

$$h_{VH} = \alpha h_H \tag{7}$$

$$h_{VM} = \alpha h_M \tag{8}$$

where α is the specific surface defined as follows:

$$\alpha = A/V \tag{9}$$

where A (m^2) is the effective surface area and V (m^3) the total volume of the material.

Different coefficients can be defined using different driving forces.

4.2. Methods of Experimental Measurement

The methods of experimental measurement of heat and mass transfer coefficients summarized in Table 8, resulted mainly from heat and mass transfer investigations in packed beds. Heat transfer techniques are either steady or unsteady state. In steady-state methods, the heat flow is measured together with the temperatures, and the heat transfer coefficient is obtained using Newton's law. Three different methods for heating are presented in Table 8. In unsteady-state techniques, the temperature of the outlet air is measured as a response to variations of the inlet air temperature. A transient model incorporating the heat transfer coefficient is used for analysis. Step, pulse, or cyclic temperature variations of the input air temperature have been used. Drying experiments during the constant drying rate period have also been used for estimating heat and mass transfer coefficients. A generalization of this method for simultaneous estimation of transport properties using drying experiments is presented in Section 7.

4.3. Data Compilation

All the data available in the literature are in the form of empirical equations, and they are examined in the next section.

4.4. Factors Affecting the Heat and Mass Transfer Coefficients

Both heat and mass transfer coefficients are influenced by thermal and flow properties of the air and, of course, by the geometry of the system. Empirical equations for various geometries have been proposed in the literature. Table 9 summarizes the most popular equations used for drying. The empirical equations incorporate dimensionless groups, which are defined in Table 10. Some nomenclature needed for understanding Table 9 is also included on Table 10.

Equations 1 to 5 in Table 9 are the most widely used equations in estimating heat and mass transfer coefficients for simple geometries (packed beds, flat plates).

Table 8 Methods for the Experimental Measurement of Heat and Mass Transfer Coefficients

Method	Reference nos.
Steady-state heating methods	
Material heating	115
Wall Heating	116
Microwave heating	117
Unsteady-state heating methods	
Step change of input air temperature	118, 119
Pulse change of input air temperature	120, 121
Cyclic temperature variation of input air	122, 123
Constant rate drying experiments	124, 125
Simultaneous estimation of transport properties using drying experiments	See Section 7

Table 9 Equations for Estimating Heat and Mass Transfer Coefficients

Equation no.	Geometry	Equation	Reference nos.
1	Packed beds (heat transfer)	$j_H = 1.06\,\mathrm{Re}^{-0.41}$ $350 < \mathrm{Re} < 4000$	126
2	Packed beds (mass transfer)	$j_M = 1.82\,\mathrm{Re}^{-0.51}$ $40 < \mathrm{Re} < 350$	127
3	Flat plate (heat transfer, parallel flow)	$j_H = 0.036\,\mathrm{Re}^{-0.2}$ $500,000 < \mathrm{Re}$	128
4	Flat plate (heat transfer, parallel flow)	$h_H = 0.0204\,G^{0.8}$ $0.68 < G < 8.1 \quad 45 < T < 150°C$	129
5	Flat plate (heat transfer, perpendicular flow)	$h_H = 1.17\,G^{0.37} \quad 1.1 < G < 5.4$	129
6	Flat plate (heat transfer, parallel flow)	$\mathrm{Nu} = 0.036(\mathrm{Re}^{0.8} - 9200)\mathrm{Pr}^{0.43}$ $1.0 \times 10^5 < \mathrm{Re} < 5.5 \times 10^6$	130
7	Packed beds (heat transfer)	$\mathrm{Nu}' = (0.5\,\mathrm{Re}'^{1/2} + 0.2\,\mathrm{Re}'^{2/3})\,\mathrm{Pr}^{1/3}$ $2 \times 10^3 < \mathrm{Re}' < 8 \times 10^3$	130
8	Rotary dryer (heat transfer)	$j_H = 1.0\,\mathrm{Re}^{-0.5}\mathrm{Pr}^{1/3}$	131
9	Rotary dryer (heat transfer)	$\mathrm{Nu} = 0.33\,\mathrm{Re}^{0.6}$	131
10	Rotary dryer (heat transfer)	$h_{VH} = 0.52\,G^{0.8}$	131
11.1	Fluidized beds (heat transfer)	$\mathrm{Nu} = 0.0133\,\mathrm{Re}^{1.6}$ $0 < \mathrm{Re} < 80$	6
11.2	Fluidized beds (heat transfer)	$\mathrm{Nu} = 0.316\,\mathrm{Re}^{0.8}$ $80 < \mathrm{Re} < 500$	6
12.1	Fluidized beds (mass transfer)	$\mathrm{Sh} = 0.374\,\mathrm{Re}^{1.18}$ $0.1 < \mathrm{Re} < 15$	6
12.2	Fluidized beds (mass transfer)	$\mathrm{Sh} = 2.01\,\mathrm{Re}^{0.5}$ $15 < \mathrm{Re} < 250$	6
13	Droplets in spray dryer (heat transfer)	$\mathrm{Nu} = 2 + 0.6\,\mathrm{Re}^{1/2}\,\mathrm{Pr}^{1/3}$ $2 < \mathrm{Re} < 200$	132
14	Droplets in spray dryer (mass transfer)	$\mathrm{Sh} = 2 + 0.6\,\mathrm{Re}^{1/2}\mathrm{Sc}^{1/3}$ $2 < \mathrm{Re} < 200$	132
15	Spouted beds (heat transfer)	$\mathrm{Nu} = 5.0 \times 10^{-4}\,\mathrm{Re}_s^{1.46}(u/u_s)^{1/3}$	6
16	Spouted beds (mass transfer)	$\mathrm{Sh} = 2.2 \times 10^{-4}\,\mathrm{Re}^{1.45}\,(D/H_o)^{1/3}$	6
17	Pneumatic dryers (heat transfer)	$\mathrm{Nu} = 2 + 1.05\,\mathrm{Re}^{1/2}\mathrm{Pr}^{1/3}\mathrm{Gu}^{0.175}$ $\mathrm{Re} < 1000$	6
18	Pneumatic dryers (mass transfer)	$\mathrm{Sh} = 2 + 1.05\,\mathrm{Re}^{1/2}\mathrm{Pr}^{1/3}\mathrm{Gu}^{0.175}$ $\mathrm{Re} < 1000$	6
19	Impingement drying	Several equations for various configurations	133–135

For nomenclature, see Table 10.

For packed beds, a lot of literature exists. In 1965, Barker reviewed 244 relevant papers (183). The equation suggested by Whitaker (130) is selected and presented in Table 9 as Eq. 7. It has been obtained by fitting to data of several investigators (cf. Refs. 126, 127). Equation 6 for flat plates comes from the same investigation (130), and it is also included on Table 9. In drying of granular materials, the equations reviewed in Reference 136 should be examined.

Rotary dryers are usually controlled by heat transfer. Thus, Eqs. 8–10 in Table 9

Table 10 Dimensionless Groups of Physical
Properties

Name	Definition
Biot for heat transfer	$Bi_H = h_H d/2k$
Biot for mass transfer	$Bi_M = h_M d/2\rho\Delta$
Gukhman number	$Gu = (T_A - T)/T_A$
Heat transfer factor	$j_H = St\ Pr^{2/3}$
Mass transfer factor	$j_M = (h_M/u_A\rho_A)Sc^{2/3}$
Nusselt number	$Nu = h_H d/k_A$
Prandtl number	$Pr = c_p\mu/k_A$
Reynolds number	$Re = u_A\rho_A d/\mu$
Schmidt number	$Sc = \mu/\rho_A D_A$
Sherwood number	$Sh = h_M d/\rho_A D_A$
Stanton number	$St = h_H/u_A\rho_A c_p$

c_p, specific heat (kJ/kg·K); d, particle diameter (m); D, diffusivity in solid (m^2/s); D_A, vapor diffusivity in air (m^{2}s); ϵ, void fraction in packed bed; G, mass flow rate of air (kg/m^{2}s); h_H, heat transfer coefficient (kW/m^{2}K); h_M, mass transfer coefficient (kg/m^{2}s); h_{VH}, volumetric heat transfer coefficient (kW/m^{3}K); h_{VM}, volumetric mass transfer coefficient (kg/m^{3}s); k, thermal conductivity of solid (kW/m·K); k_A, thermal conductivity of air (kW/m·K); μ, dynamic viscosity of air (kg/m·s); Nu', Nu' = Nu $\epsilon/(1 - \epsilon)$; ϱ_A, density of air (kg/m^3); Re', Re' = Re $(1 - \epsilon)$; Re$_s$, Re based on u_s instead of u; T_A, air temperature (°C); T, material temperature (°C); u_A, air velocity (m/s); u_S, air velocity for incipient spouting (m/s)

are proposed in Reference 131 for the estimation of the corresponding heat transfer coefficients.

Heat and mass transfer in fluidized beds have been discussed in References 6 and 137–140. The latter reviewed the most important correlations and proposed Eqs. 11 and 12 of Table 9 for the calculation of heat and mass transfer coefficients, respectively. Further information for fluidized bed drying can be found in Reference 141.

Vibration can intensify heat and mass transfer between the particles and gas. The following correction has been suggested for the heat and mass transfer coefficients when vibration occurs (6):

$$h_{H'} = h_H(A'f'/u_A)^{0.65} \qquad [10]$$

$$h_{M'} = h_M(A'f'/u_A)^{0.65} \qquad [11]$$

where u (m/s) is the air velocity, A (m) the vibration amplitude, and f (s^{-1}) the frequency of vibration. Further information on vibrated bed dryers can be found in Reference 142.

For spray dryers, the popular equation of Ranz and Marshall (132) is presented in Table 9 (Eqs. 13 and 14). They correlated data obtained for suspended drops evaporating in air.

Heat and mass transfer in a spouted bed has not been fully investigated yet because of the complex character of the flow path of the particles in a bed with zones under different aerodynamic conditions (6). However, Eqs. 15 and 16 of Table 9 can be used.

Heat transfer coefficients for pneumatic dryers have been reviewed in Reference 6.

The majority of authors examined use an equation similar to Eqs. 13 and 14 of Table 9 for spray dryers. For immobile particles, the exponent of the Re number is close to 0.5 and for free-falling particles 0.8. Equation 17 of Table 9 is proposed. The mass transfer coefficient could be estimated by the analogy Sh = Nu (6). In extensive reviews (133–135), correlations for estimating heat and mass transfer coefficients in impingement drying under various configurations are discussed.

The calculated heat and mass transfer coefficients using some of the equations presented on Table 9 are plotted versus air velocity with some simplifications in Figures 8 and 9. These figures can be used to estimate approximately the heat and mass transfer coefficients for various dryers. The simplifications made for the construction of these figures concern the drying air and material conditions. For instance, the air temperature is taken as 80°C, the air humidity as 0.010 kg/kg db, and the particle size as 10 mm (typical drying conditions). For other conditions, the equations of Table 9 should be used.

4.5. Theoretical Estimation

No theory is available for estimating the heat and mass transfer coefficients using basic thermophysical properties. The analogy of heat and mass transfer can be used to obtain mass transfer data from heat transfer data and vice versa. For this purpose, the Chilton-Colburn analogies can be used (129):

$$j_M = j_H = f/2 \tag{12}$$

where f is the well-known Fanning friction factor for the fluid, and j_H and j_M are the heat and mass transfer factors defined on Table 10. Discrepancies of the above classical analogy have been discussed in Reference 143.

In air conditioning processes, the heat and mass transfer analogy is usually expressed using the Lewis relationship:

$$h_H/h_M = c_p \tag{13}$$

where c_p (kJ/kg/K) is the specific heat of air.

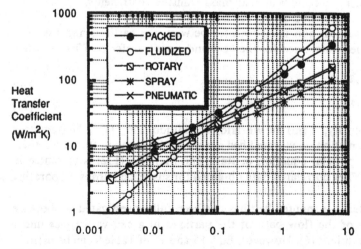

Figure 8 Heat transfer coefficients versus air velocity for some dryers (particle size 10 mm; drying conditions $T_A = 80°C$, $X_A = 10$ g/kg db.

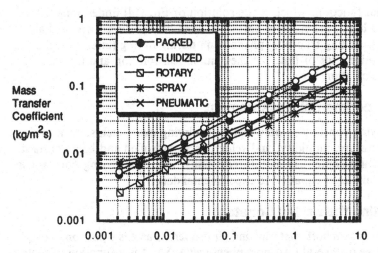

Figure 9 Mass transfer coefficients versus air velocity for some dryers (particle size 10 mm; drying conditions $= T_A = 80°C, X_A = 10$ g/kg db.

5. DRYING CONSTANT

5.1. Definition

The transport properties discussed above (moisture diffusivity, thermal conductivity, interface heat and mass transfer coefficients) describe completely the drying kinetics. However, in the literature sometimes (mainly in foods, especially in cereals) instead of the above transport properties, the drying constant K is used. The drying constant is a combination of these transport properties.

The drying constant can be defined using the so-called thin-layer equation. Lewis suggested that during the drying of porous hygroscopic materials, in the falling rate period the rate of change in material moisture content is proportional to the instantaneous difference between material moisture content and the expected material moisture content when it comes into equilibrium with the drying air (144). It is assumed that the material layer is thin enough or the air velocity is high so that the conditions of the drying air (humidity, temperature) are kept constant throughout the material. The thin-layer equation has the following form:

$$-dX/dt = K(X - Xe) \tag{14}$$

where X (kg/kg db) is the material moisture content, Xe (kg/kg db) the material moisture content in equilibrium with the drying air, and t (s) is the time. A review of several other thin-layer equations can be found in References 76 and 145.

Equation [14] constitutes an effort toward a unified description of the drying phenomena regardless of the controlling mechanism. The use of similar equations in the drying literature is ever increasing. It is claimed, for example, that they can be used to estimate the drying time as well as for the generalization of the drying curves (6).

The drying constant K is the most suitable quantity for purposes of design, optimization, and any situation in which a large number of iterative model calculations are needed. This stems from the fact that the drying constant embodies all the transport properties into a simple exponential function, which is the solution of Eq. [14] under constant air conditions. On the other hand, the classical partial differential equations, which analytically describe the four prevailing transport phenomena during drying (internal-external, heat–mass transfer), require a lot of time for their numerical solution and thus are not attractive for iterative calculations.

5.2. Methods of Experimental Measurement

Measurement of the drying constant is obtained from drying experiments. In a drying apparatus, the air temperature, humidity, and velocity are controlled and kept constant, while the material moisture content is monitored versus time. The drying constant is estimated by fitting the thin-layer equation to experimental data.

5.3. Factors Affecting the Drying Constant

The drying constant depends on both material and air properties as it is a phenomenological property representative of several transport phenomena. So, it is a function of material moisture content, temperature, and thickness, as well as air humidity, temperature, and velocity.

Some relationships describing the effect of the above factors on the drying constant are presented on Table 11. Equations 1 are Arrhenius type equations, which take into account the temperature effect only. The effect of water activity can be considered by modifying the activation energy (Eq. 2.1) on the preexponential factor (Eq. 2.2). Equations 3.1 and 3.2 consider the same factors in a different form. Equation 4 takes into account only the air velocity effect, while Eq. 5 considers all the factors affecting the drying constant. Table 12 lists parameter values for typical equations of Table 11.

Equations 3.2 and 5 were applied to shelled corn (150) and to green pepper (35), respectively, and the results are presented in Figure 10. The effects of air temperature and velocity, as well as particle dimensions, are shown for green pepper drying, while the air temperature and the small air-water activity effects are shown for the low air temperature drying of wheat.

Table 11 Effect of Various Factors on the Drying Constant

Equation no.	Materials of application	Equation	Reference nos.
1.1	Grains, barley, various tropical agricultural products	$K(T_A) = b_0 \exp[-b_1/T_A]$	75,146,147
1.2	Barley, wheat	$K(T_A) = b_0 \exp[-b_1/(b_2 + b_3 T_A)]$	148
2.1	Melon	$K(a_w, T_A) = b_0 \exp[-(b_1 + b_2 a_w)/T_A]$	149
2.2	Corn, shelled	$K(a_w, T_A) = b_0 \exp(-b_1 a_w) \exp[-b_2/(b_3 + b_4 T_A)]$	150
3.1	Rice	$K(a_w, T_A) = b_0 + b_1 T_A - b_2 a_w$	151
3.2	Wheat	$K(a_w, T_A) = b_0 + b_1 T_A^2 - b_2 a_w$	152
4	Carrot	$K(u_A) = \exp(-b_1 + b_2 \ln u_A)$	153
5	Potato, onion, carrot, pepper	$K(a_w, T_A, d, u_A) = b_0 a_w^{b_1} T_A^{b_2} d^{b_3} u_A^{b_4}$	35

K, drying constant; T_A, temperature; u_A, air velocity; a_w, water activity; d, particle diameter; b_i, parameters

Table 12 Application Examples

Material	Equation	Constants	Reference nos.
Shelled corn	$K = b_0 \exp(-b_1 a_w)\exp[-b_2/(b_3 + b_4 T_A)]$ $0.1 < a_w < 0.6,\ 23.5 < T_A < 56.9°C$	$b_0 = 170/s,\ b_1 = 1.15,$ $b_2 = 8259,\ b_3 = 492,$ $b_4 = 1.8/°C$	150
Green pepper	$K = b_0 X_A^{b_1} T_A^{b_2} d^{b_3} u_A^{b_4}$ $0.006 < X_A < 0.022\ kg/kg\ db,$ $60 < T_A < 90°C,$ $0.005 < d < 0.015\ m,$ $3 < u_A < 5\ m/s$	$b_0 = 1.11 \times 10^{-8}/s,$ $b_1 = 9.03 \times 10^{-2},$ $b_2 = 1.54,$ $b_3 = -0.982,$ $b_4 = 0.293$	35

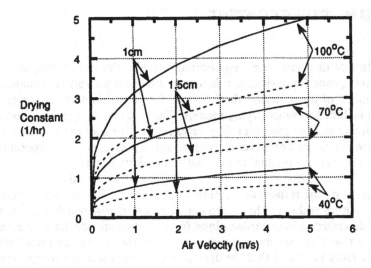

Green Pepper [Data from (35)]

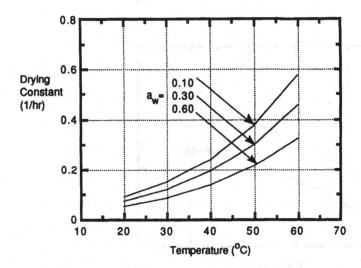

Shelled Corn [Data from (150)]

Figure 10 Effect of various factors on the drying constant.

5.4. Theoretical Estimation

It is impossible to estimate an empirical constant using theoretical arguments. The estimation of an empirical constant using theoretical arguments has little, if any, meaning. Nevertheless, if we assume that for some drying conditions the controlling mechanism is the moisture diffusion in the material, then the drying constant can be expressed as a function of moisture diffusivity. For slabs, for example, the following equation is valid:

$$K = \pi^2 D/L^2$$ [15]

where D (m²/s) is the effective diffusivity and L (m) is the thickness of the slab.

6. EQUILIBRIUM MOISTURE CONTENT

6.1. Definition

Knowledge of the state of thermodynamic equilibrium between the surrounding air and the solid is a basic prerequisite for drying, as it is for any similar mass transfer situation.

The moisture content of the material when it comes into equilibrium with drying air is a useful property included in most drying models. The relation between equilibrium material moisture content and the corresponding water activity for a given temperature is known as the *sorption isotherm*. The water activity a_w at the pressures and temperatures that usually prevail during drying is equal to the relative humidity of air.

The equilibrium moisture of a material can be attained either by adsorption or by desorption, as expressed by the respective isotherms of Figure 11. The usually observed deviation of the two curves is due to the phenomenon of hysteresis, which has not yet been quantitatively described. Many explanations for the phenomenon have been put forth that converge in that there are more active sites during the desorption than during adsorption. It is clear from Figure 11 that the desorption isotherm is the curve to use for the process of drying.

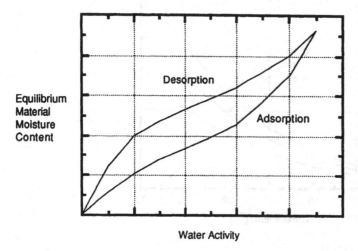

Figure 11 Hysteresis between adsorption and desorption isotherms.

In essence, the sorption isotherms express the minimum value of material moisture content that can be reached by a solid during drying in relation to the relative humidity of the drying air. On the basis of such isotherms, the equilibrium material moisture content can be calculated. Such equilibrium values are necessary for the formulation of the mass transfer driving forces.

Moreover, the isotherms determine the proper storage environment and the packaging conditions, especially for foods. Through the isotherms, the isosteric heat of sorption can be determined and, hence, an accurate prediction can be made of the energy requirements for the drying of a solid. The utility of the isotherm is extended to the determination of the moisture sorption mechanism as well as to the degree of bound water.

Brunauer et al. classified the sorption isotherms into five different types (see Figure 12) (154). The sorption isotherms of the hydrophilic polymers, such as natural fibers and foods, are of type II. The isotherms of the less hydrophilic rubbers, plastics, synthetic fibers, and foods rich in soluble components are of type III. The isotherms of certain inorganic materials (such as aluminum oxides) are of type IV. For many materials, however, the sorption isotherms cannot be properly classified since they belong to more than one type.

6.2. Methods of Experimental Measurement

A comprehensive review of existing experimental measuring methods is given in References 155 and 156. Sorption isotherms can be determined according to two basic principles, the gravimetric and hygrometric.

Gravimetric Methods

During the measurement, the air temperature and the water activity are kept constant until the moisture content of the sample attains the constant equilibrium value. The air may be circulated (dynamic methods) or stagnant (static). The material weight may be registered continuously (continuous methods) or discontinuously (discontinuous methods).

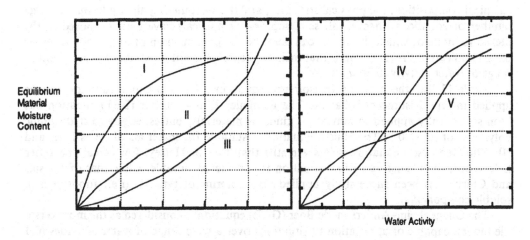

Figure 12   The five types of isotherms. (From Reference 154.)

Hygrometric Methods

During the measurement, the material moisture content is kept constant until the surrounding air attains the constant equilibrium value. The air water activity is measured via hygrometer or manometer.

The working group in the COST 90bis Project has developed a reference material (microcrystalline cellulose, MCC) and a reference method for measuring water sorption isotherms, and conducted a collaborative study to determine the precision (repeatability and reproducibility) with which the sorption isotherm of the reference material may be determined by the reference method. A detailed procedure for the resulting standardized method was presented, and the factors influencing the results of the method were discussed (157–159).

6.3. Data Compilation

A large volume of data of equilibrium moisture content appears in the literature. Data for more than 35 polymeric materials, such as natural fibers, proteins, plastics, and synthetic fibers, are given in Reference 8. Isotherms for 32 materials (organic and inorganic) are also given in Reference 92. The literature is especially rich in sorption isotherms of foods due to the fact the value of water activity is a critical parameter for food preservation safety and quality.

A bibliography on sorption isotherms of food materials is presented in Reference 160. The collection comprises 2200 references, including about 900 papers with information on equilibrium moisture content of foods in defined environments. The papers are listed alphabetically according to the names of the first author, but they are also grouped according to product.

Additional bibliographies should also be mentioned. The *Handbook of Food Isotherms* contains more than 1000 isotherms, with a mathematical description of over 800 (161). About 460 isotherms were obtained from the monograph of Reference 162. Data on sorption properties of selected pharmaceutical materials are presented in Reference 98.

6.4. Factors Affecting the Equilibrium Moisture Content

Equilibrium material moisture content depends upon many factors, among with are the chemical composition, the physical structure, and the surrounding air conditions. A large number of equations (theoretical, semiempirical, empirical) have been proposed in the literature, none of which, however, can describe the phenomenon of hysteresis. Another basic handicap of the equations is that their applicability is not satisfactory over the entire range of water activity ($0 \leq a_w \leq 1$).

Table 13 lists the best-known isotherm equations. The Langmuir equation can be applied in type I isotherm behavior. The Brunauer–Emmet–Tetter (BET) equation has been successfully applied to almost all kinds of materials, but especially to hydrophilic polymers for $a_w < 0.5$. The Halsey equation is suitable for materials of types I, II, and III. The Henderson equation is less versatile than that of Halsey. For cereal and other field crops, the Chang and Pfost equation is considered suitable, while that of Iglesias and Chirife has been successfully applied on isotherms of type III (i.e., foods rich in soluble components).

The Guggenheim–Anderson–de Boer (GAB) equation is considered as the most versatile model, capable of application to situations over a wide range of water activities ($0.1 < a_w < 0.9$) and to various materials (inorganic, foods, etc.). The GAB equation is

Table 13 Effect of Water Activity and Temperature on the Equilibrium Moisture Content

Equation name	Equation	Reference nos.
Langmuir	$a_w\left(\dfrac{1}{X} - \dfrac{1}{b_0}\right) = \dfrac{1}{b_0 b_1}$	163
Brunauer–Emmet–Tetter (BET)	$\dfrac{a_w}{(1 - a_w)X} = \dfrac{1}{b_0 b_1} + \dfrac{b_1 - 1}{b_0 b_1} a_w$	164
Halsey	$a_w = \exp\left[-\dfrac{b_1}{RT}\left(\dfrac{X}{b_2}\right)^{b_3}\right]$	165
Henderson	$1 - a_w = \exp[-b_1 \cdot T X^{b_2}]$	166
Chung and Pfost	$\ln a_w = -\dfrac{b_1}{RT}\exp(-b_2 X)$	167
Chen and Clayton	$\ln a_w = -b_1 T^{b_2}\exp(-b_3 T^{b_4} X)$	168
Iglesias and Chirife	$\ln a_w = -\exp[(b_1 T + b_2)X^{b_3}]$	169
Guggenheim–Anderson–de Boer (GAB)	$X = \dfrac{b_0 b_1 b_2 a_w}{(1 - b_1 a_w)(1 - b_1 a_w + b_1 b_2 a_w)}$ $b_1 = b_{10}\exp(b_{11}/RT),\ b_2 = b_{20}\exp(b_{21}/RT)$	170,171

X, equilibrium material moisture content; a_w, water activity; T, temperature; b_i, parameters

probably the most suitable for process analysis and design of drying because of its reliability, its simple mathematical form, and its wide use (with materials and water activity ranges). Table 14 lists parameter values of the GAB equation for some foods.

Two selected food materials are presented as an example in Figure 13. Potatoes exhibit a typical behavior. Equilibrium material moisture content is increased (172). Raisins, on the other hand, exhibit an inverse temperature effect at large water activities (173). As shown in Figure 13, potatoes and raisins exhibit sorption isotherms of type II and III, respectively.

The isotherms at 25°C for some organic and inorganic materials are presented in

Table 14 Application of the Guggenheim–Anderson–de Boer Model to Some Fruits and Vegetables

Material	b_0	$b_{10} \times 10^5$	b_{11}	b_{20}	b_{21}
Potato	8.7	1.86	34.1	5.68	6.75
Carrot	21.2	5.94	28.9	8.03	5.49
Tomato	18.2	1.99	34.5	5.52	6.70
Pepper	21.1	1.46	33.4	5.56	6.56
Onion	20.2	2.30	32.5	5.79	6.43
Raisin	12.5	0.17	22.4	1.77	−1.53
Fig	11.7	0.05	25.2	1.77	−1.55
Prune	13.3	0.07	23.9	1.82	−1.65
Apricot	15.1	0.11	21.1	2.13	−2.05

Source: Data from References 172 and 173

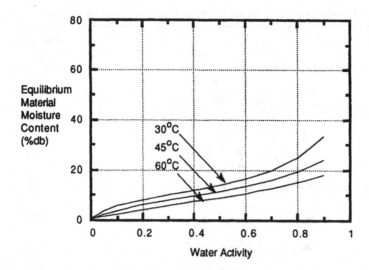

Potatoes [Data from (172)]

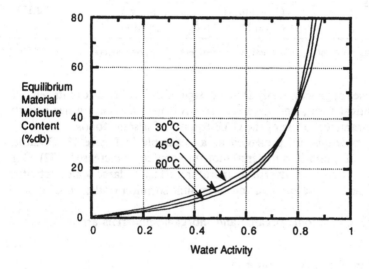

Sultana Raisins [Data from (173)]

Figure 13 Effect of air water activity and temperature on equilibrium material moisture content for two foods.

Figure 14 (92). In Figure 14, one can observe the various isotherm types, like type I for activated charcoal and silica gel, type II for leather, type III for soap, and so on.

Various regression analysis methods for fitting the above equations to experimental data have been discussed in the literature. The direct nonlinear regression exhibits several advantages over indirect nonlinear regression (173). Linear regression, on the other hand, can give highly erroneous results and should be avoided (174). When there exist differ-

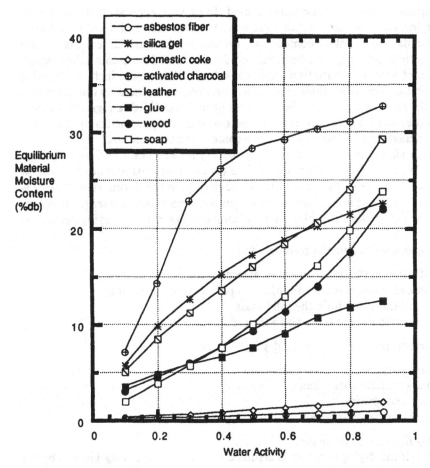

Figure 14 Equilibrium material moisture content for some organic and inorganic materials. (Data from Reference 92.)

ences in the variance of the data, the direct nonlinear weighted regression method should be used (175).

7. SIMULTANEOUS ESTIMATION OF HEAT AND MASS TRANSPORT PROPERTIES FROM DRYING EXPERIMENTS

7.1. Principles of Estimation

In the previous sections, methods of experimental determination of heat and mass transport properties have been discussed. These methods use special apparatus and are based on the equation of definition of the corresponding property. This section discusses the experimental determination of these properties from drying experiments. Some relevant techniques have been already discussed by Molnar (125). However, a generalized method based on model-building techniques is presented here. The method uses a drying experimental apparatus and estimates the heat and mass transport properties as parameters of a drying model that incorporates these properties (28,43,176–180). An outline of the method follows.

First, an experimental drying apparatus is used. In such an apparatus, the air passes through the drying material and the air humidity, temperature, and velocity are controlled, while the material moisture content and, eventually, the material temperature are monitored versus time. Second, a mathematical model that takes into account the controlling mechanisms of heat and mass transfer is considered. This model includes the heat and mass transport properties as model parameters or, even more, includes the functional dependence of the relevant factors on the transport properties. Third, a regression analysis procedure is used to obtain the transport properties as model parameters by fitting the model to experimental data of material moisture content and temperature.

Theoretically, all the properties describing the drying kinetics could be estimated simultaneously. We can define the drying kinetics (in an analogous manner to reaction kinetics) as the dependence of factors affecting the drying on the drying rate. Drying is not a chemical reaction, but it involves simultaneous heat and mass transfer phenomena. Consequently, the properties describing these phenomena describe the drying process as well.

If, for example, the phenomena considered are

The moisture diffusion in the solid toward its external surface
The vaporization and convective transfer of the vapor into the air stream
The conductive heat transfer within the solid mass
The convective heat transfer from the air to the solid's surface

then the following properties describe the drying kinetics:

Effective moisture diffusivity
Air boundary mass transfer coefficient
Effective thermal conductivity
Air boundary heat transfer coefficient

and consequently they can be estimated.

Alternatively, if the drying constant is assumed to describe the drying kinetics by the thin-layer equation, then the drying constant can be estimated using this method.

7.2. Experimental Drying Apparatus

A typical drying apparatus is shown in Figure 15. The apparatus consists of two parts, the air conditioning section and the measuring section. The air conditioning section includes the heater, the humidifier, and the fan, which are handled via a temperature, a humidity, and a flow controller, respectively. In the measuring section, the air properties, that is, temperature, humidity, and velocity, as well as the material properties (weight and temperature) are continuously recorded. The use of a computer for on-line measurement and control is preferable.

7.3. The Drying Model

An information flow diagram for a drying model appropriate for this method is shown in Figure 16. The model can calculate the material moisture content and temperature as a function of position and time whenever the air humidity, temperature, and velocity are known as a function of time, together with the model parameters. If the model takes into account the controlling mechanisms of heat and mass transfer, then the transport properties (moisture diffusivity, thermal conductivity, boundary heat and mass transfer coefficients) are included in the model as parameters. If the dependence of drying conditions

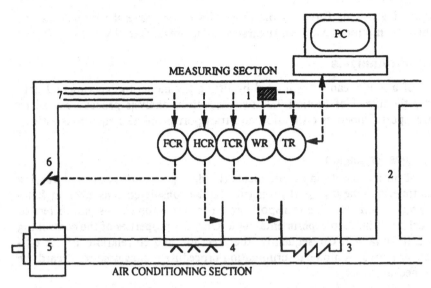

Figure 15 Typical experimental drying apparatus: (1) sample; (2) air recirculating duct; (3) heater; (4) humidifier; (5) fan; (6) valve; (7) straighteners; FCR = air flow control and recording; HCR = air humidity control and recording; TCR = air temperature control and recording; WR = sample weight recording; TR = sample temperature recording; PC = personal computer, for on-line measurement and control.

(material moisture content, temperature, and thickness, as well as air humidity, temperature, and velocity) on transport properties is also considered, then the constants of the relative empirical equations are considered as model parameters. In Figure 16 the part of the model that contains equations for the heat and mass transfer phenomena is termed the *process model*, while the equations describing the dependence of drying conditions on transport properties form the *properties model*.

In the process model, each mechanism of heat and mass transfer is expressed using a driving force and a transport property as a coefficient of proportionality between the rate

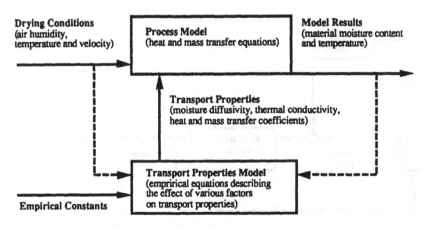

Figure 16 Model information flow diagram.

and the corresponding driving force. In the properties model, several formulas can be considered. Some assumptions have been suggested in the previous sections.

7.4. Regression Analysis

The parameters of a model can be estimated by fitting the model to experimental data (181,182). Using the model of Section 7.3, two regression analysis procedures can be applied (43): transport properties estimation and transport properties equations estimation.

Transport Properties Estimation

It is assumed that during the drying experiments the drying conditions are not varying very much with time, and the transport properties can be considered constant (not functions of the drying conditions). The transport properties are estimated as parameters of the process model by fitting it to experimental data. Only the properties of the controlling mechanisms can be obtained. Consequently, the precision and correlations of the estimates should be examined. A model discrimination procedure is suggested to discard the noncontrolling mechanisms.

Transport Properties Equations Estimation

Several empirical equations describing the dependence of transport properties on various factors are tested using a model discrimination procedure. The constants of the empirical equations are estimated as parameters of the total model (process model plus properties model) by fitting it to experimental data.

The information flow diagram for the regression analysis proposed is shown in Figure 17.

7.5. Application Example

The method described above is applied to a wide set of experimental data in potato drying (43).

Experimental Drying Apparatus

An experimental drying apparatus similar to that shown in Figure 15 was used (35). In each experiment, the air-water activity, temperature, and velocity were controlled, and

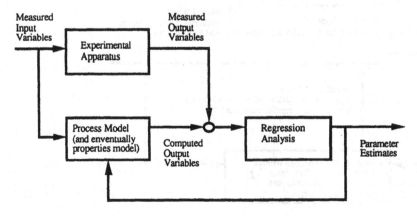

Figure 17   Regression analysis information flow diagram.

the material moisture content and temperature were monitored versus time. A total number of 100 experiments were performed for 3 different particle dimensions (5, 10, and 15 mm) at 5 air temperatures (60, 65, 70, 75, and 80°C), 3 air velocities (3, 4, and 5 m/s) and at air humidity ranging from 6 to 22 g/kg db.

Drying Model

A mathematical drying model involving simultaneous heat and mass transfer is considered for the analysis (43). The model considered has the following form:

Moisture diffusion into the solid

$$\partial(\varrho X)/\partial t = \nabla (\varrho D \nabla X) \tag{16}$$

$$D = a_0 \exp(-a_1/X) \exp(-a_2/T) \tag{17}$$

Boundary layer vapor transfer

$$-(\varrho D \nabla X) = h_M(a_{we} - a_w) \tag{18}$$

$$h_M = c_0 X_A^{c_1} T_A^{c_2} u_A^{c_3} \tag{19}$$

Heat conduction in the solid

$$\partial(\varrho\, h_s)/\partial t = \nabla (k \nabla T) \tag{20}$$

$$k = b_0 \exp(-b_1/X) \exp(-b_2/T) \tag{21}$$

Boundary layer heat transfer

$$-(k \nabla T) = h_H(T - T_A) - \Delta H_s h_M (a_{we} - a_w) \tag{22}$$

$$h_H = d_0 X_A^{d_1} T_A^{d_2} u_A^{d_3} \tag{23}$$

where X and T are the material moisture content and temperature, respectively, u_A, a_w, and T_A are the air velocity, water activity, and temperature, respectively. The thermophysical and thermodynamic properties, material density ϱ, material specific enthalpy h_s, heat of vaporization of water ΔH_s, and equilibrium air-water activity a_{we} are known functions of material moisture content and temperature.

The transport properties, moisture diffusivity D, and thermal conductivity k, are functions of material moisture content and temperature, while heat and mass transfer coefficients h_H, h_M, are functions of air velocity, water activity, and temperature.

The following adjustable constants are introduced to the relevant properties model: a_i, b_i, c_i, d_i.

Regression Analysis

If X_i and T_i are the experimental values of material moisture content and temperature and $X_{i,c}$ and $T_{i,c}$ are the corresponding calculated values using the mathematical model, then the relative deviations between experimental and calculated values (relative residuals) can be defined as follows:

$$e_{Xi} = (X_{i,c} - X_i)/X_i \tag{24}$$

$$e_{Ti} = (T_{i,c} - T_i)/T_i \tag{25}$$

The relative standard deviations between experimental and calculated values of material moisture content S_X and temperature S_T are defined as follows:

$$S_X^2 = \Sigma e_{Xi}^2/N \tag{26}$$

$$S_T^2 = \Sigma e_{Ti}^2/N \tag{27}$$

where N is the number of experimental points (including different measurements and different experiments).

A linear combination of S_X and S_T is used for parameter estimation and the resulting S_X, S_T are used for model validation (180,182). The regression analysis is performed simultaneously on all experiments.

Results

The application of the proposed method proved that

The moisture diffusivity is a function of material moisture content and temperature.
The thermal conductivity is high and cannot be estimated from these experiments.
Heat and mass transfer coefficients are constant in the region of experimentation.

More specifically, the results obtained are as follows:

$$D = 2.94 \times 10^{-7} \text{ m}^2/\text{s} \exp(-1.58 \times 10^3 \, K/T) \exp(-6.72 \times 10^{-2} \text{ kg/kg}/X) \quad [28]$$
$$h_M = 5.84 \times 10^{-7} \text{ kg/m}^2/\text{s} \quad [29]$$
$$h_H = 1.64 \times 10^{-1} \text{ W/m}^2 \quad [30]$$

The resulting model calculates the material moisture content and temperature close to experimental values and it is considered satisfactory.

ACKNOWLEDGMENT

The authors are grateful to Professor G. D. Saravacos and Dr. J. A. Palyvos for their valuable suggestions.

NOMENCLATURE

A effective surface area for heat and mass transfer, m^2
A' vibration amplitude, m
a_i constants in equations of Table 3 and in Eq. 17, various units of measure
a_w air water activity
a_{we} equilibrium air-water activity
b_i constants in equations of Table 11 and Table 12 and in Eq. 21, various units of measure
Bi_H Biot number for heat transfer (see Table 10)
Bi_M Biot number for mass transfer (see Table 10)
c_i constants in Eq. [19], various units of measure
c_p specific heat, kJ/kg·K
ΔH_s latent heat of vaporization, kJ/kg
d particle diameter, m
d_i constants in Eq. [23], various units of measure
D diffusivity in solids, m^2/s
D_A vapor diffusivity in air, m^2/s
D' diameter of spouted bed, m
db   dry base
D_O Arrhenius factor in Eq. [2], m^2/s
E activation energy in Arrhenius equation, kJ/kmol
e_{Ti} relative deviation between experimental and calculated values of material temperature, °C

e_{Xi} relative deviation between experimental and calculated values of material moisture content, kg/kg db

F constant (Table 7)

f friction factor

f' vibration frequency, 1/s

G mass flow rate of air, kg/m^{2}s

Gu   Gukhman number (see Table 10)

h_H heat transfer coefficient, kW/m^{2}K

h_M mass transfer coefficient, kg/m^{2}s

H_o static bed height for spouted beds, m

h_s specific enthalpy, kJ/kg

h_{VH} volumetric heat transfer coefficient, kW/m^{3}K

h_{VM} volumetric mass transfer coefficient kg/m^{3}s

J rate of mass transfer, kg/s

j_H heat transfer factor (see Table 10)

j_M mass transfer factor (see Table 10)

K drying constant, 1/s, 1/h

k effective thermal conductivity, kW/m·K

k_A thermal conductivity of air, kW/m·K

k_i thermal conductivity of phase i, kW/m^2·K

L slab thickness, m

N number of measurements

Nu   Nusselt number (see Table 10)

Nu'   Nu' = Nu $\epsilon/(1 - \epsilon)$

Pr   Prandtl number (see Table 10)

Q rate of heat transfer, kW/s

R gas constant, kJ/kmol·K

Re   Reynolds number (see Table 10)

Re'   Re' = Re $(1 - \epsilon)$

Re$_s$ Re based on u_s instead of u_A

Sc   Schmidt number (see Table 10)

Sh   Sherwood number (see Table 10)

St   Staton number (see Table 10)

S_T standard deviation between experimental and calculated values of material temperature, °C

S_X standard deviation between experimental and calculated values of material moisture content, kg/kg db

t time, s

T_A air temperature, °C

T material temperature, °C

T_i experimental value of material temperature during measurement i, °C

$T_{i,c}$ calculated value of material temperature during measurement i, °C

u_A air velocity, m/s

u_s air velocity for incipient spouting, m/s

V total volume of the material, m^3

wb   wet base

X material moisture content, kg/kg db

X_A air humidity, kg/kg db

X_{AS} air humidity at the solid interface, kg/kg db

X_e equilibrium material moisture content, kg/kg db
X_i experimental value of material moisture content during measurement i, kg/kg db
$X_{i,c}$ calculated value of material moisture content during measurement i, kg/kg db
Z constant (Table 7)

Greek Symbols

α specific area, m^2/m^3
ϵ void fraction (porosity)
δ constrictivity
μ dynamic viscosity of air, kg/ms
ϱ_A density of air, kg/m^3
ϱ density of material, kg/m^3
τ tortuosity

REFERENCES

1. Luikov, A. V., A Prognosis of the Development of Science of Drying Capillary-Porous Colloidal Materials, *Int. Chem. Eng.*, 10:599–604, 1970.
2. Nonhebel, G., and Moss, A. H. A., *Drying of Solids in the Chemical Industry*, Butterworths, London, 1971.
3. Reid, R. C., Prausnitz, J. M., and Poling, B. E., *The Properties of Gases and Liquids*, 4th Ed., McGraw-Hill, New York, 1987.
4. Keey, R. B., Theoretical Foundations of Drying Technology, in *Advances in Drying*, Vol. 1, Mujumdar, A. S. (ed.), Hemisphere, McGraw-Hill, New York, London, 1980, pp. 1–22.
5. Pakowski, Z., and Mujumdar, A. S., Basic Process Calculations in Drying, in *Handbook of Industrial Drying*, A. Mujumdar (ed.), 1st Ed., Marcel Dekker, New York, 1987, pp. 82–129.
6. Strumillo, C., and Kudra, T., *Drying: Principles, Applications and Design*, Gordon and Breach, New York, 1986.
7. Crank, J., *The Mathematics of Diffusion*, 2nd Ed., Oxford University Press, Oxford, 1975.
8. Crank, J., and Park, G. S., *Diffusion in Polymers*, Academic Press, New York, 1968.
9. Frisch, H. L., and Stern, S. A., Diffusion of Small Molecules in Polymers, *CRC Critical Reviews in Solid State and Materials Science*, 2(2):123–187, 1983.
10. Naesens, W., Bresseleers, G., and Tobback, P., A Method for the Determination of Diffusion Coefficients of Food Component in Low and Intermediate Moisture Systems, *J. Food Sci.*, 46:1446, 1981.
11. Hendrickx, M., Van den Abeele, C., Engels, C., and Tobback, P., Diffusion of Glucose in Carrageenan Gels, *J. Food Sci.*, 51(6):1544, 1986.
12. Gros, J. B., and Ruegg, M., Determination of Apparent Diffusion Coefficient of Sodium Chloride in Model Foods and Cheese, in *Physical Properties of Foods—2*, R. Jowitt et al. (eds.), Elsevier, London, 1987, pp. 71–108.
13. Eccles, C. D., Callaghan, P. T., and Jenner, C. F., Measurement of the Self-Diffusion Coefficient of Water as a Function of Position in Wheat Grain Using Nuclear Magnetic Resonance Imaging, *Biophys. J.*, 53:75, 1988.
14. Assink, R. A., The Concentration and Pressure Dependence of the Diffusion of Dichlorodifluoromethane in Poly(dimethyl siloxane), *J. Polymer Sci.*, 15:227, 1977.
15. Windle, J. J., An ESR Spin Probe Study of Potato Starch Gelatinization, *Starch*, 37(4):121, 1985.
16. Moyne, C., Roques, M., and Wolf, W., A Collaborative Experiment on Drying Beds of Glass Spheres, in *Physical Properties of Foods—2*, R. Jowitt et al. (eds.), Elsevier, London, 1987, pp. 71–108.

17. Schoeber, W. J. A. H., and Thijssen, H. A. C., A Short-Cut Method for the Calculation of Drying Rates of Slabs with Concentration Dependent Diffusion Coefficient, *AIChE Symp. Series*, (73) 12–24, 1977.

18. Luyben, K. C. A. M., Concentration Dependent Diffusion Coefficients Derived from Experimental Drying Curves, in *Drying, 80*, Vol. 2, A. S. Mujumdar (ed.), 1980, pp. 233–243.

19. Coumans, W. J., and Luyben, K. C. A. M., Evaluation and Prediction of Experimental Drying Curves of Slabs, paper presented at the Thijssen Memorial Symposium, Eindhoven, The Netherlands, 1987.

20. Luikov, A. V., *Transporterscheinungen in Kapillar-Porosen Korpern*, Academie, Berlin, 1958.

21. Bruin, S. and Luyben, K. C. A. M., Drying of Food Materials: A Review of Recent Developments, in *Advances in Drying*, Vol. 1, A. S. Mujumdar (ed.), Hemisphere, McGraw-Hill, New York, London, 1980, pp. 155–216.

22. Chirife, J., Fundamental of the Drying Mechanism during Air Dehydration of Foods, in *Advances in Drying*, Vol. 2, A. S. Mujumdar (ed.), Hemisphere, McGraw-Hill, 1983, New York, London, pp. 73–102.

23. Bagnall, L. O., Millier, W. F., and Scott, N. R., Drying the Alfa-Alfa Stem, *Trans. ASAE*, 13(2):232–245, 1970.

24. Saravacos, G. D., Effect of the Drying Method on the Water Sorption of Dehydrated Apple and Potato, *J. Food Sci.*, 32:81–84, 1967.

25. Rotstein, E., and Cornish, A. R. H., Influence of Cellular Membrane Permeability on Drying, *J. Food Sci.*, 43:926–939, 1978.

26. Alzamora, S. M., Chirife, J., Viollaz, P., and Vaccarezza, L. M., Heat and Mass Transfer During Air Drying of Avocado, in *Developments in Drying*, A. S. Mujumdar (ed.), Science Press, NJ, 1979.

27. Tong, C. H., and Lund, D. B., Effective Moisture Diffusivity in Porous Materials as a Function of Temperature and Moisture Content, *Biotechnol. Prog.*, 6(1):67–75, 1990.

28. Kiranoudis, C. T., Maroulis, Z. B., and Marinos-Kouris, D., Mass Transfer Model Building in Drying, *Drying Technology*, 11(6):1251–1270, 1993.

29. Syarief, A. M., Gustafson, R. J., and Morey, R. V., Moisture Diffusion Coefficients for Yellow-dent Corn Components, Winter Meeting of ASAE, December 11–14, New Orleans, paper no. 84-3551, 1984.

30. Ulku, S., and Uckan, G., Corn Drying in Fluidized Beds, in *Drying '86*, Vol. 2, A. S. Mujumdar (ed.), Hemisphere, Springer-Verlag, New York, London, 1986, pp. 531–536.

31. Jason, A. C., A Study of Evaporation and Diffusion Processes in the Drying of Fish Muscle, in *Fundamental Aspects of Dehydration of Foodstuffs*, Soc. Chem. Ind., London, 1958, pp. 103–135.

32. Pinaga, F., Carbonel, J. V., Pena, J. L., and Miquel, J. J., Experimental Simulation of Solar Drying of Garlic Using an Adsorbent Energy Storage Bed, *Journal of Food Engineering*, (3):187–208, 1984.

33. Komanowsky, M., Sinnamon, J. I., and Aceto, N. C., Mass Drying in the Cross Circulation Drying of a Foam, *Ind. Eng. Chem. Proc. Des. Dev.*, 3:193–197, 1964.

34. Ferrari, G., Meerdink, G., and Walstra, P., Drying Kinetics for a Single Droplet of Skim-Milk, *Journal of Food Engineering*, 10(3):215–230, 1989.

35. Kiranoudis, C. T., Maroulis, Z. B., and Marinos-Kouris, D., Drying Kinetics of Onion and Green Pepper, *Drying Technology*, 10(4):995–1011, 1992.

36. Okos, M., et al., *Design and Control of Energy Efficient Food Drying Processes with Specific Reference to Quality*, DOE/ID/12608-4, DE91009999, National Technical Information Service (NTIS), Springfield, VA, 1989.

37. Andrieu, J., Jallut, C., Stamatopoulos, A., and Zafiropoulos, M., Identification of Water Apparent Diffusivities for Drying of Corn Based Extruded Pasta, in *Proc. 6th Inter. Drying Symp. (IDS'88)*, France, 1988, pp. OP71–74.

38. Piazza, L., Riva, M., and Masi, P., Modeling Pasta Drying Processes, in *Engineering and Food*, Vol. 1, W. E. L. Speiss and H. Schubert (eds.), Elsevier Applied Science, 1990, pp. 592–602.

39. Palumbo, S. A., Komanowsky, M., Metzger, V., and Smith, J. L., Kinetics of Pepperoni Drying, *J. Food Sci.*, 42(4):1029–1033, 1977.

40. Saravacos, G. D., and Charm, S. E., A study of the Mechanism of Fruit and Vegetable Dehydration, *Food Technol.*, 16(1):78–81, 1962.

41. Husain, A., Chen, C. S., Clayton, J. T., and Whitney, L. F., Mathematical Simulation of Mass and Heat Transfer in High-Moisture Foods, *Trans. ASAE*, 15(4):732–736, 1972.

42. Lawrence, J. G., and Scott, R. P., Determination of the Diffusivity of Water in Biological Tissue, *Nature*, 210:301–303, 1966.

43. Kiranoudis, C. T., Maroulis, Z. B., and Marinos-Kouris, D., Heat and Mass Transfer Model Building in Drying with Multiresponse Data, *International Journal of Heat and Mass Transfer*, in press, 1995.

44. Zuritz, C. A., and Singh, R. P., Simulation of Rough Rice Drying in a Spouted-Bed, in *Drying '82*, A. S. Mujumdar (ed.), Hemisphere, McGraw-Hill, New York, London, 1982, pp. 862–867.

45. Steffe, J. F., and Singh, R. P., Parameters Required in the Analysis of Rough Rice Drying, in *Drying '80*, Vol. 2, A. S. Mujumdar (ed.), Hemisphere, McGraw-Hill, 1980, pp. 256–262.

46. Saravacos, G. D., Sorption and Diffusion of Water in Dry Soybeans, *Food Technol.*, 23: 145–147, 1969.

47. Fish, B. P., Diffusion and Thermodynamics of Water in Potato Starch Gels, in *Fundamental Aspects of Dehydration of Foodstuffs*, Soc. Chem. Ind., London, 1958, pp. 143–157.

48. Saravacos, G. D., and Raouzeos, G. S., Diffusivity of Moisture in Air Drying of Starch Gels, in *Engineering and Food*, Vol. 1, B. M. McKenna (ed.), Elsevier, London, 1984, pp. 499–507.

49. Karathanos, V. K., Water Diffusivity in Starches at Extrusion Temperatures and Pressures, Ph.D. thesis, Rutgers University, New Brunswick, NJ, 1990.

50. Vaccarezza, L., Lombardi, J., and Chirife, J., Heat Transfer Effects on Drying Rate of Food Dehydration, *Can. J. Chem. Eng.*, 52:576–579, 1974.

51. Vaccarezza, L. M., and Chirife, J., Fick's Law for the Kinetic Analysis of Air-Drying of Food, *J. Food Sci.*, 43:236–238, 1978.

52. Chirife, J., and Chachero, R. A., Through Circulation Drying of Tapioca Root, *J. Food Sci.*, 35:364–368, 1970.

53. Chirife, J., Diffusional Process in the Drying of Tapioca Root, *J. Food Sci.*, 36:327–330, 1971.

54. Margaritis, A., and King, C. J. Measurement of Rates of Moisture Transport in Porous Media, *Ind. Eng. Chem. Fund.*, 10(3):510, 1971.

55. Becker, H. A., and Sallans, H. R., A Study of Internal Moisture Movement in Drying of the Wheat Kernel, *Cereal Chem.*, 32:212, 1955.

56. Hayakawa, K. I., and Rossen, J. L., Simultaneous Heat and Moisture Transfer in Capillary-Porous Material in a Moderately Large Time Range, *Lebensm. Wiss. Technol.*, 10(2):217–224, 1977.

57. Bluestein, P. M., and Labuza, T. P., Water Sorption Kinetics in a Model Freeze-Dried Food, *AIChEJ*, 18(4):706–712, 1972.

58. Endo, A., Shishido, I., Suzuki, M., and Ohtani, S., Estimation of Critical Moisture Content, *AIChE Symp. Series*, (73):57–62, 1977.

59. Raghavan, V., and Gidaspow, D., Diffusion and Adsorption of Moisture in Dessicant Sheets, *AIChEJ*, 31(11):1791–1800, 1985.

60. Roussis, P. P., Diffusion of Water Vapor in Cellulose Acetate: 2. Permeation and Integral Sorption Kinetics, *Polymer*, 22:1058–1063, 1981.

61. Haertling, M., Prediction of Drying Rates, in *Drying '80*, Vol. 1, A. S. Mujumdar (ed.), Hemisphere, McGraw-Hill, 1980, pp. 88–98.

62. Evans, A. A., and Keey, R. B., Determination and Variation of Diffusion Coefficients when Drying Capillary Porous Materials, *Chem. Eng. J.*, 10:135–144, 1975.

63. Salas, F., and Labuza, T. P., Surface Active Agents Effects on the Drying Characteristics of Model Food Systems, *Food Technol.*, 22:1576–1580, 1968.

64. Shishido, I., and Suzuki, M., Determination of the Diffusivity of Moisture within Wet

Materials, in *Proc. 1st Int. Drying Symp.*, A. S. Mujumdar (ed.), Science Press, Princeton, NJ, 1978, pp. 30–35.

65. Chen, C. S., and Johnson, W. H., Kinetics of Moisture Movement in Hygroscopic Materials – II (An Application of Foliar Materials), *Trans. ASAE*, 12(4):478–481, 1969.

66. Edwards, W. C., and Adams, T. N., Simultaneous Heat and Mass Transfer in Wet Wood Particles, *Second Pac. Chem. Engr. Conf.*, *Heat and Mass Transfer in the Forest Products Industries*, Denver, Col., Aug. 28–31.

67. Adesanya, B. A., Nanda, A. K., and Beard, J. N., Drying Rates During High Temperature Drying of Yellow Poplar, *Drying Technology*, 6(1):95–112, 1988.

68. Saravacos, G. D., Mass Transfer Properties of Foods, in *Engineering Properties of Foods*, M. A. Rao and S. Rizvi (eds.), Marcel Dekker, New York, 1986, pp. 89–132.

69. Singh, R. P., Lund, D. B., and Buelow, F. H., An Experimental Technique Using Regular Regime Theory to Determine Moisture Diffusivity, in *Engineering and Food*, Vol. 1, B. M. McKenna (ed.), Elsevier, London, 1984, pp. 415–423.

70. Mulet, A., Berna, A., and Rossello, C., Drying of Carrots. I. Drying Models, *Drying Technology*, 7(3):536–557, 1989.

71. Sano, Y., Dry Spinning of PVA Filament, *Drying Technology*, 2(1):61–95, 1983.

72. Kiranoudis, C. T., Maroulis, Z. B., and Marinos-Kouris, D., Model Selection in Air Drying of Foods, *Drying Technology*, 10(4):1097–1106, 1992.

73. Pesaran, A. A., and Mills, A. F., Moisture Transport in Silica Gel Packed Beds – I and II. Theoretical and Experimental Study, *Int. J. Heat Mass Transfer*, 30(6):1037–1060, 1987.

74. Steffe, J. F., and Singh, R. P., Diffusion Coefficients for Predicting Rice Drying Behavior, *J. Agric. Eng. Res.*, 27:489–493, 1982.

75. Bruce, D. M., Exposed-Layer Barley Drying: Three Models Fitted to New Data up to 150°C, *J. Agric. Eng. Res.*, 32:337–347, 1985.

76. Jayas, D. S., Cenkowski, S., Pabis, S., and Muir, W. E., Review of Thin-Layer Drying and Wetting Equations, *Drying Technology*, 9(3):551–588, 1991.

77. Xiong, X., Narsimhan, G., and Okos, M. R., Effect of Composition and Pore Structure on Binding Energy and Effective Diffusivity of Moisture in Porous Food, *Journal of Food Engineering*, 15(3):187–208, 1992.

78. Marousis, S. N., Karathanos, V. T., and Saravacos, G. D., Effect of Physical Structure of Starch Materials on Water Diffusivity, *Journal of Food Processing and Preservation*, 15:183–195, 1991.

79. Van Brakel, J., and Heertjes, P. M., Analysis of Diffusion in Macroporous Media in Terms of a Porosity, A Tortuosity and a Constrictivity Factor, *Int. J. Heat Mass Transfer*, 17:1093–1103, 1974.

80. King, C. J., Rates of Moisture Sorption-Desorption in Porous Dried Food, *Food Technol.*, 22:50, 1968.

81. Rotstein, E., Prediction of Equilibrium and Transport Properties in Cellular Foods, in *Proc. 6th Inter. Drying Symp. (IDS'88)*, France, 1988.

82. ASTM Standard C-177, Thermal Conductivity of Materials by Means of the Guarded Hot Plate, *Annual ASTM Standards*, 1(14):17, 1970.

83. Mohsenin, N. N., *Thermal Properties of Foods and Agricultural Materials*, Gordon and Breach, Science Publishers, New York, 1980.

84. Fitch, W., A New Thermal Conductivity Apparatus, *American Physics Teacher*, 3(3):135–136, 1935.

85. Rahman, M. S., Evaluation of the Prediction of the Modified Fitch Method for Thermal Conductivity Measurements of Foods, *Journal of Food Engineering*, 14(1):71–82, 1991.

86. Carslaw, H. S., and Jaeger, J. C., *Conduction of Heat in Solids*, Oxford University Press, Oxford, 1959.

87. Nix, G. H., Lowery, G. W., Vachan, R. I., and Tanger, G. E., Direct Determinations of Thermal Diffusivity and Conductivity with a Refined Line-Source Technique, *Progress in Aeronautics and Astronautics: Thermophysics of Spacecraft and Planetary Bodies*, 20:865–878, 1967.

88. Sweat, V. E., Thermal Properties of Foods, in *Engineering Properties of Foods*, M. A. Rao and S. Rizvi (eds.), Marcel Dekker, New York, 1986, pp. 49–87.

89. Reidy, G. A., and Rippen, A. L., Methods for Determining Thermal Conductivity in Foods, *Trans. ASAE*, 14:248–254, 1971.

90. Murakami, E. G., and Okos, M. R., Measurement and Prediction of Thermal Properties of Foods, in *Food Properties and Computer-Aided Engineering of Food Processing Systems*, R. P. Singh and A. G. Medina, (eds.), Kluwer Academic Publishers, Boston, London, 1989, pp. 3–48.

91. Weast, R. C., *Handbook of Chemistry and Physics*, 55th Ed., CRC Press, Boca Raton, FL, 1974.

92. Perry, R. H., and Chilton, C. H., *Chemical Engineers' Handbook*, 4th and 5th Ed., Mc-Graw-Hill, New York, 1963, 1973.

93. American Society of Heating, Refrigerating and Air Conditioning Engineers, *ASHRAE Handbook of Fundamentals*, Atlanta, Ga, 1981, 1985.

94. Rohsenow, W. M., and Choi, H., *Heat Mass and Momentum Transfer*, Prentice-Hall, Englewood Cliffs, NJ, 1961.

95. Rha, C., Thermal Properties of Food Materials, in *Theory, Determination and Control of Physical Properties of Food Materials*, C. Rha (ed.), D. Reidel Publishing Company, Boston, 1975, pp. 311–355.

96. Polley, S. L., Snyder, O. P., and Kotnour, P., A Compilation of Thermal Properties of Foods, *Food Technol.*, November, 76–94, 1980.

97. Choi, Y., and Okos, M. R., Thermal Properties of Liquid Foods – Review, paper presented at the 1983 Winter Meeting of the American Engineers, Chicago, paper no. 83, p. 6516, 1983.

98. Pakowski, Z., and Mujumdar, A. S., Drying Pharmaceutical Products, in *Handbook of Industrial Drying*, A. Mujumdar (ed.), 1st Ed., 1987, pp. 605–641.

99. Miller, C. F., Effect of Moisture Content on Heat Transmission Coefficient of Grain Sorghum, ASAE Paper No. 63-80, American Society of Agricultural Engineers, St. Joseph, MI, 1963.

100. Lentz, C. P., Thermal Conductivity of Meats, Fats, Gelatin, Gels, and Ice, *Food Technol.*, 15:243–247, 1961.

101. Smith, F. G., Ede, A. J., and Game, A., The Thermal Conductivity of Frozen Foodstuffs, *Modern Refrigeration*, 55:254–259, 1952.

102. Jason, A. C., and Long, R. A., The Specific Heat and Thermal Conductivity of Fish Muscles, *Intern. Congr. Proc.*, 1:2160–2169, 1955.

103. Long, R. A., Some Thermodynamic Properties of Fish and Their Effect on the Rate of Freezing, *J. Sci. Food Agric.*, 6:621–633, 1955.

104. Kethley, T. W., Cown, W. B., and Bellinger, F., An Estimate of Thermal Conductivity Generalised Cooling Procedure and Cooling in Water, *Trans. ASAE*, 6(2):95–97, 1950.

105. Luikov, A. V., Shashkov, A. G., Vasiliev, L. L., and Fraiman, Y. E., Thermal Conductivity of Porous Systems, *Int. J. Heat Mass Transfer*, 11:117–140, 1968.

106. Parrot, J. E., and Stuckes, R. I., *Thermal Conductivity of Solids*, Pion, London, 1975.

107. Progelhof, R. C., Throne, J. L., and Ruetsch, R. R., Methods for Predicting the Thermal Conductivity of Composite Systems: A Review, *Polym. Eng. Sci.*, 16:615, 1976.

108. Mattea, M., Urbicain, M. J., and Rotstein, E., Prediction of Thermal Conductivity of Vegetable Foods by the Effective Medium Theory, *J. Food Sci.*, 51(1):113–116, 1986.

109. Maxwell, J. C., *A Treatise on Electricity and Magnetism*, Vol. 1, Dover, New York, 1954.

110. Krupiczka, R., Analysis of the Thermal Conductivity in Granular Materials, *International Chemical Engineering*, 7(1):122–144, 1967.

111. Cheng, S. C., and Vachon, R. I., A Technique for Predicting the Thermal Conductivity of Suspensions, Emulsions and Porous Materials, *Int. J. Heat Mass Transfer*, 13:537–554, 1970.

112. Wallapapan, K., Sweat, V. E., Diehl, K. C., and Engler, C. R., Thermal Properties of Porous Foods, in *Physical and Chemical Properties of Foods*, M. R. Okos (ed.), American Society of Agricultural Engineers, St. Joseph, MI, 1986, pp. 77–119.

113. Maroulis, Z. B., Druzas, A. E., and Saravacos, G. D., Modeling of Thermal Conductivity of Granular Starches, *Journal of Food Engineering*, 11(4):255–271, 1990.
114. Okazaki, M., Ito, I., Toei, R., Effective Thermal Conductivities of Wet Granular Materials, *AIChE Symp. Series*, (73):164–176, 1977.
115. Gillespie, M. B., Crandall, J. J., and Carberry, J. J., Local and Average Interphase Heat Transfer Coefficients in a Randomly Packed Bed of Spheres, *AIChEJ*, 14(3):483–490, 1968.
116. Leva, M., Heat Transfer to Gases through Packed Tubes—General Correlation for Smooth Spherical Particles, *Ind. Eng. Chem.*, 39:857, 1947.
117. Balakrishnan, A. R., and Pei, D. C. T., Heat Transfer in Gas-Solid Packed Bed Systems. 1. A Critical Review, *Ind. Eng. Chem. Proc. Des. Dev.*, 18(1):30–40, 1979.
118. Furnas, C. C., Heat Transfer from a Gas Stream to a Bed of Broken Solids, *Ind. Eng. Chem.*, 22:26, 1930.
119. Lof, G. O., and Hawley, R. W., Unsteady State Heat Transfer between Air and Loose Solids, *Ind. Eng. Chem.*, 40:1061, 1949.
120. Sagara, M., Schneider, P., and Smith, J. M., The Determination of Heat Transfer Parameters for Flow in Packed Beds Using Pulse Testing and Chromatography Theory, *Chem. Eng. J.*, (1):47, 1970.
121. Shen, J., Kaguel, S., and Wakao, N., Measurements of Particle to Gas Heat Transfer Coefficients from One Shot Thermal Response in Packed Beds, *Chem. Eng. Sci.*, 36(8):1283–1286, 1981.
122. Bell, J. C., and Katz, E. F., A Method for Measuring Surface Heat Transfer Using Cyclic Temperature Variations, Presented at Heat Transfer and Fluid Mechanics Inst. Meeting, Berkeley, CA, 243, 1949.
123. Lindauer, G. C., Heat Transfer in Packed Beds by the Method of Cyclic Temperature Variations, *AIChEJ*, 13(6):1181–1187, 1967.
124. Bradshaw, R. D., and Myers, J. E., Heat and Mass Transfer in Fixed and Fluidized Beds of Large Particles, *AIChEJ*, 9(5):590–595, 1963.
125. Molnar, K., Experimental Techniques in Drying, in *Handbook of Industrial Drying*, A. Mujumdar (ed.), 1st Ed., 1987, pp. 47–82.
126. Gamson, B. W., Thodos, G., and Hougen, O. A., Heat, Mass and Momentum Transfer in the Flow of Gases through Granular Solids, *Trans. AIChE*, 39:1–35, 1943.
127. Wilke, C. R., and Hougen, O. A., *Trans. AIChE*, 41:445–451, 1945.
128. McAdams, W. H., *Heat Transmission*, 3rd Ed., McGraw-Hill, New York, 1954.
129. Geankoplis, C. J., *Transport Processes and Unit Operations*, Allyn and Bacon, Boston, 1978.
130. Whitaker, S., Forced Convection Heat Transfer Correlations for Flow in Pipes, Past Flat Plates, Single Cylinders, Single Spheres, and for Flow in Packed Beds and Tube Bundles, *AIChEJ*, 18(2):361–371, 1972.
131. Kelly, J. J., Rotary Drying, in *Handbook of Industrial Drying*, A. Mujumdar (ed.), 1st Ed., Marcel Dekker, 1987, pp. 47–82.
132. Ranz, W. E., and Marshall, W. R., Evaporation from Drops, *Chem. Eng. Prog.*, Monograph Series, 48:141–146, 173–180, 1952.
133. Kudra, T., Mujumdar, A. S., Impingement Stream Dryers for Particles and Pastes, *Drying Technology*, 7(2):219–266, 1989.
134. Obot, N. T., Mujumdar, A. S., and Douglas, W. J. M., Design Correlations for Heat and Mass Transfer Under Various Turbulent Impinging Jet Configurations, in *Drying '80*, Vol. 1, A. S. Mujumdar (ed.), Hemisphere, McGraw-Hill, New York, London, 1980, pp. 388–402.
135. Li, Y. K., Mujumdar, A. S., and Douglas, W. J. M., Coupled Heat and Mass Transfer Under a Laminar Impinging Jet, in *Proc. 1st Int. Drying Symp.*, A. S. Mujumdar (ed.), Science Press, Princeton, NJ, 1978, pp. 175–184.
136. Sokhansanj, S., and Bruce, D. M., Heat Transfer Coefficients in Drying Granular Materials, in *Drying '86*, Vol. 2, A. S. Mujumdar (ed.), Hemisphere, Springer-Verlag, New York, London, 1986, pp. 862–867.

137. Leva, M., *Fluidization*, McGraw-Hill, New York, 1959.
138. Kunii, D., and Levenspiel, O., *Fluidization Engineering*, J. Wiley, New York, 1969.
139. Davidson, I. F., and Harrison, D., *Fluidization*, Academic Press, London, 1971.
140. Botterill, I. S. M., *Fluid Bed Heat Transfer*, Academic Press, London, 1975.
141. Gupta, R., and Mujumdar, A. S., Recent Developments in Fluidized Bed Drying, in *Advances in Drying*, Vol. 2, A. S. Mujumdar (ed.), Hemisphere, McGraw-Hill, New York, London, 1983, pp. 155–192.
142. Pakowski, Z., Mujumdar, A. S., and Strumillo, C., Theory and Application of Vibrated Beds and Vibrated Fluid Beds for Drying Processes, in *Advances in Drying*, Vol. 3, A. S. Mujumdar (ed.), Hemisphere, McGraw-Hill, New York, London, 1984, pp. 245–306.
143. Prat, M., 2D Modeling of Drying of Porous Media: Influence of Edge Effects at the Interface, *Drying Technology*, 9(5)1181–1208, 1991.
144. Lewis, W. K., The Rate of Drying of Solid Materials, *I&EC—Symposium on Drying*, 3(5): 42, 1921.
145. Sokhansanj, S., and Genkowski, S., Equipment and Methods of Thin-Layer Drying. A Review, in *Proc. 6th Inter. Drying Symp. (IDS'88)*, France, 1988, pp. OP159–170.
146. Henderson, S. M., and Pabis, S., Grain Drying Theory. I. Temperature Effect on Drying Coefficient, *J. Agric. Eng. Res.*, 16:223–244, 1961.
147. Abdullah, K., Syarief, A. M., and Sagara, Y., Thermophysical Properties of Agricultural Products as Related to Drying, in *Proc. 6th Inter. Drying Symp. (IDS'88)*, France, 1988, pp. PB27–32.
148. O'Callaghan, J. R., Menzies, D. J., and Bailey, P. H., Digital Simulation of Agricultural Dryer Performance, *J. Agric. Eng. Res.*, 16:223–244, 1971.
149. Ajibola, O. O., Thin-Layer Drying of Melon Seed, *Journal of Food Engineering*, (4):305–320, 1989.
150. Westerman, P. W., White, G. M., and Ross, I. J., Relative Humidity Effect on the High Temperature Drying of Shelled Corn., *Trans. ASAE*, 16:1136–1139, 1973.
151. Wang, C. Y., and Singh, R. P., A Single Layer Drying Equation for Rough Rice, ASAE Paper No. 78-3001, American Society of Agricultural Engineers, St. Joseph, MI, 1978.
152. Jayas, D. S., Sokhansanj, S., Thin Layer Drying of Wheat at Low Temperature, in *Drying '86*, Vol. 2, A. S. Mujumdar (ed.), Hemisphere, Springer-Verlag, New York, London, 1986, pp. 844–847.
153. Mulet, A., Berna, A., Borras, F., and Pinaga, F., Effect of Air Flow Rate on Carrot Drying, *Drying Technology*, 5(2):245–258, 1987.
154. Brunauer, S., Deming, L. S., Deming, W. E., and Teller, E., On a Theory of the van der Walls Adsorption of Gases, *Am. Chem. Soc. J.*, 62:1723–1732, 1940.
155. Gal, S., Recent Advances in Techniques for the Determination of Sorption Isothers, in *Water Relations of Foods*, R. B. Duckworth (ed.), Academic Press, London, 1975, pp. 139–154.
156. Gal, S., Recent Developments in Techniques of Obtaining Complete Sorption Isotherms, in *Water Activity: Influence on Food Quality*, L. B. Rockland and G. F. Steward (eds.), Academic Press, New York, 1981, pp. 89–11.
157. Spiess, W. E. L., and Wolf, W., Critical Evaluation of Methods to Determination of Moisture Sorption Isotherms, in *Water Activity. Theory and Applications to Food*, L. B. Rockland and L. R. Beuchat (eds.), Marcel Dekker, New York, 1981.
158. Wolf, W., Spiess, W. E. L., Jung, G., Weisser, H., Bizot, H., and Duckworth, R. B., The Water Sorption Isotherms of Microcrystalline Cellulose (MCC) and of Purified Potato Starch. Results of a Collaborative Study, *Journal of Food Engineering*, 3(1):51–73, 1984.
159. Spiess, W. E. L., and Wolf, W., The Results of the COST 90 Project on Water Activity, in *Physical Properties of Foods*, R. Jowitt et al. (eds.), Applied Science Publ., London, 1986, pp. 65–87.
160. Wolf, W., Spiess, W. E. L., and Jung, E., *Sorption Isotherms and Water Activity of Food Materials*, Science and Technology Publishers, England, Hornchurch, Essex, 1985.
161. Iglesias, H. A., and Chirife, J., *Handbook of Food Isotherms: Water Sorption Parameters for Food and Food Components*, Academic Press, New York, 1982.

162. Wolf, W., Spiess, W. E. L., and Jung, G., *Wasserdampf-Sorptionsisothermen von Lebens-mitteln Berichtsheft* 18 der Fachgemeinschaft allgemeine Lufttechnik im VDMA Frankfurt/Main, 1973.

163. Langmuir, I., The Adsorption of Gases on Plane Surfaces of Glass, Mica, and Platinum, *Am. Chem. Soc. J.*, 40:1361–1402, 1918.

164. Brunauer, S., Emmett, P. H., and Teller, E., Adsorption of Gases in Multimolecular Layers, *Am. Chem. Soc. J.*, 60:309–319, 1938.

165. Halsey, G., Physical Adsorption on Non-Uniform Surfaces, *J. Chem. Phys.*, 16:931, 1948.

166. Henderson, S. M., A Basic Concept of Equilibrium Moisture, *Agric. Eng.*, 33:29–32, 1952.

167. Chung, D. S., and Pfost, H. B., Adsorption and Desorption of Water Vapor by Cereal Grains and Their Products, *Trans. ASAE*, 10(4):552–575, 1967.

168. Chen, C. S., and Clayton, J. T., The Effect of Temperature on Sorption Isotherms of Biological Materials, *Trans. ASAE*, 14(5):927–929, 1971.

169. Iglesias, H. A., and Chirife, J., A Model for Describing the Water Sorption Behavior of Foods, *J. Food Sci.*, 41(5):984–992, 1976.

170. Bizot, H., Using the GAB Model to Construct Sorption Isotherms, in *Physical Properties of Foods*, K. Jowitt et al. (eds.), Applied Science Publ., London, 1983, pp. 43–45.

171. Van den Berg, C., Description of Water Activity of Foods for Engineering Purposes by Means of the GAB Model of Sorption, in *Engineering and Food*, Vol. 1, B. M. McKenna (ed.), Elsevier, London, 1984, pp. 311–321.

172. Kiranoudis, C. T., Maroulis, Z. B., Tsami, E., and Marinos-Kouris, D., Equilibrium Moisture Content and Heat of Desorption of Some Vegetables, *Journal of Food Engineering*, 20(1):55–74, 1992.

173. Maroulis, Z. B., Tasami, E., Marinos-Kouris, D., and Saravacos, G. D., Application of the GAB Model to the Moisture Sorption Isotherms of Dried Fruits, *Journal of Food Engineering*, 7(1):63–78, 1988.

174. Schaer, W., and Ruegg, M., The Evaluation of GAB Constants from Water Vapour Sorption Data, *Lebensm. Wiss. Technol.*, 18:225, 1985.

175. Samaniego-Esguerra, C. M., Bong, I. F., and Robertson, G. L., Comparison of Regression Methods for Fitting the GAB Model to the Moisture Isotherms of Some Drying Fruit and Vegetables, *Journal of Food Engineering*, 13(2):115–133, 1991.

176. Bertin, R., and Srour, Z. Search Methods Through Simulation for Parameter Optimization of Drying Process, in *Drying '80*, Vol. 2, A. S. Mujumdar (ed.), Hemisphere, McGraw-Hill, 1980, pp. 101–106.

177. Bertin, R., Delage, P., and Boverie, S., Estimation of Functions in Drying Equations, *Drying Technology*, 2(1):45–59, 1983.

178. Mulet, A., Berna, A., Rossello, C., and Pinaga, F., Drying of Carrots. II. Evaluation of Drying Models, *Drying Technology*, 7(4):641–661, 1989.

179. Karathanos, V. T., Villalobos, G., and Saravacos, G. D., Comparison of Two Methods of Estimation of the Effective Moisture Diffusivity from Drying Data, *J. Food Sci.*, 55(1):218–223, 1990.

180. Maroulis, Z. B., Kiranoudis, C. T., and Marinos-Kouris, D., Simultaneous Estimation of Heat and Mass Transfer Coefficients in Externally Controlled Drying, *Journal of Food Engineering*, 14(3):241–255, 1991.

181. Beck, J. V., and Arnold, K. J., *Parameter Estimation*, J. Wiley, New York, 1977.

182. Draper, N., and Smith, H., *Applied Regression Analysis*, J. Wiley, New York, 1981.

183. Barker, J. J., Heat Transfer in Packed Beds, *Industrial and Engineering Chemistry*, 57(4):43–51, 1965.

5
Rotary Drying

John J. Kelly
University College Dublin
Dublin, Ireland

1. INTRODUCTION

In common with many other unit operations of chemical engineering, drying is a mixture of "art" based on experience and "science" based on analytic approaches. The techniques for design and operation in the industrial world depend largely on the art form; the scientific approach is to be found in the studies and publications of those workers in the laboratories of university departments and research institutes. In the ideal world, there is a gradual diffusion of knowledge from the laboratories to the industrial world whereby a greater understanding of the process is obtained leading to improvement or innovation in industrial design and plant operation.

This chapter on rotary dryers is dedicated to the promotion of this "diffusional" process, for it must be admitted at the outset that there is little evidence of technology transfer, to use the contemporary expression, between the two worlds of drying research and practice. There are those in the commercial world who stoutly maintain that drying is exclusively an art form and that theory has no relevance to its design procedures whatsoever. It is true that this has been the situation in the past and indeed may be still up to the present day with the result that industrial dryers were, more often than not, somewhat oversized and inefficient. In general, however, they were mechanically sound, and it may have been that industrialists were content to purchase a well-proven design without much consideration for the capital and operating costs.

It was not possible in writing this chapter to identify a body of knowledge that could be called the industrial practice of rotary dryers. Manufacturers will disclose no design information, and each manufacturer has its own special features and claims for performance. This chapter therefore concentrates on that body of published knowledge that, although generally theoretical in character, is or should be of immediate interest and relevance to industrial designers and operators of rotary drying equipment.

For the most part, the single direct contact cascading rotary dryer is considered, in which the wet solids enter one end of a rotating drum, move down the drum length cascading from peripherally mounted lifting flights, and exit at the other end suitably

Figure 1 Top view of rotary dryer 13-ft diameter by 60-ft length, parallel flow direct heat rotary ore dryer with fired air heater for gold ore treatment process. (Courtesy Hely and Patterson, Inc., Renneburg Division, Pittsburgh.)

dried. The drying medium is either hot air or combustion gases that flow concurrently or countercurrently to the direction of the solids through the drum length. The cascading rotary dryer is the workhorse of the solids process industry and is to be found in such diverse plants as fertilizer production, pharmaceutical chemicals manufacture, seaweed drying for alginates, lead or zinc concentrate production for smelting, cement manufacture, and many more industries. They vary in shell size from 0.3-m diameter by 2-m length to 5-m diameter by 90-m length, this latter dryer weighing some 1400 t. They are generally fabricated in steel as required by the process environment (Fig. 1).

2. GENERAL DESIGN CONSIDERATIONS

The basic features to be selected in the design of a rotary dryer are as follows:

Solids feed rate F and moisture content X_{in}
Drum diameter D and length L
Drum slope α
Rotational speed N
Lifting flights number n_f and profile
Drying gases direction, cocurrent or countercurrent, and velocity V through drum

Arising from the above features, the following operational parameters will be derived:

Drum solids holdup H
Outlet moisture content X_{out}

The separate processes performed in the rotary dryer are

The dynamics or movements of the particles as they progress by cascade and kiln motion through the drum length

Heat transfer from the hot gases to the particles providing the latent heat of vaporization to the moisture within the particles

Mass transfer of the moisture from within the particles to their surface and then to the hot gases in the drum

Although the simultaneous heat and mass transfer operations are the principle objectives of the process, analyses of their respective rates require an understanding of the particle dynamics through the drum. Thus it is that investigations into the operation of rotary dryers have, for the most part, been directed toward a better understanding of the movement of particles down the drum length. Mass and heat transfer aspects of this process have been analyzed with different approaches, but until a satisfactory model of the particle dynamics is provided, the transfer phenomena theories and correlations are necessarily limited to their application. Furthermore, the movement of the particles through the drum forms the basis for the development of a residence time model, which ideally gives the average time the particles spend in the drum as well as the distribution about that average as it is not a plug flow system.

3. FLIGHT DESIGN

3.1. Particle Dynamics

Particles progress through the drum in a series of cascades and, between cascades, by a sliding or kiln motion between flights down the walls of the drum. In each cascade the average particle is "captured" by a flight in the lower half of the drum and carried stationary in a flight into the upper half until its angle of repose in the flight is greater than its equilibrium angle, when it cascades off the lip of the flight through the hot gases back into the lower half of the drum (Fig. 2). In falling through the gases, the particles are subjected to drag from the gases and are carried forward (cocurrent gas flow) or backward (countercurrent) along the drum length. When the drum is angled to the horizontal, the particles fall forward from the flight and thus move down the drum. In summary, therefore, there are three separate components making up the total movement of the particles down the drum in each cascade:

1. Forward movement of the cascading particles due to the drum angle
2. Forward or backward movement due to the drag on the particles from the drying gases
3. Kiln action between flights in the lower half of the drum

In the various residence time models that have been developed for the rotary dryer, these components have generally been incorporated in one way or another. In the following analysis, the different parameters that make up or influence the residence time of particles in the drum are examined in turn, and their inclusion into the various residence time models that have been proposed is discussed.

3.2. Kinetic Angle of Repose

All particles have a characteristic *static* or *poured angle of repose*, which is the maximum angle they will sustain with a horizontal surface when they are gently poured from a

Figure 2 Cascade pattern for staggered lifting flights. (Courtesy Hely and Patterson, Inc., Renne-burg Division, Pittsburgh.)

container. The value of this angle depends on a number of factors, principally particle
size and moisture content, which will affect their "stickiness," and a related characteristic,
the surface coefficient of friction. In the flights of a rotary drum, however, the particles
on the surface roll down the slope on top of the particles below toward the lip of the
flight. When these particles are at their *terminal rolling velocity*, that is, the forces acting
on them are in balance and the particles are moving at a constant velocity, it is possible to
predict precisely what the value of the angle of repose, known as the kinetic angle of
repose λ, will be as a function of the flight position on the drum θ, the drum speed N,
expressed in the term v, which is the ratio of the centrifugal to gravitational forces on the
particle, and a term known as the kinetic coefficient of friction of the particles on each
other, μ (1).

In balancing the three forces that are acting on the rolling particle—these are gravitational, frictional, and centrifugal (see Fig. 3)—the following important expression is
derived:

$$\lambda = \tan^{-1} \frac{\mu + v\,(\cos\theta - \mu\sin\theta)}{1 - v\,(\mu\cos\theta + \sin\theta)} \tag{1}$$

This is an important expression for the analysis or design of flights because it is then only
necessary to know the value of μ for the particles in any rotary drum process to determine
the angle of repose of the material at any angle θ on the top drum circumference. Knowing
this angle, it is a simple matter to:

1. Design a flight profile to give almost any desired cascade distribution across the drum
   or
2. Evaluate the precise distribution that any given flight shape will yield

This equation has been shown (2) to be accurate for values of v up to 0.4. This is
equivalent to a 2-m diameter drum rotating at about 20 revolutions per minute, which is
well above the speed of commercial rotary dryers. In their work, Kelly and O'Donnell (2)
devised a simple bench apparatus for the evaluation of the term μ (see Fig. 4).

Furthermore, their calculations on the distribution patterns, which would be obtained
with the more commonly used commercial flight patterns, showed them to be less than
optimum (see Fig. 5). They designed a number of "theoretical" flights designed to give a

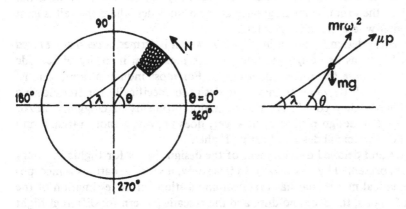

Figure 3   Forces acting on a particle prior to cascade.

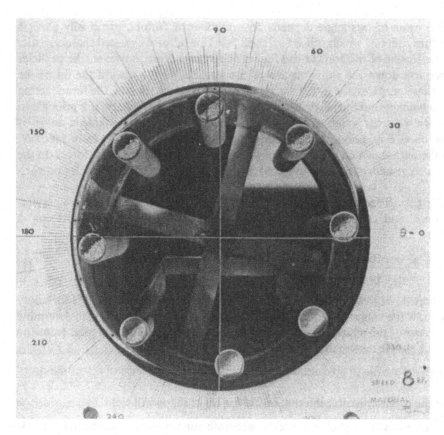

Figure 4   Rotary drum for the evaluation of μ.

precise cascade pattern of particles across the drum. An example of one of these is shown in Figure 6, which is called an *equal angular distribution* (EAD) flight as it cascades an equal amount of particles for every degree of rotation. This was not proposed as an "optimum" flight, as the intensity of cascade measured, say, by the insertion of a flat plate horizontally into the drum, is much greater at the drum sides where the fall is least than at the drum center where the fall is greatest.

A further theoretical design is shown in Figure 7, which is termed a *centrally biased distribution* (CBD) flight, as its design provides for the maximum intensity of cascade down the drum center where the fall is the greatest. For most industrial applications, these designs would be of little value as they were designed specifically for free-flowing particles, and in practice they could give rise to major problems of clogging and/or maintenance (Fig. 8). The design principle, however, does provide a more rational and efficient basis for the commercial design of lifting flights.

A comprehensive and detailed development of the design theory for flights in rotary drum dryers has been presented by Baker (28). In this study, a set of equations is incorporated into a mathematical model that has practical applications in the estimation of the optimum number of flights, the drum holdup, and the cascade pattern for different flight profiles.

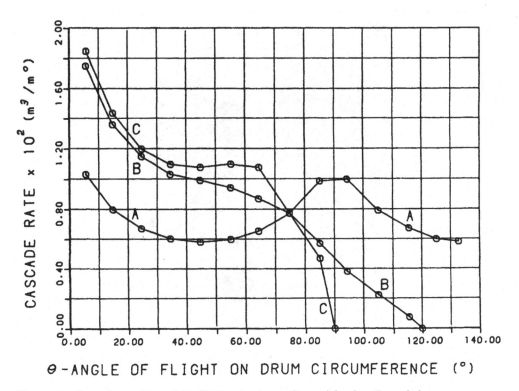

Figure 5   Cascade rate for various flights. A, square; B, semicircular; C, angled.

4.   FLIGHT AND DRUM HOLDUP

In the design or analysis of operation of a commercial rotary dryer, the quantity of particles in the drum during steady-state operation, known as drum holdup H, is a key parameter. For a fixed solid feed rate F, the mean residence time of the particles T is related to these parameters as follows:

$$T(s) = \frac{H}{F} \tag{2}$$

Where H is in m³ or kg and F is in m³/s or kg/s. A drum may be underloaded, design loaded, or overloaded, and the experienced operator will know immediately on looking into the drum which of these loading conditions applies. Underloading is a highly inefficient operation as the gas flow down the drum length, which causes the heat and/or mass transfer, will take the path of least resistance and will therefore flow down that section of the drum where the flights are not cascading because of the underloaded situation. With an overloaded drum, there is a rolling load of particles in the bottom half of the drum that bypasses the rotating flights. This will result in underdried particles or, in some instances, an unacceptable spread of moisture content in the product solids. A fully loaded drum is one in which the rotating flights are full when they pass upward through the drum's horizontal axis, thus ensuring cascade right across the drum diameter with properly designed flights and minimizing the rolling load.

The approximation of Porter (3) that the design drum holdup H^* equals the design flight holdup h^*, multiplied by half the number of flights n_f,

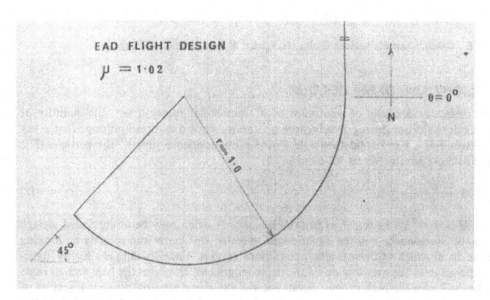

Figure 6   Equal angular distribution (EAD) flight.

$$H^* = \frac{n_f}{2} h_0^* L \tag{3}$$

has been found by Kelly and O'Donnell (2) to underestimate the true value as it ignores the particles in the air cascading. They proposed the following relationship:

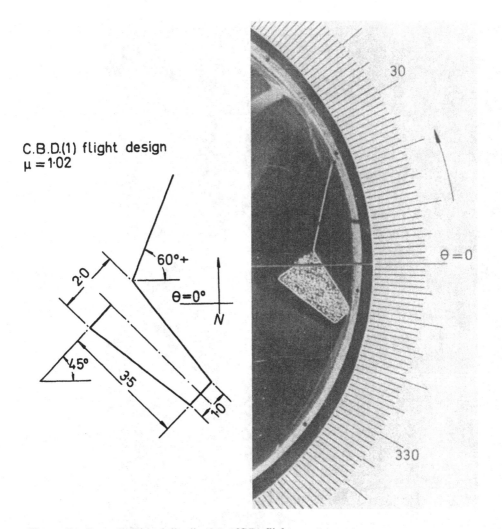

C.B.D.(1) flight design
$\mu = 1.02$

Figure 7 Centrally biased distribution (CBD) flight.

$$H^* = \frac{1}{2} (n_f + 1)h_0^* L \qquad [4]$$

It is noted from the above relationships that the key parameters determining the drum holdup are the flight holdup and the number of flights. It is understood that the practice in industrial design is to work to a drum holdup within the range 10% to 15% of the total drum volume (4). The logic behind this figure is not clear, other than experience. The procedure, then, in the design calculation is to incorporate the maximum number of flights consistent with adjacent flights not interfering with each other in collecting their load at the bottom of the drum and also consistent with the particles, which may be wet and sticky, not jamming up the works (Fig. 9). With the number of flights selected initially, and assuming 15% of drum volume or other value for the holdup, the dimensions of the flight are readily calculated for a given flight profile. In some industrial

Figure 8 View of louver dryer flights during construction. (Courtesy of Newell Dunford, Doncaster, England.)

practice, experience is the overriding factor in design, and the number and size of flights are decided without even the above simple analysis.

The actual measurement of a drum holdup is not always a simple manner. With a small pilot drum unit, the procedure is simply to stop the feed and unload the drum into containers for measurement of weight or volume. Knowing the feed rate, the calculation of average residence time may be obtained from Eq. [2].

With a large industrial drum, the above procedure is not usually possible or practical. Tracer techniques are sometimes possible when the residence time of a tracer is experimentally obtained and, knowing the feed rate, the holdup is estimated from Eq. [2].

Figure 9 Internal view of staggered lifting flight arrangement. (Courtesy Swenson Product Equipment, Inc., Harvey, Illinois.)

The radioactive tracer technique developed and described by Kelly and O'Donnell (5) is not recommended for other than strictly controlled laboratory conditions.

It may be possible in certain industrial dryers to make a reasonable estimate of the drum holdup by inspection of the flight holdup and/or cascade pattern and, working from the geometry of the drum cross-section, derive the approximate figure.

Finally, it should be noted here that it is generally preferable to operate a drum dryer in slightly overloaded conditions. This is because the cascade rate falls off rapidly at underloaded conditions, resulting in inefficient operation due to the drying gases bypassing the cascading particles. For example, it has been estimated that, when a drum is at 96% of design load, the flights are only 75% loaded as they pass upward through the horizontal drum axis (2). The optimum operating load conditions of a drum dryer may therefore be different from the "design" drum holdup, as it may correspond to a slightly overloaded condition. In general, it is advisable to experiment with a drum dryer in order to establish its optimum operating condition.

5. RESIDENCE TIME MODELS

A considerable number of studies have been done on the development of a universally acceptable residence time model, but none have really succeeded. The ideal model is one that combines all the relevant design and operating parameters to produce a calculation procedure that will accurately predict the mean residence time for a particular rotary dryer design. With such a model, the inexperienced design engineer could design the optimum dryer for a given residence time from first principles, needing only basic data on the particles to be dried, such as moisture content, size range, and specific gravity. The models that have been developed and that are summarized below go a considerable way toward this ideal model, but they should be used with caution. In all cases it is recommended that the original reference be obtained and studied before any attempt is made to use a model for design or operating purposes.

In an experimental rig, the measurement of the average residence time of particles is readily achieved from Eq. [2]. The prediction of this value to a degree of accuracy suitable for design purposes has been the principal object of most research workers in the field of rotary drum driers. In a typical design problem, the feed rate F is fixed by the process requirements and the residence time T by the heat or mass transfer requirements. The problem is then essentially to design a drum with holdup H, calculated from the above relationship, that will provide the required average residence time. Suitable allowances, when necessary, must be made for the deviation from this average residence time.

Earlier rotary drum design procedures relied entirely on scaling-up methods and on experience. Residence time relationships, such as that of van Krevelen and Hoftijzer (6), Eq. [5], were of use in predicting the effect of changing one of the design parameters, such as drum speed N or angle α, on the residence time; as design equations, they were to be applied with great caution:

$$T = \frac{kL}{ND \tan \alpha} \tag{5}$$

where k = constant = 0.13. The earlier residence time equations (Eqs. [6–9]) were, for the most part, developments of the kiln formula of Sullivan et al. (10); they related the residence time T empirically to the five principal parameters: drum diameter D, length L, slope α, drum speed N, and the gas velocity V through the drum. It was known that other factors, such as the properties of the materials being processed, the drum loading, the flight profile, and particle distribution from the flights, contributed to the residence time, and later studies (11–15) provided more analytic investigations into these and other aspects of the process. Of those latter expressions, the residence time equation of Saeman and Mitchell (11),

$$T = \frac{L}{f(H)DN(\alpha - kV)} \tag{6}$$

where $f(H) = \pi$ for heavily loaded drums and 2.0 for lightly loaded drums and k = constant, has received wide acceptance (13,16,17). In the derivation of this equation, it was assumed that the horizontal drift of the cascading particle was linearly related to the air velocity and the length of fall. The constant of proportionality k has been given values ranging from 0.001 s/ft by Porter (13) to 0.008 s/ft by Saeman (14), obtained from the data of Miller et al. (8). From a consideration of the momentum transfer associated with the gas/solid contact, Saeman proposed a theoretical method for the estimation of the value of k for different conditions. This correlation gave values considerably in excess of those realized experimentally.

A particle progresses down the drum length in a succession of cascades; each cascade is made up of a period in a flight and a period falling through the air, where the distance it moves forward is determined by the height of fall, the slope of the drum, and the gas velocity. Schofield and Glikin (1) made a theoretical analysis of this cascade motion and proposed the following residence time equation:

$$T = \frac{L}{Y_{av}\left(\sin\alpha - \dfrac{kV^2}{g}\right)}\left(t_f + \frac{1}{\delta N}\right) \qquad [7]$$

This relationship makes a clean break from the kiln-type equation and establishes a theoretically more accurate basis on which a residence time model for the process may be developed. It is not an accurate residence time as it stands:

1. It makes no allowance for kiln action of the particles in moving through the drum.
2. In the analysis of the effect of the gas velocity, the use of the drag coefficient assumed, at all times, a particle Reynolds number ≤ 20, whereas in practice values in the region 200–300 are normal.

It was found by Kehoe (17), working on a 1-ft diameter by 6-ft long drum, that this equation overpredicted the residence time by a factor of approximately 2. Kehoe concluded that kiln action of the particles, occurring at all conditions of drum loading, caused these deviations.

If it is assumed that the principal mechanism of particle movement through a drum is by cascade motion, which was the basis for the Schofield and Glikin model above, the general equation for the mean residence time may then be written as

$$T = \frac{L_{eff}}{(\text{cascade length})_{av}} \times (\text{cascade time})_{av} \qquad [8]$$

where

L_{eff} = length of the drum over which the average particle progresses by cascade

(cascade length)$_{av}$ = distance along the drum the average particle progresses with each cascade

(cascade time)$_{av}$ = time taken by the average particle for each cascade

The cascade motion is shown in Figure 10. One cascade is represented by $ABCE$, where

$AB:$ The particle is lifted stationary in a flight from the bottom half of the drum to the upper half, where is cascades downward from B.

$BC:$ Represents the actual path downward of the particle. With no gas flow, it would fall vertically BD; thus, CD is the distance the particle is blown backward because of the drag from the gases on the falling particles.

$CE:$ Represents the distance along the drum the particle moves by sliding downhill or by rolling kiln action under pressure from the particles coming behind.

The model of Kelly and O'Donnell (2) for a fully loaded drum followed this general mechanism and is

$$T = \frac{L_{eff}}{Y_{av}\sin\alpha - f(G)}\left(\frac{1}{2N} + \sqrt{\frac{2Y_{av}}{g}}\right) \qquad [9]$$

The right-hand side can be considered under the three headings:

1. L_{eff} = effective drum length = $K_c L$, where K_c is a cascade factor and represents the fraction of the drum traversed by the particles under cascade motion.
2. (Cascade length)$_{av}$ = $Y_{av}\sin\alpha - f(G)$; where Y_{av} is average height of fall off the

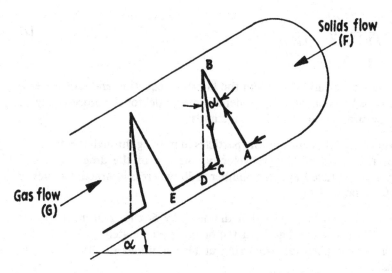

Figure 10 Cascade motion in drum: $BD = Y_{av}$; $AD = Y_{av} \sin \alpha$; $DC = f(G)$; $AC = Y_{av} \sin \alpha - f(G)$.

particles, which can be computed from the flight profile. With the drum at angle α to the horizontal, $Y_{av} \sin \alpha$ corresponds to distance AD in Figure 10. $f(G)$ is the distance the particles are blown backward (or forward for cocurrent operation) by the drying gases. Various models based on particle drag theory have been proposed for this term. Schofield and Glikin equated this to kV^2/g in Eq. [7]; Kelly and O'Donnell identified and resolved all the forces acting on the falling particle to chart its trajectory in more rigorous terms.

3. The two terms in the following expression correspond to the time taken to complete half a revolution ($1/2N$) for the time spent in the flight before cascade, and the time of fall under gravity through the distance Y_{av}, which is $(2Y_{av}/g)^{1/2}$.

$$(\text{cascade time})_{av} = \frac{1}{2N} + \sqrt{\frac{2Y_{av}}{g}} \qquad [10]$$

This model was further developed for all levels of drum loading, both underloaded and overloaded. All three parts of this model vary in value with drum loading. For the EAD (Fig. 6) flights used by Kelly and O'Donnell, Y_{av} took on the following expression for the different loading conditions:

Underloaded:

$$Y_{av} = \frac{D}{\pi m_0 \cos \alpha} (1 - \cos \pi m_0) \qquad [11]$$

Design loaded:

$$Y_{av} = \frac{2D}{\pi \cos \alpha} \qquad [12]$$

Overloaded:

$$Y_{av} = \frac{2D}{\pi M \cos \alpha} \tag{13}$$

The values of $f(G)$ are dependent on the gas velocity and the height of fall; a least-squares regression analysis on extensive experimental data yielded the expression

$$f(G) = 0.0396 V^{0.77} Y_{av}^{1.36} \tag{14}$$

The more general term for the time spent by the particle in the flight was derived as follows:

$$t_r = \frac{1}{N} (1 - 0.5 \, m_0) \tag{15}$$

In the above expressions, m_0 is the ratio of actual to design flight holdup equalling 1 at design load.

Variation in the cascade factor K_c was provided for in the expression

$$K_c = aM + b \tag{16}$$

where M is the ratio of actual H to design H^* drum loading with a and b constants. This model was further developed by O'Donnell (18) and by Thorne (19) who made more theoretical analyses of the "bouncing" of particles within the drum and of the particle drag theory.

Equation [9] for design load conditions and with no gas velocity approximates to the following expression:

$$T = \frac{L}{2.4ND \tan \alpha} \tag{17}$$

which is similar to the models proposed by other workers in this field (1,11).

More recently, studies have been carried out into the phenomenon of residence time distribution, that is, the distribution about the mean residence time, in order to identify the factors that cause some particles to move faster (and slower) through the drum than others. Hirosue and Shinohara (29) developed, and evaluated against experimental data, a velocity distribution model and a diffusion model and found good agreement, especially with the velocity distribution model. They identified particle bouncing in the lower half of the drum as the principal cause of a residence time distribution, particularly so when there was no air flow and at underloaded drum conditions. They confirmed, as other researcheers have found, that accurate analysis of the drag on particles is difficult due to the complex air-particles contact situation that occurs with the cascade action. In a radioactive tracer experimental study similar to that of O'Donnell (18), Hartman and Zeuner (30) investigated the axial motion of the particles down the drum and its relationship to the mean residence time and the residence time distribution. While equations are developed for these terms, it is not clear if their model has been evaluated experimentally.

6. INDUSTRIAL APPLICATION OF MODELS

The usefulness of the various models described above has often been questioned by industrialists and designers, who still prefer to depend on their experience developed over many years and also on pilot-plant studies carried out in their own laboratories on the solids to be dried. Who can blame them if their designs work? Nevertheless, there is evidence that the use of computer modeling in design procedure is becoming more widespread in the industry. Published studies on design methods are almost nonexistent as manufacturers understandably tend to keep their design procedures to themselves.

The principle weakness of these models lies in a number of areas:

1. *Kiln or rolling load action.* This is very difficult to quantify. Analysis of all the models indicates that this type of action can contribute over 50% to the motion of particles through a drum. If this is the case, it could be reasonably argued that this motion should be the basis for the mechanistic models rather than the cascade motion.

2. *Effect of loading.* The operation of a slightly underloaded drum is significantly different from one just overloaded. These effects are also difficult to incorporate into a model, so that most models understandably confine themselves to design or fully loaded conditions.

3. *Effect of gas velocity.* This is probably the single greatest weakness of all models. The physical situation within the dryer is quite complex, with cascading curtains moving across the drum at right angles to the gas flow. To base the $f(G)$ value on a single particle falling through the gas stream at its average velocity V is perhaps a useful, if the only, starting point for such a calculation, but the actual physical situation is far removed from such an ideal. Empirical relationships from one dryer study will not apply to another dryer, and scaleup from pilot to commercial-size plants must be handled with great care.

Perhaps the most useful role that these models play at present is their ability to predict quite accurately the specific effect of each parameter in the residence time calculation. This will have industrial application, for example, in what is generally referred to as a "debottlenecking" operation, where the rotary dryer is the constraint to plant throughout expansion. In this case the easiest variable to play around with is usually the drum speed, and its effect on the residence time is readily evaluated, as is the angle of repose. In large commercial installations these are major modifications, but they are often done with results as predicted.

It is particularly noticeable that there is a complete absence of optimization design studies for rotary dryers even though the multitude of variables involved in a design would lend themselves ideally to such a study. For example, the length-diameter (L/D) ratios for most commercial dryers fall within the narrow range 6–8. In any of the models summarized above, the same residence time T for a fixed feed rate may be obtained from an infinite variety of L/D values, varying from the long, thin tube dryer (say, $L/D = 20$) to the short, fat dryer (say, $L/D = 2$). The energy load required for these two extremes in terms of the combustion or drying gases would be very different, as indeed would the capital costs of the dryer. From a heat and mass transfer viewpoint, the parameter to be maximized would seem to be the proportion of the total drum holdup cascading through the drying gases, as the particles in the bottom of the drum or in the flights are relatively inactive. This, however, would suggest that the short, fat dryer with the longest distance of fall is optimum, but of course the compressor and fuel operating costs would be higher for this case than for the long, thin dryer. It may be, with the energy-conscious world that is fast approaching, that such studies could completely change the rotary dryer design as we know it.

7. HEAT AND MASS TRANSFER

In the rotary drying operation, the mechanism is usually described as being simultaneous heat and mass transfer, although strictly speaking they are consecutive processes, with the heat transfer preceding and controlling the mass transfer. Heat is transferred from the hot drying gases to the solids in the dryer for the evaporation and diffusion of the moisture from the solids into the gas stream.

In most industrial drying operations, the drying process is in the falling rate rather than the constant rate period so that diffusion of the moisture vapor from within the solid particles to the surface and thence into the gas phase is involved. Thus the process is essentially heat transfer from the hot drying gases to the surface of the solids, conduction of the heat from the surface to the internal moisture, evaporation of the moisture within the solid, diffusion of the moisture through the solids' pores to the surface, and, finally, diffusion from the surface into the gas stream. This is, of course, the mechanism of all solids' drying processes, but it is further complicated in the rotary drying operation because of the cascade motion of the solids through the dryer. The cascade cycle may be divided into the "cascading" and the "resting in flights" periods; in the cascading period, the solids are subjected to the heat and mass transfer process, as described above, but in the resting period, the processes are slowed down, possibly stopped altogether. Depending on the physical properties of the solid particles, in particular their thermal conductivity and porous structure, the movement toward equilibration of both its temperature and moisture concentration throughout the solid particles will take place.

The process is therefore a difficult one to describe satisfactorily in the usual mathematical equations of heat and mass transfer as it involves not only the properties of the solids and the drying gases but also the design and operating characteristics of the rotary dryer, such as the number of flights, the flight cascade pattern, and the ratio of falling to resting times, and perhaps others. The industrialist is interested in improving or optimizing rotary dryer operation either to minimize fuel costs or to maximize throughput in a given dryer, and the model that would give precise answers to all these questions simply does not exist. Nevertheless, much useful work has been done and reported in the technical literature, which in many cases will assist the manufacturer in arriving at a satisfactory if not optimum answer to a particular problem. A summary of the published work in this field is given in the following pages.

7.1. Heat Transfer

Until recently, all attempts to analyze the heat transfer mechanism in rotary dryers have been essentially of an empirical nature. Correlations have been presented in which the volumetric heat transfer coefficient U_V (defined below), or the total heat transferred Q, is related to almost every rotating parameter and/or the physical properties of the rotary drum. From the limited amount of data that has been published, it is apparent that little confidence can be placed in the use of these correlations for fundamental design problems. The wide scatter of the experimental points about the correlation curves, which is characteristic of these plots, indicates that a satisfactory analysis of this problem will only be obtained from a more fundamental approach. The statement of van Krevelen and Hotijzer (6), "Not withstanding the great spread in the data, (which unfortunately cannot be avoided in experiments like these) it appears to be quite evident that Q increases with N" is a valid observation considering the state of fundamental knowledge in this particular area, but the pessimism is hardly justified. Furthermore, the absence in most instances of heat balance data, where the nature of the apparatus is ideal for the provision of such figures, makes the specific correlations questionable.

The concept of a volumetric heat transfer coefficient was first mentioned by Marshall in the reported discussion on a paper by Miller et al. (8). It is defined by the equation

$$Q = U_V V_e \, \Delta T_{c-m} \qquad\qquad [18]$$

where U_V = volumetric coefficient (w/m^{3}K) and V_e = effective volume of the rotary drum (m^3). Miller et al. proposed the equation

$$Q = kLY_{av}n_f d_f G^a \, \Delta T_{c-m} \qquad [19]$$

where the exponent a on the gas flow G was given as 0.46 for a six-flight dryer and 0.60 for a 12-flight dryer. The operating procedures and experimental rigor of these experiments were criticized in subsequent articles, in particular by Friedman and Marshall (9). This lengthy paper of Friedman and Marshall concluded that

1. For the materials under investigation, U_V seemed to be a function of the dryer holdup to the power of 0.5 at a constant air rate.
2. From cross-plots of U_V against air rate, U_V increased with the 0.16 power of the air rate.

Thus the total effect of increasing the air velocity is to increase the holdup, thereby increasing U_V with the 0.5 power of this increase, plus a separate increase in the value of U_V to the power of 0.16 of the air rate. These workers recorded heat losses of 50%; they devised elaborate procedures to measure these heat losses. The effects on U_V of varying other operating parameters were studied and are discussed in consideragble detail in this article. Van Krevelen and Hoftijzer (6), experimenting in the drying of a single granule of wet calcium carbonate in the constant rate period, proposed the following equation to descrbe the heat transfer mechanism:

$$Nu = 1.0Re^{0.5}Pr^{0.33} \qquad [20]$$

In an article principally concerned with material transport through rotary driers, Seaman and Mitchell (11) correlated their data with the equation

$$U_V = k(0.6 + 2.5e^{-4.8d_f}) \qquad BTU/min \cdot ft^3 {}^\circ F \qquad [21]$$

where d_f = the radial flight depth (feet). It was noted that the air temperatures in the lower half of the drum were consistently higher than in the upper half and deduced that the air was entrained downward by the cascading particles. A further article by Saeman (14) analyzed the heat transfer data of previous workers in this field. Expressing agreement with the conclusions of Friedman and Marshall, it was claimed that there is an optimum loading for a rotary drum below which U_V is a function of the cascade rate, which in turn is a function of the holdup; the net result will be a first-order influence of air rate on U_V (i.e., exponent ⟶ 0.6). Above this optimum loading, further increases in the holdup will not increase the cascade rate and the effect of the air rate on U_V will be therefore reduced (i.e., exponent ⟶ 0.16). This argument was supported by the presentation of the data of Friedman and Marshall in graphic form, as shown in Figure 11. In a comprehensive analysis, McCormick (20) claimed that all previous correlations can be reduced to the general formula

$$Q = kLDG^a \, \Delta T_m \qquad [22]$$

where the exponent a has a value in the range 0.46–0.67, but where 0.67 is the most reliable. This finding would suggest that all previous experimentation was carried out on the low side of the "optimum" loading point of Saeman. In an attempt to correlate experimental results with the correlation

$$Nu = 0.33Re^{0.6} \qquad [23]$$

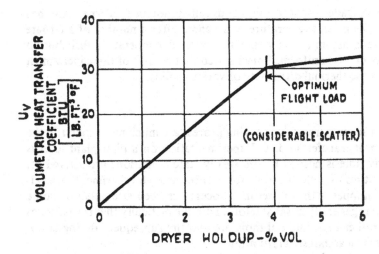

Figure 11 Effect of holdup on heat transfer. (From Reference 9.)

for values of Nu, Schofield and Glikin (1) found that their experimental values were very much smaller than calculated. It was concluded that the air-solids contact was poor or that the heat flow from the center of the granule to the surface was the controlling resistance. Mýklestad (21) investigated the simultaneous heat and mass transfer that occurs during drying, and proposed the correlation

$$U_V = 0.52G^{0.8} \qquad [24]$$

which, he noted, "is in close agreement with the work of other experimenters."

In a radically new theoretical approach to the heat transfer problem, Porter (3) criticized the use of the volumetric heat transfer coefficient U_V, stating that although it may find useful application in a scaling-up design procedure, it was not possible to use it to design a unit from first principles. He proposed an expression based on a mathematical solution of Carslaw and Jaeger (22). "This analytical expression is written in terms of the dimensionless Fourier and Nusselt or Biot groups. The particle radius and thermal conductivity appear in both these groups and the gas film coefficient of heat transfer occurs in the Nusselt group, so that this one equation takes account of the properties of the solid phase, the interfacial area, and the film resistance of the gas phase." This equation which gives the average temperature of the sphere above that of the surrounding air at time t, is written as

$$(T_{av})_t = (T_{av})_0 6Nu^2 \sum_{n=1}^{\infty} \exp(-F_0\beta_n^2) \frac{\beta_n^2 + (Nu - 1)^2}{\beta_n^2 + Nu(Nu - 1)} \frac{\sin^2 \beta_n}{\beta_n^4} \qquad [25]$$

where β_n, $n = 1, 2, \ldots$, are the roots of

$$\beta \cot \beta + (Nu - 1) = 0 \qquad [26]$$

It was claimed that the repeated application of this expression to the cascades and soaking periods of an individual particle gives the temperature of the particle after each cascade cycle; it is applicable only to those solids with thermal conductivity sufficiently high to allow the temperature within the particles to become sensibly uniform in the soaking periods between cascades. Where this condition is not met, Turner (23) has devised an expression for the mean temperature of a sphere after a number of alternate cooling and soaking periods assuming only that the original temperature distribution before the first cooling period was uniform. Porter stated that the test of these ideas must await the determination of the thermal diffusivity of various solids.

7.2. Mass Transfer

Experimental studies into the mass transfer or drying process are much more sparse in the technical literature than heat transfer studies. Experimentation on a pilot-plant level is difficult because a lengthy time is needed before steady state conditions are arrived at, thus requiring large quantitites of solid particles for a once-through experiment. Recycling the solids to the feed hopper with a rewetting process has been attempted by some workers, but the rewetting is generally unsatisfactory, with the difficulty that the particles will receive only surface rather than internal moisture and the subsequent drying is not truly representative of the typical industrial situation.

Nevertheless, a number of attempts at mathematical modeling of the combined heat and mass transfer process have been attempted. These have contained both a volumetric heat transfer coefficient U_V and a rate of drying R_D in a heat and mass balance over a section of the dryer, which extended the results over the whole dryer length. The resulting set of simultaneous differential or difference equations are then solved in a computer model solution. Typical of these are the model of Sharples et al. (24), as follows:

$$\frac{dX_s}{dL} = \frac{R_D}{V}$$

$$\frac{dX_A}{dL} = \frac{F}{G} \frac{R_D}{V_e}$$

$$\frac{dT_s}{dL} = \frac{U_V A (T_a - T_s) + F R_D / V_e - QL}{F(c_{ps} + c_s c_{pw})}$$

$$\frac{dT_a}{dL} = \frac{f c_{pw} (T_a - T_s) R_D / V_e - U_V A (T_a - T_s)}{G(c_{pa} + c_a c_{pw})} \tag{27}$$

Knowledge of the rate of drying as a function of solids' moisture content and temperature is essential for the solution of these equations, and few data on this are currently available.

A more fundamental approach, allowing for alternating periods within the cascade cycle, was proposed by Davidson et al. (25). This followed from the heat transfer model of Turner (23) and simplified the problem of alternating boundary conditions by postulating the existence of a fictitious permeability barrier at the particle surface. Unfortunately, none of these authors presented experimental data to test their models, nor did they specify how the number of cascades was to be determined. Garside et al. (26) used a finite difference algorithm to solve the equation describing drying of a spherical particle by vapor diffusion and presented data obtained using a rotary dryer simulator. The particle dynamics model of Kelly and O'Donnell (2,5), allowing as it does for the calculation of the number of cascades, enables this approach to be applied to any defined rotary dryer system. O'Donnell (18), in an extensive experimental pilot-plant program, presented data that showed good correlation with a theoretical model based on this approach. Likewise, Thorne and Kelly (27) undertook pilot-plant studies and compared their data with the prediction of a vapor diffusion model. In this case, the results were inconclusive and the difficulties of achieving steady-state operation and the validity of the solids rewetting process are given in explanation for the poor agreement. It is further noted that the theoretical predictions of the model are dependent on the value of the vapor diffusion coefficient within the pores of the solid particle and that it is difficult to determine an accurate value for this important parameter.

An extensive review of the research into the rotary dryer process is given by Baker (31) which covers flight design and loading, particle transport, heat and mass transfer and dryer simulation. A useful comparison of the various residence time equations is presented and discussed. In analyzing the various models and relationships for heat and mass transfer in rotary dryers, Baker notes that the kinetics of intermittent drying are difficult to model with any accuracy and that as a result these various studies have had very little impact on industrial dryer design—yet!

CLOSURE

In conclusion, it is believed that, up to this point in time, industrial designers and users of rotary dryer equipment have not been interested in these mathematical modeling approaches. The diversity in equipment and in solids material that is processed has not encouraged these approaches. Although it is unlikely that a universal model of the rotary drying process, combining the cascade motion with the heat and mass transfer processes, will ever be developed, it is suggested that these rate equations and mathematical models can be useful in isolating the effect of a given operating parameter to a given process. Furthermore, in the realization of the very large capital and operating expenses that are involved in industrial rotary drying, it is hoped that these more theoretical approaches will lead to a fundamental reappraisal of the current design and operating procedures of this drying process.

NOMENCLATURE

A dryer cross-sectional area, m^2
a, b constants in various equations, defined in text
c various specific heat values in Eq. [27], $J/kg \cdot K$
d diameter of particle, m
d_f radial depth of flight, m or ft
D drum diameter, m
F solids feed rate to dryer, kg/s
Fo   Fourier dimensionless number
$f(G)$ distance particle carried by gas cascade, m
$f(H)$ constant defined in Eq. [6]
g acceleration due to gravity, m/s^2
G flow rate of drying gases, $m^3/s \cdot m^2$
H solids holdup in dryer, m^3 or kg
H^* design value of H, m^3 or kg
h solids holdup in flight, m^3/m
h_0 solids holdup in flight at $\theta = 0°$, m^3/m
h^* design value of h, m^3/m
h_0^* design value of h at $\theta = 0$, m^3/m
K_c cascade factor, defined in Eq. [16]
k constants in various equations, defined in text

L drum length, m
L_{eff} effective drum length ($=KcL$), m
M ratio of actual to design drum holdup ($=H/H^*$)
m_0 ratio of actual to design flight holdup at $\theta = 0°$ ($=h_0/h_0^*$)
N drum rotational speed, rev/s
Nu   Nusselt dimensionless number
n_f number of flights
P normal force exerted by particles in flight, N/m^2
P_R Prandtl dimensionless number
Q heat transferred in drum, W
R_D drying rate, kg/kg·s
Re$'$ Reynolds dimensionless number for particle
r radius of drum, m
T average residence time of solids in drum, s
T_a temperature of gases in drum, K
T_s temperature of solids in drum, K
T_{av} average temperature of particle, K
t_f time of fall of particles in cascade, s
t_r resting time of particles in flight in a cascade, s
U_v volumetric heat transfer coefficient, W/m^3K
V velocity of drying gases, m/s
V_e effective volume (inside flights) of drum, m^3
v ratio of centrifugal to gravitational forces on particle ($=r\omega^2/g$)
X moisture content of particles, kg/kg
x horizontal axis
Y_{av} average height of fall of particle in a cascade, m

Greek Symbols

α drum slope to horizontal, degrees
β_N defined in Eq. [26]
θ angle defining position of flight on drum circumference, degrees
λ kinetic angle of repose of particles in flight, degrees
μ coefficient of particles rolling on each other in flight
Σ summation of terms in Eq. [25]
ω speed of rotation ($=2\pi N$), rad/s

REFERENCES

1. Schofield, F. R., and Glikin, P. G., *Trans. I. Ch. E.*, 40, 183, 1962.
2. Kelly, J. J., and O'Donnell, J. P., *I. Ch. E. Symp. Series*, 29, 38, 1968.
3. Porter, S. J., *Trans. I. Ch. E.*, 41, 272, 1963.
4. Kirk-Othmer, *Encyclopedia of Chemical Technology*, 3rd Ed., Vol. 8, p. 99.
5. Kelly, J. J., and O'Donnell, P., *Trans. I. Ch. E.*, 55, 243, 1977.
6. van Krevelen, D. W., and Hoftijzer, P. J., *I. Soc. Ch. Ind.*, 68, 59; 91, 1949.
7. Prutton, C. F., Miller, C. O., and Shuette, W. H., *Trans. A. I. Ch. E.*, 38, 123; 251, 1942.
8. Miller, C. O., Smith, B. A., and Shuette, W. H., *Trans. A. I. Ch. E.*, 38, 841, 1942.
9. Friedman, S. J., and Marshall, W. R., Jr., *Chem. Eng. Prog.*, 45, 482; 573, 1949.
10. Sullivan, J. D., Maier, G. C., and Ralston, O. C., U.S. Bur. Mines Tech. Paper 384, 1927.
11. Saeman, W. C., and Mitchell, J. R., Jr., *Chem. Eng. Prog.*, 50, 467, 1954.

12. Porter, S. J., and Masson, W. G., *Proc. Fert. Soc.*, p. 61, 1960.
13. Porter, S. J., *Trans. I. Ch. E.*, 41, 272, 1963.
14. Saeman, W. C., *Chem. Eng. Prog.*, 58, 49, 1962.
15. Miskell, F., and Marshall, W. R., Jr., *Chem. Eng. Prog.*, 52, 35, 1956.
16. Van Brakel, J., *Proc. First Int. Symposium on Drying* (A. S. Mujumdar, ed.), Science Press, 1978, p. 216.
17. Kehoe, J. P. G., M. Eng. Sc. thesis, National University of Ireland, 1966.
18. O'Donnell, P., Ph.D. thesis, National University of Ireland, 1975.
19. Thorne, B., Ph.D. thesis, National University of Ireland, 1979.
20. McCormick, P. Y., *Chem. Eng. Prog.*, Symp. Series, 58, 6, 1962.
21. Myklestad, O., *Chem. Eng. Prog.*, Symp. Series, 58, 41, 1962.
22. Carslaw, H. S., and Jaegar, J. C., *Conduction of Heat into Solids*, 2nd Ed., Clarendon Press, Oxford, 1959.
23. Turner, G. A., *Can. I. Ch. E.*, 44, 13, 1966.
24. Sharples, K., Glikin, P. G., and Warne, R., *Trans. I. Ch. E.*, 42, 275, 1964.
25. Davidson, J. F., Robson, M. W. L., and Rossler, F. C., *Chem. Eng. Sci.*, 24, 815, 1969.
26. Garside, J., Lord, L. W., and Reagan, R., *Chem. Eng. Sci.*, 25, 1133, 1970.
27. Thorne, B., and Kelly, J. J., *Drying '80*. Vol. 1, Developments of Drying, 1980, p. 160.
28. Baker, C. G. J., *Drying Technology*, 6(4), 754, 1988.
29. Hirosue, H., and Shinohara, H., Proc. Drying '82 (A. S. Mujumdar, ed.), Hemisphere, New York, 1982, 36–41.
30. Hartman, F. and Zeuner, A., *Zement Kalk Gips*, Edition B, 1985, 204–205.
31. Baker, C. G. J., *Advances in Drying*, Vol. 2, (A. S. Mujumdar, ed.), Hemisphere, New York, 1983, 1–51.

6
Horizontal Vacuum Rotary Dryers

James G. Moore[*]
Blaw-Knox Food and Chemical Equipment Company
Buffalo, New York

Drying operations play a significant role in many present-day processes in the food, chemical, and pharmaceutical industries. Many different types of dryers are commercially available. The various types of dryers may dry using air, direct heat, indirect heat, or atmospherically or in a vacuum. They are of all sizes, shapes, and varieties. Unfortunately, for the manufacturer and the purchaser, as well as the user, there is no universal dryer that can economically do every required or desired operation; hence, each group of materials has its own class of dryers. The type of material, its characteristics, whether it is to be a batch or a continuous operation, temperature limitations, final particle appearance, and requirements for further processing will, of course, enter into the final determination of the type of dryer to be used.

In dealing with a specific category of dryers, the vacuum rotary dryer, a batch dryer available in different forms, is usually most satisfactory. The most well-known type consists of a stationary jacket (heating) cylindrical shell, mounted horizontally, in which a set of paddle arms with blades are mounted on a revolving center tube to stir or agitate the material being dried. Generally, in this design, there may or may not be contact of the blades with the shell, and requirements are determined in the pilot plant of the equipment manufacturer unless previous experience is available. The second type, of similar design but with an extended heating surface, is also sometimes known as a steam tube dryer.

Both types, operating under vacuum, may be used when products are to be dried or desolventized and the solvent recovered. This provides economic process operation, safety when handling flammable or explosive materials, and control of temperature (see Figs. 1 and 2).

The vacuum rotary dryer is heated by condensing steam or by circulating a suitable heating medium through a jacket around the shell and, in larger dryers, through the hollow center tube and/or the added tubes on the agitator, as well as the paddle arms. These can be heated by steam, hot oil, or any suitable heating medium, but since vacuum

---

[*]Deceased.

Figure 1   Typical horizontal vacuum rotary dryer. (Blaw-Knox photograph #A-1897-1.)

Figure 2   Typical vacuum rotary dryer with agitator being installed into shell. (Blaw-Knox photograph #A-1897-2.)

rotary dryers are used for relatively low temperature operation, it is quite common for steam to be used.

The agitator design affords the maximum rigidity of the paddle arms in the event that high-power requirements during a portion of the cycle are encountered while retaining the ability to heat the arms, thereby adding to the heating surface as the arms pass through the product during the drying stage.

The smallest dryer that yields characteristic results for test purposes and data that can be easily scaled up with accuracy to the larger sizes is the 2-ft diameter by 4-ft long dryer. In this particular unit, however, the size of the agitator assembly is such that the center rube is not heated, and therefore any tests that are run in our pilot plant have a safety factor resulting from a heated center tube when the data are scaled up to a commercial unit.

Various manufacturers have slightly different types of designs, evolved from actual field installations over a period of 45–50 years of operating experience. The present designs are most flexible and inexpensive and allow the most foolproof operation from the standpoint of heating surface required, and minimization of air leakage into the system, which is always a problem. It is designed with the stuffing boxes on the outside of the dryer heads so that easy access to these is available. Also, the boxes can be sealed with a suitable liquid or grease to prevent any air leakage when low absolute pressures are used.

The actual size of the vacuum rotary dryer required for any given application or operation will depend upon the characteristics of the feed material, changes in the material that take place during the drying cycle, characteristics of the finished product, desired operating temperatures, and the capacity or batch size. Test data can be obtained in the equipment manufacturer's pilot plant for a great many different materials if a new product has not been handled previously. Obviously, as new materials are developed and new compounds are formulated, different data must be obtained at least the first time in pilot equipment.

One thing to remember if a new product is to be developed and vacuum rotary drying appears to be a suitable method of handling the product: the drying characteristics of the material and the final characteristics of the product made by the use of various solvents or a combination of solvents can result in a finished product with different characteristics. It is for this reason that a dryer test should be made on a new product before determining actual production size units and setting specifications for finished product characteristics.

In the Blaw-Knox Buflovak Division catalog (365-A), the table of sizes for vacuum rotary dryers has a column for each standard-size dryer, giving the effective heating surface at 60% full and volume capacity at 60% full (Fig. 3). Using laboratory test data together with the ratio of effective heating surface to volume, the actual size and approximate drying time can easily be determined.

The effective heating surface values shown in Figure 3 are for dryers in which the center tube is not heated. For a specific application, manufacturers of vacuum rotary dryers have developed graphs for each size of dryer showing the ratio of the effective heating surface over volume versus volume percent full for dryers with and without heating on the center tube.

The total drying cycle must also include the time to pull up vacuum to the desired absolute operating pressure and an allowance for charging and discharging material. This amount of time will normally vary with the flowability of the material (both wet and dry), the vacuum to be employed, the method of charging and discharging, and the vacuum equipment used. Generally, an allowance of one-half to one hour is given for these

TABLE OF SIZES . . . VACUUM ROTARY DRYERS

DRYER SIZE		TOTAL CAPACITY	WORKING CAPACITY 60% FULL	HEATING SURFACE 60% FULL	*OVERALL DIMENSIONS			
DIA.	LG.	CU. FT.	CU. FT.	SQ. FT.	WIDTH	LENGTH	HEIGHT	WEIGHT
2'	4'	12.2	7.3	10.6	3'	9'4"	4'	2800
2'	8'	24.2	14.5	21.2	3'	15'	4'	4300
3'	7'6"	49.1	29.4	47.4	4'	16'6"	7'	7400
3'	10'	65.6	39.3	63.0	4'	19'	7'	9500
3'	15'	98.2	58.9	96.5	4'	25'	7'	13,300
3'	20'	131.2	78.6	130.0	4'	30'	7'	16,000
4'	15'	167.6	100.6	139.6	5'	26'	8'3"	20,300
4'	20'	223.4	134.0	187.0	5'	31'	8'3"	25,100
5'	15'	246.2	147.7	183.7	6'	33'	8'	27,800
5'	20'	328.2	197.0	246.0	6'	38'	8'	34,500
5'	25'	400.2	240.1	307.2	6'	43'	8'	41,000
5'	30'	492.3	295.4	369.5	6'	48'	8'	49,000
6'4"	36'	893.6	536.2	584.4	7'	50'4"	11'	85,500

*Overall dimensions include drive, to these must be added foundation and operating space around dryer.

Figure 3 Buflovak vacuum rotary dryers' features and sizes. (From Blaw-Knox Buflovak Division catalog 365-A.)

operations, depending upon the size of the dryer and the material handled, and each individual case should be thoroughly checked by the manufacturer of the equipment to make certain that the overall capacity of the unit is not affected by this phase of the operation.

All drying can be generally divided into distinct periods. In the case of vacuum rotary drying, the drying period falls into two general categories: (1) the constant rate, or that period of time when the solvent is easily evaporated from the product's surface and initial capillary drying, which occurs when the temperature of the material is substantially at the boiling point of the evaporating liquid under the absolute pressure conditions existing in the dryer; and (2) the "falling rate" period, when the material approaches the wall temperature of the dryer, characteristic in all drying operations as diffusion drying sets in. The sensible temperature (of the material as it reduces in moisture) starts to approach the temperature of the dryer heating surfaces regardless of the indirect type of dryer used.

The overall heat transfer coefficient is almost entirely dependent upon the coefficient obtained on the material side, between the material and the inner dryer wall temperature. This coefficient is low in the normal realm of thinking, but not in terms of drying, especially when one compares this type of dryer with an air-heated dryer, for which the K factor is that obtained between a drying gas and the solid.

The coefficient will vary with the type of material being handled and the total weight bearing on the dryer surfaces, increasing with increasing bulk density, particle size, and the total load on the dryer surfaces. From test and commercial operating data obtained, the overall coefficients U range from about 5 to 55 BTU/h per square foot of contact surface per degree Fahrenheit if the dryer walls are kept reasonably clean. If caking on the walls occurs, coefficients as low as 1–5 may be encountered, and it is for this reason that there are optional types of agitators as well as direct, spring-loaded wall scrapers for those materials that have a tendency to adhere to the heated surface (Fig. 4).

Vacuum rotary dryers are usually charged to about 60% full, with smaller charges of 40–50% for very dense materials and larger charges up to 80–85% for very light, fluffy materials and those that tend to shrink a great deal while drying.

Agitator speeds usually range from 3 to 8 rpm. Faster speeds consume greater power but only slightly improve the heat transfer. The actual power requirements will depend upon the size of the dryer, the density of the material, and the capacity or number of

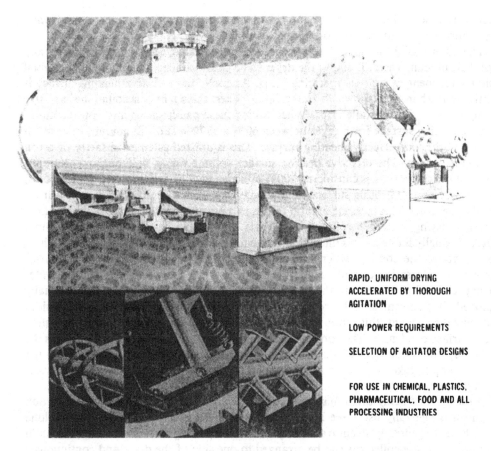

RAPID, UNIFORM DRYING
ACCELERATED BY THOROUGH
AGITATION

LOW POWER REQUIREMENTS

SELECTION OF AGITATOR DESIGNS

FOR USE IN CHEMICAL, PLASTICS,
PHARMACEUTICAL, FOOD AND ALL
PROCESSING INDUSTRIES

Figure 4   Buflovak vacuum rotary dryers. (From Blaw-Knox Buflovak Division catalog 365-A.)

batches in a given time required. Also, the maximum temperature that can be utilized without harm to the product will have a definite effect on power requirements. For feasibility studies prior to serous considerations, a good rule of thumb for power requirement is the length of the dryer in feet multiplied by the diameter of the dryer in feet times 0.2 equals horsepower. In comparing drives, it is generally found that a substantial initial cost saving can be made in most cases by using a variable-speed drive in place of a fixed-speed drive. Most of this saving is made by the elimination of a base plate, which is required for fixed-speed but not for variable-speed drives.

Another matter that must be considered in the design of the vacuum rotary dryer is the amount of air leakage anticipated since the air-handling and vacuum-producing equipment is dependent upon this. Of course, the amount of air leakage depends upon the size of the charge and discharge openings and the cleanout openings in the heads, as well as the tightness of the stuffing boxes that support the agitator. For this reason, the dryer shown in our catalog 365-A is designed to reduce the number of openings and provide quick and easy charge and discharge. Any reputable dryer manufacturer, knowing the size of the dryer, can easily select noncondensable handling equipment. Remember, vacuum equipment should be large enough to handle the air from the auxiliaries as well as the dryer, including dry dust collectors with rotary valves, wet dust collectors, and barometric condensers or surface condensers, and even the air contained in the cooling water. It is good policy to purchase any new dryer complete with vacuum system from a single supplier to avoid split responsibility.

The additional heating surface contributed by a heated center tube is sufficiently large that, in many cases, it allows the dryer to be made smaller, thereby reducing initial capital investment as well as operating costs. Actually, the effective heating surface is slightly more than that shown in our catalog, when taking into account the effective heating surface of the paddle arms. When taking these paddle arms into consideration, the extra heating surface for a dryer between 60% and 70% full will amount to as much as 5–8% of the total effective heating surface. This is utilized as an extra safety factor in the original design. The effective heating surface volume versus volume percent graphs show that the center tube is considered 100% effective at 50% full or more. At this point, the ratio of effective heating surface to volume is approximately 30–40% greater using a heated center tube. This increase in ratio results in a proportional decrease in drying time.

When drying materials containing solvents that are to be recovered, the vacuum utilized is usually limited. This limitation is due to the boiling point of the solvent, which must be above the cooling water temperature (preferably not less than 25°F above); otherwise, condensation cannot occur. If the solvent is not to be removed or if the water is being removed, a vacuum in the range of 28 to 29 inches mercury (in Hg) is usually employed. If a solvent is to be recovered, a liquid-sealed pump may be used. Besides producing the vacuum, this pump can also be considered a secondary condensing system from a surface condenser. The condensate recovered in most cases can also be used as the seal liquid, which when passed through a suitable heat exchanger can be returned to the vacuum pump to take care of the rise in temperature in the pump because of the condensate in the rotor.

If the product being dried dusts or has a tendency to dust as it dries, dust collection equipment is generally used (see Fig. 5). Dry dust collectors are of the bag or cyclone type, which are generally steam traced or jacketed for heating to prevent condensation in the collector. These collectors can be arranged to one side of the dryer and continuously feed the dust back into the dryer. Further, a wet dust collector is often used in combination with the dry dust collector. For drying most materials that are not of a dusty nature in a vacuum rotary dryer, a wet dust collector should still be recommended to remove any dust entrainment from the vapors before they pass to the condenser. In cases in which a wet dust collector is not used, it is preferable that the vapor be made to pass through the tube end of the condenser, which is furnished with removable heads to facilitate periodic cleaning. The scrubbing liquid in a wet scrubber is small in quantity and is returned to the feed of the next batch.

If the dryer is to be used to dry a number of different materials and a changeover is necessary from time to time, spray nozzles are sometimes installed to facilitate quick and easy cleaning of the interior of the dryer. A number of dryers with this arrangement have been built in the past and have provided clean-in-place advantages.

The cost of vacuum rotary dryers, as you can well imagine, varies considerably from size to size depending upon materials of construction, features, auxiliaries, and all the factors discussed above. In order to minimize the cost of this equipment, vacuum rotary dryers of the Buflovak design have been standardized in size, and the units shown in Figure 3 list the standard sizes and all pertinent engineering information. It should be kept in mind, however, that there is no reason a particular customer cannot obtain any specific length or diameter provided they are willing to pay the extra engineering charges. Usually, from an economic standpoint, one can purchase a dryer of the next larger size, which will cost less than if a special engineered unit were desired.

Also, when selecting dryers, the degree of flexibility required should be given careful

The layout and use of auxiliary equipment with a Buflovak Vacuum Rotary Dryer is dependent on the product to be processed . . . to ensure efficient and economical operation . . . and to conserve floor space and headroom.

Vacuum Rotary Dryer equipped with dry and wet dust collectors and barometric condenser. Various vacuum systems may be used. Dryer is mounted on concrete piers and arranged with drive pinion engaging bull gear mounted on the agitator shaft.

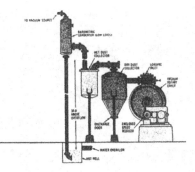

Entire contents of dryer can be quickly discharged into a receiver at completion of drying cycle. Dry material is removed from this receiver continuously, by a conveyor . . . while another batch of material is being dried in dryer.

Dry dust collector is equipped with a conveyor, which continuously returns the dust to the dryer.

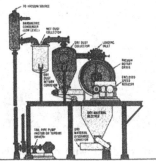

Liquid recovery and removal of non-condensable gases is accomplished through use of a surface condenser and suitable vacuum equipment. Dry material is discharged into a receiver from which it is conveyed. Dust from the dry collector is continuously returned to the dryer through a mechanically operated valve.

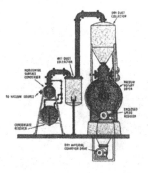

Figure 5 Vacuum rotary dryer equipment arrangements. (From Blaw-Knox Buflovak Division catalog 365-A.)

consideration. Sometimes it might be wise to pay slightly more initially for the installation of two smaller dryers rather than one large dryer. The reason for this is the flexibility afforded in the degree of capacity available as well as assurance of operation of one unit at all times.

In general, materials dried in a vacuum rotary dryer can, in most cases, be dried in a revolving double-cone dryer. However, of all the types of dryers mentioned, the vacuum rotary dryer has the best mixing action because of its continuous agitation. This accelerates the drying time by having the fresh material in contact with the effective heating surface for the greatest percentage of the drying cycle. Materials that tend to form lumps can be broken up by placing breaker bars, called Venuleth bars, in the dryer agitator to move about as the agitator rotates, keeping the heated tube of the agitator, the paddle arms, and the backside of the paddle blades clean of material. These breaker bars have been quite effective in producing a more uniform product as well as decreasing the drying

Figure 6 Vacuum rotary dryer with extended heating surfaces. (Blaw-Knox photograph #A-1072-2; retouch R-1334.)

Figure 7 Vacuum rotary dryer with extended heating surfaces. (Blaw-Knox photograph #A-1072-2; retouch R-1334.)

time since uniformity of drying takes place during the entire cycle. When using these bars in the dryer, the paddle arms generally are reinforced, and a special design is used so that rapid fatigue of the metal does not occur. Therefore, if adding such bars to an existing dryer is contemplated, the matter should be checked thoroughly with the manufacturer to make sure that undue fatigue and breaking of the paddle arms will not occur if these bars are used.

Needless to say, before using these bars, it should first be determined whether they are definitely required. It is in this instance that tests in the pilot plant of an equipment manufacturer may prove valuable. By changing components in the mixture or the drying time cycle or varying temperature during the cycle, it is often possible to achieve the same results as would be achieved with the breaker bars, and this should be determined before a final decision is made to proceed haphazardly with the unnecessary installation of inappropriate auxiliary equipment.

Vacuum rotary dryers of the type we are emphasizing have a decided advantage over other types since the materials collected from a suitable dust collector can be returned to the dryer continuously. Mostly because of this advantage, we have a number of arrangements of equipment that allow for this operation.

When headroom is a problem in a dryer installation, studies can be made and tests run to determine whether the dryer should be filled to a higher capacity than is normally the case Again, it depends upon the characteristics of the material at the start and during the drying cycle and the finished product required.

Other advantages of the vacuum rotary dryer of the type we are discussing are as follows.

1. Sampling devices can be provided in this type of dryer to allow a small quantity of material to be removed without interfering with the operation of the dryer.
2. A sight glass can easily be installed in a vacuum rotary dryer for visual inspection of the product within the dryer itself during operation.
3. The temperature of the material can be taken at all times with less elaborate and expensive equipment and can be recorded during the entire cycle if so desired.
4. The shell can be cooled more rapidly and more uniformly than any other type of unit.

The vacuum rotary dryer as shown in Figures 6 and 7, which has extended heating surfaces, can also be utilized. This unit has a special arrangement of heated tubes on the agitator that penetrates the product held in the dryer and gives a greater effective heating surface. This of course adds to the heating surface, thereby decreasing the batch time and increasing the capacity of any given dryer. This type of dryer is of limited use for special materials since any product that has a tendency to cake will quickly plug up this type of agitator and prevent free circulation of the material through the interior of the dryer shell.

In normal operation, the blades are so pitched on the agitator that during the cycle they not only give maximum agitation and tumbling effect of the product for quick drying but also circulate the material where the discharge connection is located, thereby providing shorter discharge periods and much easier discharge cycles. Figures 1 and 2 illustrate the pitch of the blades.

As mentioned initially, the vacuum rotary dryer is not a universal dryer for all applications, but it should be given careful consideration when a filter cake or a product with discrete particles is to be dried. A study of this type of dryer in a specific drying application may prove quite surprising to you and may solve in a very efficient and inexpensive manner a problem that has been long-standing and irritating.

7
Fluidized Bed Drying

Svend Hovmand
Niro Atomizer, Inc.
Columbia, Maryland

1. INTRODUCTION

The use of fluid bed drying for granular materials is now well established, and literally thousands of fluid bed dryers are operating throughout the food and chemical processing industries.

In contrast with this industrial development, the fundamental research on fluidized bed drying has not made similar progress and the design of an industrial fluid bed dryer is still very much an art based upon empirical knowledge. The advantages offered by this technology, when compared with other drying methods, are principally as follows.

The even flow of fluidized particles permits continuous, automatically controlled, large-scale operation with easy handling of feed and product.

There are no mechanical moving parts, that is, it is low maintenance.

By rapid exchange of heat and mass between gas and particles, overheating of heat-sensitive products is avoided.

Heat transfer rates between fluidized bed and immersed objects, such as heating panels, are high.

Rapid mixing of solids leads to nearly isothermal conditions throughout the fluidized bed, and thus reliable control of the drying process can be achieved easily.

When a gas is passed upward through a layer of particles supported by a grid, as shown in Figure 1, the gas will at low flow rates merely percolate through the fixed bed of particles. As the gas velocity is increased, the pressure drop across the particle layer will increase in proportion to the gas velocity until the pressure drop reaches the equivalent of the weight of the particles in the bed divided by the area of the bed. At this point, all particles are suspended in the upward-flowing gas and the frictional force between particles and gas counterbalances the weight of the particles. The layer of particles is now said to be incipiently fluidized, and although the homogeneous particle layer behaves like a liquid, only moderate particle mixing takes place. When the gas velocity is increased further above U_0, the gas velocity for the incipient fluidization, any additional fluidizing

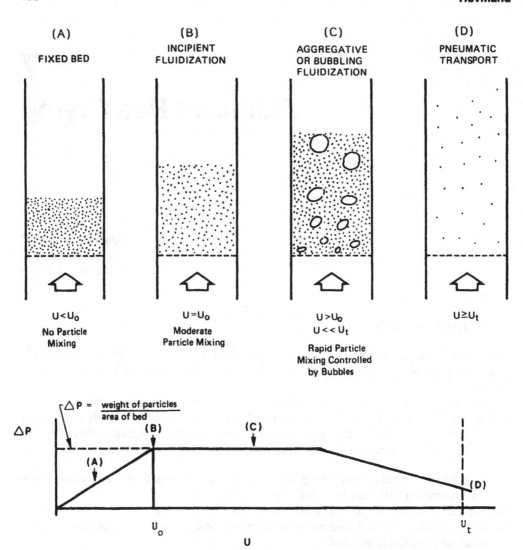

Figure 1 Regions of gas fluidization: U = gas velocity; U_0 = gas velocity for incipient fluidization; U_t = terminal fall velocity for particles; ΔP = pressure drop over particle layer.

gas will pass through the particle layer as bubbles. These gas bubbles will be small at the gas distributor; however, they coalesce rapidly and rise through the particle layer, causing vigorous mixing of the fluidized particles. At still higher gas velocities, a point is reached at which the drag forces are increased to a degree that the particles become entrained within the gas stream and are carried from the fluid bed. Finer particles are naturally entrained more easily, and the amount of entrained material is dependent on the way the gas bubbles "burst" at the surface of the fluidized layer.

In fluid bed drying of granular products, the material is dried while suspended in the upward-moving drying gas, as described above, and the fluidized layer can be compared in its behavior to that of boiling liquid, as shown in Figure 2; that is, fluidized layers assume the shape of the containing vessel and have a viscosity close to zero, which lets heavy objects sink and light objects float.

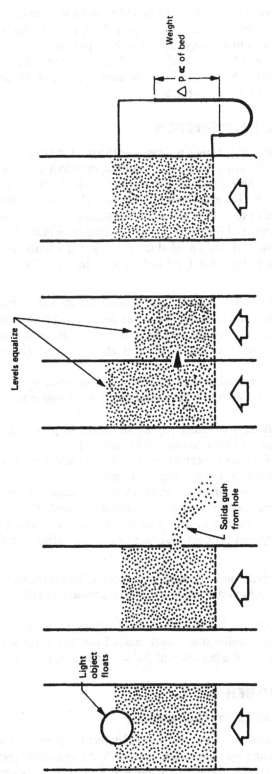

Figure 2 Gas fluidized bed compared to boiling liquid.

Several reviews have been made of fluid bed dryers and their design, including those by Vanecek et al. (1), Romankov (2), and others (3–5), and articles reviewing the recent developments in fluidized bed drying (6,7) have been published. Extensive reviews of the literature describing the behavior of vibrated fluid bed dryers (8,9) are also available, together with two books in Russian (10,11). No general application models for the design of fluid bed dryers have yet been published.

2. PARTICLE CHARACTERISTICS

The flowability of granular materials is a decisive factor in the design of a fluid bed dryer. The aeration and/or fluidization of some granular materials decrease the critical shear stress to close to zero (the critical shear stress is the value the shear stress must use to make the material flow). However, cohesion between the particles due to surface effects may make this impossible and the material is then said to be poorly fluidizable or nonfluidizable. The measurement of the angle of repose can give an indication of the flowability of the material; however this is not a reliable yardstick, particularly for sticky materials.

Materials that are suitable for fluid bed drying can generally be characterized by a number of criteria.

1. The average particle shall be somewhere between 20 μm and 10 mm in order to avoid excessive channeling and slugging. Very fine particles tend to lump together because of cohesive forces related to the very large surface areas; accordingly, these particles normally fluidize badly at gas flow rates at which excessive elutriation can be avoided.
2. The particle size distribution shall be reasonably narrow, so a fluidization gas rate can be selected to give effective fluidization without excessive particle entrainment of the fines.
3. The particles shall be reasonably regular in shape, in particular when the average particle size is large, or fibers do not fluidize properly.
4. Lumps in the wet feed shall break up readily, as they otherwise will defluidize the bed when they are retained on the gas distributor plate.
5. The particles must have sufficient strength to withstand the abrasion taking place among the fluidized particles as they are vigorously mixed by the rising gas bubbles.
6. The end product must not be sticky at the temperatures required for drying the material to the required product moisture or agglomeration and defluidization take place.

A vibrated fluid bed dryer can be applied to overcome the problems mentioned in points 1–5, as will be explained in Section 6; however, the problem in point 6 cannot be circumvented readily.

Most materials that are fluidized bed dried belong to Geldart's groups A and B (12,13). It is important to realize that models made from data from one group most likely will not accurately describe the behavior of fluidized particles from another group.

3. FLUIDIZED BED BEHAVIOR

3.1. Incipient Fluidization Gas Velocity

The incipient fluidization gas velocity is a fundamental parameter characterizing the system, and numerous attempts have been made to find a generalized equation that would predict values for the incipient fluidization gas velocity (see Refs. 14 and 15).

Using an expression for the relation between pressure drop and superficial velocity for a fixed bed, this calculated pressure drop can be made equal to the buoyant weight of the particles, assuming that ϵ_0, the voidage of the bed at the minimum fluidization velocity, is known. This voidage will depend on the shape and size range of the particles. For spherical particles, a value of 0.4 can be assumed; however, it has not been possible to make a general valid relationship between ϵ_0 and the shape factor for fluidized particles.

Pressure drop over a fluidized bed at the condition of incipient fluidization is

$$\Delta P_0 - (\rho_s - \rho_f)(1 - \epsilon_0)H_0 g \qquad [1]$$

For fine particles in a fixed layer, the pressure drop velocity relation will be given by the Carmen–Kozeny equation, which will have the following form at the incipient fluidizing point:

$$U_0 - \frac{\epsilon_0^3}{5(1 - \epsilon_0)^2} \frac{\Delta P_0}{S\mu H_0} \qquad [2]$$

Hence, from Eqs. [1] and [2] and assuming $S = 6/d$, and taking $\epsilon_0 = 0.4$, we get a first approximation for U_0:

$$U_0 = \frac{d^2(\rho_s - \rho_f)g}{1695\mu} \qquad [3]$$

This equation is only of limited use as the particles are assumed to be perfect spheres and ϵ_0 will be different from 0.4. For most systems, $0.4 < \epsilon_0 < 0.55$; however, higher values for ϵ_0 can be found in fluidized layers of fine particles.

Grace (14) and Richardson (15) recommend the following equations for calculation of U_0, again assuming $\epsilon_0 = 0.4$:

$$Re_0 = \sqrt{C_1^2 + C_2\,Ar} - C_1 \qquad [4]$$

$$Re_0 = \frac{dU_0\rho_f}{\mu} \qquad [5]$$

$$Ar = \frac{\rho_f(\rho_s - \rho_f)gd^3}{\mu^2} \qquad [6]$$

where Grace (14) recommends using

$$C_1 = 27.2 \quad \text{and} \quad C_2 = 0.0408$$

It is indicated that Eq. [4] works reasonably well over a broad range of conditions ($\pm 25\%$) if the particles do not belong to Geldart's group C (cohesive).

It should be further noted that solids of wide particle size range tend to fluidize with less gas flow than required when working with material of similar average size but of narrow size distribution because the bed voidage tends to be less. Similarly, porosities for beds of irregularly shaped material are higher than for spherical particles of similar mean size. Accordingly, U_0 can be determined by a simple bench test in a small fluid bed, measuring ΔP against the increase in gas velocity.

3.2. Two-Phase Theory for Fluidization

As already stated, when the gas velocity is increased above U_0, any additional fluidizing gas will pass through the fluidized bed as gas bubbles. The simplest theoretical concept that describes many of the properties in a gas fluidized particulate layer is often referred to as the *two-phase theory* (16).

The fluid bed will consist of two phases:

1. The particulate phase, a homogeneous mass with the voidage ϵ_0 and gas velocity U_0.
2. The bubble phase, containing all excess gas and nearly free of particles. These gas bubbles pass rapidly through the particulate phase and they have, accordingly, a major impact upon the behavior of the fluid bed, being responsible for the mixing of the particulate phase.

The gas bubbles will create at the small jets of gas coming from the holes in the gas distributor, and the gas bubbles then grow rapidly by coalescence due to the pressure decrease as the bubbles rise through the fluidized layer.

3.3. Mixing in the Fluidized Bed

Very vigorous mixing of the fluidized particles takes place just above the distributor, and heat and mass can be exchanged very effectively between particles and the fluidizing gas because of the large surface area of the particles (3). Thus, gas-particle heat and mass transfer are not limiting factors in a fluid bed drying operation.

The above has important bearing upon the rapid heat and mass transfer that takes place between the gas and the total particulate layer in a fluidized bed, just above the gas distributor plate, and explains the uniform temperature that can be achieved throughout even a large-scale fluid bed dryer. Typically, each rising bubble carries in its "wake" a mass of particles, some of which are left behind at various points; there is an additional influx of particles into the region left behind by the passage of a bubble, as illustrated by x-ray photographs made by Rowe and Partridge (17). On reaching the upper bed surface, the remaining wake material is deposited there or ejected into the space above, causing elutriation, particularly of the finer material in the fluid bed.

Because mixing is primarily caused by the rising bubbles, it depends on the total volume flow of bubbles and is increased by those conditions that produce fast-rising bubbles, such as bubbles in large equipment; accordingly, the vertical mixing rate is several times larger than the mixing rate in the horizontal direction. Further, the rate of mixing will increase as the average particle size is reduced, as finer material is fluidized more smoothly than coarse material (see Ref. 14, Sec. 8.19 for further discussion). De Groot (18) has measured the axial (vertical) diffusivity coefficient in various sizes of equipment, as shown in Figure 3, which basically supports the above remarks, and Figure 3 also shows that because the broad-range silica fluidized better than narrow-range silica the mixing was increased about 10-fold. It can further be seen from Figure 3 that the fluid bed diameter has a pronounced effect upon mixing for bed diameters less than 0.3 m. This observation points to the fact that pilot-plant data must be treated with caution when the data are used for scaleup of a fluid bed to industrial size. Botterill (19) and Potter (20) have reviewed the literature on mixing in fluid beds.

3.4. Selection of Gas Velocity: Entrainment

It is always prudent to select a gas velocity for a fluid bed that is at least two to three times larger than the incipient fluidization velocity for the particles. For industrial fluid bed dryers, it is important to have vigorously fluidized beds with sufficient particle mixing to ensure a uniform temperature throughout the particulate layer, and further, slightly oversized particles should readily be transported through the bed and not create defluidization anywhere in the particulate layer. In order to design a dryer that is as small as

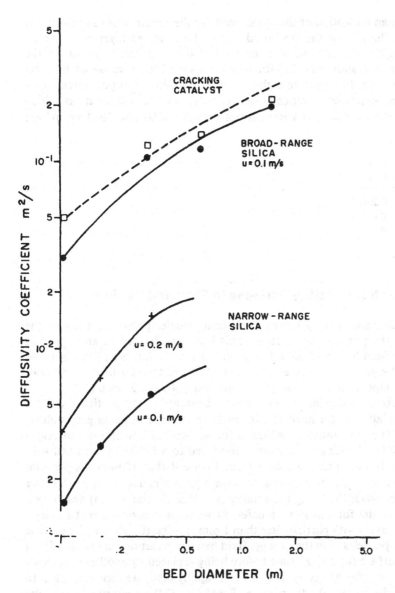

Figure 3 Variation of diffusivity coefficient.

possible, the gas velocity should be selected as large as entrainment of fine particles to the exhaust system permits.

Some material is lost from the fluidized layer when particles are thrown into the space above the bed as gas bubbles break at the surface. The finer material will then naturally have a tendency to follow the gas to the exhaust, particularly if the gas velocity is close to the so-called terminal velocity of the particles. The terminal velocity can be described when the flow is laminar by Stokes law:

$$U_t = \frac{(\rho_s - \rho_f)gd^2}{18\mu}$$

[7]

Equation [7] gives an indication of the upper limit for the selection of gas velocity. It should be noted that the size of the freeboard above the fluidized layer can have an important influence upon the amount of particles lost to the exhaust system, as the particle concentration just above the fluidized layer is very high because of bursting bubbles. The freeboard should be a little higher than the so-called transport disengaging height, which has been correlated empirically (see Ref. 21). Typically applied gas velocities for fluidizing product having particle densities between 1000 and 2000 kg/m^3 are given below.

Average particle size (μm)	Velocity (m/s)
100– 300	0.2–0.4
300– 800	0.4–0.8
800–2000	0.8–1.2
2000–5000	1.2–3.0

3.5. Heat Transfer from Heating Surfaces to Fluidized Particles

The gas fluid bed is characterized by having good heat transfer properties between the fluidized layer and heating or cooling surfaces. This heat transfer can be an important design parameter in a fluid bed, and accordingly extensive work has been done in order to develop generalized equations for the estimation of the heat transfer by Botterill (19), and Zabrodsky (22), Gelperin and Einstein (23), Schlünder (24), and Martin (25).

Heat transfer is strongly dependent on the heat capacity of the particles and the degree of particle circulation at the heat transfer surfaces because of rising gas bubbles, as shown in Figure 4. The heat transfer coefficient for wall-to-bed heat transfer increases dramatically when the bed is transferred from a fixed bed to a bubbling fluid bed with rapid particle mixing. It can further be seen from Figure 4 that 150-μm copper shot transfers the heat better than 625-μm copper shot. This is primarily due to the much better mixing that takes place when fluidizing the smaller particles. Schlünder (24) describes a so-called penetration model for this heat transfer, in which it is assumed that the heat transfer into fluidized layers with particles less than 1 mm is primarily effected by particle convection, that is, by particles with the average fluid bed temperature coming into direct contact with the hot surface for a short time before being replaced by another particle of low temperature. The particles heated in this way will rapidly dissipate their extra heat to the surrounding particles in the fluidized layer. Schlünder (24) demonstrates that the model fairly well predicts a number of experimental data for heat transfer coefficients from wall to fluid bed layers.

We may note the following:

High rates of particle displacement (via vigorous bubbling) are essential to efficient heat transfer.

Constant replacement of particles (cold) at the wall (hot) maintains maximum $\Delta t°$.

The heat transfer coefficient goes through a maximum as U increases because of the increasing preponderance of bubbles (not particles) at the wall.

Heat transfer from particles to the interior of the bed is favorable since the volumetric heat capacity of the particles is greater than that of the gas.

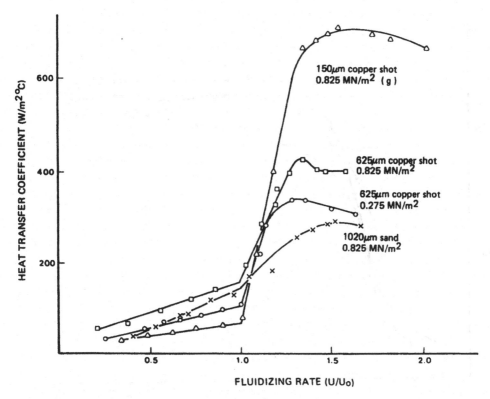

Figure 4 Variation in wall-to-bed heat transfer coefficient as a function of fluidizing gas flow rate.

Horizontal tube bundles are used extensively in fluid bed dryers as a source of heat supply to the drying process. Much heat transfer surface can be exposed to the circulation of solids induced by the rising bubbles; however, horizontal tubes have two areas with low heat transfer properties (see Ref. 19):

Defluidized particles on top of the horizontal surface will hinder effective heat transfer, and in the worse case material might be heat damaged.

The bubbles shroud the downward-facing surface of the tube, lowering the heat transfer in this area significantly.

Figure 5 shows some typical data (26) for the dependency of heat transfer coefficients for horizontal tube bundles in fluid beds with respect to gas velocity and distance between tubes. It can be seen that the heat transfer coefficient is dependent on gas velocity in the same way as described in Figure 4 and also that by placing the tubes too close together the circulation of solids is hindered and the heat transfer coefficient is decreased. Gelperin and Einstein (23) recommend a tube pitch of four to six times the tube diameter as an optimum.

Numerous equations have been suggested for predicting the heat transfer from tubes, both vertical and horizontal (see Refs. 19 and 23). However, it is recommended that the actual heat transfer coefficients in pilot plants be measured under as realistic conditions

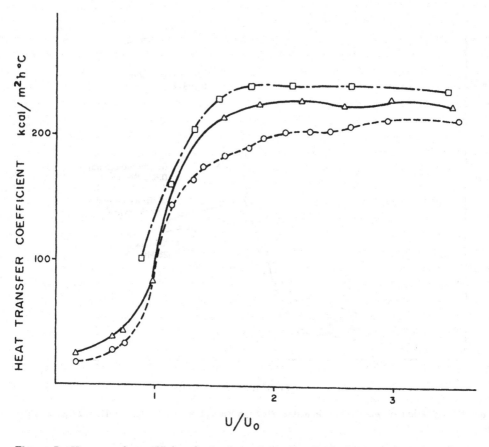

Figure 5 Heat transfer coefficient for horizontal tube bundles in fluid bed: for dependency of gas velocity and distance between tubes (26), $d_r = 25$ mm, $d_p = 0.7$ mm; □, distance between tubes $2d_r$; △, distance between tubes $4d_r$; ○, distance between tubes $6d_r$.

as possible in order to obtain reliable values for the design of the industrial-size fluid bed dryer.

3.6. Back-Mixed versus Plug Flow Fluid Bed Dryer

Continuous fluid bed units are normally characterized by the residence time distribution of the individual particles inside the unit. A broad residence time distribution is obtained in a back-mixed fluid bed in which the length-to-width ratio of the fluid bed itself is relatively small. This is indicated in Figure 6. Such a back-mixed fluid bed can, in performance, be compared to an agitated tank provided with overflow because the vigorous mixing inside the fluid bed will result in a uniform temperature; however, the particle residence time distribution is rather broad, and there may be significant variation in moisture content from particle to particle. Approximately 40% of the product stays in the fluid bed less than half the average residence time, and some particles may be over-dried when they stay too long in the fluid bed. Accordingly, the product discharge from a back-mixed fluid bed will only have a homogeneous moisture content when it is primarily surface moisture that is dried from the particles.

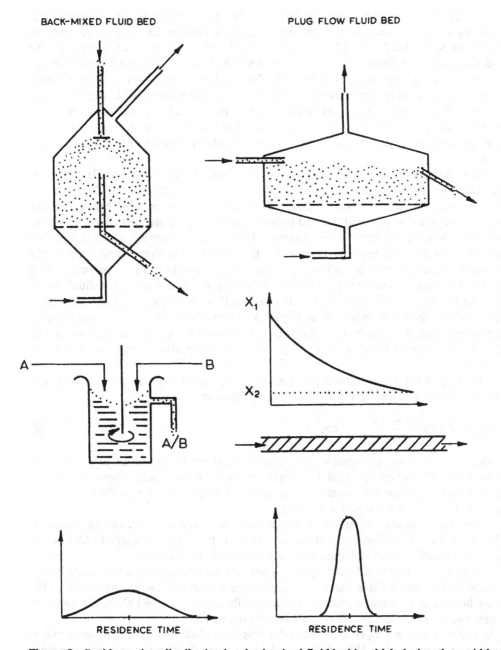

Figure 6 Residence time distribution in a back-mixed fluid bed in which the length-to-width ratio of the fluid bed is relatively small.

Because of the excellent heat and mass transfer between the fluidized particles and the drying air, equilibrium will essentially be reached between the exhaust air and the product inside the fluid bed. The back-mixed fluid bed drying concept is very suitable for drying surface moisture when residence time has little or no impact on the drying performance. As a long residence time in the fluid bed and a relatively high temperature level may be required to compensate for the broad residence time distribution inside the fluid bed unit, the back-mixed fluid bed is generally unsuitable for drying bound moisture (see Ref. 5 for a detailed discussion). Typically, for a polymer application, this would result not only in excessive heat consumption but also potential heat damage of the product.

The back-mixed fluid bed drying concept does, however, have one significant advantage compared with the plug flow concept, inasmuch as the back-mixed fluid bed can accept a feed material that is not readily fluidizable. This is possible because of the vigorous mixing inside the fluid bed and because the material inside the bed acts as a large reservoir in which the incoming feed material is immediately dispersed and the surface moisture is flashed off, making the product fluidizable. This characteristic makes the back-mixed fluid bed concept well suited as the predrying stage in many polymer drying systems. The narrow residence time distribution is obtained in a plug flow fluid bed in which the length-to-width ratio of the fluid bed itself is very large. This corresponds to a long narrow fluid bed indicated in Figure 6. Alternatively, this can be obtained by compartmentizing the fluid bed. In this way, it becomes analogous to a large number of agitated tanks in series in which every single agitated tank would represent a back-mixed fluid bed.

Reay (27) measured the lateral particle diffusivity coefficient in a shallow fluid bed (less than 10 cm), and he suggested the following equation:

$$D_{lat} = 0.17 \left(\frac{U - U_0}{U_0^{1/3}} \right) \quad cm^2 \, s^{-1} \tag{8}$$

for which a good fit was obtained with the experimental results. However, as also pointed out by Reay (27), when the fluid bed depth increased substantially above 10 cm, the observed particle diffusivity increased by an order of magnitude, due to the effect of the large bubbles in the upper layers of the fluid bed.

The residence time distribution for the particles in a plug flow bed can be calculated if the lateral particle diffusivity coefficient is known, following Levenspiel's (28) procedures for the prediction of residence time distributions in fluid flow.

The plug flow fluid bed drying concept is particularly advantageous for drying bound moisture from heat-sensitive materials because the residence time is controlled within narrow limits and a distinct moisture content profile can be obtained along the length of the unit owing to very low degree of back-mixing taking place.

In order to assure proper operation of a plug flow fluid bed, the incoming product must be readily fluidizable. If the feed material is too wet, fluidization may not take place at the feed entry point.

4. DRYING BEHAVIOR OF FLUIDIZED PARTICLES

Each product has its own drying curve characteristics. Drying curves can be determined from small-scale fluid bed drying test work by plotting residual volatiles data against time. A typical drying curve is shown in Figure 7, together with the associated product temperature curve measured during a batch fluid bed drying test.

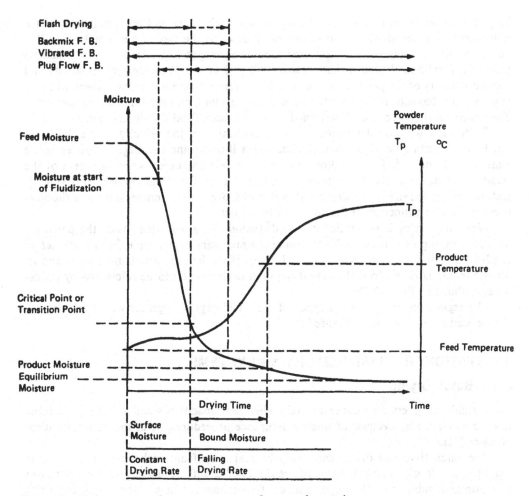

Figure 7   Drying curve and temperature curve for granular product.

It can be seen from Figure 7 that the surface moisture is rapidly evaporated into the drying gas and the so-called transition point or critical point of the drying curve is soon reached. However, at lower moisture contents, drying is controlled by the rate of diffusion of moisture inside the particles and the drying rate is decreased considerably; when very low moisture contents of the end product are desired, it may be necessary to have drying times of several hours. The temperature of the product will, after an initial period during which the evaporation of moisture keeps the temperature down, rise to close to the inlet temperature of the drying gas.

The heat and moisture exchange in a batch fluid bed is limited as the gas flow is quickly saturated, as pointed out by several authors (5,29–31). For a long time the batch fluid bed remains wetter than the continuously operating fully mixed fluid bed where the contact between wet and dry fluidized particles considerably increases the heat and moisture exchange.

However, as pointed out by Reay and Allen (30), for materials with low to moderate internal resistance to mass transfer the time required to reach a given moisture content in a batch drying test can be assumed to be proportional to the bed weight per unit area and inversely proportional to the gas velocity. Accordingly, heat and mass balances over a well-mixed fluid bed dryer can in simple cases describe the system adequately, and it is not necessary to apply such complicated models as suggested by Hoebink and Rietema (3,4). The maximum drying air temperature will be determined by the heat sensitivity and thermoplasticity of the product. If this maximum temperature is exceeded, there will be a tendency for deposits to form on the air distributor plate, and the fluidization may stop. The temperature limit can be determined by a pilot-plant fluid bed drying test.

Such a test will also determine the practical limits for the velocity of the fluidizing air. Below a certain value, uneven fluidization will occur and the larger particles in the material will not be fluidized. Above a certain air velocity, excessive entrainment of the smaller particles will take place from the fluidized layer. The total amount of drying air and, therefore, the size of the dryer will in this way, for a given temperature and fluidization velocity, be a function of the necessary heat input.

When the drying is controlled by the diffusion of the moisture inside the particles, the batch drying experiment will determine the necessary drying time for a given set of conditions in an ideal plug flow fluid bed dryer. If the lateral particle mixture is known for the industrial-size dryer, the actual required drying time can be calculated by procedures outlined by Poersch (5).

The regions in which various types of fluid bed dryers, together with flash drying, can be used are also shown in Figure 7.

5. STATIONARY FLUID BED DRYING SYSTEMS

5.1. Batch Drying

Batch fluid bed dryers are used extensively when the capacity is small and it is preferable to operate batchwise because of quality assurance procedures, as in the pharmaceutical industry (32).

The batch fluid bed dryers can easily be scaled up from such pilot plant tests, as described previously. The lower part of the fluid bed chamber, including the plenum, can be made removable from the upper part of the freeboard, which is the casing for a bag filter, allowing easy handling of material and easy cleaning of the unit.

Modern units are very flexible and can be supplied with sophisticated control equipment, fully automating the batch process. The batch fluid bed dryers are often also used as mixers, granulators, and coaters, as described in Section 7.

5.2. Totally Mixed Fluid Bed Dryer

Figure 8 shows a typical flow sheet for a totally mixed fluid bed dryer in which the wet feed is allowed to mix freely with the dried and partially dried product.

The height of the fluidized layer is controlled by a weir over which the dried material freely flows. The fluidizing air is heated and led into the plenum chamber to the fluid bed. An air distributor ensures that the fluidizing air velocity is even throughout the fluidized layer and that all material is properly fluidized. The exhaust air is cleaned by a cyclone, and entrained fines from the cyclone are returned to the dried product stream. The freeboard above the fluidized layer must be high enough to avoid excessive entrainment of particles (see Ref. 21). Because of the intimate contact between particles and fluidizing air just above the air distributor and the rapid mixing of the particles in the fluidized layer, the hot drying air is immediately quenched to the temperature of the particles and no overheating of the material will occur.

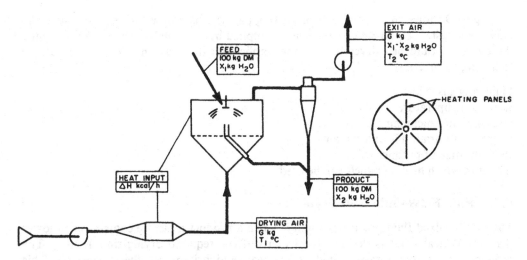

Figure 8 Totally mixed fluid bed dryer with heating panels.

A properly designed gas distributor with enough holes in order to avoid any dead areas in the fluid bed is important. Normally, pressure drops over the gas distributor are designed to be at least 30% of the pressure drop over the fluidized particle layer in order to ensure even gas distribution.

The wet feed does not need to be directly fluidizable in order to be dried in a fluid bed dryer with back-mixing. Small lumps of the feed can be fed to the top of the fluidized layer by means of a rotating product spreader. Wet lumps of feed rapidly dry when mixed with hot and dry particles in the fluidized layer and, for many products, the lumps, below a certain moisture content, will disintegrate into the originally produced particles, which are fluidizable.

The preferred feed point in most installations is above the expanded bed and in circular beds in the center, so the wet solids are not thrown out on the walls. With small feed rates and easy drying of the feed particles to moisture contents below the fluidization point, it is possible to introduce the feed without product spreader; however, it is normally advantageous to spread the feed evenly over the widest possible surface area of the fluid bed. In certain cases, it is not possible to disintegrate the lumps in the feed by this technique, and predrying of the feed (for instance, in a flash dryer) may be necessary in order to avoid lumps in the dried product.

When the material to be dried is not temperature sensitive, a very high inlet temperature for the drying gas can be applied and the fluid bed, including air distributor, will be built with refractory brick and the economy of the drying process will be good.

However, when fine particles (less than 300 μm) are dried the fluidizing gas rate must be low and often there is an upper limit to the inlet temperature. In these cases the fluid bed can be reduced considerably in size and the economy of the drying process significantly improved by the introduction of a heat source into the fluidized layer, minimizing the necessary amount of drying air.

The heat transfer coefficient between the fluidized particle layer and the heating panel must be determined by a pilot-plant drying test, simulating the conditions under which the industrial dryer operates. The maximum allowable temperature of the heating panels must also be determined.

Figure 9 shows an arrangement of heating panels that is now used extensively in the polymer industry. Up to 80% of the heat is supplied by the panels in the fluid bed. Thus, the size of a fluid bed dryer for a given application depends on one or more of the following parameters:

Fluidizing gas velocity
Inlet gas temperature
Necessary product temperature
Degree of saturation in exhaust air
Resident time of product
Extent to which heating panels can be used

5.3. Plug Flow Fluid Bed Dryer

The totally mixed fluid bed is not suitable when a product moisture content much lower than the critical point is desired, particularly if the required drying time is long. The product from a totally mixed fluid bed is near equilibrium with the exhaust air. This means the air may have to be kept at a low level of humidity and this in turn results in a large drying air requirement with accompanied increased drying costs.

Furthermore, the dried particles from a totally mixed fluid bed will have very different residence times and a large portion of the material will be either overdried or underdried. Thus, the risk of product damage will be high when drying a heat-sensitive material at near critical temperatures.

In order to overcome the above limitations of the totally mixed fluid bed dryer, a

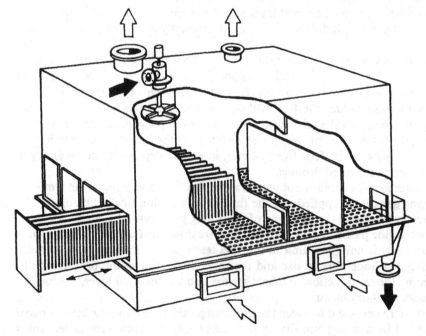

Figure 9 Totally mixed fluid bed dryer for suspension-PVC (polyvinylchloride), with heating panels.

plug flow fluid bed with a controlled particle residence time is applied. This type of fluid is shown in Figure 10.

The feed, which must be directly fluidizable, is introduced into the center of the fluid bed, and the fluidized particles are forced to follow a long narrow path (along a spiral-shaped baffle) to the periphery of the fluid bed where the dry product is discharged over a weir. Good control of the particle residence time is obtained, and the product on discharge is in near equilibrium with the hot drying gas; very low product residual moisture contents can be achieved without overheating the material.

Rectangular fluid bed dryers can also be applied. Here, the plug flow of the fluidized particles is achieved with baffles arranged transversely. However, it is difficult to avoid dead corners where the product will remain for a longer time and thus the plug flow would be less ideal.

Figure 11 shows an industrial installation for the drying of water-wet polymer. After predrying in a flash dryer, the material is dried in a plug flow bed dryer.

5.4. Two-Stage Fluid Bed Dryer

A more compact drying plant can be designed by using a totally mixed fluid bed as a predryer instead of the flash dryer, as shown for the two-stage fluid bed in Figure 12. By arranging the totally mixed fluid bed on top of the plug flow fluid bed, the product flows countercurrent to the drying air; thus, space requirements, installation costs, and heat consumption are reduced.

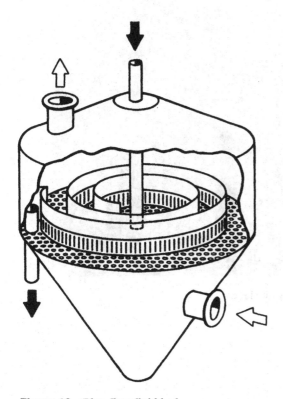

Figure 10   Plug flow fluid bed.

Figure 11   Niro Atomizer flash fluid bed drying plant.

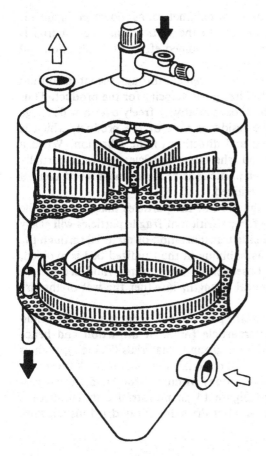

Figure 12   Two-stage fluid bed.

6.   VIBRATED FLUID BED DRYING SYSTEMS

6.1.   General Description and Advantages

Some granular products are difficult to dry in a stationary fluid bed because of one or more of the following physical properties (33):

Wide particle size distribution
Low strength of wet or dry particles
Stickiness or thermoplasticity of particles
Pasty properties of wet feed

The wide particle size distribution of some products makes the choice of fluidizing air velocity in stationary fluid beds very difficult. A high fluidizing air velocity, which is necessary for the fluidization of the larger particles, causes a large portion of the smaller particles to be entrained from the bed without being dried sufficiently; on the other hand, a low fluidizing air velocity cannot provide the fluidization, and thus the transportation of the larger particles through the dryer is hindered and defluidization may occur.

It is possible to dry products having the above-mentioned properties by using a shallow vibrated fluid bed, which normally is a long rectangular trough vibrated at a frequency of 5–25 Hz with a half-amplitude of a few millimeters, as shown in Figure 13. The vibration vector is applied at an angle (0–45°) to the vertical and the material is easily transported through the dryer because of the combined effect of fluidization and vibration.

The air velocities required to obtain good solids movement through the vibrated fluid bed may be as low as 20% of the minimum fluidizing gas velocity for the product. Thus the air rate for fluidization and drying can be chosen relatively freely over a wide range without affecting the conveying rate and the residence time of the product, which is controlled mainly by adjusting the amplitude and direction of the vibration. When a relatively low drying air rate has been selected, the larger particles are only partially fluidized; however, the entrainment of smaller particles is reduced and a good control over the residence time of smaller particles is obtained.

In a stationary fluid bed dryer, violent agitation of the particles takes place in the fluidized layer because of gas bubbling. Therefore, attrition of fragile particles will easily occur. The vibrated fluid bed provides very gentle transportation of material through the dryer. Since a low gas rate can be chosen, gas bubbling in the fluidized bed is avoided, and this greatly reduces particle attrition. Consequently, the vibrated fluid bed can be used for drying products having low mechanical strength, in either the wet or the dry state.

Considerable improvement in the fluidity of the particles will occur in a vibrated fluid bed because the vibrations break up the interparticle forces of attraction and hence improve the quality of fluidization. This is true for group C materials (Table 1), such as fluid beds of fine materials, as well as beds of moist materials that tend to stick together.

Vibrated fluid bed dryers are used extensively today within the food and dairy industry (milk, whey, cocoa, and coffee), and in Figure 14 an industrial installation of a vibrated fluid bed is shown, which is used as an after-dryer for spray-dried instant milk

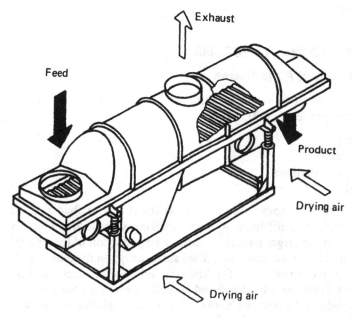

Figure 13   Vibrated fluid bed.

Table 1 Distinguishing Features of Geldart's Groups of Particles (after References 12 and 13)

Feature	Group C	Group A	Group B	Group D
Distinguishing word or phrase	Cohesive	Aeratable	Bubble readily	Spoutable
Example	Flour	Fluid cracking catalyst	Sand	Wheat
Particle size	$\leq 20\ \mu m$	$20 < d_p < 90\ \mu m$	$90 < d_p < 650\ \mu m$	$>650\ \mu m$
Channeling	Severe	Little	Negligible	Negligible
Spouting	None	None	Shallow beds only	Readily
Solids mixing	Very low	High	Medium	Low

Source: From Reference 14.

Figure 14 Niro Atomizer Vibro-Fluidizer drying plant.

products. These types of products are particularly fragile. Vibrated fluid bed dryers are also used for inorganic salts, granulated fertilizers, mining products or when a broad particle size distribution occurs (34).

A new type of multistage vibrated fluid bed has recently been developed (35) based upon gyratory motions, as generally employed for gyratory sifters. The vibrations employed are more powerful than normally used by vibrated fluid bed dryers, and accordingly very difficult fluidizable products can be successfully dried, such as smoked fish meal. This dryer easily permits the drying chambers to be built on top of one another; accordingly, a highly heat efficient and compact multistage dryer can be designed. Disassembling and washing are performed as for a sifter.

The applications of a vibrated fluid bed for the drying of tea leaves (36), sodium formate crystals (37), and granulated styrenic polymers (38) have recently been published.

Gupta and Mujumdar (9), in a review of the literature, mention that the vibrated fluid beds are a subject of intense study in the former Soviet Union; only 10% of the published articles originate in the English-speaking world in spite of the versatility and the extended use of this drying method.

6.2. Characteristics of the Vibrated Fluid Bed Dryer

Based on visual observations (9,39), three regimes of operation of a vibrated system can be defined depending on the magnitude of the vertical component of vibrational acceleration.

1. Vibrated state: When $\omega^2 a/g < 1$, the bed behaves like an ordinary fluid bed and vibrations only help improve stability and homogeneity of the fluidized layer.
2. Vibro fluidized layer: When $\omega^2 a/g \simeq 1$, both gas flow and vibrations contribute to fluidization and the bed behavior is influenced by relative magnitudes of both.
3. Vibrated fluidized layer: When $\omega^2 a/g > 1$, the bed is essentially acted on by vibrational forces only, and the air is used only as a medium for heat and mass transfer.

It should be noted that the effect of vibration is decreased with increasing bed height, and bed heights above 0.5 m are rarely used. Following Gupta and Mujumdar (9), the homogeneity of the fluid bed deteriorates when the vibratory acceleration $\omega^2 a$ exceeds 4 g.

The transition of the particles from a fixed to a suspended state is rather smooth when vibration is applied to the bed; accordingly it is impossible to define a minimum gas fluidization velocity. Gupta and Mujumdar (9,40) recommend using a new characteristic velocity, which they termed minimum solids mixing velocity U_{mm}; essentially this represents a recommended lower operating gas velocity for a vibrated fluid bed in which the solids circulation is visually strong enough.

In Figure 15 the dependency of the minimum solids mixing velocity for a vibrated fluid bed is shown with regard to variation of vibration acceleration. The curve in Figure 15 is based on data from visual inspection only, and Mujumdar points out that the curve is not generally valid; however, Figure 15 gives an indication how the gas velocity can be decreased when vibration is applied.

Pakowski and Mujumdar (41) have measured the heat transfer from a horizontal cylinder to a vibrated fluid bed, as shown in Figure 16, which illustrates how the vibration influences the heat transfer coefficient, that is, the solids mixing, in a fluidized layer of particles.

The heat transfer coefficient increases as long as the solids circulation is increased by

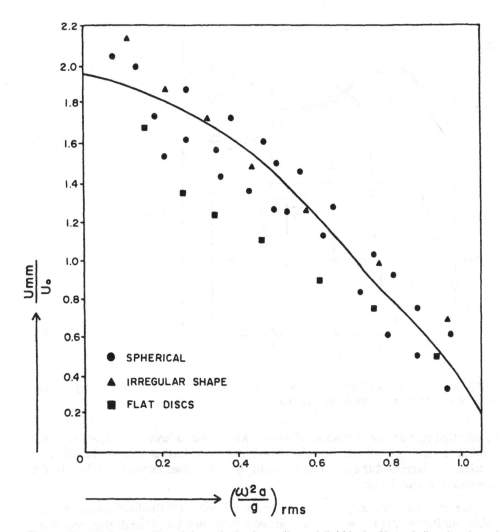

Figure 15 Minimum solids mixing velocity for a vibrated fluid bed, effect of vibrational acceleration. (From Reference 40.)

either vibration or gas flow until, when the vibrational acceleration is too high, the bed leviates. It can further be seen from Figure 16 that the effect of vibration becomes less significant as the gas flow rate is increased above the minimum fluidization gas velocity. Further measurements of heat transfer coefficients for immersed surface to particle heat transfer in vibrated fluid beds have recently been published (42).

Gupta et al. (43) studied the effect of vibration on the drying of silica gel and molecular sieve particles in a vibrated fluid bed. They found that most mass and heat transfer took place within the first few millimeters of the bed above the distributor, consistent with the behavior in stationary fluid bed dryers. Their conclusions were as follows:

The vibration did not change measurably the critical moisture content of the particles, and the mechanism of drying in the falling rate period was not significantly affected either.

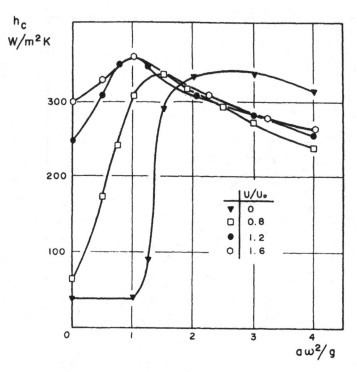

Figure 16 Heat transfer coefficient from a horizontal cylinder to a vibrated fluidized layer of dry ballontini, $d = 0.454$ mm. (From Reference 41.)

Under constant rate drying conditions vibration was found to lower the drying rate to a minimum when the vibration acceleration $\omega^2 a/g = 1$; further increase in vibration acceleration increased the drying rate again, in some cases above the value for the stationary fluidized layer.

For further discussion of the drying mechanisms in vibrated fluidized beds, see also Strumillo and Pakowski (44), who have reported a large increase in the drying rates when vibration is applied to the fluidized layer in some drying applications.

6.3. Drying of Agglomerated and Granulated Products in a Vibrated Fluid Bed

The drying of agglomerated or granulated products presents special problems, as they quite often are sticky in the wet feed state and they must be dried gently in order to assure minimum breakup of agglomerates and development of fines. A vibrated fluid bed is often the best choice for drying such granular materials.

The flow sheet in Figure 17 shows a typical agglomeration plant where a vibrated fluid bed is used as a dryer for wet agglomerated material leaving the agglomerator. The fluid bed area in the Vibro-Fluidizer is divided into sections with separate plenum chambers and with individual control of inlet air temperature and air rate. This is because a wet product often allows a higher air rate for fluidization than the end product, and it is often necessary to establish precise control of product temperature during all stages of the drying when a sticky or thermoplastic product is to be dried. After drying, the agglomerated material is sieved and a product of controlled particle size is taken out.

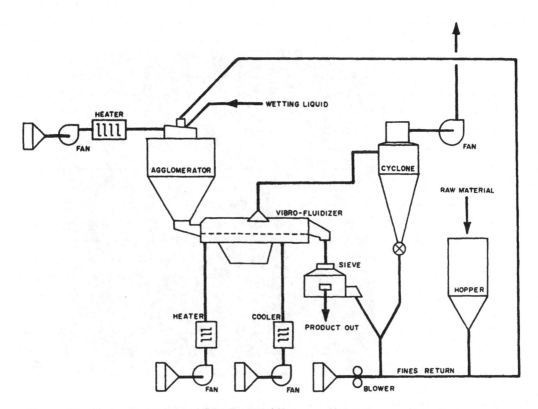

Figure 17 Agglomeration plant with Vibro-Fluidizer.

Fines from the sieve and cyclone are returned to the agglomerator by pneumatic conveying, together with the raw material. Typical agglomerated products are shown in Figure 18. A great number of food products can be made nondusty and easily dispersible in water by the above-described process, the so-called instant products. In order to dry products having a pasty consistency in the wet stage, it is necessary to granulate the feed before it is led to the Vibro-Fluidizer. In certain cases, it is not possible to directly granulate the feed into particles suitable for the Vibro-Fluidizer; thus, a certain part of the already dried material may be backmixed, as shown in Figure 19.

7. FLUID BED GRANULATION

7.1. Introduction

Fluid bed granulation can be defined as a particle-forming process by which a liquid feed containing solid is converted into a granular solid state by spraying the liquid into a fluidized layer of already formed granules. The liquid feed can be a solution, suspension, or melt, and the fluid bed granulation process involves drying, cooling, reaction, mixing, agglomeration, granulation, and coating.

The most characteristic and essential part of the process is the formation of new particles and their growth in the fluid bed. These are the main parameters in determining the size distribution and bulk density of the product. New particles are produced in two ways within the fluidized bed:

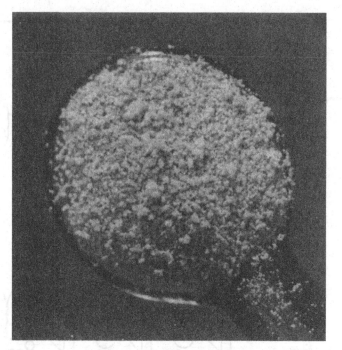

(B)

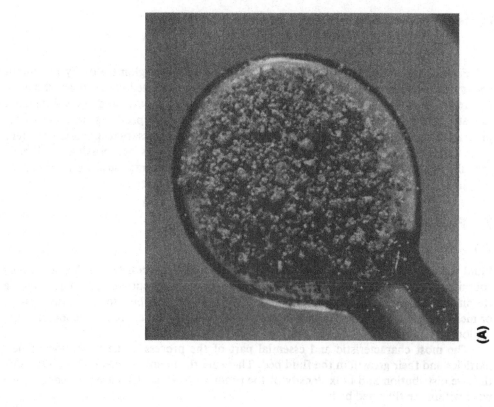

(A)

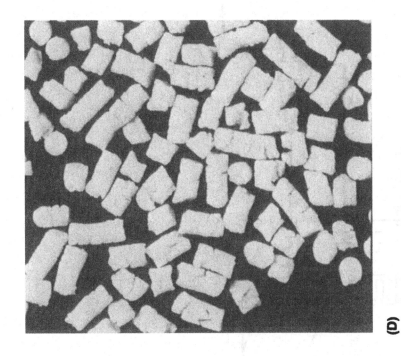

(D)

(C)

Figure 18 Some of the products suited for drying in vibrated fluid beds: (A) cocoa-sugar instant drink; (B) agglomerated food product (e.g., baby food); (C) agglomerated chicory extract solids; (D) granulated yeast.

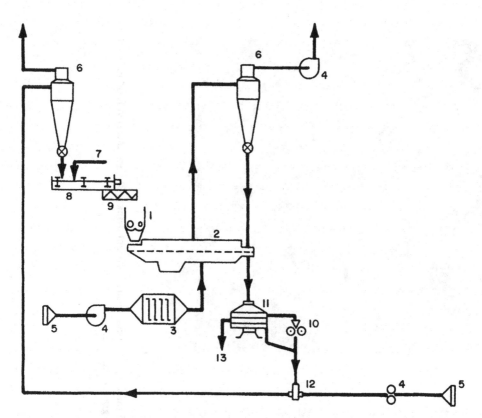

Figure 19 Plant for granulation, drying, and classification of products having pasty consistency: (1) granulator; (2) Vibro-Fluidizer; (3) heater; (4) fan; (5) air filter; (6) cyclone; (7) product inlet; (8) mixing screw; (9) feeding device; (10) mill; (11) sieve; (12) rotary valve; (13) product outlet.

When the droplets from the spray dry sufficiently prior to contact with particles in the fluidized layer
From attrition between the fluidized granules

After wetting a particle with liquid in the fluidized bed, the particles will grow in two different ways.

The wetted particle will collide with another particle creating an agglomerate that, after drying, is kept together by a bridge of solid material.
The wetted particle will dry before collision, thus creating particle growth by layering of granulation.

It is necessary to have strict control of these processes in order to produce product with constant properties.

The fluid bed granulation is conducted either continuously or in batches. A large number of industrial applications have been developed during the past decade, and the process is gaining wide acceptance. In particular, the batch operation has found widespread use in the pharmaceutical industry (32,45,46), where there is a need for nondusty, well-defined, and strong granules for the tablet presses.

The fluid bed granulation combines in one process step the drying or cooling and formation of a nondusty granular product with average particle size typically between 0.5 and 3 mm. Continuous fluid bed granulation has been developed for drying of a variety materials (7,47) and for spray cooling of urea in very large scale (48), and Romankov (2) has described a number of Russian drying processes based upon fluid bed granulation.

7.2. Batch Fluid Bed Granulation

Figure 20 shows the principal features of the batch fluid bed granulator equipment. After the fluid bed is loaded with the powder to be granulated, the powder is fluidized, the liquid binder is sprayed into the fluidized layer, and the agglomeration and granulation start. Particles are held together in the wet state by liquid bridges and in the dry state by solid bridges, which are caused by hardening binders and crystallization of dissolved substances. All powder entrained in the exhaust gas will be collected by the bag filter just above the fluid bed and recycled into the fluidized layer. After the granulation is complete the temperature of the fluid bed can be raised and the granules finally dried and possibly coated. The whole process can be fully automatic, and granules of very consistent quality can be produced.

Granulation processes in a batch fluid granulator have been studied extensively by several authors (46,49–54) in order to understand the influence of the process variables upon the characteristics of the final granules. The agglomeration-granulation process is caused by complex interaction of several variables following Schafer and Woertz (49), as shown in Table 2. In order to reliably produce granules with consistent characteristics, all the parameters listed in Table 2 must be controlled properly.

Granule growth during the batch fluid bed granulation process can be divided into three states: nucleation, transition, and ball growth regions (32,49), as shown in Figure 21. At the start, nuclei of two or more primary particles are formed and held together by liquid bridges; the size of these nuclei will depend on the size of the atomized drops, and they will continue to grow by agglomeration with fines, as it is relatively difficult to bind nuclei to each other. A decrease in content of fines therefore gives a lower growth rate; accordingly, a break in the growth curve of granule size versus quantity of binder solution is observed (see Fig. 21).

The quantity of the binder solution affects the granule size and size distribution. The highest growth rates are obtained at the largest drop size and liquid flow rate, at the

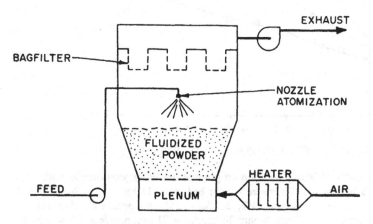

Figure 20   Principle of batch fluid bed granulation.

Table 2 Variables in Fluid Bed Granulation

Product variables	Process variables
Starting materials	Apparatus
Particle size	Bed load
Surface area	Atomization of binder
Water/absorption properties	Nozzle height
Binder	Spray angle
Binder concentration	Liquid flow rate
Quantity of binder solution	Fluidizing air velocity
Binder solution viscosity	Inlet and outlet temperature
Moisture content	Processing time

lowest inlet air temperature, and with relatively nonabsorbing particles (e.g., lactose). Growth rate is further affected by the type of binder; when using a weak binder the agglomerates formed are broken down because of the attrition taking place in the fluidized layer. When the primary particles are mostly agglomerated, the process goes into the transition region with a low growth rate and the size distribution of the granules becomes narrower. When most of the interstices in the agglomerates of primary particles are filled by further addition of liquid, a considerable and uncontrolled growth will occur through

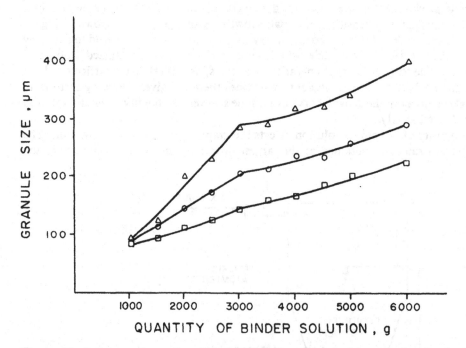

Figure 21 Influence of quantity of binder solution on granule size at varying droplet sizes (d_{50}) (49): starting materials, 80% lactose + 20% maize starch; bed load, 15 kg; ΔT_{gran}, 20°C; liquid rate flow, 150 g/min; binder solution, gelatin, 4%; Δ, d_{50} = 142 μm, nozzle air flow rate, 6 Nm^3/h; $\bigcirc$, d_{50} = 106 μm, nozzle air flow rate, 8 Nm^3/h; $\square$, d_{50} = 71 μm, nozzle air flow rate, 12 Nm^3/h.

coalescence of granules and the process is said to enter into the ball growth region. The low growth rate in the transition period produces granules with good reproducibility; the addition of liquid should be stopped during the transition period.

Granule size is directly proportional to droplet size for a given binder solution, and varying the droplet size might therefore be the most suitable way of controlling the granule size.

After granulation the granules can be dried in the fluid bed at elevated inlet gas temperatures in order to reduce the drying time. Following drying, the granules can be conveniently spray coated in the same equipment, as experience has shown that the fluidized bed is ideal for spray coating and gives constant and reproducible coatings of the granules, which in turn will have desired values, such as the slow release of active ingredients and protection against moisture.

The Wurster process (Fig. 22), a spray coating process in a fluid bed where the granules are circulated up through the center while being coated, has been specially developed for the coating of granules (50,55).

7.3. Continuous Fluid Bed Granulation

The continuous fluid bed granulation and drying process has been described in detail (47,56), and a typical flow diagram of the so-called spray fluidizer process is shown in Figure 23. The gas to the fluidized bed unit is first led into the plenum chamber, which predistributes the gases before meeting the fluidizing gas distributor, which normally consists of a 1–2-mm perforated metal plate, when the inlet gas temperature is below

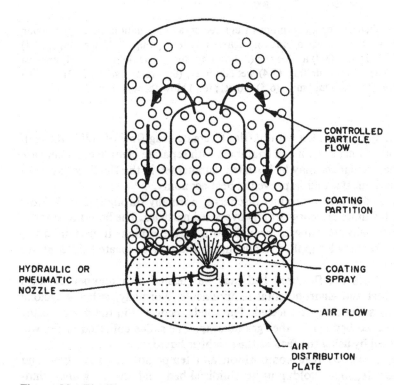

Figure 22 The Wurster process.

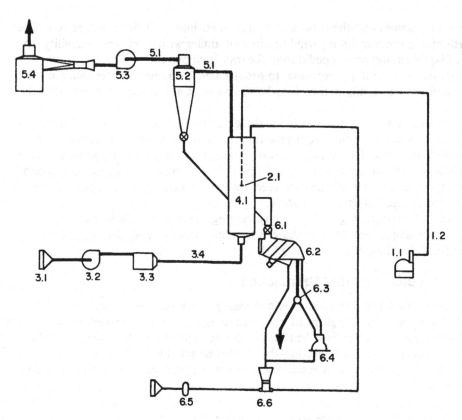

Figure 23 Typical flow diagram for a continuous fluid bed spray granulation process involving solidification of solids in solutions, suspensions, or melts: (1.1) feed pump; (1.2) feed pipe; (2.1) nozzle; (3.1) air filter; (3.2) air fan; (3.3) air heater; (3.4) hot air duct; (4.1) fluid bed; (5.1) exhaust air duct; (5.2) cyclone; (5.3) exhaust air fan; (5.4) wet scrubber; (6.1) rotary valve; (6.2) product screen; (6.3) diverter valve; (6.4) hammer mill; (6.5) blower; (6.6) blow-through valve.

600°C. The chamber may be provided with baffles to create plug flow of solids through the bed. In special applications, several bed chambers with separate plenum chambers are built in one unit. A heat exchanger may be installed, submerged in the fluidized layer, to reduce the overall equipment size and improve the heat economy.

The chamber is provided with a product discharge, normally through an underflow weir or downcomer to facilitate discharge of any oversized particles. The liquid feed spray pressure nozzles are normally introduced through ports in the ceiling. In certain cases, however, two-fluid nozzles may be applied to advantage and may be located either above or in the fluidized layer.

The exhaust gases from the fluidized bed may be cleaned in a number of ways depending on the product and environmental requirements. Normally, either a cyclone plus venturi scrubber or a bag filter is selected. The solids recovered in the dry collector are recycled to the fluidized bed for further granulating. The solids collected in the wet collector can be recovered by using the feed as the scrubber liquid.

Process parameters, such as feed composition and temperature, nozzle type and position, fluidizing gas rate, solids holdup in the fluidized bed, and screening and grinding characteristics, are adjusted initially and are normally not varied or controlled during operation. The spray fluidizer process is operated at constant feed composition in order to ensure a constant basis for control. In practice, this is achieved by maintaining a constant feed pressure at the nozzles. The temperature in the fluidized layer is kept constant by controlling the temperature of the inlet fluidizing gas. The holdup in the fluidized bed is kept constant through measurement of the bed layer pressure drop and speed control of the discharge rotary valve for the fluidized bed.

The particle size distribution in the fluidized layer may vary within certain limits given by the fluidizing gas velocity and the desired product size distribution. The process parameters are set to produce too large particles without seeding, and the desired product size distributions are then obtained by a screening operation and a crushing operation and then recycling the fines. The total solids' recycle rate in relation to the product rate, that is, the seeding ratio, is normally between 0.5 and 10, depending upon the dominating particle growth mechanism in the bed, growth by layering on a single particle or growth by agglomeration of more particles. Controlling the particle size distribution in the fluidized layer is thus selected according to the product size distribution limit and the rate of change of the distribution with varying seeding ratio. It is generally found that the lower the agglomeration tendency, the higher is the sensitivity of the process to variations in seeding ratio. When the agglomeration tendency in the bed calls for a seeding ratio of 5, it is typical that the particle size distribution remains within acceptable limits over several hours, without adjustment.

These procedures are in agreement with the observations of Grimmett (57,58), who has studied the drying and calcination of radioactive waste solution. Grimmett also found that control of the amount of seed introduced into the fluidized layer was necessary in order to control the particle size distribution.

The simultaneous drying and granulation of the radioactive waste is an example of a pure granulation process without agglomeration, and Grimmett (57) was able to describe the particle size distribution relatively simply by a partial differential equation based on a mass rate balance around a small particle size interval in the fluid bed.

The spray fluidizer process is normally capable of producing products with a particle size between 0.5 and 5 mm in diameter, and the moisture content of product can be controlled within narrow limits down to $\pm 0.25\%$.

The particle shape, sphericity, roundness, and density are functions of the product itself, but these product characteristics can also be affected by such process parameters as the degree of atomization of liquid feed, the temperature of the fluidized bed, and the amount of grinding taking place inside or outside the fluidized bed. The inherent product properties promoting irregular particle formation include hygroscopicity, low crystallization rate and thermoplasticity, and high moisture in the fluidized layer, all factors promoting the agglomeration of the particles, forming relatively low density agglomerates.

The rate of solution of the product is mainly affected by the specific surface area of the particles; thus, small and irregular particles normally have higher solution rates compared with large and regular particles, which would be formed by the granulation or layering process in the fluidized layer.

In Table 3, a number of examples of applications of the spray fluidizer process are shown, illustrating the great versatility of the process. Examples of the industrial application of the spray fluidizer are described below.

Granulation of Zinc Sulfate Monohydrate

The example of zinc sulfate monohydrate is typical for the drying and granulation of inorganic salts in general. The purpose of the plant is to convert 2600 kg/h of liquid feed containing 23% dissolved zinc sulfate into a nondusty granular product of zinc sulfate monohydrate with a particle size between 1.5 and 3.0 mm. Hot air is supplied from a

Table 3 Spray Fluidizer Applications Involving Solidification of Solids from Liquid Feed

Solidification	Solids category	Examples
Drying	Inorganic salts	Chlorides of Na, K, Ca, Mg
		Bromides of Na, K
		Sulfates of Na, NH_4, Zn, Fe, Mg, Mn
		Carbonates of Ba
		Silicates of Na
		Phosphates of Na, K
	Organic salt	Potassium sorbate
	Minerals	Clay, bauxite, kaolin
	Dyestuffs	
	Herbicides	
	Tanning products	$Na_2Cr_2O_7$, organic tanning products
	Pharmaceuticals	
	Pulping waste	Sulfite, sulfite waste liquor
Cooling	Inorganic salts	NH_4NO_3, some salt hydrates
	Organic	Urea

direct-fired air heater utilizing either ambient air or preheated air from a flue gas heat recovery system. The inlet gas temperature to the fluid bed is 550°C, and the fluidizing gas velocity is 2.3 m/s. The outlet gas, at approximately 120°C, is cleaned in a cyclone and subsequent venturi scrubber. The liquid feed may be used as scrubber liquid. Due to the specific building requirements (see Fig. 24), the discharge from the bed is lifted by a bucket elevator to a vibratory screen. The product size fraction is passed to a fluid bed cooler and then to bagging. Oversized particles, together with some product fraction, are ground in a hammer mill and then recycled by pneumatic conveying to the fluidized bed. The fuel consumption with 10°C ambient air is 1900 kW, and the total electric power consumption is 120 kW. The fuel efficiency, energy utilization over supply is 77%. By using the feed as a scrubber liquid, the efficiency is increased to nearly 90%.

Granulation of Calcium Chloride

An example of granulation of a highly hygroscopic material is a plant for production of food-grade granular calcium chloride. The plant is capable of producing 660 kg/h of a range of calcium chloride products with total moisture content ranging from 4 to 22.5%, that is, corresponding practically to anhydrous to dihydrate composition. The particle size is 0.5–1.55 mm, but a certain flexibility in the coarser and finer directions is acceptable. The process principles are identical to those mentioned above, but a number of precautions, such as extra insulation, heat tracing, and hot air purge, have been taken to prevent formation of deposits in the system. The process is also suitable for the large-scale production of road-salt-quality calcium chloride with, say 10–12% moisture.

Granulation of Organic Dyestuffs

Granulation of organic dyestuffs is typical not only for drying and granulation of dyestuffs but also for many other organic materials. The trend in drying of dyestuffs has in recent years been dominated by the requirement of nondusty products. A satisfactory product can often be achieved by spray drying through the utilization of pressure nozzle atomization. The average particle size is typically 250 μm, and the redispersibility is

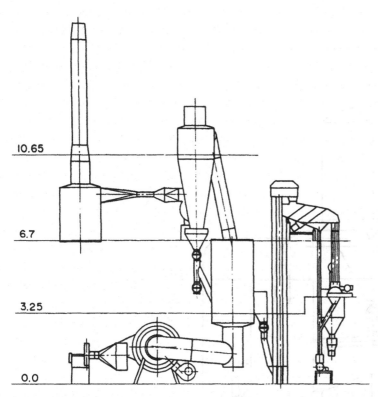

10.65

6.7

3.25

0.0

Figure 24 Arrangement of a Niro Atomizer spray fluidizer for production of about 650 kg/h granular zinc sulfate monohydrate with a particle size between 1.5 and 3 mm from 2600 kg/h solution.

excellent. In special cases in which a product is required having an even coarser average particle size, the spray fluidizer is suitable. The specific example relates to a plant for the drying and granulation of a number of dyestuff formulations. An example of an arrangement drawing for this application is shown in Figure 25. The feed rate is approximately 860 kg/h and contains 31% solids. The solids are heat sensitive and, because of fire risk, the inlet air temperature is limited to 200°C. For further protection of the plant, infrared detectors are used in connection with an automatic steam supply and emergency shutdown system. The fluidized bed temperature is approximately 75°C in order to reach 7% residual moisture in the product. The fluidizing gas velocity is 1.4 m/s. The exhaust gas is cleaned in a bag filter with cassette-mounted bags for easy product change. The solids are discharged from the fluid bed directly into a gyratory screen, and the remainder of the recycle system is identical to that described in the first examples. The fuel consumption with 10°C ambient air is 650 kW, fuel efficiency is 60%, and the total electrical power consumption is 60 kW. The overall plant dimensions are height 17 m, length 10 m, and width 8 m.

Granulation of Urea

Urea fertilizer has traditionally been granulated by prilling of a melt yielding a product with a particle size distribution of 1–2.5 mm, typically. The demand for stronger and larger size granules, in the range of 2–5 mm, typically, led to the development of pan and

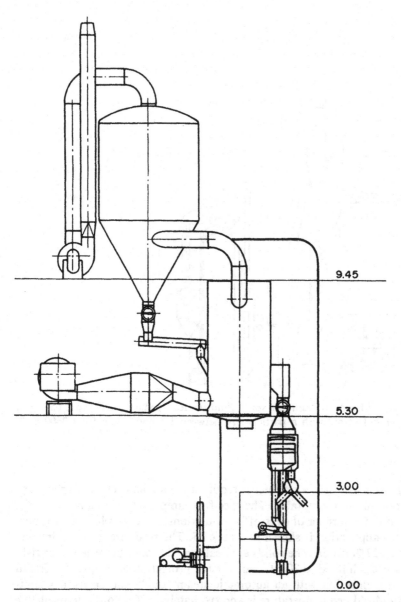

Figure 25 Arrangement of a Niro Atomizer spray fluidizer for granulation of organic dyestuffs.

drum granulation systems. Compared with these, fluid bed spray granulation offers a number of advantages, the most important being that a single unit can be built for capacities of 1500 tons/day (48). Another advantage, a consequence of the simultaneous particle formation and cooling in the fluid bed, is the relatively low rate of solids recirculation corresponding to approximately 0.5 times the production rate. In pan or drum granulation, the recirculation solids serve not only to control the particle size distribution but also to maintain temperature in the granulation zone of 100–110°C. Therefore, the recirculation rate corresponds to approximately two to three times the production rate, with the consequential increase in cost of coolers, screening, and conveying equipment.

The temperature in the granulation zone is 100–110°C, and the fluidizing gas velocity approximately 2 m/s. The biuret formation during granulation is less than 0.1%. In order to obtain granules with a high sphericity and minimize solids recirculation, it is advantageous to design the fluidized bed for solids plug flow rather than back-mix flow, as described in the previous examples. The plug flow secures the irregular seed particles that arise when the crushing is covered by a sufficiently thick layer of urea to become nearly spherical before discharge from the granulation bed.

8. MISCELLANEOUS FLUID BED DRYERS

8.1. Spouted Bed Dryer

The spouted bed is a variant of the fluid bed; instead of distributing the upward-moving gas evenly across the fluid by means of a gas distributor, the gas enters as a jet at the center of the conical base of the chamber, as shown in Figure 26. The solid particles are dispersed into the jet of gas, the spout, and the entrance of the drying gas, and a rapid exchange of heat and mass takes place similar to a flash drying operation (59). In this way a systematic cyclic movement of the solids is established: the particles travel upward in the axial jet zone and downward in the annular dense moving bed.

The spouted bed has special advantages compared with a normal fluid bed when the

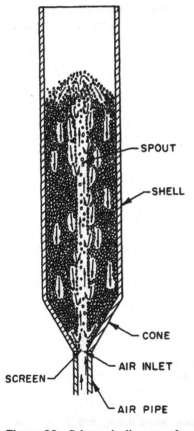

Figure 26 Schematic diagram of a spouted bed. (From Reference 62.)

particles to be dried are large, say, larger than 5 mm, and the fluidization would be sluggish. Very good solids mixing and effective gas-particle contact can be achieved in a spouted bed, and the process was developed at the National Research Council of Canada in 1954 for drying wheat. Several reviews dealing with the fluid mechanics of the spouted bed and its applications have been published (6,14,59,60) and a book by Mathur and Epstein (61) describes spouted bed technology in detail.

The spouted bed dryer is particularly suitable for drying heat-sensitive materials, such as agricultural products like wheat and peas. The excellent and controlled mixing of granular material in the spouted bed makes it possible to use a higher temperature drying gas than when applying a fluid bed dryer.

The pressure drop over an operating spouted bed is lower by about one-third than for a similar fluid bed since the chamber wall supports part of the weight of the particles; however, the pressure drop required to initiate spouting is considerably higher (60), which must be taken into account when designing the gas supply fan.

Similar to the minimum fluidization gas velocity, there is a minimum gas velocity below which a spouted bed cannot be sustained; however, the value of this gas velocity is dependent on both solid and fluid properties and bed geometry (see Refs. 14 and 60).

Further, it should be noted that there is a maximum spoutable bed depth beyond which the spouted bed becomes a fluid bed. The value of this depth is dependent on solid and fluid properties and bed geometry (see Refs. 14 and 60).

The air and solids temperatures during continuous drying of wheat in a spouted bed were recorded by Mathur and Gishler (59,62) and these profiles, shown in Figure 27, illustrate the mechanism of the drying process under spouting conditions. It can be seen

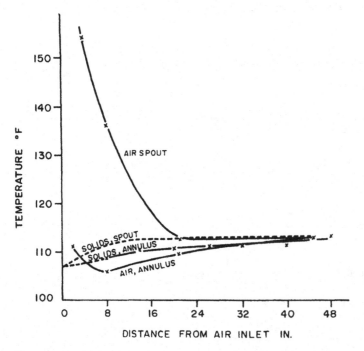

Figure 27   Air and solids temperature profiles during continuous wheat drying: bed, 12-in diameter × 48-in deep; air inlet temperature, 240°F. (From Reference 62.)

from Figure 27 that the rise in particle temperature, following the efficient heat and mass transfer in the lower part of the spout, was only a few degrees because of a large solid recirculation rate. Most of the moisture diffusion controlled drying was taking place in the annulus, where the solids moved downward, while heat was released for the drying and the temperature decreased.

Accordingly, overheating of the particles can be avoided in a spouted bed dryer and it can be used with advantage when heat-sensitive material must be dried, allowing for higher gas inlet temperatures than used for a fluid bed. However, the maximum practical equipment size seems to be limited. Application has remained confined to small-capacity dryers up to 2 tons/h, and the potential for large-scale operation using multispout beds had not been exploited (60).

The drying of a variety of granular heat-sensitive materials in a spouted bed has been studied by Romankov and Rashkovskaya (63) who in a book reviewed the extensive work carried out in the former Soviet Union.

The spouted bed dryer can also be used as a dryer for feeds in the liquid state by spraying the feed into a spouted bed, circulating either inert material or granules of the product (2) (see also Table 4 and Ref. 7).

The feed solution coats the inerts. Evaporation and drying take place on the surface of the inerts, where after abrasion the dried solids from the circulating inerts will be liberated and carried from the dryer by the exhaust air. For a detailed discussion of drying in inert beds, see the recently published review of Strumillo et al. (7) in which a table shows a number of dryers with inert materials.

Recently a method has been described (64) for drying temperature-sensitive and hygroscopic food and dairy products, in particular cattle blood products. The drying process in this spouted bed has been compared with spray drying.

The spouted bed can also be used as a granulator or coater in a similar way as the fluid bed granulator described in Section 7. The cyclic movement of the solids can be an advantage. The feed is atomized into the hot gases in the lower part of the spout containing seed granules, and the feed is rapidly dried upon the surface of the seeds. Accordingly, granules build up by layering as they cycle in the bed, and the final product is well rounded, dense, and uniform in structure. Granulated fertilizer pigments, dyes, and sulfur have been produced on an industrial scale this way.

8.2. Mechanical Agitated Fluid Bed Dryer

Increasing energy costs have increased the need for drying high-solid wet cakes in a simple one-step drying operation without external recycling and without adding water to the feed to make it suitable for spray drying. If such a paste or cake, often consisting of fine particles, does not disperse readily when introduced into a back-mix fluid bed dryer, it is

Table 4 Back-Mixed Fluid Bed Dryer With or Without Heating Panels*

Heating panels	T_{in}	T_{out}	G, kg gas/kg DM	Q, kcal/kg DM
−	100	70	24.80	594
+	100	70	19.40	467
−	130	70	12.50	382
+	130	70	6.60	287

*Feed: S-PVC, 0.2 kg H_2O/kg DM (dry mass).

possible to fluidize such powders by introducing a mechanical agitator in the feed zone of the fluid bed dryer (see Fig. 28, which shows the swirl-fluidizer process recently developed by Niro Atomizer). Gupta and Mujumdar (6) have recently reviewed the literature, and Richardt (65) and Kragh (66) have described the so-called spin flash dryer, operating after this principle in some detail.

The agitator in the bottom of the dryer rotates with approximately 100 rpm, and the agitation will disintegrate the pasty feed, disperse the lumps into the already dry material, and generally help the fluidization of the fine particles. The drying gas is introduced through air slots in the side of the lower part of the fluidized particle layer. The average residence time of the particles can be controlled by the gas velocity in the chamber up to a value of 20 min, ensuring proper drying of product.

Pastes of pigments (e.g., titanium dioxide) and dyestuffs are today dried industrially in this type of dryer, which is a combination of fluid bed and flash dryer. For drying of difficult, lump-forming pastes Ormos and Blickle (67) have recently described the application of a mechanically agitated fluidized bed of inert material into which the wet paste is fed and dried. This method relies upon the abrasion in the fluid bed to allow the dried material to disengage from the inert material.

8.3. Centrifugal Fluid Bed Dryer

The centrifugal fluid bed dryer was recently developed (6,68) for rapid constant-rate predrying and puffing of sticky, high-moisture, piece-form foods. The usable gas velocity can be as high as 15 m/s, well beyond the maximum gas velocity applied in ordinary fluid

Fluidized Bed Drying

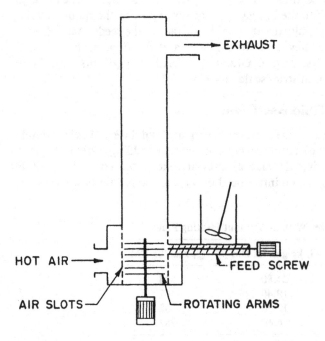

Figure 28 Niro Atomizer swirl-fluidizer, a mechanically agitated fluid bed dryer.

beds. The centrifugal fluid bed dryer is applied for diced, sliced, and shredded vegetables, which cannot otherwise be fluidized and are conventionally dried on band conveyor dryers (69).

The centrifugal fluid bed dryer shown in Figure 29 consists of a cylinder with perforated wall, rotating horizontally (centrifugal force up to 15 times gravity) in a high-velocity heat cross-flow airstream. The solids holdup in the dryer is about 10–20% of the dryer volume, and the solids are made to fluidize during part of each revolution and to form a fixed bed during the rest of the cycle. All surfaces of one piece are exposed to the hot drying air, and up to fivefold improvement in drying rates for vegetables have been observed. These high heat transfer rates help the puffing of the vegetable material, and it has recently been demonstrated (70) that high-quality, quick-cooking rice can be produced in a centrifugal fluid bed dryer.

8.4. Fluidized Spray Dryer

A large number of the food products are hygroscopic and thermoplastic, making spray drying more difficult inasmuch as deposits in the drying chamber must be avoided. Further, for such applications the inlet gas temperature for the spray dryer is limited. Recently a spray dryer with a built-in fluid bed has been developed (71,72), improving the thermal efficiency and capability of the spray dryer to handle products that are difficult to dry. The flow sheet for a fluidized spray dryer with nozzle atomizer is shown in Figure 30.

The primary hot drying gas is led into the chamber around the nozzle where the feed is atomized. Secondary drying or cooling air, amounting to approximately 25% of the total air supplied, is led to the fluid bed in the bottom of the chamber. Some spray particles will collect in the fluid bed, whereas finer particles will be entrained in exhaust gas and collected in the cyclones, from which they can be recirculated to the drying chamber for further agglomeration with the semidry particles from the spray. Agglomeration also occurs in the fluidized layer because of the impact of the partially dried particles, and a relatively high interaction between the spray and the fluidized layer will take place. This interaction is controlled by process conditions, such as the temperatures of the fluidized layer and the exhaust gas, the degree of atomization, and fluidization gas veloc-

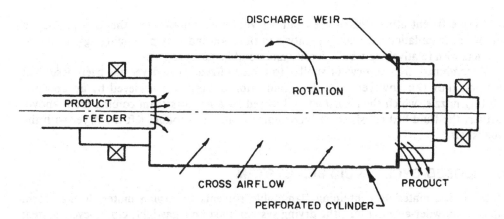

Figure 29 Modified centrifugal fluidized bed dryer (CFBD), cross-sectional view. (Courtesy APV Mitchell Driers Ltd.)

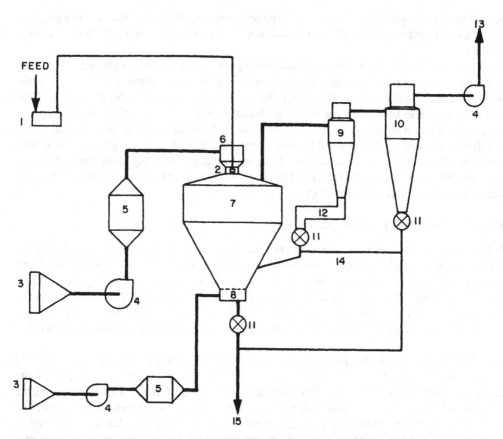

Figure 30 Fluidized spray dryer with nozzle atomizer (FSD-N): (1) feed pump; (2) nozzle atomizer; (3) air filter; (4) fan; (5) air heater; (6) roof air disperser for nozzle atomizer; (7) fluidized spray dry chamber; (8) fluidized bed air distributor; (9) primary scalping device; (10) secondary high-efficiency cyclone; (11) rotary valve; (12) conveyor; (13) exhaust or further gas cleaning; (14) fines recycle system (optional); (15) postdrying cooling (optional).

ity. Very efficient utilization of the chamber volume is obtained by this process, and a number of installations are now operating in the food industry producing agglomerated products with relatively low bulk density and good instant properties.

A combined drier somewhat similar to the fluidized spray dryer has been described by Romankov (2); however, here the atomization of feed was achieved by an upward-pointing nozzle, which then was met and mixed by a hot airstream coming from above. Further, the fluid bed consisted of inert material and all product left the dryer with the exhaust air.

9. CLOSED-CYCLE FLUID BED SYSTEM

When drying materials containing flammable solvents or drying materials that form explosive powder-air mixtures, the drying system must be a gastight, closed-cycle system with recirculation of nitrogen or other inert gas as the drying medium.

Often environmental regulations, if not simple economic considerations, will make the closed-cycle system necessary when the solvent or volatiles have to be recovered; closed-cycle plants may also be applied when products have an unpleasant odor.

The flow sheet of a closed-cycle flash fluid bed dryer (for polypropylene) is shown in Figure 31. The wet feed is led to the flash dryer, predried, and pneumatically conveyed to the cyclone above the fluid bed. After being dried in the two-stage plug flow fluid bed dryer, the product is mixed with fines from the fluid bed cyclone.

The inert gas with diluent from the flash dryer is led to the scrubber through which cooled diluent is recirculated. The diluent vapor is condensed, and any particles that have not been recovered in the cyclones are removed. The cleaned gas, saturated with diluent vapor at the temperature of the condenser, is led back via the heater to the flash dryers. In a similar way the exhaust gases from the fluid bed dryer are recirculated through a separate scrubber condenser, which is kept at a very low temperature, so the dew point of the inlet gases to the fluid bed dryer is low, enabling the dryer to dry the material to very low moisture content (73). The heat economy of such complicated drying systems will be discussed below. An industrial plant drying propylene (12 tons/h) is shown in Figure 32.

For a closed-cycle system drying aqueous-based feeds, a self-inertizing drying system, as shown in Figure 33, is used in order to eliminate powder explosion risks and undesirable oxidation of products. In such plants the formation of inert drying gas is accomplished by a direct gas heater. A controlled amount of combustion air is used, and the combustion gases are used as drying medium after being mixed with recycled gas from the scrubber, maintaining the oxygen content at low levels. Part of the recirculating gas is continuously taken out of the system in order to compensate for the combustion gases entering at the heater.

A self-inertizing drying system has the following advantages compared with the closed-cycle system shown in Figure 31.

No demand for inert gas
Higher drying gas temperatures possible because of direct heating

In comparison with the air-operated drying systems, the self-inertizing concept offers a radical reduction of air pollution from the plant.

A further development of the closed-cycle system is to use the superheated solvent as drying gas; the higher volumetric heat capacity of the solvents (hexane, benzene, and so on) compared with that of nitrogen results in reduced gas flows and thus smaller equipment size.

An industrial plant drying 50 million pounds per year polypropylene using a hexane and isopropyl alcohol azeotrope is operating in the United States (74). Further, Potter et al. (75) are developing a steam-heated, steam-fluidized, closed-cycle system for drying lignite, which offers potentially very significant energy savings. This system will be described in the next section.

10. ENERGY CONSIDERATIONS IN SELECTING FLUID BED DRYING SYSTEMS

10.1. Drying with Air

In general, a fluid bed drying system will be more energy efficient and the amount of necessary drying gas will decrease when the fluidization gas inlet temperature T_{in} is increased and when the degree of saturation of the exhaust air is increased. The maximum allowable gas inlet temperature for a heat-sensitive material must be determined by pilot-

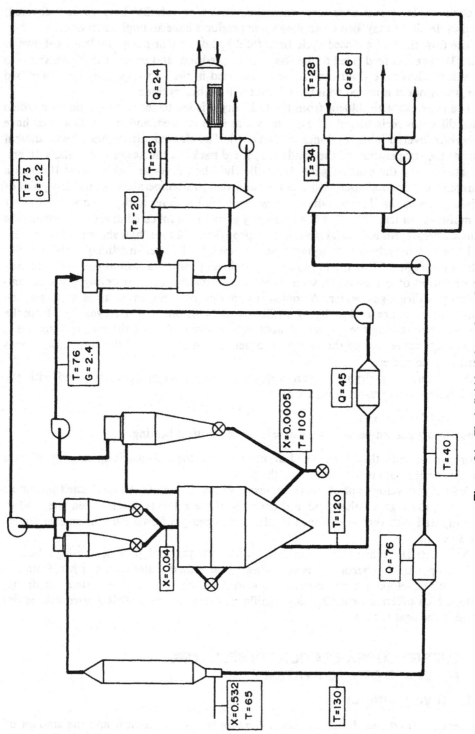

Figure 31 Flash fluid bed dryer for polypropylene.

Figure 32 Closed-cycle dryer for polypropylene, Niro Atomizer.

plant tests and can only be exceeded if a different type of dryer is selected, such as a flash dryer, which if properly designed will allow a higher gas inlet temperature than a fluid bed dryer.

Accordingly, introducing heating panels submerged in the fluidized layer is the only practical way to increase the heat input to the fluid bed dryer and increase the degree of saturation for a given amount of fluidization gas without exposing the heat-sensitive material to higher temperatures (34). In Figure 8 the flow sheet for a totally mixed fluid

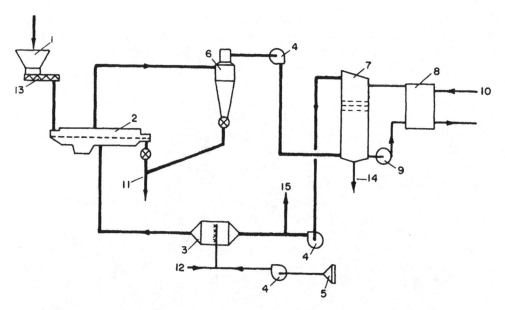

Figure 33 Self-inertizing drying plant: (1) product inlet; (2) vibro-fluidizer; (3) heater; (4) fan; (5) air filter; (6) cyclone; (7) scrubber-condenser; (8) heat exchanger; (9) pump; (10) cooling water inlet; (11) product outlet; (12) fuel gas inlet; (13) feeding device; (14) condensate outlet; (15) bleed-off.

bed dryer with heating panels is shown, and in Table 4 the savings obtainable in heat input is shown when heating panels are applied at two different inlet air temperatures typical for drying S-PVC.

If a back-mix fluid bed, drying S-PVC, is constructed with heating panels as shown in Figure 9, up to 80% of the heat for drying can be supplied by the panels in the fluid bed. It has been shown (73,76) that this type of dryer is the most economical drying system for S-PVC, offering savings up to 50% when compared with that required for a rotary dryer system or for a flash fluid bed dryer system. It should also be mentioned that the introduction of heating panels in a fluid bed can often result in better possibilities for utilization of waste heat from either a process plant or a boiler.

It is important to note that it is not possible to fully saturate the drying gas, as otherwise, uncontrollable condensation will take place in the exhaust system, in particular in the powder collection system, such as the bottom of the cyclones. The wet bulb temperature of the exhaust gas should be kept 8–15°C below the dry bulb temperature depending upon the efficiency of tracing and insulation.

10.2. Closed-Cycle Drying of Polyolefins

In processing polyolefins, polymerization normally takes place in a liquid hydrocarbon. For this reason, the centrifuge cake will contain an organic solvent rather than water, as in the case of PVC (polyvinyl chloride).

The traditional drying operation of such polymer cake would be steam stripping of solvent, drying of water-wet polymer in open cycle, and further distillation to recover the solvent. Changing to closed-cycle drying of the polymer cake represents in itself a major energy savings, also because the enthalpy for evaporation of an organic solvent normally is below 100 kcal/kg solvent.

As polypropylene normally must be dried to very low volatile levels, the final drying of the polypropylene requires that the drying gas has an extremely low dew point. For this reason, it is most economical to design the drying system as two separate loops where the evaporated solvent in the first drying state is condensed by means of cooling water while the evaporated solvent from the second state is condensed either by means of brine or in a reboiler. Figure 31 shows such a flash fluid bed drying system for polypropylene in which the solvent is hexane. In order to minimize the gas flow rate in the fluid bed loop, the fluid bed is designed as a two-stage fluid bed. Further, an economizer is introduced into the gas stream where the exhaust gas from the condenser system is preheated by means of the exhaust gas from the fluid bed dryer.

The utilization of a fluid bed with heating panels in this flow sheet serves three distinct purposes, namely, a reduction in heat requirement as previously indicated, a reduction in refrigeration requirement for the condenser, and a reduction in the gas flow through the drying loops. This results in significant savings in capital costs.

In Figure 34 the same flow sheet is presented, but the fluid bed dryer in the second drying loop is provided with heating panels, resulting in a reduction in gas flow of 38% and a similar reduction in refrigeration requirement, as well as a 20% reduction in heat consumption.

The flash dryer in the first drying loop can be replaced with a back-mixed fluid bed with heating panels, provided that the heating panels are designed in such a way that no polymer can rest on the panels for an extended period of time. As can be seen from Figure 35, this results in a significant reduction in the gas flow through the drying loop from an

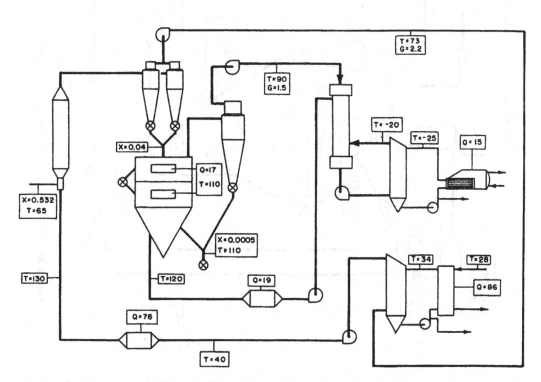

Figure 34 Flow sheet of Figure 31 with internal heating panels in the fluid bed dryer of the second drying loop.

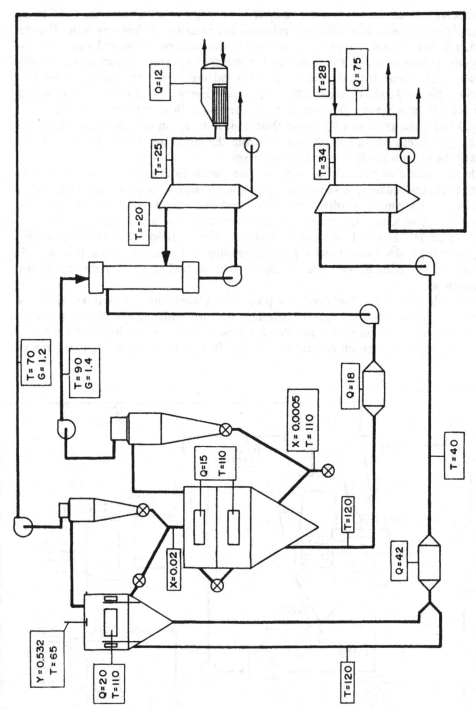

Figure 35 Replacement of the flash dryer in the first drying loop with a back-mixed fluid bed with internal drying panels.

Table 5 Closed Cycle of Polypropylene

System	Total gas G (kg/kg DM)	Total heat Q (kcal/kg DM)	Cooling load (kcal/kg)
I. Flash-two-stage fluid bed	4.6	121	24
II. Flash-two-stage fluid with heating panels	3.7	112	15
III. Back-mixed fluid bed two-stage fluid bed with heating panels	2.6	75	12
Savings from I to III	43%	38%	50%

amount of 2.2 to 1.2 kg nitrogen per kilogram dry polymer. For a smaller drying system a two-stage fluid bed dryer, such as shown in Figure 12, can be used.

Table 5 compares the consumption figures for the flow sheets in Figure 31, 34, and 35.

Mortensen (77) has studied in detail the optimization of energy consumption for drying of polypropylene in a closed-cycle fluid bed system, as shown in Figure 35, and it was found that the selection of the dew point of the inlet drying gas to the fluid bed was very important for the optimization of the system, as shown in Figure 36. As the dew point is lowered the equilibrium moisture for the product is lowered; that is, the drying force for the drying process is increased. Consequently less drying gas is necessary;

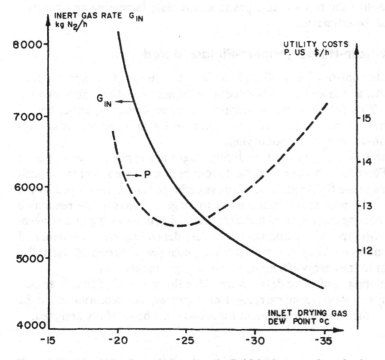

Figure 36 Optimization of closed-cycle fluid bed system by selection of dew point: solids inlet volatiles content, 0.014; solids inlet temperature, 110°C; product specification, 0.0005 kg hexane per kg DM; capacity of dryer, 10,000 kg/h.

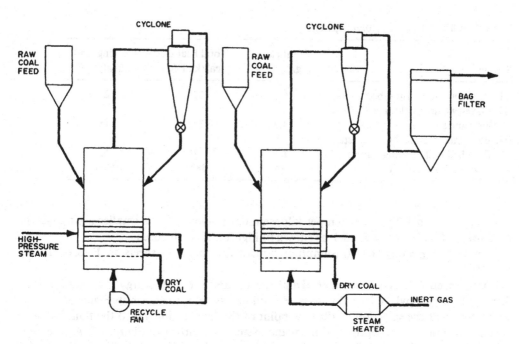

Figure 37 Two-stage atmospheric pressure dryer. First stage is heated by steam and generates steam used for heating the second stage. (From Reference 75.)

however, the cost of cooling the recirculating gas is also rapidly increasing and an optimum for utility costs can be estimated.

10.3. Drying in a Steam-Heated, Steam-Fluidized Bed

Potter et al. (75) have demonstrated in a pilot plant that heat-insensitive materials, such as lignite, can be fluidized and dried by superheated steam operating either under vacuum or at higher pressures. This points to the application of steam-fluidized, steam-heated fluid beds for the removal of moisture in order to obtain very significant energy savings when compared with conventional fluid bed drying.

In one variant of this drying process, two drying stages are applied at atmospheric pressure, as shown in Figure 37. In stage I the heat from high-pressure steam is supplied through the heating coils in the fluid bed. Part of the superheated steam developed by the drying is recycled after cleaning as fluidization gas to stage I, whereas the remaining steam is applied to the heating coils in fluid bed dryer II, which is operating as a conventional fluid bed dryer. Multiple-effect operation enables, depending on the number of effects, several kilograms water being removed from the coal per kilogram of steam fed to the first stage, similar to the energy efficiency of multistage evaporators.

The industrial development of these dryers must solve the practical problems associated with recycling superheated steam carrying fine particles; condensation must be avoided everywhere, and the handling equipment for solids must be carefully designed.

CONCLUSION

When fluid bed technology can be applied, the drying of granular products in fluid beds offers significant advantages compared with traditional drying processes.

A number of modified fluid bed dryers have been developed, such as vibrated fluid beds, spouted beds, and fluid bed granulators, because of the inherent limitations of the traditional fluid bed technology with regard to handling granular products. Accordingly, the range of products that today can be dried by fluid bed dryers is very extensive, from liquid solutions to large irregular pieces.

Reliable and highly integrated fluid bed continuous drying systems can be designed. However, the only safe basis for such designs still remains pilot-plant drying test work combined with industrial experience, as design correlations for all types of fluid beds are basically empirical and valid only within a limited range and for certain particle characteristics.

NOMENCLATURE

a	half-amplitude of vibrations
d	particle diameter
d_r	tube diameter
D_{lat}	lateral particle diffusivity coefficient
G	kg gas (free N_2) per kg dried material
g	acceleration due to gravity
H_0	bed height at incipient fluidization point
ΔP	pressure drop across fluidized bed
Q	heat supply, kcal/kg dried material
S	specific surface of particle
T	temperature, °C
U	gas velocity (superficial)
U_0	minimum fluidizing gas velocity
U_{mm}	minimum solids mixing gas velocity
U_t	terminal velocity for particle
X	kg volatile per kg dried material
ϵ_0	bed voidage at incipient fluidization
μ	viscosity of fluid
ρ_s	density of solid
ρ_f	density of gas
ω	frequency of vibrations
Ar	Archimedes number
Re_0	Reynolds number at minimum fluidization

REFERENCES

1. Vanecek, V., Drbohlar, R., and Markvard, M., *Fluidized Bed Drying*, Leonard Hill, London, 1965.
2. Romankov, P., Drying, Chapter 12 in *Fluidization* (Ed. Davidson, J. F., and Harrison, D.), Academic Press, London, 1972.
3. Hoebink, J. H. B. J., and Rietema, K., *Chem. Eng. Sci.* 35, 2135–2140, 1980.
4. Hoebink, J. H. B. J., and Rietema, K., *Chem. Eng. Sci.* 35, 2257–2265, 1980.
5. Poersch, W., *Aufbereitungs-Technik*, 4, 205–218, 1983.
6. Gupta, R., and Mujumdar, A. S., Recent Developments in Fluidized Bed Drying, Chapter 5 in *Advances in Drying*, Volume 2, 155 (Ed. Mujumdar, A. S.), Hemisphere, Washington, DC, 1983.
7. Strumillo, C., Markowski, A., and Kaminski, W., Modern Developments in Drying of Paste-

like Materials, Chapter 6 in *Advances in Drying*, Volume 2, p. 193 (Ed. Mujumdar, A. S.), Hemisphere, Washington.

8. Pakowski, Z., Mujumdar, A. S., and Sturmillo, H. C., *Advances in Drying*, Volume 3 (Ed. Mujumdar, A. S.), Hemisphere, McGraw-Hill, New York, 1983.

9. Gupta, R., and Mujumdar, A. S., *Drying '80*, Volume 1, p. 141, (Ed. Mujumdar, A. S.), Hemisphere, McGraw-Hill, New York, 1980.

10. Karmazin, V. D., *Technology and Application of Vibrated Bed*, Naukova Dumka, Kiev, 1977 (in Russian).

11. Karmazin, V. D., *Drying in Vibrated Fluid Beds*, CNITEI Legpischemash, Moscow, 1972 (in Russian).

12. Geldart, D., *Power Technol.* 6, 201–205, 1972.

13. Geldard, D., *Power Technol.* 7, 285–292, 1973.

14. Grace, J. R., Fluidized-Bed Hydrodynamics, Chapter 8.1 in *Handbook of Multiphase Systems* (Ed. Hetsroni, G.), McGraw-Hill, New York, 1982.

15. Richardson, J. F., Incipient Fluidization and Particlulate Systems, Chapter 2 in *Fluidization* (Ed. Davidson, J. F., and Harrison, D.), Academic Press, London, 1972.

16. Davidson, J. F., and Harrison, D., *Fluidized Particles*, Cambridge University Press, London, 1963.

17. Rowe, P. N., and Partridge, B. A., *Trans. Inst. Chem. Eng.* 43, 157–175, 1965.

18. de Groot, J. H., in *Drinkenburg, International Symposium on Fluidization*, p. 348, Netherlands University Press, Amsterdam, 1967.

19. Botterill, J. S. M., *Fluid Bed Heat Transfer*, Academic Press, London, 1975.

20. Potter, O. E., Mixing, Chapter 7 in *Fluidization* (Ed. Davidson, J. F., and Harrison, D.), Academic Press, London, 1972.

21. Zenz, F. A., *Encyclopedia of Chemical Technology*, Kirk-Othmer, Third Edition, Volume 10, 548, 1980.

22. Zabrodsky, S. S., *Hydrodynamics and Heat Transfer in Fluidized Beds*, MIT Press, Cambridge, MA, 1966.

23. Gelperin, N. I., and Einstein, V. G., Heat Transfer in Fluidized Beds, Chapter 10 in *Fluidization* (Ed. Davidson, J. F., and Harrison, D.), Academic Press, London, 1972.

24. Schlünder, E. V., *Verfahrenstechnik* 14(7/8), 459–467, 1980.

25. Martin, H., *Chem. Ing. Technik* 52(3), 199–209, 1980.

26. Natusch, H. J., Lokale Warmeubergangzahlen fur ein Einselrohr und fur Rohrbundel Verschiedener Anordnung in Wirbelschichten, GVT-Bericht vom 30.4, 1971.

27. Reay, D., in *Proceedings of the First International Symposium on Drying*, pp. 136–146 (Ed. Mujumdar, A. S.), Science Press, Princeton, NJ, 1978.

28. Levenspiel, O., *Chemical Reaction Engineering*, J. Wiley and Sons, New York, 1972.

29. Vanderschuren, J., and Delvosalle, C., *Chem. Eng. Sci.* 35, 1741–1748, 1980.

30. Reay, D., and Allen, R. W. K., in *Proceedings of the Third International Symposium on Drying, Birmingham, England*, pp. 130–140 (Ed. Mujumdar, A. S.), Drying Research Ltd., Birmingham, U.K., 1982.

31. Reay, D., and Allen, R. W. K., *J. Separ. Proc. Technol.* 3(4), 11–13, 1982.

32. Story, M. J., *Int. J. Pharm. Tech. Prod. Mfr.* 2(4), 19–23, 1981.

33. Danielsen, S., and Hovmand, S., *Drying '80*, Volume 1, pp. 194–199 (Ed. Mujumdar, A. S.), Hemisphere, Washington, DC, 1980.

34. Florin, G., and Hody, D., *Chem. Age India* 30(11), 1040–1044, 1979.

35. Tachibana, T., Ebisawa, T., and Okada, A., in *Proceedings of the Third International Symposium on Drying, Birmingham, England*, pp. 585–593 (Ed. Mujumdar, A. S.), 1982.

36. Shah, R. M., and Goyel, S. K., *Drying '80*, Volume 2, pp. 176–181 (Ed. Mujumdar, A. S.), Hemisphere, Washington, DC, 1980.

37. Nilsson, L., and Wimmerstedt, R., *J. Separ. Proc. Technol.* 3(4), 4–10, 1982.

38. Jinescu, G., and Balaban, G., in *Proceedings of the Third International Symposium on Drying, Birmingham, England*, pp. 124–129 (Ed. Mujumdar, A. S.), 1982.

39. Bratu, E. A., and Jinescu, G. I., *Brit. Chem. Eng.* 16(8), 691–695, 1971.

40. Mujumdar, A. S., *Lat. Am. J. Heat Mass. Transf.* 7, 99–110, 1983.
41. Pakowski, Z. and Mujumdar, A. S., in *Proceedings of the Third International Symposium on Drying, Birmingham, England*, pp. 149–155, (Ed. Mujumdar, A. S.), 1982.
42. Ringer, D. U., and Mujumdar, A. S., in *Proceedings of the Third International Symposium on Drying, Birmingham, England*, pp. 107–114, (Ed. Mujumdar, A. S.), 1982.
43. Gupta, R., Leung, P., and Mujumdar, A. S., *Drying '80*, Volume 2, pp. 201–207 (Ed. Mujumdar, A. S.), Hemisphere, Washington, DC, 1980.
44. Strumillo, C., and Pakowski, Z., *Drying '80*, Volume 1, pp. 211–225 (Ed. Mujumdar, A. S.), Hemisphere, Washington, DC, 1980.
45. Kulling, W., and Simon, E. J., *Pharmaceutical Tech. Int.* January, 29–33, 1980.
46. Aulton, M., and Banks, M., *Manufacturing Chem. Aerosol News* December, 50–56, 1978.
47. Mortensen, S., and Hovmand, S., *Chem. Eng. Prog.* April, 37, 1983.
48. Baldini, R. A., and Chan, R. C., The NSM Fluid Bed Granulation Process, Paper presented at National ACS Meeting, Washington, DC, August 1983.
49. Schaefer, T., and Woertz, O., Control of Fluidized Bed Granulation, Part I, *Arch. Pharm. Chem. Sci. Ed.* 5, 51–60, 1977; Part II, *Arch. Pharm. Chem. Sci. Ed.* 5, 178–193, 1977; Part III, *Arch. Pharm. Chem. Sci. Ed.* 6, 1–13, 1978; Part IV, *Arch. Pharm. Chem. Sci. Ed.* 6, 14–25, 1978; Part V, *Arch. Pharm. Chem. Sci. Ed.* 6, 69–82, 1978.
50. Wurster, D. E., *J. Pharm. Sci.* 49, 82, 1960.
51. Ormos, Z., and Pataki, K., *Hung. J. Ind. Chem.* 7, 237, 1979.
52. Ormos, Z., Pataki, K., and Stefko, B., *Hung. J. Ind. Chem.* 7, 31, 141, 1979.
53. Ormos, Z., *Hung. J. Ind. Chem.* 7, 153, 1979.
54. Thrun, U., Diss. No. 4511, ETH, Zurich, 1970.
55. Wurster, D. E., *J. Amer. Pharm. Assn. Sci.* 48, 451, 1959.
56. Mortensen, S., and Hovmand, S., in *Fluidization*, Volume II, (Ed. Keairns, D. L.), Hemisphere, Washington, DC, 1976.
57. Grimmett, E. S., in *Recent Advances in Fluidization*, pp. 148–159, A.I.Ch.E. Symposium Series, 1981.
58. Grimmett, E. S., *A.I.Ch.E. J.* 10, 717–722, 1964.
59. Mathur, K. B., Spouted Beds, Chapter 17 in *Fluidization* (Ed. Davidson, J. F., and Harrison, D.), Academic Press, London, 1972.
60. Epstein, N., and Mathur, K. B., Applications of Spouted Beds, Section 8.5.6 in *Handbook of Multiphase Systems* (Ed. Hetsroni, G.), McGraw-Hill, New York, 1982.
61. Mathur, K. B., and Epstein, N., *Spouted Beds*, Academic Press, New York, 1974.
62. Mathur, J. B., and Gishler, P. E., *J. Appl. Chem.* 5, 624, 1955.
63. Romankov, P. G., and Rashkovskaya, N. B., *Drying in a Fluidized Bed*, Khimiya, Leningrad, 1968.
64. Fane, A. G., Stevenson, T. R., Lloyd, C. J., and Dunn, M., The Spouted Bed Drier – An Alternative to Spray Drying, paper presented at Eighth Australian Chemical Engineering Conference, Melbourne, Australia, August 1980.
65. Richardt, K., *Drying '80*, Volume 2, pp. 379–386 (Ed. Mujumdar, A. S.), Hemisphere, Washington, DC, 1980.
66. Kragh, O. T., *Keramische Zeitschrift* 30(7), 369, 1978.
67. Ormos, Z., and Blickle, T., *Drying '80*, Volume 1, pp. 200–204 (Ed. Mujumdar, A. S.), Hemisphere, Washington, DC, 1980.
68. Lazar, M. E., and Farkas, D. F., *Drying '80*, Volume 1, pp. 242–246 (Ed. Mujumdar, A. S.), Hemisphere, Washington, DC, 1980.
69. TRADE Catalog, APV. Mitchell Dryers Ltd., U.K.
70. Carlson, R. A., Roberts, R. L., and Farkas, D. F., *J. Food Sci.* 45(5), 1177, 1976.
71. Mortensen, S., The Fluidized Spray Dryer – A New Method for Drying of Food Products, paper presented A.I.Ch.E. Summer National Meeting, Denver, Colorado, 1983.
72. Pisecky, J., *Dairy Industries, Int.* April 21, 1983.
73. Christiansen, O. B., *Chem. Eng. Prog.* 75, November, 58–64, 1979.
74. Basel, L., and Gray, E., *Chem. Eng. Prog.* 58, 67–72, 1962.

75. Potter, O. E., Beeby, C. J., Fernando, W. J. N., and Ho, P., *Drying Technology* 2(2), 219–234, 1983.
76. Herron, D., and Hummel, D., *Chem. Eng. Prog.* 76, January, 44–52, 1980.
77. Mortensen, S., Optimization of Energy Consumption in Fluidized Bed Drying Systems, paper presented at Third Engineering Foundation Conference on Fluidization, New Hampshire, August 3–8, 1980.

8
Drum Dryers

James G. Moore[*]
Blaw-Knox Food and Chemical Equipment Company
Buffalo, New York

1. INTRODUCTION

In drum drying material is dried on the surface of an internally heated revolving drum. There are various types of drum dryers and methods of operation, as discussed later. Figures 1 and 2 show a typical drum dryer, a 42-in-by-120-in atmospheric double-drum dryer viewed from opposite the drive end and from the driven end, respectively.

Major considerations and problems in the use of drum drying have to do primarily with (a) nature of the product to be dried, (b) the drum dryer as a heat transfer mechanism, (c) design and construction of the dryer, (d) conditions under which the dryer operates, and (e) economies to be affected within the drum dryer itself as well as in connection with the manufacturing process in which it is being used.

For a drum dryer to be utilized, the product to be dried is usually in fluid, slurry, or pastelike form, with the solids in either solution or suspension. The material is spread in a thin layer on the surface of the drum by various methods of application. Since the layer of the material is comparatively thin, at no time is the drying rate governed by the diffusion of the vapor through the product layer. The drying of solutions involves a change of state of the solute in addition to the evaporation of water or solvent.

The capacity of a drum dryer depends not only on the drying rate of a thin layer of the material but also on the degree of adherence of the product to the drying surface. During the operation of a drum dryer, the amount of liquid (hence, product rate) adhering to the drum surface varies with the type of feed device, as well as with steam pressure and drum speed. Feed temperature and concentration frequently can be regulated; preheating and preconcentration are desirable wherever possible, since these result in a reduction of the amount of heat to be transferred per unit weight of dry material. Concentration cannot always be carried to the limit because many times there is an optimum feed concentration above which it becomes difficult to apply a uniform coating and where a

---

[*]Deceased.

Figure 1   Typical Blaw-Knox 42-in-by-120-in atmospheric double-drum dryer, view from opposite drive end.

vapor barrier along the face of the drum at the feed point is formed, preventing uniform adherence of the product to the drum surfaces. Other properties, such as viscosity, surface tension, and wetting power, are important in governing layer formation but are generally fixed because any change in these properties may affect the quality of the dry product.

Because there are so many operating variables, as well as such a wide variation in product characteristics, it is impossible to make any definite prediction of drum dryer performance without either prior experience on the specific product or by conducting test work. Feasibility studies are most economically conducted on very small (6-in diameter by 8-in long drums) laboratory-type dryers to produce small samples of dry product for application and primary product evaluation purposes. One must be cautioned, however, that capacity, operating conditions, and relative heat transfer data obtained from such a small unit cannot be extrapolated with complete confidence for design of a large commercial unit. In order to obtain reliable information for commercial design purposes, all test and/or pilot-plant work must be performed on sufficiently large equipment to yield results that are characteristic of commercial-size dryers. Even when such data are available, it should be evaluated only by competent personnel well versed in the art and familiar with all the pertinent variables that must be considered when scaleup in size is required. Complete pilot-plant facilities with the capability for predicting processing and operating results, while requiring a minimum of material, are maintained by any compe-

Figure 2 Typical Blaw-Knox 42-in-by-120-in atmospheric double dryer, view from drive end.

tent equipment manufacturer. Through the use of such services, expert engineering knowledge, experience in application, and dependable process experience are available to resolve customer's problems for a modest fee. Also, under these conditions, the time required for product and process drying research, as well as costs, are greatly reduced.

Drum dryers are extremely flexible in operation and represent the one type of process equipment in which all variables may be changed independently without simultaneously affecting others. There are four variables involved in the operation of such a dryer on a given material: (a) steam pressure or heating medium temperature, which governs the temperature of the drum surface; (b) speed of rotation, which determines the time of contact between the film and the preheated surface; (c) thickness of the film, which may be governed by the distance between the drums, that is, gap; and (d) condition of the feed material, that is, the concentration, physical characteristics, and temperature at which the solution to be dried reaches the drum surface. In a double-drum type of operation, the feed level between the drums also determines the final concentration of the feed at the precise moment of contact with a hot drum.

2. TYPICAL DOUBLE-DRUM DRYER

With the relatively recent understanding (within the past 25 years) of the time versus temperature (especially when drying extremely heat-sensitive materials), the trend is toward the use of atmospheric double- or twin-drum dryers (Figs. 1, 2, and 3-A). This tendency results from their ability to handle a wider range of products, better economics, more efficient operations, higher production rates, and fewer operating labor requirements.

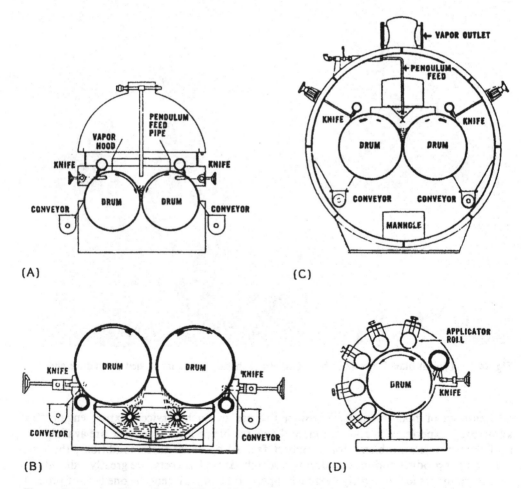

Figure 3 Types of drum dryers: (A) double drum (atmospheric); (B) twin drum; (C) double drum (vacuum); (D) single drum.

The drums of a double-drum dryer with pendulum feed rotate toward each other and the feed material passes between the preset desired drum clearance. The dry product is removed by the knives, which are located in the upper, outer quadrants. This type of drum dryer is used for a wide variety of products in both the chemical and the food industries. It handles feed materials with characteristics from dilute solutions to heavy pastes. A partial list of typical materials dried are cereals, yeast, starches, polyacrylamides, sodium benzoate, various propionates, various acetates, and many other chemical products. In some instances, because of its precise control of temperature, it is used to produce a specific hydrate of a chemical compound.

Double-drum dryers are especially applicable to drying problems in the field of food dehydration in which the preservation of the particular shape is unnecessary. Most food products of this type are characterized by being of liquid or paste form, are heat sensitive, and are ultimately to be marketed in a quickly rehydratable flake or powder form. Among this variety of products are such foodstuffs as applesauce, various fruit purees, bananas, precooked breakfast cereals, and dry soup mixtures. Since these dryers are utilized in the food industry, they must meet sanitary codes established for processing equipment. Sanitary drum dryers are available in suitable construction for all sanitary codes as well as U.S. Department of Agriculture (USDA) standards.

3. TWIN-DRUM DRYER

The twin-drum dryer, as shown in Figure 3-B, is equipped with a splash feed at the bottom. The product film applied, in some instances, depends upon drum clearance for thickness. The drums rotate away from each other, and the dry material is removed by the knives in the lower outer quadrants. This type of dryer is used for feed materials, typically slurries, saturated salt solutions, clay slips, suspensions of various insoluble solids, and materials with solids that are extremely dusty when dry. Characteristic of the type of materials dried are phosphates, chelates, aluminum oxide, sodium benzoate, metadisulfonic acid, and dusty propionates. A twin-drum dryer is also used for materials that form crystalline or sludge type of masses when concentrated and that, because of this characteristic, would cause undue concentrated point, mechanical pressure on the drums of a double-drum unit.

The twin-drum dryer may also be operated with a center top feed and can be used in combination with a second continuous dryer, many times a hot air rotary dryer. In this type of operation, the material is removed from the twin-drum dryer with a relatively high moisture content, and final dehydration is completed in the rotary dryer. This combined drying method is economical because of the increased obtainable capacity and is satisfactory because it avoids difficulties that might be inherent to either type of dryer when used separately. It has also been applied very successfully for drying a salt that was soluble in its own water of crystallization. In such an instance, the dried salt has to be completely anhydrous, and since the twin-drum dryer alone cannot produce anhydrous material, except at a great sacrifice of capacity, but can produce with high capacity a material containing so little moisture that it will not soften even when subjected to high temperatures. The partially dried material when subjected to high temperatures in a small rotary hot air dryer is completely dried and discharged in the form of dense granular particles.

Atmospheric drum dryers, equipped with enclosure (as described later), are many times used in pollution control or solvent recovery applications in which the dry material may be discarded for landfill purposes and the solvent removed is the desired product.

They are also used in drying dusty materials recovered from anti–air pollution scrubbers, as well as recovering valuable salts from waste plating solutions.

4. SINGLE-DRUM DRYER WITH APPLICATOR ROLLS

Today, the single-drum dryer, as shown in Figure 3-C, is primarily used with multiple applicator rolls. It is used to form heavy product sheets resulting in thicker and more dense flakes. It can also dry materials with feed characteristics such as pastes, mash products containing starches, and many food products. Typical products dried with this type of arrangement are potato flakes, various starches, acrylamides, and resins. The number of applicator rolls used controls the application of the material as desired and hence the final characteristics of the dry product sheet. This type of configuration is especially desirable when drying a material that has a tendency not to coat the entire hot drying surface uniformly with an even coat of product, but when repeated applications tend to fill in the differently hot areas having thinner films. Figure 4 shows two Blaw-Knox Buflovak 5-ft diameter by 16-ft long atmospheric single-drum dryers as equipped and installed for the production of dehydrated potato flakes.

5. VACUUM DOUBLE-DRUM DRYER

The double-drum and the single-drum dryer without applicator rolls are also available for vacuum operation. A typical dryer of this type is shown in Figure 3-D. These dryers are relatively expensive and are utilized only when the cost of the product will permit or the type of operation dictates its use. As mentioned previously, many applications that previously required these dryers are today being handled on special, less expensive atmospheric double-drum dryers. They will find worldwide use for very special product dehydration; for example, they are being successfully operated under sterile conditions for the drying of pharmaceutical antibiotics. They are also used for drying special products in

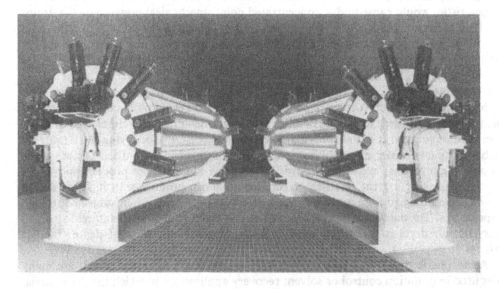

Figure 4   View of two Blaw-Knox 5-ft by 16-ft atmospheric, single-drum, potato flake dryers.

which a porous structure is desired, such as malted milk. In some instances, these dryers have found use in the recovery of valuable solvents with low vapor pressures from process waste materials. When recovering such high-boiling constituents as ethylene glycol, it is of advantage to operate under vacuum in order to reduce the boiling point of the solvent, thereby making it easier to remove it from the dry residue.

Dimensions and specifications of Blaw-Knox Buflovak drum dryers are shown in Figures 5 and 6. Figure 7 shows 6-ft by 8-ft, laboratory-size, atmospheric and vacuum drum dryers, together with enclosures for commercial-size drum dryers.

6. DRUM DRYERS WITH ENCLOSURE

Atmospheric double- and twin-drum dryers are also available with vapor-tight or dust-tight enclosures to safeguard operations during the drying of toxic, dusty, or flammable products (Fig. 7). Solvents can be recovered by suitable surface condensers. Recovery or elimination of the dust, if necessary, from the vapor can be accomplished by the addition of a scrubber. To recover solvents from within the casing while holding residual solvents in the dried material to a minimum, or to remove material that tends to sheet at the knife, a partial enclosure may be used.

In such cases, a portion of the drum extends through the enclosure, where externally mounted knives continuously remove the dry material for cooling or special handling. Other types of special enclosures may be fabricated as required.

In most cases, the dry material removed from a drum dryer by the doctor blade or knife is conveyed away from the dryer proper by screw conveyors as shown. In special cases, belt conveyors may be used or the product may be pneumatically picked up and conveyed to a central location.

Drum dryers are extremely economical in operation, usually requiring from a maximum of 1.3 lb to a minimum of 1.11 lb of steam per pound of evaporation (76–90% efficient), including losses for radiation. The steam cost should always be calculated on the basis of 1 lb of dry product, since the cost varies widely with feed concentration and acceptable dried material moisture or minimum residual solvent content.

Generally, the maximum rate of evaporation is obtained with dilute solutions, which evaporate readily. Under the most favorable conditions, this may be as high as 18 lb/h per square foot of drum surface when the water is being removed, but the rate of evaporation is not necessarily the determining factor in the economy of drum drying because the amount of acceptable dried material produced in most cases is the prime consideration.

In addition to relatively low initial cost and high steam economy, drum dryers require low horsepower and, hence, minimum electrical requirements, minimal operation personnel, low maintenance costs, and relatively low initial installation expense.

In the manufacture of dryers of this type, the drums are very vital parts. They are usually manufactured from cast iron or fabricated of steel, for whatever working pressure is required (cast iron is limited to 160 psig) and in many instances are chrome-plated. Drums are available in a variety of constructions, such as straight wall or reinforced. Based on modern-day engineering design and modern machine shop knowledge, drums that are dimensionally stable and remain concentric for their operating life are readily available. These drums, machined to extremely close tolerances for both diameter and length, allow for close clearance operation of double-drum dryers when they are to handle heat-sensitive materials.

It is important to remember that a drum to be used for drying is designed in a special

Model and Drum Dimensions (inches)	Space Required		Area (Sq. Ft.)	Approx. Weight (pounds)
	Length (inches)	Width (inches)		
24 x 24	104	48	12.5	5,000
42 x 24	124	63	22.0	10,000
42 x 36	140	63	33.0	11,250
42 x 48	154	63	44.0	12,500
42 x 60	170	63	55.0	13,750
42 x 72	186	63	66.0	15,000
60 x 120	228	118	157.1	34,000
60 x 144	252	118	188.5	36,000
60 x 192	300	118	251.3	40,000

	(mm)	(mm)	(m²)	(kg)
24 x 24	2,640	1,220	1.2	2,300
42 x 24	3,150	1,600	2.0	4,550
42 x 36	3,555	1,600	3.1	5,100
42 x 48	3,910	1,600	4.1	5,700
42 x 60	4,320	1,600	5.1	6,250
42 x 72	4,725	1,600	6.1	6,800
60 x 120	5,790	3,000	14.6	15,500
60 x 144	6,400	3,000	17.5	16,500
60 x 192	7,620	3,000	23.3	18,000

SINGLE DRUM ATMOSPHERIC

These may be dip or splash fed (not illustrated) or, as shown in the schematic and photo, equipped with applicator rolls. The latter is particularly effective for drying high viscosity liquids or pasty materials such as mashed potatoes, applesauce, fruit-starch mixtures, gelatin, dextrine type adhesives, polyacrilamide, synthetic resins and various starches. Applicator rolls eliminate void areas, permit drying between successive layers of fresh material and form the product sheet gradually. While single applications may dry to a lacy sheet or flake, the multiple layers generally result in a product of uniform thickness and density with minimum dusting tendencies.

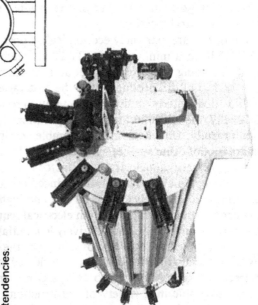

(a)

SINGLE DRUM VACUUM

Model and Drum Dimensions (inches)	Space Required			Area (Sq. Ft.)	Approx. Weight (pounds)
	Length (inches)	Width (inches)	Height (inches)		
24 x 20	108	48	72	10.5	7,700
48 x 40	132	120	156	41.9	31,000
60 x 144	228	96	108	188.5	75,000

DOUBLE DRUM VACUUM*

12 x 18	66	48	78	9.4	3,600
24 x 24	120	96	108	25.1	16,000
24 x 48	144	96	108	50.3	20,000
24 x 60	168	96	108	62.8	24,000
32 x 72	186	114	120	100.5	41,000
42 x 120	252	156	168	219.9	87,500

SINGLE DRUM VACUUM

	(mm)	(mm)	(mm)	(m²)	(kg)
24 x 20	2,745	1,220	1,830	1.0	3,500
48 x 40	3,355	3,050	3,960	3.9	14,000
60 x 144	5,790	2,440	2,745	17.5	34,000

DOUBLE DRUM VACUUM*

12 x 18	1,675	1,220	1,980	0.9	1,650
24 x 24	3,050	2,440	2,745	2.3	7,000
24 x 48	3,660	2,440	2,745	4.7	9,000
24 x 60	4,270	2,440	2,745	5.8	11,000
32 x 72	4,725	2,895	3,050	9.3	19,000
42 x 120	6,400	3,960	4,265	20.4	40,000

*Space required does not include allowance for hinged doors, working space, product receivers or condenser system.

SINGLE AND DOUBLE VACUUM

Vacuum enclosed systems are specified when products must be dried without exposure to high temperatures or reactive atmospheres. These vacuum units may be designed for sterile operation and are particularly suited for pharmaceuticals, vitamin extracts, solvent reclamation, food and fine chemicals. Continuous drying without breaking vacuum is achieved by utilizing two receivers and air locks. Feed for a single drum is usually of the pan type with pump and spreading device or a spray film for materials that are repelled by contact with heated surfaces. Double drum dryers utilize a pendulum or a perforated tube feed. Special feed devices are also available.

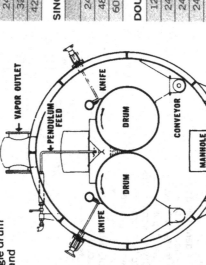

(b)

Figure 5 Buflovak drum dryers: (a) single-drum atmospheric; (b) single and double vacuum. (From *Buflovak Industrial Drum Dryers*, Catalog 405B.)

Model and Drum Dimensions (inches)	Space Required			Area (Sq. Ft.)	Approx. Weight (pounds)
	Length (inches)	Width (inches)	Height[1] (inches)		
12 x 18	60	36	56	9.4	1,650
24 x 24	138	81	90	25.1	8,500
24 x 36	150	81	96	37.7	9,200
24 x 48	162	84	108	50.3	10,000
32 x 52	196	99	110	72.6	16,800
32 x 72	216	99	110	100.5	18,400
32 x 90	234	99	110	125.7	19,600
32 x 100	246	99	110	139.6	20,500
42 x 90	252	117	120	164.9	32,500
42 x 100	264	117	120	183.3	34,000
42 x 120	284	117	120	219.9	37,000
60 x 144	312	162	168	377.0	60,000

	(mm)	(mm)	(mm)	(m²)	(kg)
12 x 18	1,525	915	1,420	0.9	750
24 x 24	3,505	2,055	2,285	2.3	3,850
24 x 36	3,810	2,055	2,440	3.5	4,200
24 x 48	4,115	2,135	2,745	4.7	4,500
32 x 52	4,980	2,515	2,795	6.7	7,600
32 x 72	5,485	2,515	2,795	9.3	8,500
32 x 90	5,945	2,515	2,795	11.7	9,000
32 x 100	6,250	2,515	2,795	13.0	9,500
42 x 90	6,400	2,970	3,050	15.3	15,000
42 x 100	6,705	2,970	3,050	17.0	15,500
42 x 120	7,215	2,970	3,050	20.4	17,000
60 x 144	7,925	4,115	4,265	35.0	27,000

[1] To top of vapor hood.

DOUBLE DRUM ATMOSPHERIC

The most versatile and widely applied — these dryers are used to dry many food, chemical and pharmaceutical materials of widely varying densities and viscosities: dilute solutions, heavy liquids, or pasty materials. They are also effective for fairly heavy sludges which become saturated and deposit salts, although splash-fed Twin Drum Drying is sometimes better. Many products sensitive to high temperatures may be dried successfully since exposure to temperature above the boiling point is restricted to only a few seconds. The movable drum permits complete control of product film thickness. Feed may be by perforated tube trough, pendulum, or special engineered devices.

(a)

Model and Drum Dimensions (inches)	Space Required			Area (Sq. Ft.)	Approx. Weight (pounds)
	Length (inches)	Width (inches)	Height[1] (inches)		
12 x 18	70	36	46	9.4	1,800
24 x 24	138	81	90	25.1	8,500
24 x 36	150	81	96	37.7	9,200
32 x 52	196	99	110	72.6	16,800
32 x 72	216	99	110	100.5	18,400
32 x 90	234	99	110	125.7	19,600
32 x 100	246	99	110	139.6	20,500
42 x 90	252	117	120	164.9	32,500
42 x 100	264	117	120	183.3	34,000
42 x 120	284	117	120	218.9	37,000
60 x 144	312	162	168	377.0	60,000
	(mm)	(mm)	(mm)	(m²)	(kg)
12 x 18	1,780	915	1,170	0.9	825
24 x 24	3,505	2,055	2,285	2.3	3,900
24 x 36	3,810	2,055	2,440	3.5	4,200
32 x 52	4,980	2,515	2,795	6.7	7,600
32 x 72	5,485	2,515	2,795	9.3	8,300
32 x 90	5,945	2,515	2,795	11.7	8,900
32 x 100	6,250	2,515	2,795	13.0	9,300
42 x 90	6,400	2,970	3,050	15.3	15,000
42 x 100	6,705	2,970	3,050	17.0	15,500
42 x 120	7,215	2,970	3,050	20.4	17,000
60 x 144	7,925	4,115	4,265	35.0	27,000

[1] To top of vapor hood.

TWIN DRUM ATMOSPHERIC

Twin drum dryers are designed for handling slurries, crystal-bearing or crystal-forming liquids and delicate, heat-sensitive products which can be exposed to high temperatures for a very limited time. These dryers may utilize pendulum or perforated tube for top feed, dip pan, or splash or spray for bottom feed. Cooling and agitation of the material in the pan may be provided to minimize pre-concentration and settling.

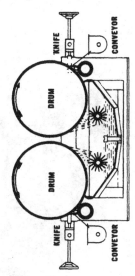

(b)

Figure 6 Buflovak drum dryers: (a) double-drum atmospheric; (b) twin-drum atmospheric. (From *Buflovak Industrial Drum Dryers*, Catalog 405D.)

LABORATORY

Experimental laboratory drum dryers have two steam heated drums of 6" diameter and 8" face with one square foot of heat transfer surface each. The drums may be rotated in either direction and one is adjustable to permit control of the product film thickness. Various feed devices are available and the dryers are furnished with adjustable knives, variable speed drive, reinforced phenolic resin endboards, dry material containers, a condensate removal syphon system, pressure gauges and other accessories.

Laboratory dryers for vacuum operation are similar to atmospheric units with the addition of a cylindrical casing of insulated double shell construction with a heating coil in the jacketed space to prevent the formation of condensate. Hinged doors at either end provide complete access to the unit and when opened permit atmospheric operation of the dryer. 6" x 8" laboratory dryers may be rented for a modest fee when the need arises.

6" x 8" Laboratory Vacuum
Double Drum Dryer.
Doors may be opened to permit
atmospheric operation.

(a)

To facilitate dried product removal while recovering solvents, a partial enclosure is used on this 60″ x 144″ Drum Dryer.

DRUM DRYERS — ENCLOSURES

To safeguard operations during the drying of toxic, dusty or flammable products, Buflovak atmospheric drum dryers may be equipped with full or partial enclosures. External controls are provided. Solvents can be recovered by connecting the vapor outlet to a condenser, and recovery or elimination of dust from the vapor can be accomplished by adding a scrubber. To recover solvents from within the casing while holding residual solvents in the dried material to a minimum, or to remove material which tends to sheet at the knife, a partial enclosure may be used. In such cases, a portion of the drum extends through the enclosure where externally mounted knives continuously remove the dry material for cooling or special handling. Other types of special enclosures may be fabricated as required.

42″ x 120″ Atmospheric Double Drum Dryer with a full dust-tight enclosure used for drying an organic sodium salt.

(b)

Figure 7 Drum dryers and enclosures: (a) laboratory-size drum dryers; (b) drum dryer enclosures. (From *Buflovak Industrial Drum Dryers*, Catalog 405B.)

manner to assure uniform surface temperature and dimensional stability, from both transient as well as permanent distortion.

For a dryer to operate correctly and produce the desired end product, the entire supporting structure (frame), knifeholders, and bearing supports must be of extremely rigid design and must be equipped with drums suitable for the specific application. Recommendations can be made only by equipment manufacturers having an extensive background of construction and operating experience, not only for the correct selection of the basic type of dryer for the particular process in question, but also the correct mechanical design.

BIBLIOGRAPHY

Perry, R. H., and Chilton, C. H., Editors, *Chemical Engineers Handbook, Fifth Edition*, McGraw-Hill Book Co., NY; 1973, Section 20.

Van't Land, C. M., *Industrial Drying Equipment*, Marcel Dekker, NY, 1992, pp. 212–215.

Williams-Gardner, A., *Industrial Drying*, Leonard Hill Books, London, 1971.

9
Industrial Spray Drying Systems

Iva Filková
IFP Process Engineering
Prague, Czech Republic

Arun S. Mujumdar
McGill University
Montreal, Quebec, Canada

1. INTRODUCTION

This chapter gives a summary of the basics of the spray drying process, its applications, principal components, and ancillaries. The contents are oriented toward the process engineer involved in the selection and specification of spray drying equipment rather than those involved in detailed design of such equipment. The information presented will also be useful to engineers and technologists responsible for operating spray drying plants in diverse industries. For in-depth information and current research and development results, the reader is referred to the literature cited.

2. PRINCIPLES OF SPRAY DRYING PROCESSES

2.1. General

The spray drying process transforms a pumpable fluid feed into a dried product in a single operation. The fluid is atomized using a rotating wheel or a nozzle, and the spray of droplets comes immediately in contact with a flow of hot drying medium, usually air. The resulting rapid evaporation maintains a low droplet temperature so that high drying air temperatures can be applied without affecting the product. The time of drying the droplets is very short in comparison with most other drying processes. Low product temperature and short drying time allow spray drying of very heat-sensitive products.

Spray drying is used to dry pharmaceutical fine chemicals, foods, dairy products, blood plasma, numerous organic and inorganic chemicals, rubber latex, ceramic powders, detergents, and other products. Some of the spray-dried products are listed in Table 1, which also includes typical inlet and outlet moisture content and temperatures together with the atomizer type and spray dryer layout used.

The principal advantages of spray drying are as follows:

1. Product properties and quality are more effectively controlled.
2. Heat-sensitive foods, biologic products, and pharmaceuticals can be dried at atmospheric pressure and low temperatures. Sometimes inert atmosphere is employed.

Table 1 Operating Parameters for Some Spray-Dried Materials

Material	Moisture Inlet (%)	Moisture Outlet (%)	Atomizing device	Liquid-air layout	Air temperature Inlet (°C)	Air temperature Outlet (°C)
Skim milk	48–55	4	Wheel		<250	95–100
($D_{3,2} \approx 60\ \mu m$)			Pressure nozzle (170–200 bar)	Cocurrent	<250	95–100
Whey	50	4	Wheel	Cocurrent	150–180	70–80
Milk	50–60	2.5	Wheel		170–200	90–100
			Pressure nozzle (100–140 bar)	Cocurrent		
Whole eggs	74–76	2–4	Wheel	Cocurrent	140–200	50–80
			Pressure nozzle			
Coffee (instant)	75–85	3–3.5	Pressure nozzle	Cocurrent	270	110
($D_{3,2} \approx 300\ \mu m$)	75–80	3–3.5	Pressure nozzle	Cocurrent	270	110
Tea (instant)	60	≈2	Pressure nozzle (27 bar)	Cocurrent	190–250	90–100
PVC emulsions:						
90% particles: >80 μm			Pressure nozzle			
5% particles: <60 μm	40–70	0.01–0.1	Rotary cup Wheel Pneumatic nozzle	Cocurrent	165–300	100
Melamine-urethane	30–50	≈0	Wheel 140–160 m/s	Cocurrent	200–275	65–75
Detergents:						
Particles: 95–100% > 60 m	35–50	8–13	Pressure nozzle (30–60 bar)	Counter-current	350–400	90–110
2–3% > 150 m						
TiO_2	Up to 60	0.5	Wheel	Cocurrent	600	120
			Pressure nozzle	Mixed flow		
Kaolin	35–40	1	Wheel	Cocurrent	600	120
Ammonium phosphate	60	3–5	Pressure nozzle	Cocurrent	400	110–195
Superphosphate			Wheel	Cocurrent	500–600	>110
Cream	52–60	4	Wheel	Cocurrent		
Processed cheese	60	3–4	Wheel	Cocurrent		
Whole eggs	74–76	2–4	Wheel			

3. Spray drying permits high-tonnage production in continuous operation and relatively simple equipment.
4. The product comes in contact with the equipment surfaces in an anhydrous condition, thus simplifying corrosion problems and selection of materials of construction.
5. Spray drying produces relatively uniform, spherical particles with nearly the same proportion of nonvolatile compounds as in the liquid feed.
6. Since the operating gas temperature may range from 150 to 600°C, the efficiency is comparable to that of other types of direct dryers.

Among the disadvantages of spray drying are the following:

1. Spray drying fails if a high bulk density product is required.
2. In general it is not flexible. A unit designed for fine atomization may not be able to produce a coarse product, and vice versa.

3. For a given capacity larger evaporation rates are generally required than with other types of dryers. The feed must be pumpable.
4. There is a high initial investment compared to other types of continuous dryers.
5. Product recovery and dust collection increase the cost of drying.

Spray drying consists of three process stages:

1. Atomization
2. Spray-air mixing and moisture evaporation
3. Separation of dry product from the exit air

Each stage is carried out according to the dryer design and operation and, together with the physical and chemical properties of the feed, determines the characteristics of the final product. A typical example of a spray drying process with the most important ancillary equipment included is shown in Figure 1.

2.2. Atomization

Atomization is the most important operation in the spray drying process. The type of atomizer not only determines the energy required to form the spray but also the size and size distribution of the drops and their trajectory and speed, on which the final particle size depends. The chamber design is also influenced by the choice of the atomizer. The drop size establishes the heat transfer surface available and thus the drying rate. A comparison of spherical droplet surface and droplet size is shown in Table 2.

Three general types of atomizers are available. The most commonly used are the rotary wheel atomizers and the pressure nozzle single-fluid atomizers. Pneumatic two-fluid nozzles are used only rarely in very special applications. Existing spray drying systems provide various forms of the dry product—from fine powders to granules. The typical ranges of the disintegrated droplets and particle sizes of various products in a spray dryer are listed in Table 3.

Drop Size and Size Distribution

The quality of dry powder is the single most important factor considerably affected by the operating conditions of the process. Powder character and quality are usually determined by further processing or by consumer requirements. To meet the required bulk density of the dry powder, it is necessary to know how the particle size and size distribution are affected by various parameters. General information about the selection of spray dryer design to meet powder specifications can be found in References 25, 26, and, for the particular case of food drying, in Reference 18.

Particle size and distribution are related to the size of the droplets and their size distribution. Hence, successful prediction of droplet size enables one to control the powder properties as desired.

The *mean size of droplet* represents a single value that characterizes the whole spray distribution. This value, together with the *size distribution*, defines the spray characteristics. Many papers have appeared on the subject of drop size prediction. The so-called Sauter mean diameter seems to be the most suitable mean value to characterize the droplet cloud together with the size distribution. This is defined as the ratio of the total droplet volume to the total droplet surface (25); that is,

$$D_{3,2} = \frac{\sum_{1}^{i} D_i^3 f_i}{\sum_{1}^{i} D_i^2 f_i} \tag{1}$$

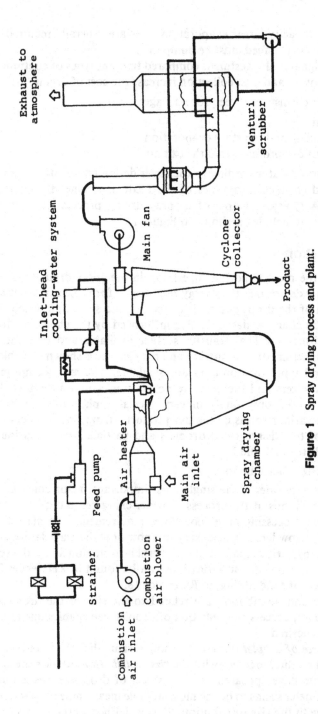

Figure 1 Spray drying process and plant.

Table 2 Spherical Droplet Surface versus Droplet Size

Total volume (m³)	Diameter of droplets		No. droplets	Surface per droplet	Total surface of droplets (m²)
1	1.234	m	1	3.14 m²	3.14
1	1	cm	1.986×10^6	3.14 cm²	623.6
1	1	mm	1.986×10^9	3.14 mm²	6,236
1	100	μm	1.986×10^{12}	31,400 μm²	62,360
1	1	μm	1.986×10^{18}	3.14 μm²	6,236,000

where f_i is the number frequency of droplet of size D_i. The Sauter mean diameter corresponds to the particle diameter with the same volume-to-surface ratio as the entire spray or powder sample. Sometimes the median diameter D_M is also used in spray drying calculations. It is that diameter above or below which lies 50% of the number or volume of droplets. It is especially useful when an excessive amount of very large or very small particles is present.

Size distribution can be represented by a frequency or cumulative distributive curve. If occurrence is given by number, a number distribution results. It may be obtained by microscopic analysis. If occurrence is given by area-volume-weights corresponding to a given diameter, then an area-volume-weight distribution results.

The lognormal and Rosin–Rammler distributions are the most common distributions used in the spray drying process calculations. The lognormal distribution has two parameters, the geometric mean size D_{GM} and the geometric standard deviation S_G. The mathematical form is as follows:

$$\frac{d(N)}{d(D)} = \frac{1}{DS_G\sqrt{2\pi}} \exp -\left[\frac{(\log D - \log D_{GM})^2}{2S_G} \right] \qquad [2]$$

where N is the number of droplets counted. The graphic form on probability paper is shown in Figure 2. The log-normal distribution is useful to represent sprays from wheel atomizers (10).

The lognormal distribution also enables us to determine the quantities for evaluation of the dispersion factor D_q, which characterizes the homogeneity of the spray (10,25).

Table 3 Range of Droplet and Particle Sizes Obtained in Spray Dryers (μm)

Rotating wheels	1–600
Pressure nozzles	10–800
Pneumatic nozzles	5–300
Sonic nozzles	5–1000
Milk	30–250
Coffee	80–400
Pigments	10–200
Ceramics	30–200
Pharmaceutics	5–50
Chemicals	10–1000

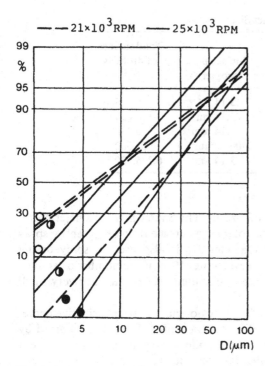

Figure 2 Log-normal distribution on probability paper, an example. (From Reference 10.)

$$D_q = \frac{D_{95} - D_5}{D_{3,2}} \tag{3}$$

Here, D_{95} and D_5 are the droplet diameters at the 95% and 5% probability, respectively.

The Rosin–Rammler distribution is used to represent sprays from nozzles. It is empirical and relates the volume percentage oversize $\overline{V}_D$ to droplet diameter D. The mathematical form is as follows (25):

$$\overline{V}_D = 100^{-[(D/\overline{D}_R)^{D_q}]} \tag{4}$$

where D_q is the dispersion factor and $\overline{D}_R$ is the so-called Rosin–Rammler mean diameter. It is the droplet diameter above which lies 36.8% of the entire spray volume. Thus it follows that the value on the ordinate corresponding to $\overline{D}_R$ equals 0.43. The graphic form of Eq. [4] on log-log paper is shown in Figure 3.

The appropriate correlations for drop size prediction will be presented along with the discussion of each individual type of atomizer in the following sections.

Wheel Atomizers

A typical wheel atomizer is shown schematically in Figure 4. Liquid is fed into the center of a rotating wheel, moves to the edge of the wheel under the centrifugal force, and is disintegrated at the wheel edge into droplets. The spray angle is about 180°C and forms a broad cloud. Because of the horizontal trajectory these atomizers require large-diameter chambers. The most common design of the wheel atomizer has radial vanes.

The linear peripheral speed ranges from 100 to 200 m/s. For the usual wheel diameter, angular speeds between 10,000 and 30,000 rpm are necessary (16) (see Fig. 5).

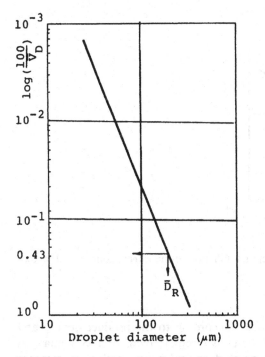

Figure 3 Rosin–Rammler distribution on log-log paper.

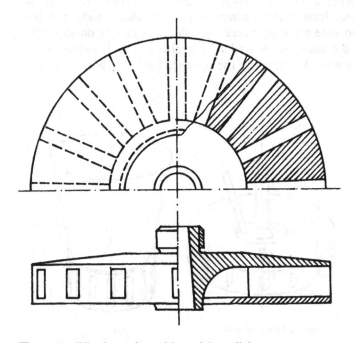

Figure 4 Wheel atomizer with straight radial vanes.

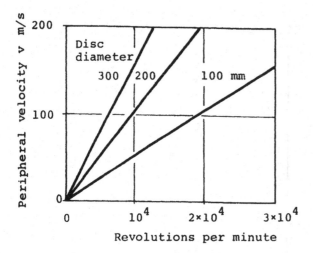

Figure 5 Peripheral velocity of a wheel atomizer as a function of the wheel diameter and revolutions.

The number and shape of the vanes differs according to the product quality and capacity requirements. The usual shape of vanes is circular, oval, or rectangular, as shown in Figure 6. The influence of vane shape on droplet size was studied in Reference 10. For high-capacity applications, where the same degree of atomization is required at higher feed rates, the number and height of the vanes are increased to maintain the same liquid-film thickness on each vane. High-capacity wheels often have two tiers of vanes (24) (see Fig. 7). The largest wheel atomizers allow feed rates up to 200 ton/h. The curved vane wheels are sometimes used instead of the standard straight radial wheels, mainly in the milk industry. The curved vane wheel produces a powder of high bulk density, up to 15% higher than the standard because of reduced air-pumping effect. The other way to achieve high bulk density of the product is to use the so-called bushing wheels, which are

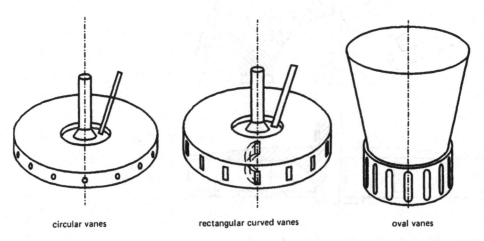

circular vanes   rectangular curved vanes   oval vanes

Figure 6 Various designs of wheel atomizers.

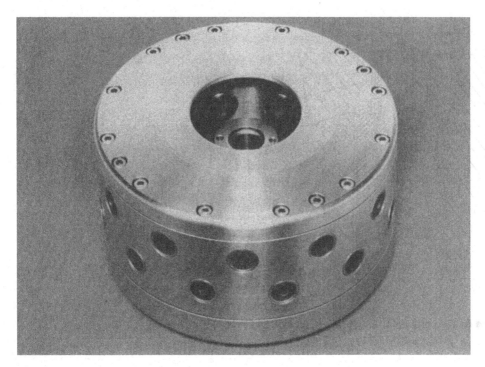

Figure 7 High-capacity wheel atomizer. (Courtesy of Niro Atomizer.)

widely used in the chemical industry. The bushings are made from very hard material, such as silicon carbide; thus the wheel can atomize even very abrasive materials (24).

Generally the wheel atomizer produces a spray of high homogeneity within a wide range of mean droplet size. The size distribution of droplets can be controlled by changing the wheel speed. Much less effect is achieved by feed rate variation. Wheel atomizers are very flexible and can handle a wide assortment of liquids with different physical properties. Factors influencing the wheel atomizer performance are specified in Reference 27, for instance.

The so-called vaneless disk atomizers should also be mentioned, although their use is limited to specialized applications in which coarse particles are required at high production rates. A typical cup-type disk atomizer is shown in Figure 8. Liquid is fed inside the cup and is pressed by centrifugal force against the cup wall, flowing down toward the cup edge, where it is disintegrated.

The wheels are conventionally driven by electric motors, either directly or by belt drives through a worm and worm wheel gearing system (lower capacity wheels) or by helical or epicyclic gearing systems (high-capacity wheels).

The power necessary to accelerate the liquid to the velocity at the periphery of the wheel is given by

$$E = \frac{\dot{M}}{2} \frac{v^2}{\epsilon} \qquad [5]$$

For example, for a mass flow rate $\dot{M} = 1$ kg/s and a peripheral velocity $v = 150$ m/s at an efficiency $\epsilon = 0.65$, the electrical power required is $E = 17.3$ kW.

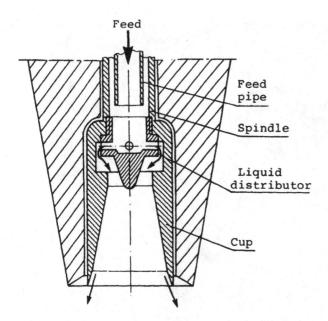

Figure 8   Cup disk atomizer. (From Reference 25.)

The drop size (Sauter mean diameter, in m) can be predicted using any one of the following correlations (25):

$$\frac{D_{3,2}}{r} = 0.4\left(\frac{\dot{M}}{N_v b\rho Nr^2}\right)^{0.6}\left(\frac{N_v \mu b}{\dot{M}}\right)^{0.2}\left(\frac{\sigma b^3 \rho N_v^3}{\dot{M}^2}\right)^{0.1} \tag{6}$$

$$D_{3,2} = 0.241\left(\frac{1}{N}\right)^{0.6}\left(\frac{1}{\rho}\right)^{0.3}\left(\frac{\mu\dot{M}}{2r\rho}\right)\left(\frac{\sigma}{N_v b}\right)^{0.1} \tag{7}$$

$$D_{3,2} = 1.62 \times 10^{-3} N^{(-0.53)} \dot{M}^{0.21} (2r)^{-0.39} \tag{8}$$

where $\dot{M}$ (kg/s) is the mass feed rate, N (rps) is rotational speed, r (m) is wheel diameter, b (m) is vane height, N_v is number of vanes, ρ (kg/m^3) is fluid density, σ (N/m) is fluid surface tension, and μ (Pa·s) is fluid viscosity. Equations [6] through [8] display deviation of $\pm 30\%$ from experimental data reported in the literature.

Most feeds are high-consistency slurries that are generally non-Newtonian, mainly pseudoplastic in nature. To these liquids, the dynamic viscosity should be substituted by the apparent viscosity, which is defined as

$$\mu_A = K\left[\left(\frac{r\omega^2\rho}{K}\right)^2 \frac{V_v}{b} \frac{2n+1}{n}\right]^{\frac{n-1}{2n+1}} \tag{9}$$

where ω is angular velocity (rad/s), V_v (m^3/s) is volumetric feed rate per vane, K (Pa·s^n) is fluid consistency, and n is the flow index (dimensionless) for power law fluids (11). The influence of non-Newtonian character on drop formation is reviewed in Reference 9.

Equation [8] has the advantage that it does not contain any physical parameters on the liquid. It can be used for a rough estimation of the drop size. For the application of more accurate correlations [6] and [7], knowledge of the feed properties is needed.

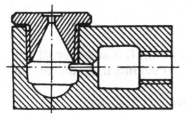

Figure 9 Pressure nozzle.

Pressure Nozzles

A *pressure nozzle*, sometimes called a *single fluid nozzle*, creates spray as a consequence of pressure to velocity energy conversion as the liquid passes through the nozzle under pressure within the usual range of 5–7 MPa (see Fig. 9). The liquid enters the nozzle core tangentially and leaves the orifice in the form of a hollow cone with an angle that varies from 40° to 140°. The orifice diameter is usually small, from 0.4 to 4 mm, and the usual capacity of one nozzle does not exceed 100 liter/h. When larger feed rate is to be processed, several nozzles are used in the drying chamber (see Fig. 10). Owing to their smaller spray angles the drying chamber can be narrower and taller. With this type of nozzle it is generally possible to produce the droplets within a narrow range of diameters, and the dried particles are usually hollow spheres. Pressure nozzles are not suitable for highly concentrated suspensions and abrasive materials because of their tendency to clog and erode the nozzle orifice.

Energy consumption of a pressure nozzle is very low in comparison with that of the wheel atomizer as well as the pneumatic nozzle.

The drop size calculation for a pressure nozzle may be done using the correlation (25)

$$D_{3,2} = 286[(2.54 \times 10^{-2})D + 0.17] \exp\left[\frac{39}{v_{AX}} - (3.13 \times 10^{-3})v_1 \right] \qquad [10]$$

where the axial velocity v_{AX} and the inlet velocity v_1 in m/s are determined as follows:

$$v_{AX} = \frac{D_1^2}{2Db} v_1 \qquad [11]$$

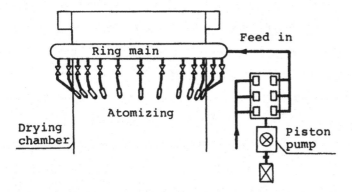

Figure 10 Multinozzle system in the drying chamber. (From Reference 16.)

$$v_1 = \frac{\dot{V}_1}{A_1} \tag{12}$$

Here, D is orifice diameter (m), D_1 is inlet channel diameter (m), A_1 is inlet channel area (m²), $\dot{V}_1$ is volumetric flow rate (m³/s), and b is thickness of fluid film in the orifice. The resulting Sauter mean diameter is in μm.

For a rough prediction, the following empirical equation may be used (6):

$$D_A = \frac{9575}{\Delta P^{1/3}} \tag{13}$$

where D_A is the average drop diameter (μm) and ΔP is the pressure drop across the nozzle (Pa).

Pneumatic Nozzles

Pneumatic nozzles are also known as two-fluid nozzles since they use compressed air or steam to atomize the fluid. Figure 11 shows the most usual type. In this case the feed is mixed with the air outside the body of the nozzle. Less frequently the mixing occurs inside the nozzle. The spray angle ranges from 20° to 60° and depends on the nozzle design. Approximately 0.5 m³ of compressed air is needed to atomize 1 kg of fluid. The capacity of a single nozzle usually does not exceed 1000 kg/h of feed. Sprays of less viscous feeds are characterized by low mean droplet sizes and a high degree of homogeneity. With highly viscous feeds, larger mean droplet sizes are produced but homogeneity is not as high. Pneumatic nozzles are very flexible and produce small or large droplets according to the air-liquid ratio. The high cost of compressed air (pressure range, 0.15–0.8 MPa) becomes important to the economics of these nozzles, which have the highest energy consumption of all three types of atomizers.

The drop size can be predicted by means of the following correlation, which yields the Sauter mean diameter in μm (25):

$$D_{3,2} = \frac{585 \times 10^3 \sqrt{\sigma}}{v_{REL}\sqrt{\rho}} + 597 \left(\frac{\mu}{\sqrt{\sigma\rho}} \right)^{0.45} \left(\frac{1000\ \dot{V}_{FL}}{\dot{V}_{AIR}} \right)^{1.5} \tag{14}$$

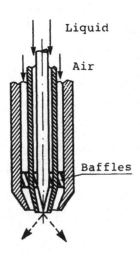

Figure 11 Pneumatic nozzle with external mixing.

where σ, ρ, and μ are the fluid surface tension (N/m), density (kg/m^3), and viscosity (cP), and $\dot{V}_{FL}$ and $\dot{V}_{AIR}$ are volumetric flow rates of fluid and air (m^3/s), respectively. Instead of relative velocity v_{REL} (m/s), the outlet velocity of air may also be substituted.

Novel Types of Atomizers

A number of liquids that cannot be atomized successfully by wheels or nozzles have generated interest in using other methods of atomization that may be more suitable for such liquids. These are, for example, highly viscous and long molecular chain structured materials and some non-Newtonian liquids, which form only filaments instead of individuals droplets from the ordinary atomizers. Attention has been paid to the use of sonic energy. The most recent development in the field of pneumatic nozzles is the sonic atomizer (34). The breakup mechanism is entirely different from that for conventional nozzles. The disintegration of a liquid occurs in the field of high-frequency sound created by a sonic resonance cup placed in front of the nozzle. However, this development has not reached the stage at which sonic nozzles can be industrially competitive with other kinds of atomizers. They have some promising aspects, such as 15% savings in energy (38) and applicability for abrasive and corrosive materials. The four main types of sonic atomizers are the Hartman monowhistle nozzle, steam jet nozzle, vortex whistle nozzle, and mechanical vibratory nozzle (25).

Selection of Atomizers

The selection of the atomizer usually means the selection between wheel atomizer or pressure nozzle because the use of the pneumatic nozzle is very limited. The selection may be based on various considerations, such as availability, flexibility, energy consumption, or particle size distribution of the final dry product. The last is the most common case. The sizes of droplets produced by various atomizers are shown in Table 3. The advantages and disadvantages of both wheel and pressure nozzle atomizers are summarized below.

Wheels: Advantages
  Can handle high feed rates in a single wheel
  Suitable even for abrasive materials
  Negligible blockage or clogging tendencies
  Simple droplet size control by changing wheel revolutions
Wheels: Disadvantages
  Higher energy consumption than pressure nozzles
  Higher capital cost than pressure nozzles
  Broad spray requires large chamber
Pressure nozzles: Advantages
  Simple and compact construction, no moving parts
  Low cost
  Low energy consumption
  Required spray characteristics can be produced by alternation of the whirl chamber
    design
Pressure nozzles: Disadvantages
  Control and regulation of spray pattern and nozzle capacity during operation not
    possible
  Swirl nozzles not suitable for suspensions because of phase separation
  Tendency to clog
  Strong corrosion and erosion effects causing enlargement of the orifice, which
    changes the spray characteristics

Table 4 Energy Consumption of Three Main Types of Atomizers

	Energy consumption for atomization of			
Atomizer type	250 kg/h	500 kg/h	1000 kg/h	2000 kg/h
Pressure nozzle, pressure 3–5 MPa	0.4	1.6	2.5	4.0
Pneumatic nozzle, air pressure 0.3				
MPa, air mass rate 0.5–0.6 m³/kg	10.0	20.0	40.0	80.0
Rotary wheel	8.0	15.0	25.0	30.0

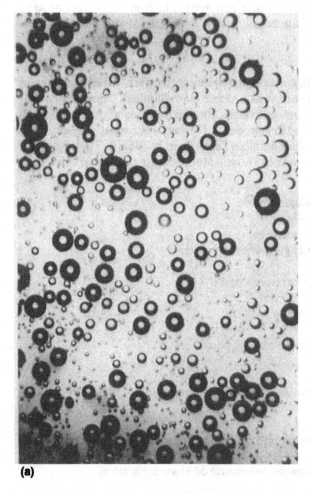

(a)

Figure 12 Spray dried powder: (a) produced by wheel atomizer, (b) produced by pressure nozzle under comparable operating conditions.

The energy consumption of three main types of atomizers is summarized in Table 4.

In many cases rotary and nozzle atomizers can be used with equal success and the choice depends entirely on the manufacturer's tradition. However, there are differences in the dry product characteristics, bulk density, and shape between the wheel and nozzle atomizers (see Fig. 12) (24). In cases in which both wheel and nozzle atomizers produce similar spray patterns, the wheel atomizer is usually preferred because of its greater flexibility.

2.3. Chamber Design

Generally the chamber design depends on the atomizer used and on the air-fluid contact system selected. The selection of atomizer and air-fluid layout is determined by the required characteristics of the dry product and production rate. The product specifications are almost always determined from small-scale tests in an experimental spray dryer. Sometimes they are available from the literature. If heat-sensitive material is involved, attention must be paid to the temperature profile of the drying air along the drying chamber. Another area that requires attention is the droplet trajectory, mainly the trajectory of the largest drops since the size of chamber must be such that the largest drop in

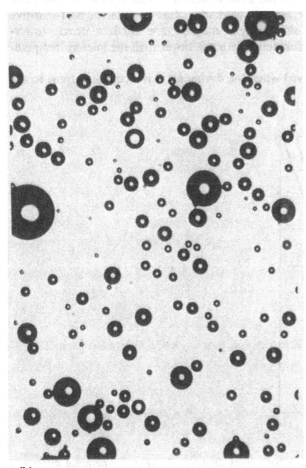

(b)

the spray is dry before it reaches the chamber wall. This requirement prevents the formation of partially dried material buildup on the chamber walls. The height H of the drying chamber as a function of droplet diameter d and temperature differences ΔT between the drying air and the particle is shown in Figure 13.

Chamber Shape

As noted earlier, the chamber shape depends on the type of atomizer employed because the spray angle determines the trajectory of the droplets and therefore the diameter and height of the drying chamber. Typical spray dryer layouts with wheel and nozzle atomization are shown in Figure 14. Correlations are available for calculating the drying chamber sizes without pilot tests. However, these equations require simplifying assumptions that make their application unreliable. The best method so far has been the scaleup from pilot to commercial size.

Air-Droplet Contact Systems

There are three basic types of air-droplet contact systems employed in spray drying processes.

1. Cocurrent contact occurs when the droplets fall down the chamber with the air flowing in the same direction. It is the most common system with both wheel and nozzle atomization. Wheel atomizers are used when fine particles of heat-sensitive material are required; heat-sensitive coarse droplets are dried in nozzle tower-chamber designs. The final product temperature is lower than the inlet air temperature.
2. Countercurrent contact is achieved when the drying air flows countercurrent to the

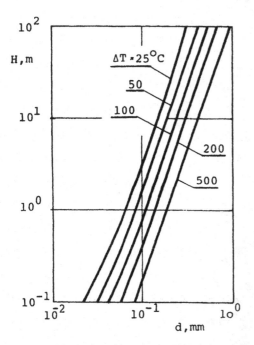

Figure 13 Height H of the drying chamber versus diameter d; ΔT = temperature difference between the drying air and the particle.

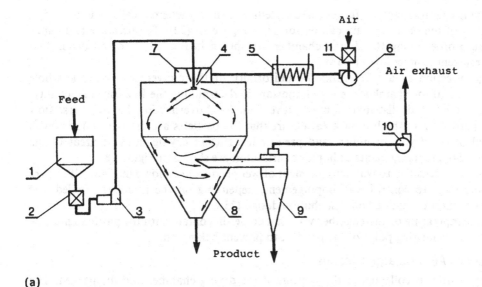

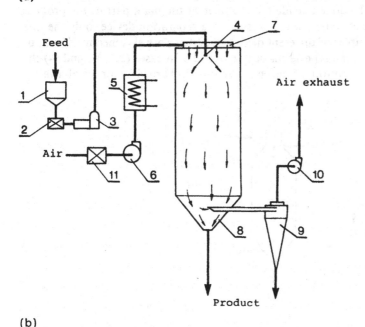

Figure 14 Spray dryer layout: (a) with wheel atomizer; (b) with nozzle atomizer: (1) feed tank; (2) filter; (3) pump; (4) atomizer; (5) air heater; (6) fan; (7) air disperser; (8) drying chamber; (9) cyclone; (10) exhaust fan.

falling droplets or particles. It is used for more heat-sensitive materials that require coarse particles, or special porosity, or high bulk density. Nozzle atomization is usually used. The final product temperature is higher than that of the exit air.

3. Mixed-flow contact is employed when a coarse product is required and the size of the drying chamber is limited. It has so far been the most economical system for a material that can withstand exposure to high temperature in dry form.

The drying chamber layouts for all three systems are shown schematically in Figure 15; a typical set of temperature data can be found in Figure 16 (22). Temperature and vapor pressure profiles along the drying chamber are shown in Figure 17 for cocurrent and countercurrent contact layouts (16).

The direction of air flow and the uniformity of the air velocity over the whole cross-section of the chamber are very important in determining the final product quality. Design of the hot air distribution must prevent the local overheating due to reverse flow of wet particles into the hot air area, ensure that the particles are dry before they reach the wall of the drying chamber, and prevent insufficient drying in some areas of the chamber. Some arrangements of hot air distribution are shown in Figure 18.

The areas sensitive to wall impingement of wet product are shown in Figure 19. There are many ways to control wall impingement, depending on the atomizer applied, the product properties, and the air distribution design (25).

The temperature of the chamber wall that comes in contact with the particles must be lower than the melting point of the product to prevent baking on.

Powder and Air Discharge Systems

The dry powder is collected at the bottom of the drying chamber and discharged. The powder-air separation can be done outside the chamber or the main part of the product can be separated inside the chamber, while outside, in a separation device, only the fine particles are collected. Examples of different discharge systems are sketched in Figure 20.

In case (1) of Figure 20, the separation occurs outside. In cases (2), (3), and (4) the product is accumulated at the bottom of the chamber and discharged by means of a valve.

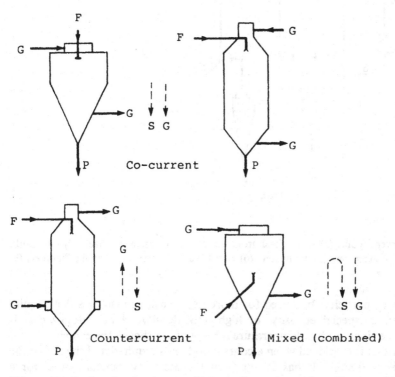

Figure 15 Drying chamber layouts: F, feed; G, gas; P, product; S, spray. (From Reference 22.)

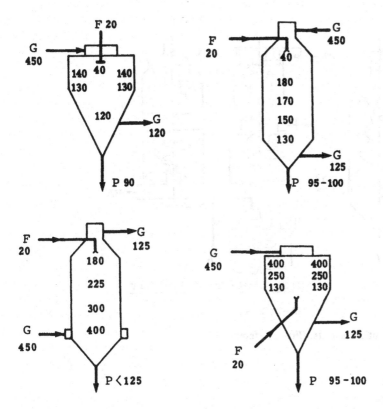

Figure 16 Temperature data in the drying chamber: F, feed; G, gas; P, product. (From Reference 22.)

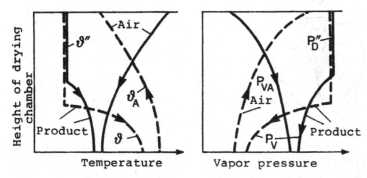

Figure 17 Temperature and vapor pressure profiles along the drying chamber; dashed lines refer to cocurrent operation, solid lines to countercurrent operation. (From Reference 16.)

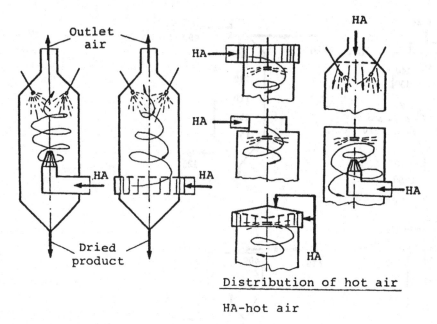

Distribution of hot air

HA-hot air

Figure 18 Hot air distribution layouts. (From Reference 16.)

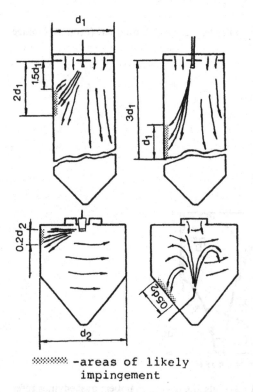

▓▓▓▓▓ -areas of likely
        impingement

Figure 19 Possible areas of impingement on the drying chamber wall. (From Reference 25.)

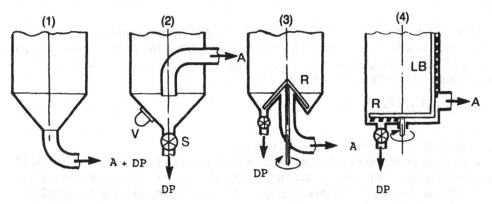

Figure 20 Examples of discharge systems: A, outlet air; DP, dry product; RS, rotating scraper; V, vibrator; AS, air sweeper.

Vibrators and rotating scrapers help convey the powder when the cone angle is too large or the chamber bottom is flat. When necessary the air sweeper can be used to discharge the dry product along the chamber wall, as in case (4).

2.4. Ancillary Equipment

The ancillary equipment used depends generally on the spray drying process layout. Nevertheless, some basic pieces of equipment must be used in any system (see Fig. 1). They are discussed below.

Air Heaters

Steam, indirect
Fuel oil, direct or indirect
Gas, direct or indirect
Electric
Thermal fluids, indirect

Direct heaters may be used if the material can come into contact with products of combustion. Otherwise, indirect air heaters must be used. The type of heater depends on the required temperature of the drying air and on the availability of the heat source; the most common air heater in the food industry is the steam-heated type. Dry saturated steam is usually used within the temperature range 150–250°C. The air temperature is about 10°C lower than the steam temperature. Steam air heaters use extended fin tubes and are relatively inexpensive. The steam rate $\dot{M}_s$ (kg/s) necessary to heat the drying air is given by (25)

$$\dot{M}_s = \frac{L_A C_{PA}(T_{Ao} - T_{Ai})}{H_s - H_c} \eta \tag{15}$$

where L_A is air flow rate (kg/s); C_{PA} is air specific heat; (J/kg·K); T_{Ao} and T_{Ai} are the air outlet and inlet temperatures (K), respectively; H_s and H_c are the enthalpies of steam and condensate (J/kg); and η is heater efficiency (about 0.95).

Indirect fuel oil heaters and gas heaters have similar features. They have separate flow passages for hot gas and drying air. The maximum air temperature at the outlet of a heater is about 400°C.

Direct air heaters are less complicated, and drying air temperatures up to 800°C can be obtained. Gas-fired direct heaters have been used more frequently in recent years because of the increasing availability of natural gas in many parts of the world. The fuel (or gas) combustion rate $\dot{M}_G$ (kg/s) is given by

$$\dot{M}_G = \frac{L_A C_{PA} (T_{Ao} - T_{Ai})}{Q_{cv}} \eta \qquad [16]$$

where Q_{cv} is the calorific value of the fuel or gas (J/kg).

Electric air heaters are used mainly with laboratory and pilot-scale dryers because of high electricity costs. Air temperatures up to 400°C can be achieved using electric heaters.

A relatively new type of air heater is one that uses a thermal fluid, a special oil that transfers heat from a boiler to the air heater. The boiler may be gas or fuel oil fired. This type of heater may be used when steam is not available and/or the temperature of inlet air up to 400°C is required and combustion gases cannot be used directly.

Fans

In a spray drying process, high flow rates of drying air are generally obtained by the use of centrifugal fans. Usually a two-fan system is used, the main fan situated after the powder recovery equipment and the supply fan located in the inlet duct to the drying chamber. Two fans enable better control of the pressure in the chamber. With a single fan after the cyclone, the whole drying system operates under a high negative pressure. The operating pressure in a drying chamber determines the amount of powder in the exhaust air and hence the capacity of the cyclones and their collection of efficiency. In special cases more fans may be used in a drying process, for example a centrifugal fan for the powder pneumatic transport or small fans for blowing cool air to potential hot spots in the drying chamber and atomizer. A typical centrifugal fan is illustrated in Figure 21.

The pressure developed by a fan depends upon the blade design. The most common type of fan in spray drying has backward-curving blades, as shown in Figure 21. Such blades are also used for the supply fan. If the powder-air ratio is high, the backward-curving blades may cause problems with deposit formation on the back side of the blades. In such cases the use of a blade profile intermediate between the backward curving and the radial is recommended. It gives better self-cleaning properties. In Figure 22 are shown the typical characteristic curves for radial and backward-curving blades for constant fan speed and air density.

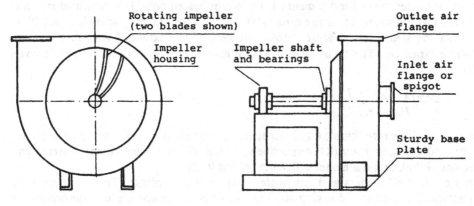

Figure 21 Typical centrifugal fan.

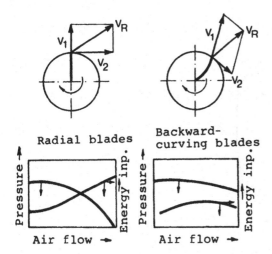

Figure 22 Typical characteristic curves for radial and backward-curving blades.

The fan energy consumption E (in W) may be calculated by means of the following relationship between the volumetric air flow rate $\dot{V}$ (m^3/s) and the total pressure drop that must be overcome, ΔP (Pa):

$$E = \frac{\Delta P \dot{V}}{\eta} \qquad [17]$$

where η is the fan efficiency, usually 0.6–0.75.

Powder Separators

Dry powder: Cyclones, bag filters, electrostatic precipitators
Wet powder: Wet scrubbers, wet cyclones, irrigated fans

Powder separators must separate dry product from the drying air at the highest possible efficiency and collect the powder. Dry separators are used for the principal dry product separation and collection; wet separators are used for the final air cleaning and hence are situated after dry collectors.

In a dry cyclone (see Fig. 23), centrifugal force is employed to move the particles toward the wall and separate them from the air core around the axis. Air and particles swirl in a spiral down the cyclone, where the particles are collected and leave the cyclone. The clean air flows upward and leaves from the top.

In a spray drying process a simple cyclone system or a multicyclone system may be used. The choice between them depends upon the following factors:

1. Size characteristics to determine the smallest particle size that can be separated in a unit
2. Overall cyclone efficiency
3. Pressure drop over the unit

Two characteristics are used to define cyclone performance. They are the *critical particle diameter* (particle size that is completely removed from the air stream) and the *cut size* (the particle diameter for which 50% collection efficiency is achieved). A typical example of theoretically and experimentally obtained efficiency curves is shown in Figure

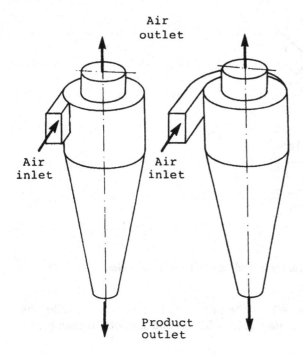

Figure 23 Dry cyclone designs with different types of air inlet.

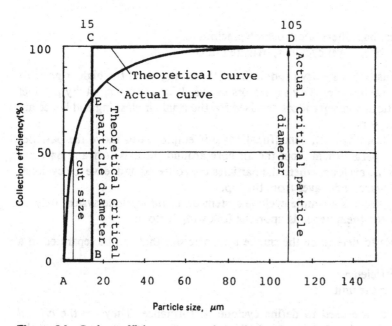

Figure 24 Cyclone efficiency curves, theoretical and actual.

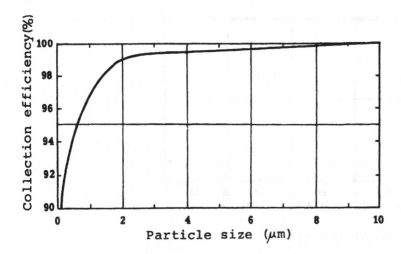

Figure 25 Bag filter efficiency curve.

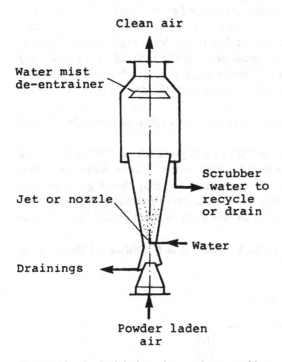

Figure 26 Typical design of venturi wet scrubber.

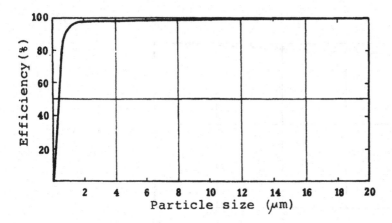

Figure 27 Venturi scrubber efficiency curve.

24. It is evident from this diagram that particles above 15 μm are removed with high efficiency in the cyclone. The pressure drop across the cyclone unit ranges between 700 and 2000 Pa.

A bag filter is widely used in spray drying processes. The air flow containing dry particles passes through a woven fabric; powder is collected on one side of the fabric and the air leaves on the other. A modern unit consists of several bags installed in a baghouse. A typical collection efficiency curve is shown in Figure 25. Very high efficiency is obtained even with 1-μm particles. Bag filters must be carefully maintained to avoid any leakage and regularly cleaned to maintain high operating efficiency. The fabric used in a bag filter is selected in accordance with the characteristics of the dry product and the air temperature.

Electrostatic precipitators are used only rarely in spray drying processes because of their high initial cost.

Wet scrubbers are very commonly used following dry collectors. The particles are separated from air by contacting it with a liquid, usually water. The well-known venturi scrubber is preferred in spray drying systems because it offers easy cleaning and maintenance. It can be used for food and pharmaceutical materials that require hygienic handling. A venturi scrubber is shown in Figure 26 and its typical efficiency curve in Figure 27.

The air carrying fine particles flows through a venturi; at the throat of the scrubber

Table 5 Heat Consumption for Evaporation of 1 kg Water (kJ)

Membrane process	140
Evaporator 1 stage	2600
Evaporator 2 stages	1300
Evaporator 6 stages	430
Evaporator 6, with thermocompression	370
Evaporator 6, with mechanical compression	220
Spray drying process	up to 6000

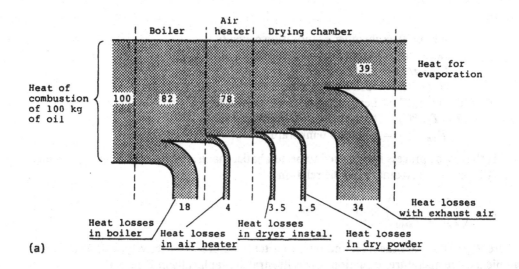

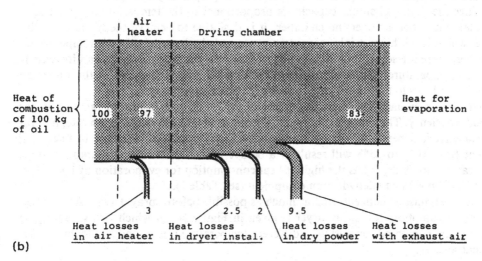

Figure 28 Thermal efficiency for two spray dryers operating with different energy sources and temperature ranges: (a) $\eta = 39\%$, steam, $T_{A\,INLET} = 180°C$, $T_{A\,OUTLET} = 90°C$; (b) $\eta = 83\%$, oil, $T_{A\,INLET} = 500°C$, $T_{A\,OUTLET} = 65°C$. (From Reference 36.)

water is injected to form a spray. The scrubbing liquid containing the product is separated out and discharged from the scrubber base either to sewage or for recirculation. The pressure drop over the scrubber is usually between 2000 and 5000 Pa.

2.5. Thermal Efficiency

The efficiency of the spray drying operation is defined as the ratio of the heat used in evaporation to the total heat input. Thus,

$$\eta = \frac{\dot{M}_{CH}\lambda}{L_A(T_A - T_{WB})C_{PA} + \dot{M}_F(T_F - T_{WB})C_{PF}} \tag{18}$$

where

$$\dot{M}_{CH} \text{ (kg H}_2\text{O/s)} = \text{chamber evaporation capacity}$$
$$\lambda \text{ (J/kg)} = \text{latent heat of evaporation}$$
$$L_A \text{ (kg/s)} = \text{air flow rate}$$
$$\dot{M}_F \text{ (kg/s)} = \text{feed flow rate}$$
$$C_{PA}, C_{PF} \text{ (J/kgK)} = \text{heat capacity of the air and feed}$$
$$T_A, T_F \text{ (°C)} = \text{temperature of the air and feed}$$
$$T_{WB} \text{ (°C)} = \text{wet bulb temperature}$$

If the drying process is supposed to be adiabatic, that is, the heat losses are negligible, Eq. [18] can be approximated to the relation

$$\eta = \left(\frac{T_{Ai} - T_{Ao}}{T_{Ai} - T_{AMB}} \right) \times 100 \qquad (\%) \qquad [19]$$

where T_{Ai} and T_{Ao} are the inlet and outlet air temperatures, respectively, and T_{AMB} is the ambient air temperature. Equation [19] is illustrated graphically in Figure 41.

Since the drying chamber capacity is proportional to the temperature difference of the inlet and outlet air over the chamber, it is desirable to achieve the highest possible value, which is the highest inlet air temperature and the lowest outlet air temperatures. However, there are some limitations to be considered. For many products an increase in the inlet temperature causes serious product damage and a decrease in the outlet air temperature leads to higher moisture content in the dried product.

On the other hand, a decrease in heat input would also cause an increase in the thermal efficiency. The heat consumption is proportional to the evaporation rate, and for a given rate, it depends on the concentration of the dryer feed. Increase in feed solid content from 10% to 25% will result in a 66.6% reduction in heat consumption for a given rate. A spray dryer has the highest heat consumption for evaporation of 1-kg water in comparison with any dehydration equipment (see Table 5).

The feed must be concentrated as much as possible before spray drying. An example of heat consumption in a spray dryer is given in Figure 28, in which two spray drying processes are compared, operating at different temperature ranges and employing different heat sources.

3. SPRAY DRYING SYSTEMS

3.1. Process Layouts and Applications

The characteristics of different process layouts are listed in Table 6 (24). The *open-cycle layout*, which represents the majority of spray dryer systems, is shown in Figure 29. The drying air is taken from the atmosphere and the exhaust air is discharged to the atmosphere. Three types of dry collectors and a wet air cleaner can be used in this layout. A relatively large amount of dry powder (in comparison with other layouts) is lost with the exhaust air. Some applications are shown in Figures 30 through 34 (36).

With reference to Figure 30, the drying medium is a combustion gas, that is, produced at a temperature of about 550°C in a combustion chamber by burning natural gas. The gas outlet temperature is about 120°C. Dryer capacity is 12 ton/h, its specific evaporation rate is in the range from 8 to 9 kg/m^{3}h.

Figure 31 shows an example of a spray dryer very commonly used in food and

Table 6   Spray Dryer Layouts

Layout	Drying medium/feed	Heating	Application
Open cycle	Air/aqueous	Direct or indirect	Exhaust air to atmosphere
Closed cycle	Inert gas/nonaqueous	Indirect (liquid phase or steam)	For evaporation, recovery of solvents; prevention of vapor emissions; elimination of explosion; fire hazards
Semiclosed (standard)	Air/aqueous	Indirect	For handling materials that cannot contact flue gases; elimination of atmospheric emissions
Semiclosed (self-inertizing)	Air with low O_2 content/aqueous	Direct	Products with explosion characteristics; elimination of powder and odor emissions
Two stage	Any of the above layouts plus fluid bed, agglomerators, flash dryers as second stage		For improved powder properties, improved heat economy
Combination	Spray/fluidizer	Direct	Improved energy consumption, reduced capital costs

pharmaceutical industries. The drying air is heated by steam. Because of the usually high cost of the final product, more complex separation equipment is justified or necessary. A spray drying system with the atomizer located in the middle of the drying chamber is shown in Figure 32. With two air inlets it is possible to control the spray characteristics very well. The secondary air entering the bottom of the chamber decreases the moisture content of the product.

The spray dryer shown in Figure 33 enables drying of high-consistency pastelike feeds, such as plastic materials, some salts, and dyestuffs. Here the feed is transported into the atomizer by means of a screw feeder and is sprayed using a two-fluid atomizer. At very high temperatures (up to 850°C) the evaporation rate is about 25 kg/m^3/h.

One of the recent applications of the spray drying process is the drying of waste sludge, which is shown in Figure 34. Solid wastes are burned in a furnace at a temperature of 800–1000°C. The heat of the exhaust gas is used in a spray dryer, where the sludge is atomized and dried. The outlet gas temperature is about 200°C.

A *closed-cycle layout* is shown schematically in Figure 35 for drying materials containing flammable organic solvents. In such cases it is not possible to discharge the solvent vapor into the atmosphere because of toxicity and odor problems. The closed-cycle layout prevents leakage of vapor or powder and minimizes explosion and fire hazards while ensuring total solvent recovery. The system is operated at a slight pressure to prevent any inward leakage of atmospheric air; open-cycle plants are usually operated at a slight vacuum. An example of a closed-cycle layout is shown in Figure 36.

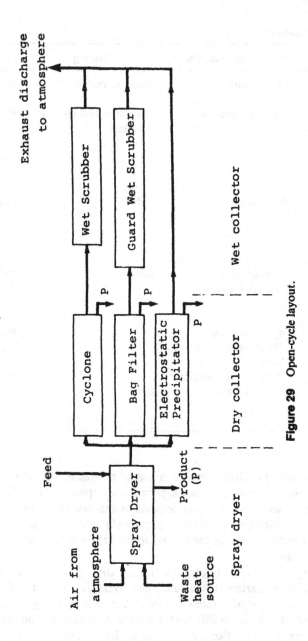

Figure 29 Open-cycle layout.

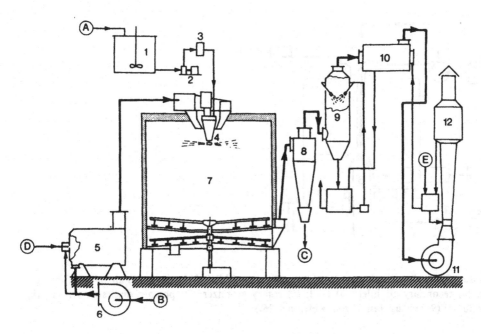

Figure 30 Spray drying of superphosphate: (1) feed tank; (2) feed pump; (3) filter; (4) atomizer; (5) hot gas generator; (6) fan; (7) drying chamber; (8) cyclone; (9) wet scrubber; (10) absorber; (11) exhaust fan; (12) gas outlet; (A) feed; (B) air; (C) product; (D) fuel; (E) water. (From Reference 36.)

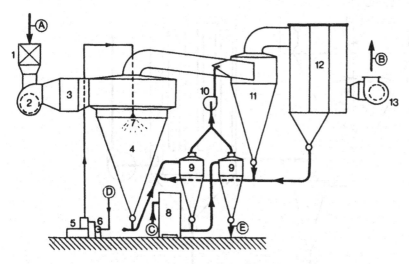

Figure 31 Spray drying of a heat-sensitive material: (1) filter; (2) fan; (3) air heater; (4) drying chamber; (5) high-pressure pump; (6) secondary pump; (7) pressure nozzle atomizer; (8) air conditioner for transporting air; (9) cyclone; (10) fan; (11) secondary cyclone; (12) bag filter; (13) exhaust fan; (A) air; (B) exhaust air; (C) secondary air; (D) feed; (E) product. (From Reference 36.)

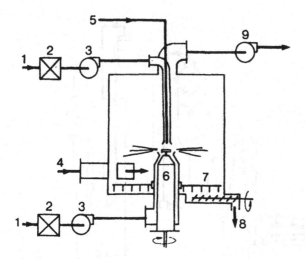

Figure 32 Spray dryer with atomizer located in the middle of chamber: (1) air inlet; (2) heater; (3) fan; (4) secondary air inlet; (5) feed; (6) rotary atomizer; (7) powder discharge system; (8) powder outlet; (9) exhaust fan. (From Reference 36.)

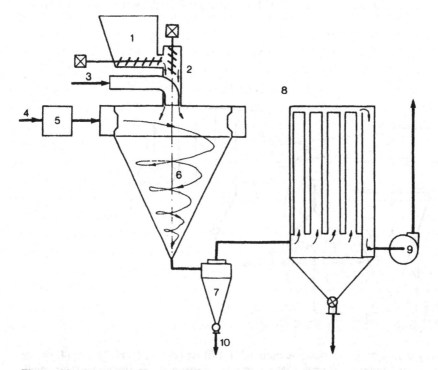

Figure 33 Spray dryer for pastelike materials: (1) feed tank; (2) atomizing device; (3) compressed air inlet; (4) drying gas; (5) heater; (6) drying chamber; (7) cyclone; (8) bag filter; (9) exhaust fan; (10) product outlet. (From Reference 36.)

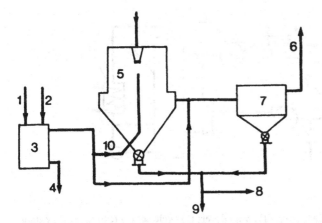

Figure 34 Drying of sludge: (1) dry material from spray dryer, see Reference 9; (2) solid wastes; (3) combustion chamber; (4) ashes discharge; (5) spray dryer with rotary atomizer; (6) exhaust gas; (7) electric filter; (8) dry product; (9) dry material to combustion chamber; (10) highly consistent sludges. (From Reference 36.)

A *semiclosed-cycle layout* is shown in Figure 37 in three types (24):

1. Partial recycle cycle
2. Vented closed cycle with indirect heater
3. Vented closed cycle with direct heater

In a semiclosed cycle, only a part of the drying air is exhausted to atmosphere, thus maintaining powder emission. Most of the drying medium is recycled. Partial recycle (Fig. 37-a) is an old idea that is now used again to utilize the heat in the outlet air and thus reduce the fuel consumption. Up to 20% reduction in fuel consumption is possible when the outlet air temperature exceeds 120°C. The amount of recycled air is about 50% of the total drying air fed to the dryer.

The vented closed cycle with indirect heating (Fig. 37-b) operates under a slight vacuum. The amount of air exhausted to atmosphere is very small and corresponds to the

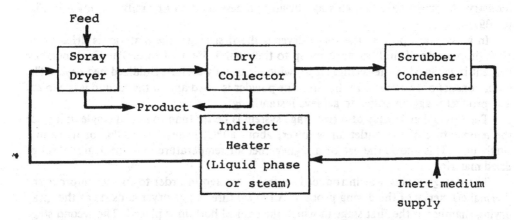

Figure 35 Closed-cycle layout.

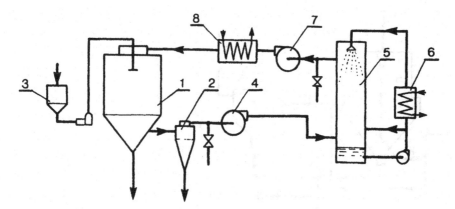

Figure 36 Example of a closed-cycle layout: (1) spray dryer; (2) cyclone; (3) feed tank; (4) exhaust fan; (5) wet scrubber; (6) scrubber cooling system; (7) fan; (8) heater. (From Reference 36.)

air that enters into the system through normal leakage because the dryer is not manufactured to be gastight. This layout can be successfully used to handle toxic materials and those having odor problems, or when contact with combustion gases must be avoided.

The direct-heater layout (Fig. 37-c) can be used for non-heat-sensitive materials. A special application of this layout is the so-called self-inertizing system for materials at high risk for fire and explosion. Any excess air entering the system by leakage is used as additional combustion air in the heater, thus passing through a flame zone where it is deactivated prior to exhaust to the atmosphere. A self-inertizing system is shown in Figure 38.

The advantages of this system are lower fuel consumption and lower cost in comparison with the closed-cycle layout, which must be gastight. It is used, for example, in the pharmaceutical industry for drying of fermentation residues.

The layouts described so far represent one-stage spray drying processes in which drying is carried out in a single unit. Developments in the last decade are oriented to achieve higher product quality at higher thermal efficiency. This has resulted in the design of a two-stage spray drying process. This system uses lower energy (~20% less) and produces a powder with "instant" characteristics, which are often required in the food industry. An example of a two-stage drying process used to dry milk is shown in Figure 39.

In a two-stage process the spray dryer is the first stage; the material is dried up to 10% (in the case of milk) instead of up to the final 3–5% moisture. The final moisture content is achieved in the second stage, which is usually a vibro fluidized bed dryer. The latter consists of two parts. In the first, the powder is dried and in the second it is cooled. The product is agglomerated to achieve instantizing.

The thermal efficiency of a two-stage system is better than that of a single unit since the temperature of the outlet air is lower, about 80°C, instead of 100°C or more in a single unit. This allows the use of a higher inlet air temperature without degradation of dried material.

Three-stage drying was introduced several years ago in order to further improve the thermal efficiency of the drying process. A typical three-stage dryer consists of the spray drying chamber as the first stage in which the conical bottom is placed. The second stage

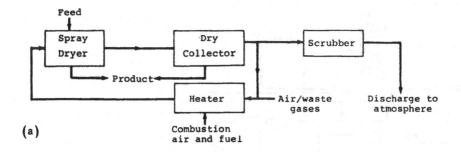

(a)

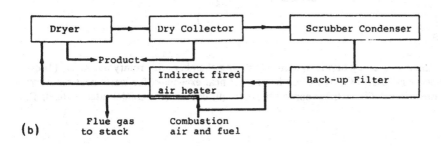

(b)

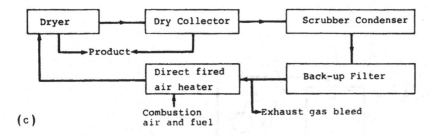

(c)

Figure 37 Semiclosed-cycle layout: (a) partial recycle cycle; (b) vented closed cycle with indirect heater; (c) vented closed cycle with direct heater. (From Reference 24.)

is a static fluid bed. The third stage represents an external fluid bed for final drying and cooling (20).

Figure 40 shows another three-stage dryer used in the food industry (33). In the first stage the feed is dried to a moisture content of 10–20%, depending on the product. The semidried powder is deposited on a moving belt (woven polyester filament) situated at the bottom of the primary spray chamber. The drying air is distributed through the belt and the powder with a relatively high velocity. This is the second stage of drying (crossflow drying). After a short stabilization period in a retention section, the powder is conveyed to the third drying stage in which drying is completed by low-temperature air. The final section is a cooling stage. The final product consists of agglomerates with "instant" properties. The temperature of the exhaust air is even lower than in two-stage systems (65–70°C), thus higher thermal efficiency is obtainable.

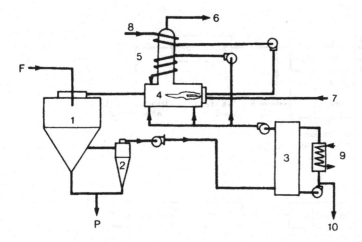

Figure 38 Self-inertizing spray drying system: (1) drying chamber with rotary atomizer; (2) cyclone; (3) scrubber condenser; (4) direct-fired heater; (5) heat exchanger for waste-heat recovery; (6) exhaust to atmosphere; (7) fuel; (8) combustion air; (9) cooler; (10) condensate; F, feed; P, product.

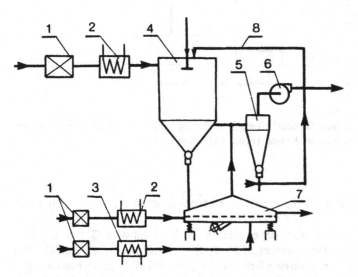

Figure 39 A two-stage spray drying process for milk: (1) air filter; (2) heater; (3) cooler; (4) spray dryer; (5) cyclone; (6) exhaust fan; (7) fluidized bed dryer; (8) return line of fine powder.

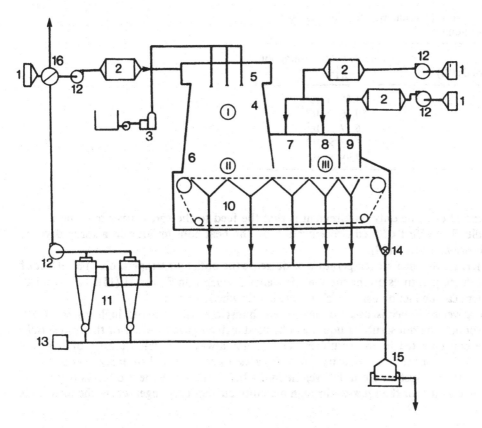

Figure 40 A three-stage spray drying process (FILTERMAT): (1) air filter; (2) heater-cooler; (3) high-pressure pump; (4) nozzle system; (5) air distributor; (6) primary drying chamber; (7) retention chamber; (8) final drying chamber; (9) cooling chamber; (10) FILTERMAT belt assembly; (11) cyclones; (12) fan; (13) fines recovery system; (14) FILTERMAT powder discharge; (15) sifting system; (16) heat recovery system; I, first drying stage, II, second drying stage; III, third drying stage.

3.2. Energy Savings

Spray drying is a very energy-intensive process. There are three main reasons for this.

1. It is necessary to supply the specific heat of evaporation in a short time.
2. The temperature difference across the drying chamber is relatively small because the heat-sensitive materials, which are mostly spray dried, do not permit the use of high-temperature inlet air. The required quality of final product does not permit the use of low-temperature outlet air, either.
3. An appreciable amount of heat is lost with the exhaust air. Nonideal performance of the powder collection system causes loss of dry product.

Hence energy savings strategies for spray drying processes correspond to the three main categories mentioned above.

1. Remove as much water from the feed as possible before drying. It is shown in Table 7 that preconcentration of the feed results in significant energy savings. The most effective is mechanical dehydration followed by membrane processes or centrifugal

Table 7 Heat Consumption for Various Feed
Concentrations

Feed solids (%)	Approximate heat consumption (kJ/kg powder)
10	23.65×10^3
20	10.46×10^3
30	6.17×10^3
40	3.97×10^3
50	2.68×10^3

separation. The only requirement is that the feed to the spray dryer must be pump-able. The effect of feed concentration on the heat consumption in a spray dryer is shown in Table 7 (24).

2. Increase the inlet air temperature or decrease the outlet air temperature. The effect of both temperatures on the thermal efficiency is evident in Figure 41 (15). This requirement can be met by using two or more stages when possible.

3. If a wet scrubber is used to clean the exhaust air and to avoid high losses of dry product, the hot scrubber liquid can be used as feed process water or the feed should be concentrated in the scrubber. This is not always possible because of hygienic problems in the food industry. In such cases a good method of energy savings is to lead the exhaust air from the cyclone into a bag filter where the product is recovered. The exhaust air then passes through a countercurrent exchanger where the inlet air is

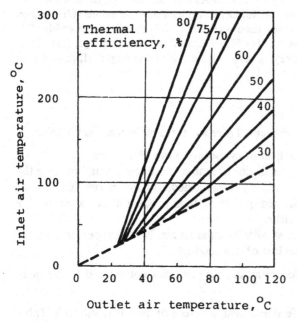

Figure 41 The effect of both inlet and outlet air temperature on thermal efficiency. (From Reference 15.)

preheated up to 75°C. The heat exchanger must be carefully designed so that it is self-cleaning since the exhaust air always contains fine particles.

3.3. Safety Aspects

In spray drying operations a potential danger of explosion and fire can exist under certain conditions.

Fire hazard exists if any of the following conditions occur:

The temperature of the air-product mixture reaches a flammability limit (see Table 8).
The oxygen content in the drying medium is high.

(See relevant chapter of this handbook for details on fire and explosion hazards.)

Dry skim milk has the lowest inflammation temperature in a layer; sugar and some detergents have the lowest temperature when in a cloud in a spray dryer. Temperature data for the drying medium in a spray dryer for three typical materials are shown in Figure 42. The temperatures are given at the following locations: T1, heater outlet; T2, drying chamber inlet; T3, drying chamber outlet; T4, exhaust air; T5, pneumatic transport location. Useful data for minimum ignition temperatures of milk products can be found in Reference 37.

It is evident that the drying air temperature nearly always exceeds the inflammation temperature in a layer. This means that ignition is possible whenever dry powder deposits are formed. In the case of detergents, ignition is also possible in the cloud.

Ignition may be initiated for any one of the following reasons:

Spontaneous combustion in product deposits
Hot solid particles entering the dryer with drying gas
Spark generation through friction

Table 8 Fire and Explosion Data for Some Spray-Dried Materials

| Powder | Inflammation temperature | | Explosion concentration (g/m³) | Explosion pressure |
	Layer (°C)	Cloud (°C)		
Wheat starch	—	410–460	7–22	High
Pudding powder	—	—	20	High
Sugar powder	—	360–410	17–77	High
Cream topping	—	—	6.3	High
Monoglyceride	290	370	16	High
Monoglyceride + skim milk	282	435	32	High
Baby food	205	450	36	High
Milk concentrate	190–203	440–450	22–32	High
Skim milk	134	460	52	High
Milk	142	420	54	High
Buttermilk	194	480	56	High
Skim milk + whey + fat	183–240	460–465	20–24	High
Coffee extract	160–170	450–460	50	
Cocoa	170	460–540	103	High
PVC	—	595	40	
Detergents	160–310	360–560	170–700	Low

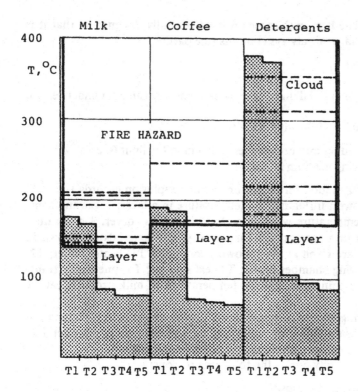

Figure 42 Temperature data for the drying air in a spray dryer. (From Reference 7.)

Electrical failure
Static electricity discharge

Fire prevention methods are based on proper operation and cleaning of the dryer. An important safety device is an automatic control system, since it can maintain the temperature within ±1°C range while the manually controlled temperature may vary ±10°C. The control system should include a detection system of parameter deviation, deposit detection, drying gas analysis, and continuous monitoring of the humidity and temperature at selected locations.

The *explosion hazard* depends on the value of the critical particle concentration, oxidation velocity, and the explosion pressure. If any of these conditions is reached, there is danger of ignition. A comparison of both explosive and process concentration for three products is shown in Figure 43. The powder concentrations are given in the following locations: C1, drying chamber; C2, cyclones; C3, exhaust fan; C4, wet scrubber; and C5, pneumatic transport. Figure 43 shows that even under normal operating conditions the danger of explosion does exist in a normal spray dryer. The explosion hazard is high above the solid line. The zone between the thick and dotted lines is potentially dangerous since the powder concentration there exceeds 50%. All process equipment in which the operating dust concentration exceeds 50% of the lower explosive concentration must be provided with a safety device. The spray drying chamber, cyclones, and pipes should be equipped with safety doors and explosion vents (21).

When highly inflammable materials are dried it is necessary to use inert atmosphere. A gastight closed-cycle spray dryer with nitrogen as the drying medium may be used in

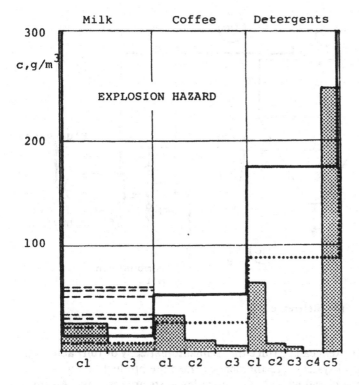

Figure 43 Powder concentration data for a spray dryer. (From Reference 23.)

such cases. When explosion hazard exists, a self-inertizing system is recommended. Good maintenance practice cannot be overemphasized.

For additional information and further references on fire and explosion hazards in drying of solids, the reader is referred to the appropriate chapter in this handbook.

3.4. Control Systems

Spray dryers can be controlled either manually (MCS) or automatically (ACS). MCS is restricted to small spray dryers or when the handled material dries easily. In other cases ACS should be adopted. The outlet air temperature from the drying chamber is the parameter that is controlled since it represents the quality of final product and is easy to monitor. The control of powder content in exhaust air is also monitored when required for environmental reasons.

Two basic types of control systems commonly employed are shown in Figures 44 and 45 (25). Control system A is used mostly with wheel atomizer processes and consists of two circuits. The first controls outlet air temperature by feed rate regulation. The inlet air temperature, controlled by the second circuit, is corrected by the fuel combustion rate. The control system is provided with a safety system that prevents any damage in case of failure in the feed system followed by rapid increase in outlet air temperature.

Control system B is used usually with nozzle dryers, where the feed rate is to be constant. The outlet air temperature is controlled by the regulation of combustion rate in the gas or oil heater or steam pressure when a steam heater is used. Similar safety arrangements, as in the previous control system, are applied here.

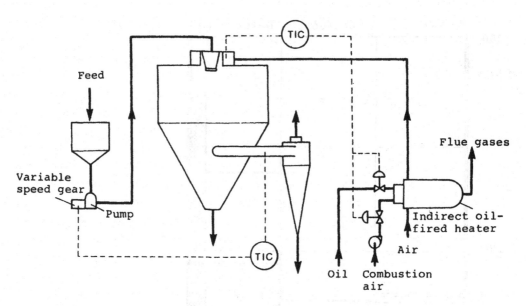

Figure 44   Control system A. (From Reference 25.)

Any spray dryer control system can be run either semiautomatically or fully automatically. A fully automatic control system is recommended when the product quality should meet very stringent requirements and when lower operating costs are essential. A timing device starts up the dryer in a predetermined sequence. In case of failure in the control equipment it is always possible to employ manual control and to continue production without interruption. The shutdown and cleaning operations are programmed and controlled by means of timing equipment.

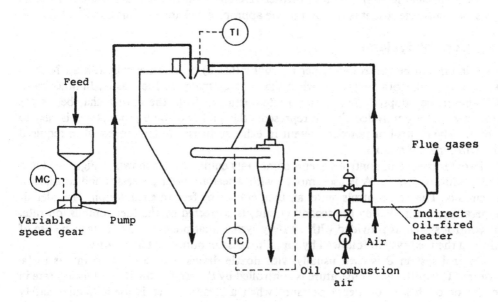

Figure 45   Control system B. (From Reference 25.)

The ACS must include a control system for fire detection, a system for fire or explosion prevention, and a programmable system for countermeasures.

3.5. Novel Concepts

A recent example of new applications for spray dryers is in flue gas desulfurization. The classic desulfurization process uses wet scrubbing techniques.

Flue gases with toxic components are passed through a spray absorber chamber where they are mixed with a fluid absorbent, usually slaked lime. A spray is formed by a rotary atomizer. Excellent contact between the gas and spray is achieved by a carefully located gas disperser. Spent absorbent is recovered from the base of the chamber and in a dry powder collector (23,1).

This technology can also be used for dry scrubbing of other toxic gases, such as HCl, HF, and HBr, which are found in exhaust gases from waste incinerators, metallurgical and allied industrial processes, and other processes. The absorption efficiency of such a process is as high as 99%.

Use of superheated steam as a drying medium is another attractive idea (2,13). Much remains to be done to test this concept industrially.

4. CONCLUSION

Within the limitation of space an attempt is made to review the overall spray drying system as used industrially. Full-scale units can be designed only with the help of pilot-plant (and in some cases experience with similar full-scale plant) results. Several efforts in recent years have been directed with partial success toward implementation of computer-aided design procedures based on some empirical information and principles of transport phenomena. The coupled heat, mass, and momentum transport processes occurring in even the simplest spray drying chamber are too complex to be modeled with confidence at the present time. However, such models may be used for preliminary estimation of the dryer size or analysis of an existing dryer installation. As of now there is no viable substitute to well-planned and carefully executed pilot-scale data for the scaleup purposes. Interested readers are referred to Refs. 3–5, 8, 12, 14, 17–19, 28–32, and 35, for further details about the spray drying and atomization processes.

ACKNOWLEDGMENT

The authors are grateful to Purnima Mujumdar for her careful typing of this manuscript.

REFERENCES

1. Abuaf, N., and Staub, F. W., Drying of liquid-solid slurry droplets, *Drying '86*, Ed. A. S. Mujumdar, Hemisphere/McGraw-Hill, New York, Vol. 1, pp. 277–284, 1986.
2. Amelot, M. P., and Gauvin, W. H., Spray drying with plasma-heated water vapor, *Drying '86*, Ed. A. S. Mujumdar, Hemisphere/McGraw-Hill, New York, Vol. 1, pp. 285–290, 1986.
3. Anderson, B. S., Low ox spray dryers, *Chemical Age of India*, Vol. 30, No. 11, 1979.
4. Crowe, C. T., Modelling spray-air contact in spray drying systems, *Advances in Drying*, Vol. 1, Ed. A. S. Mujumdar, Hemisphere/McGraw-Hill, New York, 1980, pp. 63–99.
5. Crowe, C. T., Chow, L. C., and Chung, J. N., An assessment of steam operated spray dryers, *Drying '85*, Ed. A. S. Mujumdar and R. Toei, Hemisphere, New York, 1985, pp. 221–229.
6. Dittman, F. W., and Cook, E. M., Analysing a spray dryer, *Chem. Engineering*, Jan. 17, 1977.

7. Filka, P., Safety aspects of spray drying, *Drying '84*, Ed. A. S. Mujumdar, Hemisphere/ McGraw-Hill, New York, 1984.

8. Filková, I., Nozzle atomization in spray drying, *Advances in Drying*, Vol. 3, Ed. A. S. Mujumdar, Hemisphere/Springer-Verlag, New York, 1984, pp. 181–216.

9. Filková, I., Spray drying of non-Newtonian liquids, *Drying '91*, Ed. A. S. Mujumdar and I. Filková, Elsevier Sci. Publ., Amsterdam, p. 94, 1991.

10. Filková, I., and Weberschinke, J., Effect of vane geometry on droplet size and size distribution in spray dryer, *Drying '80*, Ed. A. S. Mujumdar, Hemisphere/McGraw-Hill, New York, Vol. 2, 1980.

11. Filková, I., and Weberschinke, J., Drop size prediction of pseudoplastic fluid in a spray dryer, *Drying '84*, Ed., A. S. Mujumdar, Hemisphere/McGraw-Hill, New York, 1984.

12. Gauvin, W. H., and Katta, S., Basic concepts of spray dryer design, *A.I.Ch.E. J.*, Vol. 22, No. 4, 1976, pp. 713–724.

13. Gauvin, W. H., and Costin, M. H., Spray drying in superheated steam, *Drying '80*, Ed. A. S. Mujumdar, Hemisphere/McGraw-Hill, New York, Vol. 1, pp. 320–331, 1980.

14. Keey, R. B., *Introduction to Industrial Drying Operations*, Pergamon, Oxford, 1978.

15. Kessler, H. G., Heat conservation in concentration and spray drying of milk products, *Drying '80*, Ed. A. S. Mujumdar, Hemisphere/McGraw-Hill, New York, Vol. 1, pp. 339–342, 1980.

16. Kessler, H. G., *Food Engineering and Dairy Technology*, Verlag A. Kessler, Freising, Germany, 1981.

17. King, C. J., Control of food-quality factors in spray drying, *Drying '85*, Ed. R. Toei and A. S. Mujumdar, Hemisphere, New York, 1985, pp. 59–66.

18. King, C. J., Kieckbusch, T. G., and Greenwald, C. G., Food-quality factors in spray drying, *Advances in Drying*, Vol. 1, Ed. A. S. Mujumdar, Hemisphere, New York, 1984, pp. 71–120.

19. Kroll, K., *Trockner und Trocknungsverfahren*, Springer-Verlag, Berlin, 1978.

20. Knipschildt, M. E., Recent developments in spray drying of milk, *Concentration and Drying of Foods*, Ed. Diarmund Mac Carthy, Elsevier Sci. Publ., London, p. 235, 1986.

21. Markowski, A. S., Fire and explosion hazards in dryers, *Drying of Solids*, Ed. A. S. Mujumdar, Sarita Prabashar, India, 1990.

22. Masters, K., Recent developments in spray drying, Chemical Age of India, Vol. 30, No. 11, 1979.

23. Masters, K., Spray drying in environmental control with special reference to flue gas desulphurization, *Drying '80*, Ed. A. S. Mujumdar, Hemisphere/McGraw-Hill, New York, Vol. 2, pp. 401–404, 1980.

24. Masters, K., Spray drying, *Advances in Drying*, Ed. A. S. Mujumdar, Hemisphere/McGraw-Hill, New York, Vol. 1, pp. 269–298, 1980.

25. Masters, K., *Spray Drying*, Leonard Hill Books, London, 1979.

26. Masters, K., Impact of spray dryer design on powder properties, *Drying '91*, Ed. A. S. Mujumdar and I. Filková, Elsevier Sci. Publ., Amsterdam, p. 56, 1991.

27. Matsumoto, S., Belcher, D. W., and Crosby, E. J., Rotary atomizers: Performance understanding and prediction, *Proceedings of ICLASS-85*, London, p. IA/1/1.

28. Maurin, P. G., Peters, H. J., Petti, V. J., and Aiken, F. A., Two-fluid nozzle vs. rotary atomization for dry-scrubbing systems, *CEP*, April 1983.

29. Mujumdar, A. S., Recent developments in drying of solids, *Indian Inst. of Chem. Engineers*, Vol. 4, December 1981.

30. Oakley, D. E., and Bahu, R. E., Spray/gas mixing behaviour within spray dryers, *Drying '91*, Ed. A. S. Mujumdar and I. Filková, Elsevier Sci. Publ., Amsterdam, p. 303, 1991.

31. O'Rourke, P. J., and Wadt, W. R., A two-dimensional, two-phase model for spray dryers, *Los Alamos National Laboratory Report LA-9423-MS*, 1982.

32. Reay, D., Modelling continuous convection dryers for particulate solids—Progress and problems, *Drying '85*, Ed. R. Toei and A. S. Mujumdar, Hemisphere, New York, 1985, pp. 67–74.

33. Rheinlander, P. M., Filtermat—The 3-stage spray dryer from DEC, *Proceedings of the Third*

International Drying Symposium, Birmingham, England, Ed. J. C. Ashworth, Vol. 1, pp. 528–534, 1982.

34. Sears, J. T., and Ray, S., Acoustic spray drying of particle suspensions, *Drying '80*, Ed. A. S. Mujumdar, Hemisphere/McGraw-Hill, New York, Vol. 1, pp. 332–338, 1980.
35. Strumillo, C., Markowski, A., and Kaminski, W., Modern developments in drying of paste-like materials, *Advances in Drying*, Ed. A. S. Mujumdar, Vol. 2, pp. 193–232, 1983.
36. Strumillo, C., *Podstawy Teorii i Techniki Suszenia*, 2nd Ed., Wydawnic-two Naukowo-Techniczne, Warsaw, Poland, 1983.
37. Synnott, E. C., and Duane, T. C., Fire hazards in spray-drying of milk products, *Concentration and Drying of Foods*, Ed. Diarmund MacCarthy, Elsevier Sci. Publ., London, p. 271, 1986.
38. Upadhyaya, R. L., Some progress in atomization, *Drying '82*, Ed. A. S. Mujumdar, Hemisphere/McGraw-Hill, New York, pp. 171–173, 1982.

10
Freeze Drying

Athanasios I. Liapis and Roberto Bruttini*
University of Missouri—Rolla
Rolla, Missouri

1. INTRODUCTION

Certain biological materials, pharmaceuticals, and foodstuffs, which may not be heated even to moderate temperatures in ordinary drying, may be freeze dried. The substance to be dried is usually frozen. In freeze drying, the water or another solvent is removed as a vapor by sublimation from the frozen material in a vacuum chamber. After the solvent sublimes to a vapor, it is removed from the drying chamber where the drying process occurs.

As a rule, freeze drying produces the highest quality food product obtainable by any drying method. A prominent factor is the structural rigidity afforded by the frozen substance at the surface, where sublimation occurs. This rigidity to a large extent prevents collapse of the solid matrix remaining after drying. The result is a porous, nonshrunken structure in the dried product that facilitates rapid and almost complete rehydration when water is added to the substance at a later time.

Freeze drying of food and biological materials also has the advantage of little loss of flavor and aroma. The low processing temperatures, the relative absence of liquid water, and the rapid transition of any local region of the material being dried from a fully hydrated to a nearly completely dehydrated state minimize the degradative reactions that normally occur in ordinary drying processes, such as nonenzymatic browning, protein denaturation, and enzymatic reactions. In any food material some nonfrozen water, which is called *bound* or *sorbed water*, will almost unavoidably be present during freeze drying, but there is very often a rather sharp transition temperature for the still wet region during drying (1), below which the product quality improves markedly. This improvement shows that sufficient water is frozen to give the beneficial product characteristics of freeze drying.

However, freeze drying is an expensive form of dehydration for foods because of the slow drying rate and the use of vacuum. The cost of processing is offset to some extent by the absence of any need for refrigerated handling and storage.

Increasingly, freeze drying is used for dehydrating foods otherwise difficult to dry,

---

*****Current affiliation**: Criofarma-Freeze Drying Equipment, Turin, Italy

such as coffee, onions, soups, and certain seafoods and fruits. Freeze drying is also increasingly employed in the drying of pharmaceutical products. Many pharmaceutical products when they are in solution deactivate over a period of time; such pharmaceuticals can preserve their bioactivity by being lyophilized soon after their production so that their molecules are stabilized.

Systematic freeze drying is a procedure mainly applied to the following categories of material (1–14):

1. Nonliving matter, such as blood plasma, serum, hormone solutions, foodstuffs, pharmaceuticals (e.g., antibiotics), ceramics, superconducting materials, and materials of historical documents (e.g., archaeological wood)
2. Surgical transplants, which are made nonviable so that the host cells can grow on them as the skeleton, including arteries, bone, and skin
3. Living cells destined to remain viable for long periods of time, such as bacteria, yeasts, and viruses

Freeze drying requires very low pressures or high vacuum in order to produce a satisfactory drying rate. If the water was in a pure state, freeze drying at or near 0°C at an absolute pressure of 4.58 mm Hg could be performed. But, since the water usually exists in a combined state or a solution, the material must be cooled below 0°C to keep the water in the solid phase. Most freeze drying is done at −10°C or lower at absolute pressures of about 2 mm Hg or less.

In short, freeze drying is a multiple operation in which the material to be stabilized is

1. Frozen hard by low-temperature cooling
2. Dried by direct sublimation of the frozen solvent and by desorption of the sorbed or bound solvent (nonfrozen solvent), generally under reduced pressure
3. Stored in the dry state under controlled conditions (free of oxygen and water vapor and usually in airtight, opaque containers filled with inert dry gas)

If correctly processed, most products can be kept in such a way for an almost unlimited period of time while retaining all their initial physical, chemical, biological, and organoleptic properties, and remaining available at any time for immediate reconstitution. In most cases this is done by addition of the exact amount of solvent that has been extracted, thus giving to the reconstituted product a structure and appearance as close as possible to the original material. However, in some instances, reconstitution can be monitored in order to yield more concentrated or diluted products by controlling the amount of solvent.

Vaccines and pharmaceutical materials are very often reconstituted in physiological solutions quite different from the original but best suited for intramuscular or intravenous injections. Freeze-dried organisms, such as marine animals, plants, or tissue extracts, can also be the starting point of an extraction process (5) using nonaqueous solvents with the purpose of isolating bioactive substances. Freeze drying allows dehydration of the systems without impairing their physiological activity so that they can be prepared for appropriate organic processing.

Another example is the freeze drying of nuclear wastes, which results in the manufacture of dry powders of medium radioactivity. Mixed with appropriate chemicals, they then can be fused into glass bricks or molded to provide low-cost, high-energy radiation sources.

The freeze drying method has also been used in the synthesis of superconducting materials, and produces homogeneous, submicron superconductor powders of high purity (4).

In the chemical industry, catalyzers, adsorbing filters, and expanded plastics can be used in the dry form and placed in the path of appropriate fluids or gases. Freeze-dried dyes may also be dispersed in other media, such as oils and plastics.

These examples are not exhaustive; detailed presentations on the uses of the freeze drying process and of freeze-dried products are given in References 1-6, 8, 14, and 15.

2. FREEZE DRYING PROCESS

Freeze drying is a process by which a solvent (usually water) is removed from a frozen foodstuff or a frozen solution by sublimation of the solvent and by desorption of the sorbed solvent (nonfrozen solvent), generally under reduced pressure. The freeze drying separation method (process) involves the following three stages: (a) the freezing stage, (b) the primary drying stage, and (c) the secondary drying stage.

In the *freezing stage*, the foodstuff or solution to be processed is cooled down to a temperature at which all the material is in a frozen state.

In the *primary drying stage*, the frozen solvent is removed by sublimation; this requires that the pressure of the system (freeze dryer) at which the product is being dried must be less than or near to the equilibrium vapor pressure of the frozen solvent. If, for instance, frozen pure water (ice) is processed, then sublimation of pure water at or near 0°C and at an absolute pressure of 4.58 mm Hg could occur. But, since the water usually exists in a combined state (e.g., foodstuff) or a solution (e.g., pharmaceutical product), the material must be cooled below 0°C to keep the water in the frozen state. For this reason, during the primary drying stage the temperature of the frozen layer (see Fig. 1) is most often at −10°C or lower at absolute pressures of about 2 mm Hg or less. As the solvent (ice) sublimes, the sublimation interface (plane of sublimation), which started at the outside surface (see Fig. 1), recedes, and a porous shell of dried material remains. The heat for the latent heat of sublimation (2840 kJ/kg ice) can be conducted through the

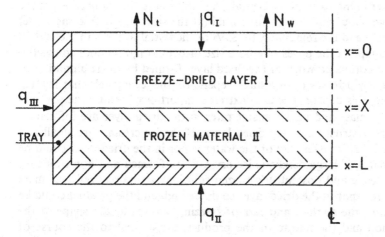

Figure 1 Diagram of a material on a tray during freeze drying. The variable X denotes the position of the sublimation interface (front) between the freeze-dried layer (layer *I*) and the frozen material (layer *II*).

layer of dried material and through the frozen layer, as shown in Figure 1. The vaporized solvent (water) vapor is transported through the porous layer of dried material. During the primary drying stage, some of the sorbed water (nonfrozen water) in the dried layer may be desorbed. The desorption process in the dried layer could affect the amount of heat that arrives at the sublimation interface and therefore it could affect the velocity of the moving sublimation front (interface). The time at which there is no more frozen layer (that is, there is no more sublimation interface) is taken to represent the end of the primary drying stage.

The secondary drying stage involves the removal of solvent (water) that did not freeze (this is termed sorbed or bound water). The secondary drying stage starts at the end of the primary drying stage, and the desorbed water vapor is transported through the pores of the material being dried.

2.1. Freezing Stage

The freezing stage represents the first separation step in the freeze drying process, and the performance of the overall freeze drying process depends significantly on this stage. The material system to be processed (e.g., gel suspension, liquid solution, or foodstuff) is cooled down to a temperature (this temperature depends on the nature of the product) that is always below the solidification temperature of the material system. For instance, if the material to be freeze-dried is a solution with an equilibrium phase diagram that presents a eutectic point (e.g., the solution of NaCl and water presents a eutectic point at $-21.6°C$), then the value of the final freezing temperature must be below the value of the eutectic temperature; in this case, the material becomes wholly crystalline.

In practice, materials display one of two different types of freezing behavior: (a) the liquid phase suddenly solidifies (eutectic formation) at a temperature that depends on the nature of solids in the sample, or (b) the liquid phase does not solidify (glass formation), but rather it just becomes more and more viscous until it finally takes the form of a very stiff, highly viscous liquid. In case (b), there is no such thing as a eutectic temperature, but a minimum freezing temperature.

At the end of the freezing step there already exists a separation between the water to be removed (frozen water in the form of ice crystals) and the solute. In many cases, at the end of the freezing stage about 65–90% of the initial (at the start of the freezing stage) water is in the frozen state and the remaining 10–35% of the initial water is in the sorbed (nonfrozen) state. The shape of the pores, the pore size distribution, and pore connectivity (6,9,11,16–18) of the porous network of the dried layer formed by the sublimation of the frozen water during the primary drying stage depend on the ice crystals that formed during the freezing stage; this dependence is of extreme importance because the parameters that characterize the mass and heat transfer rates in the dried layer are influenced significantly by the porous structure of the dried layer. If the ice crystals are small and discontinuous, then the mass transfer rate of the water vapor in the dried layer could be limited. On the other hand, if large dendritic ice crystals are formed and homogeneous dispersion of the preeutectic and posteutectic frozen solution can be realized, the mass transfer rate of the water vapor in the dried layer could be high and the product could be dried more quickly. Thus, the method and rate of freezing, as well as the shape of the container of the solution and the nature of the product, are critical to the course of lyophilization because they affect the drying rate and the quality of the product.

In industrial freeze dryers the freezing of the product is usually made in the same plant where the drying also occurs. In the vacuum-spray freeze dryer the solvent evaporation autofreezes the small particles of product (evaporative freezing), and the freezing stage begins at the same time as the drying stage. In certain food freeze dryers, the freezing of the product is also accomplished by spraying liquid nitrogen in the drying chamber in which the product is placed. In tray and pharmaceutical freeze dryers, the freezing stage is realized by contact between cooled plates and product-supporting containers.

2.2. Primary Drying Stage

After the freezing stage, the drying chamber where the product is placed is evacuated and the chamber pressure is reduced to a value that would allow the sublimation of solvent (water) to take place in the primary drying stage. When the water molecules sublime and enter the vapor phase, they also keep with them a significant amount of the latent heat of sublimation (2840 kJ/kg ice) and thus the temperature of the frozen product is again reduced. If there is no heat supplied to the product by a heat source, then the vapor pressure of the water at the temperature of the product reaches the same value as that of the partial pressure of the water vapor in the drying chamber; therefore, the system reaches equilibrium and no additional water sublimation from the product would occur. Thus, in order to have continuous sublimation of water from the product, the latent heat of sublimation must be provided to the material from a heat source. The heat is supplied to the product usually by conduction, convection, and/or radiation; conduction is realized by contact between heated plates and product-supporting containers.

The amount of heat that can be supplied to the product can not be increased freely because there are certain limiting conditions that have to be satisfied during the primary drying stage. One of the constraints has to do with the maximum temperature that the dried product (freeze-dried layer in Fig. 1) could tolerate without (a) loss of bioactivity, (b) color change, (c) the possibility for degradative chemical and biochemical reactions to occur, and (d) structural deformation in the dried layer. The maximum temperature that the dried product could tolerate without suffering any of the above-mentioned deleterious effects is denoted, for a given product, by T_{scor} (T_{scor} is often called, by convention, the temperature of the scorch point of the dried product).

Another constraint has to do with the maximum temperature the frozen layer could tolerate so that it remains a frozen layer. If the material has a eutectic form and if the temperature of the lowest eutectic is exceeded during the primary drying stage, then melting in the frozen layer (Fig. 1) can occur. The melting at the sublimation interface, or any melting that would occur in the frozen layer, can give rise to gross material faults such as puffing, shrinking, and structural topologies filled with liquid solution. When melting has occurred at some point in the frozen layer, then the solvent at that point cannot be removed by sublimation. Therefore, there is process failure in the drying of the frozen material since the frozen solvent (water) cannot be removed any more from the frozen layer (Fig. 1) only by sublimation, and there has also been, at the least, loss in structural stability.

If the material has a glass form and if the minimum freezing temperature is exceeded during the primary drying stage, then the phenomenon of collapse can occur; this makes the product collapse with a loss of rigidity in the solid matrix. Again in this case, there is process failure in the drying of the frozen material because the water cannot be removed any more from the frozen layer only by sublimation, and there has also been at least a loss in structural stability.

The *structural stability* of a material relates to its ability to go through the freeze drying process without change in size, porous structure, and shape. The maximum allowable temperature in the frozen layer is determined by both structural stability and product stability (e.g., product bioactivity) factors; that is, the maximum value of the temperature in the frozen layer during the primary drying stage must be such that the drying process is conducted without loss of product property (e.g., bioactivity) and structural stability.

Sometimes the product stability factors are related to structural stability factors (as in melting). There are systems in which the product stability factors do not depend on structural stability factors, as is the case for many vaccines, viruses, and bacteria, for which the temperature of the frozen layer during the primary drying stage must be kept well below the melting temperature so that there is a good level of bioactivity and organism survival after drying.

In general, product stability is related to the temperature of the frozen layer during the primary drying stage. The maximum allowable temperature that the frozen layer could tolerate without suffering melting, puffing, shrinking, collapse, and loss of product property or stability is denoted, for a given product, by T_m (T_m is often called, by convention, the melting temperature of the sublimation interface of the frozen layer).

The water vapor produced by the sublimation of the frozen water in the frozen layer and by the desorption of sorbed (nonfrozen) water in the dried layer during the primary drying stage travels by diffusion and convective flow through the porous structure of the dried layer and enters the drying chamber of the freeze dryer. (It should be noted that most of the water removed during the primary drying stage is produced by sublimation of the frozen water in the frozen layer.) The water vapor must be continuously removed from the drying chamber in order to maintain nonequilibrium conditions for the drying process in the system. This is usually accomplished by fitting a refrigerated trap (called an *ice condenser*) between the drying chamber and the vacuum pump; the water vapor is collected on the cooled surface of the condenser in the form of ice. The time at which there is no more frozen layer is taken to represent the end of the primary drying stage.

2.3. Secondary Drying Stage

The secondary drying stage involves the removal of water that did not freeze (sorbed or bound water). In an ideal freeze drying process, the secondary drying stage starts at the end of the primary drying stage. The word *ideal* is used here to suggest that in an ideal freeze drying process only frozen water should be removed during the primary drying stage, while the sorbed water should be removed during the secondary drying stage. But, as we discussed above, in real freeze drying systems a small amount of sorbed water could be removed by desorption from the dried layer of the product during the primary drying stage and thus there could be some secondary drying occurring in the dried layer of the product during the primary drying stage.

In real freeze drying processes, the secondary drying stage is considered to start when all the ice has been removed by sublimation (end of primary drying stage). It is then considered that during the secondary drying stage most of the water that did not freeze (bound water) is removed. The bound moisture is present due to mechanisms of (a) physical adsorption, (b) chemical adsorption, and (c) water of crystallization. While the amount of bound water is about 10–35% of the total moisture content (65–90% of the total moisture could be free water that was frozen and then removed by sublimation during the primary drying stage), its effect on the drying rate and overall drying time is very significant. The time that it takes to remove the sorbed water could be as long or longer than the time that is required for the removal of the free water.

The bound water is removed by heating the product under vacuum. But, as in the case of primary drying, the amount of heat that can be supplied to the product cannot be increased freely because there are certain constraints that have to be satisfied during the secondary drying stage. The constraints have to do with the moisture content and the temperature of the product; these two variables influence the structural stability as well as the product stability during and after drying.

For structural stability, the same phenomena, as in the case of the primary drying stage, have to be considered: collapse, melting (if temperature is increased at constant moisture), or dissolution (if moisture is increased at constant temperature) of the solid matrix can occur. Product stability (e.g., bioactivity) is a function of both moisture content and temperature in the sample, and during secondary drying the moisture concentration and temperature in the sample could vary widely with location and time. This implies that the potential for product alteration to occur in the sample will vary with time and location. The moisture concentration profile is related to the temperature profile in the dried layer; thus, the moisture content in the sample cannot be controlled independently. Since very many products are temperature sensitive, it is usual to control product stability by limiting the value of the temperature during the secondary drying process and then the final moisture content is checked before the end of the cycle (6,19).

In the secondary drying stage, the bound water is removed by heating the product under vacuum; the heat is supplied to the product usually by conduction, convection, and/or radiation. The following product temperatures are usually employed: (a) between 10°C and 35°C for heat-sensitive products and (b) 50°C or more for less-heat-sensitive products.

The residual moisture content in the dried material at the end of the secondary drying stage, as well as the temperature at which the dried material is kept in storage, are critical factors in determining product stability during its storage life. Some vaccines can remain stable for many years when they are stored at −20°C, while a significant loss of titer can be found after one year if they are stored at 37°C (20). Furthermore, certain vaccines such as live rubella and measles can be damaged by overdrying (final moisture content of about 2% is required for best titer retention), while other materials such as chemotherapeutics and antibiotics must be dried to a residual moisture content as low as 0.1% for best results.

3. MICROWAVE FREEZE DRYING

The limitations on heat transfer rates in conventionally conducted freeze drying operations have led early to the attempt to provide internal heat generation through the use of microwave power (21,22). Theoretically, the use of microwaves should result in a very accelerated rate of drying because the heat transfer does not require internal temperature gradients and the temperature of ice could be maintained close to the maximum permissible temperature for the frozen layer without the need for excessive surface temperatures.

If, for instance, it is permissible to maintain the frozen layer at −12°C, then it has been estimated (5) that the drying time for an ideal process using microwaves for a hypothetical 1-in slab would be 1.37 h. It should be noted that this drying time compares very favorably with the 8.75 h required for the case of heat input through the dry layer, 13.5 h for heat input through the frozen layer without dry layer removal, and even with the relatively short drying time of 4 h for the case in which the dry layer was continuously removed. In laboratory tests on freeze-drying of a 1-in-thick slab of beef, an actual drying time of slightly over 2 h was achieved, compared with about 15 h for conventionally dried slabs (23).

In spite of these apparent advantages, the application of microwaves to industrial freeze drying has not been successful (5,24). The major reasons for the failures are the following:

1.   Energy supplied in the form of microwaves is very expensive. It was estimated that it may cost 10–20 times more to supply 1 BTU from microwaves than it does from steam (24).
2.   A major problem in application of microwaves is the tendency to glow discharge, which can cause ionization of gases in the chamber and deleterious changes in the food, as well as loss of useful power. The tendency to glow discharge is greatest in the pressure range of 0.1–5 mm Hg and can be minimized by operating the freeze dryers at pressures below 50 μm. Operation at these low pressures, however has a double drawback: (a) it is quite expensive, primarily because of the need for condensers operating at a very low temperature, and (b) the drying rate at these low pressures is much slower.
3.   Microwave freeze drying is a process that is very difficult to control. Since water has an inherently higher dielectric loss factor than ice, any localized melting produces a rapid chain reaction, which results in "runaway" overheating.
4.   Economical microwave equipment suitable for the requirements of industrial freeze drying of foods and pharmaceuticals on a large continuous scale is not yet available.

In view of all of these limitations, microwave freeze drying is at present only a potential development (25) and is not considered in the following sections of this chapter.

4.   FREEZE DRYING PLANTS AND EQUIPMENT

In the freeze drying plant, three process sections are especially energy consuming. Process section 1 involves the freezing of the wet product. As this is normally considered one of the preparatory steps before the freeze drying proper, we will concentrate on the other two that take place in the freeze drying cabinet (1,8). Process section 2 involves the controlled supply of heat to the product to cover requirements for the sublimation and desorption processes (primary and secondary drying stages). Process section 3 involves the removal from the freeze drying chamber of the vast volumes of water vapor released during the sublimation and desorption processes. Of these three process sections, removal of the water vapor always consumes the largest amount of energy. The efficiency of water vapor removal, the vapor trap system, therefore has a decisive effect on the total energy consumption of the freeze drying plant.

The vapor trap is placed in a chamber communicating with the freeze-drying cabinet. The water vapor condenses to ice on its refrigerated surfaces. When in operation the efficiency of the vapor trip is shown by a small total temperature difference ΔT between the saturation temperature for water vapor at the pressure in the freeze drying cabinet and the evaporation temperature of the refrigerant (Fig. 2). This total temperature difference ΔT results mainly from each of the following three resistances:

1.   Pressure difference ΔP equivalent to the pressure drop caused by the resistances to the vapor flow from the freeze drying cabinet to the cold surfaces of the vapor trap.
2.   The temperature difference ΔT_{ice} through the layer of ice on the cold surface.
3.   The temperature difference ΔT_{refr} between the cold surface and the evaporating refrig-

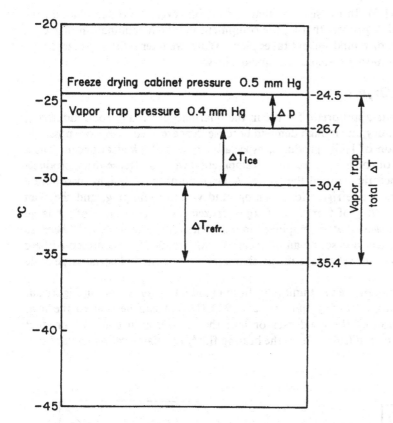

Figure 2 Graphic presentation of the variables ΔP, ΔT_{ice}, ΔT_{refr}, and ΔT.

erant. For an efficient vapor trap it is necessary to have a combination of a large cross-sectional area for the vapor flow (low ΔP), an efficient deicing system (low ΔT_{ice}), and an efficient refrigerating system (low ΔT_{refr}).

A less efficient vapor trap means a higher ΔT, thus demanding a lower evaporation temperature of the refrigerating plant to maintain the required vacuum in the freeze drying cabinet. Lower evaporation temperature means higher operation costs. In this temperature range, an evaporation temperature 10°C lower means 50% increased energy consumption.

When evaluating industrial freeze drying plants, the following characteristics are thus of prime importance:

1. Operation reliability
2. Ease and quality of process control
3. Product losses
4. Vapor trap efficiency

In the following sections, some of the commonly used types of pilot and industrial freeze drying plants are presented and their most important technical features are discussed. The sections of the plants where the product pretreatment inclusive of the freezing operation takes place are very different from one plant to another, depending on the type of products handled (1,5). In the sections that follow, however, we concentrate on the freeze drying installation proper, that is, the equipment in which sublimation of frozen solvent and desorption of bound solvent takes place. When we later refer to freeze drying plants, it will be in this more restricted meaning of the word.

4.1. Pilot Freeze Dryers

Freeze drying pilot units appropriate for use in the pharmaceutical and food industries, as well as in the laboratory, are in high demand because they are used to explore possibilities for the preservation of labile products, especially those of biological origin. These units are portable and of convenient size for developmental work on freeze-dried products in laboratories and factories around the world. A large number of designs incorporate self-contained facilities for refrigeration, heating, and vacuum pumping, and they can freeze-dry batches consisting of from 2 to 20 kg of frozen product. Because of the large variety of pilot freeze dryers that are employed in industries and laboratories and because of the limitation of space to describe all of them, a pilot freeze dryer is presented here with characteristics that are very close to the characteristics of the industrial large-scale lyophilizers.

A schematic diagram of the pilot unit (Criofarma model C5-2) is shown in Figure 3a. The unit consists of (a) a freezing fluid system (R13B1) that can be sent to the heat exchanger in the section of the condenser or into the refrigeration coils for product freezing, (b) a heating circuit (silicon oil is the heating fluid) for plate heating and defrost-

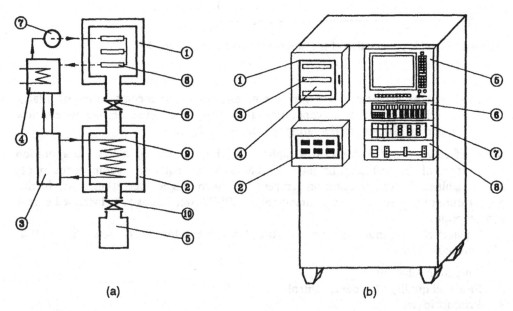

(a)   (b)

Figure 3 Pilot freeze dryer: (a) diagram of Criofarma model C5-2; 1, drying chamber; 2, ice condenser chamber; 3, refrigeration unit; 4, cooling/heating system for the plates; 5, vacuum unit; 6, isolation butterfly valve; 7, silicon oil pump; 8, cooling/heating plate; 9, refrigerated coil; 10, condenser vacuum valve; (b) frontal view of model C5-2; 1, drying chamber; 2, ice condenser chamber; 3, cooling/heating plate; 4, inspection window; 5, computer system; 6, vacuum indicator and regulator; 7, temperature control panel; 8, printer. (Model C5-2 courtesy Criofarma.)

ing of the condenser, and (c) a vacuum system for evacuating air from the apparatus before and during drying.

The rectangular drying chamber shown in Figure 3b is mounted on top of the section of the condenser and the dimensions are 0.4 m × 0.4 m with 0.6 m of depth. Viewing windows are incorporated in the sections of drying and condensation. The refrigeration and vacuum systems are in the internal part of the apparatus with complete dimensions of 1.1 m × 0.8 m × 1.8 m (the dimension 1.8 m represents the height of the apparatus).

For its pilot use, the freeze dryer offers full control of the process variables and is able to achieve conditions of pressure and temperature beyond the limits of production units. The shelf and ice condenser temperatures of the pilot unit can be −50°C and −70°C, respectively, and the pressure in the drying chamber can be as low as 1 Pa or less. The pilot freeze dryer has a control panel fully accessorized with instruments that record and display (a) the temperature inside the product, (b) the temperature on the plates, (c) the temperature of the coils of the condenser, (d) the pressure in the drying chamber, (e) the pressure in the vacuum unit, and (f) the pressure in the section of the condenser. The use of a personal computer, in this pilot unit, with programmable temperature during the freeze drying cycle and programmable input-output logic in the different freeze drying stages, offers a wide variety of drying cycles, as well as the capability for the acquisition of many data, so that process optimization could be examined and studied without the risk and cost of investigating the freeze drying system of interest in a large-production freeze dryer.

4.2. Industrial Freeze Dryers

Tray and Pharmaceutical Freeze Dryers

By far the largest number of the industrial freeze dryers in operation are of the vacuum batch type with freeze drying of the product in trays. There are two main types, depending on the type of condenser used. In one, the condenser plates are alongside the tray-heater assembly and in the same chamber; in the other the condenser is in a separate chamber joined to the first by a wide, in general, butterfly valve. This latter type of plant is always used in pharmaceutical industries, but it can also be used for the freeze drying of foods. Because of the wide variety and complexity of the problems associated with the production of pharmaceuticals by freeze drying, in the following paragraphs the principal features of an industrial tray freeze dryer for pharmaceuticals are presented.

The principal problem in the freeze-drying of pharmaceutical solutions is to operate in sterile conditions. The location of the plant must be able to warrant a sterile condition during the filling, charging before drying, and discharging after drying of the pharmaceutical product. This is realized by facing the drying chamber door in a wall separating the sterile room from the room usually called the machine or nonsterile room.

In the plant, this separation is accomplished with an isolation valve that separates the ice condenser from the drying chamber; this valve is also able to permit (a) the pressure rise test at the end of the freeze drying cycle, (b) the simultaneous discharging/loading of the product and condenser defrosting, and (c) the reduction of cross-contamination between batches to a minimum. All the internal parts of the freeze dryer are of stainless steel type AISI 304L or 316L with a finished surface of 300 mesh or more. In the modern plants, the internal sterilization of the equipment is usually made with pressurized steam at 121°C or more; in old plants, sterilization is realized with the use of certain proprietary sanitizing agents.

The product containers (vials or bottles loaded on stainless steel trays) are usually sterilized in a separate unit before the filling and charging in the freeze dryer. These operations require the presence of people in the sterile room with consequent handling of the containers and possible contamination of the batch. For this reason, the human presence in the sterile room is usually reduced to a number of people that are strictly necessary.

For this purpose, a new freeze dryer plant concept has been developed in order to reduce the risk of product contamination. The plant, as shown in Figure 4 (Criofarma model C300-7) has two doors: a small door for loading the product before drying and a full door (located in a position opposite to the small door) for discharging the product after drying. The condenser is placed on the ground floor, which is below the first floor where the drying chamber is located. The shelves of the freeze dryer are lowered to the bottom of the drying chamber and are then lifted one by one to a position in line with the loading machine. The charging of the product is made under laminar flow of sterile air; the small door is opened only for each plate loading and is then immediately closed.

If the product is unstable and must be frozen within a short time after it is filled into its container, then it is possible to load trays of product onto the precooled shelves a half plate at a time. When the product container is a bottle as shown in Figure 4, it usually has

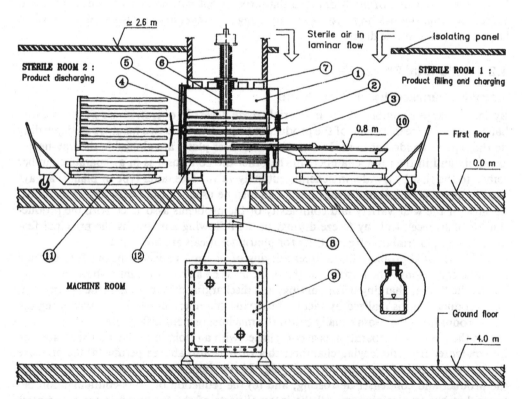

Figure 4 Layout of an industrial freeze dryer with stoppering device (Criofarma model C300-7): 1, drying chamber; 2, inspection window; 3, automatic small door opening; 4, full door; 5, hydraulic press for stoppering the bottles after drying; 6, PTFE elbows for double sterile condition inside the stoppering plug; 7, reenforcing member and cooling coils after steam sterilization; 8, isolation butterfly valve; 9, ice condenser chamber; 10, loading device; 11, discharging device; 12, unloaded shelves. (Model C300-7 courtesy Criofarma.)

on the top a silicon plug that is partially introduced into the bottle; the solvent vapor leaves the container from the free space between the inserted portion of the plug and the container. After drying and before product discharge, the bottles are stoppered in the drying chamber with the plugs that are now fully introduced into the bottles. The stoppering operation is being done (a) in vacuum conditions or (b) at atmospheric pressure by breaking the drying chamber vacuum with sterile nitrogen, which prevents successive oxidation of the product; case (b) is most often employed in practice. The silicon plug in the stoppered bottles provides a protection from contamination and it may be possible to discharge the product in a less sterile environment from the full door of the freeze dryer in only one operation. The entire process may be fully automated as the bottles are removed from the filling machine; the disadvantage of the automation is that the loading time of the freeze dryer may become as long as the time it usually takes to complete the filling step of the operation; this could reduce the theoretical freeze dryer production for a large installation.

Pharmaceutical freeze dryers are very often used to produce raw materials like ampicillin, cloxacillin, and cefazolin (usually as sodium salt), or other specialty materials like collagen. In these systems, the product is usually charged on stainless steel or polyethylene film trays and the plant is usually a medium or a large unit with a loading surface varying from 15 m^2 to 60 m^2.

If the product to be freeze dried is not particularly unstable (e.g., collagen) and can withstand a delay of some hours between being filled into its tray and being frozen, then one can usually accumulate the trays of product on a loading trolley. When the loading trolley is filled, it is placed in front of the freeze dryer and the trays are automatically pushed on the shelves without sliding contact (in order to avoid particle generation) in only one operation. This system is advantageous because it permits maximum utilization of the freeze dryer; the trolley may be loaded ahead of time when the freeze dryer is available for unloading and loading so that the loading operation can be carried out in a few minutes.

If the product is not stable in the liquid state (e.g., ampicillin sodium salt) and must be frozen within a short time after its preparation, it is common to charge the empty trays on the precooled shelves and then to fill the trays so that the freezing step is very quick and can proceed during the whole loading operation. This approach is also advantageous because it reduces the freeze drying cycle time; this happens because the cooling phase starts at the same time as the loading phase, with a consequent reduction in the total time of these two steps.

If the product is directly charged on the trays (bulk production) of the freeze drying equipment, it is found to be convenient to have an additional small ice condenser or so-called auxiliary ice condenser that is also connected with the drying chamber, together with the principal ice condenser. A typical sketch of this device (Criofarma model C1200-20) in a plant of 60 m^2 of loading surface is shown in Figure 5.

With this device, the plant is working with the principal ice condenser for the removal of free water (frozen water) during the primary drying stage (65–90% of the total moisture content is free water), while the plant is working with the unloaded auxiliary ice condenser for the removal of bound water (10–35% of the total moisture is bound water) during the secondary drying stage. The time required for the removal of bound water is usually at least as long as the time required for the removal of free water, and for this reason the auxiliary ice condenser has usually a small independent vacuum and refrigeration system with an installed power of 1/4 to 1/6 of the total refrigeration and vacuum installed power of the plant. The advantages of this device are: (a) the possibility of the

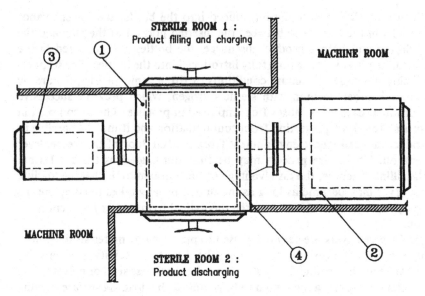

Figure 5 Top view of an industrial freeze dryer of 60-m^2 shelf area with auxiliary condenser (Criofarma model C1200-20): 1, drying chamber with double full doors; 2, principal ice condenser chamber; 3, auxiliary ice condenser chamber; 4, product support shelf. (Model C1200-20 courtesy Criofarma.)

principal ice condenser defrosting before the end of the drying cycle, (b) energy savings that can result in the reduction of the freeze drying running cost, and (c) better performance for the overall drying cycle.

When more freeze dryers than one are used in the production of raw material, then the vacuum line of each plant is connected with the vacuum lines of the other plants by a set of exclusion valves. Thus, if a failure occurs in the vacuum system of one plant, this same plant can end its drying cycle without stopping its operation by using the pumping suction of another plant. A similar device may also be used for the refrigeration units as sometimes one of them is always in a standby condition. The concept that the same vacuum and refrigeration system may operate in different plants is similar to the utilization of industrial multibatch freeze dryers where the simultaneous production of different products is made possible in a freeze drying plant built with a number of batch cabinets programmed to operate with overlapping drying cycles but served by the same central system for (a) tray heating, (b) condenser refrigeration, and (c) vacuum pumping; for each cabinet, the process is individually controlled from a separate control panel.

A user of a pharmaceutical freeze dryer must observe Good Manufacturing Practice (GMP) for processes and equipment to be validated before and during use. These can be divided into three grouped requirements about different parts and functions of the same plant as follows: (a) plant design and equipment materials, (b) control hardware and software validation, and (c) calibration of instruments. The plant design and the materials of the equipment must be such that they eliminate the potential of dirt traps and ensure successful sterilization (usually with clean pressurized steam at 121°C or more; in older plants they use proprietary sanitizing agents). Also, good cleaning access must be provided, sometimes with a clean-in-place (CIP) system (cleaning the inside part of the plant with sterile water sprayed at high pressure from internal nozzles).

The validation of control hardware and software basically requires the suitability of computer hardware assigned for the task, and that computer programs perform consistently within preestablished operational limits so that analysis of the effects of possible failures can be carried out. The calibration of instruments requires that the supplier of a freeze dryer provides a work certificate of calibration and that the user periodically verifies the performance of the instruments with an external authorized and certified instrument. In References 26–33, useful information for GMP compliance, process, and computer system validation can be found.

Multibatch Freeze Dryers

The freeze drying process in a batch plant is normally program controlled to minimize the drying time and to maximize the production of the plant. With a single-batch plant the load on the various systems will be very variable throughout the drying cycle. The material flow and the product handling operations will also be discontinuous because of the batch process characteristic. This means that optimal utilization of resources will not be possible in a single-cabinet batch plant.

To a great extent this disadvantage can be eliminated when an industrial freeze drying plant is built with a number of batch cabinets programmed to operate with staggered, overlapping drying cycles. Each of the cabinets can be charged with products from the same system, and they are served by the same central system for tray heating, for condenser refrigeration, and for vacuum pumping. But, the process is individually controlled for each cabinet from a separate control panel. This makes possible the simultaneous production of different products, which increases the operation flexibility of the plant. With only two cabinets in operation an essential part of the batch disadvantage may be eliminated; for instance, with four cabinets a very good leveling of loads will be achieved. A large number of industrial freeze drying plants operate today in this way as multicabinet batch plants (1,5,14).

Tunnel Freeze Dryers

In the tunnel type of freeze dryer (Fig. 6), the process takes place in a large vacuum cabinet into which the tray-carrying trolleys are loaded at intervals through a large vacuum lock at one end of the tunnel and discharged similarly at the other end.

The freeze dryer shown in Figure 6 consists of a tunnel with vacuum locks at each end, one for loading deep-ribbed aluminum trays containing frozen lumps of food into the tunnel and the other for discharging the freeze-dried product into an air-conditioned room where the dry product is automatically removed by machinery before being packed. The drying conditions are carefully controlled in a number of sections of the tunnel by temperature-pneumatic controllers (1). Vapor constriction plates, fitting closely inside the walls of the tunnel yet allowing the trolleys to pass through, are at two locations in the main section of the tunnel, and gate valves shut off the locks from the main section. The tunnel thus is separated into five independent process zones.

During the period when the trolley is not moving, a tray-lifting device causes all the trays in each trolley to sit on top of the heaters below. The heaters have flat top surfaces and ribs underneath through which vacuum steam circulates. They are cantilevered in pairs from both sides of the tunnel. Vacuum steam heating has several advantages, including a high latent heat of condensation and temperature control by means of pressure.

The refrigeration system consists of a large aqua-ammonia absorption refrigerator instead of a compression plant, mainly because of the ease with which the refrigeration load can be varied by controlling the oil feed to the boiler that heats the absorber.

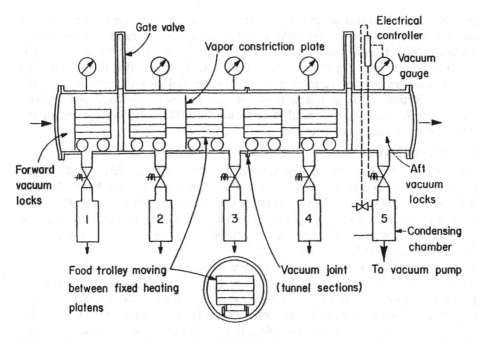

Figure 6 Schematic diagram of a typical tunnel freeze dryer. (After Reference 1, courtesy of Academic Press.)

The total capacity of a tunnel freeze dryer can be increased as the volume of business increases. Large commercial plants for processing cottage cheese and coffee have been built up in this way.

The tunnel freeze dryers have the same advantages of plant capacity utilization that can be achieved as in multibatch plants, but the flexibility for simultaneous production of different products or in switching from one product to another is lacking.

Vacuum-Spray Freeze Dryers

The vacuum-spray freeze dryer shown in Figure 7 has been developed for coffee extract, tea infusion, or milk. The product is sprayed from a single jet upward or downward in a cylindrical tower of 3.7-m diameter by 5.5-m high (1,34). The liquids solidify into small particles by evaporative freezing. In the tower a refrigerated helical condenser is coiled between the inside wall and a central hopper, the latter collecting the partially dry powder as it falls freely to the bottom of the tower, which in turn is connected to a tunnel where the drying process is completed on a stainless steel belt traveling between radiant heaters. The product passes into a hopper that feeds a vacuum lock, permitting intermittent removal of the product for packing. The whole plant operates under a vacuum of about 67 Pa. Frozen particles obtained by spraying into a vacuum are about 150 μm in diameter and lose about 15% moisture in the initial evaporation. There is no sticking of these particles.

Generally, sprayed freeze-dried coffee has less flavor than normal freeze-dried coffee and the product from this plant is no exception. However, it is hoped retention can be improved in the dried product by concentration before spraying into the tower.

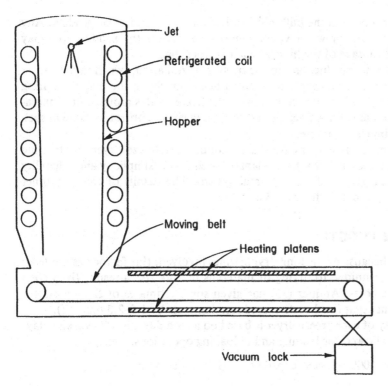

Figure 7 Layout of a vacuum-spray freeze dryer. (After Reference 1, courtesy of Academic Press.)

Continuous Freeze Dryers

Recent years have shown a growing interest in freeze drying plants operating with a continuous flow of material through the process. Particularly in industries working with a single standardized product and the preparation of the product is by a continuous process, such plants are really profitable. They give continuity in processing throughout and constant operating conditions that are easily controlled, and they require less manual operation and supervision. A particular incentive comes from the prospect of balancing the load imposed on the water vapor condensation system and the vacuum system. In a batch process the water vapor evolution rate from the foodstuff is quite high at the start of drying and becomes less as drying proceeds. The condenser system must be designed to handle the maximum water vapor removal requirement.

Continuous freeze dryers are used for freeze drying of product in trays and for freeze drying of agitated bulk materials. When handling the product in trays, the most delicate treatment of the product is achieved. The product is stationary in the tray and therefore is not exposed to abrasion, and it comes in contact only with surfaces that fully meet standards of hygiene.

When agitating a granulated product, more effective heat transfer to the single product particles can be achieved, and thus a considerable reduction of the heating surface is possible. But, both these conditions—abrasion of the product by agitation and increased water vapor production per unit heating surface—tend to carry small product particles with the vapor stream away from the bulk product bed and to cause loss of product. Any complications in the system for water vapor removal to recover the product loss may more than offset the advantage of the higher heater surface load.

The heat transfer to the product and to the trays is by radiation, which in the easiest way safeguards a correct and an evenly distributed heat transfer to the material during the process. The radiant heat is produced by horizontal heater plates grouped in temperature zones. Each tray remains for a fixed period of time in each temperature zone in such a way that the drying time is minimized.

The Conrad system (5) is a commonly used continuous freeze dryer for treating product in trays. The success of this type of plant is based on the simplicity and reliability of each component that goes to form the total system. The details of this and other continuous systems are given in References 5 and 8.

5. FREEZE DRYING COSTS

Freeze drying is costly because of the long drying times involved; this factor has hindered the application of the technique to drying of materials in bulk. As a result, fixed costs tend to exceed running costs. Anquez (35) has given some estimates of these costs for foodstuffs, and it is usual that fixed costs exceed running costs by 1.5–2.5 times (1).

The annual capacity of each freeze dryer is based on a 20-h day and 250 working days per year, with 4 h/day allowed for loading and unloading operations. Thus,

Annual capacity = 5000 × rated capacity   (kg ice per hour)

5.1. Fixed Cost

The average fixed cost is given by the ratio of the annual fixed cost to the annual capacity, assuming an annual fixed cost consisting of capital depreciation at 7.5% per year and loan charges at 8% reducible by the expression

$$C = \frac{R_1 (1 + R_1)^n}{(1 + R_1)^{n-1}} \tag{1}$$

where C equals the annual charge to repay \$1 loan and interest for n years at a rate R_1. Thus with $R_1 = 0.08$ for 10 years, $C = 0.149$ and so the loan charge per capital is \$0.049. Keey (36) points out that this expression also gives the capital recovery factor (fraction of the original capital investment set aside each year over the working life of the plant) for comparing drying systems. Mellor (1) shows that a plant would be preferred that is dearer to install but costs less to run. Other fixed capital-dependent charges at 5% of capital cost include maintenance, insurance, and taxes.

5.2. Running Cost

The running cost consists of labor and utilities costs. Data on the thermophysical properties of foods and biological materials, required for estimating utilities costs, are not always at hand and so the calculation is based only on those properties pertaining to the frozen water content of the material.

Only one person is required to operate any of the dryers, at y dollars per hour. This should increase to $2y$ per hour to cover operating supplies, supervision, payroll overhead, plant overhead, and process control. Then, average labor costs per kilogram ice equals ($2y$)24/20x (plant throughput per hour). Preparation and packing costs are not included as these will depend largely on the nature of the product. The utilities cost can be estimated from a heat and energy balance for 1-kg water undergoing freezing, sublimation, condensation, and melting.

Refrigeration. The heat extracted in freezing the water content of the material from 25°C to −30°C in 1 h is equivalent to 502 kJ/kg ice. The heat removed during condensation at −40°C is approximately 2840 kJ/kg ice. The compressor power operating on ammonia is about 0.65 kW/kW (refrigeration); thus the energy to be supplied to the compressor is 1840 kJ/kg ice.

Heating. The heat required to sublime ice at −20°C equals 2840 kJ/kg ice. The ice collected at the condenser after the completion of the drying cycle has to be melted at −40°C and requires about 419 kJ/kg ice.

Vacuum Pumping. Electricity required for two-stage vacuum pumps equals 0.36 kWh/kg ice.

Total energy	kJ/kg ice	kWh/kg ice
Refrigeration	(502 + 1840)	0.65
Vacuum pumping		0.36
		Total 1.01

If the heat of vaporization of steam is 2065 kJ/kg at a pressure of 682 kPa, then (2840 + 419)/2065 = 1.58 kg steam per kilogram ice is required for heating.

Thus, knowing the prices of 1 kg stream and 1 kWh, the utilities cost can be estimated by using the above information. It is usual to allow an increase to the utilities cost of 20% in order to cover thermal losses and other charges. A more detailed economic analysis is given in Reference 37, which has economic data, analysis, and evaluations that are based on the various operational policies considered in the research studies presented in References 6 and 37, which have considered the removal of both frozen and bound water.

6. PROCESS MODELING: PARAMETERS AND DRYING RATES

6.1. System Formulation

The goal of the process designer and of the processor is to formulate an economical drying system that gives reliably uniform and high product quality (1,2,38–40). A knowledge of the basic phenomena and mechanisms involved in freeze drying is essential for this purpose. In the following sections, a qualitative description and a mathematical model of the freeze drying process is presented; the model could be used to analyze (6,7,9–12,16–18,37,41) rates of freeze drying. The question of drying rates is all important because of the notably long cycle times or residence times that have been required for freeze drying.

In Figure 1, a material being freeze dried in a tray is shown. The thickness of the sides and bottom of the tray, as well as the material from which the tray is made, are most often in practice such that the resistance of the tray to heat transfer could be considered to be negligible (1,6,13,42). Heat q_I could be supplied to the surface of the dried layer by conduction, convection, and/or radiation from the gas phase; this heat is then transferred by conduction to the frozen layer. Heat q_{II} is supplied by a heating plate and is conducted through the bottom of the tray and through the frozen material to reach the sublimation interface or plane. The magnitude of the amount of heat q_{III} in the vertical sides of the tray is much smaller (6,13,42) than that of q_I or q_{II}; q_{III} represents the amount of heat transferred between the environment in the drying chamber and the vertical sides of the tray. Since the contribution of q_{III} is rather negligible when compared to the contributions of q_I and q_{II}, the contribution of q_{III} to the drying rate will not be considered further (6,13,42). The terms N_w and N_t in Figure 1 represent the mass flux of water vapor and the total mass flux, respectively, in the dried layer. The total mass flux is equal to the sum of the mass fluxes of water vapor and inert gas, $N_t = N_w + N_{in}$, where N_{in} denotes the mass flux of the inert gas.

Mathematical Model for the Primary Drying Stage

In the primary drying stage sublimation occurs as a result of heat being conducted to the sublimation interface through the dried (*I*) and frozen (*II*) layers. The resulting water vapor is transported by convection and diffusion through the porous dried layer, enters the vacuum chamber, and finally collects upon the condenser plate. The following assumptions are made in the development of the mathematical model: (a) only one-dimensional heat and mass flows, normal to the interface and surfaces, are considered; (b) sublimation occurs at an interface parallel to and at a distance X from the surface of the sample; (c) the thickness of the interface is taken to be infinitesimal (1,5,6,43); (d) a binary mixture of water vapor and inert gas flows through the dried layer; (e) at the interface, the concentration of water vapor is in equilibrium with the ice; (f) in the porous region, the solid matrix and the gas are in thermal equilibrium; (g) the frozen region is considered to be homogeneous, of uniform thermal conductivity, density, and specific heat, and to contain a negligible proportion of dissolved gases.

Energy balances in the dried (*I*) and frozen (*II*) layers can now be made (6,42):

$$\frac{\partial T_I}{\partial t} = \alpha_{Ie} \frac{\partial^2 T_I}{\partial x^2} - \frac{C_{pg}}{\rho_{Ie} C_{pIe}} \left(\frac{\partial (N_t T_I)}{\partial x} \right) + \frac{\Delta H_v}{\rho_{Ie} C_{pIe}} \left(\frac{\partial C_{sw}}{\partial t} \right), \quad 0 \leq x \leq X \quad [2]$$

$$\frac{\partial T_{II}}{\partial t} = \alpha_{II} \frac{\partial^2 T_{II}}{\partial x^2}, \quad X \leq x \leq L \quad [3]$$

where $\alpha_{Ie} = k_{Ie}/\rho_{Ie} C_{pIe}$, $\alpha_{II} = k_{II}/\rho_{II} C_{pII}$, and $N_t = N_w + N_{in}$. In the dried layer, effective parameters are considered that include the physical properties of both the gas and solid (6,9,11,38,42). The initial and boundary conditions are

$$\text{at } t = 0, \quad T_I = T_{II} = T_X = T^o, \quad 0 \leq x \leq L \quad [4]$$

$$\text{at } x = 0, \quad q_I = -k_{Ie} \frac{\partial T_I}{\partial x} \Big|_{x=0}, \quad t > 0 \quad [5]$$

and,

$$q_I = \sigma F(T_{up}^4 - T_I^4|_{x=0}), \quad t > 0 \quad [6]$$

for radiation heat transfer to the upper dried surface,

$$\text{at } x = X, \quad k_{II} \frac{\partial T_{II}}{\partial x} - k_{Ie} \frac{\partial T_I}{\partial x} + V(\rho_{II} C_{pII} T_{II} - \rho_I C_{pI} T_I) + N_t C_{pg} T_X$$

$$= -\Delta H_s N_w, \quad 0 < t \leq t_{X=L} \quad [7]$$

$$\text{at } x = X, \quad T_I = T_X = T_{II}, \quad t > 0 \quad [8]$$

$$\text{at } x = L, \quad q_{II} = k_{II} \frac{\partial T_{II}}{\partial x} \Big|_{x=L}, \quad t > 0 \quad [9]$$

The continuity (material balance) equations for the dried (*I*) layer are (6,42)

$$\epsilon_p \frac{\partial C_{pw}}{\partial t} + \frac{\partial C_{sw}}{\partial t} + \frac{\partial N_w}{\partial x} = 0 \tag{10}$$

$$\epsilon_p \frac{\partial C_{pin}}{\partial t} + \frac{\partial N_{in}}{\partial x} = 0 \tag{11}$$

where N_w and N_{in} represent the mass fluxes of water vapor and inert gas, respectively, in the dried layer. The term $\partial C_{sw}/\partial t$ in Eqs. [2] and [10] accounts for the change in the concentration of sorbed or bound water with time. The mass fluxes N_w and N_{in} can be obtained from the following constitutive equations:

$$N_w = -D_{win,e} \frac{\partial C_{pw}}{\partial x} + \left(\frac{C_{pw}}{C_{pw} + C_{pin}} \right) N_t \tag{12}$$

$$N_{in} = -D_{inw,e} \frac{\partial C_{pin}}{\partial x} + \left(\frac{C_{pin}}{C_{pw} + C_{pin}} \right) N_t \tag{13}$$

The total mass flux N_t ($N_t = N_w + N_{in}$) is given by

$$N_t = v_p (C_{pw} + C_{pin}) \tag{14}$$

where v_p represents the convective velocity of the gas (water vapor and inerts) in the porous dried (I) layer. The convective velocity v_p of the gas in the dried layer is obtained from Darcy's equation as follows:

$$v_p = -\left(\frac{\kappa}{\mu} \right) \frac{\partial P}{\partial x} \tag{15}$$

In Eq. [15], κ is the permeability of the porous dried (I) layer and μ is the viscosity of the gas. By combining Eqs. [14] and [15], the following expression is obtained for N_t:

$$N_t = -(C_{pw} + C_{pin})\left(\frac{\kappa}{\mu} \right) \frac{\partial P}{\partial x} \tag{16}$$

Equations [12] and [13] are then substituted into Eqs. [10] and [11]. The term $\partial C_{sw}/\partial t$ in Eqs. [2] and [10] can be quantified if a thermodynamically consistent mathematical model could be constructed that could describe the change in the concentration of bound water with time. Different rate mechanisms may be considered (6,11,39,44–47). One of the rate mechanisms could be given by the following expression:

$$\frac{\partial C_{sw}}{\partial t} = k_1 C_{pw} (C_T - C_{sw}) - k_2 C_{sw} \tag{17}$$

where C_T denotes the maximum equilibrium concentration of sorbed water, and k_1 and k_2 represent the rate constants of the adsorption and desorption steps, respectively. The parameters k_1 and k_2 can be functions of temperature (39,44–48). In Eq. [17], the term $\partial C_{sw}/\partial t$ is negative if $k_2 C_{sw}$ is greater than $k_1 C_{pw}(C_T - C_{sw})$. Of course, if $k_2 C_{sw} \gg k_1 C_{pw}(C_T - C_{sw})$ for all times and everywhere in the dried layer, then the term $\partial C_{sw}/\partial t$ could be the set equal to $-k_2 C_{sw}$ without introducing any significant error in the calculations of the drying rate and time.

The initial and boundary conditions of Eqs. [10], [11], and [14]–[17] are as follows:

at $t = 0$, $C_{pw} = 0$ for $x > 0$ [18]

at $t = 0$, $C_{pin} = 0$ for $x > 0$ [19]

at $t = 0$, $C_{sw} = C_{sw}^o$ for $0 \le x \le L$ [20]

at $x = 0$, $C_{pw} = C^o_{pw} = M_w\left(\dfrac{p^o_w}{RT_I|_{x=0}}\right)$, $t \geq 0$ [21]

at $x = 0$, $C_{pin} = C^o_{pin} = M_{in}\left(\dfrac{p^o_{in}}{RT_I|_{x=0}}\right)$, $t \geq 0$ [22]

at $x = X$, $C_{pw} = C_{pwX} = M_w\left(\dfrac{p_{wX}}{RT_X}\right) = M_w\left(\dfrac{g(T_X)}{RT_X}\right)$,

$\qquad\qquad\qquad 0 < t \leq t_{X=L}$ [23]

at $x = X$, $\dfrac{\partial C_{pin}}{\partial x}\Big|_{x=X} = 0$, $0 < t \leq t_{X=L}$ [24]

at $x = 0$, $P = P^o = p^o_{in} + p^o_w$, $t \geq 0$ [25]

at $x = X$, $N_t = -((C_{pw} + C_{pin})|_{x=X})\left(\dfrac{\kappa}{\mu|_{x=X}}\right)\left(\dfrac{\partial P}{\partial x}\Big|_{x=X}\right)$,

$\qquad\qquad\qquad 0 < t \leq t_{X=L}$ [26]

The total pressure at $x = X$ is given by $P_X = p_{wX} + p_{inX}$, where $p_{wX} = g(T_X)$. The variable p^o_w is the chamber water vapor pressure determined by the condenser design, and the function $g(T_X)$ represents the thermodynamic equilibrium between the frozen product and water vapor (1,6,11).

The mathematical model is completely specified by a material balance at the interface that defines its velocity as

$$V = \frac{dX}{dt} = -\frac{N_w}{\rho_{II} - \rho_{Ie}} \tag{27}$$

where the variable X (position of the interface) is a function of time, $X = X(t)$.

Equations [2]–[27] represent the mathematical model that could be used to describe the dynamic behavior of the primary drying stage of the freeze drying process (6,11,42). This model involves a moving boundary (the position of the sublimation interface) and accounts for the removal of frozen water by sublimation, as well as for the removal of bound water by secondary drying in the dried layer during the primary drying stage. External transport resistances can be easily incorporated into this model by including the expressions developed by Liapis and Litchfield (9).

However, in a well-designed freeze dryer the external mass and heat transfer resistances should not be controlling in determining the drying time. A point to be stressed is that in any freeze drying process it will be desirable to fix the design and operating conditions so that the process is not rate limited by external resistances to either heat or mass transfer. The internal heat and mass transfer resistances are characteristic of the material being dried, but the external resistances are characteristic of the equipment. The design conditions refer to having appropriate capacities for the vacuum pump, the water vapor condenser, and the heaters, and that the spacings between trays are such that the external heat and mass transfer resistances are not significant.

Two limits may possibly be reached during the primary drying stage. First, the surface temperature $T_I(t,o)$ must not become too high because of the risk of thermal damage. Second, the temperature of the interface T_X must be kept well below the melting point. If the outer surface temperature limit (T_{scor}) is encountered first as $T_I(t,o)$ is raised, the process is considered to be heat transfer controlled; to increase the drying rate further, the thermal conductivity k_{Ie} of the dried layer must be raised. Many commercial freeze drying processes are heat transfer controlled (1). If the melting point temperature T_m is

Table 1 Frozen Layer and Maximum Dry Surface Temperatures in Typical Freeze Drying Operation Conducted with Heat Input through the Dry Layer

Food material	Chamber pressure (mm Hg)	Maximum surface temperature (°C)	Frozen layer temperature (°C)
Chicken dice	0.95	60	−20
Strawberry slices	0.45	70	−15
Orange juice	0.05–0.1	49	−43
Guava juice	0.05–0.1	43	−37
Shrimp	0.1	52	−29
Shrimp	0.1	79	−18
Salmon steaks	0.1	79	−29
Beef, quick frozen	0.5	60	−14
Beef, slow frozen	0.5	60	−17

Source: From Reference 5

encountered first, then the process is considered to be mass transfer limited and, in order to increase the drying rate, the effective diffusivity of water vapor in the dried layer $D_{win,e}$ and the total mass flux N_t must be raised (an increase in N_t implies that the convective velocity of the vapor in the pores of the dried layer is increased); the values of $D_{win,e}$ and N_t could be raised by decreasing the pressure in the drying chamber.

The frozen layer temperature must be maintained below the melting point, which may in some cases be 10°C or more below the melting point of ice for the reasons discussed previously. Typical ice temperatures existing in the freeze drying of foods under conditions in which the total pressure was primarily due to water vapor and the heat transfer took place via the dried (I) layer are shown in Table 1 (5). Typical ice temperatures existing in the freeze drying of pharmaceuticals under conditions in which the total pressure was primarily due to water vapor and the heat transfer took place via the dried (I) and frozen (II) layers are shown in Table 2.

It should be noted that in the model presented above, diffusion of sorbed water on the surface of the pores of the dried layer (surface diffusion) and diffusion of sorbed water in the solid material of the dried layer (solid diffusion) were not considered. The data of Pikal et al. (10) suggest that the contribution of solid diffusion in the removal of

Table 2 Frozen Layer and Maximum Dry Surface Temperatures in Typical Freeze Drying Operation Conducted with Heat Input through the Dry and Frozen Layers

Pharmaceutical material	Chamber pressure (mm Hg)	Maximum surface temperature (°C)	Frozen layer temperature (°C)
Ampicillin sodium salt	0.15	40	−24
Cloxacillin sodium salt	0.20	45	−20
Cefalosporin sodium salt	0.15	40	−25
Collagen	0.30	70	−20

bound water is not significant. The surface diffusion mass flux of bound water could be incorporated in the continuity equation for the bound water, and this would increase the complexity of the model; furthermore, the value of the surface diffusion coefficient has to be estimated. The data of Pikal et al. (10) appear to indicate that the desorption (evaporation) of bound water represents the rate-limiting mass transfer process in secondary drying. The model presented above accounts for the desorption of bound water in secondary drying.

The equations of the model presented above can be solved by the numerical method developed by Liapis and Litchfield (49) and Millman (42). This method immobilizes the moving boundary and transforms the problem of the freeze drying process into a problem of fixed extent; then, the numerical solution of the partial differential equations is obtained by the method of orthogonal collocation (39,42,49). This model has been found (11,16) to provide theoretical predictions that agree well with the experimental freeze drying rate and time data. It should be mentioned at this point that, in certain pharmaceutical products, because of their processing origin or for freezing processing purposes, solvents other than water are used together with water. In this case, the mass flux of the solvent, the material balance equation for the solvent, and the rate expression for the removal of bound solvent have to be introduced in the structure of the mathematical model presented above.

For the lyophilization of a pharmaceutical product in vials, the mathematical model of Tang et al. (7) could be used. This model accounts for the removal of frozen and bound water, and the temperature and concentration variables vary with time and with two space variables (one space variable is along the length of the vial and the other is along the radial coordinate of the vial because a cylindrical coordinate system is used).

Mathematical Model for the Secondary Drying Stage

In the secondary drying stage, there is no frozen (II) layer, and thus there is no moving sublimation interface. The secondary drying stage involves the removal of bound water. The thickness of the dried (I) layer is L, and the energy balance in this layer (it has the same form as Eq. [2]) is as follows (6,42):

$$\frac{\partial T_I}{\partial t} = \alpha_{Ie} \frac{\partial^2 T_I}{\partial x^2} - \frac{C_{pg}}{\rho_{Ie} C_{pIe}} \left(\frac{\partial (N_t T_I)}{\partial x} \right) + \frac{\Delta H_v}{\rho_{Ie} C_{pIe}} \left(\frac{\partial C_{sw}}{\partial t} \right), \qquad 0 \le x \le L \qquad [28]$$

The initial and boundary conditions of Eq. [28] are:

$$\text{at } t_s = 0, \qquad T_I = \Psi(x), \qquad 0 \le x \le L \qquad [29]$$

$$\text{at } x = 0, \qquad q_I = -k_{Ie} \frac{\partial T_I}{\partial x} \Big|_{x=0}, \qquad t_s > 0 \qquad [30]$$

and,

$$q_I = \sigma F(T_{up}^4 - T_I^4 |_{x=0}), \qquad t_s > 0 \qquad [31]$$

for radiation heat transfer to the upper dried surface,

$$\text{at } x = L, \qquad q_{II} = k_{Ie} \frac{\partial T_I}{\partial x} \Big|_{x=L}, \qquad t_s > 0 \qquad [32]$$

The material balance equations for the water vapor and the inert gas are given by Eqs. [10] and [11], and the constitutive expressions for N_w and N_{in} are obtained from Eqs. [12] and [13]. Equation [17] represents one possible form for the rate expression of the removal of bound water (as discussed in the primary drying stage), and this equation (or its simpler form, as discussed above) could be used to describe the change in the concentration of sorbed water with time. The total mass flux N_t through the porous dried layer is obtained from Darcy's law, and is given by Eq. [16].

The initial and boundary conditions of Eqs. [10], [11], [16], and [17] in the secondary drying stage, are given by the following expressions:

$$\text{at } t_s = 0, \quad C_{pw} = \gamma(x), \quad 0 \leq x \leq L \tag{33}$$

$$\text{at } t_s = 0, \quad C_{pin} = \delta(x), \quad 0 \leq x \leq L \tag{34}$$

$$\text{at } t_s = 0, \quad C_{sw} = \theta(x), \quad 0 \leq x \leq L \tag{35}$$

$$\text{at } x = 0, \quad C_{pw} = C_{pw}^o = M_w\left(\frac{p_w^o}{RT_I|_{x=0}}\right), \quad t_s > 0 \tag{36}$$

$$\text{at } x = 0, \quad C_{pin} = C_{pin}^o = M_{in}\left(\frac{p_{in}^o}{RT_I|_{x=0}}\right), \quad t_s > 0 \tag{37}$$

$$\text{at } x = 0, \quad P = P^o = p_{in}^o + p_w^o, \quad t_s \geq 0 \tag{38}$$

$$\text{at } x = L, \quad \frac{\partial C_{pw}}{\partial x}\bigg|_{x=L} = 0, \quad t_s > 0 \tag{39}$$

$$\text{at } x = L, \quad \frac{\partial C_{pin}}{\partial x}\bigg|_{x=L} = 0, \quad t_s > 0 \tag{40}$$

The functions $\Psi(x)$, $\gamma(x)$, $\delta(x)$, and $\theta(x)$ provide the profiles of T_I, C_{pw}, C_{pin}, and C_{sw} at the end of the primary drying stage or at the beginning of the secondary drying stage; these profiles are obtained by the solution of the model equations for the primary drying stage. The total pressure at $x = L$ is given by $P_L = p_{wL} + p_{inL}$.

Equations 10–17 and 28–40 represent the mathematical model that could be used to describe the dynamic behavior of the secondary drying stage of the freeze drying process. The numerical solution of the partial differential equations of this model can be obtained by the method of orthogonal collocation (6,39,42,49). External transport resistances can be easily incorporated into this model by including the expressions developed by Liapis and Litchfield (9). But, as it was discussed above (for the model of the primary drying stage), in a well-designed freeze dryer the external resistances should not be controlling in determining the drying time.

For the lyophilization of a pharmaceutical product in vials, the mathematical model of Tang et al. (7) could be used to describe the dynamic behavior of the secondary drying stage (as was discussed above for the case of the primary drying stage).

Effect of Chamber Pressure on the Heat and Mass Transfer Parameters of the Dried Layer

The effective thermal conductivity k_{Ie} in the dried material has been found to vary significantly with the total pressure and with the type of gas present. At very low pressures the thermal conductivity reaches a lower asymptotic value independent of the surrounding gas. This asymptotic conductivity reflects the geometric structure of the solid matrix itself, with no contribution from the gas in the voids of the material since the gas pressure is so low.

At high pressures the thermal conductivity levels out again at a higher asymptotic value. This higher asymptote is characteristic of the heterogeneous matrix composed of solid material and the gas in the voids. Consequently, the high-pressure thermal conductivity is dependent upon the nature of the gas present and specifically increases as the thermal conductivity of the gas increases and, hence, as the molecular weight of the gas decreases.

When the thermal conductivity attains the high-pressure asymptotic value, the mean free path of the gas molecules within the void spaces of the dried layer has become substantially less than the dimensions of the void spaces. During the transition in thermal conductivity from the low-pressure asymptote to the high-pressure asymptote, the mean free path of the gas molecules rivals the void space dimensions in magnitude but, once the mean free path is reduced to the point at which the gas phase within the solid matrix obeys simple kinetic theory, the thermal conductivity stops rising.

This reflects the fact that the thermal conductivity of a gas obeying simple kinetic theory is independent of the pressure. The transition in thermal conductivity between asymptotes usually occurs between 0.1 and 100 mm Hg, which includes the pressures characteristic of freeze drying processes. The pressure range over which the transition in thermal conductivity between asymptotes occurs is characteristic of the pore size distribution of the void spaces within the freeze-dried material (50). A smaller pore dimension means that the gas must achieve a higher pressure in order for the mean free path of the gas to become comparable to the pore spacing and, hence, means that the transition between asymptotes will occur at higher pressures.

Since fast freezing before freeze drying leads to smaller pore spacing after freeze drying (51,52), it follows that faster freezing should lead to lower thermal conductivities at a given pressure. If a freeze drying process is rate limited by internal heat transfer, the rate of freeze drying for fast-frozen material should then be less than that of a slowly frozen material. Slower freeze drying rates for food pieces frozen more rapidly have been reported (51,52). Also, Triebes and King (53) and Saravacos and Pilsworth (54) have found that the thermal conductivity of freeze-dried materials is higher at higher relative humidities of the surrounding gas, in rough proportion to the volume fraction of sorbed water present, and weighted in proportion to the thermal conductivity of liquid water.

Tables 3 and 4 summarize thermal conductivity measurements of freeze-dried food substances and pharmaceuticals, respectively. The surrounding gases and the pressure range are indicated, along with the range of thermal conductivities encountered. Thermal conductivities can be measured by the use of a thermopile apparatus or may be inferred from actual freeze drying rate measurements (8). It will probably be helpful in many cases to make use of the thermal conductivity models for porous media (8,53,55,56) in order to extrapolate and interpolate data to different conditions.

As shown in Tables 3 and 4, the thermal conductivities of dry layers of foods and pharmaceuticals are extremely low compared with the conductivities of insulators, such as cork and styrofoam. As a consequence, the temperature drop across the dry layer is large, and with surface temperatures often limited to values below 65°C because of danger of discoloration and in some cases to values below 38°C because of the danger of denaturation, the resultant ice temperature is usually well below −18°C. Except for materials with very low melting points, it is the surface temperature that limits the drying rate.

As a consequence of this limitation, drying rates attainable in practice are very much below the maximum rates attainable with ice. Thus, for materials loaded into the freeze dryer at about 8–18 kg/m² of tray surface, which corresponds to industrial practice (1), average drying rates are of the order of 1.5 kg of water removed per square meter. The corresponding drying times are 6–10 h. A much more rapid rate of course can be achieved by decreasing the particle size and loading rates (6,37). This corresponds to reduction of the average thickness of the dry layer and thus of mass and heat transfer resistances. This

Table 3 Thermal Conductivities of Freeze-Dried Food Substances

Food substance	Surrounding gas	Range of pressures (mm Hg)	Range of thermal conductivities (BTU/h·ft·°F)
Beef	Water vapor	0.5–2.4	0.020–0.032
Mushrooms	Air	0.3–760	0.006–0.021
Cornstarch solutions	Water vapor and air	0.1–2.0	0.008–0.019
Beef	Air	0.001–760	0.022–0.038
Apple	Air	0.001–760	0.009–0.024
Peach	Air	0.001–760	0.009–0.025
Pear	Freon-12, carbon dioxide, nitrogen, neon, hydrogen	0.02–760	0.013–0.108
Apple	Same	0.02–760	0.013–0.115
Beef	Same	0.02–760	0.022–0.116
Apple	Water vapor	0.01–0.3	0.020–0.067
Milk	Water vapor	0.01–0.3	0.013–0.047
Salmon	Water vapor	0.15	0.024–0.077
Haddock	Water vapor	0.08	0.011–0.015
Perch	Water vapor	0.08	0.013–0.020
Beef	Water vapor and air	0.007–80	0.020–0.037
Beef	Water vapor	0.2–3.0	0.030–0.042
Potato starch	Water vapor and air	0.03–760	0.005–0.024
Gelatin	Same	0.03–760	0.009–0.024
Cellulose gum	Same	0.03–760	0.011–0.032
Egg albumin	Same	0.03–760	0.008–0.024
Pectin	Same	0.03–760	0.007–0.024
Tomato juice	Water vapor	0.4–1.5	0.020–0.100
Turkey	Air, water vapor, Freon-12, carbon dioxide, helium	0.01–760	0.008–0.112

Source: From Reference 8

Table 4 Thermal Conductivities of Freeze-Dried Pharmaceutical Materials

Pharmaceutical material	Surrounding gas	Range of pressures (mm Hg)	Range of thermal conductivities (BTU/h·ft·°F)
Ampicillin sodium salt	Water vapor	0.01–0.15	0.012–0.040
Cloxacillin sodium salt	Water vapor	0.01–0.20	0.012–0.040
Cefalosporin sodium salt	Water vapor	0.01–0.15	0.012–0.017
Collagen	Water vapor	0.1–0.4	0.015–0.040

approach, however, is limited to selected products, since efficient operation requires specialized equipment, such as continuous freeze dryers.

The effective diffusivities $D_{win,e}$ and $D_{inw,e}$ in the dried material are functions of the structure of the material, Knudsen diffusivity, and molecular diffusivity. Simplified expressions for $D_{win,e}$ and $D_{inw,e}$ are given as (38)

$$D_{win,e} = \frac{\epsilon_p}{\tau} \left(\frac{1}{(1 - \alpha y_w)/D_{win} + 1/D_{Kw}} \right) \qquad [41]$$

$$D_{inw,e} = \frac{\epsilon_p}{\tau} \left(\frac{1}{(1 - \beta y_{in})/D_{inw} + 1/D_{Kin}} \right) \qquad [42]$$

where τ is the tortuosity factor of the porous dried layer and y_w and y_{in} are the mole fractions of water vapor and inerts, respectively. The expressions for α and β in Eqs. [41] and [42] are as follows: $\alpha = 1 + (N_{in}M_w/N_wM_{in})$ and $\beta = 1 + (N_wM_{in}/N_{in}M_w)$. The Knudsen and the molecular diffusivities can be obtained from the following expressions (38,57):

$$D_{Kw} = 97.0\bar{r}(T_I/M_w)^{1/2} \qquad [43]$$

$$D_{Kin} = 97.0\bar{r}(T_I/M_{in})^{1/2} \qquad [44]$$

$$D_{win} = \frac{1.8583 \times 10^{-7}T_I^{3/2}}{P\sigma_{win}^2\Omega_{win}} \left(\frac{1}{M_w} + \frac{1}{M_{in}} \right)^{1/2} \qquad [45]$$

$$D_{inw} = \frac{1.8583 \times 10^{-7}T_I^{3/2}}{P\sigma_{inw}^2\Omega_{inw}} \left(\frac{1}{M_{in}} + \frac{1}{M_w} \right)^{1/2} \qquad [46]$$

In Eqs. [43] and [44], the term $\bar{r}$ is the average pore radius in the dried layer in m(meter). In Eqs. [45] and [46], the terms σ_{win} and σ_{inw} are the average collision diameters, and Ω_{win} and Ω_{inw} are the collision integrals.

From Eqs. [45] and [46], it can be observed that the magnitude of D_{win} and D_{inw} would decrease if the total pressure P is increased. If the value of D_{win} is decreased because of increased total pressure, then Eq. [41] indicates that the value of the effective pore diffusivity $D_{win,e}$ could decrease as the total pressure is increased. Therefore, if the total pressure in the drying chamber is increased, then the effective diffusivity $D_{win,e}$ could decrease, and thus the diffusional mass flux of water vapor in the dried layer could decrease.

Furthermore, when the total pressure in the drying chamber is increased, the gradient of the total pressure $\partial P/\partial x$ in the dried layer could be reduced, and this could decrease the convective velocity v_p and the total mass flux N_t (Eqs. [14]–[16]). Since the freeze drying process will become internal mass transfer controlled above a certain pressure ($D_{win,e}$ and N_t decrease with increasing pressure, and k_{Ie} increases with pressure), the highest rate under mass transfer control will occur at the pressure of transition from heat transfer control to mass transfer control and the attainable drying rate will decrease at higher pressures.

More elaborate expressions of $D_{win,e}$ and $D_{inw,e}$ can be found in References 6, 9, 11, 16, 17, and 58. These expressions are more complex than Eqs. [41] and [42]. In all cases, $D_{win,e}$ and $D_{inw,e}$ decrease with increasing total pressure.

In general, operating conditions in freeze drying of foods include maximum surface temperatures of 38-82°C and chamber pressures of 0.1-2 mm Hg. Freeze drying of biological specimens, vaccines, and microorganisms is usually conducted with maximum surface temperatures of 20–32°C and chamber pressures below 0.1 mm Hg. It is possible to conduct freeze drying at atmospheric pressure, provided the gas in which the drying is conducted is very dry. In this case, the heat transfer is improved but the external and internal mass transfer rates deteriorate; mass transfer becomes limiting in atmospheric freeze drying. As a consequence, for all but very small particles, the drying rates are very slow.

7. CONTROL VARIABLES AND POLICIES IN FREEZE DRYING

Much interest has been focused on ways to reduce the drying times of the freeze-drying process so that the proportional amortization and operating costs are minimized. Thus, there has been considerable interest in investigating the factors affecting the batch time of freeze dryers, since this variable is most amenable to control, and efforts have been made to minimize the batch time (1,6,9,12,37,42,59).

The heat variables q_I and q_{II} from the energy sources and the drying chamber pressure P_{ch}, $P_{ch} \cong P^o = p_w^o + p_{in}^o$ ($P_{ch} \cong P^o$ when the external mass transfer resistance is insignificant, as would be the case with a well-designed freeze dryer) are natural control variables. It should be emphasized at this point that (as the equations of the mathematical models for the primary and secondary drying stages and the material in the preceding section indicate) the effects on the heat and mass transfer rates resulting from changes in the values of q_I, q_{II}, and P_{ch} are coupled.

The variable p_w^o is taken to represent the water vapor pressure in the drying chamber (the external mass transfer resistance is taken to be insignificant) and its value is determined by the design and the operational temperature of the ice condenser. Thus, P_{ch} may be changed by changes in p_w^o (p_w^o could be changed by changes in the temperature of the ice condenser), and by increasing or decreasing p_{in}^o. Therefore, changes in the temperature of the ice condenser affect the pressure P_{ch} (through p_w^o) in the drying chamber, and thus the mass transfer rate in the dried layer. The controls q_I, q_{II}, and P_{ch} (the temperature of the ice condenser and the value of p_{in}^o can increase or decrease the value of P_{ch}) must be selected from a set of admissible controllers

$$q_{I^*} \leq q_I \leq q_I^*$$

$$q_{II^*} \leq q_{II} \leq q_{II}^*$$

$$P_{ch^*} \leq P_{ch} \leq P_{ch}^* \qquad [47]$$

This set of controllers excludes those that would produce an unacceptable product quality. Two important constraints on the product state are that the surface temperature $T_I(t,o)$ must not exceed the scorch point of the dried product

$$T_I(t,o) \leq T_{scor} \qquad [48]$$

and that the frozen interface and, in general, the frozen layer must not melt

$$T_X \leq T_m \qquad [49]$$

and

$$T_{II}(t,x) \leq T_m, \qquad X \leq x \leq L \qquad [50]$$

If during the drying run $T_I(t,o) = T_{scor}$ and $T_X < T_m$, then the process is called heat transfer controlled, and if $T_I(t,o) < T_{scor}$ and $T_X = T_m$, then the process is considered to be mass transfer controlled.

The objective is to minimize the total batch time t_b, equivalent to defining a performance index of the form

$$\overline{Q} = \min_{q_I, q_{II}, P_{ch}} Q = \min_{q_I, q_{II}, P_{ch}} \int_o^{t_b} dt \qquad\qquad [51]$$

where t_b is defined as the time when a fixed amount of water remains in the product. The problem given in Eqs. [47] and [51] along with a mathematical model of the freeze drying process (see the sections on the mathematical models for the primary and secondary drying stages) is the standard time optimal control problem (9,12).

Liapis and Litchfield (9) performed a quasi-steady-state analysis for a system where $q_I \neq 0$ and $q_{II} = 0$ and obtained general guidelines about the optimal control policy at the beginning of the drying process (when neither of the state constraints is active), as well as during operation, when the process may be heat and/or mass transfer limited (6,9).

The complete unsteady-state optimal control problem has been studied by Litchfield and Liapis (12) for a system where $q_I \neq 0$ and $q_{II} = 0$ using turkey meat and non-fat-reconstituted milk as model foodstuffs. The results of the dynamic analysis for non-fat-reconstituted milk confirm the suggested control policies of the quasi-steady-state analysis. At low chamber pressures the dynamic analysis with turkey meat showed control results similar to those obtained by the quasi-steady-state analysis. However, at higher pressures the assumed control policy based on the quasi-steady-state analysis was not optimal. The optimal control dynamic study of Litchfield and Liapis (12) suggested that the policies of the quasi-steady-state analysis may be useful guidelines but they should be interpreted with some caution. To obtain accurate optimal control policies on the heat input and chamber pressure of the freeze drying process, the complete unsteady-state optimal control problem should be solved.

Millman et al. (6) studied the freeze drying of skim milk under various operational policies that included the case where $q_I \neq 0$ and $q_{II} \neq 0$. They found that the control policy that produced the shortest primary drying stage was also the policy that provided the shortest overall drying time. Their results show that at least 80% of the heat used during the primary drying stage was transferred through the frozen layer of the sample. They also showed that the type of criterion used in terminating the secondary drying stage is of extreme importance, especially for samples of large thicknesses, since it may lead to an undesirable sorbed (bound) water profile that may deteriorate the quality of the dried product.

It is important to report at this point that since a batch of a pharmaceutical product in an industrial freeze dryer can easily be worth significant amounts of money, it is of paramount importance that the units of the plant and the control systems of the process should always operate under conditions at which there is insignificant loss in the quality of the product being freeze-dried. For this purpose, the freeze dryer usually has one additional refrigeration and/or vacuum unit in a standby condition and furthermore the control instrumentation is designed in such a way that the control policies can be implemented either automatically by computer or by manual override.

Mellor (1) has suggested that periodically time-varied chamber pressure will produce improvements in drying time when compared with conventional steady-state pressure operation. The basis of the argument was that heat must be transmitted through an insulating dried layer to the ice interface in order to provide energy for sublimation and that the average effective thermal conductivity of the dried layer can be enhanced by cycling the pressure. Several industrial and pilot-scale cyclic pressure plants have been constructed, mainly in Australia by CSIRO, and substantial reductions in drying time have been reported (1).

It should be noted that any analysis and/or evaluation of the cyclic pressure freeze drying process should involve non-steady-state heat and mass transfer equations like those presented in the mathematical models sections. The effectiveness of cyclic pressure freeze drying and the effect of cycle period and shape on drying times have been the subject of a number of investigations (1,60,61). Litchfield and Liapis (12) found that optimal policies with respect to pressure could be closely approximated by a constant pressure policy over the entire period of the primary drying stage. This near-optimal constant pressure policy formed the basis for comparison with the cyclic pressure process (1,60). The results for turkey meat showed that all the cyclical policies tried were inferior (although only slightly in some cases) to a near-optimal constant pressure policy developed by Liapis and Litchfield (9) and Litchfield and Liapis (12). Since the capital cost of a cyclic pressure process is considerably greater than that of a constant pressure process (1,8), it appears that the latter process would be preferred.

It should be noted that an increase in pressure will increase k_{Ie} but at the expense of resistance to mass transfer, which is also increased. Hence, the mass flux is reduced and consequently the temperature of the sublimation interface and of the frozen layer is increased. Thus, the thermal conductivity increases but the temperature driving force is decreased.

A simple analysis (2,60) has shown that an increase in pressure (which will increase k_{Ie}) will not guarantee an increase in heat transfer and that an optimum pressure may exist that will maximize the heat flux. If, furthermore, such an optimum does exist, then a cyclic policy will have no beneficial effect since perturbations in either direction away from the optimum will be detrimental.

An optimum chamber pressure for turkey meat has been found experimentally by Sandall et al. (62) and has also been established through theoretical analysis (12). Litchfield et al. (61) compared cyclic pressure and near-optimal constant pressure freeze drying processes in a situation in which operation at a pressure that would minimize drying time was not possible because of an interface temperature constraint; at no time during the entire run did the cycled pressure process prove superior. In view of these considerations, it may be inferred that for materials exhibiting an attainable optimum with respect to pressure, there will be no advantage in cycling the chamber pressure when compared with near-optimal constant pressure operation.

8. CLOSURE

The evolution of freeze drying in the past 40 years indicates that this separation process (unit operation) is a convenient method for drying those decomposable products (mostly pharmaceuticals, e.g., plasma, vaccines, antibiotics, sera, and growth hormones) that cannot be stabilized in any other way or that show markedly improved quality for a rather high average cost (coffee, mushrooms, diced chicken, and others); however, most food products are still dried by conventional means for obvious economic reasons. The absence of interfacial forces during freeze drying has been exploited to produce highly dispersed, homogeneous, free-flowing, and very reactive powders (15). This separation method has found uses in the engineering ceramics area (3) and in the synthesis of superconducting powders (4).

The economics of the process indicate that freeze drying can be suitable for high-value products with specific biological and/or physicochemical properties. The highest cost advantages would be obtained from the processing of concentrated solutions of expensive materials; in this respect, freeze drying could represent a viable alternative to filtration and crystallization. It is certain that freeze drying has a future, and it is likely that this will essentially be in the fields of food, chemistry, materials science, biological sciences, and biotechnology. Its potential evolution is still great and will, of course, depend upon progress in basic research and upon the level of creativity in the design and operating conditions of plants and instruments.

NOMENCLATURE

C annual charge to repay \$1 loan and interest

C_{pg} heat capacity of gas in the dried layer, kJ/kg·K

C_{pin} concentration of inert gas in the dried layer, kg/m^3

C_{pin}^o concentration of inert gas at $x = o$, kg/m^3

C_{pw} concentration of water vapor in the dried layer, kg/m^3

C_{pw}^o concentration of water vapor at $x = o$, kg/m^3

C_{pwX} concentration of water vapor at $x = X$, kg/m^3

C_{ple} effective heat capacity of dried layer, kJ/kg·K

C_{pII} heat capacity of frozen layer, kJ/kg·K

C_{sw} concentration of bound water, kg/m^3 dried layer

C_{sw}^o initial concentration of bound water, kg/m^3 dried layer

C_T maximum equilibrium concentration of bound water, kg/m^3 dried layer

D_{inw} molecular diffusivity of a binary mixture of inert gas and water vapor (Eq. [46]), m^2/s

$D_{inw,e}$ effective pore diffusivity of a binary mixture of inert gas and water vapor in the dried layer (Eq. [42]), m^2/s

D_{Kin} Knudsen diffusivity for inert gas (Eq. [44]), m^2/s

D_{Kw} Knudsen diffusivity for water vapor (Eq. [43]), m^2/s

D_{win} molecular diffusivity of a binary mixture of water vapor and inert gas (Eq. [45]), m^2/s

$D_{win,e}$ effective pore diffusivity of a binary mixture of water vapor and inert gas in the dried layer (Eq. [41]), m^2/s

$g(T_X)$ functional form of the thermodynamic equilibrium between the water vapor and the frozen layer at the temperature of the sublimation interface, T_X ($p_{wX} = g(T_X)$), N/m^2

k_1 rate constant in Eq. [17], m^3/kg·s

k_2 rate constant in Eq. [17], s^{-1}

k_{Ie} effective thermal conductivity in the dried layer, kW/m·K

k_{II} thermal conductivity in the frozen layer, kW/m·K

L sample thickness, m

M_{in} molecular weight of inert gas, kg/kg mole

M_w molecular weight of water vapor, kg/kg mole

N_{in} mass flux of inert gas in the dried layer, kg/m^2·s

N_t total mass flux in the dried layer ($N_t = N_{in} + N_w$), kg/m^2·s

N_w mass flux of water vapor in the dried layer, kg/m^2·s

P_{inX} partial pressure of inert gas at $x = X$, N/m^2

p_{in}^o partial pressure of inert gas at $x = 0$, N/m^2

p_w^o partial pressure of water vapor at $x = 0$, N/m^2

p_{wX} partial pressure of water vapor in equilibrium with the sublimation front ($p_{wX} = g(T_X)$), N/m^2

P total pressure ($P = p_{in} + p_w$) in the dried layer, N/m^2

P^o	total pressure at $x = o$, N/m^2
P_{ch}	total pressure in the drying chamber, N/m^2
q_I	heat flux at $x = o$, kW/m^2
q_{II}	heat flux at the bottom of the tray, kW/m^2
Q	performance index
$\underline{Q}$	optimum performance index
R	gas law constant
R_1	interest rate
t	time, s
t_b	batch time, s
t_s	time for secondary drying stage, s
$t_{X=L}$	time at which the sublimation front arrives at $x = L$, s
T_I	temperature in the dried layer, K
T_{II}	temperature in the frozen layer, K
T^o	initial temperature, K
T_m	melting temperature, K
T_{scor}	scorch temperature, K
T_X	temperature of the sublimation front, K
T_{up}	temperature of upper plate, K
v_p	convective velocity of the binary mixture of water vapor and inert gas in the porous dried layer (Eq. [15]), m/s
V	velocity of the sublimation front (Eq. [27]), m/s
x	space coordinate, m
X	position of sublimation front (interface), m

Greek Symbols

ΔH_s	heat of sublimation of ice, kJ/kg
ΔH_v	heat of vaporization of bound water, kJ/kg
ΔP	pressure drop, N/m^2
ΔT	total temperature difference in Figure 2, K
ΔT_{ice}	temperature difference through the layer of ice on the cold surface, K
ΔT_{refr}	temperature difference between the cold surface and the evaporating refrigerant, K
ϵ_p	void fraction in the dried layer
κ	permeability of the porous dried layer, m^2
μ	viscosity of the binary mixture of water vapor and inert gas in the porous dried layer, kg/m·s
ρ_{Ie}	effective density of the dried layer, kg/m^3
ρ_{II}	density of the frozen layer, kg/m^3
σ	Stefan–Boltzmann constant
τ	tortuosity factor of the porous dried layer

Superscripts

*	maximum value

Subscripts

*	minimum value
I	dried layer
II	frozen layer

REFERENCES

1. Mellor, J. D., *Fundamentals of Freeze Drying*, Academic Press, London, 1978.
2. Liapis, A. I., Freeze Drying, in *Handbook of Industrial Drying*, 1st Edition (A. S. Mujumdar, Editor), Marcel Dekker, New York, pp. 295–326, 1987.
3. Dogan, F., and Hausner, H., The Role of Freeze-Drying in Ceramic Powder Processing, in *Ceramic Transactions*, Volume 1 (Ceramic Powder Science II) (G. L. Messing, E. R. Fuller, Jr., and H. Hausner, Editors), American Ceramic Society, Westerville, Ohio, pp. 127–134, 1988.
4. Johnson, S. M., Gusman, M. I., and Hildenbrand, D. L., Synthesis of Superconducting Powders by Freeze-Drying, in *Materials Research Society Symposium Proceedings*, Volume 121 (Better Ceramics Through Chemistry III) (C. J. Brinker, D. E. Clark, and D. R. Ulrich, Editors), Materials Research Society, Pittsburgh, Pennsylvania, pp. 413–420, 1988.
5. Goldblith, S. A., Rey, L., and Rothmayr, W. W., *Freeze Drying and Advanced Food Technology*, Academic Press, London, 1975.
6. Millman, M. J., Liapis, A. I., and Marchello, J. M., *American Institute of Chemical Engineers Journal (AIChE Journal)*, *31*, 1594, 1985.
7. Tang, M. M., Liapis, A. I., and Marchello, J. M., A Multi-Dimensional Model Describing the Lyophilization of a Pharmaceutical Product in a Vial, in *Proceedings of the Fifth International Drying Symposium*, Volume 1 (A. S. Mujumdar, Editor), Hemisphere Publishing Corporation, New York, pp. 57–65, 1986.
8. King, C. J., *Freeze-Drying of Foods*, Chemical Rubber Co. Press, Cleveland, Ohio, 1971.
9. Liapis, A. I., and Litchfield, R. J., *Chemical Engineering Science*, *34*, 975, 1979.
10. Pikal, M. J., Shah, S., Roy, M. L., and Putman, R., *International Journal of Pharmaceutics*, *60*, 203, 1990.
11. Litchfield, R. J., and Liapis, A. I., *Chemical Engineering Science*, *34*, 1085, 1979.
12. Litchfield, R. J., and Liapis, A. I., *Chemical Engineering Science*, *37*, 45, 1982.
13. Bruttini, R., Rovero, G., and Baldi, G., *Chemical Engineering Journal*, *45*, B67, 1991.
14. Snowman, J. W., Lyophilization Techniques, Equipment, and Practice, in *Downstream Processes: Equipment and Techniques*, Alan R. Liss, New York, pp. 315–351, 1988.
15. van Zyl, A., *ChemSA*, pp. 182–185, June 1988.
16. Liapis, A. I., and Marchello, J. M., A Modified Sorption-Sublimation Model for Freeze-Dryers, in *Proceedings of the Third International Drying Symposium*, Volume 2 (J. C. Ashworth, Editor), Drying Research Limited, Wolverhampton, England, pp. 479–486, 1982.
17. Petropoulos, J. H., Petrou, J. K., and Liapis, A. I., *Industrial and Engineering Chemistry Research*, *30*, 1281, 1991.
18. Petropoulos, J. H., Liapis, A. I., Kolliopoulos, N. P., Petrou, J. K., and Kanellopoulos, N. K., *Bioseparation*, *1*, 69, 1990.
19. Roy, M. L., and Pikal, M. J., *Journal of Parenteral Science and Technology*, *43*, 60, 1989.
20. Freyrichs, C. C., and Herbert, C. N., *J. Biol. Stand.*, *2*(1), 59, 1974.
21. Burke and Decareau, *Advances in Food Research*, *13*, 1, 1966.
22. Copson, D. A., *Microwave Heating*, Avi Publishing Company, Westport, Connecticut, 1962.
23. Hoover, M. W., Markantonatos, A., and Parker, W. N., *Food Technology*, *20*, 807, 1966.
24. Bouldoires, J. P., and LeViet, T., Microwave Freeze-Drying of Granulated Coffee, Second International Symposium on Drying, McGill University, Montreal, Canada, July 7–9, 1980.
25. Sunderland, J. E., *Food Technology*, pp. 50–56, February 1982.
26. Lee, J. Y., *Pharmaceutical Technology*, pp. 54–60, October 1988.
27. Food and Drug Administration, *Lyophilization of Parenterals, Inspection Technical Guide*, U.S. Department of Health and Human Services, Washington, D.C., 1986.
28. Food and Drug Administration, *Guideline on General Principles of Process Validation*, U.S. Department of Health and Human Services, Washington D.C., 1987.
29. Food and Drug Administration, *Guide to Inspection of Bulk Pharmaceutical Chemical Manufacturing*, U.S. Department of Health and Human Services, Washington D.C., 1987.
30. Trappler, E. H., *Pharmaceutical Technology*, pp. 56–60, January 1989.

31. Chapman, K. G., and Harris, J. R., *Pharmaceutical Technology International*, pp. 54–58, May/June 1989.
32. Alford, J. S., and Cline, F. L., *Pharmaceutical Technology*, *14*(9), 88, 1990.
33. Masterson, P. M., *Pharmaceutical Technology*, pp. 48–54, January 1989.
34. Thuse, E., Ginnette, L. F., and Derby, R., U.S. Patent 3,362,835, 1968.
35. Anquez, M., *Fifth International Course on Freeze-Drying*, Lyon and Dijon, France, 1966.
36. Keey, R. B., *Drying: Principles and Practice*, Pergamon Press, Oxford, England, 1972.
37. Millman, M. J., Liapis, A. I., and Marchello, J. M., *Journal of Food Technology*, *20*, 541, 1985.
38. Geankoplis, C. J., *Transport Processes and Unit Operations*, Allyn and Bacon, Boston, 1983.
39. Holland, C. D., and Liapis, A. I., *Computer Methods for Solving Dynamic Separation Problems*, McGraw-Hill, New York, 1983.
40. Belter, P. A., Cussler, E. L., and Hu, W.-S., *Bioseparations—Downstream Processing for Biotechnology*, Wiley-Interscience, New York, 1988.
41. Millman, M. J., Liapis, A. I., and Marchello, J. M., *Journal of Food Technology*, *19*, 725, 1984.
42. Millman, M. J., The Modeling and Control of Freeze Dryers, Ph.D. dissertation, University of Missouri — Rolla, Rolla, Missouri, 1984.
43. Harper, J. C., and El Sahrigi, A. F., *Industrial and Engineering Chemistry Fundamentals*, *3*, 318, 1964.
44. Liapis, A. I., *Separation and Purification Methods*, *19*(2), 133, 1990.
45. Ruthven, D. M., *Principles of Adsorption and Adsorption Processes*, Wiley-Interscience, New York, 1984.
46. McCoy, M. A., and Liapis, A. I., *Journal of Chromatography*, *548*, 25, 1991.
47. McCoy, M. A., B. J. Hearn, and A. I. Liapis, *Chemical Engineering Communications*, *108*, 225, 1991.
48. Liapis, A. I., and McCoy, M. A., *Journal of Chromatography*, *599*, 87–104, 1992.
49. Liapis, A. I., and Litchfield, R. J., *Computers and Chemical Engineering*, *3*, 615, 1979.
50. King, C. J., Lam, W. K., and Sandall, O. C., *Food Technology*, *22*, 1302, 1968.
51. Haugh, C. G., Huber, C. S., Stadelman, W. J., and Peart, R. M., *Transactions ASAE*, *11*, 877, 1968.
52. Lusk, G., Karel, M., and Goldblith, S. A., *Food Technology*, *19*, 620, 1965.
53. Triebes, T. A., and King, C. J., *Industrial Engineering Chemistry Process Design and Development*, *5*, 430, 1966.
54. Saravacos, G. D., and Pilsworth, M. N., *Journal of Food Science*, *30*, 773, 1965.
55. Luikov, A. V., Shashkov, A. G., Vasiliev, L. L., and Fraiman, Y. E., *International Journal of Heat and Mass Transfer*, *11*, 117, 1968.
56. Marcussen, L., Mathematical Models for Effective Thermal Conductivity, in *Thermal Conductivity*, Volume 18 (T. Ashworth and D. R. Smith, Editors), Plenum Press, New York, pp. 585–598, 1985.
57. Bird, R. B., Stewart, W. E., and Lightfoot, E. N., *Transport Phenomena*, John Wiley and Sons, New York, 1960.
58. Mason, E. A., and Malinauskas, A. P., *Gas Transport in Porous Media—The Dusty-Gas Model*, Elsevier, New York, 1983.
59. Lombrana, J. I., and Diaz, J. M., *Vacuum*, *37*, 473, 1987.
60. Litchfield, R. J., Farhadpour, F. A., and Liapis, A. I., *Chemical Engineering Science*, *36*, 1233, 1981.
61. Litchfield, R. J., Liapis, A. I., and Farhadpour, F. A., *Journal of Food Technology*, *16*, 637, 1981.
62. Sandall, O. C., King, C. J., and Wilke, C. R., *American Institute of Chemical Engineers Journal (AIChE Journal)*, *13*, 428, 1967.

11
Microwave and Dielectric Drying

Robert F. Schiffmann
R. F. Schiffmann Associates, Inc.
New York, New York

It is sometimes surprising to realize that dielectric and microwave heating have been in use for quite some time. It appears to many engineers that these are new forms of heating when in fact practical applications began during World War II and the home microwave oven was invented shortly after that war. Yet, these remain small industries, and for the most part, the equipment manufacturers are likewise small companies. The older of the two, dielectric heating, is a "workhorse" heating method used in many industries, including plastics, wood, ceramics, furniture, textiles, and paper. It is also by far the larger of the two industries; however, it is also not very glamorous, and the industrial microwave heating industry has glamour, but limited sales. To try to quantify the relationship, there are probably only 50–60 MW of microwave power in use globally for industrial heating purposes, whereas a single large dielectric heating system may employ as much as 2 or 3 MW of power. The annual worldwide sales of industrial microwave heating systems probably amounts to only 15–25 million dollars, but the sales of the home microwave ovens in the United States is of the order of 1.5–2.0 billion dollars. The reasons for the relatively small size of these markets are several, but two stand out: first, the heating mechanisms are not familiar to most engineers, and second, they often represent a radical departure from conventional systems and there is generally a tendency to resist real innovation in most industries.

In the past few years there has been a surge of interest in the applications of microwave and dielectric heating for industrial purposes. This is due primarily to the worldwide energy crisis and the growing acceptance of and familiarity with microwave ovens. The unique heating mechanisms of microwaves and dielectrics permit dramatic energy savings in many instances, as well as providing other benefits. This is nowhere better seen than in some of the applications in drying. It is the purpose of this chapter to provide background into these heating methods and their applications and, it is hoped, thereby stimulate their consideration in new drying systems.

1. BACKGROUND

The terms *dielectric* and *microwave* are somewhat confusing and must be defined as best we can. The term *dielectric heating* can be applied logically to all electromagnetic frequen-

Table 1 Frequencies Designated by the International Telecommunication Union for Use as Fundamental Industrial, Scientific, and Medical Frequencies[a]

Center frequency (MHz)	Frequency range (MHz)	Maximum radiation limit[b]	Number of appropriate footnote to the table of frequency allocation to the ITU Radio Regulations
6.780	6.765–6.795	Under consideration	524[c]
13.560	13.553–13.567	Unrestricted	534
27.120	26.957–27.283	Unrestricted	546
40.680	40.66–40.70	Unrestricted	548
433.920	433.05–434.79	Under consideration	661[c], 662 (Region 1 only)
915.000	902–928	Unrestricted	707 (Region 2 only)
2450	2400–2500	Unrestricted	752
5800	5725–5875	Unrestricted	806
24,125	24,000–24,250	Unrestricted	881
61,250	61,000–61,500	Under consideration	911[c]
122,500	122,000–123,000	Under consideration	916[c]
245,000	244,000–246,000	Under consideration	922[c]

Source: From Reference 1.
[a]Resolution No. 63 of the ITU Radio Regulations applies.
[b]The term *unrestricted* applies to the fundamental and all other frequency components fallling within the designated band. Special measures to achieve compatibility may be necessary where other equipment satisfying immunity requirements (e.g., EN 55020) is placed close to ISM equipment.
[c]Use of these frequency bands is subject to special authorization by administrations concerned in agreement with other administrations with radio communication services that might be affected.

cies up to and including at least the infrared spectrum. The lower frequency systems operate at frequencies through at least two bands: HF (3–30 MHz) and VHF (30–300 MHz). Thus the names high frequency (HF), dielectric, radiofrequency, and RF heating can often be used interchangeably. However, it is generally accepted that dielectric heating is done at frequencies between 1 and 100 MHz, whereas microwave heating occurs between 300 MHz and 300 GHz. This makes the wavelengths in dielectric heating extend

Table 2 Frequencies Designated on a National Basis in CENELEC Countries for Use as Fundamental Industrial, Scientific, and Medical Frequencies

Frequency (MHz)	Maximum radiation limit[a]	Notes
0.009–0.010	Unlimited	Germany only
3.370–3.410	Unlimited	Netherlands only
13.533–13.553	110 dB(μV/m) at 100 m	United Kingdom only
13.567–13.587	110 dB(μV/m) at 100 m	United Kingdom only
83.996–84.004	130 dB(μV/m) at 30 m	United Kingdom only
167.992–168.008	130 dB(μV/m) at 30 m	United Kingdom only
886.000–906.000	120 dB(μV/m) at 30 m	United Kingdom only

Source: From Reference 2.
[a]Distance measured from the exterior wall outside the building in which the equipment is situated.

to many meters. Microwave wavelengths range from 1 mm to 1 m. Tables 1 and 2 show the industrial, scientific, and medical (ISM) bands established by international agreement (1–3). Note that these frequency allocations are made by the International Telecommunication Union (ITU) and some frequencies are specific to certain countries. For example, 915 MHz is allowed in the United States but not in European countries.

Practical heating applications, including those for drying, are done at 13.56, 27.12, 40.68, 896, 915, and 2450 MHz. It should be noted, however, that a great deal of dielectric industrial heating is done at frequencies other than these ISM bands, as explained below.

Although the basic principles of heating and drying at dielectric and microwave frequencies are the same, the methods of generation and equipment are different. These will be described separately later in the chapter. In other cases, the two terms may be used interchangeably in the text.

2. FUNDAMENTALS OF MICROWAVE AND DIELECTRIC HEATING

2.1. Electromagnetic Waves

We are surrounded by electromagnetic waves at all times. Light, x-irradiation, TV, AM and FM radio waves, ultraviolet, infrared, and microwaves are some of the common manifestations of these waves. All bodies in the universe, above absolute zero temperature, emit electromagnetic waves. The relationship of these waves is found in the electromagnetic spectrum (see Fig. 1). All electromagnetic waves are characterized by their wavelength and frequency, and an illustration of a plane monochromatic electromagnetic wave is seen in Figure 2. It is seen that an electromagnetic wave is a blend of an electric component E and a magnetic component H. Note that E and H are perpendicular to each other and both are perpendicular to the direction of travel. This is what makes this a

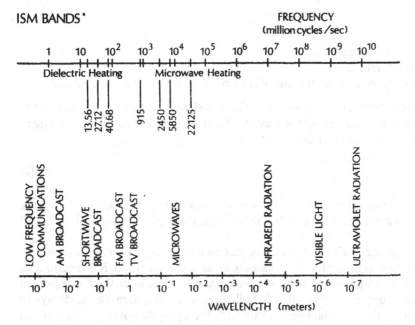

Figure 1 The electromagnetic specturm.

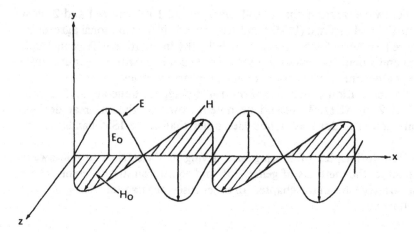

Figure 2 Diagrammatic illustration of a plane electromagnetic wave. E and H represent the electrical and magnetic components of the wave; E_o and H_o are their respective amplitudes.

"plane" wave. Note further that the field strength at any point may be represented by a sine or cosine function, which is what makes it "monochromatic." Further, it is "linearly polarized" since the electric and magnetic field vectors E and H lie in one direction only. The plane of polarization is YX for the E vector and ZX for the H vector.

Figure 1 also indicates that the wave is traveling in the X direction at the velocity C, which is equivalent to the speed of light in air or vacuum but slows as it passes through another medium, as indicated in Eq. [1]:

$$V_p = \frac{C}{\sqrt{\epsilon'}} \qquad [1]$$

where

    V_p = the velocity of propagation
    C = the speed of light in air
    ϵ' = the dielectric constant of the material through which the wave is propagated

We further note the distance λ, which is the wavelength. Equation [2] shows the relationship between the frequency of the wave f, that is, how many times it goes through a full cycle per second, and the wavelength.

$$f = \frac{V_p}{\lambda} \qquad [2]$$

Note that as an electromagnetic wave passes through a material its frequency remains the same; therefore, its wavelength changes, and this affects the depth of penetration, as will be seen later.

A further examination of Figure 1 indicates that an electromagnetic wave is an energy wave that changes its energy content and amplitude as it travels through a medium, as seen by a change in the amplitude of the wave. For example, if we trace the E component we see that at some point it is zero; then it builds up to a maximum value, decays to zero, and again builds up to a maximum value with the opposite polarity before again decaying to zero. The same thing happens to the H component. The amplitude of the wave at any point along the X axis represents the electrical (E) or magnetic (H) field strength, which are measured as volts or amperes per unit distance, respectively. It is this periodic flip-flopping of the wave's polarity and its decay through zero that causes the stress upon ions, atoms, and molecules, which is converted to heat, and the greater the field strength, the greater will be the whole effect.

2.2. Heating Mechanism

A crucial fact to keep in mind at all times is that microwaves and dielectrics are not forms of heat but rather forms of energy that are manifested as heat through their interaction with materials. It is as if they cause materials to heat themselves. There are many mechanisms for this energy conversion, as can be seen in only a partial listing in Table 3 (3).

2.3. Ionic Conduction

Since ions are charged units they are accelerated by electric fields. In a solution of salt in water, for example, there are sodium, chloride, hydronium, and hydroxyl ions, all of which will be caused to move in the direction opposite to their own polarity by the electric field. In so doing they collide with un-ionized water molecules, giving up kinetic energy and causing them to accelerate and collide with other water molecules in billiard ball fashion, and when the polarity changes the ions accelerate in the opposite fashion. Since this occurs many millions of times per second, large numbers of collisions and transfers of energy occur. Therefore, there is a two-step energy conversion: electric field energy is converted to induced ordered kinetic energy, which in turn is converted to disordered kinetic energy, at which point it may be regarded as heat. This type of heating is not dependent to any great extent upon either temperature or frequency. The power developed per unit volume (P_v) through ionic conduction is shown as

$$P_v = E^2 q n \mu \qquad [3]$$

Table 3 Partial List of Energy Conversion Mechanisms[a]

Ionic conduction
Dipole rotation
Entire molecule quantized
Twist
Bend
Interface polarization
Dipole stretching
Ferroelectric hysteresis
Electric domain wall resonance
Electrostriction
Piezoelectricity
Nuclear magnetic resonance
Ferromagnetic resonance
Ferrimagnetic resonance

[a]It is the first two with which we are primarily interested in dielectric heating phenomena.

where

    q = amount of electrical charge on each of the ions

    n = ion density, the number of ions per unit volume

    μ = level of mobility of the ions

2.4. Dipolar Rotation

Many molecules, such as water, are dipolar in nature; that is, they possess an asymmetric charge center. Water is typical of such a molecule. Other molecules may become "induced dipoles" because of the stresses caused by the electric field. Dipoles are influenced by the rapidly changing polarity of the electric field. Although they are normally randomly oriented, the electric field attempts to pull them into alignment. However, as the field decays to zero (relaxes), the dopoles return to their random orientation only to be pulled toward alignment again as the electric field builds up to its opposite polarity. This buildup and decay of the field, occurring at a frequency of many millions of times per second, causes the dipoles similarly to align and relax millions of times per second. This causes an energy conversion from electrical field energy to stored potential energy in the material and then to stored random kinetic or thermal energy in the material. This temperature-dependent, molecular size-dependent time for buildup and decay defines a frequency known as the *relaxation frequency*. For small molecules, such as water and monomers, the relaxation frequency is already higher than the microwave frequency and rises further as the temperature increases, causing a slowing of energy conversion. On the other hand, large molecules, such as polymers, have a relaxation frequency at room temperature that is much lower than the microwave frequency but that increases and approaches it as the temperature rises, resulting in better energy conversion into heat. This may lead to runaway heating in materials that at room temperature are very transparent to the microwave field. This must be superimposed upon the fact that such liquids as water and monomers are better absorbers of microwave energy than polymers. Since in drying or curing applications it is the liquids and monomers that require heating, not the existing polymeric substrate, it is possible to execute the process well, often at a lower temperature. In fact, it is even possible to dry such materials as foods and medicinals at cold or subfreezing temperatures.

    The power formula for dipolar roatation is

$$P_v = kE^2 f\epsilon' \tan \delta$$

or

$$P_v = kE^2 f\epsilon'' \tag{4}$$

where

    k = constant dependent upon the units of measurement used

    E = electric field strength, in volts per unit distance

    f = frequency

    ϵ' = relative dielectric constant, or relative permitivity

   tan δ = loss tangent or dissipation factor

    ϵ'' = loss factor

The relative dielectric constant expresses the degree to which an electric field may build up within a material when a dielectric field is applied to the material. The loss tangent is a measure of how much of that electric field will be converted into heat.

    A further examination of Eq. (4) reveals that E and f are functions of the equipment, whereas ϵ', ϵ'', and tan δ are factors related to the material being heated. Another important point is that as frequency f is changed it is necessary to increase the electric field strength E in order to maintain a particular power level P_v. Since dielectric heating frequencies are much lower than microwave frequencies, this requires that the field strengths be much higher for comparable power output in a dielectric system. This may lead to voltage breakdown of air (arcing) or in the process material. The sparking threshold for air is about 30,000 V/cm (75,000 V/in).

2.5. Interaction of Electromagnetic Fields with Materials

We may divide materials and the way they interact with electromagnetic fields into four categories:

1. Conductors. Materials with free electrons, such as metals, are materials that reflect electromagnetic waves just as light is reflected by a mirror. These materials are used to contain and direct electromagnetic waves in the form of applicators and waveguides.
2. Insulators. Electrically nonconductive materials, such as glass, ceramics, and air, act as insulators which reflect and absorb electromagnetic waves to a negligible extent and primarily transmit them (that is, they are transparent to the waves). They are therefore useful to support or contain materials to be heated by the electromagnetic field and may take the form of conveyor belts, support trays, dishes, or others. These materials may also be considered "nonlossy dielectrics."
3. Dielectrics. These are materials with properties that range from conductors to insulators. There is within this broad class of materials a group referred to as "lossy dielectrics," and it is this group that absorbs electromagnetic energy and converts it to heat. Examples of lossy dielectrics are water, oils, wood, food, and other materials containing moisture, and the like.
4. Magnetic compounds. These are materials, such as ferrites, that interact with the magnetic component of the electromagnetic wave and as such will heat. They are often used as shielding or choking devices that prevent leakage of electromagnetic energy. They may also be used for heating in special devices.

As indicated earlier, those properties that govern whether a material may be successfully heated by a dielectric or microwave field are the dielectric properties: relative dielectric constant ϵ', loss tangent or dissipation factor (tan δ), and the loss factor ϵ''.

Note that the complex dielectric constant ϵ may be expressed as

$$\epsilon = \epsilon' - j\epsilon'' \qquad [5]$$

where $j = \sqrt{-1}$, which indicates a 90° phase shift between the real (ϵ') and imaginary (ϵ'') parts of the complex dielectric constant. The loss tangent is defined as the ratio of dielectric loss to dielectric constant:

$$\tan \delta = \frac{\epsilon''}{\epsilon'} \qquad [6]$$

These various factors are affected by several parameters.

Moisture Content

The amount of free moisture in a substance greatly affects its dielectric constant since water has a high dielectric constant, approximately 78 at room temperature; that of base materials is of the order of 2. Thus, with a larger percentage of water the dielectric constant generally increases, usually proportionally. It should be emphasized that very complex phenomena occur when different dielectrics are mixed. However, a few rules of thumb may be applied.

1. The higher the moisture content, usually the higher is the dielectric constant.
2. The dielectric loss usually increases with increasing moisture content but levels off at values in the range of 20% to 30% and may decrease at still higher moisture.
3. The dielectric constant of a mixture usually lies between that of its components.

   Various materials, including alcohols and some organic solvents, also exhibit dielectric properties that make them suitable for heating with microwave and dielectric energy and, so, behave similarly to water. Table 4 indicates the heating properties of various classes of materials.

   Since drying is concerned with removal of water or a solvent, it is interesting to note that as these liquids are removed the dielectric loss decreases and hence, the material heats less well. In many cases this leads to self-limitation of the heating as the material becomes relatively transparent at low moisture content. This has great value in obtaining moisture leveling, especially in sheet materials, in which the electromagnetic energy is likely to preferentially dry the wetter areas. Figure 3 shows a general graph of the variation in loss factor with moisture content (4). Water exists in materials in different states, for example, bound or free, and these states may be ascribed to different regions on the graph, as indicated by the change of slope ($d\epsilon''/dm$). Thus at low moisture contents, below the critical moisture content, we are dealing primarily with bound water; above it we encounter primarily free water. (Note that the dielectric loss of bound water is very low since it is not free to rotate under the influence of the electromagnetic field. This is seen in an analogous situation with ice, which has a dielectric loss factor of approximately 0.003; that of water is approximately 12.) The change in the slope may be quite gradual for some materials, making positive identification fairly difficult. The critical moisture content for highly hygroscopic materials occurs between 10% and 40% (dry basis); for nonhygroscopic materials it is in the region of about 1%. It is obvious that moisture leveling will be

Table 4   Heating Properties of Various Materials

Heat well	Heat poorly
Water	Hydrocarbons
Acid anhydrides	Halogenated hydrocarbons
Alcohols	(symmetrical)
Aldehydes	Alkali halides (e.g., salt)
Ketones	Inorganic oxides (e.g.,
Amides	alumina)
Amines	Some elements (e.g., sulfur)
Nitrates	Boron nitride
Cyanides	Mica
Proteins	
Halogenated hydrocarbons	
(unsymmetrical)	
Ferrites	
Ferroelectrics	
Ionic solutions	

Source: From Reference 3

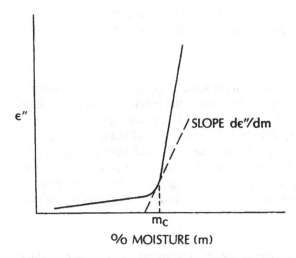

Figure 3 The critical moisture content m_c. The dielectric loss factor is ε''. The region below m_c is indicative of bound water, whereas above free water is more easily removed.

quite effective above the critical moisture content but not so effective below it. Although some materials become quite transparent below the critical moisture content, there are others, such as wood and textiles, that will continue to heat and may scorch or burn.

Density

The dielectric constant of air is 1.0, and it is, for all practical purposes, transparent to electromagnetic waves at industrial frequencies. Therefore, its inclusion in materials reduces the dielectric constants, and as density decreases so do the dielectric properties, and heating is reduced.

Temperature

The temperature dependence of a dielectric constant is quite complex, and it may increase or decrease with temperature depending upon the material. (See discussion concerning water and polymers in Section 2.4.) In general, however, a material below its freezing point exhibits lowered dielectric constant and dielectric loss. Above freezing the situation is not clear-cut, and since moisture and temperature are important to both drying and dielectric properties, it is important to understand the functional relationships in materials to be dried. Wood, for example, has a positive temperature coefficient at low moisture content (5); that is, its dielectric loss increases with temperature. This may lead to runaway heating, which in turn will cause the wood to burn internally if heating continues once the wood is dried.

Frequency

Dielectric properties are affected by the frequency of the applied electromagnetic field. However, since industrial heating is restricted to the ISM frequencies, the engineer is limited in making use of this phenomenon. It may, however, be useful in measuring moisture content.

Conductivity

Conductivity refers to the ability of a material to conduct electric currents by the displacement of electrons and ions; this effect is described in detail in Sec. 2.3. Suffice it to say that these charged units can have a major effect in heating, and in a drying situation in which the ion concentration increases as the water is removed, this effect can be very complex.

Thermal Conductivity

Thermal conductivity often plays a lesser role in microwave and dielectric heating than in conventional heating because of the great speed with which the former heat thus reducing the time in which thermal conductivity can be effective. There are cases, however, in which it has a major role. For example, when penetration depth of the electromagnetic energy is small in comparison with the volume being heated, thermal conductivity may be depended upon to transfer the heat to the interior. Another important case is to even out the nonuniformities of heating that may occur with electromagnetic fields. Sometimes the microwave or dielectric power is pulsed on and off to allow for this evening out of temperature, as in microwave thawing.

Specific Heat

The specific heat parameter is often neglected by the researcher or engineer dealing with electromagnetic heating who focuses attention only upon the dielectric properties. However, specific heat can have profound effects and may, in fact, be the overriding parameter, causing materials to heat much faster than one would predict by looking only at their dielectric properties.

Penetration Depth

Although not a property of a material but rather a result of its various properties, penetration depth is of utmost importance. Since electromagnetic heating is, in effect, bulk heating, it is important that the energy penetrate as deeply as possible. If it does not, then the heating is limited to the surface. Those parameters affecting the depth of the field into the material are the wavelength, the dielectric constant, and the loss factor, as shown in Eq. [7]:

$$D = \frac{\lambda_0 \sqrt{2}}{2\pi} \left[\epsilon' \left(\sqrt{1 + (\epsilon''/\epsilon')^2} - 1 \right) \right]^{-\frac{1}{2}} \qquad [7]$$

where D is the penetration depth at which the available power in the material has dropped to about 37% ($1/e$) of its value at the surface and λ_0 is the free space wavelength. If ϵ'' is low, Eq. [7] may be simplified:

$$D = \frac{\lambda_0 \sqrt{\epsilon'}}{2\pi\epsilon''} \qquad [8]$$

This equation is reasonably accurate for most foods even though many have relatively high ϵ'' values.

From these equations it is obvious that materials with high dielectric constants and loss factors will have smaller depths of penetration than those with lower values. It is also apparent that the depth of penetration is greatly affected by the wavelength (and hence the frequency) of the applied field. This is illustrated in Table 5, in which penetration depth into Douglas fir is shown for various wavelengths (5). This demonstrates the very clear superiority of dielectric heating of very large materials with substantial dielectric properties.

A peculiarity of this type of heating is the unusual temperature gradients that may be generated. This is due to a number of factors. First, unless an auxiliary heat form is

Table 5 Wavelength and Depth of Penetration in Douglas Fir at Various Frequencies

	Frequency (MHz)					
	5.0	13.56	27.12	40.0	915	2450
Wavelength						
Meters	60.0	22.1	11.1	7.5	0.328	0.122
Feet	196.9	72.6	36.3	24.6	1.07	0.400
Depth of penetration						
Meters	23.9	8.8	4.4	3.0	0.130	0.049
Feet	78.4	28.9	14.4	9.8	0.425	0.158

Source: From Reference 5

applied, the air in the system remains cold. Hence, the surface will be cooler than a zone somewhat below the surface. This is especially true in a drying system in which evaporative cooling of the surface will occur.

Another circumstance concerns the depth of penetration as it relates to the size of the piece being heated. If the piece is several times larger than the depth of penetration, then the temperature gradient will resemble conventional gradients, with a cooler interior and a warmer exterior. However if the piece is small in comparison with the penetration depth, for example only one or two times greater, then there may be a focused accumulation of the electromagnetic field in the center of the piece due to the multiple passes of the waves and internal reflections. In this case, the center may be the hottest place, and in fact, if it is overheated, the center may burn while the surface remains cool.

3. PROCESS ADVANTAGES OF MICROWAVE AND DIELECTRIC SYSTEMS

3.1 Advantages of Microwave and Dielectric Heating

Heating and drying with microwave and dielectric energy is distinctly different from conventional means. Whereas conventional methods depend upon the slow march of heat from the surface of the material to the interior as determined by differential in temperature from a hot outside to a cool inside, heating with dielectric and microwave energy is, in effect, bulk heating in which the electromagnetic field interacts with the material as a whole. The heating occurs nearly instantaneously and can be very fast, although it does not have to be. However, the speed of heating can be an advantage, and it is often possible to accomplish in seconds or minutes what could take minutes, hours, and even days with conventional heating methods. The fastest industrial heating system of which this author is aware heats fine plastic thread at the rate of about 30,000°C/s (the material was actually heated about 100°C in about 3 ms) (3). On the other hand, one can heat at the rate of 1°C per century, if desired. The governing parameters here are the mass of the material, its specific heat, dielectric properties, geometry, heat loss mechanisms, and coupling efficiency, the power generated in the material, and the output power of the microwave-dielectric heating system. All other things being equal, the speed may be doubled by doubling the output power.

A list of advantages of microwave and dielectric heating includes the following:

Process speed is increased, as described above.

Uniform heating may occur throughout the material. Although not always true, often the bulk heating effect does produce uniform heating, avoiding the large temperature gradients that occur in conventional heating systems.

Efficiency of energy conversion. In this type of heating, the energy couples directly to the material being heated. It is not expended in heating the air, walls of the oven, conveyor or other parts. This can lead to significant energy savings. Also, the energy source is not hot and plant cooling savings may be realized.

Better and more rapid process control. The instantaneous on-off nature of the heating and the ability to change the degree of heating by controlling the output power of the generator means fast, efficient, and accurate control of heating.

Floor space requirements are usually less. This is due to the more rapid heating.

Selective heating may occur. The electromagnetic field generally couples into the solvent, not the substrate. Hence, it is the moisture that is heated and removed, whereas the carrier or substrate is heated primarily by conduction. This also avoids heating of the air, oven walls, conveyor, or other parts.

Product quality may be improved. Since high surface temperatures are not usually generated, overheating of the surface and case hardening, which are common with conventional heating methods are eliminated. This often leads to less rejected product.

Desirable chemical and physical effects may result. Many chemicals and physical reactions are promoted by the heat generated by this method, leading to puffing, drying, melting, protein denaturation, starch gelatinization, and the like.

3.2. Advantages of Microwave and Dielectric Drying

The mechanism for drying with microwave and dielectric energy is quite different from that for ordinary drying. In conventional drying, moisture is initially flashed off from the surface and the remaining water diffuses slowly to the surface. Although the potential of energy transfer for heating is the temperature gradient, which results in energy transfer to the interior of the material, the potential for mass transfer is the mass concentration gradient existing between the wet interior and the drier surface. This is often a slow process, diffusion rate limited, which requires high external temperatures to generate the temperature differences required.

With internal heat generation, in microwave and dielectric systems, mass transfer is primarily due to the total pressure gradient established because of the rapid vapor generation within the material (6). Most of the moisture is vaporized before leaving the sample. If the sample is initially very wet and the pressure inside the sample rises very rapidly, liquid may be removed from the sample under the influence of a total pressure gradient. The higher the initial moisture, the greater is the influence of the pressure gradient on the total mass removal. Thus there is in effect a sort of "pumping" action, forcing liquid to the surface, often as a vapor. This lead to very rapid drying without the need to overheat the atmosphere and perhaps cause case hardening or other surface overheating phenomena. Table 6 summarizes the advantages of microwave and dielectric drying.

Of great interest today are the potential energy savings achievable from such a system. This is due to speed of drying, the direct coupling of energy into the solvent, possible lower drying temperatures, far more effective use of conventional heating in combination with the dielectric methods, and less overall heat loss.

A word of caution must be expressed here. These systems can heat and dry quickly, but too rapid heating can be destructive. Care must be taken not to heat so fast that the

Table 6 Advantages of Microwave and Dielectric Drying

Efficiency: in most cases, the energy couples into the solvent, not the substrate

Nondestructive: drying can be done at low ambient temperatures; no need to maintain high surface temperatures, leading to lower thermal profiles

Reduction of migration: solvent often mobilized as a vapor; therefore does not transfer other materials to the surface

Leveling effects: coupling tends toward the wetter areas

Speed: drying times can be shortened by 50% or more

Uniformity of drying: by a combination of more uniform thermal profiles and leveling

Conveyorized systems: less floor space, reduced handling

Product improvement in some cases: eliminates case hardening, internal stresses, and other problems

---

material may scorch, burn, or be otherwise damaged or dry so quickly that the steam or other vapors cannot escape quickly enough, leading to internal pressure buildup, which can lead to rupture of the piece or an explosion.

When drying with dielectric heating it is usual to combine hot air with the system, particularly with microwave systems. This is because it usually improves the efficiency and the economics of the drying process (7). Hot air is, by itself, relatively efficient at removing free water at or near the surface, whereas the unique pumping action of dielectric heating provides an efficient way of removing internal free water as well as bound water. By combining these properly, it is possible to draw on the benefits of each and maximize efficiency and keep the costs of drying down. Note that drying with microwaves or dielectrics alone can be very expensive in terms of both equipment and operating costs.

There are three ways in which microwave and dielectric energy may be combined with conventional drying methods, as illustrated in Figure 4.

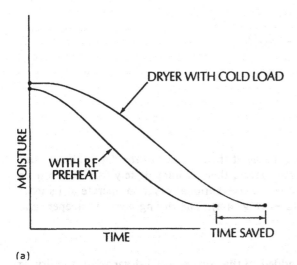

(a)

Figure 4 Typical drying curves for microwave and dielectric drying systems: (a) preheating with microwave or dielectrics; (b) booster drying; (c) finish drying.

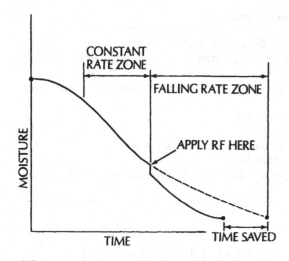

(b)

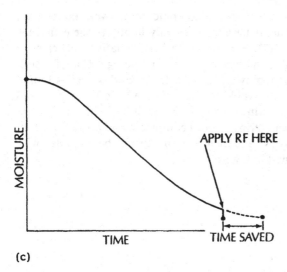

(c)

Figure 4 Continued

Preheating

By applying the microwave or dielectric energy at the entrance to the dryer, the interior of the load is heated to evaporation temperature, thereby immediately forcing moisture to the surface and immediately permitting the conventional dryer to operate at its most efficient condition, at higher temperatures (Fig. 4-a). The drying curve is steeper, and drying time is shortened.

Booster Drying

The microwave or dielectric energy is added to the conventional dryer when the drying rate begins to fall off (Fig. 4-b). The surface of the material is dry, and moisture is concentrated in the center. The added electromagnetic energy generates internal heat and vapor pressure, forcing the moisture to the surface, where it is readily removed. The drying is sharply increased with a leverage of 6 : 1 or 8 : 1 in terms of increased drying capacity for each unit of electromagnetic energy added. This is most effective on thick, hard to heat materials.

Finish Drying

The least efficient portion of a conventional drying system is near the end, when two-thirds of the time may be spent removing the last one-third of the water (Fig. 4-c). By adding a microwave or dielectric dryer at the exit of the conventional dryer, this replaces the inefficiency of hot air drying with internal heat generation. The conventional dryer may also be speeded up, thereby increasing the throughput of the dryer while presenting the dryer with a wetter load, thus increasing the efficiency. This method also provides close control of the terminal moisture and moisture leveling while avoiding overdrying.

The most common methods of application are booster and finish drying, and in spite of the greater cost for electrical energy than gas, the overall increase in drying efficiency and throughput can bring about large economic savings.

4. EQUIPMENT FOR MICROWAVE AND DIELECTRIC HEATING AND DRYING

The heating mechanisms for microwave and dielectric heating are similar, but the means of achieving them are somewhat different. The basic components of these systems are a means of generating the high-frequency energy—the generator—and a means of applying it to the workpiece—the applicator. These are described below.

4.1. Generators

The basic function of the generator is to convert the alternating current of 50 or 60 Hz to the high frequencies desired. The means of doing this are quite different for dielectric and microwave systems.

Dielectric systems usually employ negative grid triode tubes, although some systems operating in the 50–100-MHz range use beamed power types. RF circuits are usually simple, self-excited oscillators of the Hartley, Colpitts, or tuned plate-tuned grid type. These circuits usually consist of coils and capacitors and/or coaxial lines. The load to be heated may actually form part of the tank circuit capacitance, it may be separately tuned or inductively or capacitively coupled to the oscillator, or it may be a combination of these, being directly connected to the oscillator and also partially tuned to the oscillator circuit. This last option has a distinct advantage in drying applications, in which the electrodes may be partially tuned to the oscillator circuit on the side of the tuning curve, where they will be detuned when a dry load with a lower dielectric constant is present. Thus a wet load will cause an increase in electrode voltage, and a dry load will cause a power reduction. In this way there is a self-limiting or leveling effect (8). Note that energy not developed in the load can be lost in the coupled circuits, causing less efficiency and wasted power.

A negative grid tube is a variable power device that draws current from the supply line only in sufficient quantity to supply circuit losses plus the power consumed in the load. Thus, most dielectric systems control power to the load by varying the RF electrode voltage, often by means of a variable capacitor. In many instances this system is self-limiting and self-regulating, supplying power only as demanded by changes in size or electrical characteristics of the load (8).

Dielectric heating systems utilize a wide range in frequencies, from 3 MHz to more than 150 MHz. Many of these are not confined to the ISM bands and may vary frequency to improve the efficiency of heating. In so doing, however, they must be properly shielded and filtered to comply with Part 18, Subpart D of the FCC Rules and Regulations, to prevent out-of-band radiation leakage.

Microwave systems operate on a nomimal fixed frequency of 915 or 2450 MHz, with the frequency controlled by the tube dimensions and geometry. They must also be shielded to prevent excessive radiation of harmonics, as well as for safety.

A microwave generator consists of a dc power supply and a tube—either a magnetron or a klystron. These tubes are constant output power devices, and power to the load may be controlled by sensing the load requirements and controlling the input power accordingly, usually by indirectly varying the dc anode voltage. Although magnetrons and klystrons are capable of withstanding a reasonable degree of mismatching, manifested as power reflected back to the tube, precautions must be taken to avoid overheating or in other ways damaging these tubes (see ferrite circulators in the section, "Protective Devices").

4.2. Applicators

The means of applying microwave and dielectric energy to a work piece differ in a very significant manner. Microwave energy may be transported through free space and must be focused upon the load. On the other hand, dielectric energy is usually applied by means of electrodes, in which the field oscillates through the load, which is placed between the electrodes. This is described in detail below.

Dielectric Systems

There are many types of electrodes, but they may be described in three basic categories, which are illustrated in Figure 5.

The platen type usually consists of flat plates in pairs, between which the work piece may be held in a batch system or pass on a conveyor belt. Often, a conveyor belt may represent one of the plate electrodes. This is especially useful for bulky objects. A drawback is that as the plates become widely separated the high field strengths required may cause voltage discharge, which can burn the load.

With the stray field type, the load, usually in the form of a thin web, passes over the electrodes of alternating polarity. Since the load represents the path of least electrical resistance, the dielectric field passes through it, causing heating.

The staggered type is usually for sheet materials and thick webs. The distance between the electrodes is kept to a minimum in order to achieve heating without arcing.

As for other types, electrodes may also be shaped to conform to the geometry of the load or may be part of the conveyor or hydraulic or pneumatic press. The entire system is confined within a metal housing to prevent leakage of radiation. At the same time, hot air of controlled temperature, humidity, and/or velocity may be passed through the applicator.

Microwave Systems

After generating the microwave energy it must be transported to the applicator. This is usually accomplished by means of waveguides, although coaxial cable is also useful for lower power.

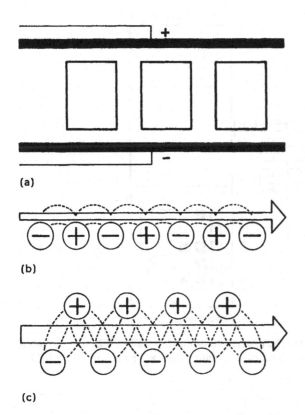

Figure 5 Electrode configurations for dielectric heating systems: (a) platen type for bulky objects; (b) stray field type for thin webs; (c) staggered type for thick webs or board.

The waveguide is ordinarily a hollow rectangular metallic conduit, usually made of brass or aluminum. Its interior dimensions are carefully chosen to control the nature of the microwave field presented to the applicator. Applicators are of several major types and are always constructed of metal.

Waveguides themselves may be used as applicators. Since the electric field may be maximum in the center of a waveguide, it is possible to pass a material through this intense field to obtain very efficient heating. A good example is the heating of filamentary materials.

Traveling wave applicators are also known as slotted, folded, or serpentine waveguides. A slot is cut into the narrow sides of the waveguide, and several waveguides are joined together as shown in Figure 6. A thin sheet material, such as paper or textile, may be passed through the slots. The microwave energy makes several passes through the load, heating it as it travels. These are highly efficient heating systems, although they may cause some side-to-side nonuniformity.

Cavity applicators are a large class of applicators, but they are probably the most common type. Home microwave ovens are a typical example. They consist of metal boxes, which may be used for conveyorized systems or in batch operations. The microwave energy may be coupled into this applicator by means of waveguide or coaxial cable through a single port or multiple ports (see Figure 7). There is an industrial system in which over 100 magnetrons are separately introduced into the cavity. In this type of

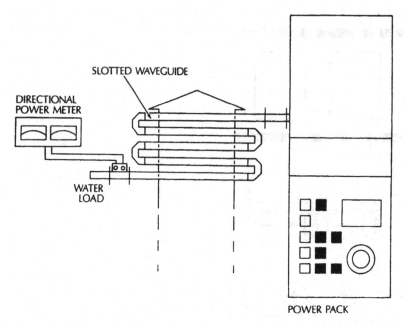

Figure 6 Slotted or serpentine waveguide. The material to be dried, usually a thin web, is passed through slots in the sides of the waveguide and exposed by multiple passes to the microwave field.

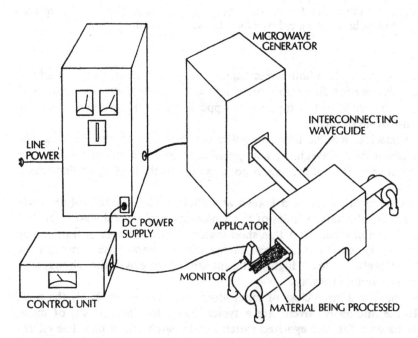

Figure 7 A typical microwave heating system utilizing a conveyorized cavity applicator. A feedback system monitors the heated material and automatically adjusts the output power of the magnetron to control the final moisture.

applicator, the load usually represents only a small fraction of the volume of the applicator and is subjected to the microwave field being reflected from the sides of the applicator and passing through it from all sides. This causes a three-dimensional bulk heating effect that is unique and of great use.

A major problem with cavity applicators is uniformity of the microwave field in the load. In order to ensure uniformity of heating, a number of steps may be taken, usually in combination: moving or turning the load in the applicator by means of conveyors or turntables; providing mode stirrers, which often resemble slowly rotating fans and increase the number of modes in the oven, causing reflective scattering of energy; using multiple inputs for the microwave energy; using multiple microwave sources with slight differences in frequency that cause different mode patterns; and choosing the cavity dimensions to support the maximum number of modes, the so-called resonant multimode cavity.

4.3. Other Devices

A number of auxiliary devices and systems should be mentioned at this point.

Control Systems

Since the output power of the microwave or dielectric heating system is governed by electrical energy, unique control systems can be designed utilizing feedback loops that monitor some function of the load, such as moisture, and automatically control the output power to give better and faster control of the moisture content.

Leakage and Safety Control Systems

As mentioned earlier, the amount of radiation leaking from a microwave or dielectric heating system must be controlled, both to contain radiofrequency interference within acceptable limits and for personnel safety. Numerous devices, often called *chokes* or *attenuating tunnels*, are used for this purpose at conveyor openings, around doors and windows, at seams, and the like. Good engineering design should make it possible to keep leakage radiation well below the limits and guidelines set by the various controlling governmental organizations.

Protective Devices

Several protective devices are used in microwave systems to prevent high levels of reflected microwave energy from damaging the magnetron or klystron. The simplest of these are thermal switches that sense overheating of the tube and shut off the power. These may not be sufficient to protect the tube, however. Another method is the use of directional power sensors that discriminate between forward and reflected power and can shut off the systems when the latter becomes excessive. By far the most sophisticated system is the ferrite circulator or isolator, which by influencing the magnetic field passes microwave energy only in the forward direction, causing the reflected power to be shunted off into a dummy load. This system is highly efficient and especially recommended for high-power applications.

5. INDUSTRIAL APPLICATIONS OF MICROWAVE AND DIELECTRIC DRYING

It has been estimated that in Western Europe and the United Kingdom, RF equipment with a total of 30 MW is manufactured annually compared with about 2 MW of microwave industrial equipment (9). Roughly half of the RF is for plastic welding, with the rest given over to diverse systems. The industrial applications of microwave and dielectric heating are many and varied. In some cases, the application is unique to one form of energy or the other; in other cases, either form may be used. Although it is not possible to give any hard and fast rules for selecting one over the other, there are some guidelines that may be followed (8).

5.1. Guidelines for the Selection of Microwaves or Dielectrics

Size of Load

If the load is very large or very wide, dielectric heating may be preferred. The depth of penetration is also directly proportional to wavelength, and in dielectric heating this may be measured in meters; in microwaves it is in centimeters. On the other hand, if the piece is small, microwave heating is preferred.

Watt Density

If the watt density requirement is very high, microwaves may be preferred to avoid arcing and burning of the material. For example, a bulky product consisting of loosely packed particles with a loss factor less than 0.05 favors microwaves.

Power

If the power requirement is high, over 50 kW, economics favor the dielectric system.

Geometry

If the product has an irregular shape with no rectangular cross-section, the multimode microwave cavity will provide more uniform heating.

System Compatibility

If the system requires the use of pneumatic or hydraulic presses, metal conveyors, metal dies, or tenter frames, dielectric heating may be the only choice.

Self-Regulation

If the load fluctuates rapidly or goes through drastic changes in dielectric constant or dissipation factor during its heat cycle, partially tuned electrodes and instantly variable power from dielectric heating are advantageous.

Self-Limiting

If the load has a low-frequency loss tangent greater than 1.0, it is more resistive than capacitative, and in many cases this element disappears as the load dries or cures. RF will heat the resistive elements, and it will be more self-limiting in these areas and will not overheat or overdry. However, if the load has a low-frequency dissipation factor of less than 0.5, then it is mostly capacitive, and reduction of the dielectric constant as it dries or cures will be the predominant change and will permit better self-limiting at microwave frequencies. For example, dielectric heating has very little leveling effect on the moisture content of paper below about 5% because it is self-limiting in that region, but microwave energy could be used to dry the paper to near zero moisture content.

5.2. Dielectric Drying Systems

There are numerous systems in the lumber, furniture, textile, paper, food, tire, and ceramics industries, to name but a few. A brief description of these follows. In only a few cases, such as the postdrying of crackers, cookies, and biscuits or the drying of foundry sand cores, are there many systems utilizing the same basic equipment. In most cases, the systems are customized or "one of a kind" so the number of actual applications is much larger. The same holds true for microwave drying.

Lumber

Dielectric heating is used for both drying and gluing lumber. It is used in the manufacture of plywood for drying of the veneer in order to remove pockets of moisture and provide moisture leveling. Otherwise, during hot pressing, steam pockets would form and delaminate the board. It is also used to cure the glue in plywood, medium-density fiberboard, and particle board. In all these cases the dielectric system can utilize the plates of the presses as electrodes to give fast, efficient heating. These systems range in output power from 250 kW to over 1500 kW.

The rapid drying of lumber is also made use of in the furniture industry, in which precut furniture parts may be dried in minutes rather than days or weeks and the shrinkage is well controlled. Dielectrics are also used to dry the glue. Golf clubs are also dried in this manner (10,11).

Textiles

A large number of dielectric drying systems are used in the textile industry for drying of textile packages, hanks, skeins, tops, and loose stock. Speed, prevention of surface overdry, and leveling effects are all benefits of this technique, which results in superior product quality. Another benefit is more even distribution of dyes due to the diffusion of water vapor rather than liquid water during drying. Textile systems are in the 50–100 kW range primarily, although some as large as 250 kW and as small as 3 kW have been built (12). A recent RF application is for the drying of loose fibers, especially for high-grade animal hair such as cashmere in which a loss of quality is unacceptable. The lower the temperature of the fiber mass the better and, by drawing air heated by the waste heat from the RF tube through the bed, a low fiber temperature can be maintained (9).

Paper

Dielectric heating is used to dry printing inks, adhesives, and coating materials on paper, as well as to dry the paper itself. This has been combined with hot air, infrared, and other heating media to achieve optimum results. Some examples of commercial installations include (9):

1. Business forms, in which the RF units are used in conjunction with both presses and collators, with energy savings up to 75%. Up to 14-part forms can be dried as quickly as 180 meters per minute.
2. Direct mail line speeds up to 10,000 sheets/hour or 30 meters per minute for web-feed systems.
3. Envelopes, for which the selectivity of RF energy heats and dries only the adhesive lines, leaving the bulk of the paper cool and flat.
4. Book binding, in which the RF dries the adhesive, usually low-cost polyvinyl acetate.
5. Varnishes and water-based coatings for book covers, record sleeves, confectionary boxes, publicity brochures, and more.
6. Film laminates, in which nontoxic water-based adhesives are used in conjunction with polyester and polypropylene films.

For paper making, machines are in use with 50 to 500 kilowatts of RF at 27.12 or 13.56 MHz. The RF is able to overcome the most common problems encountered at the dry end of the process: low efficiency due to moisture distribution through the paper thickness, uneven moisture distribution across the width of the web, temporary unevenness or streaks of moisture due to some failure of the equipment, and cyclic variations in the machine direction.

Automobile Tires

An unusual application is the drying of the latex coating on fiberglass for fiberglass cord automobile tires. The coating is necessary to prevent abrasion of the fiberglass. Air drying of the coating must be slow to prevent surface skinning, which results in rupture of the coating when the internal steam pressure becomes sufficiently large. The dielectric drying of the coating results in uniform moisture loss in as little as 2.5 s, thereby increasing the solids level of the coating from 25% to 98%. Superior bonding of the latex to the fiberglass results from this process as well. PPG Industries utilizes such equipment with a capacity of several megawatts (13).

Food

Dielectric drying has several uses in the food industry, including drying of breakfast cereal; postbake drying of crackers, cookies, and biscuits; and postbake drying of dog biscuits. There are now several hundred such postbaking systems in operation in North America and Europe. These usually consist of a short postbaking conveyor, 3 to 4 meters in length, immediately following the fuel-fired oven. Thus, the product may exit the baking oven at a much higher moisture content, which is then rapidly removed by the RF dryer. Oven speeds may thus be increased by 30% to 50% or more, yielding higher quality product through the avoidance of case hardening (9).

Ceramics

Many drying systems of hundreds of kilowatts in power have been installed in the last 10 years for drying of ceramic monoliths, which must be done to permit firing of the ceramic. Because ceramics are good insulators, normal drying times are 24 h or more at high oven temperatures. The dielectric drying system accomplishes this in 20–22 min, which results in substantial energy savings as well as high product quality due to the uniformity of drying.

5.3. Microwave Drying Systems

Numerous industries use microwaves for drying, many of which are the same as those that use dielectric drying. However, there are several unique drying systems.

Food

Microwaves are used to dry pasta products, and there are over a dozen operational industrial systems. The systems utilize microwaves and hot air of controlled humidity to dry pasta and macaroni products in less than 1 h instead of the conventional 8-h drying time. These systems handle approximately 3000 lb of product per hour with 60 kW of microwave energy at 915 MHz. These systems offer substantial savings in energy, operation, and maintenance. They also provide bacteriologically more acceptable product, with reductions in microbial contamination and insect infestation (14).

Other food industry drying applications include drying of onions, seaweed, and potato chips (15). The drying of onions is particularly interesting in providing substantial benefits in terms of moisture leveling, a 30% reduction in energy costs in the final drying, and a reduction in bacterial count of 90%. Here, hot air reduces the moisture level from 80% to 10% and the microwaves from 10% to 5% (16). This is an ideal example of combining the two forms of energy in the most economical form.

Lumber

An unusual microwave application was in the drying of lumber for the manufacture of baseball bats from wood from the tanoak tree. This wood normally takes up to 2 years to dry. A microwave system at 2450 MHz heated the wood for 4 h, after which it was allowed to dry for another 2 weeks at ambient conditions. This was done on precut billets, which were later shaped into the bats, but is no longer being used (17).

Laboratory Analysis

Several microwave systems have been developed for analytical laboratory drying to determine solids and moisture content. These systems have great advantages of speed with good precision and accuracy. It is often possible to do a complete moisture determination in 2 or 3 min that might otherwise take several hours.

Microwave Freeze Drying and Vacuum Drying

There has been great interest for many years in the possibility of utilizing microwaves for freeze drying. A problem in freeze drying is that, as the moisture front recedes, the product becomes harder to dry because of the reduced thermal conductivity of the material. Microwave radiation could be ideal to provide the required heat to the receding moisture. Unfortunately, at the high vacuum pressures involved in freeze drying there is increased opportunity for ionization of the gases, causing plasma discharge that can burn the product. There is indication that these and other problems have been overcome, and commercial microwave freeze drying systems are now feasible and may, in fact, already be operational for coffee (18).

The microwave vacuum dryer offers an interesting alternative to freeze drying, and several systems are in commercial production, manufacturing fruit juice concentrates, tea powder, and enzymes. Pilot-plant tests have also been successfully performed for drying such vegetables as mushrooms, onions, and asparagus. Still another pilot system is being used for the drying of soya beans. The operational cost of microwave vacuum drying is said to be midway between spray and freeze drying (19).

In these systems, the material, often in paste form, is spread on a conveyor belt and passed through the specially built tunnel at a vacuum of 1–20 torr. This causes formation of a foam that, when dried, has excellent rehydration properties. An advantage of this method is that it allows materials of much higher solids content than in spray and freeze drying (19), which reduces the cost since less energy must be expended.

There has also been great interest in a new system aimed at drying grain with a combination of microwaves and vacuum. By pulling a vacuum of about 20 torr, moisture in the grain can be evaporated at approximately 125°F rather than the 200°F air temperature currently used (20).

A recent overall review of microwave applications in the food industry covers these and other systems and discusses the industry's problems in adopting microwave technology (21).

Pharmaceuticals

The pharmaceutical industry has become very interested in microwave vacuum drying, particularly for the manufacture of tablet granulations (22). These are blends that are then formed into tablets. During the course of manufacture they may be mixed with water, ethanol, or acetone and must, subsequently, be dried. These microwave systems are gaining use and may combine mixing, granulating, lubricating, and dry sizing in a single step. Systems as large as 1200 liters, employing 36 kW of microwave power, are in use. They demonstrate advantages in operator safety, cleaning, pollution control, and energy savings at costs often comparable to conventional systems.

Industrial Coating

There are a number of microwave drying systems for drying coatings on plastics and paper. (Included in these is the drying of silver halide on photographic film.) These combine high-speed drying with moisture leveling effects for high efficiency.

Ceramics

The ceramics industry has, for many years, examined the use of microwaves for drying purposes. Today, several uses are operating successfully. One such system, MCB Ceramics in Toronto, Canada, uses microwaves at two stages to replace a slow, hand-operated batch system with a continuous process. A 27-kW microwave oven is used to speed up the initial drying in the mold to 20 min from its previous 1 h, during which the microwaves are applied for only 2 min. The final drying used to take 24 h, but now is done with microwaves in only 8 min, after which the piece is glazed and fired. The process is used to produce small bathroom accessories such as towel bar holders and soap dishes (23).

In another process, ceramic filters, which are used to clean the slag in foundries prior to pouring the liquid metals, are uniquely produced with microwaves. These filters, which may be as large as 12-in square and 2-in thick, are made by coating, both internally and externally, an expanded polyurethane or rubber foam with a ceramic slurry. This is then dried in about 25 min for even heating, and then the filter is placed into a kiln that burns away the foam leaving the porous ceramic structure (Krieger, 1994, private communication).

Casting Molds

The use of microwaves in the foundry industry for drying and polymerizing the sand molds used for casting is very important. This allows the complete recovery of the sand and provides a great increase in speed in making the mold, which otherwise must be slowly dried and cured with hot air. An example of this application is the manufacture of the internal castings for automobile engines. Many RF systems are also used for this process.

A new method of casting, the lost foam process, makes excellent use of the unique heating mechanisms of microwaves. Intricate castings, such as automobile engines and marine parts, are made of polystyrene foam and coated with a ceramic that must then be dried. The foam is an excellent insulator so hot air drying may take many hours, while microwave alone or in combination with hot air shortens the time dramatically. The system for automobile engines for the Ford Motor Company utilizes foam clusters hanging from a monorail traveling through a microwave oven 45 ft × 25 ft × 7 ft and utilizes 48 kW of microwave power. In this case, the clusters are first dried with warm air for about 70 min and then finish dried with microwaves for 20 min, a much shorter and efficient process than the 4 h required if only hot air is used.

The demands on this process are very stringent since the ceramic coating must be 100% dry, have a smooth, evenly coated exterior and interior, have no brittleness nor be overdried and browned, and have no flaking or separation of the coating. The equipment must also deal with varying process rates and so the number of units in the oven may vary from full capacity to none and with a large opening at the entrance and exit with no significant leakage so as to maintain personnel safety. The system has been operating flawlessly since 1985.

5.4 Criteria for Successful Microwave and Dielectric Drying Systems

There are several criteria for successful microwave and dielectric drying systems. Cost is reduced. This is often a major factor. Cost savings may be realized through energy savings, increased throughput, labor reduction, reduction in heat load in the plant, speedup of the process, operational efficiencies, and reduced maintenance costs.

Quality is improved. Two examples are the drying of the latex-coated fiberglass cord and the drying of onions. In the first, there is prevention of rupture of the coating and, in the second, a reduction of bacterial contamination.

Yield is higher. The avoidance of high surface temperatures prevents overheating of the material being dried and may lead to a lower level of rejects. The near instantaneous control of temperature in these systems allows better control of drying to closer tolerances. Moisture leveling effects also avoid over- or underdrying of products.

Product cannot be produced any other way. Again, the careful control of temperature combined with the unique manner in which microwave and dielectric energy couple into materials allows the drying of extremely thermolabile materials with no damage to the product. Occasionally a unique beneficial effect may be obtained, such as the slight puffing of the pasta noodles when they are microwave dried. This allows them to be cooked more quickly.

It is usual that two or more of these attributes may be combined. However, the bottom line is economics, and if a process does not produce a sufficient return on investment, it will not meet with success.

6. ECONOMICS OF MICROWAVE AND DIELECTRIC DRYING SYSTEMS

In analyzing the economics of a microwave or dielectric drying system, the costs can be divided into capital and operating costs, and this latter may be further broken down to tube replacement, general maintenance, energy, and floor space costs. We could also add costs for cooling water for the tubes or for materials specific to a system; however, we will concentrate on only the following: capital equipment cost, tube replacement cost, and energy costs.

6.1. Capital Equipment Costs

Capital equipment costs refers to the cost of the equipment and is sometimes a bit difficult to define. For example, a microwave drying system may increase the throughput of a product dramatically and thereby necessitate the purchase of additional packaging equipment, conveyors, feed systems, and the like. In that case, the overall capital outlay would be much higher than for the microwave dryer alone.

Another problem in trying to compare the costs of microwave and dielectric dryers is that the latter often include hydraulic or pneumatic presses, materials-handling devices, and the like, built into the system, whereas this is unusual for a microwave system (8). Also, suppliers express their costs in many different ways.

In general, though, we can limit ourselves to looking only at the basic drying system, consisting of the generator, tube, applicator, control system, and conveyor, if there is one. In that case, some rough figures are available. Dielectric systems are usually the less costly of the two, especially at high power output, and recent estimates given by Zimmerly (24) of $2500–$5000 per kilowatt, with the higher cost associated with lower power equipment. Microwave systems vary between $4000 and $7000 per kilowatt, again with the lower cost associated with higher power equipment. Higher costs are also related to higher degrees of sophistication and automation of these systems. Therefore, for drying systems, by estimating that 1 kW of microwave or dielectric energy will remove 3.0 lb of water per hour, we begin to get a feeling for the economic feasibility of applying these energy systems to a specific drying process. If it appears that over 100 kW is required to remove the water, then the capital cost may be prohibitive, particularly for the microwave system. However, it is imperative to remember that it is unlikely that all the water will be removed by microwaves, but they should be employed discretely along with a conventional form of heat energy. In that case, they may only be required to remove the last few percentages of moisture from a material, at a much reduced capital cost.

As an example, consider the drying of bread crumbs from 27% to 5% moisture, at a rate of 1000 lb of wet bread crumbs per hour. In such a system, it would be necessary to evaporate 231.6 lb of water per hour (1000 lb of bread crumbs contain 270 lb of water, but when dried to 5% contain only 38.4 lb of water). This would require about 77 kW to dry (231.6 ÷ 3.0), plus an additional 20 kW to heat to the drying temperature, for a total of 97 kW, neglecting all heat losses. If we assume this system to have a coupling efficiency of 75% (that is, the efficiency of coupling microwaves into the product), then a system of 130 kW is required, which would cost in excess of $650,000. On the other hand, if a conventional hot air dryer is used to reduce the moisture from 27% to 12% and the microwave dryer to finish drying it to 5%, then the amount of water to be evaporated by the microwave system is only 61.6 lbs, which requires 20.4 kW (61.1 ÷ 3.0). Since the product is already hot, we need only increase the output to 27 kW to account for the 75% coupling efficiency (20.4 ÷ 0.75). Such a system would cost of the order of $160,000 plus the cost of the less expensive conventional dryer.

Note, however, that there are times when such a microwave or dielectric dryer can be less costly than the conventional, especially when drastic improvements are made in the drying rate such that 25% or less time is required to dry a product.

6.2. Tube Replacement Costs

The cost of tubes varies greatly, depending upon the output power. The least expensive tubes by far are the microwave oven tubes, the output power of which is of the order of 750 W and may be purchased, in quantity, for under $15 each. However, higher power tubes are far more expensive. The approximate replacement costs and the approximate tube lives are shown in Table 7.

6.3. Energy Cost

There was a time when electrically powered systems were considered too costly to operate for high-power applications. However, today, with the rising costs of oil and gas, this is no longer necessarily true, especially when far greater heating and drying efficiencies are possible with these systems. It is not practical to try to compare gas, oil, and electricity prices at this point since these are so variable. However, we can look at the conversion efficiencies of the various systems. In general, the conversion efficiency of electricity to microwave energy is estimated to be of the order of 45–50%, which includes losses in conversion from ac to dc (about 4%), from dc to microwaves (about 40%), and waveguide and applicator losses (about 10%). For dielectrics there is an approximately 60% overall efficiency, based on a combined filament, control, and dc supply efficiency of 92%, times a tube efficiency of 73%, times a circuit efficiency of 90% (8). Thus the energy cost per hour may be calculated as

Energy cost/hour = (utility rate/kWh)(kW of system)(efficiency)   [9]

Table 7 Tubes of Approximate Replacement Costs

Size (kW)	Type	Frequency (MHz)	Life (h)[a]	Actual cost ($)	($) per hour of operation per kW
Microwave tubes					
2.5	Magnetron	2450	4,000	1,500	0.15
6.0	Magnetron	2450	6,000	3,300	0.09
50	Klystron	2450	25,000	69,000	0.06
30	Magnetron	915	8–10,000	5,300	0.02
50	Magnetron	915	6– 8,000	6,000	0.02
Dielectric tubes					
5		Up to 100	5–10,000	1,600	0.04
10		Up to 100	5–10,000	2,200	0.02
50		Up to 100	5–10,000	3,600	0.01
100		Up to 30	5–10,000	8,400	0.01
200		Up to 30	5–10,000	16,800	0.01

[a]Approximate

6.4. Other Costs

Other costs may be considered, such as the cost for floor space for the system and the cost of maintenance. These may vary for individual situations and cannot be considered here. We may, however, consider the cooling water cost. Many of the tubes are water cooled, and the cost for the cooling water should be considered. The water requirements are directly proportional to the output power of the tube; for example, a 6-kW magnetron at 2450 MHz requires 1.5 gal/min; a 30-kW magnetron at 915 MHz requires 5 gal/min.

An article by Jones and Metaxas (25) describes a case study on providing additional drying capacity in a paper making operation. The paper compares the present conventional drying system to microwave, RF, and infrared systems and provides the analysis that led to the final decision. A detailed cost benefit analysis is provided that is a good summary of the procedures that should be followed in such circumstances.

Another good summary of the economic factors to be considered in choosing a microwave or dielectric drying system is given in an early paper by Jolly (26). Here a detailed comparison is given for the economic considerations that enter into the choice of a new microwave system over the standard conventional procedure.

7. CONCLUSION

The application of microwave and dielectric heating to industrial drying systems is of increasing interest, particularly because of the increased energy and operational efficiencies they afford. Their unique heating means provide benefits not obtainable from other, more conventional methods. However, their high capital costs generally require that they be used judiciously in conjunction with more conventional heat forms. Their unique properties also require the systems be designed by engineers thoroughly familiar with the art and science of microwave and dielectric heating applications.

Considerable research and development has been devoted to the understanding of the fundamentals of dielectric drying and to the development of industrial applications in the past decade. For more recent, authoritative, in-depth reviews of this subject the interested reader is referred to the papers by Schmidt et al. (27) and Jones (28). Turner and Rudolph

(29) have presented a model for combined microwave and convective drying. Cohen et al. (30) have demonstrated the benefits of microwave-assisted freeze drying of peas. A number of other applications can be found in the proceedings of the biennial International Drying Symposium (IDS) series. Further, a special issue of *Drying Technology—An International Journal* dealt with all aspects of dielectric drying (31).

REFERENCES

1. IEC CISPR Publication 11, Second Ed., 1990-09, *Limits and Methods of Measurement of Electromagnetic Disturbance; Characteristics of Industrial, Scientific, and Medical (ISM) Radio-Frequency Equipment.*
2. Cenelec European Standard, CISPR 11, modified, *Limits and Methods of Measurement of Radio Disturbance Characteristics of Industrial, Scientific, and Medical (ISM) Radio-Frequency Equipment.* 1991.
3. J. White, *Transactions of the International Microwave Power Institute*, 1:40–61 (1973), Manassas, Virginia.
4. A. C. Metaxas, *Transactions of the International Microwave Power Institute*, 2:19–47 (1974), Manassas, Virginia.
5. W. R. Tinga, *Proceedings of the International Microwave Power Institute Short Course for Users of Microwave Power*, 19–29 (1970), Manassas, Virginia.
6. D. W. Lyons, J. D. Hatcher, and J. E. Sunderland, *J. Heat Mass Transfer*, 15:897–905 (1972).
7. P. Bhartia, S. S. Stuchly, and M. Hamid, *J. Microwave Power*, 8:243–252 (1973).
8. M. Preston, *Theory and Applications of Microwave Power in Industry*, 65–85 (1971), International Microwave Power Institute, Manassas, Virginia.
9. P. L. Jones, *J. Microwave Power*, 22(3):143–153 (1987).
10. D. Ward and R. C. Anderson, *Woodworking Digest*, August 1964.
11. D. Ward and R. C. Anderson, *Woodworking Digest*, September 1964.
12. K. W. Peterson, *Proceedings of the Industrial Short Course* (1982), International Microwave Power Institute, Manassas, Virginia.
13. M. D. Preston, Technical Bulletin of Fitchburg Dryer Division of SPECO, Inc., Schiller Park, IL.
14. R. Maurer, M. Tremblay, and E. Chadwick, *Food Processing*, January 1972.
15. R. F. Schiffmann, *J. Microwave Power*, 8:137–142 (1973).
16. F. J. Smith, *Microwave Energy Applications Newsletter*, 12(6):6–12 (1979).
17. *Varian Associates Magazine*, April 1969.
18. J. E. Sunderland, *Food Technology*, 36(2):50–56 (1982).
19. N. Meisel, *Microwave Energy Applications Newsletter*, 12(6):3–6 (1979).
20. J. Forwalter, *Food Processing*, November 1978.
21. R. F. Schiffmann, *Food Technology*, 46(12):50–52, 56 (1992).
22. R. Poska, *Pharmaceutical Engineering*, 11(1):9–13 (1991).
23. *Initiatives and Payback No. 10*, Ontario Hydro, Toronto, Canada (1988).
24. J. Zimmerly, private communication (1993).
25. P. L. Jones and A. C. Metaxas, *J. Microwave Power*, 23(4):203–210 (1988).
26. J. A. Jolly, *J. Microwave Power*, 11(3):233–245 (1976).
27. P. S. Schmidt, T. L. Bergman, J. A. Pearce, and P.-S. Chen, *Drying '92*, Pt. A, A. S. Mujumdar, ed., Elsevier, Amsterdam, pp. 137–160 (1992).
28. P. L. Jones, *Drying '92*, Pt. A, A. S. Mujumdar, ed., Elsevier, Amsterdam, pp. 114–136 (1992).
29. I. W. Turner and V. Rudolph, *Drying '92*, Pt. A, A. S. Mujumdar, ed., Elsevier, Amsterdam, pp. 553–570 (1992).
30. J. S. Cohen, J. A. Ayoub, and T. C. S. Yang, *Drying '92*, Pt. A, A. S. Mujumdar, ed., Elsevier, Amsterdam, pp. 585–594 (1992).
31. *Drying Technology—An International Journal*, Special Issue on Dielectric Drying, 8(5) (1991).

12
Solar Drying

László Imre
Technical University of Budapest
Budapest, Hungary

1. INTRODUCTION

Open-air sun drying has been used since time immemorial to dry plants, seeds, fruit, meat, fish, wood and other agricultural or forest products as a means of preservation. However, for large-scale production the limitations of open-air drying are well known. Among these are high labor costs, large area requirement, lack of ability to control the drying process, possible degradation due to biochemical or microbiological reactions, insect infestation, and so on. In order to benefit from the free and renewable energy source provided by the sun numerous attempts have been made in recent years to develop solar drying mainly for preserving agricultural and forest products.

2. ASPECTS AND LIMITATIONS OF SOLAR DRYING

2.1 General Considerations

Among the advantages of solar drying one may cite a free, nonpolluting, renewable, abundant energy source that cannot be monopolized (1). At the same time, in using solar radiation for planned drying, several difficulties must be overcome. There is the very basic problem of the periodic character of solar radiation, a problem that gave rise to the idea of storing part of the energy gained during radiation periods. This difficulty can be eliminated, aside from employing heat storage devices, only with the use of an auxiliary energy source. Even the radiation periods may produce certain difficulties. First, the intensity of incident radiation is a function of time. This is a circumstance that demands adequate control strategy and the means necessary for the control. Another problem is caused by the low energy density of solar radiation, which requires use of large energy-collecting surfaces (collectors).

Thus, the nature of solar radiation has innate problems that require means (heat stores, auxiliary energy source, control system, and large-surface solar collectors) for their solution, and so the investment costs are considerable. Obviously, a prerequisite to utilizing solar energy is economics and the need to achieve an acceptable rate of return.

Examination of the technoeconomics of solar drying has led to the knowledge of the main factors, their roles, and influencing mechanisms. The first obvious discovery was that solar energy can be economically used for drying only if the purpose can be coordinated with the specific characteristics of solar radiation. Thus, geographic circumstances deciding the number of sunny days yearly and the incident radiation intensity give different energy gain in various areas of the earth.

The relatively small energy flux density of solar radiation implies that it is particularly suited to drying processes with small energy demands.

Seasonal changes of solar radiation suggest use of solar drying in the maximum radiation intensity season (e.g., part of the agricultural products should be dried during this period).

Matching of the drying process and the specific characteristics of solar radiation is also important in governing the investment costs. Because of the small flux density of solar radiation, a high-temperature drying medium can only be produced with concentrating collectors. Such collectors are generally very expensive. Cheaper, flat-plate collectors, on the other hand, can be applied only for producing a moderate-temperature medium (usually under 60°C), and their efficiency improves with decrease in operation temperature. So dryers with flat-plate collectors can be used for products requiring low-temperature drying.

One way to reduce the costs of solar collectors is to strive for cheap and simple construction. This can mean a decrease in operation life and efficiency, so that the task must be handled as an optimization problem. Another possibility is multipurpose construction, for instance, building the collector into the roof structure as an integrated part.

At this point technical development can proceed in two directions: simple, low-power, short-life, and comparatively low-efficiency dryers in one direction, and high-efficiency, high-power, long-life, expensive dryers in the other direction. The latter are characterized not only by an integrated structure but also by integration in an energy system involving processes other than drying. The aim is twofold: coupling a solar energy dryer to a farm's energy system gives the possibility of using solar collectors practically throughout the whole year, for example to produce hot water when the dryer is not in use; also, the hot water tank of the farm can be used as a heat store of the solar system.

2.2.   Role and Importance of Solar Drying in the Developing Countries

Due to the lack of adequate preservation methods, direct open-air drying is still a widely used means of food preservation in most parts of the developing world (147,152). This traditional practice has its inherent disadvantages:

Damage to the crop by rodents, birds, and animals
Degradation through exposure to direct irradiation of the sun and to rain, storm, and dew
Contamination by dirt, dust, wind-blown debris, and environmental pollution
Splitting of the grain bleaching and loss of germination capability due to overdrying
Insect infestation
Growth of microorganisms
Additional losses during storage due to insufficient or nonuniform drying

Postharvest losses can be estimated at more than 30% (148,149) and it could be reduced to a great extent by adequate drying of crops (150).

Table 1 Total Horizontal Solar Insolation and
Sunshine Hours for Some Developing Countries

Country	Average insolation (kWh/m² day)	Sunshine hours (h/day)
Cameroon	3.8–5.5	4.5–8.0
Egypt (Cairo)	6	9.6
Guatemala	5–5.3	–
India	5.8	8–10
Indonesia	4.24	–
Kenya	5.25–5.6	6–7
Malaysia	4.41	–
Mali	4.34	8.4
Mauritius	4.5	7
Mexico (Jalapa, Veracruz)	4.65	–
Nicaragua	5.43	–
Nigeria	3.8–7.15	5–7
Papua New Guinea	4.6–9.6	4.5–8
Philippines (Metro Manila)	4.55	–
Sierra Leone	3.4–5.3	3–7.5
Thailand	4.25–5.66	–
Togo	4.4	5.5–7.2

There are two possible ways for the proper preservation by drying: by using fossil fuels and by the use of solar energy. Not considering the disadvantageous environmental pollution effects caused by the CO_2, SO_2, and NO_x emission, the use of fossil fuel fired/electrically powered dryers is limited and inappropriate for most of the farmers in developing countries. The main reasons are as follows:

Expensive investment and high energy costs.
Lack of skilled personnel for operation and maintenance.
Conventional sources of energy are either unavailable or unreliable (151).

During the past two decades, several developing countries have started to change their energy policies toward further reduction of petroleum import and to alter their energy use toward the utilization of renewable energies (152).

With very few exceptions, the developing countries are situated in climatic zones of the world where the insolation is considerably higher than the world average of 3.82 kWh/m² day (153). In Table 1 daily average horizontal insolation data and sunshine hours are given for some developing countries (152,154).

An alternative to traditional drying techniques and a contribution toward the solution of the open-air drying problems is the use of solar drying. The main reasons are as follows:

1. Solar drying provides the desired reduction of losses together with improved quality of the dried products.
2. The time of drying can be significantly reduced.
3. The harvesting period can be shortened, which enables the soil to be prepared for the cultivation of another crop.

4. The drying season can be lengthened by successive harvests and by using solar dryers in which various types of products can be preserved.
5. Farmers may have a greater income by the production of marketable crops.
6. The additional costs involved in installing solar dryers can be returned by the increased profits.

Accordingly, the availability of solar energy and the operational marketing and economy reasons offer a good opportunity for using solar drying in the developing countries. A great number of successful practical applications have already been reported (152,155–174).

3. CONSTRUCTION PRINCIPLES OF SOLAR DRYERS

3.1 Main Parts of Solar Dryers

Solar dryers have the following main parts:

1. Drying space, where the material to be dried is placed and where the drying takes place
2. Collector to convert solar radiation into heat
3. Auxiliary energy source (optional)
4. Heat transfer equipment transferring heat to the drying air and/or to the material
5. Means for keeping the drying air in flow
6. Heat storage unit (optional)
7. Measuring and control equipment (optional)
8. Ducts, pipes, and other appliances

3.2. Classification of Solar Dryers

The structure of solar dryers is adjusted to the quantity, character, and designation of the material to be dried as well as to the energy sources used. Accordingly, a great variety of solar dryers have been developed and are in use. The following classification suggests three main groups for solar dryers on the basis of the energy sources used (175):

1. *Solar natural dryers* using ambient energy sources only
2. *Semiartificial solar dryers* with a fan driven by an electric motor for keeping a continuous air flow through the drying space
3. *Solar-assisted artificial dryers* able to operate by using a conventional (auxiliary) energy source if needed

3.3 Solar Natural Dryers

In the main group of solar natural dryers two subgroups are included: the subgroup of the *passive, natural convection* solar dryers (cabinet, tenth type, greenhouse type, chimney type dryers) and the subgroup of *active, partly forced convection* solar dryers having a fan driven by electric energy converted by photovoltaic solar cells or driven by a small wind turbine.

Cabinet Dryers

The simplest solar dryers are the *cabinet dryers* (Fig. 1). Their main characteristic is that the heat needed for drying gets into the material through direct radiation and through a south-oriented, transparent (glass or foil) wall 1. Other walls of the dryer are opaque and

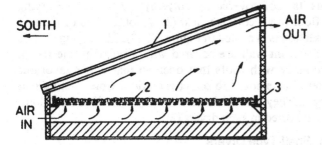

Figure 1 Structure of a cabinet dryer. (From References 2 and 3.)

well insulated. The drying material 2 is spread in a thin layer on a tray 3. The bottom plate of the tray is perforated. Air flows through the holes by natural convection through the material and finally leaves through the upper part of the cabinet (2,3). The design of the dryer is simple, and its cost is low. It is suitable for drying small quantities (10–20 kg) of granular materials (e.g., for individual farmers). The products dried in cabinet dryers are mainly agricultural products—vegetables, fruits, spices, and herbs. Drying of the material can be made more even by periodic turning over of the material. It is employed chiefly in tropical countries, but during the warm months it can be used in the temperate climates as well. The usual size of the drying area is 1–2 m^2.

A variation of the cabinet dryer is the *tent dryer*, which consists of a triangular framework covered with a thin sheet (Fig. 2-a). The south-oriented front wall 1 is transparent; the back wall is covered by a black sheet. The material is spread on a tall tray made of netting or wire mesh.

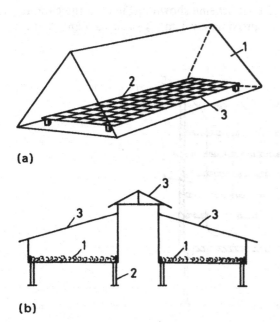

(a)

(b)

Figure 2 Cabinet dryer variations: (a) tent-type dryer; (b) its terrace-type solution. (From Reference 2.)

Another type of tent dryer has its roof covered by polyethylene sheet. The drying material is spread over a concrete floor (e.g., coffee beans) (2). Another type of the tent dryer is the *terrace dryer*. Its cross-section is sketched in Figure 2-b (2). The drying shelves 1 stand on posts 2; the roof and the front wall are covered with a polyethylene foil 3. Certain types of terrace dryers are made with roofs that open so that under favorable weather conditions the drying material is exposed to direct radiation. Constructing tent and terrace dryers is cheap and simple. They are widely used for drying coffee. In Colombia, about 70% of the coffee beans are dried in such dryers (2).

Natural Convection, Static Bed, or Shelf-Type Dryers

The capacity per unit area of cabinet dryers is limited by two conditions: need for direct radiation on the drying material and small airflow rate. To dry larger quantities of material, the basic area of the dryer has to be increased. To avoid this problem it is preferable to place the material in several independent layers; the necessary heat transfer is thus accomplished by convection. The increase in the mass flow rate of air can be achieved by increasing the effects that produce natural convection. These effects must also be increased if the air is to be circulated through a material laid in several layers one over the other, or through a thick layer, as in the case of the static bed type. To keep up the natural pressure difference without using a ventilator (for instance, in a field), the "chimney effect" must be exploited. For this purpose the vertical flow of hot air in the dryer must be increased.

In Figure 3 a scheme of the so-called *shelf dryer* can be seen (4). The material to be dried is placed on perforated shelves 1 built one above the other. The front wall of the case faces south, its top and sides 2 are covered by transparent walls (glass or sheet), and the back wall 3 is heat insulated. The back wall and the floor are covered with a coating of black paint. The ambient air is warmed in a flat-plate collector 4 joined to the bottom of the case, and it flows up to the space under the lowest shelf. Moist air exits to the open through the upper opening of the case 5. In the scheme shown in Figure 3 the chimney effect is ensured by the increased height (approximately 1 m). Test measurements have

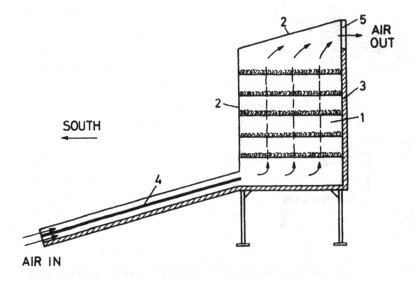

Figure 3   Shelf-type dryer with separate collector. (From Reference 4.)

shown that over a period of 5 years the best results were obtained with the aid of a glass-covered absorber plate placed in the middle of the air opening. Costs of utilized energy (Thailand climate, January–April) amount to about U.S. $.03/kWh. The experiments indicated that separation of the collector is only justified with a high-efficiency collector. The dryer is suitable for drying fruit and vegetable goods. (For a theoretical analysis of shelf-type solar dryers, see Reference 5.)

For large amounts of material an appropriately high chimney has to be connected to the dryer housing. Figure 4 gives the cross-section of a chimney-type dryer designed and built for drying 1000-kg rice (2). Rice is placed in a static bed 1 in a 0.1-m thick layer. The collector 2 consists of a plastic covering and roasted rice shell, the latter playing the role of absorber. The front surface 3 over the layer of rice is also transparent. The wall of the chimney 5 is made of black plastic foil. The framework of the dryer is wood and wire; manufacture is inexpensive and simple. The air needed for drying amounts to 5.7 m³/min per m³ rice. The chimney is 5 m high. Drying is not uniform, so the rice in the static bed must be turned over at intervals. The duration of drying is 3–4 days in the case of 15 MJ/m², day mean global sun radiation, and 23 m² collector surface. With the application of a larger (36 m²) collector surface, drying time can be reduced to 1–2 days in good weather. As a rule of thumb, the solar collector surface must be approximately three times the surface of the bed.

Preliminary steps (6) have been made for the development of heat-storing chimney-type dryers with the purpose of extending the drying process over radiation-free periods. A schematic of the dryer is shown in Figure 5. Air gets through the collector 1 to the heat-storing space 2. The collector is foil covered; its angle of inclination can be adjusted to a small degree. A reflection panel 3 is placed near the air entry, which serves for warming the entering air to a small degree. Water-filled vessels 4 serve to store heat. The walls of the heat-storing space are insulated by reflecting panels that can be turned down for the night. During night operation the outside air can be let into the heat-storing bodies through openings 5 made in the bottom of the heat-storing space. The front and side walls of the heat-storing space are covered with transparent foils like the southern side wall of the chimney 6. Its back wall and bottom plate, as well as the drying space, are well insulated. In the drying space 7, the drying material is spread on trays with perforated bottoms. Test measurements indicate about 10% drying efficiency as related to the input solar energy.

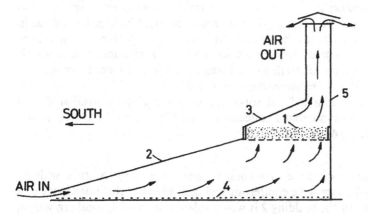

Figure 4 Static bed type solar dryer with chimney. (From Reference 2.)

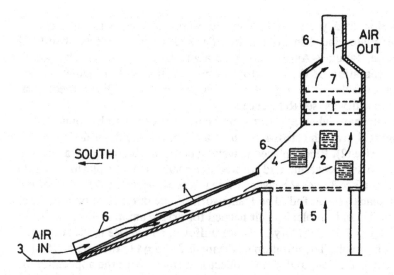

Figure 5   Tray- and chimney-type solar dryer with heat storage. (From Reference 6.)

3.4.   Semiartificial Solar Dryers

The greatest advantage of chimney-equipped natural convection dryers is that no auxiliary energy source is needed and thus they can be operated far from populated areas. The disadvantage is that the height of inexpensive-finish chimneys (without special stiffeners and foundation) is limited mainly because of the increased wind loading. Limitation of the chimney height means a limitation on the hydrostatic pressure difference and also that of maintainable air flow rate. Another disadvantage of natural convection dryers is that the air entering the collector flows through the drying space and then into the atmosphere. During drying the temperature of the material approaches the dry bulb temperature and the enthalpy increase of the air taken up from the collector is used for drying in decreasing quantity while an ever-increasing part leaks into the atmosphere through the chimney as exit heat loss.

In view of this, solar dryer variations have been developed in which a small power ventilator is fitted for maintaining air flow while recirculation is controlled by simple flaps built at suitable spots to improve thermal efficiency. The construction of such dryers is relatively simple and inexpensive. In a high-performance tent-type dryer variation designed for drying 3–4 tons of peanuts (7), the material to be dried is placed and dried in a drying drum located in a closed chamber with perforated walls, which plays the role of a solar collector as well. A ventilator delivers outside air into the chamber and as the air is warmed it comes into the drum through its perforated mantle and from there into the open air; a part of the exit air may be recirculated if desired.

An effective solution of semiartificial solar dryers is the directly irradiated, foil covered solar tunnel dryer with integrated collector section (183). Scheme and operation of that dryer is given in Figure 46 in Section 7.3.

Room Dryers

Figure 6 shows the schematic structure of a solar timber dryer (8,9). A stack with air clearances 1 is made of 30–65-mm wide coniferous and oak timber in the inner space of a building. The northern wall of the building 2 is well insulated; its roof 3, southern wall 4, and side walls are made of special two-layer transparent synthetic plates. The dryer is

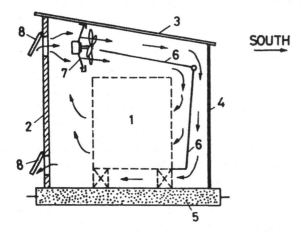

Figure 6   Forced convection solar dryer for timber. (From References 8 and 9.)

built on a concrete base 5. Solar radiation coming in through the transparent walls warms the black-painted aluminum absorber 6. The airflow delivered by an axial ventilator 7 flows along the two sides of absorber 6; one part of it enters the stack at the back side, the other from the bottom. The adjustable angle of inclination of the upper (top, roof) part of the absorber 6 makes it possible to control the quantity of air directly led into and circulated in the stack. The proportion of fresh and recirculated air can be changed by simple flap valves 8. The flow volume of the ventilator is 2.5 m³/s, with 180 Pa. Depending on the width of the building, one or several ventilators may be used. (For example, two ventilators are needed for a 5.64 m wide building. The height of the southern wall is 2.50 m; the width of the building is 3.05 m. Stack volume is 5–9 m³. For tests made with a dryer of similar construction, see Refs. 10 and 11; with rock heat storage placed on the floor, see Ref. 12.)

The application of forced airflow is necessary for drying products in static beds, which form a comparatively large flow resistance in the bed. Such products include grains and hay. One solution is to build a solar room dryer (13) as shown in Figure 7-a. The grains to be dried are placed as a bed 1 on a perforated flooring. Collectors 2 are located on the southern wall and the roof of the building. The air warmed in the collectors is kept in circulation through channel 3 by fan 4 and ventilated through the bed across the lower distributing space 5. Wet air exits to the open air from the roof space through side wall openings. The collector area was chosen to be 4 m² for 1 m³ wheat of 20–24% initial moisture content (double covering, 200 W/m² long-term mean collector power). Measurements showed that a maximum of 55°C entering air temperature could be reached without influencing germination ability. In the lower layers there is the possibility of overdrying, which may be avoided economically by employing separate heat storage.

The role of the dryer housing can be played by a grain bin (8,14). In this case the collectors are integrated with the wall of the cylindrical or square-shaped bin. Other variations use a separate plane collector system (15).

Figure 7-b shows the schematic construction of a solar rough fodder dryer (16). The material is placed in a static bed 1. The collector system 2 is placed, as in Figure 7-a, on the southern wall and the roof, but with airflow in the opposite direction. The collecting channel is placed at the bottom, joined to the housing of a fan 3. The fan is able to draw in outside air directly 4. For bad weather there is the possibility of using an auxiliary

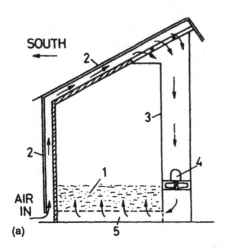

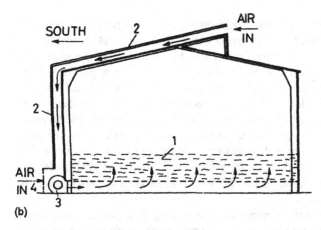

Figure 7 Solar room dryers: (a) room solar grain dryer (from Reference 13); (b) solar fodder dryer (from Reference 16).

energy source (e.g., gas-heated heat exchanger) on the suction side of the ventilator. (For further applications of collectors integrated into the roof, see References 17 and 18.)

Solar Dryers with Physical Heat Storage

The application of heat storage in solar drying systems is justified by three circumstances:

1. Drying period can be extended by the stored energy.
2. The surplus energy appearing at the radiation peaks can be stored to avoid local overdrying.
3. The temperature of the drying air can be controlled to avoid damage to material.

   In any case, when dimensioning the collector surface of the dryer with heat storage, attention must be paid to the fact that the energy getting into the storage unit forms a part of the energy gained by the collector system. Also, use of heat storage will necessarily involve a decrease in the temperature level of the energy obtained. In the case of directly

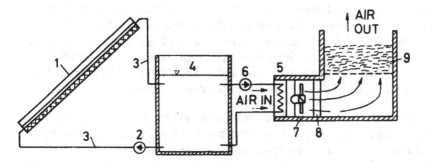

Figure 8 Solar dryer equipped with water-type heat storage. (From Reference 19.)

radiated heat storage (formed, e.g., as the absorber of the collector) this effect is less important.

One pays for the advantage of using heat storage with higher investment and operating costs. Careful technoeconomic evaluation must be made before use of solar energy storage in solar drying.

Natural or artificial materials may be employed for heat storage. Natural materials (water, pebble bed, and rock bed) are usually cheaper than synthetic materials (e.g., latent heat-storing salt solutions and adsorbents). Detailed discussion of heat stores is beyond the scope of this chapter.

Sensible heat storage of high capacity calls for water as the working medium (indirect heating system). Accordingly, the collectors are more expensive, and the application of a water-air heat exchanger also involves further cost (see Sec. 5.2).

In Figure 8 the construction of a solar dryer with water storage is shown (19). The dryer is an indirect system. Pump 2 circulates the working medium of the collectors 1 along a pipe 3 and warms the fluid in a storage tank 4. The dryer uses outside air drawn by ventilator 7 and heat exchanger 5. The primary medium of the heat exchanger is the fluid from tank 4 circulated by pump 6. Air can be warmed to the necessary temperature by the heater 8. The material to be dried is placed in a static bed 9. Measurements have proved that 50–60% of the energy needed for drying can be gained from solar energy.

Figure 9 shows the arrangement of a solar dryer with rock-bed storage (20). The dryer

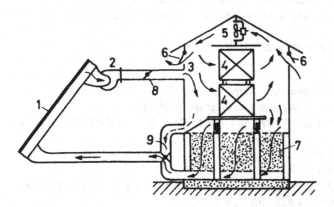

Figure 9 Solar timber dryer equipped with rock-bed heat storage. (From Reference 20.)

is a direct system; the collectors 1 are located on the ground and have an area of 193 m^2. The air warmed in the collectors is forwarded into drying space 3 by a fan 2. The dryer has room for a maximum of 6.5 m^3 timber 4. In the upper roof space two ventilators are placed that are used for continuous circulation of the air. Vents 6 are placed in the sidewalls of the roof space for allowing inflow and outflow of air. The rock bed 7 is about 22 tons of 19-mm crushed basalt. The dryer operates as follows.

1. During the warming period, a fan 2 revolves, first slowly, then at full rotation; a damper 8 opens gradually; damper 9 is in the position marked in the figure by the dotted line. Air flows from the dryer space back into the collector.
2. During drying and charging of heat storage, damper 9 is in the position shown in the figure by a solid line, air flows from the dryer space into the rock bed and back to the collector.
3. During operation when there is no solar radiation, damper 9 is in the medium position, air flows from the dryer into the rock bed, is warmed, and flows again into the dryer. The operation of dampers 6 is controlled by the wet bulb temperature measured in the dryer.

3.5. Solar-Assisted Artificial Dryers

Solar-Assisted Dryer for Seeds

In Figure 10 the scheme of a solar-assisted seed dryer is presented. Figure 10-a shows the cross-section of the dryer with layer-type arrangement and Figure 10-b a containerized construction for sensitive materials (185,197). The drying space is divided into two cells 1, 2 for the better direction of the drying process (see also Sec. 7.3). Each cell has an individual fan 5 of two RPM stages. Fans are arranged in separate spaces 3. Among the two fan spaces the space 4 of the auxiliary heater 6 operating with natural gas is situated (Fig. 10-c).

As solar energy converters, uncovered flat-plate air collectors 7 are used integrated into the roof structure of the building. Air ducts of the collector field are connected to a collecting-distributing air channel 10. By moving sliding plates 11 the collecting channel can be opened and connected to the fan spaces to divide the total preheated air flow, in a proper ratio, between the cells. For seed grains of small dimensions a layer-type static bed is preferred (Fig. 10-a). For seed grains of larger dimensions (e.g., beans) use of containers with perforated bottoms is recommended to avoid the possible damages during transportation and feeding in and out. Moist air leaves the drying cells through the openings 20.

The main technical data of the dryer are as follows:

Number of cells: 2
Effective surface area of the bed for one cell: 56 m^2
Dry mass of seed for one cell/seed grains of meadow grass/: 5600 kg
Mass flow rate of air of one fan: at RPM 1090 min^{-1}, 41,000 m^{3}h^{-1}; at RPM 475 min^{-1}, 12,500 m^{3}h^{-1}
Surface of the collector field: 191 m^2
Average effectiveness of the collector: 0.3
Average temperature increase of the air preheated by solar energy (July, Hungary): at RPM 1090 min^{-1}, 2.9°C; at RPM 475 min^{-1}, 9.86°C
Output of the auxiliary air heater: 93 kW

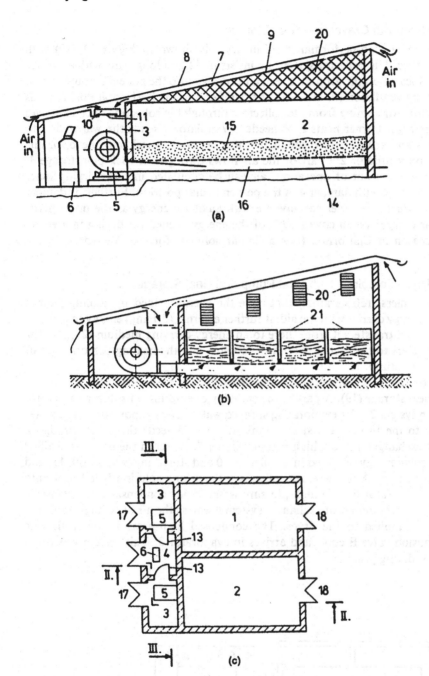

Figure 10   Solar-assisted dryer for seeds: (a) cross-section of the dryer in layer-type arrangement; (b) containerized construction; (c) ground plan of the dryer.

Solar-Assisted Dryer with Gravel Bed Heat Storage

The construction of a high-performance raisin dryer is shown in Figure 11 (19) with rock-bed heat storage. The collector system 1 consists of 42.7-m long units with a surface area of 1812 m², located on the ground. Fresh air is drawn into the system through a heat recovery wheel 9 by ventilator 10, and with the damper 4 in the horizontal position it is sent to the collector. Air coming from the collectors through the collecting duct 11 arrives in space 6, where a gas burner heats it as needed. Ventilator 7 sends the warm air into a drying tunnel 8 and from there through the heat recovery wheel into the open air. When switching on ventilator 2 a part of the air flows through the rock pile storage 3. If the collector system is out of operation, air can be circulated into the drying space through the heat storage with damper 4 in the perpendicular position. Further, if damper 5 closes the upper duct, the dryer can operate with auxiliary energy as the only energy source. The solar energy system covers 69% of the energy needed for drying in a yearly 214-day sunny season in California. (For a similar solution for crop dehydration, see Reference 21.)

Solar-Assisted Dryer Combined with Heat Pump and Heat Storage

Heat pumps are coolers (refrigerators) that raise the energy gained by cooling from a low-temperature energy carrier with the aid of further external (driving) energy to a higher temperature level and transfers it from there to an energy-carrying medium (22,23). The term *heat pump* refers to the fact that both the cooling and the heating performance of the refrigerator are utilized.

Figure 12 shows the schematic arrangement of a solar dryer equipped with absorption heat pump and heat storage (19). A part of the enthalpy of entering outside air 1 is used — interposing pump system 2 — for evaporating sprayed water in an evaporator 3. The water vapor goes over to the brine sprayed into tank 4. Pump 5 feeds the brine through a regenerator heat exchanger 6 into a high-pressure boiler 7. Water in the boiler is distilled with the help of solar energy obtained in a collector 10 and stored in a water tank 11, and by using auxiliary energy A to the extent necessary the strong solution is led back into tank 4 through regenerator 6. The high-pressure water vapor condenses in condenser 8 and with the help of the pump heat exchanger system 9 warms the air of reduced moisture content, which is supplied to the dryer. The condensed high-pressure water flowing through an expansion valve E cools and arrives in evaporator 3. This system was originally designed for drying peanuts.

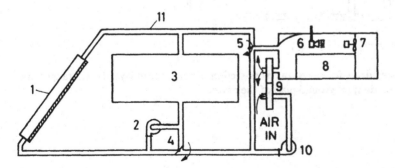

Figure 11   Solar raisin dryer with gravel-bed heat storage. (From Reference 19.)

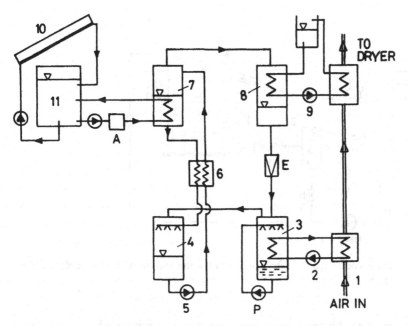

Figure 12 Solar dryer for peanuts equipped with absorption heat pump and heat storage. (From Reference 19.)

Solar-Assisted Dryer Integrated into a Complex Energy System

One economic factor in the use of solar dryers is the amount of solar energy over the year and the yearly drying period. In any case, even in the drying season there are unavoidable breaks, and during the rest of the year the collector system cannot be used for utilizing solar energy. For year-round utilization it is desirable to look for other possibilities for using solar heat. Such a possibility is given, for instance, by satisfying the hot water needs of a stock-breeding farm. With the integration of the solar system into the hot water system of the farm, investment costs can be saved. The storage tank of the hot water system can be used as water storage for the solar system. In addition, the complex system solution may permit increasing the efficiency of the solar system and thus the amount of solar energy obtainable. The efficiency of flat-plate collectors (see more details in Sec. 5.2) improves with decrease in operational temperature. So it is more advantageous if the collectors are used for warming cold water from wells or the water supply system than if the working medium returning from the fluid-air heat exchanger of the dryer is led into the collector at a temperature higher than the ambient.

Figure 13 shows the scheme of a solar alfalfa dryer joined to the hot water system of a stock-breeding farm (24,25). The fluid medium collector system 5 built on top of the dryer building is connected to a closed circuit. The system can have different operating modes. When valves 2 and 3 are closed, the collector system works on the fluid-air heat exchanger 6 and serves dryer 7. With values 1 and 3 closed, the water heat storage 8 is warmed. In the transition position of valves 1 and 2 (valve 3 is closed), the two modes can partially operate simultaneously. If valves 1 and 4 are closed, the drying air is warmed in heat exchanger 6 by using the hot water reserves of heat storage 8.

The air leaving the dryer has almost the same enthalpy it had on entering the dryer. A considerable part of the enthalpy used on drying can be regained by condensing the

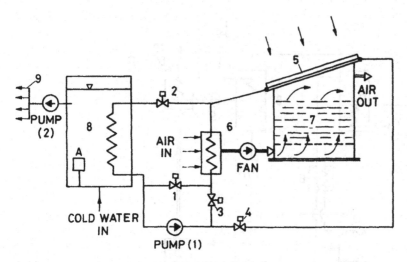

Figure 13 Solar hay dryer connected to technological hot water system of stock-breeding farm. (From References 24 and 25.)

absorbed water vapor. For this purpose a heat pump may be inserted in the energy system. Figure 14 illustrates the scheme of a system complete with a heat pump (24). Part of the moist air leaving the dryer flows through the evaporating heat exchanger 9 of the heat pump, and a proportional part of its moisture content is condensed. The heat input to the working medium of the heat pump (complemented by the input energy of compressor 10 and with the aid of the condenser heat exchanger 11) can be taken into the hot water system. Depending on the ambient state, the air leaving heat exchanger 9 can be returned to heat exchanger 6 of the dryer. (Other labels in the figure are the same as those

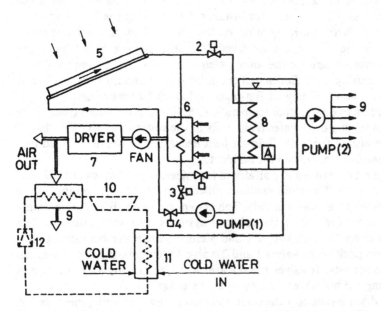

Figure 14 Complex solar dryer system combined with heat pump. (From Reference 24.)

in Figure 13.) In the case of a dryer connected to the energy system of a cattle-raising farm, a heat pump can be also used for cooling milk and producing hot water at the same time.

Inasmuch as the stock-breeding farm possesses a biogas-producing system, the hot water produced can be utilized for heating the gas-producing containers in place of biogas, which can be utilized in other ways. In this case, naturally, biogas can be used as an auxiliary energy source for the dryer during periods of bad weather.

When solar dryers are integrated into the complex energy system of a farm, adsorbent beds can also be utilized as auxiliary units. Adsorbent materials have to be regenerated for exploiting their dynamic adsorption capacities. The regeneration temperature is typically over 150°C, depending on the adsorbent (26). A condition of economical application is that a considerable part of the energy used for regeneration should be used for producing hot water, for example. During breaks in solar radiation, drying can be continued at the expense of the adsorption capacity of the adsorbents. Therefore, the advantage of applying adsorbents is that a considerable part of the energy used for regenerating the adsorbent can be utilized more cheaply for other purposes than using the same energy for heating the drying air.

Solar-Assisted Adsorption Dryer

Figure 15 shows the scheme of a complex solar system complemented with adsorbent units (24). Air warmed in the air-type collector system is delivered to the dryer by fan 1. The ducts of the adsorbent units are joined to the duct section before the fan at points 2 and 3. Unit A2 in the state drawn in the figure (thick lines) is under regeneration; active unit A1 is undergoing adsorption. At joint 3 dry air flows to the airflow of fan 1. Fan 2 draws air through filter F; then the airflow divides into two parts. In the open position of 8 and closed position of 9 a part of the airflow goes through A1 to the suction side of fan 1; the other part is warmed in heat exchanger H1 and serves the purpose of regenerating unit A2 and then moving into heat exchanger H2. The primary medium of heat exchanger H1 is air heated by energy source E (e.g., biogas), and driven by fan 3. The medium leaving heat exchangers H1 and A2 flows into the second part of heat exchanger H2. Heat exchanger H2 is connected to the water system of tank T of the hot water system. Water is circulated by pump 1. Pump 2 supplies hot water for 15 consumer lines. (A is the auxiliary energy source of the water tank.)

The energy conditions of solar dryers integrated into a complex energy system are favorable. At the same time, the number of necessary auxiliary equipment is greater, investment costs are higher, and control of the system is more complicated. The economics of such a system depend on local conditions.

Solar dryers can be used with an adsorption bed for energy storage (12,26–29). The storage of energy accordingly occurs, partly in the form of physical heat and partly in the form of moisture adsorption capacity. The total energy storage capacity of the adsorbent heat store per unit mass is about 10 times greater than that of a physical heat store. An optimally designed timber dryer fitted with adsorbent energy storage has been compared with rock-bed storage and proven competitive (28,29).

4. ECONOMICS OF SOLAR DRYERS

4.1 Main Economic Factors

The economics of solar drying depend on the costs involved and benefits gained. Interpretation of the benefits gained by solar dryers is less unambiguous than that of other solar

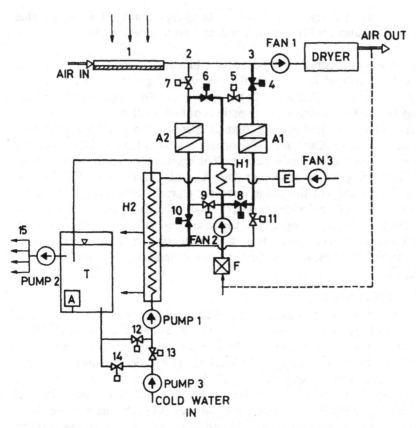

Figure 15 Arrangement of solar-assisted dryer combined with adsorbent units. (From Reference 24.)

systems (196). The main reason can be found in the great variety of solar dryers. For evaluation of the gain a correct basis should be interpreted by finding the extra income (i.e., the *savings*) of the solar dryers as compared to a basic solution (175,176).

Natural solar dryers should be compared to open-air drying. Savings are realized by reducing the losses of the open-air drying and no energy savings can be considered. *Semiartificial solar dryers* should be compared to an artificial dryer with the same performance. Their advantages are in reducing the first costs by an unsophisticated construction and by the energy substituted by solar energy. With *solar-assisted artificial dryers*, savings should be interpreted by the substituted energy. As costs, the investment of the solar energy converting system should be taken into account only.

The savings depend on the lifetime of the dryer and, for the last two main groups of solar dryers, on the cost of the substituted fuel or energy carrier. The lifetime of the dryer can be estimated in advance; it is in any case related to the maintenance costs as well. An error made in the estimation of lifetime may cause significant uncertainty in the economic evaluation.

The price of the dryed products and/or of the substituted energy is not stable. These prices may change during the lifetime of any dryer. For changes in the prices, predictions can be used; however, these must be taken as estimates, again causing some further uncertainty in economy calculations.

The sum of overall installation costs is composed of investment costs, interest and maintenance, service, tax, and insurance charges. Inflation modifies the total cost of installation.

In view of the uncertainties mentioned it is expedient to make some estimations for the calculations.

4.2. Dynamic Method of Economic Evaluation

Economic evaluation of solar dryers usually aims at determining the payback time. The dynamic method of calculations takes the influence of inflation into consideration. The following considerations summarize this method, according to Böer (30).

Payback occurs when the accumulated savings S equals the sum of investment capital I plus yearly interest and the accumulated costs E:

$$S = I + E \tag{1}$$

The annual accumulated savings can be calculated from the net income D by reducing the mass and quality losses and thus by increasing the marketing price of the product in case of natural solar dryers. For semiartificial and solar-assisted artificial dryers savings can be calculated from the price of the substituted conventional energy D. Considering the annual interest rate r and the yearly inflation rate e for the prices, for n years the annual accumulated savings can be calculated as follows:

$$S = \frac{(1 + r)^n - (1 + e)^n}{r - e} D \tag{2}$$

The sum of first investment cost C with interest will be, during n years,

$$I = C(1 + r)^n \tag{3}$$

Accumulated yearly costs, taking the annual fixed charge rate mC and inflation rate i into consideration, will be

$$E = \frac{mC(1 + r)^n - mC(1 + i)^n}{r - i} \tag{4}$$

Knowing C and D, diagrams can be made for the determination of payback time n, referring to values of r, m, i, and e. From these diagrams the requirements for the expected payback time can be easily seen. Figure 16 shows the payback time as a function of D/C for various values of r, m, i, and e, following Böer (30). One can see from the diagram which D/C values can bring about the desired payback time when the various other parameters are at given values. With parameters differing from the above, the calculation must be made separately following Eqs. [1] through [4]. A comparison of curves 1 and 2 indicates the influence of interest rate r in cases with no inflation; curves 3 and 6, when compared, lead to the effect of energy prices. As can be seen in all the cases examined, the D/C ratio needed for a 10-year payback time falls in the 0.12–0.23 range. Since payback time is a function of D/C, it is obvious that cheaper (smaller C) and less efficient (smaller D) installations are justified insofar as realization of less expense does not mean a significant decrease in the durability (lifetime) of the system.

Payback calculations refer to the whole solar energy drying system (31,32). With appropriate division of the costs, there is of course nothing in the way of making the calculation only for the collector system (33,34). Construction of the collectors can be planned on the basis of the economic optimum (35,36).

When the application of a solar dryer results in improved quality of the dried prod-

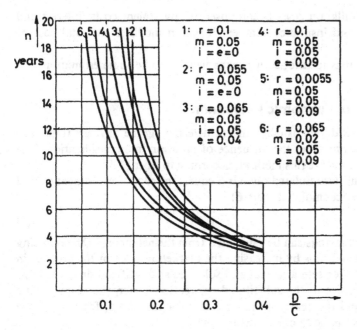

Figure 16   Payback time in years as a function of D/C, with different parameters r, m, i, and e. (From Reference 30.)

uct, the value of D is savings S has to be increased in relation to the value of the quality enhancement.

5.   KEY ELEMENTS OF SOLAR DRYERS

5.1.   Solar Collectors

Construction of Solar Collectors

The solar collector plays the part of primary energy source for a solar dryer. Essentially it has functions of energy conversion and energy transfer.

As energy converter the collector converts the direct and diffuse radiation coming from the sun into heat. This energy transformation takes place in the so-called absorber of the collector (Fig. 17). The absorber is made of a material of high absorption coefficient for the radiation of the sun or has a coating with such a character. The radiation absorbed causes the inner energy of the absorber to grow and its temperature to rise.

The energy-transferring function is to transfer the radiation energy transformed into heat in the absorber to the working medium of the collector. The working medium is, with direct system solar dryers, the drying air itself; with indirect systems this is an appropriately chosen liquid (distilled water or, in winter operation, a fluid with low freezing point, oil, and nonaqueous liquids).

Heat transfer between absorber and the medium flowing through the collector occurs by convection. Only part of the heat coming from the incident radiation gets into the working medium.

The part of the radiant energy irradiated that causes in increase of enthalpy of the working medium flowing through the collector is considered utilized heat; the rest is heat

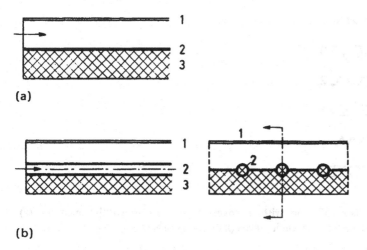

(a)

(b)

Figure 17 Setup of flat-plate collectors: (a) air; (b) liquid as working medium (1, covering; 2, absorber; 3, heat insulation).

loss. For attaining a realistic rate of heat loss most collectors are covered with transparent materials to solar radiation (glass, plastic foil, and others). If the absorption of the covering material and its reflection is high for the absorber's own long-wave radiation, it will reduce the radiation loss of the absorber. For reducing the radiation loss, a coating that selectively reflects long-wave radiation can also be applied to the covering. Since the temperature of the covering is considerably lower than that of the absorber, the coating will also reduce the convective heat loss from the structure to the ambient air. The number of coverings is usually not more than two. For the reduction of further heat loss it is desirable to insulate the nontransparent parts of the collectors. Efficient means for reducing convective heat loss are collectors with the space between covering and absorber evacuated (vacuum collectors); this makes the collectors expensive.

The final form of the collector is a problem of reading a technoeconomic optimum (see for instance, the D/C ratio in Sec. 4). In general, beyond a certain limit the reduction of heat loss is no longer economical.

The simplest types of solar dryers (e.g., cabinet dryers, tent dryers, and certain chimney shelf dryers) (see Figs. 1, 2, and 4), do not employ a separate absorber; the role of the absorber is played by the irradiated material itself. The majority of high-performance solar dryers are equipped with flat-plate collectors.

Figure 18 presents flat-plate collectors without covering using air as the working medium. Air flows in the channel between the absorber and the heat insulation. The absorber is a commercially available, rolled metal sheet, usually with a surface coating. Absorbers made of zinc or steel galvanized with zinc can be applied without a coating. The application of collectors without a covering is justified for low-performance dryers.

Figure 19 shows some variations of air-type collectors with one covering. These structures can also be made with two coverings. Flow under the absorber reduces the convective heat loss of the air from the covering. There are designs in which air flows on both sides of the absorber. A corrugated or finned absorber surface improves heat transfer between air and absorber. With the latter, flow direction is usually parallel to the fins.

Improvement of heat transfer between air and absorber is aimed at collector designs with absorbers of divided surface (Fig. 20). The air is forced to flow through the gaps.

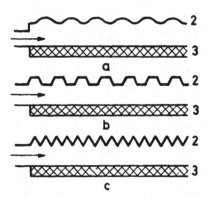

Figure 18 Scheme of flat-plate collectors without covering, with air as working medium: (a) corrugated plate; (b) trapezoid plate; (c) triangle waved plate (as absorber).

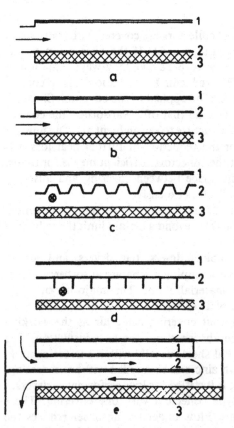

Figure 19 Scheme of air-type flat-plate collectors with covering: (a) plane absorber, flow over absorber; (b) plane absorber, flow under absorber; (c) absorber with corrugated surface; (d) finned absorber (in c and d, flow is under the absorber, perpendicular to the plane of the paper); (e) plane absorber, flow over and under absorber (two way). (From Reference 35.)

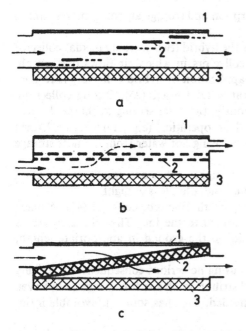

Figure 20 Air-type flat-plate collector constructed with divided surface absorber: (a) stepwise divided absorber made of overlapped plates; (b) perforated double-plane absorber; (c) matrix-type absorber.

There are a large number of matrix-type and porous absorbers. The matrix-type absorber, for instance, can be made of wire bundles and of slit and expanded aluminum foil (Fig. 20-c).

An air collector with integrated latent heat storage is shown schematically in Figure 21-a. The casing of the heat storage material acts as the absorber (37). Here advantage over air-warmed storage is that heat store warms through direct irradiation; thus heat-storing materials having a phase change temperature (38,39) higher than the temperature

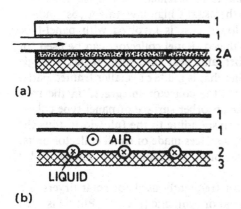

Figure 21 Integrated collector types: (a) latent heat storage filling, as absorber (2A); (b) hybrid (air-liquid) collector.

of the air can be used. An integrated rock absorption and storage air collector system has also been developed (40).

Another variation of combined collectors is the hybrid (two-working media) collector (Fig. 21-b). Hybrid collectors are liquid-type collectors in which air flows over an absorber that is common for air and liquid. The application of hybrid collectors is reasonable if the dryer is connected with sensible heat water storage (25). During collection, hybrid collectors warm air and water simultaneously, the latter serving to charge the heat storage tank. During the radiation-free period of operation (e.g., night), the hybrid collector works as a water-air heat exchanger and, using hot water from the heat storage tank, preheats the drying air.

Both integrated collector types can operate at night as they are usually made with a double (glass-glass or glass-foil) covering to reduce radiation loss at night.

A very simple design of air collectors is the polyethylene tube collector (41). A black absorber tube is placed inside a clear tube connected to the fan. The tube assumes its cylindrical form on overpressure. It can be laid on the ground in appropriate length, chiefly for agricultural drying purposes.

Air-type collectors have the advantage of causing no serious consequences if leaking occurs (42). Difficulties may arise in uniform distribution of air: for advantageous even distribution, considerable fan performance is needed. This has some unfavorable influence on operating costs.

Collectors with a liquid working medium are used for indirect-type solar dryers, usually with water heat storage. Their application for high-performance installations is justified because no large and costly distributing and collecting air channel system is needed as for the air-type collectors. On the other hand, a water-air heat exchanger must be employed.

A drawback of the liquid-type collectors is the danger of leakage and freezing. The former can be averted by appropriate junctions that permit dilatation, the latter by using antifreeze liquids as working media, for example, by integration into the hot water system of the farm for year-round performance.

Owing to the widespread application of solar hot water systems, a great number of variations of fluid collectors have been developed and commercialized circulation. It is mostly the 1–2 m^2 surface-mounted units that are available commercially; these, joined in an appropriate number, form a full collector system. The advantage of commercially available collector surfaces is the quick and simple replacement of the elements and guaranteed thermal efficiency. A disadvantage is the usually high investment cost and the long payback time. For indirect solar dryers, lower cost is involved with panel-type collectors, if the costs of collectors need to be reduced. Panel collectors can be used for both air and fluid working media. The absorber and heat insulation forms a single continuous surface, and it is only the glass covering that is made of smaller framed parts. If placed on the roof of an agricultural building, the collector integrated in the roof structure can play the role of roofing as well. The absorber surface of panel-type collectors can be mounted on 5–10-m long elements corresponding to the full width; thus the number of joints is considerably less than those of surfaces made of small collector units. In this way not only the construction cost is lower, but the probability of leakage is also reduced.

The absorber of the liquid-type collector most frequently used for solar dryers is a surface formed of ribbed pipes. Its material is metal or synthetic (plastic). Plastic is used for producing carpetlike collectors containing appropriate fluid ducts. The collector carpet usually brought into circulation can be spread over a very large, contiguous surface and can be used with panel collectors, too, with or without covering. Its advantage is simple mounting; its disadvantage is its wear. The ultraviolet (UV)-stabilized construction has a comparatively long lifespan. Plastic is also used for making absorbers of pipelines laid side by side (43). Their disadvantage — apart from that already mentioned — is that a great number of connections and very careful mounting are required. Another disadvantage of plastics is their sensitivity to high temperature. Even on using a cover a temperature rise in the absorber of over 100°C may occur if there is no heat removal. The temperature of collector types without a covering does not rise to a dangerous level.

Most liquid collectors used with high-performance solar dryers are made with a finned metal tube absorber. The absorbers applied to solar hot water producing systems are often made of sheet halves with stamped passages bonded by seam welding or by rolling them together (Fig. 22-a).

Flat-plate collectors with finned tube absorber (shown in Fig. 22-b) can be built of extruded elements. This is proposed for integrated or panel collectors. The absorber elements perpendicular to the plane of the paper can be ordered from the manufacturer by length of the panel. The structure must be designed so that dilatational movement of the elements is possible. Collectors built of absorbers from pipes soldered or welded to a sheet are shown in Figures 22-c through f. The type in Fig. 22-f ensures great strength (rigidity) even with long panel collectors.

The materials commonly used for finned tube absorbers are copper, aluminum, or steel (44). Copper is rather expensive for dryer collectors. Aluminum gives a long operation life with nonaqueous working media. The corrosion of steel can be reduced with the application of inhibitors.

Efficiency of Flat-Plate Collectors

The surface required for the collectors of solar dryers can be determined from the energy demand of the dryer. In most solar dryers, drying takes place in stages and only a small part of a dryer is used for drying of continuous material flow.

In the case of drying in stages, the energy demand is not constant: it is greater at the beginning of the drying process and decreases as drying proceeds. For dimensioning the collectors, the starting point must be the drying requirements and thus the drying characteristics of the material. Drying requirements specify the planned drying time and the permissible material temperature, among others. The drying characteristics of the material serve to determine, with the help of simulation or laboratory drying experiments (or both), the necessary inlet characteristics of drying air (temperature and relative humidity) and the necessary air mass flow rate $\dot{m}_a$ for the drying period with the highest energy demand.

The air is usually taken from the surroundings. In the case of a direct system, the air collector, and with an indirect system the liquid-type collector and liquid-air heat exchanger, serve to heat the air. If heat storage is also employed, the temperature of the medium leaving the collector has to be set to a value that can ensure the prescribed temperature of the drying air even when air heated by the heat storage is being used.

In the case of drying in continuous material flow or of preliminary drying, the drying energy demand for a given material is nearly constant over time. The standard energy demand of the dryer ϕ_d is covered by the enthalpy increase of the drying air. When using a direct system, the temperature of the air entering the dryer $T_{d,\text{in}}$ is equal to the temperature of the air leaving the collector ($T_{c,\text{out}} = T_{d,\text{in}}$). The necessary enthalpy increase of the air in the collector is

$$\phi = \dot{m}_a c_{p,a} \left(T_{c,\text{out}} - T_{c,\text{in}} \right) \tag{5a}$$

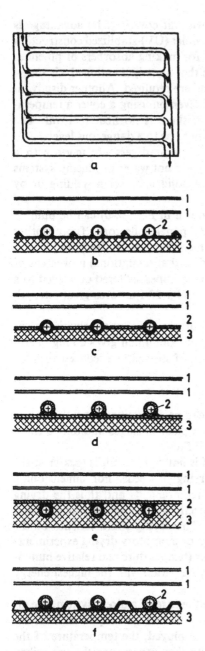

Figure 22 Some designs of liquid-type collectors: (a) absorber plate made of stamped sheets; (b) collector with extended finned tube absorber elements; (c) through (f) different tube-sheet, flat-plate collectors.

With air-type collectors, if recirculation from the dryer is not employed the temperature of air entering the collector is equal to the ambient temperature ($T_{c,\text{in}} = T_o$).

With liquid-type collectors, Eq. [5a] is valid for the air flowing through the fluid-air heat exchanger ($T_c = T_H$). The necessary temperature T'_f of the fluid entering the heat exchanger from the collector ($T'_f = T_{c,\text{out}}$) can be determined by the efficiency of the heat exchanger H. For $\dot{m}_a c_{p,a} > \dot{m}_f c_{p,f}$, the energy balance for adiabatic heat exchanger gives:

$$\phi_d = \dot{m}_f c_{p,f} H (T'_f - T_o) \tag{5b}$$

If the heat loss of the flow duct system ϕ_l is not negligible (45), the energy demand for the collector is $\phi_u = \phi_d + \phi_l$.

Therefore from [5b]

$$T'_f = T_{c,\text{out}} = \frac{\phi_u}{\dot{m}_f c_{p,f} H} + T_o \tag{5c}$$

The necessary enthalpy increase for the fluid flowing through the liquid-type collector is

$$\phi_u = \dot{m}_f c_{p,f} (T_{c,\text{out}} - T_{c,\text{in}}) \tag{6}$$

When using air collectors, the full airflow demand by the dryer is generally led through the collector (the value of $\dot{m}_{a,c}$ will be equal to the energy demand of the dryer, $\dot{m}_a$). The air heated in the collector can be mixed with the air sucked in directly from outside in a mixing space formed on the suction side of the fan.

The $\dot{m}_f$ mass flow rate in the collectors of indirect system dryers can be chosen, within certain limits. However, $\dot{m}_f$ is interdependent with the thermal efficiency η of the collector. After clearing up the necessary energy flow rate to be utilized in the collector, the required collector surface must be determined. On the basis of utilized ϕ_u heat flow rate and the energy flux of the incident radiation (energy flow rate per surface unit: irradiance), the efficiency of the collector can be expressed as

$$\eta = \frac{\phi_u}{A_c I} \tag{7}$$

where A_c is the necessary collector surface. Correlation [7] can be interpreted only as a transient value owing to the time dependence of the irradiance.

For a definite period, the so-called long-term efficiency of the collector can be expressed with the time integral of utilized and input energy flow rates

$$\eta = \frac{\displaystyle\int_{\tau_o}^{\tau} \phi_u \, d\tau}{A_c \displaystyle\int_{\tau_o}^{\tau} I \, d\tau} \tag{8}$$

The duration for averaging can be chosen in accordance with the operating time of the collector and the purpose of calculation (daily, monthly, or yearly long-term efficiency).

The efficiency of the collector can be determined by calculation and measurements. For design purposes, different calculation methods can be used (34,35,37,46–52). The efficiency data for commercially available collectors are determined by standard measurements (53,54). (For "second law efficiency," see Ref. 55.)

Simplified Calculation of Collector Efficiency

In correlation [7] of the instantaneous efficiency, utilized heat flow rate ϕ_u is the difference between the heat flow rate absorbed by the absorber ϕ_a and the heat flow rate lost ϕ_l to the ambient air:

$$\phi_u = \phi_a - \phi_l \qquad [9]$$

where

$$\phi_a = t\alpha I A_c \qquad [10]$$

is the heat flow rate absorbed by the absorber from the irradiation getting through the covering, and

$$\phi_l = A_c U_l (T_a - T_o) \qquad [11]$$

is the heat flow rate transferred to the ambient air from an absorber at T_a temperature. In Eq. [11], U_l is the overall heat transfer coefficient of the collector to the ambient air. Substituting into Eq. [7], the instantaneous efficiency of the collector is (46–50)

$$\eta = t\alpha - U_l \frac{T_a - T_o}{I} \qquad [21]$$

If t, α, and U_l are taken as constant values, instantaneous efficiency in the function

$$f = \frac{T_a - T_o}{I}$$

(efficiency function, an independent variable) can be plotted as shown in Figure 23. At a given operating point, the utilized energy flow rate from the collector is $\phi_u = \eta A_c I$.

These considerations can be appropriately applied according to Eq. [8] for expressing the long-term efficiency by substituting time averages $(I)_{av}$, $(T_a)_{av}$, and $(T_o)_{av}$:

$$\eta = t\alpha - U_l \frac{(T_a)_{av} - (T_o)_{av}}{(I)_{av}} \qquad [13]$$

From correlations [12] and [13] the threshold value of incident radiation flux can be determined with which the absorbed energy flow rate and the loss heat flow rate are equal and thus the efficiency is zero:

$$I_{th} = \frac{U_l[(T_a)_{av} - (T_o)_{av}]}{t\alpha} \qquad [14]$$

From I_{th}, using the appropriate meteorologic data, the possible operation time of the collector can be stated.

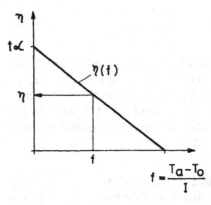

Figure 23 Instantaneous efficiency diagram of a flat-plate collector.

The instantaneous efficiency of the collector can also be expressed from the known inlet temperature T_{in} of the working medium with the aid of the heat removal factor F_R (56):

$$\eta = F_R \left[t\alpha - \frac{U_l (T_{in} - T_o)}{I} \right] \tag{15}$$

Coefficient F_R takes into consideration the relative decrease of efficiency caused by the increase of T_a absorber mean temperature compared with T_{in} inlet temperature of the working medium.

Further, instantaneous efficiency can be expressed directly as a ratio of useful heat flow rate coming into the working medium and the incident heat flow rate on the absorber:

$$\eta = \frac{\dot{m}c_p (T_{c,\text{out}} - T_{c,\text{in}})}{A_c I} \tag{16}$$

In practice, $\eta(T_{in} - T_o)$ or $\eta(T_{c,\text{out}} - T_{c,\text{in}})$ diagrams are often used in place of the $\eta(f)$ efficiency diagram. For representation of the thermal behavior of collectors, besides those above, other practical diagrams, such as $\eta(\dot{m}c_p)$ and $\eta(T_{c,\text{in}})$, function curves, can also be used. In these cases other factors in the η equation appear as the parameters of the efficiency curves.

The simplified calculation method has several weak points. One is that the value of T_a must be known to perform the calculation. The temperature of the absorber changes in the flow direction of the working medium, and T_a can be interpreted only as a mean temperature and can be determined only with knowledge of the absorber temperature distribution.

The greatest error appears in the application of the overall heat transfer coefficient U_l and its use as a constant value. U_l models the overall effect of complex and nonlinear heat transfer processes. Its value for a given collector depends on the local values of T_a, on the sky temperature T_s in view of radiation, on the mass flow rate of the working medium, and on the weather (e.g., wind) conditions. In the value of U_l, the temperature dependence of the heat transfer from the covering is strong. One can interpret the value of U_l as the sum of three coefficients: heat transfer from top covering (U_t), from the bottom plate (U_b), and from the edges (U_e):

$$U_l = U_t + U_b + U_e \tag{17}$$

For a simple determination of the top loss coefficient U_t, a set of diagrams is presented by Duffie and Beckman (56).

Simulation of Flat-Plate Collectors

For computer simulation of the performance of flat-plate solar collectors, several methods and computer programs have been used. Finite-difference (57), network (26,37,54, 57), stochastic (58), dynamic (59), and simplified models (60) and methods (see Reference 99) have been elaborated.

In the following a short description of a simulation method based on a heat flow network model is given. This method was applied for optimum design and control of roof panel-type collectors of solar dryers (26,37).

The essential point of a heat flow network model is the division of the structure of the collector into discrete parts with temperatures that can approximately be characterized by a single value. In the network the discrete parts are represented by nodes. The heat capacities and heat sources and the so-called temperature sources modeling the boundary condition reference temperatures are connected to the nodes. In this way the ambient air is also represented in the network by a node. The nodes are connected to a network by heat transfer resistances characterizing the thermal interactions among the discrete parts.

Identification of network elements proceeds with constant values at the beginning of the calculation. The identification of temperature-dependent and time-changing network elements takes place in subroutines in the computer program, over the course of calculation, with time increments.

The system of equations for the network can be written from the node and branch equations (37). For its solution, finite difference or finite time element schemes can be applied. An advantage of the network model is its flexibility and multipurpose applicability for collectors of different construction. Refinement of discretization is simple to accomplish.

The construction of a network model is shown for a hybrid collector in Figure 24-a (double covering, finned tube absorber) (57). The main model conditions are as follows:

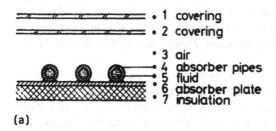

- 1 covering
- 2 covering
- 3 air
- 4 absorber pipes
- 5 fluid
- 6 absorber plate
- 7 insulation

(a)

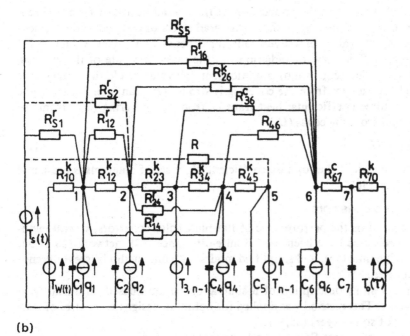

(b)

Figure 24 Heat flow network model of a hybrid collector: (a) setup of the collector; (b) the scheme of the HFN model. (From Reference 37.)

M1: In the collector plane, the temperature distribution is uniform in the direction perpendicular to the flow of the medium.

M2: For collectors with pipe ducts, the temperature nonuniformity of the absorber perpendicular to the flow direction of the medium is taken into account by an average temperature using the fin efficiency (43,56).

M3: The frame of the collector is assumed well insulated.

M4: The flow distribution between the collector tubes is uniform (for the effect of nonuniform distribution, see Reference 51).

M5: The spectral variations of absorption and transmission relative to heat radiation are taken into account by average values weighted with the solar spectrum and with the distribution of the low-temperature Planck radiator.

The collector of length L is divided in the flow direction of the working medium in z number of discrete sections of length ΔL ($L = z \, \Delta L$). The length of the sections is not necessarily equal:

$$L = \sum_{n=1}^{z} \Delta L_n$$

Further, if it is assumed that the thermal relation between each discrete section is established only by the flowing medium, the full network model of the collector can be separated into individual network models of discrete sections z. Therefore the solution for the whole collector will be produced through a sequence of the solutions of the individual (in this case, seven-node) part networks proceeding by time increments along z sections.

In Figure 24-b, the connection of the heat flow network of the nth discrete part is sketched. The numbering of the nodes in Figure 24-b corresponds to the numbers of the discrete parts in Figure 24-a.

In the network model, heat flow sources ϕ_1 through ϕ_7 represent the heat flow rates absorbed from incident radiation; C_1 through C_7 are the heat capacities of the discrete parts; $T_s(\tau)$ and $T_o(\tau)$ are the temperature sources giving the temperature of the sky and the ambient air, respectively; and $R^r_{i,j}$, $R^c_{i,j}$, and $R^k_{i,j}$ are radiation, conduction, and convection heat transfer resistances between the discrete parts, respectively. Network element $T_{i,n-1}$ is the temperature source representing the inlet temperature of the working medium flowing into the nth section from the $n-1$ discrete section. From the heat balance equation,

$$T_{i,n-1} = T_{i,n-2} + \frac{\phi_{n-1}}{\dot{m}c_p} \qquad [18]$$

In the case of liquid-type collector simulation, $T_{3,n-1}$ is omitted (see Fig. 24-b).

To use the network model for an air-type collector, branches 3–4 and 4–5 of the network are disconnected.

For liquid-air hybrid collectors with counterflow movement of the media, the heat flow network of the collector cannot be separated into z number of part networks with m nodes. In this case the part networks of the sections must be connected into a single zm node network with the aid of the relationship $R_{i(n,n-1)} = -\dot{C}^{-1}$ formal resistances inserted between the nodes of the working media. (In writing the nodal equations, the formal resistances will be taken into account only in the direction of the medium flow [61].) Taking the lengthwise heat conduction in the absorber into account will lead to a network of the same type. In the following, the correlations for calculating the network elements are described.

The heat flux incident on a given element can be written as

$$\phi_i(\tau) = \alpha_i I(\tau) \prod_{m=1}^{i-1} t_m \qquad [19]$$

The incident solar flux density I and its directional distribution vary in time. In $I(\tau)$ time function the geographic position, the relative position of the collector to the sun, is present as a regular, periodic variation. The degree of cloudiness, the humidity content of the air, and the degree of pollution of the atmosphere cause a stochastic change in the $I(\tau)$ value, which is superimposed on the regular variation. For the calculation, the solar radiation and meteorologic data relating to the given geographic position are necessary. Such data are usually available (43,56,62-78). Direct and diffuse components of solar radiation for simulation are characterized by an intensity-time function. There is a significant body of knowledge for determining the effect of cloudiness, the relationships of direct and diffuse radiation intensities (43,56,72,79-84). The time dependence of absorbance and transmittance of the materials due to direction can be taken into account with an average value (37) (see also Reference 104). (For the determination of typical weather for use in solar energy simulations, see also Reference 100.)

Assuming that the diffuse radiation is isotropic, the density of energy flow rate from incident solar radiation for a collector in a given location can be calculated from the following equation (37,43,56,84-89):

$$I(\tau) = I_{H-T}(\tau)(\beta(\tau)\chi(\tau) - \chi(\tau) + 1) \qquad [20]$$

where $I_{H-T}(\tau)$ is the time dependent total flux density incident on a horizontal surface;

$$\chi = \frac{I_d}{I_t} \qquad [21]$$

is the instantaneous ratio of direct and total flux density, a function of time, weather, and the location;

$$\beta(\tau) = \frac{\cos (\phi - \psi) \cos \delta \cos \omega + \sin (\phi - \psi) \sin \delta}{\cos \phi \cos \delta \cos \omega + \sin \phi \sin \delta} \qquad [22]$$

is a factor characterizing the relative position of the collector and the sun, where ϕ is the latitude, ψ is the inclination angle of the collector, δ is the angle of declination, and ω is the hour angle.

The daily changes of ambient air temperature T_o are known from meteorologic data. The standard sky temperature T_s in clear weather is influenced by air temperature and humidity. A good approximation for determining T_s is given by Swinbank's formula (90):

$$T_s = 0.0552(T_o)^{1.5} \qquad [23a]$$

Bliss's correlation (91) also takes air humidity content into account:

$$T_s = T_o \left(0.8 + \frac{T_{dp} - 273}{250}\right)^{0.25} \qquad [23b]$$

where T_{dp} is the local dew-point temperature of the air (temperatures in Kelvin). (For examining night conditions, see References 92-94.)

Heat transfer resistances can be interpreted by the branch equation of the heat flow network:

$$\phi_{i,j} = \frac{\Delta T_{i,j}}{R_{i,j}} = K_{i,j}\Delta T_{i,j} \qquad [24]$$

where $K_{i,j}$ is the conductivity of the branch. Radiation conductivities for surface units are

$$K_{s,i}^{(r)} = \varepsilon_i \sigma (T_i + T_s)(T_i^2 + T_s^2) \qquad [25]$$

$$K_{i,j}^{(r)} = \varepsilon_{i,j} \sigma (T_i + T_j)(T_i^2 + T_j^2) \qquad [26]$$

$$\varepsilon_{i,j}^{-1} = \varepsilon_i^{-1} + \varepsilon_j^{-1} - 1 \qquad [27]$$

where

ε_i = emissivity of the glass
ε_j = emissivity of the absorber
σ = Stefan-Boltzmann constant

Heat conductivity in solid elements of δ thickness for surface unit is

$$K_{i,j}^{(c)} = k\delta^{-1} \qquad [28]$$

Conductivities (convection) for the outer surface units are

$$K_{i,o}^{(k)} = h \qquad [29]$$

where h is the heat transfer coefficient that depends on wind velocity. Approximate correlations for h as a function of wind velocity w are

$$h = 5.7 + 3.8w \qquad \text{for } w < 5 \text{ m/s}$$

$$h = 7.6w \qquad \text{for } w \geq 5 \text{ m/s}$$

(See Reference 95 for further details on wind effects.)

Heat transfer coefficients for forced and natural convection in air gaps can be calculated from correlations (43,56).

Substituting, on the basis of the branch equations [23], the $\phi_{i,j}$ values into the nodal equations, we get the following system of equations for the network:

$$\underline{C}\dot{\underline{T}}(\tau) + \underline{K}(\underline{T}, \tau)\underline{T} = \underline{\phi}(I, T_s, T_o, \tau) \qquad [30]$$

where

$\underline{C}$ = diagonal matrix of the nodal heat capacities
$\dot{\underline{T}}$ = column vector of the time derivates of the nodal temperatures
$\underline{T}$ = column vector of the nodal temperatures
$\underline{\phi}$ = source vector

The solution of the system [30] can be obtained numerically by time discretization. Because of the nonlinearities, the integrated mean value of the conduction matrix $\underline{\hat{K}} = \underline{\hat{K}}(\underline{T}(\tau))$ for a given period of time can be created and a numerical scheme of one or two time levels can be applied (96). Favorable results have been obtained by the linear Galerkin scheme (97):

$$\left[\frac{1}{3}\underline{\hat{K}} + \frac{1}{2\Delta\tau}\underline{C}\right]\underline{T}^{(k)} = \left[-\frac{1}{6}\underline{\hat{K}} + \frac{1}{2\Delta\tau}\underline{C}\right]\underline{T}^{(k-1)} + \frac{1}{6}[\underline{\phi}^{(k)} + 2\underline{\phi}^{(k-1)}] \qquad [31]$$

In this equation, k is the number of discrete time sections in the step-by-step solution ($\tau = k\,\Delta\tau$). Meteorologic data are at one's disposal hourly, so in the calculation a time step of 1 hour can be applied. For stability it is advantageous if the system of equations is well conditioned. Accordingly, heat capacities of insignificant influence are best neglected. As a result of calculation, the time dependence of the medium outlet temperature, as well as the time variation of the approximate temperature distribution of the collector, is obtained. The results can be used for instantaneous and steady-state efficiency diagrams, for collector design and optimization, and for the solution of process control problems (33,34,43,50,56,98). It is of course desirable to verify the calculation results with experimental measurements where possible.

Thermal Performance of Flat-Plate Collectors

Figure 25 shows the characteristics of a single-covering, air-type collector (black absorber) obtained by calculations (37). In the figure, ΔT is the rise in temperature of air in the collector ($\Delta T = T_{c,\text{out}} - T_{c,\text{in}}$), η is the instantaneous efficiency, and $\dot{m}$ is the mass flow rate of air. The parameter in the figure is the surface area per unit collector length (series connection).

In Figure 26, the instantaneous efficiency and the outlet temperature of a liquid-type collector (single-covering, steel finned tubes, black absorber) are illustrated as a function of liquid heat capacity flow rate (37). The parameter is the irradiation. As the variations of I are accompanied by nonlinear heat transfer resistance variations, the curves for different I values deviate. The entry temperature of the medium is equal to the outside temperature ($T_{c,\text{in}} = T_o$).

Figure 27 illustrates the instantaneous efficiency of a liquid-type collector as a function of the inlet temperature with single (Fig. 27-a) and double (Fig. 27-b) covering (37). The parameter is the outside temperature. With an increase in the inlet temperature, the mean temperature (T_a)$_{av}$ and the heat loss of the absorber increase while the efficiency

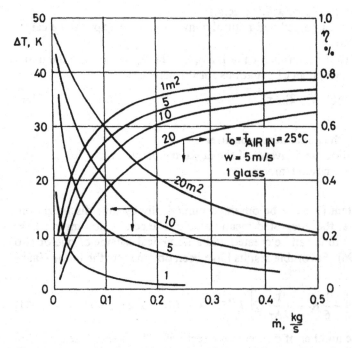

Figure 25   Main characteristics of an air-type collector (From Reference 37.)

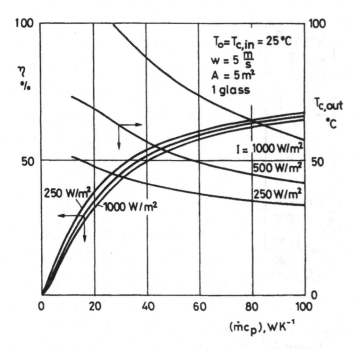

Figure 26 Main characteristics of a liquid-type collector. (From Reference 37.)

deteriorates. With double covering, the efficiency with a higher inlet temperature is greater than that with a single covering.

Figure 28 shows the instantaneous efficiency diagrams for different collectors. It can be seen from this figure that the instantaneous efficiency of collectors now in general use can be as high as 50–60%. The value of daily long-term efficiency amounts to approximately 25–30%.

On the basis of the long-term collector efficiency, the collector surface necessary for the operation of the dryer can be determined. In view of the fact that the instantaneous efficiency of the collector is over one part of operation time greater than the long-term efficiency, the energy utilized by the collector over this period is greater than the necessary value. The surplus energy can be stored for the period when there is no solar energy available. In cheap, simple construction dryers without heat storage, this surplus energy can be used for some temporary enhancement of the drying process if the material to be dried can withstand it.

5.2. Heat Storage for Solar Dryers

From a thermal viewpoint heat storage of solar dryers can be classified into two main groups:

1. Directly irradiated heat storage
2. Heat storage charged by the working medium of the collector

The storage temperature of directly irradiated heat storage is not limited by the collector outlet temperature of the medium. However, in heat stores warmed by the working medium the maximum temperature cannot exceed the collector outlet temperature.

The aim of heat storage is to store surplus energy appearing in strong radiation

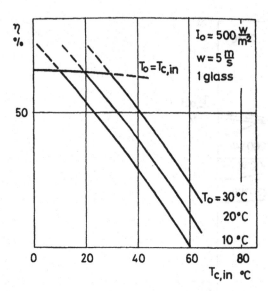

(a)

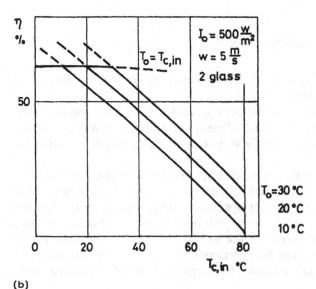

(b)

Figure 27 Instantaneous efficiency of a liquid-type collector as a function of inlet temperature; parameter is the ambient temperature: (a) single covering; (b) double covering. (From Reference 37.)

periods; however, the aim may also be to store enough energy for full-scale drying operation at night as well.

When determining the necessary surface area of the collector, the amount of energy to be stored must also be taken into account.

Directly Irradiated Heat Storage

Directly irradiated heat storage of solar dryers can first be used in direct air systems. The collector sketched in Figure 21 is built with phase-change storage. The phase-change

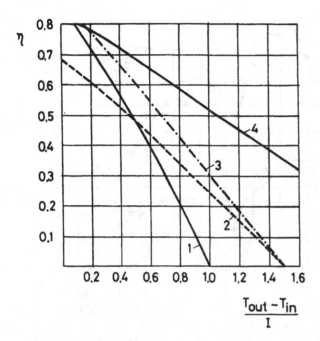

Figure 28 Efficiency diagrams of collectors of different construction: (1) black absorber, one covering; (2) black absorber, two coverings; (3) selective absorber, one covering; (4) vacuum collector.

material is placed in a plastic honeycomb matrix casing. Thermal expansion is made possible by additives. The outer surface of the plastic cells containing the phase-change material take over the function of the absorber. The thickness of the cells is limited by phenomena occurring in the course of phase change (38,39,101,102). Therefore the mass to be placed on 1 m^2 is also limited, and so is the overall amount of storage heat.

The primary purpose of applying directly irradiated latent heat storage is to attain an equalizing effect during cloudy periods over the day as well as to lengthen the daily drying time. To show the thermal behavior of a collector integrated with latent heat storage, Figure 29 is presented. The temperature-time function was determined by calculation using the simulation model described above and checked by measurements. The heat capacities of other elements of a heat storage collector are negligible when compared with that of the heat storage. The effect of latent heat was built in the volumetric heat capacity of the absorber and simulated (38,103,104) (data: material, $CaCl_2 \times 6H_2O$; phase-change temperature, 29°C; latent heat, $\lambda = 209$ MJ/m^3; size of cells, 9 mm $\times$ 10 mm; mass/m^2, 6.3 kg).

The absorber may be a solid (e.g., concrete) wall with a black coating. In the case of solar dryers, this solution is justified if the wall is the dryer housing wall. Its operating principle is the same as that of the Trombe wall used in passive solar heating (56). (For collector-integrated solar water heaters, see References 105 and 108.)

Heat Storage Charged by the Medium of the Collector

In heat storage warmed by the working medium, the maximum temperature cannot exceed the maximum temperature of the medium leaving the collector. As storage material, various solids, water, various phase-change materials, as well as chemical storage materials may be used.

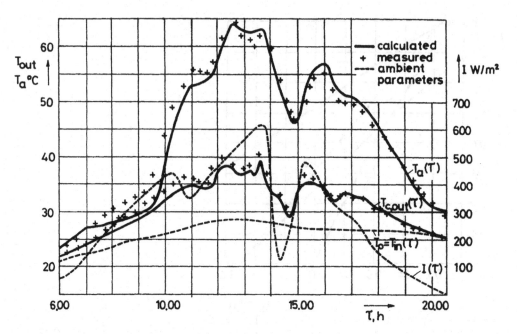

Figure 29 Behavior of air-type collector integrated with phase-changing filling (functions of T_{out} and T_a determined by simulation and measurements).

The temperature of the medium entering the storage is equal to the temperature of the medium leaving the collector if the heat loss of the duct is negligible. The medium in liquid systems after leaving heat storage flows back into the collector. With air systems sometimes the air leaving the heat storage unit is exhausted into the atmosphere.

Water-type heat storage may be a direct or indirect system. In direct systems the working medium flows in a closed circuit (Fig. 30-a). In indirect systems, the working medium can be liquid (antifreeze liquids in some cases). The closed collector circuit and the closed storage circuit are connected through a heat exchanger (Fig. 30-b). In water heat storage, two types of storage systems are employed: stratified and well-mixed storage.

In stratified water storage the warm water from the collector enters near the top of the tank; the fluid led back to the collector is drawn from the bottom of the tank by a pump. Thus in the upper layers of the tank there is always warm water, and the lower layers contain cold water. The advantage of this method is that the collector receives cold water as long as cold layer exists near the bottom of the tank (56,106,107); accordingly, the collector works with approximately constant efficiency. The thickness of the transient temperature zone is determined by the time boundaries of the temperature changes of the water coming from the collector. In operation periods of reduced radiation, the temperature of the water from the collector is lower than that of the temperature in the top layer. This water descends and causes mixing in the tank.

When using stored hot water (Fig. 31), the water is led into the air-water heat exchanger from the top of the tank. The returning water enters at the bottom of the tank. The mass of water in the tank therefore makes on charging a slow downward motion and on discharge a slow upward motion. The rate of this motion depends on the mass flow rate of the water (and on other effects, see below). The arrangement shown in Figure 31

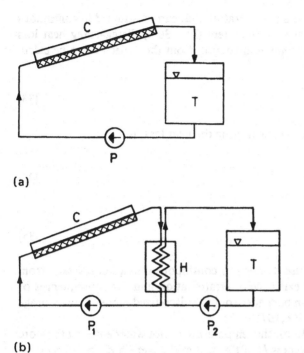

(a)

(b)

Figure 30 Water-type heat storage: (a) direct; (b) indirect (C, collector; T, tank; P, pump; H, heat exchanger).

allows regulation of the water flow rate, the change of operational mode, and a simultaneous drying-charging operation.

The amount of heat to be stored Q can be determined from the heat demand of the dryer:

$$Q = \int_0^\tau \phi(\tau) \, d\tau \tag{32}$$

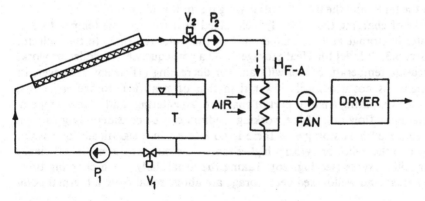

Figure 31 Connection of water-type heat storage to the dryer (H_{F-A}, fluid-air heat exchanger; V_1 and V_2, valves).

Storage temperature is limited by the fluid temperature that can be attained in continuous operation of the collector. In a direct storage system (Fig. 30-a) disregarding heat loss from the pipes, the temperature of the medium coming from the collector can be calculated by using Eq. [16]:

$$(T_{c,\text{out}})_{av} = (T_{c,\text{in}})_{av} + \frac{A_c I \eta}{\dot{m} c_p} \qquad [33]$$

The heat to be stored in τ charging time T, neglecting the heat loss, is

$$\begin{aligned} Q &= M_T c_T [(T_{c,\text{out}})_{av} - (T_{c,\text{in}})_{av}] \\ &= \dot{m} c_p [(T_{c,\text{out}})_{av} - (T_{c,\text{in}})_{av}] \tau \end{aligned} \qquad [34]$$

Hence,

$$M_T = \frac{\dot{m} c_p}{c_T} \tau \qquad [35]$$

This calculation is approximate because it refers to continuous operation assessed from time averages and does not consider existence of stratification and the consequences of heat loss. A more exact calculation can be made from the discretized model of the storage tank with due regard to heat loss (56,106,107).

In indirect storage systems (Fig. 30-b), the temperature of hot water entering the store T_T can be calculated from the effectiveness H of the heat exchanger ($\dot{m}_c c_{pc} \geq \dot{m}_T c_{PT}$):

$$T_{T,\text{in}} = T_{T,\text{out}} + H(T_{c,\text{out}} - T_{T,\text{out}}) \qquad [36]$$

The relationship corresponding to Eq. [34] can be written with the respective values of $T_{T,\text{in}}$ and $T_{T,\text{out}}$.

In the case of stratified heat storage the mass flow rate of the collector medium is held during charging at a relatively low value for the purpose of getting as high a value of $T_{c,\text{out}}$ exit temperature as possible. A disadvantageous consequence (see Fig. 25) is only moderate efficiency.

With a *well-mixed heat storage system*, the temperature of the water in storage is practically uniform. Mixing of the water can be produced by several (eventually simultaneous) effects. These are a large mass flow rate and inlet velocity, horizontal location of the tank, in an indirect system the heat exchanger located at the bottom of the tank, baffle plates in the tank, and the use of a circulating pump for mixing.

In the course of charging the perfectly well-mixed heat storage, the temperature of the mass of water in storage and the temperature of the water returning to the collector rises (see also Sec. 6.3, "Model for Heat Storage"). As a consequence, the collector works at an ever-increasing temperature level with an ever-decreasing efficiency. However, its long-term efficiency is not necessarily lower than that of stratified storage collectors, although these receive, during a major part of charging, an entering fluid of low temperature. That is, the mass flow rate of the working medium can be considerably greater in collectors of well-mixed heat storages as there is no interest in a significant increase of fluid temperature in the collector. With a higher mass flow rate, however, the efficiency of the collector will increase (see Fig. 26). Taking the usual daily 6–8 h charging time, stratified heat storage and well-mixed heat storage are about equal from a thermal point of view.

In the operation of a solar dryer, technological interest is attached to a sufficiently high temperature of the fluid leaving the collector for preheating the drying air. At the same time, during certain periods of operation, simultaneous actions must be carried out for charging the store and for drying (see Fig. 31). Consequently, the mass flow rate of working medium of the collector must be limited to reach the necessary exit temperature of the collector. Therefore, in the operation of solar dryers, the stratified heat store is preferred.

In solar dryers with air-type collectors *rock-bed storage* (crushed stone or pebbles of 2–4 cm size) is used most commonly. When selecting the material for the rock or pebble bed, among other considerations, there is the question of the pressure drop across the bed. The flow resistance of a pebble is usually smaller than that of crushed rock. Uniform pebble or rock size must be chosen to obtain uniform air distribution in the bed. The necessary mass of rock-bed or pebble-bed heat storage is typically about threefold that of water-type heat storage.

Because of the point contact between the particles in the rock bed, the heat conduction is negligible. Therefore, rock beds work practically as stratified heat stores no matter the arrangement for air inlet and exhaust.

During charging of the rock-bed heat store, the cooling of the air at the entry spot takes place within a layer of a certain thickness. If the entry temperature of the air was constant, this layer of changing temperature would push down deeper and deeper (Fig. 32). However, because during the day periods of varying radiation intensity occur, it is possible that during certain periods of the day the temperature of entering air is lower than that of the heat-storing material at the place of entry. In this case the air in the entry layer continues to get warmer and delivers the earlier stored heat into deeper layers of the storage, to a lower level of temperature, and enlarges the width of the layer of changing temperature. The principle of approximate calculation of rock- or pebble-bed heat storage is similar to that of a stratified water-type heat storage. Figures 10 and 11 present the arrangement of solar dryers equipped with pebble-bed heat storage.

Between water and solid-bed heat storages there is a so-called transitional option heat storage with the aid of fluid-filled cans. An example of this is shown in Figure 5.

In heat storage devices warmed by a working fluid, phase-change materials can also be used (38,39,101,102). Phase-change energy storage has two main advantages. One advantage is that the energy stored per unit volume of the store is significantly greater than that for pebble-bed heat storage, for example. One of the cheapest materials with good working characteristics (109) is $Na_2SO_4 \cdot 10H_2O$; with this the proportion of stored

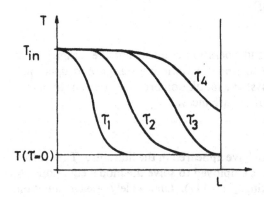

Figure 32 Passing of temperature wave in a gravel-bed heat storage with depth L (T_{in} = constant).

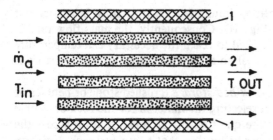

Figure 33 Example for placing phase-changing fillings in the heat storage: (1, insulated wall; 2, containers; $\dot{m}_a$ = constant).

energy is about fourfold (latent heat λ = 251 kJ/kg; phase-changing temperature, 32°C). Also, the fixed phase-change temperature reduces temperature variations.

The application of phase-changing materials (PCM) also involves certain problems. Eutectic salts are susceptible to phase separation (110,111). To avoid this (38,112) it is desirable to make the PCM containers thin (Fig. 33). Another phase-change material is paraffin wax (λ = 209 kJ/kg). Its disadvantage is inflammability.

A general problem with PCM stores is the fabrication of an inexpensive casing resistant to corrosion. For some PCMs, degradation has been experienced over the course of repeated cycling. The choice between phase change and sensible heat storage must be made on economic grounds.

Heat storage by chemical reactions has also been studied (56,112,113). These methods have not progressed beyond the experimental stage and do not appear applicable for solar drying applications.

Energy storage in adsorbent beds can be applied to solar dryers because air of reduced humidity content leaks from the adsorbent. Silica gel, activated aluminum oxide gel, zeolites, and various molecular sieves can be used as adsorbents (27,29,114). Detailed analyses and optimization procedures are known for energy storage in timber dryers (29) and garlic dryers (12). The applications of adsorbent energy storage are also found in complex systems (25). Although the advantages are undisputed, the costs of adsorbents and the auxiliary equipment are considerable.

6. SIMULATION OF SOLAR DRYERS

6.1. Purpose of Simulation

Simulation is an important tool for design and operation control. For the designer of a drying system simulation makes it possible to find the optimum design and operating parameters. For the designer of the control system, simulation provides a means to devise control strategies and to analyze the effects of disturbances.

6.2. Methods of Simulation

Various simulation models for solar processes have appeared in the literature. They differ mainly in the assumptions made and strategies employed to solve the model equations. A majority of the models refer to solar heating (115–117). One widely used simulation program is the TRNSYS (56,118). The *f*-chart method for solar heating of buildings is also well known (56,119,120). Other relevant references are design of closed-loop solar systems (121), control and optimization (117), and process simulation based on a stochastic model (58) and for a resistance network (56,122).

Selcuk et al. describe the simulation of a shelf-type dryer (5,123). Close (124) worked out the simulation of an air-type solar drying system equipped with gravel-bed heat storage using a 10-node discretized model for the heat storage and forward finite difference technique for solving the equation system. Imre et al. (25,125) presented a simulation of solar dryer for alfalfa.

Daguenet (187,188) elaborated a simplified methodology for the calculation of solar-assisted convective dryers considering the meteorological conditions. Norton and Hobson (189) suggested a finite difference numerical analysis for chimney-type solar dryers for drying crops. Tiguert and Puiggali published (190,191) a performance model for a solar natural chimney dryer given in Figure 5. The one-dimensional, concentrated parameter model consists of four elementary models (i.e., for the collector, for the heat storage in daytime and in nighttime, and for the material situated in thin layers in the drying space).

Steinfeld and Segal (192) proposed a simulation model for a solar thin-layer drying process including the technique of estimating the solar radiation, the procedure of obtaining the thermal performance of a solar air-heater, and an analysis of the drying process based on the Lewis analogy and the Equilibrium Moisture Content (EMC) concept. Hasnaoui et al. (186) elaborated a one-dimensional model for a solar dryer of static thick bed.

Patil and Ward (194) prepared a simulation model for a solar-assisted bin dryer with and without mixing in atmospheric air. The model consists of the part model of the solar collector and the part model of the thick throughflow layer divided into thin layers.

Weitz et al. (195) proposed a simulation model for a multishelf type semiartificial solar dryer for drying processes in thin layer based on Luikov's theory by considering the shrinkage of the material.

6.3. Simulation Model of Solar Dryers

Solar dryers are thermohydraulic systems composed of various units such as collector, heat storage, heat exchanger, pump, ventilator, tubes, valves, closing and controlling devices, and the dryer. A simulation model of solar dryers is therefore made up of three main subsystems:

1. Model of the flow subsystem
2. Model of the thermal subsystem
3. Model of the drying volume

Model of the Flow Subsystem

The units of the solar dryer are joined into a system by flowing media; therefore a model of the system is best built on the flow model of these media (126). The number of flow models is necessarily equal to the number of working media in the solar dryer.

The flow model serves to determine the mass flow rates of the flowing media. In the following section the mass flow network (MFN) modeling method is presented. The MFN model of solar dryers essentially divides the flow system into discrete parts in which the effects causing pressure changes are modeled by network elements; the network elements are joined according to the flow path of the medium.

Pressure changes can be produced by external as well as internal effects. An example for an external effect is due to the pump or the fan. Internal effects can be the consequences of cross-sectional changes, heat transfer, acceleration or deceleration, and flow resistances. Some of the most important mass flow network elements are given in Table 2 (26).

In the MFN model the flow resistances and pressure sources are dependent on heat flow or temperature. Discretization has to be made so that over an individual discrete section the temperature-dependent characteristics can be expressed by lumped values with sufficient accuracy. Pipes, which can be taken as isothermal and have a constant cross-section can be modeled as a single flow resistance.

Flow resistances are characterized for the jth section on the basis of the pressure drop Δp_j and the mass flow rate $\dot{m}_j$ (see Table 2):

$$R_j(\dot{m}_j, T_m) = \Delta p_j \dot{m}_j^{-1} \qquad [37]$$

For the simulation it is necessary to know the state functions $\Delta p(\dot{m}, T, n)$ of the pumps and fans used.

Figure 34-a shows the arrangement of an indirect system solar alfalfa dryer with water-type heat storage serving simultaneously as the hot water supply (25). (For a detailed description, see legend, Fig. 13.)

The system has two working media. The primary flow circuit is closed; its medium is water. The secondary flow circuit is open; the working medium is air. Thermal interaction of the working media of the two circuits is carried out by a heat exchanger.

Figure 34-b and 34-c outline the reduced mass flow network model of the primary

Table 2 Identification of the Elements of the Mass Flow Network Model

Element	Relationship	Note
Flow resistances	$R = \Delta p\, \dot{m}^{-1}$	
Tube friction resistances Laminar flow	$R = \dfrac{c\eta L}{\rho\, d_h^2 A}$	From Poiseuille equation $(\dot{m} = \rho A w)$
Turbulent flow	$R = \dfrac{2fL}{\rho\, d_h A^2}\dot{m}$	From Fanning equation
Shape resistance	$R = \dfrac{\zeta}{2\rho A^2}$	From: $\Delta p = \zeta\dfrac{\rho}{2}w^2$
Jumplike change of cross-section	$R = \xi_B \dfrac{\dot{m}}{2\rho}\left(\dfrac{1}{A_1^2} - \dfrac{1}{A_2^2}\right)$	ξ_B: Borda–Carnot coefficient
Pressure sources Centrifugal pump and fan	$\Delta p = A + B\dot{m} + C\dot{m}^n$	Equation of the characteristic
Change of cross-section $(A_1 \lessgtr A_2)$	$\Delta p = \dfrac{\dot{m}}{2\rho}\left(\dfrac{1}{A_2^2} - \dfrac{1}{A_1^2}\right)$	From Bernoulli equation
Hydrostatic pressure source	$\Delta p = \rho g(z_1 - z_2)$	z_1, z_2 level heights
Closed thermosyphon loop	$\Delta p = g\phi\rho(T(z))\,dz$	$T(z)$: temperature as a function of the level height z
Heat input pressure source between points 1 and 2 of a pipe $(A = \text{const})$	$\Delta p = \dfrac{R}{(c_p)_{av}}\left[\rho_{av}\dfrac{\phi_h}{\dot{m}} - \dfrac{\dot{m}^2}{2A^2}\left(\dfrac{1}{\rho_1} - \dfrac{1}{\rho_2}\right)\right]$	From enthalpy balance equation
Acceleration pressure source	$\Delta p = \rho a L$	$\rho = \text{constant}$
Mass flow source (independent)	$\dot{m}\rho^{-1} = \text{constant}$	Volumetric pump

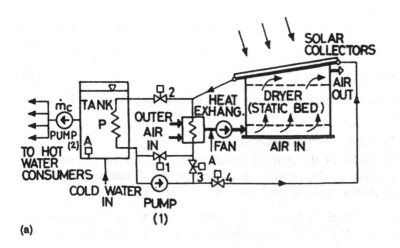

(a)

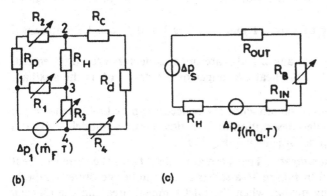

(b)   (c)

Figure 34 Solar-assisted indirect drying system for alfalfa: (a) scheme of system; (b) reduced mass flow network model of the fluid; (c) reduced mass flow network model of the airflow system.

and secondary flow circuits, respectively. In the reduced model of the primary flow circuit (Fig. 34-b), R_c and R_H are the resultant flow resistances of the collector and the heat exchanger, respectively, including the resistances of the pertinent pipe sections. (In the detailed model there are several partial resistances corresponding to the discretization.) In Figure 34-b, R_d is the flow resistance of the forward pipe section, R_1 through R_4 are the variable shape resistances of the control valves, and Δp is the pressure source due to the pump.

In the model of the secondary flow circuit, Δp_s is fixed by the atmospheric pressure, Δp_f is the pressure source of the fan, R_H is due to the heat exchanger, and R_{in} and R_{out}, respectively, are the inlet and outlet flow resistances of the dryer itself. R_B refers to the varying flow resistance of the bed with thickness changing over time.

The system of equations of the mass flow network model consists of the nodal equations [38], the loop equations [39], and the branch equations [40]. If the network contains m number of independent nodal points, i number of branches for each node, l number of independent loops, and j number of branches for each loop, the equation system can be written in the form

$$(\Sigma \dot{m}_i)_m = 0 \qquad [38]$$
$$\underset{i}{}$$

$$\Sigma[\Delta p_j + \Delta p(\dot{m}, T, n)]_l = 0 \qquad [39]$$
$$\underset{j}{}$$

$$\Delta p_j = R_j(\dot{m}_j, T)\dot{m} \qquad [40]$$

The temperature-dependent elements of the mass flow network can be identified from the thermal subsystem model.

Thermal Subsystem Model of the Primary Circuit

The thermal subsystem model of the primary circuit consists of the models M of the collector M_c, the heat exchanger M_H, and the heat storage M_T, in accordance with the mode of operation. The modes of operation are as follows (Fig. 35):

Mode I: Values 2 and 3 are closed, and 1 and 4 are open; the air passing through the heat exchanger can be preheated by the water coming from the collector (see also Fig. 34-a).

Mode II: Valves 1 and 3 are closed, and 2 and 4 are open; the water in the tank can be heated by the spiral pipe P.

Mode I + II: By partial closing of valves 1 and 2 modes I and II can be maintained simultaneously.

Mode III: Valves 1 and 4 are closed, and 2 and 3 are open; collectors are not operated. The water of the tank is led to the heat exchanger of the dryer. (A is the auxiliary heater; see Fig. 34-a.)

Thermal Model for the Collector. The model of the collector can be built according to different model concepts. A detailed description of heat flow network model is given in Sec. 5.1. The equations of this model are given in Eq. [30].

Thermal Model for the Heat Exchanger. The thermal model M_H of the water-air heat exchanger of the system outlined in Figure 34-a serves to determine the outlet temperatures T_F'' and T_A'' of the working media, when the inlet temperatures and the thermal capacity flows $\dot{C} = \dot{m}c_p$ in modes I, I + II, and III are known. Because eventual transients are slow, the effect of the thermal capacities is disregarded.

Effectiveness H of the heat exchanger as a function of the ratio $\dot{C}_A/\dot{C}_F$ must be known for the calculation. For an adiabatic heat exchanger,

$$T_A'' = T_A' + H(T_F' - T_A') \qquad [41]$$

$$T_F'' = T_F' - H\frac{\dot{C}_A}{\dot{C}_F}(T_F' - T_A') \qquad [42]$$

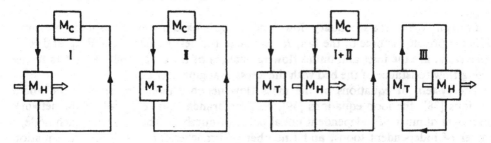

Figure 35 Thermal subsystem models according to the operating modes of the solar dryer.

Model for Heat Storage.   For the dryer in Figure 34-a, the tank of the warm water system of the farm is used as heat storage. The emergency energy source A is built into the tank. The heat storage is of the well-mixed type. The model for heat storage serves for determining the outlet temperature T_{F2} of the collector working medium and the temperature T_T of the water in store. The equations of the storage with $C_T = M_T c_p$ at any time, for the case of $\dot{C}_c = \dot{m}_c d_p$ hot water consumption, are

$$C_T \frac{dT_T}{d\tau} + \dot{C}_c (T_T - T_{w,\text{in}}) = \dot{C}_F (T_{F1} - T_{F2}) + \phi_A + \phi_T \qquad [43]$$

$$T_{F2} = T_T + (T_{F1} - T_T) \exp\left(-\frac{hA}{\dot{C}_F}\right) \qquad [44]$$

where
$\quad T_{w,\text{in}}$ = temperature of entering water
$\quad\;\; \phi_A$ = heat flow rate of the auxiliary energy source
$\quad\;\; \phi_T$ = rate of heat loss from the storage tank
$\quad\; T_{F1}$ = temperature of water from the collector
$\qquad A$ = heat transfer surface of coil heater

Model for the Control System.   The control strategy of dryer operation must be known for the simulation of the system, and the operating model for the control system has to be built into the simulation model. The model contains the condition system of the operation mode change, with the values of resistances $R_1, \ldots, R_4$ (see Fig. 34-b) ordered to the corresponding points of time or to the limit values of the state characteristics of the working media of the material to be dried and of the atmospheric states.

An advantageous solution is optimum control by sampling, using microprocessors (for optimum temperature control, see, e.g., Reference 127).

Coupling Equations.   Insofar as heat losses from the pipes between the units are disregarded, the temperature of the medium leaving the preceding unit is equal to the inlet temperature of the adjoining unit:

$$T_{c,\text{out}} = T'_F \qquad T_{c,\text{out}} = T_{F1} \qquad T'_F = T_{F1} \qquad\qquad [45a]$$
$$T_{c,\text{in}} = T''_F \qquad T_{c,\text{in}} = T_{F2} \qquad T''_F = T_{F2} \qquad\qquad [45b]$$

If the heat losses from the connecting pipes are taken into consideration (45), further component models must be developed for determination of the losses. In this case the coupling equations are interpreted for the relation of loss part models and the models of the units. If the loss heat flow rate of the collector-storage pipe is ϕ_{C-T}, that of the storage-collector duct is ϕ_{T-C}, and those of the collector-heat exchanger-collector pipes accordingly are ϕ_{C-H} and ϕ_{H-C}, respectively, and the coupling equations between the units can be written for the different modes of operation on the basis of the enthalpy balance equations

$$T'_F = T_{c,\text{out}} - \frac{\phi_{C-H}}{\dot{C}_F} \qquad\qquad [46a]$$

$$T_{F1} = T_{c,\text{out}} - \frac{\phi_{C-T}}{\dot{C}_F} \qquad\qquad [46b]$$

$$T_{c,\text{in}} = T''_F - \frac{\phi_{H-C}}{\dot{C}_F} \qquad\qquad [46c]$$

$$T_{c,\text{in}} = T_{F2} - \frac{\phi_{T-C}}{\dot{C}_F} \qquad\qquad [46d]$$

In the combined mode I + II of operation, the inlet temperatures can be determined from the mixing enthalpy balance equation.

Thermal Subsystem Model of the Secondary Circuit

The airflow circuit of the system in Figure 34-a is open; thus the operation of the dryer does not react to the operation of the primary circuit. In the case of solar dryers, with partial recirculation of the drying air (25), the condition of the air entering the dryer depends on the operation of the dryer itself. Accordingly, with recirculating systems' dryer operation reacts on the primary circuit and the calculation of the primary and secondary circuit must be coupled.

Model for the Dryer. In this system the thermal model of the secondary circuit contains the heat exchanger part model M_H and the dryer part model M_D. Using the heat exchanger Eq. [41], the inlet air temperature of the dryer can be determined; the solution of the mass flow network model equation system in Figure 34-b gives the mass flow rate of the drying air $\dot{m}_a$. Thickness and initial state of the material in the dryer can be considered as given. For the simulation of static bed dryers, different methods are used (128–130).

In the drying space of the system in Figure 34-a, the drying of alfalfa takes place in a static bed so that a new wet layer is laid on that already dried until the maximum layer thickness (approximately 6 m) is reached. Thus the thickness of the bed grows; consequently the value of the air mass flow rate belongs only to the thickness of the given layer of material (to R_B; Fig. 34-c). Owing to the net weight of the layers laid one above the other, the porosity of the bed is not constant, either. The bed porosity, the heap density of the alfalfa, and the specific phase contact surface also depend on the thickness of the bed.

The model assumptions regarding alfalfa drying are as follows:

1. The drying of leaves and stems are modeled separately (two-component model: $k = 1, 2$)
2. The internal moisture-conduction resistance in the leaves and in the crushed stems is disregarded.
3. The alfalfa bed is divided into discrete layers along the height z, and within these layers the temperature T_k and material moisture content X_k are characterized by lumped values.
4. In the discrete layers the airflow rate is divided proportionally to the drying surfaces of the components.
5. The mixed mean state of the air leaving the components is regarded as standard for the state of the air entering successive layers (x_{av}, T_{av}, enthalpy).

The equation system describing the dryer model for a given elementary layer of the bed (25,125,131) is as follows: moisture mass balance for the material,

$$\frac{\partial X_k}{\partial \tau} = -\beta_k a_k (X_k - X_{e,k}) \tag{47}$$

enthalpy balance for the material,

$$\frac{\partial T_k}{\partial \tau} = \frac{h_k a_k}{c_k \rho_k}(T_a - T_k) - \frac{\beta_k a_k r}{c_k}(X_k - X_{e,k}) \tag{48}$$

moisture mass balance for the air,

$$\frac{\partial x_k}{\partial z} = A \frac{\sigma_k a_k}{\dot{m}_{a,k}} (x_{e,k} - x_k) - A \frac{\rho_a \xi_k}{\dot{m}_{a,k}} \frac{\partial x_k}{\partial \tau} \tag{49}$$

enthalpy balance for the air,

$$\frac{\partial T_{a,k}}{\partial z} = \frac{a_k A}{\dot{m}_{a,k} c_{pa}} [\sigma_k (x_{e,k} - x_k) c_{pw} + h_k](T_k - T_a)$$

$$- \frac{A \rho_a \xi_k}{\dot{m}_{a,k}} \frac{\partial T_{a,k}}{\partial \tau} + \frac{(1 - \xi_k) b_k A \rho_k}{\dot{m}_{a,k} c_{p,a}} \tag{50}$$

equation of desorption isotherms,

$$f(X_{e,k}, x_{e,k}, T_a, p_w, p_b) = 0 \tag{51}$$

The values of drying and material characteristics (σ_k, β_k, a_k, b_k, ξ_k, c_k, ρ_k, and sorption isotherms) in the equation system of the dryer model must be determined by experiment (132).

6.4. Strategy of Solution

Numerical solution of the system of equations above can be obtained by discretization in time (25). The main steps of the solution process are as follows:

1. Data input, calculation of constant network elements and characteristics.
2. Selection of time step.
3. Solution of the working model of the process control system on the basis of the initial state; determination of mode of operation.
4. On the basis of the initial state $\underline{T}^{(k)}$, calculation of the dependent network elements and characteristics.
5. Generation of the mass flow network models corresponding to the mode of operation.
6. Solution of the mass flow network models for the subsequent period; determination of $\dot{m}_F^{(k+1)}$ and $\dot{m}_a^{(k+1)}$, for example by the Newton–Raphson method.
7. Solution of thermal component models on the basis of $\dot{m}_F^{(k+1)}$, $\dot{m}_a^{(k+1)}$, and the initial state $\underline{T}^{(k)}$ (for the collector, the finite time-element scheme see Reference 31; for the storage, the finite difference scheme can be applied); determination of the temperature of air entering the dryer.
8. Solution of the equations for the dryer by applying an implicit finite difference scheme; determination of material temperature and moisture distribution in the bed.

6.5. Results of Simulation

As an example, the results of simulation of the dryer in Figure 34-a are presented (25): the thickness of the fresh alfalfa layer in the bed, $z = 0.3$ m; the discretized layer thickness, $\Delta z = 0.075$ m; time step for collector $\Delta \tau_c^* = 0.5$ h, for the dryer, $\Delta \tau_d = 0.1$ h; air mass flow rate density, $\dot{m}_a = 0.171$ kg/s·m^2. After live wilting, $\dot{m}_c$ (dry basis) for leaf is $x_1(0) = 1.47$ kg/kg; for stem, $x_2(0) = 1.62$ kg/kg. Air inlet temperature $T_{a,\text{in}}$ as a function of time and the drying curves are given in Figure 36. As can be seen from the figure, the drying rates for the leaf and stem are substantially different. In the top layer of the bed, the stem reaches $x_2 = 0.14$ kg/kg moisture content in about 42 h. Under the conditions given, drying can be carried out in 2-day cycles.

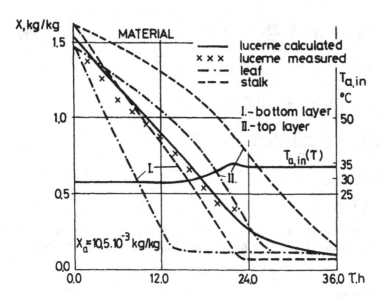

Figure 36   Drying curves of alfalfa.

7.   DIRECTION AND CONTROL OF SOLAR DRYERS

7.1.   Aims of the Direction and Control

Direction and control of solar dryers aim to ensure the economical operation of the dryer in every stage of the drying process considering the actual state of the material under drying as well as the actual meteorological conditions (177).

Economy Aspects

The economic analysis of solar drying is presented in Section 4. In this section economy aspects are interpreted in connection with the direction and control only. To fulfill the requirements of a good economy the savings should be increased and the costs reduced.

Utilizing solar energy for drying is not simply a method for saving conventional energy carriers but a technology for producing dried materials of high quality. This aspect is especially important when drying materials sensitive for quality deterioration. Products serving human and/or animal foods are generally very sensitive and their main characteristics (i.e., color, smell, taste, shape, nutritive and other internal substances) are highly dependent on the thermal and sorption history of the material from the harvesting to the preserved state and on the time interval of drying. In Table 3 drying data are given for some agricultural products. Cereal grains and grain legumes need to be dried from an initial moisture content of about 30% (wet basis) at harvest to a level of 12%. Leafy green vegetables and fruits have an initial moisture content of about 60–80% to be reduced to the range of 10–25% for safe storage. Safe drying air temperatures are of 35–60°C and for some products these temperatures are higher at the end phase of the drying.

Quality of the dried products has an effect on the economy by influencing the marketing capability and income of the products because a higher price can be achieved by better quality. To ensure the required preconditions for drying of such sensitive materials a technological direction and process control is needed. The loss in quality of the dried product should be considered as a saving of negative value (176).

One of the main components of savings for semiartificial and solar-assisted dryers is

Table 3 Drying Data for Some Agricultural Products

| Product | Moisture percent (wb) | | Drying air temperature (°C) |
	Initial	Final	
Bananas	80	15	70
Barley	18–20	11–13	40–82
Beets	75–85	10–14	–
Cardamom	80	10	45–50
Cassava	62	17	70
Chilies	90	20	35–40
Coffee seeds	65	11	45–50
Copra	75	5	35–40
Corn	28–32	10–13	43–82
Cotton	25–35	5–7	–
French beans	70	5	75
Garlic	80	4	55
Grapes	74–78	18	50–60
Green forages	80–90	10–14	
Hay	30–60	12–16	35–45
Longan	75	20	
Medicinal plants	85	11	35–50
Oats	20–25	12–13	43–82
Onions	80–85	8	50
Peanuts	45–50	13	35
Pepper	80	10	55
Potato	75–85	10–14	70
Pyrethrum	70	10–13	
Rice	25	12	43
Rye	16–20	11–13	
Sorghum	30–35	10–13	43–82
Soybeans	20–25	11	61–67
Spinach leaves	80	10	
Sweet potato	75	7	75
Tea	75	5	50[a]
Virginia tobacco	85	12	35–70
Wheat	18–20	11–14	43–82

Source: From Reference 152
[a] At the end of drying, for 2–3 hours 100°C

the price of the conventional energy carriers substituted by solar energy. It should be emphasized that the solar energy utilized by the solar dryer is not equal to the energy collected by the solar collector and transferred into the drying air but the energy effectively used in the drying process. The energy effectiveness of solar dryers depends also on the exit energy losses, which can be reduced by applying a proper direction and control strategy. In the case when the energy effectiveness of a given solar dryer is lower than that of a conventional one a negative value in savings should also be considered.

The possible savings and the investment and maintenance costs of solar dryers are interdependent. Well-directed solar dryers need higher costs. Inexpensive, simple, and unsophisticated solar dryers, generally, have no appropriate devices for direction and control. These types of solar dryers are of great importance first of all for country use, for substituting open-air drying of nonsensitive materials and eliminating the well-known disadvantages of natural drying.

Solar dryers of high performance or for drying of quality-sensitive materials should be well directed. The higher investment and maintenance costs can be balanced by the better quality of the dried product and by the longer annual operation time. The annual operation time can be extended when drying materials of long growing time and several harvests in a year (e.g., meadow grass, alfalfa) or in cases of solar dryers applicable for drying of various materials having differing ripening times, one after the other (e.g., herbs, medicinal plants, spices and aromatic plants, seeds). Another way for year-round utilization is the multipurpose application of the solar energy converter of the dryer when, in the idle periods of drying, the solar energy collected is utilized for other technological purposes (e.g., for satisfying technological hot water demands of an agricultural farm). These complex systems should have appropriate direction and control devices to realize an economical operation strategy.

Strategy of Direction

The strategy of the direction should be elaborated by taking into consideration the drying characteristics of the material and the regulation possibilities of the solar dryer to be used. In the knowledge of the drying characteristics of the materials to be dried (i.e., sorption isotherms, drying curves) the appropriate schedule of the drying operation should be elaborated and, using it as a basis, the possible methods for interventions should be determined. Even in case of the most simple solar dryers some methods in the direction are recommended.

7.2. Direction and Control Actions

Direction of Drying Operation

Direction includes actions that are required for realization of an appropriate drying process in the dryer (i.e., to follow with attention the actual state of the material under drying) and determine and execute the necessary interactions by applying a direction strategy.

The first phase of the drying to be directed is the feeding of fresh material into the dryer. It should be emphasized that the good quality of the fresh material is a precondition of the good quality of the dried product. This action should be in harmony with the ripening state of the material and the point of views of the drying should be asserted in the harvesting technology.

The main direction actions of the drying process can be summarized as follows:

1.   Feeding fresh material into the dryer.
2.   Turning or tedding the layer of the material under drying occasionally in the case of unsophisticated (e.g., tent type) dryers.
3.   Regulating the airflow rate.
4.   Regulating the recirculation of the air.
5.   Regulating the intermittent drying process (determining the beginning and the interval of the break).
6.   Separating the solar dryer from the atmosphere in the night and in rainy weather when no auxiliary energy source exists.
7.   Regulating the operation of the auxiliary energy source.

8. Distribution of solar energy collected inside the drying space and between the cells in the case of multicell solar dryers.
9. Determining the mode of operation in the case of complex and multipurpose solar dryers and ensuring the optimal distribution of solar energy collected between the dryer, the storage, and the other heat consumers.
10. Regulating the operation of storage in the case of solar dryers with heat storage.
11. Determining the appropriate inlet temperature of the drying air in the different stages of the drying process.

Control of Drying Operation

Control actions of the drying operation are concerned with holding the given values of some operational parameters determined by the direction strategy. The main control actions are as follows:

1. Temperature control of working mediums.
2. Relative humidity control.
3. Mass flow rate control of flowing mediums.
4. Switch in and out devices (e.g., fans, humidifiers, valves or dampers, auxiliary heaters) when the limit values of some parameters occur.
5. Control of charging the thermal storage of the system.
6. Control of the rate of drying.
7. Control of the recirculation.
8. Control of the intermittent drying process.

7.3. Principles of the Direction

Direction Strategy of Static Bed Solar Dryers

For the drying process with forced convection in a static bed, the static bed is generally arranged in a drying chamber and the heat is transferred from the drying medium to the material by convection. In through-flow dryers air is led below the layer and flows upward through the bed.

The state of the material to be dried in a static bed can be approximately characterized by the change of state of the drying air flowing through the bed. The change of state of the air can be followed in the enthalpy (h) and absolute moisture content (x) chart of Mollier.

In Figure 37 the h-x diagram for an open-cycle drying process is presented. The actual state of the atmospheric air is represented by 0. Supposing that the mass flow rate of the air is $\dot{m}$, into the air is ϕ_h, the temperature increase of the air can be calculated

$$\Delta t = t_1 - t_0 = \frac{\phi_h}{c_p \dot{m}},\qquad [52]$$

where c_p is the specific heat capacity of the air. The temperature of the air entering the dryer is t_1. Supposing further that the drying by convection is nearly adiabatic, the change of state of the air in the bed is approximately of $h_1 =$ constant. The air will approximate the equilibrium relative humidity U_e of the material and the absolute moisture content of the air will increase by Δx (in Fig. 37, point 2). The mass flow rate of evaporation N from the material can be expressed indirectly

$$N = \dot{m}\Delta x\qquad [53]$$

and the energy flux practically consumed for drying is

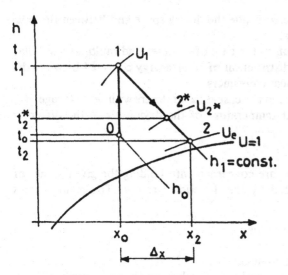

Figure 37 Change of state of the air in an open-cycle drying process in *h-x* chart.

$$\phi_D = \dot{m} \, \Delta x r = Nr, \tag{54}$$

where *r* is the total heat of evaporation of the water from the material under drying. The ϕ_D value can also be expressed by the temperature difference of the air

$$\phi_D = \dot{m} c_p (t_1 - t_2). \tag{55}$$

When the material to be dried has a quasi-constant rate period between the initial integral moisture content W_0 and the first critical moisture content $W_{cr,1}$ the drying rate curve *NW* has the shape given in Figure 38. In the figure the temperature curve $t(W)$ is also given. In the constant rate period the temperature of the material approximates the wet bulb temperature t_{wb} and remains practically constant ($t_m \cong t_{wb}$ = const). In the falling rate period the temperature of the material will approximate the dry bulb temperature t_1.

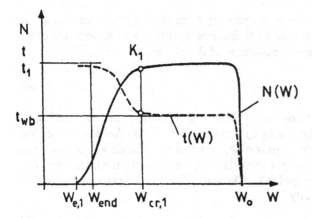

Figure 38 Typical drying rate and temperature curves of materials having constant drying rate period.

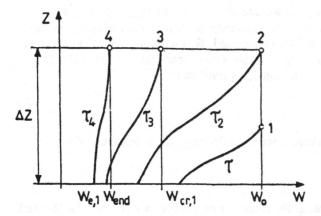

Figure 39 Moisture distribution in a static bed at the different stages of drying.

The moisture distribution in the bed of a thickness Δz can qualitatively be characterized by the curves in Figure 39 given for the different stages of drying. For time $\tau = \tau_1$ from the beginning, the upper layer of the bed above point 1 is almost of the initial moisture content. Drying of the surface layer starts at time $\tau = \tau_2$ and will be continued with approximately constant rate until the moisture content of the surface layer will be reduced to the first critical ($W_{cr,1}$) value $\tau = \tau_3$. In time interval $\tau < \tau_3$, ϕ_D remains practically constant, not considering a short starting period. When $\tau > \tau_3$ (i.e., in the falling rate period of drying), t_m and t_2 will increase $t_2 \rightarrow t_2^{\bar{x}}$ (see Fig. 37) and, according to Eq. [55] the exit energy loss of the dryer $\phi_L = \dot{m}\, c_p(t_2^x - t_2)$ will increase. As a consequence of the increasing energy loss the effectiveness of the dryer will successively decrease in the falling rate period ($\tau_3 < \tau < \tau_4$).

Direction of Operation of Static Bed Dryers. Considering the drying process in a static bed described qualitatively before, conclusions for the direction can be summarized as follows.

1. In Figure 40 the $\phi_D(\tau)$ function is presented at constant inlet parameters t_1, U_1

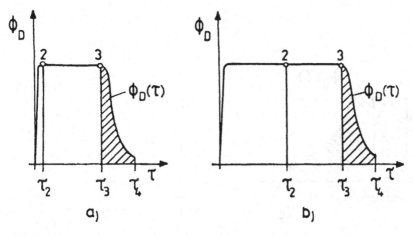

Figure 40 Energy flux consumed for drying in the function of time: (a) thin bed; (b) increased bed.

for two different bed thicknesses. In the case of a thin bed (Fig. 40-a) τ_2 is almost negligible (for τ_2 see also Fig. 39), $\tau_3 - \tau_2$ is the time interval of the practically constant rate period, $\tau_4 - \tau_3$ is that of the falling rate period. By increasing the thickness of the bed τ_2 will increase, $\tau_3 - \tau_2$ and $\tau_4 - \tau_2$ intervals remain practically constant (Fig. 40-b). The energy effectiveness of the dryer (taking ϕ_h as constant)

$$e = \frac{\phi_D}{\phi_h} \qquad\qquad [56]$$

is the function of the time. For the time interval of drying τ_4 the average value of e is

$$e_{av} = \frac{1}{\tau_4 \phi_h} \int_0^{\tau_4} \phi_D(\tau) d\tau. \qquad\qquad [57]$$

The e_{av} can be increased by increasing the thickness of the bed (see the ratio of the dark and white areas of Figs. 40-a and b).

2.   Though by increasing the layer thickness the energy effectiveness of the dryer can be improved, another effect has also to be taken into consideration. Since the time interval τ_2 will be longer, this method is not advantageous for materials sensitive to quality deterioration because the loss of internal substances is directly proportional with the time passed from the harvesting to the preserved state. For the material situated in the upper layer of the bed, drying will start after a longer time and some deterioration may occur. To solve this contradiction in the economy requirements, drying should be started with a fairly thin layer thickness. The thickness of the bed should be increased from time to time by feeding successively a new fresh layer in at $\tau = \tau_3$. This multilayer feeding method can be applied when the harvesting can also be fulfilled successively (e.g., in case of drying meadow grass or alfalfa in a solar drying-storing barn). In this case the task of the direction is to determine the time points of τ_3. Observation of τ_3 is possible with a fairly good approximation by measuring continuously the temperature t_2 and the relative humidity U_2 of the air leaving the bed.

3.   Exit energy loss can be reduced by the multilayer feeding also in case of materials not having a constant drying rate period at all. As it can be seen from Figure 41, by feeding new layers in after the time intervals of $\Delta\tau$, the value of e_{av} will be much higher than that of the single processes fulfilled one after the other.

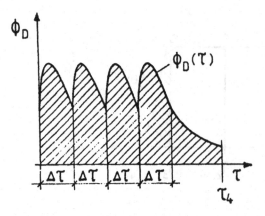

Figure 41 $\phi_D(\tau)$ function with multifeeding in for drying materials not having any constant rate period.

4. Another well-known direction method is the application of intermittent operation in the falling rate period of drying (Fig. 42). The first drying interval of $\Delta\tau_{1d}$ is followed by a break of $\Delta\tau_{1b}$ and so on ($\Delta\tau_{2d} - \Delta\tau_{2b}, \ldots$). During the break the uneven moisture distribution inside the material tends to equalize and, in the next drying interval, the drying rate will be higher again. This way e_{av} can be increased and the electrical energy consumed for driving and fan can be reduced. To find the best combinations of the drying periods and the breaks, an optimization problem has to be solved. For optimization the drying curves of the material, the time function of the internal moisture equalization process at different integral moisture contents, and the technical data of the dryer itself are needed. The direction strategy can be elaborated by computer simulation and proved by experiments.

For solar dryers not having any heat storage the utilization of the solar energy collected in the breaks is a problem. One possibility is to time the break intervals in the night. This condition in some cases will not permit fulfillment of the optimal strategy.

5. Another possibility is to divide the drying space into individual cells (see Fig. 11). Each cell should have its own fan and the construction should permit the distribution of the solar energy collected between the cells in an optional ratio. The intermittent drying processes in the different cells are shifted in time making it possible to utilize more solar energy in one cell, while in another cell a break is in progress.

6. In the falling rate period of drying the energy effectiveness of the solar dryer can be improved by recirculating one part of the air leaving the material. This well-known method can be recommended, first of all, for drying materials of long drying time.

A simplified scheme of a solar dryer having a separated collector is presented in Figure 43. Part of the recirculated air can be regulated by valves 4 moving together built in the air ducts serving for the outlet air 5 and for the recirculating air 6. Air flow of the air duct 6 is mixed with fresh air before entering the collector 1. The mixed air will be preheated in the collector and transported by the fan 2 to the drying chamber 3. The recirculated air can also be mixed with the fresh air preheated by the collector before the fan (dotted line 6*).

Principles of the direction can be followed in an h-x diagram (Fig. 44). The state of the fresh air is represented by 0, the air preheated in the collector (without recirculation)

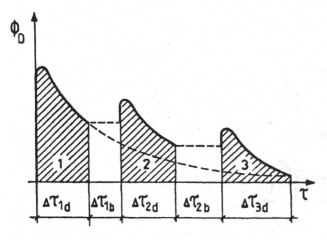

Figure 42 $\phi_D(\tau)$ function for intermittent operation.

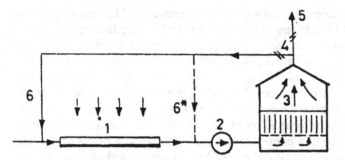

Figure 43 Simplified scheme of a solar dryer with recirculation.

by 1 and, the state of the air leaving the dryer by 2. In case of applying recirculation in a proportion of $(b/a)\dot{m}$, the state of the mixed air will be of M. Flowing through the collector the air will be preheated to t_1^* (point 1*). Supposing that the temperature of the material is t_2 (point 2 represents the state of air in equilibrium with the material), the direction of the change of state of the air will be of 1*2. As a result of the increased vapor pressure of the air ($p_{v_1}^*$) the rate of drying will temporarily be decreased and the temperature of the material will increase. The state of the air leaving the material will tend to 2*.

Applying recirculation the material can be heated to a higher temperature level than that without recirculation, where the rate of drying will be higher. The operation can be directed by regulating the b/a ratio. In the course of drying, temperature of material tends toward 1*. This fact has to be considered when regulating the recirculation.

In the case when the recirculated air is mixed with the air leaving the collector (see duct 6* in Fig. 43) the airflow rate in the collector will decrease and the temperature t_1^* will increase. Air temperature depends on the b/a ratio (see Fig. 45). The mixing point M will represent the state of the air entering the dryer and it will be situated between 1* and 2. The position of M will change in the function of the material temperature (2 tends to 2*). The possibilities for drying will be similar to the previous version. In both cases

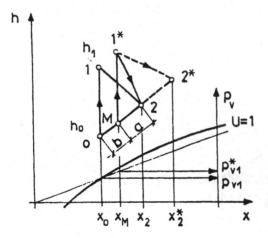

Figure 44 Change of state of the air with recirculation and mixing before the collector.

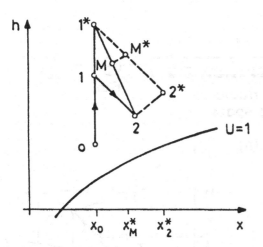

Figure 45 Recirculation with mixing after the collector.

the energy effectiveness of the dryer will be higher in spite of the greater heat losses in the collector and the drying chamber. A disadvantage of this system is the additional cost of the air duct 6. By applying a special construction duct 6 can be eliminated. As an example the solar timber dryer presented in Figure 6 can be considered.

Direction of the operation of such a dryer is, principally, the same as that written before. Representing the process in an *h-x* diagram, some differences arose from the facts that the stack is partially irradiated by the absorber and the airflow contacts a differing area of the absorber surface before entering the stack. To elaborate the direction strategy of solar dryers operating by recirculation, experiments are recommended.

Direction of Solar Dryers with Heat Transfer by Convection and Direct Irradiation

In Figure 46-a, a simplified scheme of a solar dryer operating with convection and direct irradiation (184) is presented. A single covered solar collector having a transparent foil covering is connected to the drying space. In the drying space the material to be dryed is arranged in a fairly thin layer and, through the transparent covering, it is directly irradiated. Drying air is transported through the collector and the drying space by a fan.

The operation of the dryer can be followed in an *h-x* psychrometric chart (Fig. 46-b). In the collector the atmospheric air is heated to the drying temperature t_1. In the drying space temperature t_1 is approximately constant if the heat demand of evaporation is satisfied by direct irradiation. The state of the air leaving the dryer is represented by point 2. The rate of drying is $N = \dot{m} \, \Delta x$, where $\dot{m}$ is the mass flow rate of the air. During the sunny hours of the day the state of the atmospheric air 0 and the solar irradiation is changing over time. As a consequence, t_1 and t_2 temperatures are varying, too. Nevertheless, the character of the change of state process of the drying air can be represented by the line 012 in Figure 46-b.

Let us suppose that the temperature of the material—and of the air in equilibrium with it—is t_2 when, in the evening, the solar irradiation stops. If the fan will be stopped, too, the material having a higher partial vapor pressure will keep evaporating. Without ventilation the water evaporated from the material will partially condense on the internal surface of the covering. In the night the atmospheric temperature will decrease (t_0^x, see Fig. 46-c) and the covering will cool down by convection and radiation. To avoid the condensation and the possible damage to the quality of the wet material it may seem to be

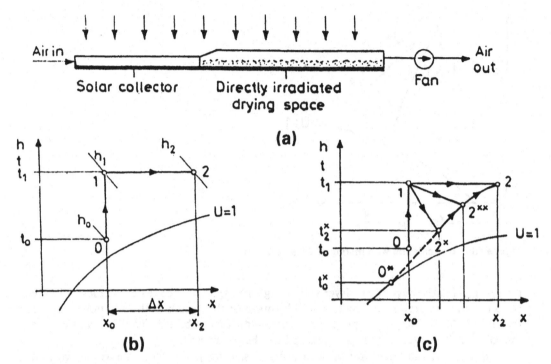

Figure 46 Solar dryer with heat transfer by convection and direct irradiation: (a) simplified scheme of the dryer; (b) change of state of the air in h-x chart; (c) night operation.

advisable—at least in the first stage of drying—to keep the fan in operation also in the night. As a consequence of the evaporation and of the cooling effects the temperature of the material will decrease. If the temperature decreases below the atmospheric temperature that existed in daytime ($t_2^x < t_0$), next morning the partial vapor pressure of the preheated air will be higher than that of the material and, for a time interval, the material will absorb some water vapor from the air (*rewetting effect*). The temperature of the material will increase by the convective heat transfer from the air as well as by the direct irradiation of the sun (from t_2^x to t_2).

Possible Methods for Direction.   During drying the rate of evaporation on the surface of the layer will be higher than that in the bottom of the layer. In the falling rate period the temperature of the surface layer will increase over the temperature of the air stepping in the drying space from the collector. Heat transfer from the surface into the layer will increase and, at the same time, a part of the energy gained by direct radiation will be transferred into the air by convection. As a consequence, the energy effectiveness of the dryer will decrease. The disadvantageous effects can be avoided by turning the layer over from time to time. This operation needs some handwork and can be applied only for materials not sensitive to the mechanical effects of turning or mixing. Without turning, the overheating of the surface should be prevented by the appropriate design of the dryer.

In the night, the cooling down effects can be moderated by reducing the mass flow rate of the air. It can be realized by a throttling or by using a driving motor of variable speed. Heat losses can be reduced by shading the transparent covering. A further possibility is the application of a solar collector with heat storage. A heat storage pebble bed or latent heat storage can be used. The surface of the storage layer may serve as absorber. Application of heat storage results in more sophisticated construction and higher investment costs. Economy aspects should be determined in each case separately.

Direction of the Operation of Tent, Greenhouse, and Cabinet-Type Solar Dryers

Tent, greenhouse, and cabinet-type dryers have simple and unsophisticated structure. The material to be dried is partially or totally irradiated during drying. Problems that may arise during the operation of such dryers are similar to those discussed in the previous section. The possible means of direction of the operation are also the same: turning the material over from time to time and to close and shadow the drying space during the night or, in some cases, to reduce the air flow in night operation.

In Figure 47 a scheme of a cabinet dryer for household use is presented to dry fruits, herbs, and vegetables. Material to be dried is arranged in a thin layer on a tray 1 with perforated bottom inside the cabinet 2 with a single glass covering 3. The air is introduced to the dryer by free convection using a black-painted metal sheet 4 of the same surface area. The metal sheet serves as a secondary uncovered collector by increasing the inlet air temperature. The airflow can be controlled by the throttling device 5 at the air outlet. During the day the position of the dryer can be changed according to the position of the sun. At the first stage of the drying the slider 5 is open. The outlet cross-section area of the air should be decreased according to the progress of the drying process. Thus, the airflow rate will also be decreased and the temperature of the air increased. In the night, the material should be shaded by posing the black-painted sheet 4 on the transparent covering of the cabinet 2 and having the slider 5 closed.

Direction of the Operation of Chimney-Type Solar Dryers

In chimney-type solar dryers the driving force of the airflow is the hydrostatic pressure difference caused by the decreasing density of the preheated air (*chimney effect*). Since no conventional energy sources are needed, chimney-type solar dryers can effectively be used as country dryers. For keeping the airflow on in the night, chimney-type dryers usually have some kind of heat storage.

In Figure 5, the construction of a chimney-type solar dryer (6) is presented. In night operation the transparent walls 6 can be insulated by reflecting panels. The air duct of the collector should be closed and air ducts 5 below the drying chamber should be opened. The atmospheric air will be preheated by convection when flowing through the heat storage space and contacting with the water containers.

Another construction is presented in Figure 48 (178). This dryer has a collector 2 with a latent heat storage material ($CaCl \cdot 6H_2O$) as absorber 4. Two additional latent heat storage plates are applied 5 that are pulled out from the covering of the collector and are directly irradiated in daytime. The southern walls, including the wall of the chimney, are made of transparent material; the northern wall and the bottom 6 are insulated. The material to be dryed is arranged on trays 7 in the drying chamber 1 and partially irradiated. On the northern wall of the chimney 3 a layer of latent heat storage is built in 4 that produces additional preheating effects for the air in night operation.

As to direction of the operation the latent heat plates 5 should be pushed below the covering of the collector for night operation (see the figure). Since the heat storers are directly irradiated, they can be charged to a higher temperature level than they could be when heating by the air. The airflow through such dryers can be assisted by a windmill driving a fan built in the chimney.

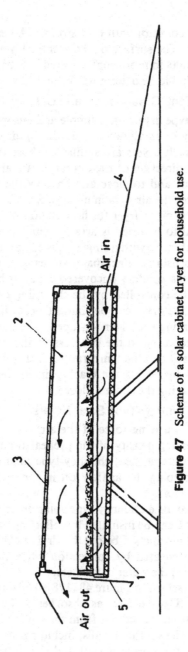

Figure 47 Scheme of a solar cabinet dryer for household use.

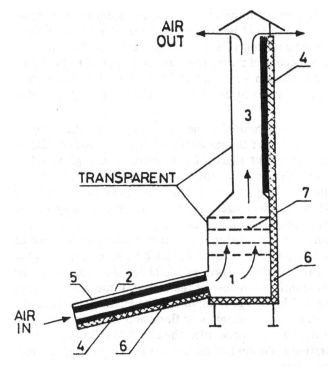

AIR
OUT

TRANSPARENT

AIR
IN

Figure 48   Simplified scheme of a chimney-type solar dryer with latent heat storage.

7.4.   Basic Principles of Control

Application of Automatic Control

Economy Aspects.   Values of the operational parameters required by the direction strategy can be effected automatically by using controllers. Application of automatic control can be justified by the possible reduction of manpower as well as by the higher reliability compared to manual control. Though the instrument cost is high, its effect on economics can be balanced by the better quality, by the better energy effectiveness, and, in case of sophisticated solar dryers of high performance, by the improved security of the drying operation. Economy analysis has to be done in each case separately.

Aspects for Designing the Control.   Control systems generally consist of three main elements. A sensing element (*sensor*) is used for measuring the actual value of the parameter to be controlled. The control device serves for forming commands for intervention, if necessary. The element for intervention (e.g., a motoric valve) executes the command of the controller.

Process control systems generally operate with a constant set point. It should be emphasized that exact control is not always necessary and overinstrumentation should be avoided. Solar dryers generally have some parts of large capacities that reduce the effects of disturbances.

For the accurate design of the control system a detailed operational analysis is needed. Requirements for the design can be determined on the basis of the information produced by the analysis. In simple cases, the appropriate controller can be selected by using rules of thumb. Solar dryers having various input parameters can often be controlled by individual controllers. In cases when the control aims at optimizing the drying operation, commands for interventions should be formed by considering several input and output variables. For this purpose a microprocessing control device can be applied.

The detailed description of process control theory is beyond the scope of this chapter. In actual designing problems a consultation with experts in process control is recommended. A brief overview of the main types of controllers is offered here (for further details, see Refs. 179–181).

Main Types of Control Systems.   *On-off control* is the most simple and the cheapest method and it is widely used. If the value of the measured variable is less than the set point value, the controller is on. The output signal of the controller is a given value. When the measured variable is above the set point, the controller is off. In solar dryers on-off temperature controllers are used (e.g., for control of the operation of auxiliary heaters in a water storage tank). A disadvantage of this control system may be the uncertain operation and, actually, some overshoot may occur.

Closed loop or feedback control systems operate by adjusting automatically one of the input variables of the process by comparing a signal fed back from the output of the process with a reference input. The difference serves as signal for the controller. The system can be characterized by the transient response of the output of the process due to some specific variations in the input. The change in input may be either a change in the set point or in one of the load variables (e.g., uncontrolled flows and temperatures). Two different operations can be realized. With servo operation the aim is to follow changes in the set point. With regulator operation the output of the process should be kept constant in spite of some changes in load variables.

Open-loop control systems are used when every input variable of the process should be constant. Open-loop control can effectively be used when a closed control is not needed, when the change in inputs is not strong, or in cases when the feedback control is not good enough. The open-loop control is called *feed-forward control* when one of the input variables is measured and used for adjusting of another input variable.

Main Types of Control Actions.   In Figure 49, a block scheme of a control system is presented. In the case when the set point value x_0 is constant the control is called value keeping. If the set point is a function of the time, that is, $x_0(\tau)$, a signal is needed for operating the set point device. Various principles can be used to form commands for

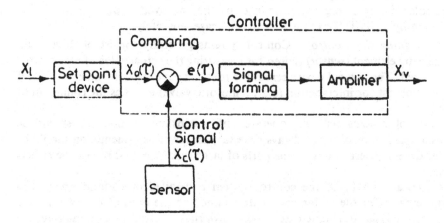

Figure 49   Block scheme of a control system.

interventions by the controller. The basis of the methods is the error e, which is the difference between the controlled parameter (control signal x_c) and the set point x_0: $e = x_0 - x_c$. Controllers have a signal-forming unit that produces the command signal as output x_i.

The *proportional controller* (P) produces its output proportional to the error

$$x_i = K/x_0 - x_c/ = Ke, \qquad [58]$$

where K is the gain of the controller. At a given working point, the change in the output related to a differential change in input is called the gain of the controller.

With *integral control* (I) output of the controller is proportional to the time integral average value of the error

$$x_i = \frac{1}{\tau_R} \int_0^{\tau_R} e d\tau, \qquad [59]$$

where τ_R is the reset time.

The two-mode *proportional integral* (PI) controller produces its output by addition

$$x_i = K\left(e + \frac{1}{\tau_R} \int_0^{\tau_R} e d\tau\right). \qquad [60]$$

Derivative control action (D) can improve the response of slow systems when coupling parallel to proportional control by adding an effect proportional to the time derivative of the error. This way some disadvantageous effects of large load changes and the maximum error can be reduced.

The *three-mode controller* has a proportional and an integral character with derivative action (PID). The output signal of a PID controller is

$$x_i = Ke + \frac{1}{\tau_R} \int_0^{\tau_R} e d\tau + \tau_D \frac{de}{d\tau}, \qquad [61]$$

where τ_D is the derivative time.

The output signal x_i of the signal-forming unit will be amplified and modified. The output of the controller is the signal for intervention x_v. As an example the scheme of the automatic temperature control of a liquid-air heat exchanger is given in Figure 50.

Selection of Control Systems. In the operation of control systems stability is required. Operation is stable when continuous cycling will not occur. Instability could be the consequence of the increase in the overall gain of the controller above a maximum value. The overall gain of the controller is the product of gain terms in a closed loop. Role of the time lag may also be considered. Different stability criteria have been elaborated and various rules developed. Integral control and derivative control action can improve the stability of systems added to proportional control. The final performance of a system is affected by the characteristics of the process to be controlled, by the operational character of the controller used, and by the nature of the disturbances to be expected.

Operational characteristics can be described by the response function of the system $x(\tau)$ to a step change in load. The system consists of the process to be controlled and of the controller. The mathematical relationship between input and output is called a *transfer function*. The time τ_c necessary to approximate the required value within a given difference Δx (mostly, $\Delta x = \pm 0.02, \ldots, 0.05$) is called *control time*. The response function of the controlled process (not having any integrating character) can be characterized by the time lag τ_t and by the time constant τ_t.

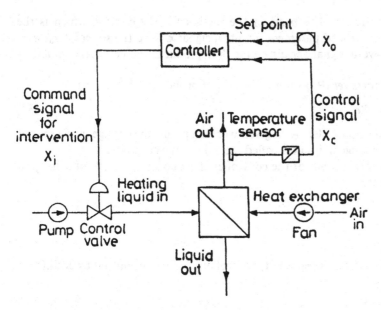

Figure 50　Control scheme of a liquid-air heat exchanger.

For the selection of the appropriate control system a dynamic analysis is needed that can inform the designer about the type of controllers really needed. Some general recommendations follow. In cases when the expected disturbances are strong and the control time permitted is long enough, proportional (P) controller can be used. PI controllers can be applied with $\tau_c \tau_t > 6$. PID controllers are recommended in the range of 4 $\tau_t < \tau_c < 6$. For flow rate control and for level control P and PI controllers are used; for temperature control, P, PI, and PID controllers are mostly applied.

Control of Solar Dryer Operating with Recirculation

With solar dryers of multicell construction the collector field serves for more than one drying chamber. Recirculation of the drying air can be directed for keeping the inlet air temperature constant. The solar energy saved by recirculation can be utilized for preheating the air of another drying chamber (or for other technological purposes). In these cases the aim of applying recirculation is the improvement of the energy effectiveness of each dryer and, by this way, of the system. As it can be seen from Figure 44, constant inlet air temperature can be realized by the appropriate variation of the b/a ratio. A precondition of such operation is that the point 0 (i.e., the state of the atmospheric air) should be of lower absolute water content than that of the outlet air. The scheme of the control is presented in Figure 51 for a multicell solar dryer having a liquid-type collector and a liquid-air heat exchanger. (In the figure one drying chamber is indicated.)

Operation by recirculation can be realized by controlling the b/a ratio and the heat input to the heat exchanger. By controlling the air valves moving together (see also 4 in Fig. 43) the absolute moisture content of the air x_1 can be ensured (according to the position of the mixing point M). The controlling signal for the recirculation is the wet bulb temperature T_{wb} of the inlet air. Inlet temperature of the drying air is controlled by the valves V of the liquid working medium of the collector serving as heating (primary) medium for the heat exchanger. The controlling signal is the dry bulb temperature ($T_{db} = T_1$). The control action can be realized automatically or manually.

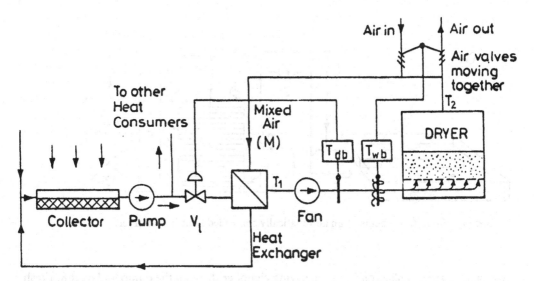

Figure 51   Control of a solar dryer with recirculation.

Direction and Control of Solar Dryers with Rock Bed Heat Storage

For dryers having separated rock-bed heat storage (see Fig. 10) three main modes of operation should be applied:

1. Drying with air preheated by the collector
2. Drying and simultaneously charging of the heat storage with the air preheated by the collector
3. Drying with air preheated by the heat storage (discharging period) when no solar radiation exists

In mode of operation 1, damper 8 in Figure 10 is open while damper 9 closes the air duct below the heat storage 7. In mode of operation 2, damper 9 closes the upper air duct and the air flows from the drying space into the rock bed. In mode of operation 3, damper 9 is in a medium position, fan 2 is out of operation, and the air flows from the dryer into the rock bed; there it will be preheated and flow back into the drying space in the upper air duct.

Regulation of the mode of operation can be realized manually or automatically. As a signal for regulation the temperatures of the drying space, of the outlet air of the collector, and of the rock bed can be used. As a controlling signal for the operation of the dampers 6, the wet bulb temperature or the relative humidity of the drying space can be applied.

Automatically Controlled Solar Dryer with Auxiliary Heater

Construction of the Dryer.   The simplified scheme of an automatically controlled solar lumber kiln dryer with an auxiliary energy source of wood residue burner is shown in Figure 52 (182). Collector 1 has a charcoal absorber and a gravel bed storage is arranged below it. Airflow through the collector 1 is induced by two blowers 2. Four collectors are coupled in parallel to the kiln. The preheated air is distributed by the manifold duct 3 behind the four fans serving for the internal circulation in the drying chamber 4. Four blowers 6 exhaust humid air from the kiln through the stack 7. A part of the air is

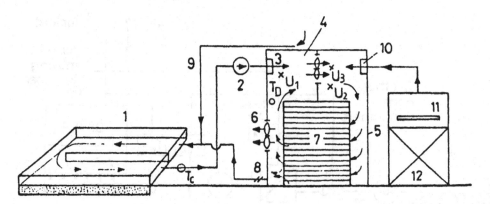

Figure 52 Simplified scheme of an automatically controlled lumber kiln dryer.

recirculated to the collector from the drying chamber through the dampered duct 8. Fresh air, slightly preheated by the black-painted roof surface, is led into the collector through duct 9. The wood residual burner 12 produces hot air for the drying space that is distributed by the manifold 10. A humidifier 11 is coupled to the burner for use when the humidity is below the minimum level.

Direction and Control of the Dryer. Operation of the dryer is directed and controlled as follows. Solar blowers 2 start when the temperature in the drying space T_D is lower than the collector outlet temperature T_c, dampers 8 are open. Blowers are activated by an on-off temperature control device. As a precondition of the operation of blowers the relative humidity U_1 should be lower than the set point value U_{1s}. Set point selection is manual. If $U_1 < U_{1s}$ the internal fans 4 are on. The operation of the exhaust blowers 6 is controlled by the signal of the relative humidity sensor U_2 situated behind the internal fans. The set point U_{2s} is high initially and should be reduced in the progress of drying. Exhaust blowers are actived if $U_2 > U_{2s}$. In the drying space the relative humidity should also be above a minimum level U_{3s}. When $U_3 < U_{3s}$ humidifier 11 will be activated by the control signal of the U_3 sensor. By applying water spray into the furnace chamber, humidification of the air led to manifold 10 will be effected. Operation of the burner is directed by U_3 or manually. The solar blowers 2, the internal fans, the exhaust vents, the humidifier 11, and the dampers 8 can also be controlled manually by using bypass switches. When solar blowers 2 are off, the internal fans 4 and the exhaust blowers 6 can be in operation if the state of the air in the dryer satisfies the requirements.

Drying time can be influenced by the operation time of the burner. For the sake of electric energy saving the number of internal fans in operation can be reduced, generally, in the final period of drying. In this period the humidifier 11 can be used for relief of drying stresses. Control of the drying process can be performed according to a schedule by applying a timer that opens or closes the control relay at a determined time point. It can be bypassed manually. When the dryer is out of operation, dampers 8 are closed and the drying space can be isolated from the collectors (e.g., in the night).

Direction and Control of Solar Dryers with Water Storage

Solar dryers applying a water tank for heat storage have an indirect heat transfer system: the working medium of the collector is liquid (e.g., water) and the drying air should be preheated in a liquid-air heat exchanger. Two different constructions are discussed below.

Construction and Control of a Dryer with Water Storage. The scheme of a simple system is presented in Figure 53. In this system three flow loops are applied. The first is that of the collector-tank loop in which the flow of the liquid working medium is maintained by pump P_1. The second loop is that of the tank-heat exchanger with pump P_2. The third one is the open loop of the air that is transported by the fan through the heat exchanger toward the drying space. Drying air is preheated by the heat exchanger using water from the heat storage tank.

Temperature required for drying T_D is controlled by control device CV using temperature sensor T_D and valve V. Fan is in operation when T_D is higher than the lowest temperature limit as set point value T_{DS}. The fan is turned on by thermoswitch SD. The minimal temperature level of the water needed for ensuring T_D is T_s, which is the set point for the temperature sensor in the tank T_T. Operation of pump P2 is induced by the thermal switch SP2. Pump P1 and the collector is in operation if the liquid outlet temperature $T_L \geq T_T$. Operation of the pump P1 is induced by the control device CP1. The controlling signal is the temperature difference $T_L - T_T$.

When T_T is lower than the set point value, the auxiliary energy source in the tank (A.H) will turn on. It is operated by on-off control device S.A.H. Each of the control operations can also be realized manually using bypass switches.

Construction, Operation, and Control of a Complex Solar Drying and Hot Water Supply System. Application of a heat storage tank permits the year-round utilization of solar energy collected. In the idle periods of drying the heat storage can be charged by the collector and a hot water supply can be ensured for other heat consumers. The simplified block scheme of a complex solar drying and hot water supply system is presented in Figure 34. The system can operate in five different modes of operation.

1. Solar-only operation. Valves 2 and 3 are closed, 1 and 4 are open; the collector serves for the heat exchanger of the dryer.
2. Simultaneous operation for charging the heat storage and drying. Valve 3 is closed, 4 is open, and 1 and 2 are in a partially opened position.
3. Drying operation by using stored energy. Valves 1 and 4 are closed, 2 and 3 are open; heat exchanger operates by using stored energy.
4. Charging the heat storage by the collector. Valves 1 and 3 are closed, 2 and 4 are open. Heat exchanger is out of operation.
5. Technological hot water supply. Hot water flow is induced by pump 2. This operation can be realized simultaneously with other modes of operation.

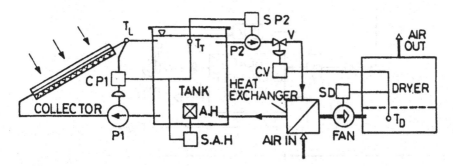

Figure 53 Control scheme of a solar dryer with water storage.

Tasks of the direction and control system are the selection of the mode of operation, the control of the input temperatures of the heat exchanger, and the control of charging of the heat storage tank. A scheme of the control system is presented in Figure 54. Four control units are applied (CU1–CU4).

CU1 and CU4 are value-keeping controllers for the control of the outlet temperatures of the collector and of the HWS heat exchanger, respectively. Control signals are produced by the temperature sensors T_{OC} and T_W. Control is realized by changing the mass flow rates with one-way motor valves V_C and V_W, respectively.

CU3 serves for controlling the charging of the stratified heat storage tank T by the collector. As a control signal the temperature difference $(T_L - T_S)$ is applied between the primary liquid (T_L) and the stored water in the upper layer (T_S). Valve system V_s serves to direct the flow in the layer of appropriate temperature into the tank.

CU2 process controller serves for automatic controlling of the inlet air temperature of the dryer T_{OD}. The aim is to direct the rate of drying and realize intermittent drying

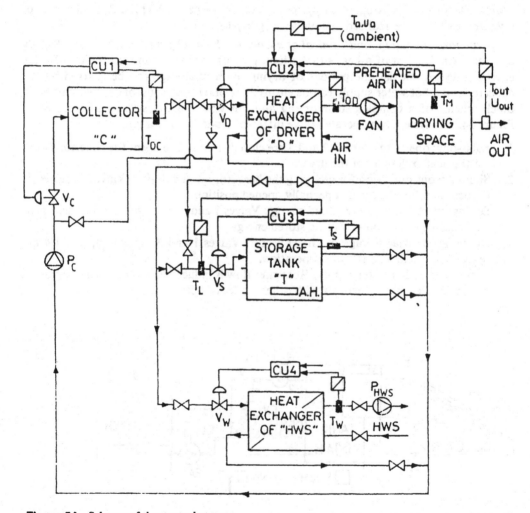

Figure 54 Scheme of the control system.

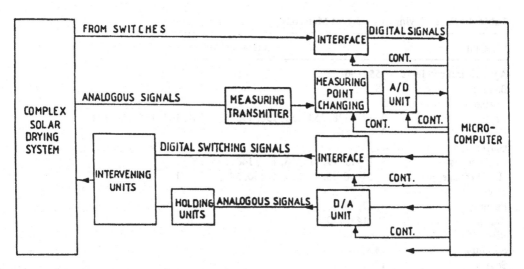

Figure 55 Block scheme of the microprocessor.

operation in the falling rate period of drying. As controlling signals, the surface temperature of the material under drying T_M, the temperature of the outlet T_{out}, and of the ambient air T_a, the relative humidity of the outlet air U_{out}, and the ambient air U_a are used. Control is realized by adjusting the mass flow rate of the liquid stepping into the heat exchanger of the dryer by applying the motoric valve V_D. The other valves indicated in the block scheme serve for realizing the different modes of operation.

Control of drying and selection of the appropriate mode of operation is directed by a microprocessor. The simplified block scheme of the microprocessor is presented in Figure 55. As control signals the following parameters are used: temperature and relative humidity of the ambient air T_a and U_a, respectively, and that of the air step in and out of the dryer (T_{OD}, T_{out}, U_{out}), the actual inlet and outlet temperatures of the liquid working medium of the collector, temperature distribution of the water in the storage tank, temperature of the hot water produced for consumers. Direction of the intermittent drying process is realized by computing parameters serving for the indication of the beginning and the time interval of the break. A microprocessor forms commands for the execution of the direction and control actions, indicates the main information about the actual state of the system, and calls attention to manual interventions when needed.

8. PROSPECTS FOR SOLAR DRYING

Further research and development of various components of solar drying systems continues to proceed internationally. Major current applications are confined to drying of agricultural and forest products. Table 4 summarizes selected references on solar drying of various materials.

Some significantly good payback times (1–7 years) have been achieved (14,17,133), mainly with simple and cheap dryers. Chances for the extensive use of high-performance systems may be improved by integrated construction and multipurpose operation. Using modern process control techniques, the efficiency of solar dryers can be increased. Due attention must be paid to system maintenance and training of the operating personnel.

In the design of solar dryers modern methods of modeling and simulation can play

Table 4 Solar Drying of Different Materials

Material	Reference nos.
Agricultural products	155, 166
Banana	18
Coffee	3
Crop	3, 129, 138, 153, 155, 156, 158, 159, 161, 163, 164, 168, 173
Fruits	3, 6, 11, 41
Garlic	12
Grain crops	3, 13–15, 19, 31, 128, 135, 148, 155, 167
Hay, herbage, grass	3, 13, 16, 132, 133, 135, 136, 141, 142, 171
Jute	3
Peanuts	3, 17, 19
Raisins	19, 143, 170
Rice	146
Sorghum	140
Soybeans	19
Timber	3, 8–10, 19, 20, 29, 124, 133, 134
Tobacco	19, 137, 144, 145, 169
Tomato	139
Vegetables	3, 193, 194, 198
Lumber	162, 171, 172

an important role in both the design and optimum operation. The renewable nature of solar energy is a definite asset in most parts of the world.

NOMENCLATURE

a	specific transfer surface by volume basis, m^{-1}
a	acceleration, $m \cdot s^{-2}$
A	area, m^2
b	biologic heat source, $J \cdot kg^{-1}$
c	specific heat capacity, $J \cdot kg^{-1} K^{-1}$
C	heat capacity, $J \cdot K^{-1}$
CU	control unit
$\dot{c}$	heat capacity flow rate, $W \cdot K^{-1}$
d_h	hydraulic diameter, m
e	effectiveness
e	error
E	irradiance, $W \cdot m^{-2}$
f	efficiency function
f	Fanning coefficient
g	acceleration of gravity, $m \cdot s^{-2}$
h	heat transfer coefficient, $W \cdot m^{-2} K^{-1}$
H	heat exchanger effectiveness
I	global irradiance, $W \cdot m^{-2}$
k	heat conductivity, $W \cdot m^{-1} K^{-1}$
K	heat conduction, $W \cdot K^{-1}$
K	gain of controller

L	length, m
M	mass, kg
$\dot{m}$	mass flow rate, $\text{kg} \cdot \text{s}^{-1}$
N	mass flow rate of evaporation, $\text{kg} \cdot \text{s}^{-1}$
n	revolution, s^{-1}
p	pressure, Pa
q	radiant flux density, $\text{W} \cdot \text{m}^{-2}$
Q	radiant energy, J
r	heat of evaporation, $\text{J} \cdot \text{kg}^{-1}$
P	pump
R	resistance in a network
R	gas constant, $\text{J} \cdot \text{kg}^{-1} \, \text{K}^{-1}$
S	switch
w	velocity, $\text{m} \cdot \text{s}^{-1}$
W	integral moisture content, $\text{kg} \cdot \text{kg}^{-1}$
W	water
z	height, m
t	transmittance
T	temperature, K
U	overall heat transfer coefficient, $\text{W} \cdot \text{m}^{-2} \, \text{K}^{-1}$
U	relative humidity of the air, %
V	value
x	absolute water content of the air by dry basis, $\text{kg} \cdot \text{kg}^{-1}$
X	moisture content of the material, dry basis, $\text{kg} \cdot \text{kg}^{-1}$

Greek Symbols

α	absorbance
β	moisture transfer coefficient, $\text{m} \cdot \text{s}^{-1}$
ε	emittance
ϕ	heat flux, W
η	efficiency
η	dynamic viscosity, $\text{kg} \cdot \text{s}^{-1} \, \text{m}^{-1}$
ζ	shape flow resistance coefficient
ξ	porosity
ρ	reflectance
ρ	mass density, $\text{kg} \cdot \text{m}^{-3}$
σ	Stephan–Boltzmann constant
σ	evaporation coefficient, $\text{kg} \cdot \text{m}^{-2} \, \text{s}^{-1}$
λ	latent heat, $\text{J} \cdot \text{kg}^{-1}$
λ	wavelength, m
ν	frequency, s^{-1}
τ	time, s

Subscripts

A	auxiliary energy source
a, A	air
a	ambient
av	average

b	beam (direct)
b	bottom
b	break
c	collector
c	control
cr	critical
d, DIR	direct
db	dry bulb
D	drying
e	equilibrium
f, *F*	fluid
h	heat
H-T	horizontal-total
i	incident
i, *j*	nodal points in a network
in	inlet
k	component
l	lag
l, *L*	loss
L	liquid
lt	long term
m	number of nodes in a network
m, *M*	material
n	number of the discrete parts
n	normal
o	outside atmosphere
0	set point
out	outlet
p	at constant pressure
r	reflected
R	reset
s	space
s	sky
S	set point
t, *TOT*	total
T	temperature
th	threshold
u	useful
v	vapor
w	water
wb	wet bulb
pw	for humid air
(*k*)	time period
'	inlet
"	outlet

REFERENCES

1. P. Barbet, *Panorama de l'Énergie,* *20*:33 (1980).
2. Special issue, *Sunworld,* 4:179 (1980).

3. H. P. Garg, in *Proc. Third Int. Drying Symp.* (J. C. Ashworth, ed.): Drying Res. Ltd., Wolverhampton, England, 1982, p. 353.
4. P. Wibulswas and C. Niyomkarn, *Reg. Workshop on Solar Drying, CNED-UNESCO,* Manila:1 (1980).
5. M. K. Selcuk, Ö. Ersay, and M. Akyurt, *Solar Energy, 16*:81 (1974).
6. J. R. Puiggali and M. A. Lara, in *Proc. Third Int. Drying Symp.* (J. C. Ashworth, ed.), Drying Res. Ltd., Wolverhampton, England, 1982, p. 390.
7. M. N. Özisik, B. K. Huang, and M. Toksoy, *Solar Energy, 24*:397 (1980).
8. Résumé de l'étude en cours on CTB sur l'utilization de l'énergie solaire, *Centre Technique du Bois,* Paris, 1978.
9. K. C. Yang, *Forest Product Journal, 30*:37 (1980).
10. A. Schneider, F. Engelhardt, and L. Wagner, *Holz als Roh- und Werkstoff, 37*:427 (1979).
11. C. Rosello, A. Berna, and M. Mulet, *Drying Technology Int. J., 8*(2):305 (1990).
12. F. Pinaga, J. V. Carbonell, and J. L. Pena, in *7th Int. Congress of Chem. Eng. Praha*:C5.8 (1981).
13. W. Wieneke, *Agricultural Mechanization in Asia, Autumn*:11 (1980).
14. G. J. Schoenau and R. W. Besant, in *Proc. of Sharing the Sun, Solar Technology in the Seventies* (K. W. Böer, ed.), Winnipeg, Canada, 7:33 (1976).
15. G. Roa and I. C. Macedo, *Solar Energy, 18*:445 (1976).
16. W. Dernedde and H. Peters, *Landtechnik*: 29 (1978).
17. D. H. Vaughan and A. J. Lambert, *Trans. ASAE*: 218 (1980).
18. R. G. Bowrey, K. A. Buckle, I. Hamey, and P. Pavenayotin, *Food Technology in Australia, 32*:290 (1980).
19. W. W. Auer, in *Drying '80* (A. S. Mujumdar, ed.), Hemisphere, New York, 1980, p. 292.
20. W. R. Read, A. Choda, and P. I. Cooper, *Solar Energy, 15*:309 (1974).
21. P. W. Niles, E. J. Carnegie, J. G. Pohl, and J. M. Cherne, *Solar Energy, 20*:19 (1978).
22. T. L. Freeman, J. W. Mitchell, and T. E. Audit, *Solar Energy, 22*:125 (1979).
23. J. A. Duffie and N. R. Sheridan, *Mech. and Chem. Eng. Trans. of the Inst. of Engs., Australia*: MC1(1), 1965.
24. L. Imre, L. I. Kiss, and K. Molnár, in *Proc. Third Int. Drying Symp.* (J. C. Ashworth, ed.), Drying Res. Ltd., Wolverhampton, England, 1982, p. 370.
25. L. Imre, I. Farkas, L. I. Kiss, and K. Molnár, *3rd Int. Conf. on Num. Methods in Thermal Problems,* Seattle, 1983.
26. L. Imre, in *Handbook of Drying* (in Hungarian), Müszaki KK, Budapest, 1974.
27. D. J. Close and L. L. Pryer, *Solar Energy, 18*:287 (1976).
28. D. J. Close and R. V. Dunkle, *Solar Energy, 19*:233 (1977).
29. N. A. Duffie and D. J. Close, *Solar Energy, 20*:505 (1978).
30. K. W. Böer, *Solar Energy, 20*:225 (1978).
31. O. Y. Kwon and P. J. Catania, in *Drying '80* (A. S. Mujumdar, ed.), Hemisphere, New York, 1980, p. 445.
32. J. M. Gordon and A. Rabl, *Solar Energy, 28*:519 (1982).
33. J. S. Vaishya, S. Subrahmaniyam, and V. G. Bhide, *Solar Energy, 26*:367 (1981).
34. M. Kovarik, *Solar Energy, 21*:477 (1978).
35. N. E. Vijeysundera, L. L. Ah, and L. E. Tijoe, *Solar Energy, 28*:363 (1982).
36. J. P. Chiou, M. M. EI-Wakil, and J. A. Duffie, *Solar Energy, 9*:73 (1965).
37. L. Imre and L. I. Kiss, in *Num. Methods in Heat Transfer* (R. W. Lewis, K. Morgan, and B. A. Schrefler, eds.), Wiley, Chichester, England, Vol. 2, Chap. 15, 1983.
38. D. J. Morrison and S. I. Abdel-Khalik, *Solar Energy, 20*:57 (1978).
39. M. Telkes and R. P. Mozzer, in *Proc. Annual Meeting Am. Section ISES,* Denver, 1978.
40. J. H. Schlag, D. C. Ray, A. P. Sheppard, and J. M. Wood, *Solar Energy, 20*:89 (1978).
41. H. R. Bolin, A. E. Stafford, and C. C. Huxsoll, *Solar Energy, 20*:289 (1978).
42. D. J. Close and M. B. Yusoff, *Solar Energy, 20*:459 (1978).
43. P. J. Lunde, *Solar Thermal Engineering,* Wiley, New York, 1980.

44. S. H. Butt, J. W. Popplewell, W. S. Lymann, P. Anderson, P. Kruger, and S. C. Byone, *Solar Age*, *2*:5 (1977).
45. W. A. Beckman, *Solar Engineering*, *21*:531 (1978).
46. H. C. Hottel and B. B. Woertz, *Trans. ASME*, *64*:91 (1942).
47. H. C. Hottel and A. Willier, *Trans. Conf. on Use of Solar Energy*, Univ. of Arizona, Vol. 11, 1955, pp. 74–104.
48. R. W. Bliss, *Solar Energy*, *3*:55 (1959).
49. E. Marschall and G. Adams, *Solar Energy*, *20*:413 (1978).
50. A. Whillier, in *Low Temp. Eng. Appl. of Solar Energy* (R. C. Jordan, ed.), ASHRAE, 1967.
51. J. P. Chiou, *Solar Energy*, *29*:487 (1982).
52. H. S. Robertson and R. P. Patera, *Solar Energy*, *29*:331 (1982).
53. ASHRAE STANDARD, ASHRAE Inc., New York, 1977.
54. J. E. Hill, J. P. Jenkins, and D. E. Jones, *NBS Building Science Series*, 117, USDC, January 1979.
55. C. D. Adler, J. W. Byrd, and B. L. Coulter, *Solar Energy*, *26*:553 (1981).
56. J. A. Duffie and W. A. Beckman, *Solar Engineering of Thermal Processes*, Wiley, New York, 1980.
57. K. S. Ong, *Solar Energy*, *16*:137 (1974).
58. A. A. Sfeir, *Solar Energy*, *25*:149 (1980).
59. A. J. De Rou, *Solar Energy*, *24*:117 (1980).
60. W. F. Phillips, *Solar Energy*, *29*:77 (1982).
61. L. Imre, *Heat Transfer of Composite Devices* (in Hungarian), Acad., Budapest, 1983.
62. G. O. G. Löf, J. A. Duffie, and C. O. Smith: *Solar Energy Lab. Univ. of Wisconsin*: Rep. No. 21, 1966.
63. J. I. Yellott, in *NAS Conf. Proc.*, 1976.
64. *Climatic Atlas of the United States*, U.S. Gov. Printing Office, 1968.
65. C. R. Attwater et al., Canadian Solar Rad. Base: *ASHRAE Arm. Meeting*, Detroit, 1979.
66. Catalogue of Solar Rad. Data Australia: *Aust. Government Publ. S.*, Canberra, 1979.
67. Beuzeman and Cook, *N. Zealand J. Sci.*, *12*:698 (1969).
68. N. K. O. Choudhury, *Solar Energy*, *4*:44 (1963).
69. Anderson et al., *Extract from Meteorological Data: Thermal Insulation Lab.*, Techn. Univ., Denmark, 1974.
70. K. R. Rao and T. N. Sechadri, *Ind. 7. Meteor Geophys.*, *12*:267 (1961).
71. A. A. Sefir, *Solar Energy*, *26*:497 (1981).
72. J. W. Spencer, *Solar Energy*, *29*:19 (1982).
73. M. Iqbal, *Solar Energy*, *22*:81 (1979).
74. G. Stanhill, *Solar Energy*, *10*:96 (1966).
75. P. Berdahl and R. Fromberg, *Solar Energy*, *29*:299 (1982).
76. A. I. Kudish, D. Wolf, and Y. Machlav, *Solar Energy*, *30*:33 (1983).
77. E. Scerri, *Solar Energy*, *28*:353 (1982).
78. W. Palz (ed.), Atlas über die Sonnenstrahlung in Europa, *Komm. Eur. Gemeinsch.*, Grösschen V1, Dortmund, 1979.
79. G. Haurwitz, *J. Meteorology*, *5*:110 (1948).
80. M. A. Atwater, P. J. Lunde, and G. D. Robinson, *ISES Silver Jubilee Congress*, Atlanta, 1979.
81. B. Choudhury, *Solar Energy*, *29*:479 (1982).
82. C. Castagnoli et al., *Solar Energy*, *28*:289 (1982).
83. D. G. Erbs, S. A. Klein, and J. A. Duffie, *Solar Energy*, *28*:393 (1982).
84. R. Walraven, *Solar Energy*, *20*:393 (1978).
85. W. D. Dickinson, *Solar Energy*, *21*:249 (1978).
86. P. Doratio and M. Jamshidi, *Solar Energy*, *29*:351 (1982).
87. K. J. A. Revfeim, *Solar Energy*, *28*:509 (1982).
88. K. Scharp, *Solar Energy*, *2*:531 (1982).
89. I. M. Gordon and Y. Zarmi, *Solar Energy*, *28*:483 (1982).
90. W. C. Swinbank, *Quarterly J. Royal Meteor. Soc.*, *89*: 1963.

91. R. W. Bliss, *Solar Energy*, 5:103 (1961).

92. J. Yellot and Kokoropsulos, in *Proc. U.N. Conf. on New Sources of Energy*, Paper No. 5/34, Rome, 1961.

93. K. G. Picha and J. Villanueva, *Solar Energy*, 6:151 (1962).

94. M. Centeno, *Solar Energy*, 28:489 (1982).

95. E. M. Sparrow, J. S. Nelson, and W. Q. Tao, *Solar Energy*, 29:33 (1982).

96. L. Imre, in *Num. Meth. in Heat Transfer*, Vol. 1 (R. W. Lewis, K. Morgan, and O. C. Zienkiewicz, eds.), Wiley, Chichester, England, 1981, p. 51.

97. D. C. Zienkiewicz, *Finite Element Analysis in Eng. Science*, McGraw-Hill, New York, 1978.

98. B. J. Huang and J. H. Lu, *Solar Energy*, 28:413 (1982).

99. D. L. Evans, T. T. Rule, and B. Wood, *Solar Energy*, 28:13 (1982).

100. W. R. Petrie and M. McClintock, *Solar Energy*, 21:55 (1978).

101. S. B. Marks, *Solar Energy*, 30:45 (1983).

102. M. Telkes, in *Critical Materials in Energy Production*, Chap. 14, Academic Press, New York, 1978.

103. G. Comini, S. Del Guidice, R. W. Lewis, and O. C. Zienkiewicz, *Int. J. Num. Meth. Eng.*, 8:613 (1974).

104. I. Benkö and L. I. Kiss, in *Energy Conservation in Heating, Cooling and Ventilating Buildings*, Vol. 1, (C. J. Hoogendorn and W. H. Afgan, eds.), Hemisphere, Washington, D.C., 1978, p. 251.

105. I. Tanishita, in *Melbourne Int. Solar En. Soc. Conf.*, 1970.

106. Y. Jaluria and S. K. Gupta, *Solar Energy*, 28:137 (1982).

107. W. F. Phillips and R. N. Dave, *Solar Energy*, 29:111 (1982).

108. M. Sokolov and M. Vaxmann, *Solar Energy*, 30:237 (1983).

109. C. S. Herrick, *Solar Energy*, 28:99 (1982).

110. D. Y. S. Lou, *Solar Energy*, 30:115 (1983).

111. C. Vaccarino and T. Fioravanti, *Solar Energy*, 30:123 (1983).

112. A. Abhat and T. Q. Huy, *Solar Energy*, 30:93 (1983).

113. R. F. Childs, D. L. Mulholland, M. Zeya, and A. K. Goyal, *Solar Energy*, 30:155 (1983).

114. R. Gopal, B. R. Hollebone, C. H. Langford, and R. A. Shigeishi, *Solar Energy*, 28:421 (1982).

115. D. M. Brooks and R. N. Dave, *Solar Energy*, 29:129 (1982).

116. S. C. Klein et al., *Solar Energy*, 17:29 (1975).

117. A. H. Eltimsahy and C. H. Copass, *Mathematics and Computers in Simulation*, XX:114 (1978).

118. S. A. Klein and W. A. Beckmann, *ASHRAE Trans.*, 82:623 (1976).

119. W. A. Beckmann, S. A. Klein, and J. A. Duffie, *Solar Heating Design by f-Chart Method*, John Wiley and Sons, New York, 1977.

120. M. E. McBabe, *ASHRAE Trans.*, 86:420 (1980).

121. S. A. Klein and W. A. Beckmann, *Solar Energy*, 22:269 (1979).

122. H. F. W. DeVries and J. C. Franken, *Solar Energy*, 25:279 (1980).

123. M. K. Selcuk et al., *Int. Solar Energy*, 16:81 (1975).

124. D. J. Close, *Int. Solar En. Congress*, Los Angeles, 1975.

125. L. Imre, I. L. Kiss, T. Környey, and K. Molnár, in *DRYING '80* (A. S. Mujumdar, ed.), Hemisphere, New York, 1980, p. 446.

126. L. Imre and I. Szabó, *Proc. 1st Int. Drying Symposium* (A. S. Mujumdar ed.), Science Press, Princeton, 1978, p. 76.

127. P. Doratio, *Solar Energy*, 30:147 (1983).

128. S. Pabis, *Proc. Agric. Eng. Symposium*, Silsoe, England, 1967.

129. T. L. Thomson, R. M. Peart, and G. H. Foster, *Trans. Am. Soc. Agric. Engs.*, 11:582 (1968).

130. G. W. Ingram, *J. Agric. Eng. Res.*, 21:263 (1976).

131. L. Imre and K. Molnár, *Proc. 3rd Int. Drying Symp.*, Birmingham, England (J. C. Ashworth, ed.), Vol. 2, 1982, p. 73.

132. L. Imre, K. Molnár, and S. Szentgyörgyi, in *DRYING '83* (A. S. Mujumdar, ed.), Hemisphere, Washington, D.C., 1984.
133. JAKRAP SOLARKILN, Technical Inf. Sheet, Fuels Ltd., Silverburn, Penicuik, U.K.
134. H. N. Rosen and P. Y. S. Chen, *AIChE, 76*(200):82 (1980).
135. A. Kangro, *Sveriges Landbruksuniversitet, Specielmeddelande, 111*, Lund, 1981.
136. W. E. Ferguson and P. A. Bailey, *Proc. Int. Solar En. Society Congress, Brighton, England, 1981*, Pergamon Press, Oxford, 1982, p. 1006.
137. L. Andreotti et al., *Solar Energy Int. Progress*, Vol. 3 (T. N. Veziroglu, ed.), Pergamon Press, New York, 1980, p. 1645.
138. P. W. Niles et al., *Solar Energy, 20*:19 (1978).
139. A. M. A. El-Bassononi and A. M. Tayeb, *Proc. 3rd Int. Drying Symp., Birmingham, England*, Vol. 1 (J. C. Ashworth, ed.), Drying. Res. Ltd., Wolverhampton, England, 1982, p. 385.
140. A. P. Soponronnarit, *Proc. 3rd Int. Drying Symp., Birmingham, England*, Vol. 1 (J. C. Ashworth, ed.), Drying Res. Ltd., Wolverhampton, England, 1982, p. 375.
141. J. P. Ratschow and H. G. Claus, *Landtechnik, 1*, (1978).
142. U. Facchini and G. Frosi, *Sunworld, 3*:160 (1979).
143. B. W. Wilson, *Australian J. Agric. Res., 13*:662 (1962).
144. B. Sadikov, A. Vardjasvili, and T. Sadikov, *Geliotechnika, 5*:77 (1976).
145. C. Artikov and T. M. Maksudov, *Geliotechnika, 1*:72 (1978).
146. T. Mackawa, K. Toyoda, and K. Matsumoto, in *DRYING '82* (A. S. Mujumdar, ed.), Hemisphere, Washington, D.C., 1982.
147. A. K. Mahapatra and L. Imre, *Int. J. of Ambient Energy, 4*:205 (1990).
148. C. G. Odunukwe, *Proc. of Biennial Congress of ISES*, Vol. 3. Pergamon Press, Hamburg, FRG, 1987, p. 2515.
149. C. Wereko-Brobby, *Proc. of Workshop on Solar Drying in Africa*, Dakar, Senegal, 1986, p. 147.
150. N. K. Bansal et al., *Sun World, 8*(1):9 (1984).
151. B. Norton and O. V. Ekechukwu, *Solar Thermal and Photovoltaic Conversion*, Workshop, Dhaka, Bangladesh, 1987.
152. A. K. Mahapatra and L. Imre, *Int. J. of Ambient Energy, 10*(3):163 (1989).
153. D. K. McDaniels, *The Sun: Our Future Energy Source*, 2nd ed., John Wiley and Sons, New York, 1984, p. 138.
154. J. C. McVeigh, *Sun Power*, Pergamon Press, 1977, p. 17.
155. J. Nagirju et al., *Proc. of Biennial Congress of ISES*, Vol. 3, Pergamon Press, Hamburg, FRG, 1987, p. 2538.
156. N. Sitthiphong and P. Terdtoon, *Proc. of Biennial Congress of ISES*, Vol. 3, Pergamon Press, Hamburg, FRG, 1987, p. 2528.
157. F. B. Sebbowa, *Proc. of Workshop on Solar Drying in Africa*, Dakar, Senegal, 1986, p. 60.
158. W. O. N. Harvey et al., *Proc. of Biennial Congress of ISES*, Vol. 2, Pergamon Press, Montreal, Canada, 1985, p. 1082.
159. C. J. Minka, *Proc. of Workshop on Solar Drying in Africa*, Dakar, Senegal, 1986, p. 11.
160. M. Dicko, *Proc. of Workshop on Solar Drying in Africa*, Dakar, Senegal, 1986, p. 75.
161. Y. K. L. Yu Wai Man, *Proc. of Workshop on Solar Drying in Africa*, Dakar, Senegal, 1986, p. 92.
162. R. Martinez et al., *Solar and Wind Technology, 4*:223 (1984).
163. E. A. Arinze, *Proc. of Workshop on Solar Drying in Africa*, Dakar, Senegal, 1986, p. 128.
164. M. W. Bassey et al., *Proc. of Workshop on Solar Drying in Africa*, Dakar, Senegal, 1986, p. 207.
165. K. Amonzon et al., *Proc. of Workshop on Solar Drying in Africa*, Dakar, Senegal, 1986, p. 252.
166. I. Segal and M. Reuss, *Proc. of Biennial Congress of ISES*, Vol. 3, Pergamon Press, Hamburg, FRG, 1987, p. 2368.

167. V. Muthuveerappan et al., *Proc. of Biennial Congress of ISES*, Vol. 2, Pergamon Press, Montreal, Canada, 1985, p. 1077.
168. V. Asiedu-Bondzie and A. Ayensu, *Proc. of ANSTII Symposium on Renewable Energy for Development*, Kumasi, Ghana, Vol. 2, 1986, p. 160.
169. S. Janjai, V. Guevezov, and M. Daguenet, *Drying Technology Int. Journal*, 4(4):605 (1986).
170. G. S. Rauzeous and G. D. Saravacos, *Drying Technology Int. J.*, 4(4):633 (1986).
171. J. Muller et al., *Solar and Wind Technology*, 5:523 (1989).
172. M. A. Sattar, *Proc. of 1st World Renewable Energy Congress*, Reading, U.K., Vol. 2 (A. A. M. Sayigh, ed.), Pergamon Press, Oxford, 1990, p. 59.
173. H. P. Garg, *Proc. of 1st World Renewable Energy Congress*, Reading, U.K., Vol. 2 (A. A. M. Sayigh, ed.), Pergamon Press, Oxford, 1990, p. 618.
174. Futal Huang et al., *Proc. of 1st World Renewable Energy Congress*, Reading, U.K., Vol. 2 (A. A. M. Sayigh, ed.), Pergamon Press, Oxford, 1990, p. 633.
175. L. Imre, *Proc. of X. Congress on Energy*, Opatija, Yugoslavija, 1988, p. 23.
176. L. Imre, *Proc. of ASRE 86 Symposium*, Vol. 2, Cairo, Egypt, 1986, p. 1105.
177. L. Imre, *Direction and Control of Solar Dryers*, in *Manual of Industrial Solar Dryers*, UNESCO Report (M. Daguenet, ed.), Chapter 2, 1988.
178. L. Imre, *Proc. 4th Int. Drying Symposium*, Kyoto, Japan, Vol. 1 (R. Toei and A. S. Mujumdar, eds.), 1984, p. 43–50.
179. P. Harriott, *Process Control*, McGraw-Hill, New York, 1964.
180. M. Gopal, *Modern Control Systems Theory*, John Wiley and Sons, Chichester, 1984.
181. K. Ogata, *Modern Control Engineering*, Prentice-Hall, Englewood Cliffs, N.J., 1970.
182. J. L. Tsernitz and W. T. Simpson, *Drying Technology Int. J.*, 4(4):651 (1986).
183. L. Imre, I. Farkas, and L. Gémes, in *DRYING '86* (A. S. Mujumdar, ed.), Vol. 2, 1986, p. 678.
184. K. Lutz and W. Mühlbauer, *Drying Technology Int. J.*, 4(4):583 (1986).
185. L. Imre et al., *Drying Technology Int. J.*, 8(2):343 (1990).
186. M. Hasnaoui, G. Le Palec, and M. Daguenet, *Revue Général de Thermique*, XXIII:(265):7 (1984).
187. M. Daguenet, *Solar Dryers. Theory and Practice* (in French), UNESCO, Paris, 1985.
188. M. Daguenet, Methodology for Calculation of Solar-Assisted Convective Dryers, in *Manual of Industrial Solar Dryers*, UNESCO Report, (M. Daguenet, ed.), Chapter 1, 1988.
189. B. Norton and P. A. Hobson, Thermal Analysis of Thermosyphon Solar Energy Crop Dryers, in *Manual of Industrial Solar Dryers*, UNESCO Report, (M. Daguenet, ed.), 1988.
190. J. R. Puiggali and A. Tiguert, *Drying Technology Int. J*, 4(4):555 (1986).
191. A. Tiguert and J. R. Puiggali, A Modelling Approach to the Performance of a Country Solar Dryer, in *Manual of Industrial Solar Dryers*, UNESCO Report, (M. Daguenet, ed.), 1988.
192. A. Steinfeld and I. Segal, *Drying Technology Int. J.*, 4(4):535 (1986).
193. D. Stehli and F. Escher, *Drying Technology Int. J*, 8(2):241 (1990).
194. B. G. Patil and G. T. Ward, *Solar Energy*. 43(5):305 (1989).
195. D. A. Weitz, E. A. Lugue, and R. D. Piacentini, *Drying Technology Int. J.*, 8(2):287 (1990).
196. L. Imre, *Drying Technology Int. J.*, 4(4):503 (1986).
197. L. Imre, *Proc. of ASRE 89 Symposium*, Cairo, Egypt, Vol. 2. 1989, p. 1029.
198. M. Tsamparlis, *Drying Technology Int. J.*, 8(2):261 (1990).

13
Spouted Bed Drying

Elizabeth Pallai and Tibor Szentmarjay
Hungarian Academy of Sciences
Veszprém, Hungary

Arun S. Mujumdar
McGill University
Montreal, Quebec, Canada

1. INTRODUCTION

The applicability of the spouted bed technique (1–5) to drying of granular products that are too coarse to be readily fluidized (e.g., grains) was recognized in the early 1950s. Interest in this area received appreciable impetus two decades later as the energy-intensive drying processes were reexamined with renewed vigor. Spouted bed dryers (SBDs) display numerous advantages and some limitations over competing conventional dryers. Because of the short dwell time in the spout, SBDs can be used to dry heat-sensitive solids, such as foods, pharmaceuticals, and plastics. With simple modification the so-called modified spouted beds can be designed to ensure good mixing, controlled residence time, minimum attrition, and other desirable features. Also, the operations of coating, granulation agglomeration, and cooling, among others, can be carried out by the same apparatus by varying the operating parameters. SBDs can be used for solids with constant as well as falling rate drying periods. Using inert solids as the bed material, SBDs have been used successfully to dry pastes and slurries.

This chapter is devoted mainly to the generally less accessible results on spouted bed drying obtained at the Research Institute of Chemical Engineering of the Hungarian Academy of Sciences (MTA MÜKKI). Other results are readily available in literature cited.

2. EXPERIMENTAL DEVICES AND PROCEDURES

The classic or conventional spouted bed (CSB) is a cylindrical vessel with a conical bottom fitted with an inlet nozzle for introduction of the spouting air (drying medium). This device suffers from limited capacity due to the maximum spoutable bed height and inability to scale up the apparatus beyond 1 m diameter. Introduction of a hollow, tall vertical tube (draft tube) some distance above the nozzle eliminates the former restriction by acting as a pneumatic conveyor.

Figure 1 displays a draft-tube SBD, which is a tremendous improvement over the

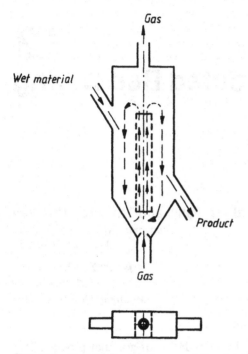

Figure 1   Longitudinal and cross-sectional schematics of the spouted bed predryer.

classical SBD (6,7). The tube may be impermeable, porous, or partly porous. It may be cylindrical or slightly tapered. The bed height can be increased severalfold by inserting a suitable draft tube with solid walls. The solids in the annulus flow in a plugflow fashion, guaranteeing a uniform residence time distribution (RTD) in the SBD. Experiments in two-dimensional beds showed that the RTD is more uniform in two-dimensional rather than circular cross-sectional SBDs. Tables 1 and 2 show the geometries of the two SBDs studied.

For the laboratory-scale predryers, the drying capacity for corn is about 80 kg/h of product, with a 10% decrease in moisture content. In the course of this calculation it was presumed that the particles on the average made two cycles within the bed.

The second part of the two-stage SBD is a conical-cylindrical device of traditional shape (Table 2). The drying capacity of this device is about 80 kg/h of dried corn with 150°C air at a flow rate of about 200 m³/h, which results in a moisture removal of about 5% in the falling rate period.

Efforts have been made recently to find some means of introduction of air, which will avoid the high pressure loss caused by the central nozzle. The problem was solved by

Table 1   Dimensions of the Angular Spouted Bed Predryer

Longer dimension (mm)	Shorter dimension (mm)	Height (mm)	Nozzle diameter (mm)	Cross-section of the annulus (m²)	Mean sliding velocity (m/s)
350	40	1000	20–40	0.0124	0.01

Table 2 Dimensions of the Cylindrical Spouted Bed Afterdryer

Diameter (mm)	Height (mm)	Nozzle diameter (mm)
170	1500	17–30

tangential air feeding with the use of horizontal slits (see Fig. 2) (8) and by so-called swirling rings (see Fig. 3) (9). The first type of equipment is referred to as a *vortex bed dryer*; the latter device ensures the maintenance of uniform particle circulation in the spouted bed, favoring the development of the spout channel. Figure 3 shows this device, a novel type of spouted bed dryer equipped with an inner cylinder, a draft tube, and an attachment permeable to air. This type of dryer can operate with flow rates three to five times higher than those used in conventional SBDs and the start-up of the dryer also becomes simpler.

To select the optimal geometry of the draft tube it was necessary to carry out experiments in semicircular spouted beds for visualization of the flow patterns. It was then also possible to measure the particle velocities in the annulus as well as the spout by introduction of marked (tagged) particles. High-speed photography was employed for this purpose (10).

Since the velocity of particle recirculation depends mainly upon the velocity of the slow sliding in the annulus, the mean residence time of the particles is about $\bar{\tau} = H/w_a$. In order to investigate the possibility of controlling the circulation of particles, various sets of measurements were carried out with glass beads, with ground-activated carbon, and with plastics as model particles. It was found that the sliding velocity of the particles varies according to the following empirical correlations. It is affected by the velocity of the entering air, the bed height, and the nozzle diameter (10).

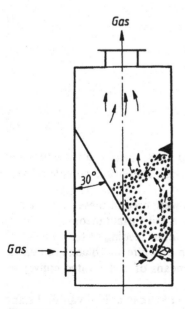

Figure 2 Dryer with a slit for gas introduction.

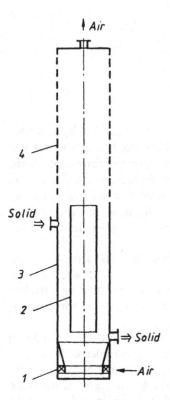

Figure 3 Spouted bed dryer with a draft tube and an air-permeable attachment: 1, tangential air feed; 2, draft tube; 3, dryer body; 4, air-permeable attachment.

$$w_a' = w_a'' \left(\frac{v'}{v''} \right)^{3/2} \tag{1}$$

$$w_{aH} = w_{aH_{max}} \left(\frac{H}{H_{max}} \right)^{1/3} \tag{2}$$

and

$$w_a' = w_a'' \left(\frac{D_i''}{D_i'} \right)^{2/3} \tag{3}$$

However, the operation of the dryer depends not only upon the mean residence time but also upon the residence time distribution of the particles. Therefore, complementary measurements were carried out to study this variable.

Polyvinyl chloride (PVC) granules of various colors were used in these experiments. To a white particle layer some black PVC particles were fed instantaneously onto the upper level of the annulus at the right side. During continuous feeding of the white PVC granules, the composition of the mixture (the proportion of white and black granulates) leaving the device could be continuously measured by means of a calibrated conveying belt and a movie camera.

The proportion of the particles of different color was recorded at intervals of 3 s and 5 s. For example, the distribution of residence times is presented in Figure 4, in which the

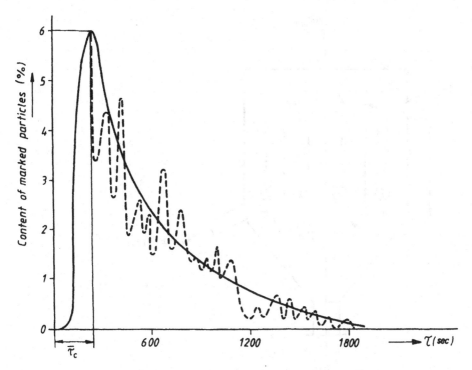

Figure 4 A typical distribution of the residence times of particles; $\bar{\tau}_c = 2\tau_1 + \tau_2$.

broken line denotes measured values and the solid line corresponds to the mean values of the distribution. The local peak values recurring periodically in the density curve may be explained by the fact that the marked particles are entering the outlet in different periods of the recirculating movement of particles, and by the fact that they are occasionally swallowed up from the annulus into the spout channel and thus fractions of the full recirculation period may occur.

On the basis of our experiments it appears that correlations that assume perfect mixing, as found in the literature (11,12), can be applied for the calculation of the distribution of residence times only under conditions that ensure intense circulation of particles in the bed. Even in these cases, the agreement is better at short dimensionless times ($\Theta = \tau/\bar{\tau}$) corresponding to the initial section of the curve. According to the above, some modification of the physical model of the distribution of residence times appears desirable. Better agreement between the measured and calculated data was obtained with the plug-flow model, simultaneously taking into account the internal recirculation (13,14).

Using the symbols of Figure 5, the RTD function can be expressed by

$$\rho(\tau) = \left(\frac{1}{\tau_{\text{in}}}\right)(1 - \xi_a - \xi_b + \xi_a\xi_b\sum_{n=0}^{\infty} (\xi_a + \xi_b - \xi_a\xi_b)^n\Theta[\tau - 2\tau_1 - \tau_2$$
$$- n(\tau_1 + \tau_2)] - \Theta[\tau - \tau_{\text{in}} - 2\tau_1 - 2\tau_1 - \tau_2 - n(\tau_1 + \tau_2)] \tag{4}$$

Here,

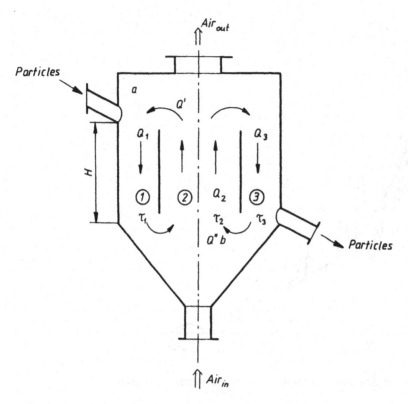

Figure 5 Proposed model of particle circulation.

$$\tau_1 = \frac{S_a H}{2Q_1} \qquad \tau_2 = \frac{S_s H}{Q_2} \qquad \tau_3 = \frac{S_a H}{2Q_3}$$

and the recirculation fractions are $\xi_a = Q'/Q_2$ and $\xi_b = Q''/Q_3$.

From Eq. [4] it can be estimated that these impulses appear at attenuated amplitudes with periods of $\tau_1 + \tau_2$. Because of the irregular pattern of recirculation of particles, the peak value does not appear as a sharp peak (6).

A spouted bed dryer with tangential air inlet equipped with an inner conveyor screw has also been developed (15) (see Fig. 6). According to the developed and patented solution, air is injected into the bed through specially designed "whirling" rings. Along the vertical axis of the device is a houseless open conveyor screw capable of ensuring, independently of the airflow rate, the typical spouted circulating motion even with materials of small particle size ($D_p/d_p > 500$) for which conventional spouted beds are not suited. The diameter of the screw is nearly equal to the diameter of the gas channel (spout) around which a similar dense sliding layer (annulus) is formed and a circulation motion (similar to convectional beds) can be visible.

Pressure drop is lower by 25–30% in comparison with air injection through a nozzle of a conventional spouted bed. Another advantage is that gas velocity can be regulated in a wide range due to the possibility of choice of proper size and number of slots. Due to mechanical particle circulation fan energy consumption can be reduced by 15–20% (16). Moreover, using the inner screw, air volume rate can be chosen only from the point of view of drying requirements, resulting in optimum drying conditions. This mechanically

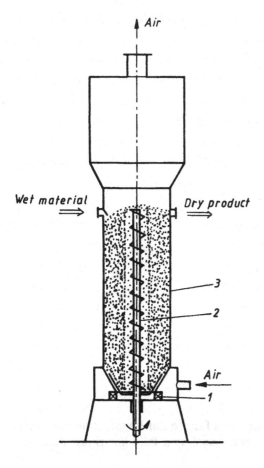

Figure 6 Spouted bed dryer with a tangential air inlet and a central conveyor screw: 1, tangential air inlet; 2, conveyor screw; 3, dryer body.

spouted bed (MSB) construction offers further advantages in solving scaleup problems. Bed volume and diameter to height ratio can be selected quite freely, materials of wide particle size ranges can be circulated and particle circulation time and rate can be controlled within wide limits (17).

Scaleup data of conventional spouted bed dryers are well known from the literature, but the relations cannot be applied to MSB dryers with inert packings due to different hydrodynamical and drying characteristics. Therefore the drying mechanism itself, effects of various process and operational parameters, and the relevant relationship had to be investigated in more detail. Important dimensions as well as geometric, physical, and hydrodynamical characteristics of equipment and of the inert particles used are summarized in Table 3. Measurements of the pressure drop as a function of airflow rate across the static bed as well as the mechanically spouted bed have been performed on both the laboratory- (Fig. 7) and the industrial-scale equipment (18).

The speed of rotation of the screw of the laboratory-scale equipment was set to 600 rpm. Taking the nearly identical peripheral speed, this corresponds to 195 rpm on the pilot-plant dryer. Relevant parameters are summarized in Table 4.

Table 3 Characteristics of Laboratory and Pilot-Scale Mechanically Spouted Bed Dryers

	Laboratory	Pilot
D_c (m)	0.138	0.380
Number of slots	2	6
h_i (m)	2×10^{-3}	4×10^{-3}
A_i (m^2)	1.28×10^{-3}	2.1×10^{-3}
A_c/A_i	11.7	5.5
D_{sc} (m)	0.04	0.12
d_{sc} (m)	0.016	0.05
$s(m)$	0.028	0.09
s/D_{sc}	0.70	0.72
D_{sc}/D_c	0.29	0.31
	Ceramic spheres	
Inert particles		
d_p (m)	6.6×10^{-3}	7.4×10^{-3}
ρ_p (kg/m^3)	3640	3520
ϵ	0.36	0.36
a (m^2/kg)	0.30	0.23
Re_{mf}	1100	1285
u_{mf} (m/s)	2.6	2.7

Pressure drop across the beds was measured as a function of Reynolds number, and the results are shown in Figure 8. The curve represented by a solid line is calculated by Ergun's equation:

$$\Delta p_E = (H/d_p)(1 - \epsilon)/\epsilon^3 \, \rho_f v''^2 \, [150 \, (1 - \epsilon)/Re_p + 1.75] \qquad [5]$$

In order to show the range investigated, the dotted line represents the pressure drops Δp_{mf} corresponding to minimum fluidization velocities, calculated as follows:

$$\Delta p_{mf} = (1 - \epsilon)Hg(\rho_p - \rho_f) \qquad [6]$$

It can be concluded from Figure 8 that pressure drop on the spouted bed is always smaller than on the static bed, probably due to the bed-loosening and particle transporting effect of the conveyor screw. Curves calculated from Eq. [5] at Re < 600 fit well-measured values.

3. DRYING RESULTS

The inventors of the original patent concerning the spouted bed method and the construction of the apparatus (1) proposed primarily to solve the problem of the drying of cereals, such as wheat and corn, in a continuous and efficient way. However, the demand for an up-to-date drying apparatus that emerged in the course of the manufacture of products in various industries resulted in widening the field of application of the spouted bed dryer. Operational experience and frequent emergence of novel problems of drying initiated an activity to develop newer devices. As a result of this activity, various modified spouted bed dryers proved to be suitable for energy-efficient drying in the agricultural and food industry, the chemical industry, and many other industries, for drying of powdered

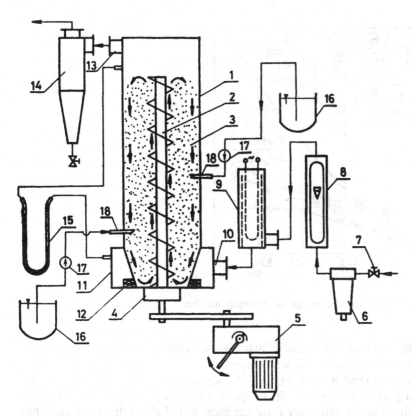

Figure 7 Schematic of the laboratory-scale mechanically spouted bed (MSB) dryer: 1, column; 2, screw; 3, inert particles; 4, bearing; 5, drive; 6, air filter; 7, valve; 8, rotameter; 9, heater; 10, air inlet; 11, air distributor chamber; 12, air slots; 13, air outlet; 14, cyclon; 15, manometer; 16, suspension tank; 17, pump; 18, feeding pipe.

Table 4 Data for Hydrodynamical Experiments

		Lab 1	Lab 2	Pilot
H	(m)	0.34	0.2	0.45
H/D_c	–	2.5	1.4	1.2
m_p	(kg)	9.2	5.2	55
A_p	(m^2)	2.8	1.6	12.7
n_{sc}	(rpm)	600	600	195
v_{ph}	(m/s)	1.25	1.25	1.22
q	(kg/s)	0.48	0.48	4.30
τ_p	(s)	10.8	19.2	12.8
v_a	(m/s)	1.5×10^{-2}	1.5×10^{-2}	1.9×10^{-2}
V''	(Nm^3/h)	20–140	20–140	360–1000
v''	(m/s)	0.4–2.6	0.4–2.6	0.9–2.4
v_s	(m/s)	4.3–30.3	4.3–30.3	4.8–13.0
Re$_p$	–	160–1100	160–1100	420–1140

It should be noted that hydrodynamical comparability at $H = 0.2$ m is given by nearly identical values of geometric ratios H/D_c of beds, of sliding velocities v_a of particles, of superficial air velocities v, and of Reynolds numbers.

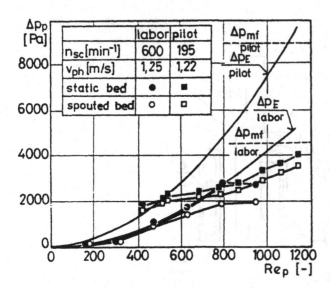

Figure 8   Pressure drop across the bed against Reynolds number.

materials ($d_p < 100$ μm) and of centrifuged materials with high moisture content, as well as coagulating and adhesive pulps, and suspensions.

   In the following we shall discuss typical SBD applications. We shall also describe briefly the operating conditions and results of spouted bed (SB) drying in laboratory- or industrial-scale units. These results may form the basis of industrial dryer designs when coupled with appropriate engineering judgment.

3.1.   Drying of Agricultural Products

The optimum parameters for drying of granular agricultural products are determined not only by their chemical composition and physical properties but also by their potential use. In the case of corn and oats, for example, as cattle feed, when the valuable nutrients (including proteins and vitamins) must be saved, the temperature of drying air may be higher (the seed temperature can be as high as 60–75°C) than is the case for seeds for sowing. In this latter case, in order to conserve the ability of germination, the temperature must not exceed 40–45°C. However, the actual level depends on the seed type and on the residence time of the particle. In the case of some seeds a maximum seed temperature of 30–35°C is required. The quality of the dried product is determined, furthermore, by the rate of drying, the temperature, and the flow rate of the drying agent, mainly in the initial period of the drying.

   Cereal seeds and, generally, also most other seed varieties are colloidal materials with capillary pores. This means that the walls of their capillaries are elastic and change their shape on absorbing moisture. They are composed of a skin, an endocarp, a so-called aleuron layer between these two parts, and the embryo. The hygroscopic properties of these parts are different. At equilibrium the embryo has the highest moisture content, then the aleuron cells, and last the fiber cells. Accordingly, a greater part of moisture is present in the external layer of the seeds, from which it can be removed relatively easily. On drying the seed is crumbled and becomes hard. At a drying temperature below 50°C, the drying rate of the seed decreases at first abruptly, then much more slowly. In the temperature range between 46°C and 50°C, the aleuron cells form an almost closed, elastic skin in which the diffusion of moisture is very slow. At a drying temperature above 50°C, the structures of various parts of the seed change. The thermally labile cell content is hardened, the cell walls are split, and the moisture moves quickly through these splits toward the surface of the seed.

In the case of seeds for sowing, that is, when the temperature of seeds must definitely be maintained below 45°C, the drying rate is lower. Consequently, it is practical to increase, instead of the temperature, the velocity of the drying agent, obviously only to the optimal limit. At this limiting value of air velocity the moisture migrating from the interior of the seed to the seed surface is removed immediately. Beyond this limit it is not reasonable to increase air velocity. The drying curves of corn and wheat are shown in Figures 9 and 10, respectively, at various drying temperatures; the critical moisture contents are also indicated.

The drying rate is affected by the manner of drying and the flow rate and temperature of the drying agent. Two grass varieties and two herbs (scarlet clover and coleseed) were dried in a flow-through oven and in a laboratory-size spouted bed dryer with a central screw conveyor. Significant deviations are observed in the times required for attaining the final moisture content of 14% prescribed for storage (see Table 5).

For determination of the drying rates, airflow rates used in the spouted bed dryer experiments varied from 0.6 to 1.0 m/s for the various seeds. The typical drying rate curves also present some information concerning the structure of the material. The curve labeled "grass seed I" in Figure 11 is used as an illustration.

It can be seen in Figure 11 that, for oven drying, constant rate drying extends from a moisture content of 32.0% to 26.5% (by weight), characterizing the removal of moisture from the surface layers. The section of falling drying rates points to a complex structure

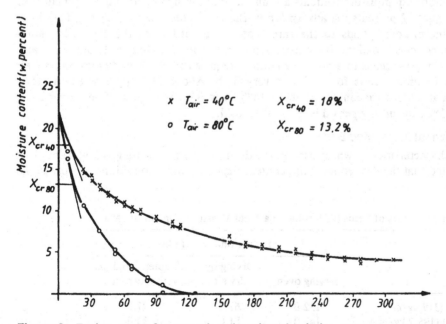

Figure 9 Drying curves for corn plotted against the drying temperature measured in a flow-through drying oven.

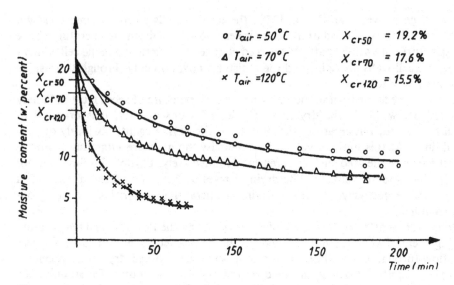

Figure 10 Drying curves of wheat plotted against the drying temperature measured in a flow-through drying oven.

of the internal pore. However, it is quite striking that in dryings carried out in the spouted bed the shape of the rate curve is different. The protracted shape of the constant drying rate (i.e., from 32% to 13.5% by weight) allows the conclusion that in the inside of the seed there are open pores—spaces between the fibers—from which the moisture can be removed more quickly by increasing the flow rate of the drying agent. The critical moisture contents of the seed varieties listed in Table 5 are given in Table 6.

The equilibrium moisture contents are important in design of drying equipment. For example, Figure 12 presents the adsorption isotherm of grass seed I at 20°C. The shape of this isotherm corresponds to the rare type III, according to the BET (Brunauer-Emmett-Teller) classification. It appears from this isotherm that, although it is also possible to dry grass seeds to a moisture content required for storage by atmospheric air of 70–80% humidity, time for drying is very high. About 24 h are needed to attain equilibrium at a moisture content of about 14% by weight. In a spouted bed dryer using hot air at 45°C, the drying period is reduced 10-fold.

Drying of Wheat in Spouted Bed

The main characteristics of wheat used in the drying experiments are given in Table 7. It must be noted that the desired moisture content depends on the conditions of storage. At

Table 5 Drying Times of Seeds for Sowing to a Final Moisture Content of 14%

Seed Variety	Drying time, $t_{14\%}$ (min)		
	Drying oven	Drying by dry air	Spouted bed dryer with inner screw
Grass seed I (319 kg/m³)	132.0	61.2	27.0
Grass seed II (195.2 kg/m³)	147.0	81.0	39.0
Scarlet cover (604.4 kg/m³)	132.0	55.5	33.0

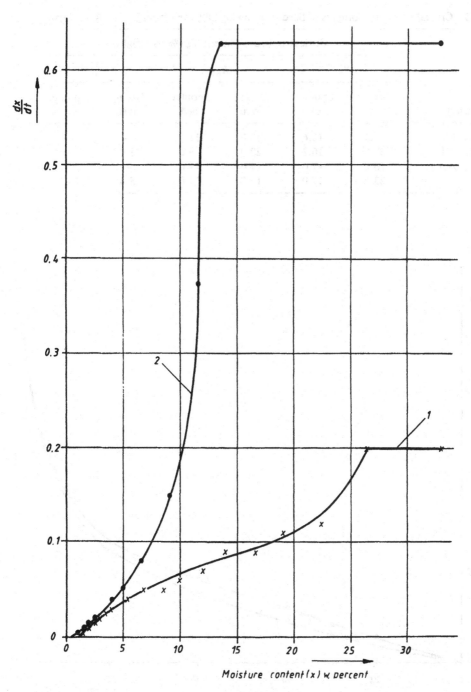

Figure 11 Flow rate curve of the drying of grass seed I on the basis of drying experiments carried out in a drying oven (curve 1) and in a spouted bed dryer (curve 2).

Table 6　Critical Moisture Contents of Seeds for Sowing Obtained from Drying Rate Curves

	Critical moisture content X_{cr} (% by weight)					
	X_{cr1}		X_{cr2}		X_{cr3}	
Seed variety	Drying oven	Spouted bed	Drying oven	Spouted bed	Drying oven	Spouted bed
Grass seed I	26.5	13.6	16.0	11.5	—	—
Grass seed II	27.0	16.2	20.0	9.6	8.0	—
Scarlet clover	26.0	17.0	14.0	11.0	—	—
Coleseed	23.5	17.0	17.0	11.0	5.4	—

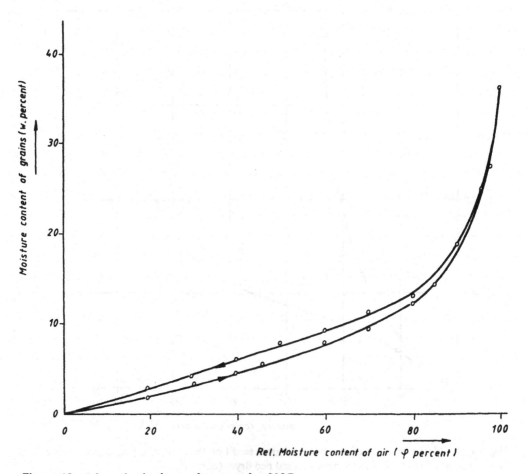

Figure 12　Adsorption isotherm of grass seed at 20°C.

Table 7 Main Characteristics of Wheat

Mean particle size $(\bar{d}_p)$	6.5 mm (longer diameter), 3.5 mm (shorter diameter)
Density	798 kg/m³
Initial moisture content	22–25.5% m/m
Moisture content of the product	11.5–19% m/m
Minimum fluidizing gas velocity	1.22 m/s
Operating gas velocity	3.19 m/s
Terminal gas velocity	10.2 m/s

a moisture content of 14% by weight, wheat can be stored for a long time in bulk in large quantities, at 15% in bags for about 1 year, and at 16–18% in bags for some weeks, whereas at 19% by weight at most for only a few days.

For drying wheat to be used as seed for sowing, for example, removal of the superficial moisture can be attained by means of air at 50–70°C (see the values of critical moisture content given in Figure 10); by parameters to achieve a particle residence time of about 25 min, the moisture content is reduced to 19% and 17.5% m/m with air at 50°C and 70°C, respectively. Such highly efficient predrying may be needed to prevent damage of seeds prior to shipping to warehouses or processing plants. However, wheat to be stored over long periods in bulk must be at the prescribed moisture content of 15% m/m.

It is practical to dry wheat for processing in mills in one step in a spouted bed dryer by air at 100–140°C without any damage to its composition. According to Figure 10, the value of x_{crit} is 15.5% by weight, which exceeds only marginally the desired final moisture content of 14% by weight. Thus, much of the drying may be carried out essentially at a constant drying rate.

The wheat drying studies were carried out in a spouted bed predryer of rectangular cross-section equipped with a draft tube, and also in a cylindrical afterdryer (Tables 1 and 2). The experimental conditions and the data of the batch and continuous drying experiments are presented in Tables 8–11.

It appears from the data of Table 8 that wheat can be dried without any damage to the embryo to 17.0% m/m of moisture content in a spouted bed within 20 min by air at 100–102°C. In the afterdryer, the moisture content of wheat was decreased from 25.5% m/m to 11.7% m/m (see Table 9) at a mean temperature of 105°C of the drying air within 35 min without any damage to the embryo.

Table 8 Drying Experiments for Wheat in a Spouted Bed Predryer

Time (min)	Moisture content of wheat (% m/m)	Air temperature		Embryo number (%)
		T_{in} (°C)	T_{out} (°C)	
0	25.5	102	98	93
5	23.0	101	91	94
9	22.5	101	75	92
14	21.4	102	76	94
18	19.1	100	77	94
20	17.0	102	78	94

Table 9 Drying Experiments for Wheat in a Cylindrical Spouted Bed Afterdryer[a]

Time (min)	Moisture content of wheat (% m/m)	Air temperature		Grain temperature (°C)	Quality[b] (%)
		T_{in} (°C)	T_{out} (°C)		
0	25.5	105	99.5	20.4	93
15	20.7	105	73.0	40.1	94
25	15.9	103	80.0	58.5	96
30	14.0	104	81.0	70.2	94
35	11.7	105	88.0	72.8	96

[a]Weight of charge, 8 kg of moist wheat; volume of drying air, 120 m^3/h (at 20°C)
[b]Percentage of viable seeds

Table 10 Continuous Wheat Drying in a Spouted Bed Predryer

Experimental conditions	
Initial charge	15 kg wheat of a moisture content of 17% m/m
Feed rate	15 kg/h wheat of a moisture content of 25% m/m
Mean residence time of grains	20 min
Air volume rate	50 m^3/h at 20°C
Temperature of drying air	105°C
Drying results	
Mean temperature of air leaving the dryer	68°C
Mean moisture content of the dried product	16.9% m/m
Amount of removed moisture	1.53 kg/h
Water evaporating capacity referred to the dryer cross-section	110 kg water/m^{2}h
Specific energy consumption	3360 kJ/kg water

Table 11 Continuous Drying of Wheat in a Spouted Bed Afterdryer

Experimental conditions	
Initial charge	7 kg wheat of a moisture content of 14% m/m
Feed rate	40 kg/h wheat of a moisture content of 17% m/m
Mean residence time of grains	10 min
Air volume rate	120 m^3/h at 20°C
Temperature of drying air	100°C
Drying results	
Mean temperature of air leaving the dryer	72.5°C
Mean moisture content of the dried product	13.8% m/m
Amount of removed moisture	1.48 kg/h
Water evaporating capacity referred to the dryer cross-section	74 kg water/m^{2}h
Specific energy consumption	7683 kJ/kg water

Table 12 Main Characteristics of Shelled Corn

Mean grain size ($\bar{d}_p$)	10.5 (longer distance), 7.5 mm (shorter diameter)
Bulk density	677.6 kg/m^3
Initial moisture content	22–30% m/m
Desired final moisture content	14% m/m
Minimum fluidizing gas velocity	1.58 m/s
Terminal gas velocity	14.6 m/s

Drying of Corn in a Spouted Bed

A number of prototype spouted bed dryers have been used to dry shelled corn. One pilot dryer has a designed capacity of 100-kg water per hour, based on 10% moisture removal. This dryer is suitable for drying at two levels and for carrying out cooling after the drying. Accordingly, the equipment consists of a unit of smaller capacity (300 × 400 mm) and is equipped with a gas inlet nozzle and a draft tube, and of a spouted bed dryer of a cross-section of 0.32 m^2 (800 × 400 mm) developed with two gas inlet nozzles (19). The main characteristics of the shelled corn investigated in these drying experiments are presented in Table 12, whereas the results of the laboratory and pilot-plant experiments are listed in Tables 13–15.

The bed height was 1200 mm in order to utilize more completely the drying medium. The total pressure drop across the dryer was 1260 mm of water column (12.4 kPa); that is, drying that may be considered very good from the aspect of drying efficiency and of specific energy consumption represented in fact a significant energy requirement for aeration. It must be noted further that the products of the drying experiments 1 and 2 were also satisfactory from the aspect of the capability of germination; the bed temperature did not exceed 80°C even at the highest temperature used.

The aims of the corn-drying experiments carried out in a large laboratory-size spouted bed dryer (see Fig. 6) were to decrease the pumping power requirement and to study the effects of various modifications on SBD performance. Table 15 summarizes the relevant results. This novel SBD has improved drying efficiency and favorable specific energy

Table 13 Drying of Shelled Corn in a Continuous Operation in a Laboratory-Scale Spouted Bed

Experimental conditions	
Charge	5 kg of shelled corn of a moisture content of 14% m/m
Feed rate	18 kg/h of corn of a moisture content of 23.7% m/m
Mean residence time of grains	16.6 min
Air volume rate	70 m^3/h at 20°C
Temperature of drying air	90°C
Drying results	
Mean temperature of air leaving the dryer	42°C
Mean moisture content of the dried product	14.2% m/m
Amount of removed moisture	2.0 kg/h
Evaporational rate referred to the cross-section of the dryer	143 kg water/m^{2}h
Specific energy consumption	3290 kJ/kg water

Table 14 Drying of Shelled Corn in a Continuous Operation in a Pilot Plant-Scale Spouted Bed Dryer of 0.16 m² Cross Section[a]

Drying results	Experiments			
	1	2	3	4
Temperature of the inlet air (°C)	150.0	120.5	178.5	190.0
Temperature of the outlet air (°C)	68.0	46.6	67.5	76.3
Mean residence time of grain (min)	20.0	14.6	11.1	10.0
Mean moisture of product (m/m)	14.8	17.2	16.8	16.0
Amount of removed moisture (kg/h)	65.6	25.4	42.3	48.6
Capacity referred to the cross-section of the dryer (kg water/m²h)	396.0	132.3	220.3	252.1
Specific energy consumption (kJ/kg water)	2884	3868	3811	3485

[a]Experimental conditions: Charge, 120 kg shelled corn of a moisture content of 14% m/m; feed rate, 360–700 kg/h corn of a moisture content of 29% m/m; mean residence time of grains, 10–20 min; amount of drying air, 1100 m³/h at 20°C; temperature of drying air, 120.5–190°C.

Table 15 Drying of Shelled Corn in a Continuous Spouted Bed Dryer Equipped with a Draft Tube and an Inner Conveying Screw

Experimental conditions (diameter of the dryer, 138 mm)	
Drying air flow through a swirling ring	
Feeding of the wet material through the feeding tube moving circularly above the bed surface, distributing the material uniformly on the surface of the sliding bed (annulus)	
Draft tube	
Diameter	70 mm
Length	1000 mm
Conveyor screw	
Diameter	32 mm
Length	1200 mm
Speed of rotation of the screw	700 rpm
Weight of bed	13 kg of shelled corn with a moisture content of 14%
Mean residence time of particles	20 min
Air volume rate	120 m³/h at 20°C
Temperature of drying air	160°C
Drying results	
Temperature of the outlet air	67.5°C
Mean moisture content of product	13.8% m/m
Amount of removed moisture	5.64 kg/h
Evaporational rate referred to the cross-sectional area of the dryer	376 kg water/m²/h
Specific energy consumption	3535 kJ/kg water

consumption; the bed pressure loss is reduced significantly from the earlier pressure drop of 1000–1400 mm of water column (9.8–13.7 kPa) to as low as 300–400 mm of water column (2.9–3.9 kPa).

3.2. Drying of Other Food Products

In this section the solution of some special drying tasks will be presented, such as the drying of preground red paprika, of various cut, chopped vegetables containing significant moisture (about 50% by weight) (e.g., cubed carrots), and of various pulpy materials (e.g., potato pulp).

Drying of Washed and Centrifuged Tomato Seeds

Drying of agglomerated granular materials of high (45–50% m/m) moisture content causes difficulties. A further problem emerges in the treatment of the washed and centrifuged seeds with a moisture content of about 50% m/m when a large amount of water must be removed at low temperatures in order to avoid damage at higher temperatures.

The drying of washed and centrifuged tomato seeds for sowing presents a problem of this type. On the basis of drying experiments performed in the laboratory-scale dryer, an industrial SBD (with a capacity of 25 kg water/h) has been designed, manufactured, and put into operation (20).

Tomato seeds for sowing are one of the end products of the tomato processing technology, containing significant amounts of fibrous pulp and skin fragments. These are removed by washing combined with repeated sedimentation. The seed for sowing obtained in this way is adjusted by centrifugation to a moisture content of 45–50% m/m. Tomato seeds for sowing treated in this way are still very moist to the touch and in lump form. Uniform drying of this material can be carried out only by ensuring a mobile state of the particles. A further task is abrasion of the dry seeds, that is, the removal of the undesirable residue on the seed surface to permit the use of a mechanical technique of sowing single seeds. This operation can be carried out practically in combination with the drying process in an SBD.

For this purpose, an SBD with a draft tube was chosen (enabling the application of large amounts of air), together with swirling-ring air entry (tangential air feeding). In order to prevent external damage to seeds, the removal of residue was carried out simultaneously with the drying procedure. The velocity of particles was increased to about three times the usual level. The particle velocity in the annulus was 0.05–0.07 m/s. The upper part of the equipment was made from linen permeable to air (see Fig. 13).

Table 16 presents the calculated fundamental parameters of design based on laboratory measurements and also the main dimensions of a dryer of industrial scale. The operation of this type of SBD is as follows. Air for the dryer is supplied by a fan, heated by an air heater, and then led to the swirling ring along the periphery and fed tangentially. The cylindrical draft tube stretches about 20–30 mm to the linen part, which retains the dried residue. Figure 13 shows this dryer schematically. Table 17 gives other pertinent characteristics of this dryer. Quality tests indicated that the dried product displayed no damage to the seed embryo.

Drying of Paprika

Dried paprika pods are ground in a hammer mill. According to the practice followed so far, breaks of a moisture content of 10–16% were transported directly to a paprika mill to be passed through several special rolls and runs in order to obtain grits suitable for use

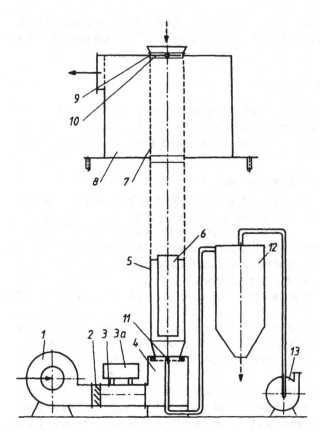

Figure 13 Spouted bed dryer of batch-type operation for drying of tomato seeds: 1, fan; 2, air feed control; 3, heater; 4, air chamber; 5, drying tank; 6, draft tube; 7, sieve; 8, dust chamber; 9, feed of wet material; 10, slap spindle; 11, discharge of the dried product; 12, cyclon; 13, ventilator.

as a spice. If the initial moisture of paprika breaks is no more than 6–8%, the paprika mills can double their production capacity.

Continuous drying of coarse breaks of paprika presents some difficulties. It is not possible to ensure a uniform motion of particles in a fluidized bed dryer, nor can the breaks be treated in a conventional spouted bed apparatus. Spouted bed devices proved to be suitable only for treating materials of a particle size above a few millimeters (e.g., 3 mm) and in a rather limited range. However, paprika breaks do not meet either of these conditions, as shown by the data in Figure 14.

A spouted bed apparatus suitable for maintaining the recirculation of paprika breaks of a wide range of particle sizes with minimal clogging tendencies was developed for this purpose. Its efficiency was improved by using an air-fed annulus (swirling ring) of special design and by simultaneous application of a mixer and a draft tube. The spouted bed dryer is shown in Figure 3. The main parts of the device are the base, the drying column, and the separating column. An electric heater, a stirrer, and the air inlet swirling ring are located in the base. The drying column includes the draft tube controlling the recirculation of particles. The device is completed by a wet feed hopper, an overflow chamber, a cyclone, and an air blower. The drying column has a cross-section of 0.015 m^2.

Table 16 Main Dimensions of Spouted Bed Dryers and Experimental Parameters

Dimensions of the dryers (mm)	Laboratory scale	Industrial scale
Air dashpot		
Diameter	128	633
Length	650	900
Cylindrical part		
Diameter	140	538
Length	450	1830
Draft tube		
Diameter	75	317
Length	740	1220
Linen cloth permeable to air		
Diameter	200	538
Length	1500	1460
Bronze sieve cloth		
Diameter	—	538
Length	—	1500
Size of air inlet slits	10	5 rows of rings; each has 12 slits of diameter 8 mm (slit height: 8mm)
Characteristic air velocity data		
Minimum fluidization velocity	0.94 m/s (dry seed)	
Velocity	9.60 m/s	
Gas velocity in the draft tube	11.90 m/s	
Gas velocity in the annulus	1.30 m/s	

In Table 18 note that the dryer capacity was in the range 1000–1400 kg/m^{2}h with a water evaporating capacity of 105–115 kg/m^{2}h. Although the drying air temperature was rather high, the temperature of outlet air and the dried product was in the range of 39°C to 62°C, and the product quality was good at a moisture level of 6–7%. The amount of air used here is several times that used in a conventional SBD. In test 4 of Table 18 carried out with a bed of 1600 mm height (approaching industrial dimensions), the specific heat

Table 17 Experimental Results Obtained with Tomato Seeds for Sowing in a Dryer of Industrial Scale

Drying air flow rate	3500 m^3/h at 20°C
Temperature of drying air	45–60°C
Temperature of outlet air	26–41°C
Pressure drop across the dryer	1.76 kPa
Drying period (total time of drying and abrasion)	85 min
Weight of the charge	200 kg of wet seeds
Initial moisture content	50–52% m/m
Evaporational rate	200 kg water/m^{2}h
Specific energy consumption	3156 kJ/kg water

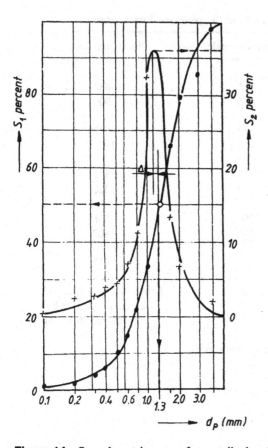

Figure 14 Granulometric curves for paprika breaks.

Table 18 Drying of Paprika Breaks in a Spouted Bed

	Measurement			
Conditions	1	2	3	4
Initial moisture content (% m/m)	9.63	14.4	15.0	18.8
Bed height (mm)	490	490	490	1665
Mean residence time of particles (min)	12	13.5	11.5	48
Air feeding rate (m³/h)	56	61	78.8	63
Air temperature (°C)	120	147	144	122
Water evaporating efficiency (kg/h)	1.64	1.57	1.68	1.73
Specific rate of evaporation (kg water/m²h)	109	104.6	112	115.3
Specific air utilization (kg air/kg water)	37.9	43.9	53.0	40.7
Specific moist-material efficiency (kg/m²h)	1400	1116.8	1288.6	1001.6
Specific energy consumption (kJ/kg water)	4318	5644	6158	4091

Table 19 Characteristics of Raw Carrot Cubes

Dimensions of the cubes	$10 \times 10 \times 10$ mm
Initial moisture content	85–89% m/m
Bulk density	491.5 kg/m^3
Desired final moisture content	20–22% m/m
Terminal velocity	4.3 m/s

consumption was found to be 4091 kJ/kg water. The pressure drop is 4.4–4.9 kPa, which is significantly lower than that for an SBD using a nozzle.

Drying of Carrot Cubes

The main characteristics of the raw carrot cubes tested are presented in Table 19. Figure 15 shows typical drying curves. In drying of cut carrots (and cut vegetables, in general) one must reckon with a significant deformation and also with a decrease in specific gravity. With respect to the latter, the flow rate of the drying medium must be adequately decreased. Furthermore, the temperature of the drying agent must be lowered in order to prevent caramelization. These aspects suggest drying in two stages. The experimental conditions and results for the drying of carrot cubes in a two-stage SBD are presented in Tables 20 and 21.

These results are favorable in terms of both the quality of the product and drying efficiency. This can be ascribed to the fact that the drying takes place in both stages almost completely in corresponding to the constant drying rate periods. It must be noted that the shape of the end product changed due to the effect of shrinkage, resulting in smaller granules of a nearly spherical shape.

Pastelike materials occur in many technological processes in the food processing, chemical industry, and so on. They are involved in the production of foodstuffs, organic intermediate products, pigments, pharmaceuticals, inorganic salts, and the like. Due to the variety of the occurrence of these materials in engineering, the process of drying the pastelike materials can be considered an important stage of process technology. The

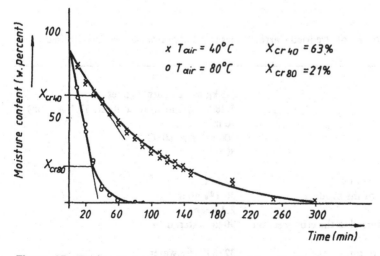

Figure 15 Drying curve of raw carrot cubes determined in flow-through drying oven.

Table 20   Drying of Carrot Cubes in a Laboratory-Scale Continuous Spouted Bed Predryer

Experimental conditions	
Charge	2.5 kg of raw carrot cubes
Feed rate	6 kg/h (moisture content, 80.5% m/m)
Mean residence time	25 min
Drying airflow rate	115 m³/h at 20°C
Temperature of air	91°C
Drying results	
Mean temperature of the outlet air	35°C
Mean moisture content of the product	57.1% m/m
Removed moisture	3.26 kg/h
Rate of evaporation referred to the cross-section of the dryer	233 kg water/m²h
Specific energy consumption	2885 kJ/kg water

drying process may greatly influence the quality of the product. The specific properties of the pastelike material affect the drying and general process requirements. Thus, in organizing the drying process it is necessary to take into account the properties of the materials being dried. A knowledge of these properties and the laws governing the changes in parameters during drying is the basis for choosing the most practicable method of drying and the optimum conditions for carrying out the processes.

In view of the high moisture content of suspensions economic drying can be performed only in dryers with intensive heat/mass transfer. The inert bed dryers provide good conditions for this purpose since the drying process is performed

On a large and continuously renewed surface
In a thin layer formed on the surface of inert particles
With intensive contact between the wet material and the drying agent of high flow rate

Drying of pastelike materials of high thermal sensitivity such as animal blood and liquid vegetable extracts in a spouted bed of inert bodies is technically feasible (21).

Table 21   Continuous Drying of Predried Carrot Cubes in a Laboratory-Scale Spouted Bed Afterdryer

Experimental conditions	
Charge	3.6 kg of raw carrot cubes
Feed rate	5.48 kg/h (moisture content, 57.1% m/m)
Mean residence time	40 min
Drying airflow rate	100 m³/h at 20°C
Temperature of air	80°C
Drying results	
Mean temperature of the outlet air	42.5°C
Mean moisture content of the product	30.1% m/m
Removed moisture	2.1 kg/h
Rate of evaporation referred to the cross-section of the dryer	90 kg water/m²h
Specific energy consumption	3276 kJ/kg water

Experimental investigations were carried out in a drying chamber consisting of a 60° conical base followed by a 0.14-m diameter cylindrical section. The chamber contained polypropylene beads as inert particles (ρ = 820 kg/m^3 and d = 3.9 × 10^{-3} m). It has been observed during animal blood drying that when the air outlet temperature was kept at 74°C a dry product with soluble protein content of about 85% of its original amount can be obtained. At this temperature the process thermal efficiency is about 55%.

The dried vegetable products have shown the same organoleptic properties and moisture content as commercial vegetable powders for pharmaceutical use. Therefore, their active compounds do not lose their desirable characteristics after drying.

The product moisture can be adopted as a quality control parameter. It is recommended to keep it around 5% for dried vegetable extracts and around 6% for dried animal blood. Drying of agricultural products and by-products has received a great deal of attention in Brazil in recent years. The spouted bed with inert particles has also been investigated as a paste dryer for bovine blood and tomato paste, as well as for solutions of various pharmaceutical products. Another by-product that has been underutilized is the yeast produced in large quantities in the alcohol industry.

Since yeast is a living organism, great care must be taken in its drying to ensure a viable product for use in the baking industry. Owing to this heat sensitivity, the SBD with inert particles, in which the particle temperature stays well below the gas temperature, is considered to be a suitable dryer (22). The SBD consisted of a conical base with 60° angle, followed by a cylindrical section with a diameter of 500 mm. The diameter of the gas inlet was 50 mm. Drying was carried out on inert glass beads (d = 1.8 mm and ρ = 2.5 g/cm^3). The inlet air temperature values were 60°C, 79°C, and 80°C. The airflow rate was kept at twice the minimum spouting value. It was found that the viability of the product was in the 50–70% range, and that a feed concentration of 10% was satisfactory.

Drying of Potato Pulp

Potato is a colloidally dispersed material with a capillary porous structure. The mean composition of potato flour is presented in Table 22. Both raw and boiled potatoes are used in the production of potato flour. With potato flour prepared from boiled potatoes, the peeled, washed potatoes are treated with sulfite and boiled for 30 min under pressure. Pulping may be carried out in a hammer mill before the material is dried on a cylindrical dryer at a temperature of 60–70°C to a final moisture content of 7–9% by weight.

Table 22 Composition of Potato Flour

Components	Content (% m/m)
Water	75.00
Total carbohydrate	21.00
Sugar	1.50
Starch	18.00
Protein	2.00
Fibrous material	1.00
Ash	1.00
Vitamin C (mg/100g)	1.00
Vitamin B$_1$–thiamine (mg/100g)	0.11

Production of Potato Grits

Drying of the potato pulp has been carried out in a large, laboratory-size SBD fitted with a central screw conveyor, with an inert charge and tangential air feeding. A schematic of such an SBD, which can be used to dry pulp materials, pastes, and suspensions, is presented in Figure 16.

Potato pulp (moisture content, 75–80%) is fed by means of a suitable feeding system into the sliding part of the inert charge, which recirculates into the lower third part of the bed. (In the laboratory-scale SBD, the diameter of the pipe feeding the pulp was 3 mm.) The size of this diameter plays an important role in the development of the product grain size. The uniform distribution of the pulp fed along the entire cross-section is ensured by stirring elements attached to the inner screw axis at appropriate distances above and below the feeding site.

The recirculating motion of the inert particles, which are thinly coated with pulp, dries, grinds, and backmixes the dried powder with pulp to obtain good texture. The fine, dry product is collected in a cyclone or in a bag filter. The grain size of the product can be controlled by the location of the discharge pipe and by the flow of air. The conditions and results of drying potato pulp in an SBD are presented in Table 23.

In some areas the spouted bed dryers for solutions and suspensions can be an economically interesting alternative to the spray dryer.

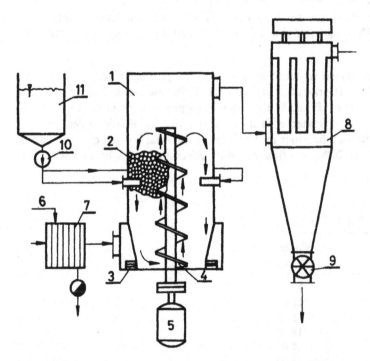

Figure 16 Flow scheme of mechanically spouted bed (MSB) dryer with inert particles for drying of materials of high moisture content: 1, dryer; 2, inert packing; 3, slots for air inlet; 4, vertical conveyor screw; 5, drive; 6, air inlet; 7, heat exchanger; 8, bag filter; 9, rotary valve; 10, pump; 11, suspension tank.

Table 23   Drying of Potato Pulp in a Mechanically Spouted Bed Dryer with an Inert Charge

Experimental conditions	
Wet solid	potato pulp of a moisture content 78% m/m
Feed rate	2.88 kg/h
Drying airflow rate	110 m^3/h at 20°C
Mean temperature of the inlet air	114.5°C
Temperature of the outlet air	63°C
Speed of rotation of the screw	240 rpm
Drying results	
Removed moisture	2.87 kg/h
Mean temperature of the outlet air	42.5°C
Rate of evaporation referred to the cross-section of the dryer	191 kg water/m^{2}h
Specific energy consumption	3025 kJ/kg water
Mean moisture content of the product	0.2% m/m

Drying of Brewery Yeast

The dried and inactivated brewery yeast with its vitamin and protein content preserved is widely used as a filling material in the food industry and as an ingredient of medicative preparations. Drying has been carried out under suitable circumstances in the MSB dryer (cf. Fig. 17) with inert particles (polyformaldehyde), the dimensions and operational parameters of which are summarized in Table 24.

3.3.   Drying of Some Chemical Products

An SBD for PVC tested in the Technological Institute in St. Petersburg, Russia has the following dimensions: $D = 400$ mm, $D_i = 70$ mm, and $\alpha = 40°$. The cross-section of the dryer was 1 m^2, with a capacity of 400 kg/h of moisture and 1600 kg/h of dry material. Additional data for this operation are given in Table 25 (23).

Some of the examples described below relate to products of high moisture content; they range from pasty or pulpy products to granular, crystalline substances.

Drying of Calcium Carbonate of High Purity and Fine Grain Size

The moisture content of limestone produced in the desired grain size of 2–3 μm and in desired crystalline form, upon filtration or centrifugation, is 45–50% m/m. This pulpy product can be dried in an SBD with inert charge. The desired final moisture of the material during the drying was 100–110°C (see Fig. 16). Here, the conveying screw was made of Teflon. Metallic moving elements would cause abrasion and contamination of product.

The diameter of the drying chamber was $D = 140$ mm; its height was 450 mm. The upper part had a disengagement zone with a threefold increase of area. The conveying screw, diameter of 75 mm and 650 mm long, extended into the upper enlarged duct. The surface of the bed, consisting of inert grains, must be lower by about one screw profile than the end of the conveying screw. This way larger bits of grains do not accumulate at the top but are transported back into the bed.

The results of these drying experiments are presented in Table 26. Note that here the specific energy consumption is rather high. The result is considered favorable since one must take into account also that the grinding and the screening operations are carried out simultaneously in such an apparatus.

Figure 17   Industrial-scale mechanically spouted bed (MSB) dryer with inert particles.

Drying of Potassium Permanganate

Large-grain crystalline potassium permanganate after centrifuging can be dried in an SBD (24). The 0.5–1.2-mm grain size fraction varied between 75% and 80%, but the fraction under 0.5 mm was around 20–25%. To dry this highly sticky material, 50 Nm^3/h air of 260°C inlet (100°C outlet temperature) was used, bringing down its original 4% moisture content to 0.22%. Gas velocity in the gas inlet nozzles was 16.1 m/s; in the annular part of the dryer it was 0.66 m/s. Specific air and heat consumption values were 16.6 Nm^3 air/ kg water and 7975 kJ/kg water. Results of similar drying experiments conducted in a fluidized bed dryer were 50–55 Nm^3 air/kg water and 7350 kJ/kg water, that is, a much higher air consumption with a marginally lower heat requirement.

Dehydration of Salts

The dehydration of moist inorganic salts with water of crystallization content is, in general, a complex task. The loss of crystalline water usually consists of several compo- nents with different temperature or residence time requirements for each case. In the

Table 24   Drying of Brewery Yeast in a Mechanically
Spouted Bed Dryer with Inert Particles

Technical parameters	
Diameter	1.0 m
Height	1.8 m
Revolution of the screw	160 rpm
Diameter of the screw	0.25 m
Diameter of particles	12 mm
Weight of bed	380 kg
Bed height	0.8 m
Drying conditions	
Moisture content	
Initial	5.7 kg/kg db
Final	0.05 kg/kg db
Air flow rate	6000 Nm^3/h
Temperature	
Inlet	110°C
Outlet	60°C
Rate of evaporation	90 kg water/h
Specific drying rate	115 kg water/m^{2}h
Specific energy consumption	3500 kJ/kg water

Table 25   Drying of Polyvinyl Chloride in a Spouted
Bed Dryer

Initial moisture content	60% m/m
Temperature of drying air	
Inlet	120°C
Outlet	60°C
Specific energy consumption	9000 kJ/kg
Specific rate of evaporation	6.4 kg/m^{3}h
Final moisture content	0.5% m/m

Table 26   Drying Fine-Grain Calcium Carbonate in a Mechanically Spouted Bed Dryer
with Inert Particles

Experimental conditions	
Inert charge	glass beads (d = 4 mm)
Feeding	with a pump
Feed rate	3 kg/h of suspension of 50% m/m
Drying airflow rate	110 m^3/h at 20°C
Temperature of drying air	135°C
Drying results	
Temperature of the outlet air	75°C
Mean moisture content of the product	0.4% m/m
Removed moisture	1.49 kg/h
Rate of evaporation referred to the cross- section of the dryer	95 kg water/m^{2}h
Specific energy consumption	9715 kJ/kg water

course of dehydration, the material to be processed (e.g., $ZnCl_2$ or $MnCl_2$) may form undesirable compounds. In numerous cases, as with magnesium chloride and iron chloride, oxygen impedes the full loss of crystal water as a result of heat effects. On the basis of these problems, publications dealing with successful dehydration experiments conducted in spouted bed equipment are of particular interest.

A diluted solution of manganese sulfate forms granular particles free of water if the solution is sprayed into the bottom of the spouted bed of manganese sulfate particles (24). The SBD dimensions were $D = 196$ mm, $D_i = 60$ mm, and $\alpha = 40°$. The operating data for this experiment are given in Table 27. The output of the dryer was 50 kg solution per h at a concentration of 20–25%. At concentration levels of 40–45%, an insufficient number of nuclei can be formed and the spouting motion stops (24).

Rabinovich (25) dried a diluted solution of $MnCl_2$ by spraying the 24–25% solution onto a 3.5–5 mm aluminum silicate catalyst acting as an inert packing bed or, alternatively, anhydrous $MnCl_2$ was the bed material.

Ermakova (26) developed an SBD for the washing medium drained from an electrolysis bath containing manganese chloride and iron chloride. The inert packing bed was an aluminum silicate catalyst. To prevent hydrolysis, aluminum chloride was added to the washing medium (30–50 kg/m^3). The inlet gas temperature was 500–550°C; the outlet temperature was 140–150°C. Gas velocity in the annular part of the spouted bed dryer was in the range 3.2 to 3.6 m/s.

The importance of the residence time was observed by Mitev in the course of vortex bed dehydration of plaster of paris. At a bed temperature of 128–136°C, a residence time of 130 min was required to remove 1.5 molecules of water of crystallization. The average residence time of 30 min even at an inlet gas temperature of 800°C was not sufficient to attain the final desired moisture level. The cross-section of the vortex bed dryer was 3.9 m^2, with an output of 6000 kg/h and a specific evaporating capacity of 1500 kg/m^{3}h. The inlet temperature of the 2000 m^3/h drying medium was 650°C; the outlet temperature was 150°C.

Drying of Pigments and Dyes

The treatise by Romankov and Rashkovskaya (23) outlines SBD procedures for a number of inorganic pigments and organic dyes. The authors detail methods of drying pastelike materials with 40–70% moisture content. No deterioration in the quality of the product was observed at relatively high (150–390°C) inlet temperatures. Some significant specific data about these dryers of 200, 300, and 870 mm diameter are as follows: 14.2–24 kg gas/ kg water; 3250–4310 kJ/kg water; capacity 55–90 kg water/m^{2}h.

Similar values were attained in a spouted bed of industrial scale. For example indus-

Table 27   Drying of Solution of Manganese Sulfate

Air flow rate	180 Nm^3/h
Gas velocity in the nozzle	80.6 m/s
Gas velocity in the cylindrical part	2.49 m/s
Temperature	
Inlet	600°C
Outlet	140°C
Specific requirement of gas	4.76 Nm^3/kg water
Specific requirement of heat	3840 kJ/kg water
Specific evaporation	1260 kg water/m^{2}h

trial-scale drying of 200 kg/h "Z-type black" dye resulted in 11.8 kg gas/kg water and 4600 kJ/kg water; the diameter of the dryer was 1.6 m. The inlet temperature of 330°C in the spouted bed decreased to 100°C in the process of drying. The initial moisture content of the wet material was 5%. Gas velocity in the gas inlet nozzle was 30 m/s and 0.75 m/s in the annulus. Pressure drop over the 400-mm bed at this velocity was 5 kPa.

Drying of NaCl solution and alumina suspension were investigated (27) in laboratory-size SBD inert packing. Experiments have been carried out in a cylindrical chamber 150 mm in diameter with a conical base and an air inlet of 20 mm. The inert particles were mainly glass and plastic beads of 2 to 5 mm in size and of different shapes.

To conclude, solutions and suspensions can be dried by spraying onto the outside of large inert particles. The drying proceeds at a constant rate until nearly all the water has evaporated. The drying rate then falls rapidly. The dried product is not attrited from the surface of the particles until the drying passes into the falling rate period.

Drying of Cobalt Carbonate

For the drying of cobalt carbonate, bed height to column diameter ratio H/D has been changed at constant airflow rate and temperature, comparing specific drying performances related to equipment cross-section and total inert particle surface, also calculating drying efficiency from theoretical and actual heat consumption necessary for the evaporation of water. Measured and calculated data are summarized in Table 28 (18).

It can be seen from the data that specific drying performance P increases and the heat consumption r_r necessary for the evaporation of water decreases with reducing of the ratio H/D_c. If the ratio H/D_c is large, only a definite fraction of the inert particles participate in the drying process, that is, their surface is coated by the wet material, which dries on it and subsequently abrades. This means that the fraction of particles not participating in the drying process is relatively high. This fraction can be reduced with a decreasing H/D_c ratio, and the amount of heat relieved in such a way can be used for removing the moisture. Drying efficiency was highest at the lowest bed height to diameter

Table 28 Drying Experiments in the Laboratory-Scale Dryer ($D = 0.135$ m)

		Lab 1	Lab 2	Lab 3
m_p	(kg)	9.2	7.0	5.2
A_p	(m^2)	2.8	2.1	1.6
H/D_c		2.5	2.0	1.4
V''	(Nm^3/h)	80	80	80
v''	(m/s)	1.5	1.5	1.5
x_{in}	(kg/kg)	5.5	5.3	5.3
x_{out}	(kg/kg)	0.05	0.05	0.03
G	(kg/h)	3.7	4.0	4.7
T_{in}	(°C)	174	178	178
T_{out}	(°C)	64	59	61
W	(kg water/h)	3.1	3.4	3.9
P_w	(kg water/m^{2}h)	209	227	264
$P_{w,p}$	(kg water/m^{2}h)	1.1	1.6	2.5
P_d	(kg dry/m^{2}h)	40	40	53
$P_{d,p}$	(kg dry/m^{2}h)	0.2	0.3	0.5
r_{ra}	(kJ/kg)	3350	3350	2950

ratio ($H/D_c = 1.4$). The experiment performed on the pilot-plant dryer ($D = 0.38$ m) proved that drying can be performed at a low H/D_c ratio with the same good efficiency as under laboratory conditions (cf. Table 29).

The experimental results are interesting from the point of economic operation, that is, that one can obtain very high specific drying performance and low specific energy consumption at packing heights much lower than with conventional spouted bed dryers. Energy consumption for airflow can be reduced considerably by decreasing the volume of the inert particles of relatively high density.

Moreover, the comparison of specific drying performances related to equipment cross-sectional area P_d and the total inert packing surface $P_{d,p}$ also show good agreement.

Drying of Sludge from Metal Finishing Industries' Wastewater Treatment Plants (28)

Metal finishing industries, in particular the plating industries, typically produce large volumes of metal hydroxide sludges from the treatment of rinse waters and wastewaters. These sludges typically contain 20–25% of solid after dewatering in a filter press. The spouted bed is potentially an excellent alternative for drying such materials. Tests were carried out in a 154 mm diameter, half-cylindrical spouted bed designed for operation at high temperature. The bed has a total height of 1156 mm and a conocylindrical base with an included angle of 60°. The sludge used for the trials is a mixture of ferric and zinc hydroxides containing 41% iron, 7.5% zinc, and 4.0% lead on dry basis with the balance represented by oxide and hydroxide. The nonbound moisture percentage of the sludge was 83.4%.

In all cases sand was used as the bed material. Spouting air preheated to a controlled temperature up to 500°C was introduced through a single orifice of 19-mm diameter. After steady spouting had been attained at a bed temperature of 350°C and a superficial gas velocity of 1.4 m/s, sludge was fed approximately 3 mm into the spout. Particle size distribution of material was between 15 and 30 mm. The product had a moisture content

Table 29 Comparison of the Laboratory and Pilot-Plant Drying

		Lab 3	Pilot
m_p	(kg)	5.2	55
A_p	(m^2)	1.6	12.7
H/D_c		1.4	1.2
V''	(Nm^3/h)	80	620
u''	(m/s)	1.5	1.5
x_{in}	(kg/kg)	5.3	5.0
x_{out}	(kg/kg)	0.03	0.02
G	(kg/h)	4.7	37.0
T_{in}	(°C)	178	178
T_{out}	(°C)	61	61
W	(kg water/h)	3.9	30.7
P_w	(kg water/m^{2}h)	265	270
P_{wp}	(kg water/m^{2}h)	2.5	2.4
P_d	(kg dry/m^{2}h)	53	56
P_{dp}	(kg dry/m^{2}h)	0.5	0.5
r_r	(kJ/kg)	2950	2900

of less than 3%. The sand discharged from the spout was almost uniformly coated with a thin layer of dried sludge.

A conservative capital cost estimate was made for a 610 mm diameter SBD designed to dry 220 kg/h of wet sludge with 20% by weight solids. The cost compared favorably with available kiln-type technologies and shows that the spouted bed is cost competitive with conventional technology.

4. ASSESSMENT OF DRYING RESULTS

On the basis of extensive results obtained at the Research Institute of Chemical Engineering of the Hungarian Academy of Sciences on SBD techniques, it is established that with proper design and selection of proper operating parameters the SBD lends itself to a very wide range of applications in various industries, such as drying granular, pastelike, or pulpy materials with a wide range of possible particle sizes. SBDs are especially suited for drying of heat-sensitive materials with surface moisture, as well as those with bound moisture, such as seeds, foodstuffs, pharmaceuticals, and synthetic products, in one or two stages. Among the advantages of the SBD are

Intense particle motion. Good particle mixing prevents localized overheating and ensures the product uniform moisture content.

The recirculating motion of particles ensures that during the residence time the drying particles contact the inlet warm air at regular intervals. The velocity of this recirculatory particle motion can be adjusted as required by varying the operating parameters, such as gas velocity and bed height, and geometry, such as size of the gas inlet nozzle, the use of an inner transport screw, and draft tube.

The residence time of particles may be changed and regulated within very wide limits, for example, by changing bed height or by the use of suitable internal elements, such as partitions or draft tubes.

To dry materials with bound moisture (e.g., plastics), the tangential air inlet and an inner transport screw are highly recommended since the volumetric gas velocity can be adjusted as required by the drying process regardless of the gas velocity requirement for particle motion.

5. CONCLUSION

Spouted beds for drying of granular materials, pastes, and slurries are an emerging technology. Although there are few large-scale industrial applications reported in the literature, the modified spouted beds display significant potential for future applications. Scaling up of axisymmetric and three-dimensional spouted beds remains a difficult area. Two-dimensional spouted beds, such as those proposed by Mujumdar (29), have a decided advantage as far as scaleup and modular design are concerned. Also, spout-fluid beds (which combine the advantageous features of spouting with fluidization) with internal heat exchange surfaces are expected to find special applications. For a more comprehensive survey of the literature along with a discussion of the design of conventional as well as several modified spouted beds for drying, the reader is referred to Passos et al. (30). This chapter is based to a great extent on results obtained at the Hungarian Academy of Sciences; other investigations reported in the literature (see Reference 30) are consistent with the findings and conclusions presented here.

ACKNOWLEDGMENTS

Acknowledgment is due to the Hungarian Academy of Sciences, the National Committee for Technical Development (OMFB), and various industrial firms and institutes (CHEMI-MAS, the Paprika Processing Works of Szeged, the Pét Nitrogen Works, and the TECH-NOVA Engineering Co., BiYo-Product Co.) for the financial support and grants they provided with the aim of assisting research work and promoting development of spouted bed dryers. Purnima Mujumdar, Brossard, Canada, and Agota Barta, Hungary, retyped revised versions of the original draft; their assistance is gratefully acknowledged.

NOMENCLATURE

A_c	cross-sectional area of column, m²
A_i	surface of inlet slots, m²
A_p	surface of particle, m²
a	specific surface, m²/kg
D_c	column diameter (diameter of the dryer), m
D_i	diameter of the air inlet nozzle, m
D_{sc}	diameter of the conveyor screw, m
d_p	diameter of particles, m
G	wet feeding rate, kg/h
H	bed depth, m
h	height of slots, m
m_p	bed weight, kg
n_{sc}	speed of rotation of the conveyor screw, rpm
P_w	rate of evaporation related to the cross-section of the dryer, kg/m²h
P_{wp}	rate of evaporation related to the particle surface, kg/m²h
P_d	specific drying performance (dry product) related to the cross-section of the dryer, kg/m²h
P_{dp}	specific drying performance related to the particle surface, kg/m²h
Q_1, Q_2, Q_3	bulk velocities in spaces 1, 2, and 3, respectively (see Fig. 5), m³/s
Q', Q''	bulk velocities in points a and b, respectively, m³/s
q	conveying rate of the screw, kg/s
Re_{mf}	Reynolds number at minimum fluidization velocity
Re_p	particle Reynolds number
r_r	real heat of consumption, kJ/kg
s	pitch of the conveyor screw, m
T_{in}	inlet air temperature, °C
T_{out}	outlet air temperature, °C
v', v''	inlet air velocity, m/s
v_{ph}	peripheral speed of the screw, m/s
w_a	sliding velocity of particles in the annulus, m/s
w_a', w_a''	sliding velocity of particles in the annulus at inlet air velocity v' and v'', Eq. [1], and air inlet nozzle D_i' and D_i'', respectively (Eq. [3]), m/s
$w_{aH}, w_{aH_{max}}$	sliding of particles in the annulus at bed depth h and H_{max}, respectively (Eq. [2]), m/s
W	evaporated water, kg/h
x_{in}	initial moisture content, kg/kg d.b.
x_{out}	final moisture content, kg/hg d.b.

Δp	pressure drop
ϵ	voidage of bed
ρ_f	density of fluid
ρ_p	density of particles
τ	time, residence time
τ_1, τ_2, τ_3	residence time in spaces 1, 2, and 3, respectively (see Fig. 5)
τ_{in}	time of feed of material
$\bar{\tau}$	average residence time of particles

REFERENCES

1. P. E. Gishler and K. B. Mathur, U.S. Patent 2,736,280 to Nat. Res. Council of Can (1957), filed 1954.
2. W. S. Peterson, *Can. J. Chem. Eng.*, 9:111 (1974).
3. K. B. Mathur and N. Epstein, *Spouted Beds*, Academic Press, New York, 1974.
4. D. V. Vukovich, F. K. Zdanski, and H. Littman, Present Status of the Theory and Application of Spouted Bed Technique, Int. Congr. Chem. Eng. CHISA, Prague (1975).
5. K. B. Mathur and P. E. Gishler, *J. Appl. Chem.*, 5:624 (1955).
6. J. Németh, thesis, Budapest, Hungary (1978).
7. J. Németh, E. Pallai, T. Blickle, and J. Györy, Hungarian Patent 160.333 (1973).
8. E. O. Sulg, D. T. Mitev, N. B. Rashkovskaya, and P. G. Romankov, *J. Appl. Chim.* (USSR), 43:2204 (1970).
9. E. Aradi, E. Pallai, T. Blickle, E. Monostori, and J. Németh, Hungarian Patent 176.030 (1976).
10. E. Pallai, postgraduate degree thesis, Hung. Acad. Sciences, Budapest (1970).
11. H. A. Becker and H. R. Sallans, *Chem. Eng. Sci.*, 13:245 (1961).
12. M. Kugo, N. Watanabe, O. Uemaki, and T. Shibata, *Bull. Hokkaido Univ. Sapporo*, Jap., 39:95 (1965).
13. E. Pallai and J. Németh, Int. Conf. Drying, 3rd Budapest Conf., Paper D. 7 (1971).
14. E. Pallai and J. Németh, Int. Congr. Chem. Eng. CHISA, Prague, Paper C3.11 (1972).
15. T. Blickle, E. Aradi, and E. Pallai, *Drying '80*, (A. S. Mujumdar, ed.) Vol. 1, Hemisphere, New York, 1980, p. 265.
16. T. Szentmarjay and E. Pallai, Drying of Suspensions in a Modified Spouted Bed Dryer with an Inert Packing. *Drying Technology*, 7(3):523–536 (1989).
17. J. Németh, E. Pallai, and E. Aradi, Scale-up Examination of Spouted Bed Dryers, *Can. J. Chem. Eng.*, 61:419–425 (1983).
18. T. Szentmarjay, A. Szalay, and E. Pallai, Scale-up Aspects of the Mechanically Spouted Bed Dryer with Inert Particles, *Drying Technology*, 12(1&2):341–350 (1994).
19. T. Blickle, E. Pallai, J. Németh, E. Aradi, and J. Varga, Monograph of Acad. Group at Veszprém of the Hung. Acad. Sciences (1978).
20. E. Aradi, E. Pallai, and M. Péter, Int. Congr. Chem. Eng. CHISA, Prague, Paper C4.25 (1981).
21. H. J. Re and J. Teixeira Freire, Drying of Pastelike Materials in Spouted Bed, *6th IDS, Versailles*, 1:OP119–125 (1988).
22. J. Texeira Freire and J. A. Morris, Drying Yeast in a Spouted Bed Dryer, *IDS '90 Prague*, Vol. 2, Paper E5.46.
23. P. G. Romankov and N. B. Rashkovskaya, *Sushka v vzveshennom sloye* (*Drying in Fluidized Bed*), Chimiya, Leningrad, 1979.
24. M. J. Rabinovich, *Teplociye processi fontaniruyuschem sloye*, (*Thermal Processes in Spouted Beds*), Nauka Dumka, Kiev, 1977.
25. M. J. Rabinovich, *Chim. Prom. Ukraini*, 6:60 (1970).

26. T. T. Ermakova, Poluchenie i svoystva in Pramoye Poluchenie Zheleza (Direct Method of Obtaining Iron), *Metallurgia*, Moscow, 134–142 (1974).

27. T. Schneider and J. Bridgwater, Drying of Solutions and Suspensions in Spouted Beds, *6th IDS Versailles 1988*, 1:OP113–117.

28. C. Brereton and C. J. Lim, Spouted Bed Drying of Sludge from Metals Finishing Industries Wastewater Treatment Plant, *Drying Technology*, 11(2):389–399 (1993).

29. A. S. Mujumdar, *Drying '85*, (A. S. Mujumdar, ed.), Hemisphere/Springer-Verlag, New York, 1984.

30. M. L. Passos, A. S. Mujumdar, and G. S. V. Raghavan, *Advances in Drying*, Vol. 4, Hemisphere, New York, 1986.

14
Impingement Drying

Arun S. Mujumdar and Bing Huang
McGill University
Montreal, Quebec, Canada

1. INTRODUCTION

Impinging jets of various configurations are commonly used in numerous industrial drying operations involving rapid drying of materials in the form of continuous sheets (e.g., tissue paper, photographic film, coated paper, nonwovens, and textiles) or relatively large, thin sheets (e.g., veneer, lumber, and carpets), or even beds of coarse granules (e.g., cat or dog food). In this chapter we will not examine the last-mentioned application, which is a novel operation in which hot jets are directed normally onto thin beds of pellets transported on a slow-moving conveyor. The jets "pseudofluidize" the bed to ensure good gas-solid contact needed for effective drying.

Specific, large-scale applications of impingement drying, such as drying of tissue paper on Yankee dryers, combined impingement and through-drying of newsprint (the so-called Papridryer process), and drying of wood, are covered elsewhere in this handbook. The objective of this chapter is to review the subject in more general terms and to provide empirical guidelines for the designer of impingement systems on the basis of recent published literature on this subject.

Since impingement yields very high heat or mass transfer rates, it is a popular system for convective drying when rapid drying or small equipment is desired. High production capacities are attained at the expense of increased capital and operating expenses because of the more complex fabrication and increased air-handling requirements. Impinging jet drying is recommended only if a major fraction of the moisture to be removed is unbound. If the drying rate is internal diffusion controlled, the high heat transfer rates of the impingement system can often result in product degradation if the product is heat sensitive. For rapid drying of very thin sheets (e.g., tissue), high-velocity impingement of hot air jets is very effective. On the other hand, for drying of heavier grades of paper and textiles, for example, impingement drying is effective only to remove surface moisture. Another important industrial application of impingement drying is in the printing, packaging, and converting industry, in which printing techniques are used to deposit a thin film of coating onto a moving substrate.

It should be noted parenthetically that although most current applications use hot air jets for drying, the use of superheated steam in impinging jet configuration is being seriously considered. At least one firm in India markets stenter dryers for textiles using steam jets. A combined impingement and through-drying scheme has also been proposed by the author for drying of permeable paper grades and textiles. Potential for use of steam drying of veneer has also been explored and found to be favorable. Much developmental work needs to be done in the area of drying with steam jets.

It should be pointed out at the outset that since impinging jets are recommended only to remove surface or unbound moisture, the process calculation of such dryers is based essentially on the external heat or mass transfer rates, which can be estimated empirically for a very wide variety of jet configurations. Much of the empirical information available in the literature was motivated by other than drying applications, such as heating or cooling of glass, metal sheets, turbine vanes, and furnace walls. This information can be readily adapted for dryer calculations if the geometric configurations are similar. It should be stressed that when the moisture removal is internal diffusion controlled, the design can be based only on experimentally determined drying rate data or on a validated mathematical model of the falling rate drying process.

If the internal resistance is large, designing a dryer to operate in a combination mode takes advantage of the high external transfer rates achievable with jets. Thus, for drying of foodstuffs extruded in the form of thin sheets, one may combine impingement drying with a microwave heat source that heats the solid volumetrically without depending on the development of large thermal gradients required by Fourier's law of heat conduction. Impingement may be combined with a through-flow to dry thin packed beds or permeable sheets. In this case, application of suction has a synergistic effect: aside from through-drying, suction also augments the impingement heat or mass transfer rate.

In a few cases, the jet velocity is governed by the fluid mechanics of the impacting jet rather than by the heat or mass transfer rates. In drying of coated papers (films), very high shear stresses generated by jet impingement can cause the high-velocity ink or coating to flow, causing undesirable streaks (railroading) on the product. The jet velocity and spacing of the nozzle from the target surface should be adjusted to avoid such problems. Extremely high drying rates obtained by use of high-velocity jets at elevated temperatures may cause such problems as case hardening or cracking due to rapid shrinkage. For combustible products there is a significant fire hazard unless due care is taken in design and operation. Appropriate controls must be built in to avoid fire or explosion hazard.

In most industrial applications the jet temperatures are such that the radiative heat transfer contribution is very small (a small percentage) of the convective. Thus, a design based purely on convective heat/mass transfer is conservative. This chapter will therefore consider only convective heat or mass transfer correlations for design. The reader is referred to the literature for full details.

2. DESIGN OF IMPINGEMENT DRYERS

2.1. General Observations

An impingement dryer system should be chosen only if the product is in a form that is amenable to being subjected to a multiplicity of hot jets directed normally (or nearly normally) onto one or both of its surfaces. In some cases (e.g., pulp or double-coated paper), it is possible to support the web to be dried entirely aerodynamically by strategically locating the blow boxes from which the jets emerge. In most other cases, the material being dried is supported on a conveyor or roll, which may be solid or permeable. Because of their complex fabrication and high air-handling costs (due to the high nozzle pressure drop and high recycle ratio needed to achieve reasonable thermal efficiencies), impingement dryers are recommended only (or primarily) to remove unbound moisture. These should not be used to remove moisture well below the critical moisture level. Simpler parallel airflow dryers are just as effective and more economical under such conditions (e.g., wood and carpet).

Typically, the jet velocity and temperature may range from 10 to 100 m/s and 100°C to 350°C, respectively, depending on the product. Because of the extremely high evaporative cooling in the constant rate period, the product surface temperature can be held well below its degradation or ignition temperature. Care must be exercised in drying bound moisture under such conditions since the product temperature can rise rapidly. If the feed moisture or rate is susceptible to random variation or different products are to be dried in the same dryer, a suitable control system sensitive to the product surface temperature rise is recommended strongly.

If very high jet temperatures are needed, combustion gases may be used directly. Otherwise, steam-heated air may be used. When using high-velocity, high-temperature jets, up to 90% of the exhaust air may be recycled after reheating or directly mixed with the inlet jet air. Under these conditions, the jet contact time is too short to utilize the total moisture uptake capacity of the jet. Hence the need for such large recycle ratios, which result in expensive air-handling systems. For low-performance impingement dryers the recycle ratios are typically low (less than 50%). In impingement dryers consisting of distinct compartments, each with its own operating temperature and air velocity, exhaust from the upstream compartment can be introduced directly into the downstream compartment(s), provided the exhaust air is still far from dew point, and can accomplish drying. In any event, fresh makeup air is needed to account for the moisture pickup; otherwise the drying medium will quickly become saturated.

The dryer along with its ducting must be well insulated to obtain good thermal efficiency.

2.2. Design Parameters

Design of impingement dryers is both simplified and aggravated by the excessive number of design variables or parameters that can be specified or chosen arbitrarily; at least a dozen of these are important and have been studied systematically in recent literature. The designer's task is simplified by the fact that almost any arbitrary design can be made to work by adjusting one or several of the other operating parameters (e.g., jet velocity or temperature). The difficult task for the design is to choose a system that will give optimal system design (in terms of energy consumption and throughput and product quality) without undue adjustments during operation. It is also desirable to be able to control the dryer readily and be able to predict its performance when one or several of the operating variables are changed randomly.

Thus it is apparent that the design problems may be posed in an infinite number of ways. The following is just one approach suggested by the author; no pretense is made that it is the most desirable approach. Most industrial designs are based on more arbitrary and trial-and-error procedures. However, field experience can be used to advantage in improving the design.

1. Select nozzle configuration (e.g., multiple slot and round jet arrays or exhaust port location) (see Figs. 1 and 2).
2. Select nozzle geometry (at least 50 different designs are in use, most chosen arbitrarily).
3. Select jet velocity, temperature, and nozzle-target spacing. (These are interrelated. A minimum geometric spacing is needed for fabrication purposes, and ease of access, for example. Jet temperature and velocity are dictated by product thermal sensitivity: the higher the temperature, the better the thermal efficiency will be in general. Jet Reynolds number must exceed 1000, probably 2000, to achieve fully turbulent regime.)
4. Calculate drying rates (constant rate only) using pertinent empirical correlations and applying various corrections for surface motion, high transfer rates, large temperature differences between the air jet and the web surface, and so on. Compute product surface temperature. Make a parametric study to determine quantitatively the influence of various parameters.
5. Determine air recycle ratio (by mass and enthalpy balance).
6. Redo steps 4 and 5, accounting for changes in jet temperature and humidity due to recycle. (*Note:* The recirculated exhaust may be heated in some instances.)

2.3. Nozzle Configuration

Although not popularly recognized until recently, selection of the nozzle geometry and multinozzle configuration have important bearing on the initial capital cost and operating costs, as well as the product quality (e.g., nonuniform moisture distribution). It appears that most of the old designs utilize nozzle shapes and configurations chosen arbitrarily. Once selected for ease of fabrication and economy of scale, the same nozzle box is retained in all subsequent designs. The dryer is typically constructed of a series of similar nozzle boxes arranged in a modular fashion with strategically located exhaust ports. The

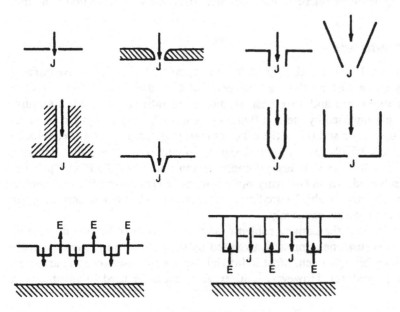

Figure 1 Various nozzle types studied in literature.

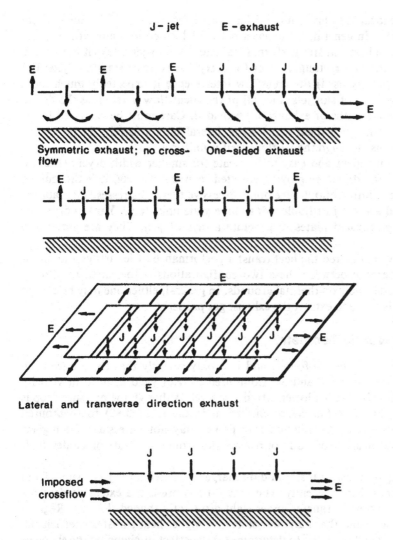

J – jet   E – exhaust

Symmetric exhaust; no cross-flow

One-sided exhaust

Lateral and transverse direction exhaust

Imposed crossflow

Figure 2 Various arrangements for exhaust of spent jets.

air extraction systems must be chosen to minimize any nonuniformities in the heat transfer distribution in the cross-machine direction. The seriousness of this factor depends on the nozzle configuration used; for example, it is more important for multiple slot arrangement than for round jet arrays. If the target surface moves rapidly with respect to the stationary nozzle boxes (i.e., moves at a speed at least 10% of the jet velocity at impact), the nonuniformities in heat-mass transfer rate in machine direction tend to even out. On the other hand, such smoothing does not occur in the case of cross-machine direction nonuniformities if they persist throughout the dryer length.

2.4. Selection of Nozzle Configuration

The two basic multiple jet configurations for nozzles are arrays of round jets (issuing out of orifices or perforated plates) and rows of slot jets. In both configurations, suitable exhaust ports must be located to prevent the deleterious effects of crossflow due to spent flow from upstream jets. In general, the exhaust ports are located to remove the exhaust flow before it affects the heat transfer performance seriously. For jet arrays it is common to provide a circular port or a rectangular slot for every five rows of staggered jets. For in-line arrays (not popularly used) the crossflow interference is expectedly much more significant. This is also true of slot jets since all of the spent flow must cross the downstream jet unless alternate paths of exhaust are provided. Care must be exercised in the design of extraction ports since these can lead to lateral nonuniformity in heat/mass transfer rate distributions although the jet flow itself is uniform.

Slot nozzles are convenient and easy to fabricate for smaller width dryers (1-2 m). For larger widths and for dryers employing elevated temperature jets, it is difficult to maintain a uniform jet width without developing excessive thermal stresses. The fabrication costs are high and are not justifiable. Most large-scale units (e.g., Yankee dryer for tissue) therefore use perforated plates to generate arrays of jets. They are easier and cheaper to fabricate.

Obot et al. (1) have compared the heat transfer performance of jet arrays and multiple slot jets. The difference between these two configurations is too small to allow a logical choice to be made solely on thermal grounds. In general, either one may be chosen if adequate provision is made to extract the exhaust gas properly.

2.5. Selection of Nozzle Geometry

Nozzle geometry effects can be profound, and it is only recently that these have been recognized (e.g., see References 2 and 3). Contoured nozzles yield a uniform velocity profile and the exit plane but with a lower turbulence level. Although the pressure drop is reduced at the given flow rate (discharge coefficient in excess of 0.90) for contoured nozzles, the heat transferred per unit pumping power may not necessarily be higher. Indeed, the cost of fabrication of countoured nozzles (round or slot) precludes their industrial application.

Obot (2) has compared heat transfer and discharge coefficients for jets issuing from round nozzles and tubes. More recently, Hardisty (3) has presented extensive data for discharge coefficients and heat transfer under eight differently shaped slot jets. Regardless of nozzle shape, he found that narrower slots give higher heat transfer coefficients. He carried out tests for single slot jets to determine (a) the effect of changing the shape of the nozzle holding the slot width constant, and (b) the effect of changing the nozzle width while maintaining the same nozzle shape. These data are valid for single slots, but the relative trends may be expected to hold even for multiple jets. He presented correlations for the average heat transfer coefficient using the effective nozzle width w' (which is less than the geometric nozzle width w), which is related to the geometric nozzle width w by

$$w' = wC_D$$

where C_D = discharge coefficient of the nozzle defined as C_D = (actual flow rate)/$(\rho w L V_j)$. Here, w = nozzle width, L = length of nozzle, ρ = air density, V_j = jet centerline velocity. Note that C_D can only be measured empirically. Florscheutz et al. (4) have shown that C_D for punched plates depends not only on the orifice geometry and Reynolds number but also on the array pattern and presence of crossflow. For small H, C_D may also depend on H.

It is impossible to give definitive recommendations for nozzle design. Simplicity of fabrication suggests the use of punched orifices or slots despite their low discharge coefficient values. More complex nozzle shapes are found in practice although their value in augmenting heat transfer rates is debatable. Often patent considerations dictate selection of the nozzle geometry and configuration. Regardless of the choice of the nozzle, it is important to provide for proper exhaust of the spent flow to ensure that it does not cross the neighboring jets.

2.6. Product Quality Considerations

Depending on the type and quality of product, the dryer designer will be limited in the choice of one or more of the design variables, namely, the type of nozzle (round or slot), jet temperature, and jet velocity. If the surface being dried is sensitive to mechanical stresses (e.g., coated or printed sheet), the velocity at impact is limited to a value determined experimentally in laboratory tests. If cross-machine drying uniformity is critical (e.g., coated papers), slot nozzles are preferred over arrays of round holes although the latter may yield higher drying rates at a given open area; indeed, according to Pinter and Greimel (5), the heat transfer rates for round hole arrays may be 50–100% higher than those for slot arrays of the same open area (1–4%).

The energy consumption for an impingement dryer consists of two components: (a) electrical power for fan drives and (b) heat for the air jets. The electrical power needed to achieve a given heat transfer coefficient depends on the design of the nozzle, plenum chamber, jet velocity, recycle ratio, and pressure losses in the ducting. In general, an optimum value for the internozzle spacing exists for a given slot width. (See Reference 5 for data on different slot widths, open area ratios, Reynolds numbers, and so on.) With good control of supply and exhaust air and good insulation, a well-designed dryer should yield a thermal efficiency of the order of 75–90%. Specific power consumption of 3100 kJ/kg of water evaporated have been reported for air floater dryers. When jets are used for pneumatic conveyance of the web or sheet to be dried, additional aerodynamic considerations need to be taken into account.

3. HEAT TRANSFER CORRELATIONS

3.1. Recommended Correlations

Obot et al. (1), Obot (2), Saad (6), Das (7), Martin (8,9), Dyban and Mazur (10), and several others have tabulated the numerous empirical correlations for stagnation and space-averaged heat transfer under single or multiple, round or slot turbulent jets impinging on stationary, impermeable, plane surfaces. It is not the intention of this chapter to list and review all these correlations. The reader is referred to the literature cited for details.

Since single slot or round jets are of no practical interest, correlation for this case will not be listed here. Only arrays of round and slot nozzles will be considered. Martin (9) has recommended the following correlations for these two configurations in the Hemisphere *Heat Exchanger Design Handbook*. Appropriate caution must be exercised by the user to account for various extraneous effects already noted and those discussed in subsequent chapters.

Figure 2 shows the spatial arrangement of round and slot nozzles in arrays and the averaging area appropriate to each. Let f = nozzle exit area/area of square or hexagon attached to it. The following table defines f for the triangular or square pitch round nozzles and parallel slot arrays.

ARN$_\Delta$:

$$f = \frac{\pi}{2\sqrt{3}} \left(\frac{D}{L}\right)^2$$

ARN$_\square$:

$$\frac{\pi}{4} \left(\frac{D}{L}\right)^2$$

ASN:

$$\frac{w}{L}$$

where

ARN$_\Delta$ = array of round nozzles, triangular pitch
ARN$_\square$ = array of round nozzles, square pitch
ASN = array of slot nozzles
L = spacing between nozzles

Martin gives the following correlations for these major configurations. For arrays of round nozzles (ARN),

$$\left(\frac{\overline{Sh}}{Sc^{0.42}}\right)_{ARN} = \frac{Nu}{Pr^{0.42}} = \left(1 + \left(\frac{H/d}{0.6\sqrt{f}}\right)^6\right)^{-0.05} \sqrt{f} \left(\frac{1 - 2.2\sqrt{f}}{1 + 0.2(H/d - 6)\sqrt{f}}\right) Re^{2/3}$$

Range of application is $2000 \leq Re \leq 100{,}000$; $0.004 \leq f \leq 0.04$; and $2 \leq H/D \leq 12$; the accuracy is $\pm 15\%$.

For arrays of slot nozzles (ASN),

$$\left(\frac{\overline{Sh}}{Sc^{0.42}}\right) = \frac{Nu}{Pr^{0.42}} = \frac{2}{6} f_0^{3/4} \left(\frac{4\,Re}{f/f_0 + f_0/f}\right)^{2/3} \quad \text{where } f_0 = (60 + 4(H/2w - 2)^2)^{-1/2}$$

valid in the range $1500 \leq Re \leq 40{,}000$; $0.008 \leq f \leq 2.5 f_0$; $1 \leq H/S \leq 40$; the accuracy is $\pm 15\%$.

Martin (9) has also provided a detailed discussion on optimal spatial arrangements of nozzles. Here the term *optimal* means a combination of geometric variables that results in maximal average $\overline{Nu}$ (or $\overline{Sh}$) for a given blower rating per unit area of the heat transfer or drying surface. Within engineering approximation, his optimization analysis yields the following results for both round and slot nozzles: optimal spacing for slots and optimal pitch for ARN are both $\approx 1.40H$; with $S_{opt} = 0.2H$ and $D_{opt} = 0.18H$. These optima should be taken only roughly. Other effects may overshadow these and shift the optima significantly.

3.2. Effect of Various Parameters

Following is only a brief outline of the various extraneous effects that have been investigated to varying degrees. For details the reader is referred to the literature cited. The effects of crossflow, movement of the impingement target, large temperature differences, high drying rate, and artificial turbulence are among the most important to be considered in the design of impingement dryers. With large temperature differences between the jet and the surface, some radiant transfer may also be present. All dryers must be enclosed:

the effect of confinement must be accounted for if the correlations used are for uncon-
fined impingement. The effect of entrainment of low-temperature ambient air is always
to reduce the temperature driving force and thus worsen the thermal performance. Infil-
tration of ambient air should be scrupulously avoided in impingement dryers, as in any
other types.

Effect of Crossflow

Spent flow from individual jets or clusters of jets must be exhausted properly or the
performance of adjoining jets can fall even by a factor of 2. For slot jets and rows or
arrays of round jets, experimental studies have measured the effect of induced crossflow
(i.e., crossflow due to upstream rows of jets made to cross the downstream jets to
exhaust), as well as imposed crossflow (i.e., a flow parallel to the impingement surface
that is not due to upstream jets). For ASN the effect of crossflow is severe (6). No more
than two jet rows should be allowed prior to exhaust. For staggered ARN arrangements
the crossflow effect is less severe; an exhaust port may be provided for every three to five
rows. After five rows the crossflow effect drops the average Nu rapidly in the downstream
direction (11).

If the drying performance of an existing dryer is to be improved by adding extra
nozzles, care must be taken that the extra jets do not impair the performance of existing
jets.

Another undesirable effect of crossflow is that, depending upon how it is directed
with respect to the motion of the surface being dried, one may encounter a nonuniform
moisture profile.

Recently, Galant and Martinez (12) have provided an empirical correlation for heat
transfer under ARN subjected to known crossflow. As a rule the designer should avoid
the development of strong crossflows in the dryer by providing well-located exhaust ports.
The work of Nagpal (13) and Kercher and Tabakoff (14) is also relevant when evaluating
the effect of spent flow on ARN impingement.

Effect of Semiconfinement

Saad et al. (11), Obot et al. (15), Folayan (16), and others have observed that, because of
the favorable pressure gradient induced by confining walls (parallel to the impingement
surface), the heat transfer rates are reduced somewhat (10–20%) compared with free jet
impingement. The reduction is more significant at lower spacings; at larger spacings the
effect may be neglected.

Most of the recent work, as opposed to the earlier work, is concerned with semicon-
fined impinging jets. The results are then applicable without correction for confinement.

In the absence of confinement, ambient air entrainment effects may cause unpredictable effects; for hot air impingement the heat transfer performance may worsen by 20–
50% depending upon the flow rate, difference between the jet and the ambient temperature, and nozzle-surface spacing.

Effect of Mass Transfer

Large mass fluxes normal to the surface are caused by high evaporation rates; this is
especially true of the high-velocity, hot air jet impingement employed in paper drying.
The presence of normal flux causes thickening of the thermal/concentration boundary
layer and hence a reduction in the heat and mass transfer rates. Since most correlations
used in practice are for no net mass flux, corrections must be applied to the empirically
determined $\overline{Nu}$ or $\overline{Sh}$. Keey (17) has discussed the correction factors developed on the
basis of film theory and boundary layer theory. For spray drying the reduction due to

high evaporation flux in the nozzle zone may be as high as 15%. Similar corrections ($\sim 10\%$) may apply in an intense impingement drying of paper.

More recently, Kast (18) has presented modified correction factors for high mass fluxes in laminar and turbulent flows. Although more complex in form, their effect on the calculated drying rates for superheated steam drying of paper was found by Loo and Mujumdar to be negligible (19). The simpler forms given by Keey are therefore recommended for practical calculations.

Effect of Large Temperature Differences

Very little work exists on the effect of large temperature differences between the jet and the target surface. Das (7) has presented empirical correlations of the following form for a single slot jet; the correlation may be applied to ASN for large internozzle spacings and for exhaust slots located midway between adjoining slots.

$$\overline{\mathrm{Nu}}_j = K \, \mathrm{Re}_j^a \left(\frac{H}{w} \right)^b \left(\frac{T_j}{T_s} \right)^c \mathrm{Pr}_j^{1/3}$$

His data and correlation apply over $5000 < \mathrm{Re} < 20{,}000$, $8 < H/w / 12$, and $1.18 \leq T_j/T_s \leq 2.06$. Values of a, b, and c are given in Reference 7. Because of higher viscosities of gases at higher temperatures, impingement dryers often operate with low jet Reynolds numbers (100–2000) for which no correlation exists for large temperature differences.

No correlations exist for other configurations. Correction factors recommended for tube flows may be applied for impinging jets in the absence of suitable correlations.

Effect of Suction

It is observed that a synergistic effect occurs when impingement drying is combined with through-drying, as in combined drying of permeable paper or textile. Application of suction increases the impingement heat transfer rate. However, it is not clear if the added capital and operating costs of through-drying are justified in a general application. For design purposes a 10–15% increase in the average impingement heat or mass transfer rate may account for the suction effect within engineering accuracy. Very limited experimental or analytic information is currently available to permit prediction of the suction effect.

It is worth noting that the product permeability in general will change with moisture content as drying proceeds. Thus, for a fixed applied pressure drop across the permeable target, the suction velocity will in general increase in the downstream direction.

Effect of Surface Motion

Van Heiningen et al. (20) have reported local Nu values for a confined single turbulent slot jet impinging on a rotating drum using a specially designed rig. Earlier, Fechner (21) made measurements of average Nu on rotating drums subjected to single and multiple slot jet impingement. On the basis of the limited data at hand, it appears it is safe to neglect the effect of surface motion if the surface linear velocity is less than 20% of the jet velocity at impact. The local Nu is altered significantly, but the average value is little affected. Work currently underway at McGill University will shed more light on the effects of surface motion with or without suction and oblique impingement (22,23).

Effect of Oblique Impingement

For rapidly moving impingement surfaces it is likely that an optimum angle of impingement (other than normal) may exist. Korger and Krizek (24) found that, with oblique impingement, although the local Nu changed the average value changed very little compared to normal impingement for a stationary surface. According to Baines and Keffer (25), the surface-averaged shear stress for a single turbulent slot jet impinging obliquely on a rotating drum was maximum for an angle of 60° between the target and the jet axis. A similar effect may be present for heat or mass transfer, although no analogy can be made between momentum and heat or mass transfer in the impingement region.

Miscellaneous

The effects of surface roughness, curvature of impingement surface, and artificially induced turbulence have been studied primarily with regard to their application in turbine vane cooling rather than drying (see Refs. 26–28).

To tailor heat/mass transfer rates and their distribution on the impingement surface, the designer has numerous possible choices. One that is not commonly exploited is the use of nonuniform arrays (e.g., varying D with fixed or variable pitch for ARN or variable S or W for ASN), which can be tailored to meet the requirements of varying drying rates. This is particularly useful for drying in the falling rate period of heat-sensitive products.

The effect of swirl in the jet flow has been studied by Ward and Mahmood (29). Since the $\overline{Nu}$ decreases with increasing swirl number, it is clearly undesirable to introduce swirl in the jet flow.

On the other hand, an artificial increase in jet turbulence in the region of the target surface helps augment the average heat or mass transfer rates significantly. Ali Khan et al. (30,31) have studied, for round jets, the effect of placing a perforated plate just upstream of the impingement surface. It is thus possible to manipulate the magnitude and distribution of Nu by varying few parameters. They used punched plates (diameter of holes = 3–10 mm, pitch = 4–36 mm with open area 0.63–51%). For a punched plate hole diameter 0.06 times the jet diameter, they found that maximal augmentation occurred for the punched plate placed three times the punched hole diameter. Although optimal, this spacing is too close for practical utility. Further work is needed to exploit this technique of augmentation of impingement heat/mass transfer.

The use of turbulence generation placed in the nozzle or upstream is wasteful since the jet turbulence decays rapidly as it approaches the impingement surface. Thus the added pumping power leads to little or no increase in the impingement transfer rate.

4. CONCLUSION

More recently Viskanta (32) and Polat (33) among others have reviewed the relevant literature on impingement heat and mass transfer. The theses of Bond (34), Chen (35), and Seyedein (36) contain extensive discussion of the empirical as well as computational fluid dynamic modeling of complex impingement-type flows. Effects of coupled surface motion and throughflow are discussed by Polat et al. (37). Bond (34) has presented a calculation procedure for evaluating impinging jet heat transfer under arrays of superheated steam jets, including the effect of evaporation and large temperature differences.

Seyedein (36) examined the effect of crossflow on multiple slot jet impingement, including the effect of inclining the confinement surface so as to accelerate the flow in the direction of exhaust. He showed that it is possible to maintain a nearly uniform heat transfer distribution by accelerating the exhaust flow to counteract the deleterious effect of crossflow. Saad et al. (38) have reported on experimental studies of multiple slot jet impingement with exhaust slots located midway between adjacent jets so as to eliminate crossflow effects.

ACKNOWLEDGMENT

The author appreciates with gratitude the assistance of Purnima Mujumdar in the preparation of this typescript.

REFERENCES

1. Obot, N. T., Mujumdar, A. S., and Douglas, W. J. M., *Drying '80*, Vol. 1 (A. S. Mujumdar, ed.), Hemisphere, New York, 1980.
2. Obot, N. T., Ph.D. thesis, McGill University, Montreal, Canada, 1981.
3. Hardisty, H., *Proc. Inst. Mech-Eng.*, *197C*: 7–15 (1983).
4. Florschuetz, L. W., Metzger, D. E., et al., *NRSA Cr. Rept.* 3630 (1982).
5. Pinter, R., and Greimel, R., *Proc. 3rd Int. Drying Symp.*, Birmingham, AL, U.S., September 1982.
6. Saad, N. R., Ph.D. thesis, McGill University, Montreal, Canada, 1981.
7. Das, D., M.Eng. thesis, McGill University, Montreal, Canada, 1982.
8. Martin, H., *Advances in Heat Transfer*, Vol. 13, Academic Press, New York, 1977, pp. 1–60.
9. Martin, H., *Heat Exchanger Design Handbook*, Hemisphere, New York, 1982.
10. Dyban, Y. P., and Mazur, I. A., *Impinging Jet Heat Transfer*, Kiev, Nakova, USSR, 1983 (in Russian).
11. Saad, N. R., Mujumdar, A. S., Abdel-Messeh, W., and Douglas, W. J. M., ASME Paper, 1980.
12. Galant, S., and Martinez, G., *Proc. 7th Int. Heat Transfer Conf. Munich*, FC60, Hemisphere, New York, 1982.
13. Nagpal, S. C., Ph.D. thesis, Colorado State University, Ft. Collins, CO, 1974.
14. Kercher, D. M., and Tabakoff, W. J., *J. Eng. Power Trans. ASME*, 73–82 (January 1970).
15. Obot, N. T., Mujumdar, A. S., and Douglas, W. J. M., *Proc. 7th Int. Heat Transfer Conf.*, Munich, Hemisphere, New York, 1982.
16. Folayan, C., Ph.D. thesis, Imperial College, London, England, 1977.
17. Keey, R. B., *Introduction to Industrial Drying Operations*, Pergamon Press, London, 1978.
18. Kast, W., *Proc. 7th Int. Heat Transfer Conf.*, Vol. 3, Munich, FC47, Hemisphere, New York, 1982.
19. Loo, E., and Mujumdar, A. S., *Drying '84*, (A. S. Mujumdar, ed.), Hemisphere, New York, 1984.
20. Van Heiningen, A. R. P., Mujumdar, A. S., and Douglas, W. J. M., *Proc. 1st Int. Symp. Turbulent Shear Flows*, Pennsylvania State University, College Park, PA, 1977.
21. Fechner, G., Dr.Eng. dissertation, Technical University, Munich, 1972.
22. Huang, B., Ph.D. thesis, McGill University, Montreal, Canada, 1987.
23. Polat, S., Ph.D. thesis, McGill University, Montreal, Canada, 1987.
24. Korger and Krizek, *Verfahrenstechnik (Mainz)*, 6:223 (1972).
25. Baines, W. D., and Keffer, J., *Drying '80*, Vol. 1 (A. S. Mujumdar, ed.), Hemisphere, New York, 1980.
26. Sakipov, Z. B., Kozhakhmetov, D. B., and Zubareva, L. I., *Heat Transfer—Soviet Research*, 7(4) (July/August 1975).
27. Mori, Y., and Daikoku, T., *Bull. JSME*, 15(90) (1972).
28. Mujumdar, A. S., McGill University, unpublished work, 1981.
29. Ward, J., and Mahmood, M., *Proc. 7th Int. Heat Transfer Conference, Munich*, Hemisphere, New York, 1982.
30. Ali Khan, M., Ph.D. thesis, University of Tokyo, 1980.
31. Ali Khan, M., Kasagi, N., Hirata, M., and Nishikawa, N., *Proc. 7th Int. Heat Transfer Conf.*, *Munich*, CFC63, Vol. 3, Hemisphere, New York, 1982, pp. 363–368.
32. Viskanta, R., *Experimental Thermal and Fluid Science*, 6:111–134 (1993).
33. Polat, S., *Drying Technology*, 11(6) (1993).

34. Bond, J.-F., Ph.D. thesis, McGill University, Montreal, Canada, 1990.
35. Chen, G., M.Eng. thesis, McGill University, Montreal, Canada, 1989.
36. Seyedein, S., M.Eng. thesis, McGill University, Montreal, Canada, 1993.
37. Polat, S., Mujumdar, A. S., and Douglas, W. J. M., *Can. J. Chem. Eng.*, 69:266–274 (1991).
38. Saad, N. R., Polat, S., and Douglas, W. J. M., *Int. J. Heat and Fluid Flow*, 13(1):2–14 (1992).

15
Flash Drying

Bilgin Kisakürek
Ankara University
Ankara, Turkey

1. INTRODUCTION

Drying is a thermal process in which a solid material is stripped of its moisture content by means of the heat supplied to it. It is one of the most important unit operations used in the chemical process industry. Since most manufactured solid products are dried at least once at some stage of their manufacture, drying has a very wide application in the chemical industry. In addition, the tendency of industrial processes to use more economical methods based on technical progress has also had its effect in one of the most common industrial steps, that is, in the drying process.

Fulford (1), in his major survey of research on the drying of solids, comments that drying remains largely an art even though the number of publications on this area increases very rapidly every year. It is called an art because most of the design methods available for various dryer types are generally empirical in nature. Flash drying is a very good example.

There are several reasons that drying has received low priority in terms of academic and industrial research in the past. The first is that the other process steps, such as the reaction kinetics and reactor design, are generally assumed relatively more critical than drying. Another reason for the slow development of drying science is the breadth of the subject. During drying, simultaneous heat and mass transfer occur both at the surface and inside the solid structure. The effect of other external forces, such as gravity and radiation, should also be included in the analysis. Finally, in addition to these, the dryer used for a specific purpose also imposes very important geometric forces on the controlling drying mechanism. Because of these complicated effects, one can easily say that no other unit operation covers the handling of such a wide range of materials and of different types of equipment in which the choice of equipment and operating conditions can have such a strong impact upon the product's properties.

1.1. Flash Drying

Flash dryers are typical examples of suspended bed dryers. Fulford's remark seems to apply to suspended bed dryers to a greater extent than to any other type of dryer. To

503

carry out a suspended bed drying operation, one must be able to form a suspension in a gaseous drying medium at an elevated temperature. Experience comes into effect at this stage. Other examples of such type of dryers are spray and fluid bed dryers. It should be noted that these three processes do not tend to compete for the same spot in an overall process scheme because they handle different types of feed and differ widely in operation. In a published survey of pneumatic drying equipment (2,3), it was suggested that flash dryers are more efficient than rotary dryers. However, this statement was not supported by any technical data.

Drying Phenomena

Flash drying of fine solid materials is based on the phenomenon of the homogenous suspension of these materials by the hydrodynamic effect due to the upward flow of the drying and conveying medium. Apart from certain other disadvantages, the main drawbacks of these processes are the limitation of particle sizes to fine powders and control problems for both balanced suspension and entrainment at desired hydrodynamic levels. On the other hand, the main advantage is the large contacting surface area for heat and mass transfer, which is very important mainly in diffusion-controlled drying processes. High heat and mass transfer rates are possible because of good contact between particles and the gas. Another advantage is the great circulation intensity of the solid material, which approaches ideal mixing, thus giving uniform product moisture concentration. Not the least important advantage is the easy transportation of free-flowing materials, combined with several other specific advantages. Intensive solids mixing, because of the presence of bubbles, gives almost homogenous bed temperature control and operation of the bed to the highest temperature acceptable in relation to the thermal degradation of solids.

Flash dryers are used to dry efficiently those materials that can be transported pneumatically in large volume. During this conveying operation, small solid particles can be dried by preheating the carrier air or gas to such an extent that it acts as a drying medium. These types of dryers always operate in a cocurrent manner, and the drying takes place in a very short period of time.

Dryer Operation

The operation of flash dryers is very simple. Fine, powdered, wet feed material is introduced into the dryer together with the hot drying air or gas that conveys the drying material to the dry product collection system. The name *flash* comes from the very short residence time, say, 0.5–3.5 s. A typical flash dryer unit is illustrated in Figure 1. Such a simple system consists of an air heater, a material feeder, a drying duct, a separator, and an exhaust fan. A modification of this system could include a dispersion unit, which is required to break up the dried product into an acceptable particle size. The air heater, product collection system, and air moving device all have much the same requirements as for other drying systems. Drying ducts and feed systems are very important in terms of successful flash dryer design. Feed to a flash dryer must be in particle form. Because of the very short residence times, the feeding mechanism should supply nonlumpy wet solid into the stream of drying air with a very controlled flow rate.

Applications

Granular materials that are relatively free flowing in the wet state can be dispersed merely by dropping them into the heated airstream. Sludges, filter-press cakes, and similar nongranular or lump-tendency materials, however, must be disintegrated by a disintegrator. It must be remembered that all of the feed must be immediately accelerated from a static

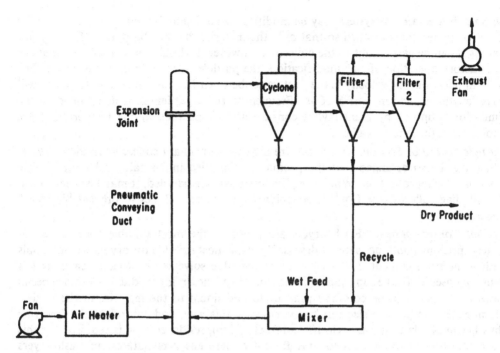

Figure 1   A typical flash dryer system.

condition to the conveying velocity within the dryer. This acceleration can be done in the disintegrator. In some cases, materials that have a tendency to form lumps in the wet state must be mixed with the dry product coming from the recycle line to permit suitable disintegration. Typical applications are with distiller's spent grain, sludge, corn gluten, gypsum, and pigments. Mixing the wet feed with recycled dry product also prevents blockages, fires, and other undesired effects because the recycle stream in a dry condition and at high temperature somewhat impinges first on the dryer walls before the wet material hits and sticks there. However, great care must be taken with the geometry of the dryer in the feed area. In addition, the feed system must be properly controlled so that it supplies the feed material at a very uniform rate into the dryer. Otherwise, the adjusted residence time in the dryer will be changed, resulting in off-specification of the product quality.

Equipment

Feed Section.   For proper vertical pneumatic conveying, gas velocities must exceed a certain minimum because of the phenomenon of choking. For a given gas flow rate, any discontinuities in the solid flow rate causes the entire suspension to collapse into a slug flow in which the drying rate reduces to very low values. Slug flow must be avoided for large-diameter ducts because of the violent vibrations that may be generated. In conventional feeding systems, finely pulverized materials with proper size distribution are compressed by a screw pump and aerated in a mixing chamber. Screw conveyors, vibratory conveyors, rotary air locks with rotating table feeders, ribbon mixers, and venturi feeders are used singly or in combination.

Dryer Section.   The dryer section of the system may be a square or a circular vertical duct in which the solids are conveyed by the hot airstream. To increase the time of contact with the heated air, the system may be modified by adding an inverted cone in which the air velocity decreases upward so that only the material that has become sufficiently dry and light can be carried out to the collector. However, it should be remembered that for satisfactory operation of this modification, the particles should be uniform in size. For most applications the velocity in the duct varies between 15 and 30 m/s, and the dryer will have lengths in the range of 5–50 m. Frequently, the dryer length is determined by the limitations imposed by plant layout considerations, and then drying temperatures are chosen to fit the length available.

Particle Separator Section.   Cyclones are usually the primary collectors in these dryers. They may be so arranged that the product is classified into small- and large-particle fractions during collection. When very fine materials are handled, it may be necessary to provide bag collectors or Cottrell precipitators after the cyclones to prevent high yield losses.

Special Considerations.   Flash dryers are probably the most suitable form of atmospheric pressure continuous direct dryers. They are most suitable for drying wet materials with a moisture content in the form of recoverable solvents. A stream heater unit is generally used as the heating source if air is used for the drying medium. In some special cases (4), it has become common practice to add liquid to the material in the feeder. Homogeneity in particle size can be increased in this way, and control of the feed rate thus becomes relatively easy. On the contrary, a long residence time is required for this extra moisture in the system to dry. Recently there has been interest in flash dryers handling liquid feed only. Such systems are commercially attractive if the excessive recycle rates and long residence times are not affecting the product quality. Larger product sizes can be obtained in this way owing to agglomeration.

2. DRYER TYPES

The flash dryers used in industry can be analyzed in two groups, conventional and alternate designs. Williams-Gardner gives a comprehensive picture of the flash dryers commonly used in the chemical industry (5). Only a short summary will be given here.

2.1. Conventional Design

A.P.V.-Kestner Thermo Venturi Dryer

The A.P.V.-Kestner thermo venturi dryer has a vertical drying duct. The duct has a convergent-divergent section similar to a venturi shape for establishing the optimum air velocity. The diameter of the venturi is adjusted in such a way that the largest particles in the wet feed are in complete aerodynamic balance. The modifications to this simple dryer are contraflow and S-T-V dryers. With the help of this modification it is possible to extend the particle residence time for drying of larger particles with high surface moisture contents. Therefore, excessive column heights are prevented to a large extent. This is achieved by using double columns with an expansion chamber between them.

Rietz Air-Lift and Proctor-Mark Dispersion Dryers

The Rietz air-lift and Proctor–Mark dispersion dryers operate on the principle that the material is first pneumatically entrained at the bottom of the dryer column. Then, it is passed through several expansion chambers in which the gas velocities are alternately reduced and increased so that larger materials with a diffusion-controlled drying mechanism can be properly dried, because with this alternate type of operation, longer residence times can be obtained. In the Proctor-Mark dryer, the vertical tube contains several expanded sections in which the drying air is spun at high velocity by introducing secondary air tangentially into the chambers. Material circulation may be modified for larger particle sizes and secondary hot gases may be introduced at the expansion sections so that the residence times for such large particles can be further increased.

Raymond Flash Dryer

The Raymond flash dryer has conventional design features, namely, the drying air, after being heated, is sent to the bottom of the vertical conveying and drying duct. Moisture removal is accomplished by dispersing the material to be dried in a hot air zone followed by conveying at high velocities. The wet material enters the vertical duct at the bottom of the column. After drying is finished, the dry product is removed from the airstream in a product collection system. Drying, disintegration, grinding, pneumatic conveying, and classification can be accomplished simultaneously in this dryer by proper design of the component parts. The possible modifications to this dryer can be introduction of a disintegrator, a dry product recycling system by a back-mixer and a pulverizer. When furnished with a disintegrator, the wet material and hot air are usually both fed into the disintegrator so that initial drying occurs within it. In the recycle system, the dryer is maintained under suction, which is produced by the exhaust fan. This fan is usually located after the dust recovery equipment. Performance data (6) are given in Table 1.

Buttner–Rosin Flash Dryer

In the Buttner–Rosin flash dryer, drying air together with entrained feed material first passes through a vertical drying duct. Then, they pass through a pneumatic sifter and a classifier. The larger and heavier particles are separated here and sent back to the disintegrator for further drying. It is possible to modify this dryer system by designing an alternate arrangement so that the system is composed of two-stage dryers. The larger and

Table 1 Manufacturer's Performance Data for Flash Dryers

Data	Filter cake	Clay (dry)	Coal
Dryer size, evaporator capacity, lb/h	3500	2060	7.480
Inlet moisture, & wet basis	80	27	9
Outlet moisture, & wet basis	10	5	3
Inlet gas temperature, °C	700	525	650
Outlet gas temperature, °C	120	75	80
Air rate, ft^3/min	7600	5000	32,000
Production rate, lb/h	1000	6700	113,300
Air-solids ratio in duct	0.004	0.34	1.0
Material temperature in, °C	15	15	15
Material temperature out, °C	70	50	60
Type of fuel	Gas	Oil	Coal
Fuel consumption, Btu/lb water evaporation	1725	1700	1,600
Drying without disintegration	1600–3750		
Drying with disintegration	1600–2800		
Power consumption, kWh/lb water evaporation	0.012	0.037	0.010
Drying without disintegration	0.008–0.015		
Drying with disintegration	0.010–0.020		

heavier particles from the classifier after the first dryer are sent to the second dryer for further processing.

Stork–Bowen Dryer System

The Stork-Bowen dryer system is a modification of the Buttner–Rosin dryer and is generally composed of two stages. Buttner–Rosin, Stork–Bowen, and Barr and Murphy flash dryers are typical examples of such dryers; the residence time for larger particles is increased by performing the material classification at the dryer outlet and recycling over-size wet material. These units have the advantage of being able to mill the oversize material if required. These dryers offer greater versatility in operation than the other types (4).

2.2. Alternate Design

Dispersion and Stephanoff Dryers

The J. S. Dispersion dryer, Stephanoff dryer (6), Berk Ring dryer, Alfred Herbert Atritor-dryer pulverizer, and the Buttner jet dryer can be considered alternative designs with respect to formal conventional flash dryer designs. In the J. S. Dispersion dryer, the wet material entrained in the hot airstream is injected at high velocity upward into a free venturi. Thus, suction is produced at the venturi entrance point. Fine and easily dried particles are ejected from the venturi outlet. Larger and less easily dried particles, on the other hand, stay in the dryer by internal recirculation. The Stephanoff dryer is a combination dryer and fine grinder in a closed-circuit system. The wet particles are introduced through hot air or stream jets at sonic velocities. The wet material is circulated in the closed system and simultaneously dried and ground.

Berk Ring, Alfred Herbert Atritor, and Buttner Jet Dryers

In the Berk Ring dryer design, the drying duct is a ring duct through which the wet particles are circulated pneumatically by the hot gases. Humid exhaust air is continuously extracted and replaced by the hot gases. The dried product leaves the duct at the center of the duct ring. The Atritor-dryer pulverizer pulverizes the wet material by pure attrition. In this zone the material is repetitively carried forward toward the center and outward to the periphery until the desired final degree of drying and pulverization is achieved. The Buttner jet dryer was originally developed for the airborne drying of wood chips in the particleboard industry. The wet material enters the horizontal dryer, which is composed of a cylinder having a rotating agitator, and leaves the dryer by passing to a cyclone. Heated gases are introduced tangentially at high velocities through lateral jets. This tangential movement of the drying gases maintains the desired drying suspension. A possible modification, again, is an alternate design for recirculation of the less well dried products.

3. PROPERTIES OF FEED

The feed to a flash dryer must satisfy the following requirements:

1. It must be in fine and homogeneous particle form.
2. It should flow freely in the system, especially while it is in a wet condition.
3. It should not be so large that clogging of the venturi-type narrow joints occurs.
4. It should easily be dispersed and entrained by the turbulence generated by the hot airstream.

With these properties, one can form a suspension in the drying medium either directly or with back-mixing. In this way, high heat and mass transfer rates are possible because of good contact between feed particles and the drying medium. Intensive solid mixing, due to the presence of turbulence generated, gives almost homogeneous duct temperature, allowing easy temperature control and operation of the dryer up to the highest temperature acceptable in relation to thermal degradation of the solids. Moreover, the long residence times needed to overcome possible diffusion limitation can be achieved in an apparatus that is relatively simple and has no moving parts.

In addition to the feed specifications already mentioned, the following requirements should also be considered.

1. It should easily be conveyed so that power consumption is minimized.
2. It should not be very heat sensitive because very high temperatures are usually employed in flash dryers.
3. It should have good drying characteristics; the moisture migration by capillary movement controls the drying mechanism instead of the diffusion mechanism so that very high drying rates are maintained in a very short period of drying.
4. The material should enter the dryer with a relatively low moisture content, so that excessive lumping of the feed is prevented.
5. It should not be abrasive, scarred, or broken up.

3.1. Drying Rate

Final product temperature is critical in most dryers because that temperature determines product moisture more than any other. Since the temperature of a wet solid in hot air depends on the rate of evaporation and drying, the key to analyzing all these dryers is the psychrometric relations modified for high-temperature surroundings (7). Psychrometric charts simplify the crucial calculations of how much heat must be added and how much moisture will be removed from the system. Thus, they are an indirect indication of the drying rate.

3.2. Design Criteria

With the help of psychrometric charts and detailed drying rate calculations (8), one can easily illustrate the opposing factors of interest in the design of flash dryers as follows:

1. A high air inlet velocity leads to a very high drying rate in the acceleration zone because of the high relative velocity. However, the residence time in the dryer is lower.
2. Large product particles have longer residence times in the dryer than smaller particles for the same air velocity because of the smaller relative velocity of the latter. However, the drying rate is lower in this case.

The material that enters the dryer with a given initial velocity first gains acceleration. During its stay in the dryer, its temperature rises from its initial value up to the wet bulb temperature of the heating medium. Drying occurs under a constant rate period of drying and reaches its maximum in this period. For the case of fine particles, the critical moisture content is easily reached and drying extends into the falling rate period. The temperature of the material increases further, and in the extreme case, the air and product temperatures become equal. This condition is unlikely to be reached for relatively large particles.

Because of the increase in product temperature and the drying process, which begins as soon as the product is fed into the dryer, the air temperature falls relatively quickly, which leads to a reduction in velocity. The air velocity may increase by a relatively small amount due to the decrease in density caused by the pressure drop. The curve of the product velocity, after the acceleration section, runs parallel to the air velocity curve. However, the distance between the two curves, the relative velocity, must reduce with time owing to the loss in weight and possible shrinkage of the product during the drying process, which is taking place at a high temperature.

3.3. Factors Affecting the Drying Rate

The drying rate of a particle in such a system depends on many factors, which can be summarized as follows.

1. Particle diameter
2. Type of moisture transfer mechanism, whether it is diffusion or capillary mechanism controlling
3. Length of diffusion path for moisture transfer within the particle, if it is diffusion mechanism controlled
4. Length and size of particles, if it is capillary mechanism controlled
5. Size uniformity
6. Physical properties of the wet and dry solid particles
7. Product mass flow rate
8. Hot air mass flow rate
9. Diameter of the dryer section
10. Critical moisture content of the material
11. The characteristic drying rate curve

3.4. Observations

For small particles, it can be assumed that the moisture transfer within the solid structure takes place by capillary forces; thus the resistance to heat and mass transfer is relatively lower with respect to diffusion-controlled drying. In most dryers, the particle sizes are not uniform but have a reasonably wide distribution. This effect is enhanced when agglomerates are taken into consideration. As a result, the dryer contains both fine particles and coarse particles. The fine particles, having relatively small resistance to drying because of the capillary flow controlling mechanism, can be overdried, whereas the coarser ones, in which the moisture migration by diffusion is predominant, are underdried. To overcome this undesirable result, most of the flash dryers available in industry have recycle design arrangements. With the help of such modifications, the coarser particles are allowed to remain longer in the dryer and thus more uniform product distribution can be obtained. The uniformity in the product moisture distribution also lowers the required driving force, giving a lower energy consumption. Operation at lower temperatures can also be maintained, which causes less thermal damage to the heat-sensitive materials and, to a certain degree, to the equipment itself.

4. FEED MATERIALS

Flash dryers can be used to dry most chemicals, food products, polymers, several by-products, and minerals. Because of the short residence time in the dryer, even temperature-sensitive products can be dried, using high air inlet temperatures, without suffering a loss of quality. Magnesium sulfate, magnesium carbonate, copper sulfate, dicalcium phosphate, ammonium sulfate and phosphate, calcium carbonate and phosphate, and boric and adipic acids are common examples of chemicals and by-products. Antibiotics, salt, blood clot, bonemeal, bread crumbs, cornstarch, corn gluten, casein, gravy powder, soup powder, vegetable protein, spent tea, wheat starch, soybean protein, meat residue, and flour are examples of food products. Cement, aniline dyes, blowing agents, chlorinated rubber, coal dust, copper oxide, gypsum, iron oxide, and silica gel catalyst are typical by-products and minerals that can be dried in a very efficient way in flash dryers. In the plastic and polymer industry, flash dryers can be used for drying of rubbers, polyethylene, polypropylene, polystyrene, and polyvinyl chloride.

4.1. Drying Metallic Stearates, Starch, Acids, and Cakes

The Buttner–Rosin flash dryer is generally used for drying metallic stearates to reduce the initial moisture content from 40% wet basis to 1% (6) with the help of material recirculation. Other applications for this dryer are the drying of starch, adipic acid, fibers, and coal filter cakes. Almost all types of starch can be dried, but corn, potato, and wheat starches are more common. Care should be taken in controlling temperatures and avoiding gelatinization. In starch drying, this type of dryer is generally heated by steam although, providing proper attention is paid to the explosive nature of the dried product, direct firing with natural gas is also possible.

4.2. Drying Minerals, Grain, and Organics

The Raymond flash dryer is used for drying fine minerals, spent grain, organic chemicals, and fine coal filter cakes. Relatively higher product rates can be achieved (2000–25,000 lb/h), and moisture can be reduced from 40–60% to 10% on a wet basis in these dryers. Direct oil heating is utilized in Raymond flash dryers, and material can be recirculated if necessary.

4.3. Drying Clays, Cement, and Coal

Even though the Alfred Herbert Attritor dryer was originally designed to pulverize and dry the coal generally used in firing cement kilns, these dryers can now be used for drying limestone, chalk, clay, and filter cakes in addition to coal. Because of the limitations in pulverization operation, the material size should be somewhere between 100 and 300 mesh. Very dry products are obtained.

4.4. Drying Polymers

Suspension-grade polymer materials and copolymers are frequently dried in a two-stage system (4). The first stage is the flash dryer and the second a fluid bed dryer. It is possible to arrive at the required product final moisture content by flash drying alone, but because of the residence time available in the flash dryer, the high temperatures required give an unsatisfactory product. There is a wide range of copolymers varying in molecular weight, particle size, and other properties, and they all have different drying characteristics. For these reasons the dryer velocities are generally left low, around 15 m/s, and special care is exercised in handling the dried product.

The following two factors should be considered in polymer drying:

1. The limitations on dryer temperature to avoid softening of the polymers
2. The overall operating partial pressure level of the solvent in the system

In polypropylene drying, the feed enters the dryer wet with hexane. Longer residence times are necessary, but this introduces danger with the flammable solvent. Therefore, closed-cycle drying in a nitrogen atmosphere is used, and the solvent is recovered by a scrubber-condenser.

5.   TRANSPORT PHENOMENA IN FLASH DRYERS

It can easily be said that the available literature on the transport mechanism in flash dryers is relatively scant with respect to the other dryer types. Because of this and other complications present in flash dryer operations, there are still no generally reliable theoretically accepted principles for calculating momentum, heat and mass transfer coefficients, and their corresponding solution for gas suspension systems. In some cases the problem with heat and mass transfer both for an individual particle and for a gas suspension may be exactly formulated theoretically by the corresponding system of transport equations. However, when heat-sensitive materials and high-temperature surroundings are involved, this problem cannot be theoretically solved owing to its complexity. In order to obtain a method of calculation, one must definitely refer to the available experimental data and semiempirical relations. Unfortunately, even at this stage, several difficulties should also be faced. This is because of the complexity of these transport processes in gas suspensions, which easily restricts the use of experimental data because there is no reliable model for such a mechanism (9). In addition, the available experimental data contradict each other and generally are not always reliable.

In this section, we attempt to provide the user with some design correlations that can be used to give an idea about the transport mechanism. A thorough survey of the available literature indicates that it is possible to use well-established correlations for the range of existence and the other particle-gas heat and mass transfer equations (10–13).

5.1.   Problems in Using Transport Correlations

In order to have a detailed design of flash dryers, one must consider the factors that should be used in (a) momentum transfer, (b) heat transfer, (c) mass transfer, and (d) drying rate equations.

Factors Involved

The factors involved in selecting the available transport correlations are summarized below.

1.   Shape and sphericity of particles
2.   Diameter of the average-size particles
3.   Diameter of the dryer duct
4.   Density and heat capacity of the feed
5.   Enthalpy of desorption, if any
6.   Diffusion constants
7.   Inlet drying air temperature and humidity
8.   Air or gas flow rate
9.   Critical moisture content of the feed
10. Moisture levels desired at end points of the dryer

11. Thermal damage that may occur to heat-sensitive materials and to the equipment itself

Design Parameters

With these conditions and using the available semiempirical correlations, it may be possible to get some information on the following design parameters:

1. Characteristic drying rate (14,15).
2. Residence time in the dryer. It decreases with an increase in the flow rates of both phases. The effect of solid load is not important here (13).
3. Flow pattern and its dependence on the geometry of the dryer.
4. Particle velocity and Reynolds number. Note that the average particle velocity at the very dilute feeding rate is several feet per second times higher than at the denser feeding rate and that the larger particles travel faster than small particles having the same density. The dependence of gas velocity on particle velocity is in general less with the pneumatic riser (13,16,17).
5. Gas-particle mean slip velocity, which is defined as the difference between gas and solid velocities. Note that a high slip velocity occurs in relatively short duct segments. Figures 2 and 3 illustrate this effect (2).
6. Drag coefficient (18,19).
7. Friction factor and pressure drop. Kmiec and Mielczarski (13) have found that the pressure drops in the spouting chambers of a flash dryer depend on the flow rates of both phases and they increase with the ratio of the chamber diameter to the riser diameter.
8. Heat transfer coefficient and the Nusselt number.
9. Mass transfer coefficient and the Sherwood number.

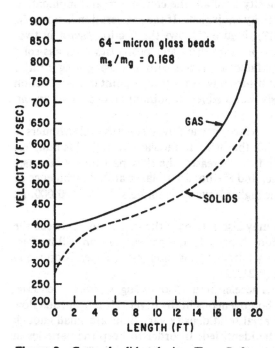

Figure 2 Gas and solids velocity. (From Reference 2.)

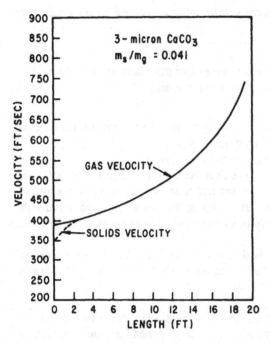

Figure 3 Gas and solids velocity. (From Reference 2.)

Other Considerations

In addition to these design parameters, one must also consider the following complications during process design. If the slip velocity is small, the continuum and momentum summation mathematical models are essentially equivalent. If the acceleration is slow, the choice of drag coefficient is not critical (19). Figure 4 shows the friction factor in two-phase flow using available correlations for pressure drop and properties of a gas-solid mixture. The rate of decrease in the friction factor decreases with increasing weight ratio. This is consistent with the observation that if the converse is true, a point of zero friction would be approached, which is obviously impossible. A minimum occurs at $m_s/m_g = 0.8$.

The character of gas motion near the surface of stationary spheres and cylinders is another complication. It has been shown (20) that with an increase in Reynolds number, the eddy formation at the rear of the body flow increases, the flow past the frontal part of a sphere remains by its nature invariable, and the boundary layer at low turbulence of an incoming flow remains laminar up to the higher Reynolds numbers, Re = 100,000. At this point turbulence starts to occur.

The presence of the adjacent particles may also influence the interaction between the gas and a particle owing to particle collision. A particle in a gas suspension will interact with a gas flow differently from a stationary particle. This interaction depends on many factors that are very peculiar and complex (9,21,22).

The available literature (2,23–25) on the acceleration of solids suggests that the radial effect may be omitted when solids flow is homogeneous in the dryer. Homogeneous flow is defined as flow in which the radial and axial solid density variations are small enough that clusters of solids (slug flow) cannot be identified. In order to keep the particles in uniform suspension, the turbulent fluctuations must offset gravitational effects. There is

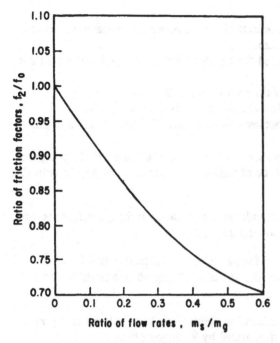

Figure 4   Two-phase friction factor. (From Reference 2.)

a substantial interaction of solids with the wall when there is a pressure drop for gas-solid suspension flow (26).

The concentration gradient of solids is another important parameter in the description of flash dryer design. However, it should be noted that no data have been presented for these fluctuation estimations at high Reynolds numbers (20).

5.2.   Pressure Drop

The effect of solids in gas-solid flow in flash dryers can be described by measuring the velocity profile of the solids, their concentration distribution, and their effect upon eddy dissipation. The experiment should be performed in such a way that it yields homogeneous flow of solids in the pilot-scale dryer. Recall that the homogeneous flow requires high Reynolds numbers.

The semiempirical correlations available in the related literature (19,20,27–29) can only be used with great caution for predicting the momentum transport properties in these dryers. In general, most of the correlations and theoretical approaches are based on several simplifying assumptions.

Assumptions

In order to simplify the mathematical models so that this complicated dryer behavior can at least be formulated, one must assume the following:

1. The flow is steady, and the motion is one dimensional. The effect of turbulence enters only in a characteristic parameter.
2. The solid particles are uniform in diameter and are distributed over each cross-section even though it is understood that they are suspended by turbulence and interactions exist between them.

3. The physical properties are constant.
4. The drag on the particles is mainly due to differences between the mean velocities of particles and the surrounding gas stream.
5. Slip flow, if it occurs, can be accounted for by an appropriate characteristic parameter.
6. The velocity and density variations of the solids are small.
7. The volume occupied by solid particles is neglected, as is the gravity effect.
8. The driving force for heat transfer between the gas and the solid is basically due to their mean temperature difference.
9. The solid particles, owing to the small size and high thermal conductivity compared with those of the gas, are assumed to be at uniform temperatures at any level in the dryer.

The acceleration of solid particles is calculated using standard drag coefficients, and the use of this correlation involves the following questions:

1. Does the standard drag coefficient hold for particles in accelerating flow?
2. In dilute systems, is it possible to consider the drag acting on one particle and then sum over the other particles?

Measurements of drag coefficients in accelerating fields have been attempted by many individuals, and their efforts have been summarized by Rudinger (29).

Drag Coefficients

In the movement of nonspherical particles in the dryer, the rotation of particles is observed (30). Turbulence plays a significant role here. The distribution of particle concentration, mass flow rate, and velocities in a vertical pipe are the parameters that should be taken into account. For a microsize particle, the concentration increases near the wall of the pipe, whereas the mass flow decreases.

The drag coefficient appears to depend primarily on the magnitude of the relative turbulence intensity and on the particle Re. The determination of the particle velocity and the concentration distribution represents a complicated problem for flash dryers, especially for nonspherical particles.

A simple approximating equation for the drag coefficient of a single sphere, which is correlated from experimental data (2,20), is

$$C_D = \frac{1}{3}\left[\left(\frac{72}{\mathrm{Re}}\right)^{1/2} + 1\right]^2 \tag{1}$$

which describes with sufficient accuracy the theoretical and experimental curve over the entire pertinent range of $0 < \mathrm{Re} < 10^5$. This is illustrated in Figure 5.

The drag coefficient for a swarm of particles is less than that of a single sphere (2,31) and is given as

$$F = \frac{M_s V_s}{g_c} \frac{dV_s}{dx} = \frac{C_R \rho_f (V_g - V_s)^2 A_p}{2g_c} \tag{2}$$

where F is drag force (lb), C_D is drag coefficient for particles, V_g and V_s are gas and particle velocities (ft/s), ρ_f is fluid density (lb/ft^3), and A_p is area (ft^2). Equation [2] gives the solid velocity as

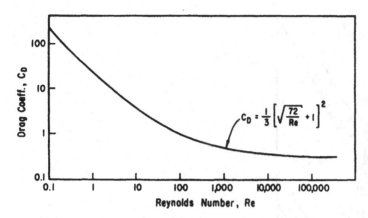

Figure 5 Drag coefficient of a single sphere(s).

$$\frac{dV_s}{dx} = \frac{3C_D\rho_f(V_g - V_s)^2}{4\rho_p D_p V_s}$$ [3]

In design, the minimum flow rate for dilute phase solid transportation in a gas stream for a flash dryer can be calculated from (32,33)

$$\log \text{Fr} = \frac{V_e - 2}{28} + 0.25 \log F_1$$ [4]

where Fr is the Froude number, F_1 is feed rate, and V_e is entrainment velocity. This correlation seems to be valid for larger dryer diameters. The entrainment velocity is related to the drag coefficient as

$$V_e = \frac{4}{3} \frac{D_p}{C_D} \frac{\rho_s - \rho_g}{\rho_g} g$$ [5]

Figure 6 shows the pressure gradients, in feet of water per foot line (18), as a function of air velocity for three different levels of solids flow rate. Note that in these curves a typical shape for this type of plot, in which pressure gradient decreases at first with increasing air flow, goes through a minimum, and then increases again. With air velocity just below the minimum pressure gradient, slugging may occur. The same tendency is observed with empty and packed lines.

Applying the continuity equation and the ideal gas laws and assuming constant temperature, McCarty and Olsan (2) obtained the following expression for pressure drops for gas-solids mixtures:

$$P = \left(P_0 + \frac{4fm_m^2RTL}{D_p g_c} + \frac{2m_m^2RT}{g_c} \ln \frac{P}{P_0}\right)^{1/2}$$ [6]

where P is pressure, P_0 is stagnation pressure, f is friction factor, m_m is mass flow rate (lb/ft^2/s), T is temperature ($^\circ$R), and D and L are dryer diameter and length (ft), respectively. Equation [6] is implicit in pressure and needs only the initial or final system pressures, mass flow rate, and stagnation temperatures to solve for the pressure.

Hallström and Wimmerstedt found that a simple model (34) for the constant rate period of drying and uniform solid temperature can explain the evaluation along the dryer of the main process variables. They agree with Thorpe (12) and Schlunder et al.

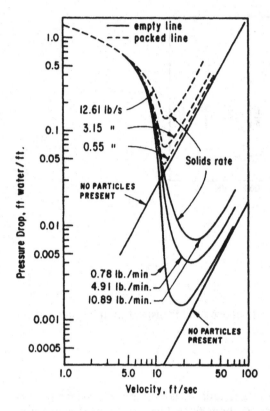

Figure 6   Pressure drop versus velocity. (From Reference 18.)

(15), who state that the exact modeling must take into account elementary momentum and heat and mass transfer between the particles and the surrounding air.

5.3. Heat and Mass Transfer

The application of flash dryers to heat-sensitive material is limited by several factors. The main controlling factor for drying should be related to the external convective heat transfer taking place on the wet surface of the particle, not to internal heat transfer. For larger particles this may lead to some errors because in this case the internal heat and mass transfer within the particle structure may be controlling even in the very short residence time. If the convective heat flux between the gas and particles is too high, the rate of moisture removed from the particle surface would be higher than the rate of internal moisture migration to the surface from the interior structure of particle because of the resistances encountered against diffusion and capillary flow phenomena. Dry spots appear on the surface of the particles, which changes the solubility of water in the skin layer at the surface. Other changes as a result of thermal degradation could also alter the internal heat and mass transfer mechanism. For this range of conditions the observed rate of drying may be smaller than for conditions in which a lower entrance gas temperature is applied.

The heat transferred during the flash drying process may be considered the result of the following four stages:

1. Drying in the disintegrator if dry recycle steam is used
2. Drying during high air-particle slip velocity, which occurs after the feed point and after each elbow
3. Drying during a nearly steady particle velocity after the acceleration point
4. Drying in the cyclones (35)

Since convective heat transfer is always influenced by flow phenomena, we must first concern ourselves with the fluid dynamics of dispersed two-phase systems. The important design considerations are (a) velocity of the particles and their swarms, (b) the rate of rise of these swarms, (c) the agglomerations taking place, and (d) the loading rate with respect to the phase in which it is introduced. It can be observed that the heat transfer coefficient increases sharply after the feed entrance point, and it may reach 10–100 times the value for stationary beds because dispersed two-phase systems differ from a single-phase flow because they contain solid particles that have a difference in density from the suspension phase (36). The relative velocity of these particles characterizes the fluid behavior. This difference in the mechanism (26) necessarily influences the heat transfer.

The experimental studies of many authors (37–39) have demonstrated that the fluid-particle heat and mass transfer coefficients in suspension beds fit quite well between the corresponding values for fixed beds on the one hand, and for individual particles in flow on the other. The following correlations are suggested. Assuming that the coefficients of heat and mass transfer stay constant throughout the flash dryer, it is only necessary to calculate particle Nusselt and Sherwood numbers. For single spheres, one can use

$$Nu_{single} = 2 + (Nu_{lam}^2 + Nu_{turb}^2)^{1/2} \tag{7}$$

$$Nu_{lam} = 0.664 \, (Pr)^{1/3} \, (Re)^{1/2} \tag{8a}$$

$$Nu_{turb} = \frac{0.037 \, Re^{0.8} Pr}{1 + 2.443 \, Re^{-0.1}(Pr^{2/3} - 1)} \tag{8b}$$

$$Nu = [1 + 1.5(1 - e)]Nu_{single} \tag{9}$$

where e is the porosity. It is known that the heat and mass transfer coefficients are proportional to the Reynolds number, but the exponent for Re depends on many parameters.

5.4. Generalized Drying Curves

In this section generalized drying curves that can be used in the prediction of drying times of fine solids are presented. The model assumes that the movement of moisture in a solid body is by capillary motion and that the major resistance to heat and mass transfer is on the boundary layer at the solid surface. As discussed in previous sections, flash drying easily fits this model.

Mathematical Model

A short description of the mathematical model will be given here. For more detailed presentation, the reader should refer to previous publications in this area (40).

Theory. Many solids can be considered as containing a very large number of capillaries extending in all directions. During the constant rate period of drying, these capillaries are full and the entire surface area is covered with a film of water. As drying proceeds this film evaporates and dry surfaces begin to appear. Owing to very high drying rates, the process takes place in flash dryers even though the residence time is very short.

Transfer Coefficients. The convective and radiative heat transfer coefficients are found from the following relations:

$$Nu = 2.0 + 0.236 \, (Re)^{0.6} \, (Pr)^{0.33} \qquad [10]$$

$$h_r = k_1 e_1 (T_R^2 + T_S^2)(T_R + T_S) \qquad [11]$$

where k_1 and e_1 are constants. The radiative heat transfer coefficient should always be included in flash dryer calculations. Total heat transfer coefficient is equal to the sum of h_r and h_c.

Assuming that the effect of mass transfer on heat transfer is negligible, then the following correlation for psychrometric ratio is used (7):

$$B = \left(\frac{Sc}{Pr} \right)^{-0.567} \qquad [12]$$

where $B \, (=29 c_p K_G P_{BM}/18 h_c)$ is the psychrometric ratio, K_G is the mass transfer coefficient, and P_{BM} is the log-mean pressure difference.

For thin materials the surface moisture content is approximately equal to the average particle moisture content. Then the mass transfer area can be related to the heat transfer area during the falling rate period of drying as

$$\frac{A_m}{A_p} = \left(\frac{W}{W_c} \right)^{2/3} \qquad [13]$$

where A is area and W is moisture content.

Drying Rate Equations. The following rate equations govern the drying rate in both the constant rate and falling rate periods of drying:

$$\frac{-dW}{d\theta} = \frac{h_T A_p}{\lambda} (T_R - T_S) \qquad [14]$$

$$\frac{-dW}{d\theta} = K_G A_m (P_S - P_R) \qquad [15]$$

where $-dW/d\theta$ is drying rate. When the vapor pressure on the particle surface is related to the surface temperature, the following final relation can be obtained:

$$\frac{P_s}{P_0} = A - b\phi \, \frac{T_R - T_S}{T_R} \qquad [16]$$

where A and b are constants and $\phi = 18\lambda/RT_S$.

The rate equations reduced to Eqs. [19] and [20], which upon integration give the generalized drying curves, with the help of the following dimensionless defined quantities:

$$C = \frac{W}{W_c} \qquad F = \frac{h_T A_p T_R \phi}{W_c \lambda} \qquad [17]$$

$$G = \frac{K_G P_0 \lambda}{h_T T_R} \qquad E = h_T A_p \theta \, \frac{T_R - T_{WB}}{W_c \lambda} \qquad [18]$$

and

$$F_2 - F_1 = \frac{3(C_1^{1/3} - C_2^{1/3}) + Gb\phi(C_1 - C_2)}{G(A - H_R)} \qquad [19]$$

with

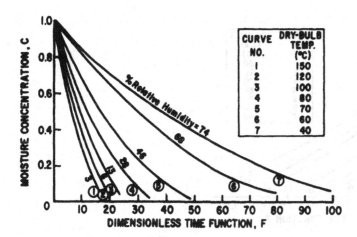

Figure 7 Generalized drying curves.

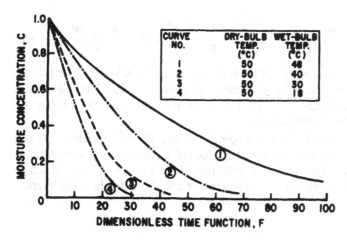

Figure 8 Effect of humidity on drying rate.

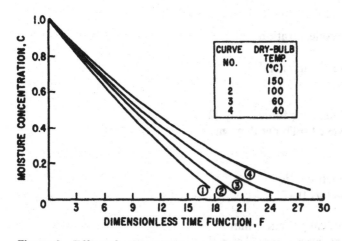

Figure 9 Effect of gas temperature on drying rate.

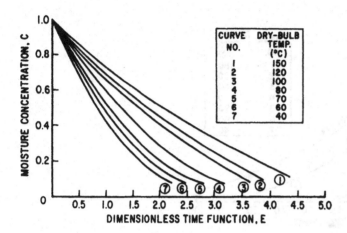

Figure 10   Generalized drying curves.

$$\frac{(h_T A_p \phi_F)(T_R - T_{WB})}{W_c \lambda} = \frac{W_c - W_F}{W_c} + e_2 \left(\frac{W_c - W_F}{W_c}\right)^2 + e_3 \left(\frac{W_c - W_f}{W_c}\right)^3 \qquad [20]$$

where the e's are constants and $e_1 = 1$; for flash dryers e_3 can be taken as 0. Equation [20] is derived for the case in which the feed moisture content is equal to the critical moisture content. The dimensionless concentration, air temperature, and drying time are represented by C, G, and F (or E), respectively.

Generalized Drying Curves.   These equations contain G and H_R as parameters. G is a function of K_G, h_T, and T_R. The first two can be predicted from empirical correlations, and both are functions of T_R. Hence G is a function of T. The results are shown in Figures 7–10. The effects of humidity and air temperature are illustrated in these figures.

NOMENCLATURE

A	particle area, cm^2, ft^2 in Eq. [1]
A	mass transfer area, cm^2
B	psychrometric ratio
C	dimensionless moisture concentration
C	drag coefficient
D	average particle diameter in dryer, ft^2
E	dimensionless drying time
e	porosity
F	dimensionless drying time, drag force in Eq. [2], lb
F	loading rate, lb of conveyed solid per lb of gas
Fr	Froude number
f	friction factor
G	dimensionless psychrometric ratio
g	Newton's conversion factor, ft s^{-2}
H	relative humidity
h	heat transfer coefficient, cal g^{-1} cm^{-2} min^{-1}
K	mass transfer coefficient, g cm^{-2} min^{-1} atm^{-1}
L	length of pipe, ft

m mass flow rate, lb ft^{-2} s^{-1}

Nu   Nusselt number

P pressure drop in Eq. [6], vapor pressure, atm

P log mean of $(1 - P_0)$ and $(1 - P_s)$, atm

Pr   Prandtl number

R universal gas constant, ft R^{-1}

Re   Reynolds number

Sc   Schmidt number

Sh   Sherwood number

T temperature, °R in Eq. [6], °C

V velocity, ft s^{-1}

W moisture content

Subscripts

c convection, critical point in Eqs. [17] to [20]

e entrainment

F final

f fluid

0   initial

p particle

R dryer

r radiative

s solid surface

T total

wb wet bulb

Greek Symbols

θ drying time

λ latent heat of vaporization, cal/g

ρ density, g/cm^3

REFERENCES

1. Fulford, G. D., *Can. J. Chem. Eng.* 47, 378, 1969.
2. McCarty, H. E., and J. H. Olsan, *IEC Fund*, 7, 471, 1968.
3. McCormick, P., *Brit. Chem. Eng. Equip. Suppl.*, 14, 1225, 1969.
4. Lang, R. W., Stork-Bowen Ltd. (1982).
5. Williams-Gardner, A., *Industrial Drying*, George Godwin, London, 1976.
6. Perry, J. H., *Chemical Engineers Handbook*, 5th ed., McGraw-Hill, New York, 1963.
7. Henry, H., and N. Epstein, *Can. J. Chem. Eng.*, 48, 595, 1970.
8. Kisakürek, B., Generalized Drying Curves for Porous Solids. Ph.D. thesis, Illinois Institute of Technology, Chicago, 1972.
9. Chukhanov, Z. F., *Int. J. Heat Mass Transfer*, 13, 805, 1970.
10. Lee, K., and H. Barrow, *Int. J. Heat Mass Transfer*, 11, 1013, 1968.
11. Thorpe, G. R., *Trans. Inst. Chem. Eng.*, 51, 339, 1973.
12. Thorpe, G. R., *Chem. Ind. Dev.*, Nov. 13, 1975.
13. Kmiec, A., and S. Mielczarski, *Drying '80*, vol. 2, (A. S. Mujumdar, ed.), Hemisphere, New York, 1980, p. 213.
14. Kisakürek, B., and A. Akit, *Drying '80*, vol. 1 (A. S. Mujumdar, ed.), Hemisphere, New York, 1980, p. 486.

15. Schlunder, E. U., H. Martin, and P. Gummel, Drying Fundamentals and Technology, Lecture Notes, McGill University, Montreal, 1977.
16. Soo, S. L., *A.I.Ch.E. J.*, 7, 384, 1961.
17. Soo, S. L., *Fluid Dynamics of Multiphase Systems*, Blaisdell, Waltham, Massachusetts, 1967.
18. Capes, C. E., *Can. J. Chem. Eng.*, 49, 182, 1971.
19. Capes, C. E., and K. Nakamura, *Can. J. Chem. Eng.*, 51, 31, 1973.
20. Schlicting, H., *Boundary Layer Theory*, 6th ed. McGraw-Hill, New York, 1968.
21. Botterill, J. S. M., *Fluid-Bed Heat Transfer*, Academic Press, New York, 1975.
22. Petrovic, L. J., and G. Thodos, Proc. Intern. Symp. on Fluidization, Eindhoven, 1967.
23. Broothroyd, R. G., *Trans. Inst. Chem. Eng.*, 45, 297, 1967.
24. Rao, J. S., and A. E. Dukler, *IEC Fund*, 10, 520, 1971.
25. Van Swaaij, W. P. M., C. Buurman, and J. W. Van Breugel, *Chem. Eng. Sci.*, 25, 1818, 1970.
26. Soo, S. L., and G. L. Trezek, *IEC Fund*, 5, 388, 1966.
27. Van Breugel, J. W., J. J. M. Stein, and R. J. de Vries, *Proc. Inst. Mech. Eng.*, 148, 18, 70, 1969.
28. Richardson, J. F., and W. N. Zaki, *Trans. Inst. Chem. Eng.*, 32, 35, 1954.
29. Rudinger, G., Multiphase Symposium, ASME, 1963.
30. Martin, H., *Int. Chem. Eng.*, 22, 30, 1982.
31. Richardson, J. F., M. N. Romani, and K. J. Shakiri, *Chem. Eng. Sci.*, 31, 619, 1976.
32. Debrand, S., *IEC Proc. Des. Dev.*, 13, 396, 1974.
33. Doig, J. D., and G. H. Roper, *Aust. Chem. Eng.*, 4, 9, 1963.
34. Hallström, A., and R. Wimmerstedt, *Chem. Eng. Sci.*, 38, 1507, 1983.
35. Brook, R. C., and F. W. Bakker-Arkema, *Drying '80*, vol. 2 (A. S. Mujumdar, ed.), Hemisphere, New York, 1980, p. 263.
36. Mersmann, A., H. Noth, D. Ringer, and R. Wunder, *Int. Chem. Eng.*, 22, 16, 1982.
37. Gel'perin, N. I., P. D. Lebedev, G. N. Napalkov, and V. G. Ainshtein, *Int. Chem. Eng.*, 6, 4, 1966.
38. Moncman, E., *Int. Chem. Eng.*, 9, 471, 1969.
39. Hoffman, T. W., and L. L. Ross, *Int. J. Heat Mass Transfer*, 15, 599, 1972.
40. Kisakürek, B., R. E. Peck, and T. Cakaloz, *Can. J. Chem. Eng.*, 53, 53, 1975.

16
Conveyor Dryers

Lloyd F. Sturgeon
S. T. Hudson Engineers, Inc.
Camden, New Jersey

1. INTRODUCTION

One kilogram of air at 20°C and 50% relative humidity occupies about 0.84 cubic meters of volume at sea level pressure. That volume also contains about 7.2 grams of water vapor—about half the amount 20°C air can hold at saturation. If this same kilogram of air (including the 7.2 grams of water vapor) is now heated to 40°C, the mixture will expand to a volume of 0.90 cubic meters. It still includes 7.2 grams of water vapor, but its relative humidity has dropped to 16%, and it could hold almost seven times as much water vapor, or about 48 grams, at saturation.

Continuing to heat the mixture to 75°C results in a further increase in volume to one cubic meter. The relative humidity, however, has now dropped to about 3.1%, and it could hold about 380 grams of water at saturation, or over 50 times the original amount. This enormous increase in the ability of air to hold water vapor with increasing temperature is the key to the effectiveness of air as a drying medium.

The majority of air-dried products are processed in rotary dryers of various types, in which the product is continuously tumbled in a rotating cylinder through which a heated airstream is passed. When the product is robust and bulk moisture-content reduction is the principal objective, rotary dryers usually offer the most economical approach. Other dryer types, such as the fluid bed and flash dryer, are also employed, which again depend upon more or less violent motion of the drying particles in the heated airstream.

There is a class of products, however, that requires gentle handling—and in some cases close control of process conditions—during drying. Such products are usually dried in a through-circulation dryer in which the (typically) granular product is evenly spread in a bed on a perforated plate, and dried by the passage of air up or down through the bed. This is accomplished in small batch or experimental operations in a tray dryer, but for large-scale production the conveyor dryer—also called an apron or band dryer—is employed. Truck and tunnel dryers are also used in some production operations, but employ much less efficient cross-bed (vs. through-bed) circulation, and are highly labor-intensive in operation.

In the conveyer dryer, the product is distributed on a moving belt, typically of perforated plate, that passes through a tunnellike structure in which vertical airflow is closely controlled. With the exception of transfer operations (or occasionally deliberate stirring of the bed), the individual particles of product in a conveyor dryer remain fixed in position with respect to each other, in contrast to rotary or fluid bed dryers in which the product particles are continuously tumbling. Each particle, therefore, has essentially the same residence time, a particularly important factor where color changes occur during drying or uniformity of moisture is important for downstream operations such as cereal puffing.

The flow of heated air in a conveyor dryer can be adjusted so that each area of the bed is exposed to the process conditions of air velocity, air temperature, and air humidity most suitable for that phase of the drying process. The bed depth can also be adjusted as appropriate by transfer to a section of conveyor moving at a different speed.

As production air dryer types, truck, tunnel, and conveyor dryers all offer the two key features of (a) gentle physical handling of the product and (b) close control of process conditions. The conveyor dryer, however, offers the additional advantages of a continuous process and the greatly increased drying efficiency obtained with positive through-the-bed air circulation. An important factor in the application of the through-circulation conveyor dryer is the preforming or other preparation of the product to permit forming a bed permeable to airflow. This is discussed further below.

2. MACHINE DESCRIPTION

Figure 1 is a cutaway view of a typical single-conveyor dryer (other conveyor dryer types are discussed below). Product in granular form is here being distributed onto a perforated-plate conveyor by an oscillating feeder conveyor. The dryer conveyor, de-

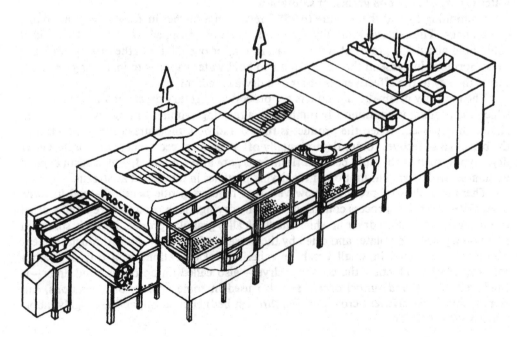

Figure 1 Cutaway view of single-conveyor dryer. (Courtesy Proctor & Schwartz, Ltd.)

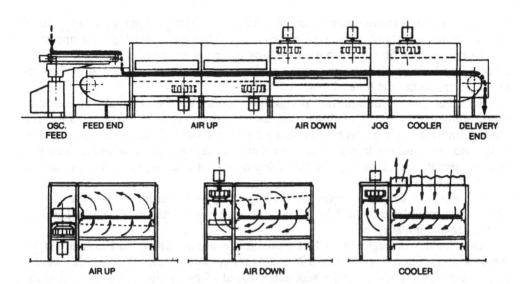

Figure 2 Side elevation and cross-section of typical single-conveyor dryer. (Courtesy Proctor & Schwartz, Ltd.)

scribed below, can be from 1.0- to 4.5-m wide, with 2.5 to 3.0 m most typical. The product is carried through an insulated housing within which heated air is forced through the material. Housing lengths can be from 3 m (in narrow widths) to 60 m or more.

In Figure 1 the product passes first into an "air-up" zone, where heated air is forced up through the moving bed of material by backward-blade turbine fans, chosen for their efficiency at high static pressures and their nonoverloading characteristics. The return air is reheated by steam heat exchangers (finned coils). Other possible heat sources are discussed below. Uniform delivery of the heated air to the bed is ensured by passing it first through perforated air distribution plates installed across the full width of the conveyor.

The conveyor in this illustration now moves into an "air-down" zone in which the flow direction is reversed. Air passing through a bed of drying material is giving up heat and picking up moisture as it goes, thus the product particles on the leaving side of the bed are not subjected to as effective drying conditions as those on the entering side. Air direction reversals are often employed, therefore, to enhance moisture uniformity throughout the bed.

Finally the product passes through an air-down cooling zone within which ambient or chilled air is drawn through the bed by another turbine fan discharging to an exhaust duct. The product is then dumped from the dryer conveyor, typically onto a cross-conveyor for transport to the next stage in the process.

Figure 2 portrays the machine of Figure 1 in side elevation and cross-section, showing more clearly the directions of airflow described above. Note particularly the positions of the perforated air distribution plates or baffles. Note also that a neutral zone or "jog space" with no air circulation is employed to isolate the cooler from the preceding dryer zone.

Dryer housings are typically built in standard-length modules (here, 1 m), sized to optimize utilization of commercial-size materials. This standardization permits efficiencies in engineering and manufacture as well as in on-site assembly. Each of the airflow direction zones in a conveyor dryer can be further subdivided by partitions into one or more temperature control zones to permit controlled changes in entering air temperature as drying progresses. Similarly, compartment-by-compartment control of exhaust (extraction) and fresh air makeup airflows can be used to maintain wet bulb temperature (humidity) at the optimum level for each phase of drying.

Note that during constant-rate drying the product will be essentially at wet bulb temperature, so that control of air humidity can effectively control the temperature of the product. In some applications high humidity during the early phases of drying is used to inhibit too rapid surface drying that could result in surface cracks. In other instances it is used to prevent *case hardening*, the formation of an outer sealing layer that can inhibit the loss of internal moisture.

Bed depths in conveyor dryers are most typically of the order of 100 mm but can range from a monolayer of particles to over a meter in some instances. Conveyor loadings range up to about 600 kg/m^2.

Air velocities through the bed of material are usually about 1.25 m/s in single-conveyor dryers but can range from 0.25 to 2.5 m/s in special applications. Static pressure drops across the bed are normally kept below about 25-mm water gauge to avoid problems with zone-to-zone leakage and blowout at the feed and discharge ends.

Operating temperatures can be up to about 400°C for the type of machines described here, but most applications are within the range of 10°C to 200°C. It should be noted that conveyor dryers are often used for processes other than, or in combination with, drying. These include toasting, puffing, curing, baking, and calcining, among many others.

3. CONVEYOR DESIGN

The details of construction of a typical conveyor are shown in Figure 3. The individual conveyor plates are typically made of perforated stainless steel sheet, perhaps 1-mm thick. Widths in the machine direction are from 50 to 300 mm, with a 200-mm pitch most common. The conveyor plates are perforated to from about 6% to 45% of net open area, with perforation patterns to suit the product size, cleaning requirements, and other factors.

For example, a 1.5 × 6 mm slot running in the machine direction at about 25% open area might be used where it is necessary to brush the conveyor on the return run to remove residue. The very low open areas are used when the conveyor itself is employed to maintain uniformity of air distribution, as in applications in which uneven product loading is unavoidable. For products with a high fines content (e.g., bread crumbs), the conveyor plates can be overlaid with a cover screen of 30- to 50-mesh polyester or similar material.

As shown in Figure 3, the individual plates are hinged together, with a vertical reinforcing girt perhaps 100 mm deep running parallel with the hinge at each joint to minimize deflection under load. The plates are supported at their ends by attachments on a specially designed roller chain. Traveling guards are attached at the same points, designed to overlap in such a way as to minimize side-spilling of the product, especially as the conveyor passes around the sprocket at the dump point. Sheet metal stationary guards fastened to the main body of the dryer and fitted with a resilient rubbing seal between them overlap the outside of the traveling guard, ensuring both containment of the product and minimal bypassing of circulating air (see also the cross-sectional drawings in Fig. 2).

Scrapers, doffers, and pin breakers are often employed at the dump point to aid in

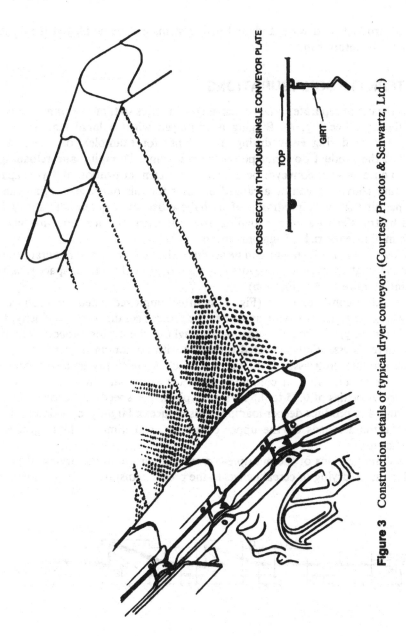

Figure 3 Construction details of typical dryer conveyor. (Courtesy Proctor & Schwartz, Ltd.)

transfer of product, and wash and/or brush systems can be employed for cleaning the conveyor on the return run.

4. ALTERNATE CONFIGURATIONS

Staging is the use of separate conveyors arranged in series so that one transfers its product load to the next (see Fig. 4). Staging is employed when a large amount of product shrinkage occurs during early drying (in onions, for example), or when the "green strength" of the product does not permit deep loading. Dumping and reloading with a deeper bed onto another conveyor permits more efficient utilization of dryer capacity. As many as four conveyors can be arranged in series for this purpose. Staging can also be used to permit the quick predrying of a shallow bed of a potentially sticky product, followed by transfer to a deeper loading after sufficient surface moisture has been removed to minimize the risk of agglomeration.

As shown in Figure 4, staging can be accomplished either by a horizontal arrangement of successive single-conveyor conveyors (Fig. 4-a) or, to reduce floor space requirements, by stacking them vertically (Fig. 4-b).

The multiple-conveyor dryer (Fig. 5) is often employed when the product requires relatively long-time, gentle drying and when the process conditions (air velocity, temperature, and humidity) can remain constant throughout the drying process. The multiple-conveyor dryer is essentially a heated enclosure with no zoning or partitioning through which the product progresses from one (of 2 to 15) vertically arranged conveyors to the next. Air circulation is a combination of crossflow and through-circulation, and through-bed velocities of 0.25 to 0.5 m/s are typical. The speed of the lowest conveyor(s) can be reduced to produce deeper loading. This induces a larger proportion of air to flow through the thinner beds on the upper conveyors, sometimes yielding greater overall drying efficiency.

Typical applications of the multiple-conveyor dryer are in the drying of pasta products and in the predrying of cereal pellets to the proper moisture content prior to puffing.

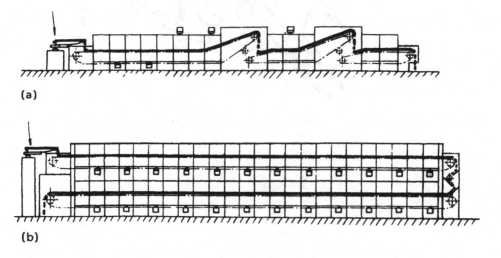

(a)

(b)

Figure 4   Examples of staged conveyor dryers. (Courtesy of Proctor & Schwartz, Ltd.)

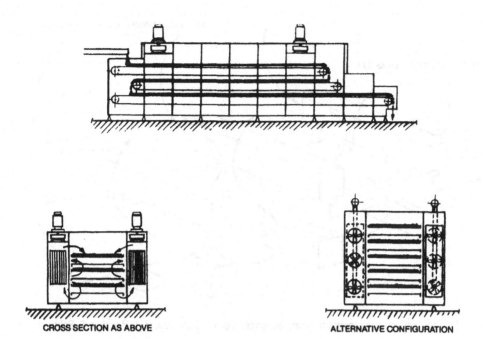

CROSS SECTION AS ABOVE   ALTERNATIVE CONFIGURATION

Figure 5 Multiple-conveyor dryer. (Courtesy Proctor & Schwartz, Ltd.)

Figure 5 illustrates two of several possible configurations: the side elevation shows product loaded onto the top conveyor and then dumped successively onto lower conveyors, each of which moves in a direction opposite to that of the one immediately above. As noted above, up to 15 conveyors are used, but 3 to 5 are most common.

The cross-section on the left in Figure 5, corresponding to the side elevation above it, shows turbine fans arranged to force return air through vertically mounted steam coils and then through baffles into the conveyor space. The baffles are arranged in this case so that air passes down through the top conveyor, up through the second conveyor, and down again through the bottom conveyor before returning to the fan intake. This alteration of airflows, along with the mixing that occurs at the dump points, results in excellent uniformity of product treatment.

Turbine fans in a multiple-conveyor dryer offer the advantages of simplicity and efficiency. As the number of conveyors increases, however, considerations of air distribution and overall height favor a vertical array of propeller fans (as shown in the right cross-section of Figure 5) that can move large volumes of air at the low static pressures normally encountered in machines of this type.

The multiple-conveyor dryer offers a large conveyor holding area for the floor space employed. The product, however, must be able to withstand repeated dumps, which are prone to generate fines even with the aid of transfer chutes between conveyors. Sticky products or those requiring frequent washdown are difficult to deal with unless a built-in clean-in-place (CIP) system is employed. The precise process control possible with the zoned single-conveyor dryer is also not available in a multiple-conveyor dryer.

The multiple-conveyor dryer is not inherently more energy efficient, as is sometime claimed, but, owing to its compact configuration and functional simplicity, it lends itself

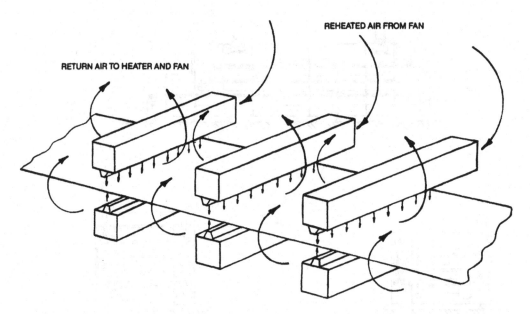

REHEATED AIR FROM FAN

RETURN AIR TO HEATER AND FAN

Figure 6 Typical airflow arrangement in an impingement dryer. (Courtesy Proctor & Schwartz, Ltd.)

more readily to the installation of energy-saving devices such as automatic controls and heat recovery discussed below. Sometimes the best choice is a combination of types: for example, a single-conveyor predryer to remove surface moisture (or stickiness), followed by a multiple conveyor for final drying.

A special class of single-conveyor dryer is the impingement dryer, in which the drying air does not pass through the product bed but instead is blown against the product by special nozzles. A typical airflow arrangement is shown in Figure 6. The product is usually in sheet form but can also be a monolayer of individual pieces arranged on a conveyor. The conveyor itself can be perforated sheet metal or woven wire, or can be a solid sheet, in which case heat transfer from the bottom nozzles is wholly by conduction.

The nozzle boxes in Figure 6 are fed from a plenum from which they receive heated air under fan pressure. After impinging on the product from the nozzle slots, the cooled and moisture-laden air returns around and between the nozzles to the heater and fan, where some is exhausted and the rest recycled.

Nozzle exit velocities range from 5 to 20 m/s. The jet of air from the nozzle slot impinges first upon the product directly below the nozzle, then sweeps along the product surface for some distance before joining the return airflow. The average heat transfer coefficient across the product surface is a function of nozzle exit air velocity, slot width, and nozzle-to-nozzle and nozzle-to-product spacing. A typical value is 50 W/m$^2 \cdot$°C. Other nozzle types include arrays of round holes and jet tubes, which sometimes offer higher heat transfer coefficients with less disturbance of the product compared with slot nozzles. Nozzle design is a careful trade-off between requirements for uniformity versus cost and fan horsepower.

Impingement dryers are used for drying textiles, sheet tobacco, and other materials in sheet form. Another application is discussed in the next section.

5. ANCILLARY EQUIPMENT

Various feeding devices are used to provide a uniform bed across the width of the conveyor. Probably the most common is the oscillating boom carrying either a passive chute or a vibrating or moving-belt conveyor, as shown in Figures 1 and 2. The simplest type of feed is the full-width hopper with adjustable discharge gate (Figure 7-a), often used for nuts and other free-flowing products.

As noted above, efficient through-circulation drying depends upon establishing and maintaining a structural configuration that will permit the free passage of air through the bed of material. As an example, the rolling extruder (Fig. 7-b) is used to convert heavy pastelike products (e.g., filter cake) into noodlelike extrusions, say, 6 to 8 mm in diameter, greatly increasing the surface area.

The rolling extruder shown in Figure 7-b consists of a pair of rubber-covered rolls that are driven back and forth on a curved perforated plate mounted across the full width of the conveyor. Product in cake form is distributed into a hopper supported between the oscillating rollers from which it is forced by the rollers through holes in the plate. The extrusions fall freely to form an open bed on the conveyor. The maintenance of this structure depends in turn upon the gentle handling characteristics of the conveyor dryer; many extrusions are quite delicate and will slump into an impermeable mass if disturbed before an appreciable amount of moisture has been removed. Other common preforming devices include auger extruders, briquetters, and pelletizers. Many natural products, such as vegetables, are sliced or diced prior to drying.

Slurried products can also be spread onto a solid (unperforated) conveyor and passed through the impingement-type predryer. After the product has dried sufficiently to form

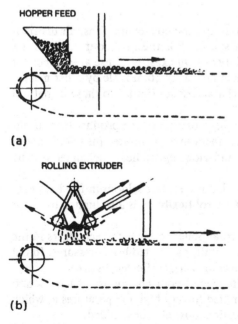

(a)

(b)

Figure 7 Typical conveyor feed devices. (Courtesy Proctor & Schwartz, Ltd.)

a solid cake, it is scored, then broken into pieces as it is transferred to a perforated conveyor for through-bed drying in the conventional manner.

6. SIZING

Sizing of conveyor dryers is usually accomplished by scaleup from laboratory tests run on specially designed tray dryers. Such dryers are designed to permit accurate simulation and measurement of process conditions at each stage of drying so that the configuration of the commercial-size machine can be optimized for the product characteristics desired.

Tests currently are being run with increasingly sophisticated equipment, including weigh-in-place devices for determining product moisture loss, automatic recording of process conditions, and computer analysis of results. Every effort is made to simulate as exactly as possible the conditions that will be met in commercial services, from preparation of the product to dumping from the discharge conveyor. Test planning will include analysis of the probable commercial-scale machine configuration, required preparation and/or preforming of the product, and allowable temperatures.

Test observations will include – in addition to drying characteristics – product shrinkage, fines generation, fume or smoke generation, and the tendency of the product to stick to conveyor plates. Test tray bottoms are fabricated using the same material (including nonstick coatings) and perforations as will be used in the proposed commercial-scale machine.

A scaleup factor is employed, in converting from laboratory data to commercial-scale equipment, which is based on a combination of analysis plus laboratory and field experience; it can vary from less than 1.0 to 2 or more.

7. HEAT SOURCES

The next step in the design process is to select components: conveyors, fans, air distribution devices, ancillary equipment, and the heat source. For the heat source, direct gas firing with natural gas offers the greatest flexibility and quickest response at least cost; it also permits operation at higher temperatures than are easily obtainable by other means. Safety requirements are stringent. Propane has characteristics similar to those of natural gas but is more costly.

Most dryers currently are steam heated, offering isolation of the product from products of combustion, along with simple control. Temperatures are limited (usually to about 150°C) by available steam pressure, however, and clogging of heat exchanger fins by fines is frequently a maintenance problem.

Direct oil firing is sometimes used in process industries, but seldom in food applications due to the risk of product contamination. Control flexibility is also limited by lower burner turndown ratios.

Hot oil (transfer fluid pumped from an external boiler to air heat exchangers in the dryer) offers higher temperatures than steam without the associated high pressures. Initial costs are higher, but without condensate losses overall energy efficiency is greater.

Electric heating is seldom used because of its overall poor energy-use efficiency and resulting high cost. Applications are generally limited to very high temperatures at which product contamination by productions of combustion must also be avoided.

Indirect oil (or gas) fired into radiant tubes is seldom employed due to low efficiency, sluggish response, and high engineering and materials costs.

8. ENERGY CONSIDERATIONS

The following tabulation shows a typical conveyor dryer heat load calculation, indicating how input energy is distributed. The first three components of the heat load are established by operational requirements of the machine; the last two are losses to be minimized. The loss levels shown are not necessarily typical, but represent what can be achieved with good design and careful maintenance.

Component	Watts	%
Evaporation (+ moisture, vapor)	528,000	55
Product heatup	117,000	12
Exhaust makeup	264,000	28
Product energy/exit losses	29,000	3
Radiation/convection losses	15,000	2
Total	953,000	100

Each component of the heat load should be examined during design for factors that could reduce energy consumption. Following are some guidelines.

The evaporative heat load can be reduced by reducing initial moisture (e.g., by mechanical dewatering) and by closer approach to the allowable final moisture through automatic product moisture control. Overall drying efficiency is critically dependent upon uniformity of product loading; uneven loading will allow bypassing of air, causing over-drying of thin spots in the bed and underdrying elsewhere.

The product heatup load may be reduced by conserving heat already in the product from upstream processes, and can be recovered by recycling air from a downstream cooler. Heatup losses into the conveyor itself (included under the same head) can be reduced by enclosing return runs and feed or delivery end extensions and also by transferring product to a separate conveyor for cooling.

The selection of gas burner type can have a critical effect on exhaust (and resulting fresh air makeup) volumes. Burners requiring secondary (tempering) airflows beyond requirements for drying should be avoided. Manufacturers offer a variety of burner gas-air mix arrangements, from stoichiometric upward, and a wide range of turndown capabilities.

Decreasing overall exhaust levels may be possible without loss of capacity, but insulation of ductwork may be necessary to avoid condensation at lower dew points. Losses due to unnecessary exhaust during product interruptions and drops in initial moisture can be minimized by an automatic exhaust humidity control system. Sensible heat can often be recovered from the exhaust, and sometimes latent heat as well.

Entry and exit losses can be minimized through proper seal design; careful adjustment and maintenance are also required to avoid bypass leakage. Special sealing devices and airlocks are sometimes employed, along with enclosed feed and delivery end extensions. In food dryers such enclosed extensions require careful attention to temperature control to avoid creating biologic incubation sites. Automatic exhaust pressure balance controls are also used.

Typical dryer housing insulation consists of removable panels with inner and outer skins of aluminized or galvanized steel, stainless steel or aluminum, with an inner core of mineral wool or fiberglass insulation. Through-metal at the edges of panels has a significant effect on overall radiation and convection losses, and special designs must be employed to minimize such losses.

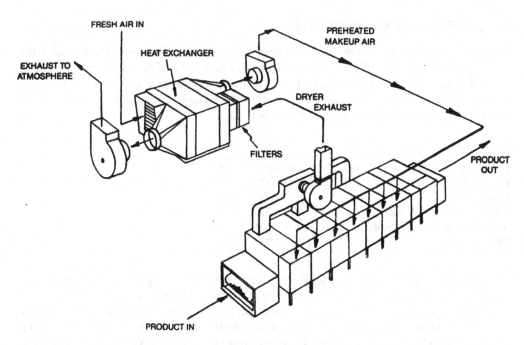

Figure 8   Typical conveyor dryer heat recovery system. (Courtesy Proctor & Schwartz, Ltd.)

Fan horsepower requirements can be minimized by optimizing dryer air velocity (along with temperature and humidity) to conform to the drying characteristics of the product. High fan horsepower may have an overall beneficial effect on dryer energy efficiency during constant rate drying but can be largely wasted when drying is diffusion limited.

Exhaust heat recovery should be considered at the time of new dryer design, and retrofitting is sometimes justified for older units. Figure 8 shows the basic parts of a typical heat recovery system; the hot exhaust air from the dryer is passed through one side of an air-to-air heat exchanger in which 50% to 70% of the available sensible heat is transferred to the incoming fresh airstream, which is thus preheated for injection as makeup air. Other system elements include booster fans, exhaust air filters as required, ductwork, and bypass dampers. It should be noted that the cost of the heat exchanger itself—plate, wheel, or heat pipe—is only a fraction of total system cost. Careful consideration must be given to protecting the heat recovery element from fouling due to contaminants (particulates and condensibles) in the dryer exhaust, and to provision for periodic (and effective) cleaning.

The economic feasibility of such an installation can be checked quickly. If an acceptable payback (1 to 3 years) is indicated, a more detailed design, cost, and justification analysis can be carried out, which should include probable escalation in the price of fuel.

CONCLUSION

In summary, the conveyor dryer provides gentle physical handling of products, which must in general be preformed in some manner to permit free passage of air through the bed. The conveyor dryer also offers the opportunity to closely control process conditions at each stage of the product drying cycle to optimize both product quality and energy utilization. The various configurations available—single, staged, and multiconveyor, used singly or in combination—allow a great deal of flexibility in process design.

The well-engineered conveyor dryer is outwardly a simple machine, robustly constructed, with easy access for cleaning and maintenance. Most of its technical sophistication lies in subtleties of air handling and process control. To obtain maximum benefits in capacity and efficiency the potential user must be sure the dryer manufacturer clearly understands present and projected process requirements, and must follow the manufacturer's recommendations with regard to operation and maintenance.

17
Impinging Stream Dryers

Tadeusz Kudra* and Arun S. Mujumdar
McGill University
Montreal, Quebec, Canada

Valentin Meltser
Belorussian Academy of Sciences
Minsk, Belorussia

1. INTRODUCTION

Impinging stream dryers (ISDs) is a generic term for an emerging class of dryers in which moisture evaporation from wet particles or liquid droplets occurs in an impingement zone that develops as a result of "collision" of two oppositely directed high velocity gas streams, at least one of which contains the dispersed material to be dried (Fig. 1). At the outset a distinction should be made from the well-known impingement dryers in which gas jets are directed onto the weblike or slablike materials, or "jet-zone" dryers in which a layer of particulates is "pseudofluidized" by a multiplicity of high-velocity airstreams (1). Due to the hydrodynamic characteristics of the impingement zone, and the large inertia of the solid/liquid phase, the particles flowing originally with a gas stream oscillate about the impingement plane with damped amplitude until their velocity drops to the terminal velocity to be entrained with the outgoing gas stream. The high intensity of turbulence in the impingement zone and the rapid, unsteady particle motion enhance significantly the heat and mass transfer processes and hence reduce the drying times (2). A typical set of operating data for impingement stream dryers in coaxial configuration is listed in Table 1.

The main features of the impinging streams' configuration that result directly from their special hydrodynamics are

High intensity of drying, especially when surface or weakly bound water is to be removed
High product quality
Simple design and operation (in general, no moving or rotating parts)
Compactness
Possibility of combining drying with other operations (e.g., granulation, disintegration, heating, cooling, chemical reactions, etc.)

Because of the high gas velocities, solids loading ratios, and momentum loss in the collision zone, the pressure loss in ISDs is much greater than in pneumatic dryers but it is

---

*Current affiliation: CANMET, Energy Diversification Research Laboratory, Varennes, Quebec, Canada

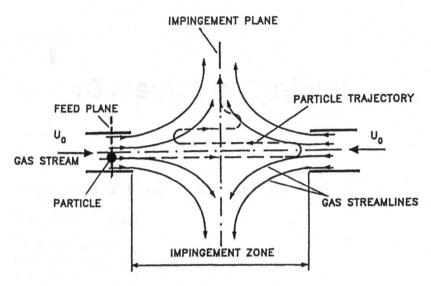

Figure 1   The principle of impinging stream dryers (ISDs).

comparable with that of fluidized and spouted bed dryers (2,3). The impinging stream configurations can, however, compete in various aspects with the classical systems for drying of particulates and pastes (Table 2).

The unique hydrodynamic conditions of impinging streams also offer special advantages such as intermittent heating, control of the processing time, possibility of particle disintegration not only for drying or thermal processing but also for other unit operations such as absorption, chemical reactions, combustion, mixing, and so on (4,5).

2. CLASSIFICATION OF IMPINGING STREAM SYSTEMS

Among the number of criteria that can be considered for impinging stream classification, the most obvious is the flow direction (countercurrent or co-current), number of impinging streams (two or multistreams), flow characteristics of the streams (swirling, nonswirl-

Table 1   Range of Operating Parameters for Coaxial Impinging Stream Dryers

Gas velocity, m/s	10–150
Particle diameter, mm	0.2–3.0
Volumetric concentration of solids, m^3/m^3	Up to 0.0035
Particle residence time, s	0.5–15.0
Production rate, kg/h	
Of dry material	50–2000
Of water evaporated	Up to 5000
Volumetric throughput, $kg/(m^3 h)$	
Of wet material	1500–30,000
Evaporative capacity, $kg/(m^3 h)$	Up to 1700
Pressure drop, kPa	1–10

Table 2 Performance of Selected Dryers for Granules of Aluminum Alloy[a]

Parameter	Vibrofluidized bed	Fluidized bed	ISD
Drying time[a], s	300–480	600	1–20
Unit gas consumption[b], Nm^3/kg	8	13	1.2–2
Unit energy consumption[b], kJ/kg	1380	3240	850
Product quality[c]	Decreased	Decreased	Unchanged

[a]Reference productivity, 250 kg/h; $d_p = 0.65$–2 mm; $X_1 = 0.17$ kg/kg
[b]To reach $X_2 = 0.01$ kg/kg
[c]Per kg of dry solid
[d]Estimated from oxidation tests

ing, and direction of swirl of streams that may co- or counterrotate), and type of material processed (e.g., the material may be dried with or without inert particulates).

Depending on the geometry and flow direction the following basic variants of impinging streams can be identified by

Type of flow
    Coaxial
    Rotating (swirling)
    Curvilinear
Flow direction
    Countercurrent, in which gas (or solid suspension) streams flow in opposite directions
    Co-current, in which gas (or solid suspension) streams flow in the same direction
Streamline direction
    Counterrotating
    Co-rotating
Type of impingement zone
    Stationary, in which the position of the impingement plane does not change in time
    Mobile, in which the position of the impingement plane changes periodically or
        continuously
Geometry of impingement plane
    Planar with radial flow, in which streamlines diverge radially toward the impinge-
        ment plane
    Planar with circumferential flow, in which streamlines form concentric circles in the
        impingement plane
    Tubular
    Annular

Examples of the basic types of impinging stream systems as well as the fundamental configurations of impinging planes are presented in Figures 2 and 3, respectively.

3. CHARACTERISTICS OF IMPINGING STREAM SYSTEMS

When two gas streams collide, the initial velocity profiles characteristic of a free flow deform in the vicinity of the impingement plane and additional components of velocity (radial, axial, or circumferential, depending on the impinging streams' configuration) appear as a result of this deformation.

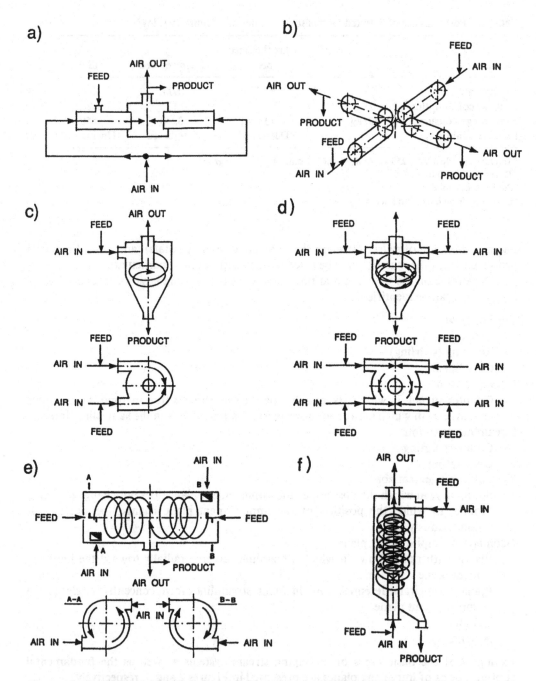

Figure 2 Basic types of impinging stream flows: (a) coaxial; (b) mutually perpendicular X configuration; (c) curvilinear countercurrent; (d) four impinging streams; (e) counterrotating countercurrent; (f) co-rotating countercurrent.

a)

c)

b)

d)

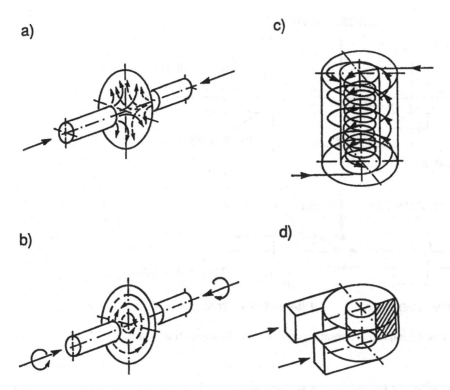

Figure 3 Impinging plane geometries: (a) planar with radial flow; (b) planar with circumferential flow; (c) tubular; (d) annular.

If a solid or liquid particle flows with one of the impinging streams then in the impingement zone, the particle penetrates into the opposite stream due to its inertia and decelerates to a full stop some distance of penetration within the domain of the opposing jet flow (cf. Fig. 1). Thereafter the particle accelerates in the opposite direction and penetrates the original gas stream. Thus, the process of deceleration and acceleration repeats itself. Because of energy dissipation, after several damped oscillations, the particle leaves the impingement zone and flows with an outlet gas stream into the discharge chamber.

Because of such oscillatory motion, the residence time of a single particle in the impingement zone is longer than that of the gas stream. This residence time is usually reduced for a number of particles because of interparticle collisions that lead to enhanced energy dissipation. When solids are present in both gas streams, the rate of collision and the resulting loss of momentum is much higher than that for a single stream feed, which might result in such significant decrease of the residence time and penetration depth that the beneficial effects of impingement are dramatically reduced.

The above disadvantage is practically eliminated in an impinging stream system with mobile impingement zones (6). In this arrangement, the impingement plane is made to move between the two locations I and II by alternate switching of the gas flows from left to right and then from right to left (Fig. 4). The particles fed into the accelerating flow duct pass through the central (reverse flow) tube that links the two impingement chambers. Flowing with the original gas stream into the first impingement chamber the parti-

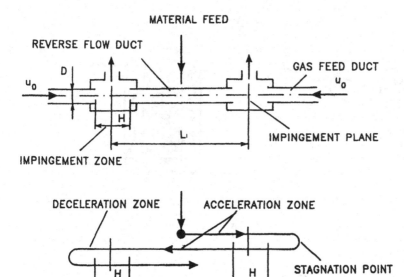

Figure 4   Coaxial impinging stream dryer with a mobile impingement zone.

cles collide with the secondary gas stream, penetrate it up to the stagnation point, and then begin to accelerate in the opposite direction. At this moment the gas outlet of the first impingement chamber is closed while the outlet of the second impingement chamber is open. This procedure results in the flow of a secondary gas stream with accelerating particles toward the second impingement chamber where the process of jet collision, penetration of particles into the original gas stream, and the following acceleration toward the first impingement zone is repeated. The period of oscillatory motion of the particles is controlled by decreasing the inertia force due to reduction of particle size in the course of processing, by reduction of particle mass due to moisture evaporation, or by placing a suitable limiting grid at the outlet ducts, which restricts particle entrainment.

Another modification of ISDs with mobile impingement zones contains inert spherical particles undergoing oscillatory motion between the impingement zones. This allows drying of slurries or suspensions sprayed on the surface of inert particles. Drying occurs by combined convective and conductive heat transfer. Also, drying with simultaneous grinding can be performed if metallic beads are used as the inert particles.

Identical transport phenomena occur in an impinging stream system consisting of four mutually perpendicular ducts of the same diameter (X configuration). There, the primary gas streams mixed in an impingement zone are split into two secondary streams carrying the particles off the impingement zone. Single impinging stream units can be combined in series and/or in parallel to form a system allowing extension of the residence time and development of different flow or temperature regimes (Fig. 5).

Figure 6 shows an interesting semicircular duct with built-in impinging stream zones. Flows in this case are affected by centrifugal forces.

Another variant of impinging streams can be formed when a secondary gas-solid suspension stream is brought into collision with a primary flow, as is the case of spouted bed, jet, or cyclone dryers. Such configurations offer better conditions for processing of

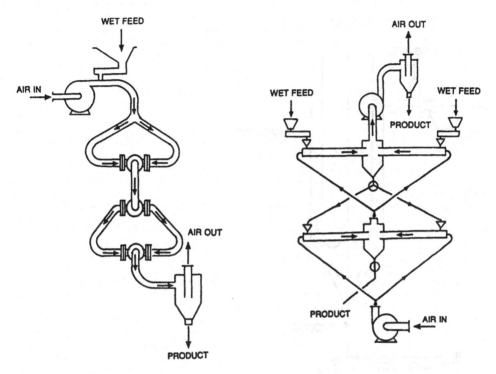

Figure 5 Two-stage impinging stream dryer: (a) co-current; (b) countercurrent.

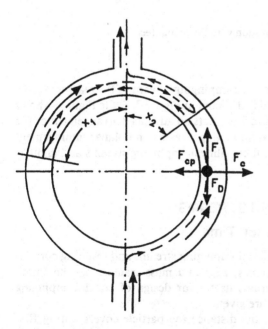

Figure 6 Particle trajectory for double penetration in semicircular impinging streams (F_c, centrifugal force; F_{cp}, centripetal force; F_D, drag force; F, lift force).

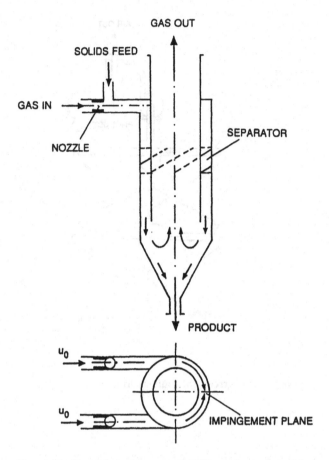

Figure 7   Two-impinging-stream (TIS) configuration with tangential feed.

dispersed systems, and therefore are termed two-impinging-stream (TIS) reactors (3–5). A single-stage TIS dryer with tangential feed is depicted in Figure 7, while Figure 8 shows a schematic design of a multistage unit. Figure 9 presents an advanced arrangement of the TIS that doubles the combination of coaxial impinging streams and planar impingement plane (type a in Figure 3) with countercurrent curvilinear impinging streams and annular impingement plane (type d in Figure 3) (7).

4.   HYDRODYNAMICS OF IMPINGING STREAMS

4.1.   Penetration Depth and Oscillation Time

Detailed studies of the fluid-particle dynamics in impinging streams and resulting correlations allowing modeling of the hydrodynamics, heat and mass transfer can be found elsewhere (2,8–10). Here, only the correlations useful for design of coaxial impinging streams (which are commercially available) are given.

The maximum penetration depth (i.e., the distance the particle covers during first penetration into the opposing stream) and penetration time (i.e., time spent by a particle in the penetration distance) are of prime importance in the apparatus design. Taking

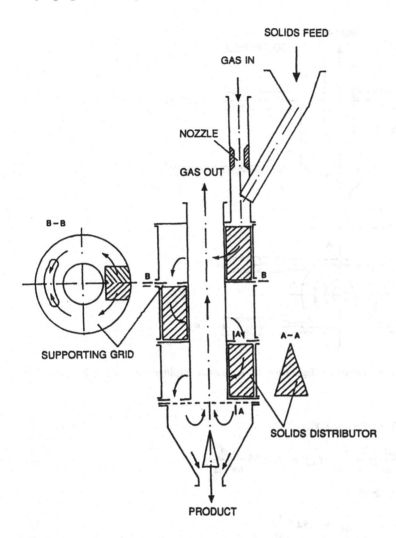

SOLIDS FEED

GAS IN

NOZZLE

GAS OUT

B - B

SUPPORTING GRID

A - A

SOLIDS DISTRIBUTOR

PRODUCT

Figure 8 The structure of a three-stage two-impinging-stream dryer.

initial particle velocity at the impingement plane equal to the gas velocity in the accelerating flow duct and the final particle velocity equal to 0, the following correlations for the maximum penetration depth and for the maximum penetration time have been proposed (9):

Laminar region ($Re_r \leq 1$)

$$x_{\max} = 0.016 \frac{ud_p^2\rho_p}{\nu\rho} ; \qquad t_{\max} = 0.036 \frac{d_p^2\rho_p}{\nu\rho} \tag{1}$$

Transient region I ($1 \leq Re_r \leq 13$)

$$x_{\max} = 0.01415 \frac{\rho_p u^{0.8} d_p^{1.8}}{\rho\nu^{0.8}} ; \qquad t_{\max} = 0.328 \frac{\rho_p d_p^{1.8}}{\rho\nu^{0.8} u^{0.2}} \tag{2}$$

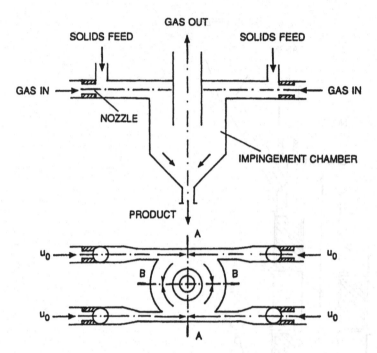

Figure 9 Four-impinging-streams configuration (A-A, planar impingement plane; B-B, annular impingement plane).

Transient region II ($13 \leq \mathrm{Re}_r \leq 800$)

$$x_{\max} = 0.02675 \frac{\rho_p u^{0.5} d_p^{1.5}}{\rho \nu^{0.5}} ; \qquad t_{\max} = 0.0635 \frac{\rho_p d_p^{1.5}}{\rho \nu^{0.5} u^{0.5}} \tag{3}$$

Self-similar region ($\mathrm{Re}_r \geq 800$)

$$x_{\max} = 0.598 \frac{\rho_p d_p}{\rho} ; \qquad t_{\max} = 1.54 \frac{\rho_p d_p}{\rho u} \tag{4}$$

where the Reynolds number for the accelerating/decelerating particle is defined in terms of the gas-particle relative velocity ($\mathrm{Re}_r = d_p(u \pm u_p)/\nu$).

The penetration depth and oscillation time for subsequent periods of oscillatory motion useful for process calculation and dryer design can be determined from the following empirical correlations valid over $2.5 \times 10^{-4} \leq d_p \leq 4 \times 10^{-3}$ m, $10 \leq u \leq 150$ m/s, $0.2 \leq x_0 \leq 4.8$ m, $3 \leq \mathrm{Re}_p \leq 800$, and $30 \leq T \leq 300°C$ (11):

$$t^* = 1.42 \mathrm{Re}_p^{0.28} L^{-0.51} ; \qquad x^* = 0.09 \mathrm{Re}_p^{0.21} L^{-0.44} \tag{5}$$

with

$$t^* = \frac{t_{\max} u}{x_0} ; \qquad x^* = \frac{x_{\max}}{x_0} \tag{6}$$

where

$$\mathrm{Re}_p = \frac{u d_p \rho}{\mu} \qquad \text{and} \qquad L = \frac{2\rho x_0}{\rho_p d_p} \tag{7}$$

The dimensionless time and distance are related to their maximum values, which appear in the first period of oscillation.

For $Re_p > 800$ Eq. (5) can be simplified to the form

$$t^* = 9.23L^{-0.51}; \qquad x^* = 0.37L^{-0.44} \tag{8}$$

In a given impinging streams' configuration, the following inequalities should be satisfied for oscillatory motion of a single particle about the impingement zone of the length H (12):

$$\frac{8u}{H} \geq \frac{3\mu}{4d_p^2 \rho_p}; \qquad H \leq \frac{9ud_p^2 \rho_p}{4\nu\rho} \tag{9}$$

Thus, the minimum gas velocity in the flow ducts and minimum particle diameter are given by

$$u \geq \frac{4\nu\rho H}{9d_p^2 \rho_p}; \qquad d_p = \sqrt{\frac{9\nu\rho H}{4\rho_p u}} \tag{10}$$

Since the particle velocity decreases during consecutive passage through the impingement plane, during some oscillation it approaches the gas velocity necessary to suspend the particle in a horizontally flowing gas stream. This critical gas velocity at which the particle is carried away from the impingement zone can be calculated from the following equation (12):

$$u_{cr} = 0.9(D/d_p)^{1/7}\sqrt{\frac{\rho_p - \rho}{\rho}\, g d_p} \tag{11}$$

or

$$u_{cr} = 5.6D^{0.34}d_p^{0.36}\left(\frac{\rho_p}{\rho}\right)^{0.5}\beta^{0.25} \tag{12}$$

4.2. Pressure Drop

As expected, the pressure drop depends on the geometry of the ISD, gas flow rates, particle properties, and the loading ratio. It generally varies from 1 to 10 kPa for the range of parameters that result in a stable operation. Referring to the pressure drop in a TIS configuration, the pressure drop in impinging streams is about 20 times lower than that for a fluidized bed and 50 times lower than that for a spouted bed under the same operating conditions (13).

The pressure drop in the two-phase flow ΔP_s in the impingement zone can be calculated from

$$\Delta P_s = \Delta P + \zeta_s \frac{u^2 \rho}{2} \tag{13}$$

where ΔP is the pressure drop in the single phase flow and ζ_s is the friction coefficient resulting from the solid phase present in a gas stream. This coefficient is related to the geometry of the impingement zone, gas velocity, and the solids' concentration and can be estimated using one of the following equations (14,15):

$$\zeta_s = 3420Re_t^{-0.18}\beta \tag{14}$$

which is valid for an ISD with a cylindrical impingement zone for $45 \leq \text{Re}_t \leq 1150$ and $\beta \leq 0.0006 \text{ m}^3/\text{m}^3$. Also,

$$\zeta_s = 1580(H/D)^{-0.47}\beta \qquad [15]$$

which was obtained for the X configuration for $H/D = 0.25\text{--}1.0$ and $\beta \leq 0.00045$ m^3/m^3.

Note that the term *geometry of impingement zone* refers only to the "active" fraction of the impinging chamber volume that contributes to the transfer processes. On the basis of numerous experiments, the limiting value of H/D below which the impingement zone can be considered to be active is 1.2–1.5.

The friction coefficient for horizontal gas-solid duct flow can be calculated from the following formula (16):

$$\zeta_s = 7.85 \cdot 10^{-5}(D/d_p)\text{Re}_t^{0.32}\text{Fr}^{0.5} \qquad [16]$$

The pressure drop for single-phase flow (ΔP in Eq. [13]) can be calculated from the following correlations derived for impinging streams with cylindrical impingement zones for $H/D = 0.25\text{--}1.125$ (14,15):

$$\text{Eu} = 8.3(H/D)^{-1.3}\text{Re}^{-0.25} \qquad [17]$$

for $1.2 \cdot 10^4 \leq \text{Re} \leq 4 \cdot 10^4$, and

$$\text{Eu} = 0.6(H/D)^{-1.3} \qquad [18]$$

for $\text{Re} > 4 \cdot 10^4$.

For single-phase impinging streams of X-type impingement zone the relevant correlations are as follows:

$$\text{Eu} = 14.2\text{Re}^{-0.25} \qquad [19]$$

for $\text{Re} \leq 2 \cdot 10^4$, and

$$\text{Eu} = 1.16 \qquad [20]$$

for $\text{Re} > 2 \cdot 10^4$.

It should be noted that for the X-type ISD with rounded edges in impingement zone $(r/d) \sim 1$ the pressure drop is approximately half of that for sharp-edged duct outlets, thus the Euler number is equal to 0.54 (15).

For the TIS configuration, the following correlation for pressure drop valid over the range of particle-to-gas mass flow rate ratio from 0 to 4 and $\text{Re} = 5000\text{--}45000$ is recommended (13):

$$\frac{\text{Eu}_s}{\text{Eu}} = 1 + A\frac{W_p}{W} \qquad [21]$$

where $A = 0.037$ for a coaxial vertical TIS (17), $A = 0.15$ for a coaxial horizontal TIS (18), $A = 0.33$ for a coaxial horizontal TIS with cyclone-type impingement chamber (13), and $A = 0.0961$ for a countercurrent curvilinear TIS (19).

Although the relation for calculation of the Euler number for pure gas flow in TIS has not been given it seems reasonable to use Eqs. [17] and [18], which have been derived for the similar geometry of impingement zone.

4.3. Heat Transfer

Due to the inherent hydrodynamics of ISDs the heat transfer rates from gas to particle depend on time as well as space. Reported data indicate, however, that the average heat transfer coefficient for ISD differs by less than 10% from its local value, which is sufficient for dryer design (20,21).

In general, the average gas-to-particle heat transfer coefficients for ISDs are much higher than those in classical dryers that operate under similar hydrodynamic regimes. For example, when the heat transfer coefficient in a coaxial ISD is 850 W/(m²K), the values for pneumatic dryers calculated for the same operating conditions from several equations compiled by Strumiłło and Kudra (22) ranged from 300 to 520 W/(m²K). For the TIS dryer with coaxial impinging streams the heat transfer coefficient is 1.1 to 1.8 times higher than that for the geometrically similar spouted bed under the same operating conditions (17). The volumetric heat transfer coefficients for ISDs are also higher, reaching 125,000 W/(m³K), which is 2.5-3 times higher than the values for spouted bed dryers (17) and 15-100 times higher than those for spray dryers (2).

The average heat transfer coefficient in an impingement zone can be calculated within ±18% from the following correlations (20):

$$Nu = 0.173 Re_r^{0.55} \beta^{-0.61} \tag{22}$$

for $300 < Re_r < 3500$ and $\beta \leq 0.0009 \text{ m}^3/\text{m}^3$, or

$$Nu = 1.59 Re_r^{0.55} \tag{23}$$

for $300 < Re_r < 3500$ and $0.0009 < \beta < 0.0021 \text{ m}^3/\text{m}^3$, where the Reynolds number is based on the relative gas-particle velocity averaged over the length of the impingement zone.

For semicircular impinging streams the effect of β on heat transfer occurs only for $\beta \geq 0.0009 \text{ m}^3/\text{m}^3$. Beyond this solids' concentration the following correlations are recommended (21):

$$Nu = 0.186 Re_p^{0.8} \tag{24}$$

for $80 < Re_p < 480$, or

$$Nu = 1.14 Re_p^{0.5} \tag{25}$$

for $480 \leq Re_p \leq 2000$.

The heat transfer coefficient in impinging streams with mobile impingement zones can be calculated from the standard equation for gas flowing past a sphere with particle Reynolds number based on the gas velocity in an accelerating flow duct:

$$Nu = 2 + 1.05 Re_p^{0.5} Pr^{0.33} Gu^{0.175}. \tag{26}$$

For a TIS dryer with tangential feed the following correlation for calculation of the heat transfer coefficient has been developed, assuming that the W_p/W (particle mean flow rate/gas mean flow rate) at the feed remains constant along the dryer (23):

$$Nu = 1.386 \cdot 10^{-8} Re_p^{3.46}. \tag{27}$$

The volumetric heat transfer coefficient (recalculated from the reported data in Ref. 23 for $r = 2500.8 \text{ kJ/kg}$) in TIS dryers depends on the mass concentration of solids defined as the mass holdup m_h to the reactor volume V_r ratio, and has been correlated by

$$h_V = K(m_h/V_r)^n. \tag{28}$$

The parameters K and n depend on the impinging stream configuration and vary from 1270 to 3800 for K, and from 0.90 to 1.70 for n (23). An equation for the volumetric heat transfer coefficient in coaxial and curvilinear TIS configurations valid for $0.00037 < V_r < 0.00615$ m^3, $0.0016 < W_p < 0.0265$ kg/s, $0.0053 < W < 0.0212$ kg/s, $0.0015 < m_h < 0.022$ kg is

$$h_V = 2.19.10^4 (m_h/V_r)^{1.13} W^{0.626} \tag{29}$$

where

$$m_h = 0.967\, W_p - 0.000334. \tag{30}$$

For any impinging stream system, the heat transfer coefficient, both local and averaged over the total residence time of particles, increases as the gas velocity increases and decreases as the gas temperature increases. Such a temperature influence on heat transfer results not only from the variation in gas properties but also from an appreciable change in the hydrodynamic conditions; with increase in the gas temperature the penetration depth into the opposing gas stream increases, which extends the period of unsteady particle motion thus enhancing the total heat transfer.

4.4. Drying

Drying in impinging streams is purely convective. Thus, typically both constant and falling drying rate periods can be observed if the processing materials have both internal and external resistances to mass flow. Generally, the period of constant drying rate is short because of the high heat and mass transfer rates.

Figure 10 shows the time variation of moisture content, temperature, and location

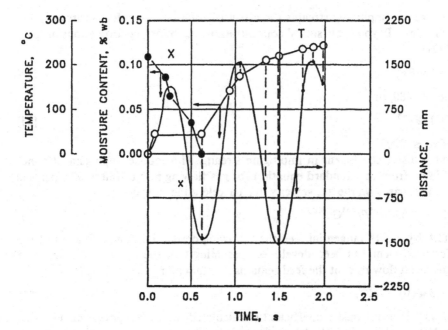

Figure 10 Time variation of axial position, moisture content, and particle temperature in an impinging stream dryer with a mobile impingement zone ($T_{g1} = 250°C$; $u_0 = 30$ m/s; $X_1 = 0.11$ kg/kg).

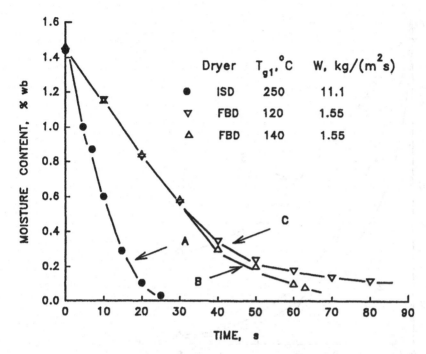

Figure 11 Drying curves for sucrose in a fluidized bed (FB) and an impinging streams (UVS-2) dryer.

for 1-mm aluminum beads covered with a thin layer of surface water, and dried in an ISD with mobile impingement zones (cf. Fig. 4) (24). It can be seen that the heat transfer rate differs significantly in the period of particle deceleration and acceleration due to the different gas-particle relative velocities. Despite the stabilizing effect of oscillatory motion this difference can be as high as 35–40%. Because of surface water evaporation, the drying rate is controlled by the rate of external heat transfer. Thus it is higher in the deceleration period than in the acceleration period. The particles are completely dry within 0.6 s, that is, within less than one cycle of oscillation.

Figure 11 compares the drying kinetics of sugar with average crystal size of 1 mm dried in an ISD with mobile impingement zones with the corresponding drying kinetics in a fluidized bed dryer. It is clear that even for materials with intracrystalline moisture, which in the course of drying results in recrystallization of sucrose, the drying rate in the ISD is much higher than in a fluidized bed. In particular, in the period of surface water evaporation (down to 0.5–0.6%), the drying rate in the ISD is 3 to 4 times higher than in a fluidized bed. In the second phase of drying when water is removed from a saturated sucrose solution, the corresponding drying rate is still one order of magnitude higher. Moreover, the drying time in the ISD is about 10–12 s while in a fluidized bed it is greater than 60 s under similar drying air conditions.

Figure 12 presents the drying kinetics of crystalline lysine with an initial moisture content of 15.2% in an ISD with mobile impingement zones using hot air at 120°C and flow velocity of 20–23 m/s; the frequency of reversing motion is 1.0–1.2 Hz (25). Curve 1 represents drying of monodisperse crystals (0.4 mm mean diameter) at mass concentrations of 0.2–0.5 kg/kg of air. In this case the surface water is removed within one period of motion (2–3 seconds). The desired moisture content level of 1% can be achieved within

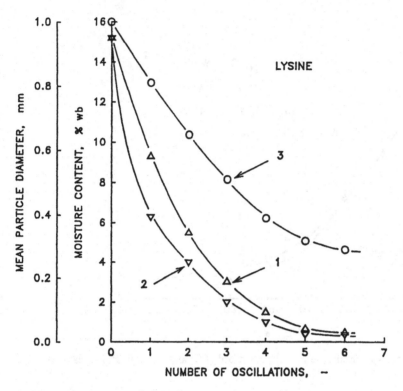

Figure 12 Drying and grinding characteristics for lysine: *1*, drying kinetics for monodisperse crystals; *2*, drying kinetics for polydisperse crystals; *3*, grinding curve for polydisperse crystals.

5 periods of oscillation. Curve 2 shows the drying kinetics for polydisperse lysine with a mean diameter $(d = \Sigma\, x_i d_i)$ equal to 1 mm with simultaneous grinding due to the oscillatory motion of the inert material (mixture in 1 : 1 mass ratio of 2 mm steel and 3 mm aluminum beads) at a mass concentration of 1.0–1.5 kg of inert per kg of drying material. Compared with curve 1 the drying rate with simultaneous disintegration is appreciably higher in the first period of reverse motion mainly due to a reduction of the heat transfer resistance inside the material and increase of the mass transfer area due to particle disintegration. After a certain number of oscillations the drying rate decreases and follows a similar decrease in the intensity of crystal disintegration (curve 3).

5. SCALEUP AND DESIGN CONSIDERATIONS

Analysis of the reported data indicates that pressure drop in impinging streams depends greatly on the Reynolds number, solids concentration, and the length of the impingement zone if $H/D < 1$. Assuming the solids concentration must be determined from the process requirements, the following conditions should be satisfied for the scaleup of impinging streams:

$$\frac{\mathrm{Eu}^s}{\mathrm{Eu}^l} = 1, \qquad \frac{\mathrm{Re}^s}{\mathrm{Re}^l} = 1, \qquad \text{and} \qquad \frac{G^s}{G^l} = 1 \tag{31}$$

where G is the scale factor (e.g., the H/D ratio for coaxial impingement streams); the indices s and l refer to the small- and large-scale devices, respectively.

Since for $H/D > 1.5$ the effect of impinging streams' geometry in coaxial configuration on the Euler number is negligible, and at high Reynolds numbers the Euler number does not depend on the hydrodynamic conditions (18), these two dimensionless groups in Eq. [31] can be neglected. Hence the scaleup of the hydrodynamics of impinging streams reduces to only the Euler's number ratio. This has been confirmed by Tamir and Shalmon (26) for countercurrent, curvilinear impinging streams. Based on experiments in small- and large-scale equipment with effective volumes equal to 0.00038 m^3 and 0.00334 m^3 respectively, they proved that the pressure drop data for both sizes fit well a plot of the Euler number ratio against the mass flow ratio. Hence, the pressure drop for a large-scale apparatus operating at Re > 5000 for any prescribed solid-to-gas mass flow ratio can be obtained from the respective Euler number ratios determined experimentally (e.g., the plot in Fig. 13), or computed if the relevant correlations are available. Assuming that for high Reynolds numbers the ratio of the Euler numbers for small- and large-scale apparatus (for gas flow only) is inversely proportional to the geometry scaleup factor, the Euler number for gas flow in a large-scale apparatus can be computed from the known Euler number for the small-scale, laboratory-size apparatus. For Re < 5000, the Euler number

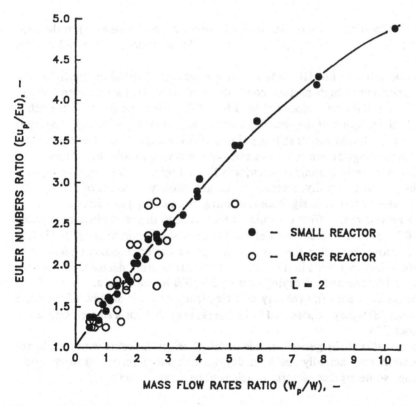

Figure 13 Euler number ratio versus Reynolds number ratio for two-impinging-stream reactor. (Reprinted with permission from *Ind. Chem. Res.*, 27(2):238–242, 1988. Copyright 1992 American Chemical Society.)

as a function of the Reynolds number must be determined experimentally for gas flow or calculated from available correlations (e.g., Eqs. [17]–[20]).

For the residence time and mass holdup, the scaleup criteria can be written as

$$\frac{t^s}{t^l} = \frac{m_h^s}{m_h^l} = \frac{G^s}{G^l} \qquad [32]$$

where t is the mean residence time.

According to Tamir and Shalmon (27) the mean gas and particle residence times can be calculated from the following equation:

$$\frac{t_p^s}{t_p^l} = \frac{L^s}{L^l}, \qquad \text{and} \qquad \frac{t^s}{t^l} = \left(\frac{L^s}{L^l}\right)^3 \qquad [33]$$

while the mass holdup of the gas and solid phases can be determined from

$$\frac{m_{h,p}^s}{m_{h,p}^l} = \frac{L^s}{L^l}, \qquad \text{and} \qquad \frac{m_h^s}{m_h^l} = \left(\frac{L^s}{L^l}\right)^3 \qquad [34]$$

where L is a characteristic linear dimension of the impinging stream configuration such as distance between gas feed duct outlets in a coaxial arrangement.

The following factors should also be considered in the design of impinging stream dryers:

The maximum solids loading (mass of wet feed per unit cross-sectional area) in the feed duct is in the order of 250–300 kg/m^2. In practice, a loading rate of 20 to 80 kg/m^2 is recommended.

The gas velocity calculated by Eq. [10] may not be adequate to suspend the particles in a gas stream, especially at high particle concentrations or if the particles tend to agglomerate. Therefore the gas velocity should be 1.8–2.2 times greater than the minimum value. At the location of the feed this velocity should be 20–30% higher because of the high particle concentration at this point, relatively large mass of wet particles, as well as because of agglomerates that can be formed owing to cohesion forces.

The maximum gas velocity in impinging streams is limited by the mechanical properties of material being dried. Usually, increase of the gas velocity in excess of 3.5–4 times the minimum velocity results in significant disintegration of the particles.

Interparticle collisions have no effect on solids' concentration in the impingement plane up to $\beta = 10^{-5}$ m^3/m^3. The collision effect becomes significant when $\beta > (1$–$1.5)$ 10^{-3} m^3/m^3. Above this value the particle holdup at the impingement plane rises gradually, which further results in increase of the pressure drop, decrease of the heat transfer rate, and increase of the drying time by 30–50% in some cases.

In the absence of relevant data the velocity and frequency of reversal flow of "particle cloud" in coaxial ISDs may be assumed to be the same as that for a single particle for design purposes (21).

At a certain limiting ratio of the gas-to-particle terminal velocity, drying with simultaneous disintegration may actually result in drying with simultaneous agglomeration when processing some organic materials such as bone-meat extract (27).

6. SOME INDUSTRIAL IMPINGING STREAM DRYERS

Figure 14 shows a schematic of an industrial ISD for drying of dewatered sewage sludge in a two-stage configuration. The first stage is a coaxial ISD in which the feed is dispersed

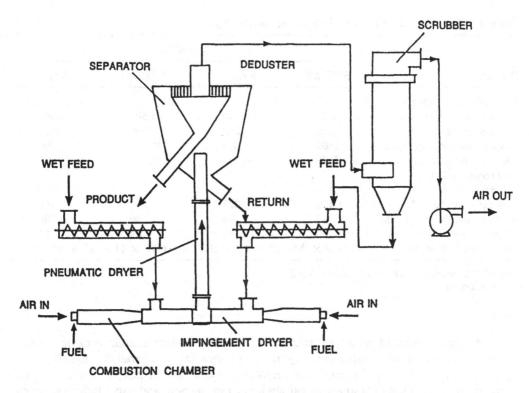

Figure 14   Impinging stream dryer for waste sludges (SVS type, Mashkhimizdat, Russia).

and a significant amount of surface moisture removed. The second stage is comprised of flash drying in the pneumatic duct and drying in a swirling stream during segregation in the centrifugal separator. The mixture of air and furnace gases from combustion of liquid or gaseous fuels (oil or natural gas) is used as the drying agent at inlet temperatures of 560–700°C. The outlet gas temperature is in the range of 90°C to 140°C. The gas velocity in ISD tubes and the pneumatic duct is in the order of 25–50 m/s, which ensures stable transport of the dispersed material without settling or sticking to the tube walls. The gas velocity in the profiled nozzle of the ejector-type feed is typically 200–250 m/s, which is sufficient to discharge a screw feeder, disintegrate lumps, and disperse particles well in the feed section. The mechanically predewatered sludge at initial moisture content 72% to 83% wb is blended in the ratio 3 : 1 with under- and oversize solids from the centrifugal separator. This back-mixing with the initial sludge allows the moisture content to be adjusted to the feed moisture content to about 50%, which facilitates operation of the ISD (lower gas velocities can be employed with a relatively dry feed) and also reduces the energy consumption from 3300–3800 kJ/kg to about 2900 kJ/kg evaporated water. The specific air consumption per kg of water evaporated is 4.5–5 Nm^3, whereas the electric energy consumption varies from 0.02 to 0.03 kWh/kg. The final moisture content depends on the particle size in product, and ranges from 2% for the 0.25-mm particles to 55% wb for 5-mm particles. The average moisture content of the 1–2-mm granules is 19.5% wb.

Table 3 presents the key operating parameters for a unified series of the industrial ISDs for drying of waste sludges with an initial moisture content of 80% wb.

Table 3 Characteristics of Industrial Impinging Stream Dryers

Parameter	Model[a]			
	SVS, 2.5	SVS, 5	SVS, 10	SVS, 20
Production rate, kg/h				
Wet sludge	3120	6250	12,500	25,000
Dry product	620	1250	2500	5000
Evaporation capacity, kg/h	2500	5000	10,000	20,000
Air consumption, Nm³/h	5000	10,000	20,000	40,000
Fuel consumption				
Fuel oil, kg/h	130	255	510	1020
Natural gas, Nm³/h	140	225	550	1100
Water consumption[b], m³/h	10	20	40	80
Electric power, kW	30	40	44	65
Overall dimensions, m	19 × 9 × 9.5	23 × 11 × 10.5	24 × 13 × 11.5	25 × 18 × 13.5

[a]Model SVS produced by Khimmash, Glazov, Russia.
[b]For wet scrubber.

A large number of municipal and industrial waste sludges contain organic components that can be used as supplementary fuel thus increasing the overall thermal efficiency of an ISD plant. Figure 15 presents the performance and fuel consumption of an impinging stream system used for drying waste sludges from the pulp and paper industry, which contain up to 50% cellulose fibers. Table 4 compares the performance characteristics of a standard impinging stream dryer at reference evaporative capacity of 10,000 kg/h with and without combustion of the dried sludge.

Coaxial ISDs with mobile impingement zone as presented in Figure 4 are especially suitable for drying quartz and foundry sand, metal granules, polymers, and like materials with surface or loosely bound moisture. Table 5 gives the characteristics of an industrial unit used for drying of metal granules. The same type of ISD with mobile impingement zones but furnished additionally with constraining grids at the outlets from the impingement chambers has been used in the former Soviet Union for drying with simultaneous grinding of otherwise hard-to-dry products containing tightly bound moisture.

An example of such a product is crystalline lysine, a highly heat-sensitive material with tendency to lump and with a large amount of bound water (up to 17% wb). The crystal size after centrifugation is 0.2–2.5 mm. These conditions along with the additional requirements for final moisture content below 1–1.5% wb with crystal dimension to be less 0.5–0.6 mm are met by the ISD dryer shown in Figure 16. The dryer consists of air conveying ducts with two impingement chambers situated at the ends of a central drying duct 1.5 m in length. Both chambers are connected via outlet ducts and cyclones to a flow-switching device. The outlet ducts are constrained by grids with a grid opening of 1.3 mm. A mixture of steel (2 mm) and aluminum (3 mm) beads in 1 : 1 mass ratio is used to disintegrate the crystals and break lumps. The mass concentration of the inert beads in the drying duct is maintained at 1–1.5 kg/kg. The frequency of reverse motion between the impingement chambers is 1–1.2 cycles per second. Hot air at 135°C and 20–23 m/s enters both ends of the impingement chambers. A switching device moves the impingement zone periodically between the two impingement chambers. This results in an oscillatory motion of both the inert beads and the wet particles, which are continuously fed to

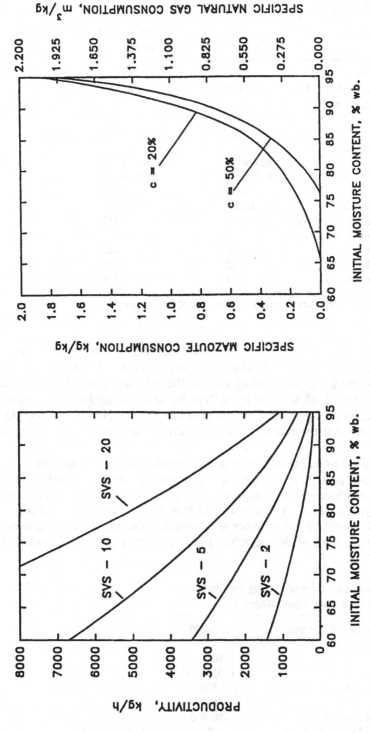

Figure 15 Drying performance and unit fuel consumption of impinging stream dryers (SVS type) versus initial moisture content and mass concentration of cellulose fibers.

Table 4 Performance of an Industrial Impinging Stream Dryer[a]

Parameter	Drying	Drying and combustion
Evaporation capacity, kg/h	10,000	10,000
Production rate, kg sludge/h	14,000	12,860
Moisture content, % wb		
Initial	80	80
Final	30	10
Gas temperature, °C		
Inlet	800 ± 30	800 ± 30
Outlet	150 ± 30	150 ± 30
Furnace exit	1400	1400
Air consumption, Nm^3/h	33,500	32,100
Dry sludge consumption, kg/h	—	3012
Fuel consumption		
Fuel oil, kg/h	1160	362
Natural gas, Nm^3/h	1025	320
Electrical power, kW	478	478

[a]Model TT8-01 SV 0.5–1.2 VC-01 produced by Khimmash, Glazov, Russia.

one end of the duct only. After performing 4–5 oscillations, which are the equivalent of 10–15 seconds of drying time, the material is well ground and dry (cf. Fig. 12) so it is carried away through the constraining grids to the cyclones. The volumetric evaporative capacity of the dryer is about 1700-kg H_2O/(m^3h) while air and heat consumption (without heat recovery) per kg of water evaporated are 200–250 Nm^3 and 3–3.6 MJ, respectively. This shows that the performance of the ISD is superior to that of a fluidized bed dryer (25).

Figure 17 presents an industrial setup with semicircular impinging stream ducts developed for thermal processing of grains (e.g., drying, puffing, and carrying out certain thermally induced biochemical reactions). In this case superheated stream is used as the carrier and drying medium. A multiplicity of semicircular tubes (type a ISD as in Fig. 2) are placed within a cylindrical chamber with a conical bottom. The product is collected at the bottom of the cone while the exhaust gas is led to a cyclone to collect the fines.

Table 5 Performance of an Impinging Stream Dryer[a]
for 0.5–2.5-mm Aluminum Alloy Granules

Production rate, kg wet material/h	150–200
Moisture content, % wb	
Initial	10–12
Final	0.01–0.02
Inlet air temperature, °C	300
Drying time, s	5–20
Air consumption, Nm^3/h	250–300
Electric power, kW	70
Overall dimensions, m	3 × 2.2 × 3.5

[a]Model USV produced by ITMO, Minsk, Belorussia.

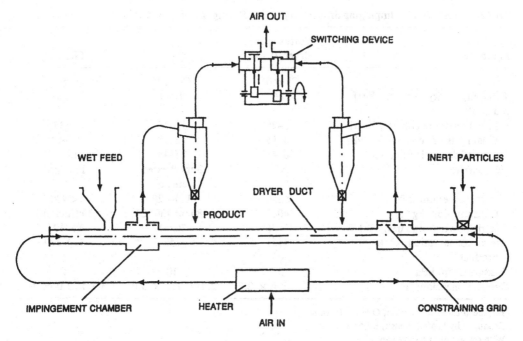

Figure 16   Impinging stream dryer with mobile impingement zone for drying of crystalline lysine.

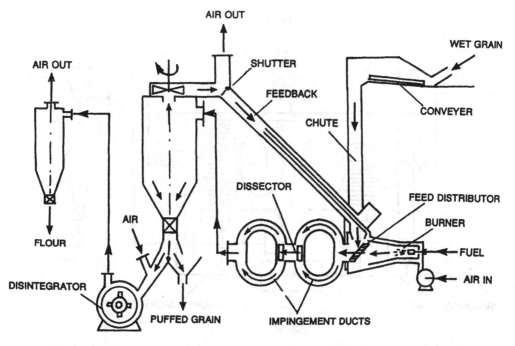

Figure 17   Semicircular impinging stream dryer (Vortex spray dryer, 0.8/1.2 type) for thermal processing of grains in superheated steam.

Table 6 Semicircular Impinging Stream Dryers for Drying/Puffing of Grains

Parameter	Vortex spray dryer, 0.15[a]	TD, 3[a]	TRD, 3[b]
System	Closed[c]	Closed[c]	Open
Throughput (dry product), kg/h	260	1000	1200
Material[d]			
Initial moisture content, % wb	5–25	12–16	<35
Final moisture content, % wb	3–15	6–12	6–12
Degree of dextrinization, %	30–45	25–40	35–45
Heat carrier	Air	Superheated steam	Flue gas
Inlet temperature, °C	<380	380–420	<420
Consumption, kg/h	400	180 (at 130°C)	30 (fuel oil)
Electric power (total/heater), kW	43/26	70/81	85/103
Energy consumption, kJ/kg dry product	455	252	310–370
Pressure drop, kPa	5	10	5
Overall dimensions, m	2.4 × 1.7 × 2.7	2.4 × 1.7 × 3.0[e]	3.5 × 2.0 × 1.0

[a]Produced by Kislorodmash, Odessa, Ukraine
[b]Produced by ITMO, Minsk, Belorussia
[c]With release of excessive vapor
[d]Corn, rye, barley, wheat, rice
[e]Without steam heater

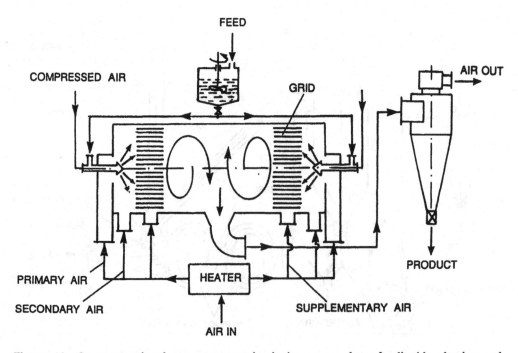

Figure 18 Counterrotational, countercurrent impinging stream dryer for liquids, slurries, and pastes.

Table 7 Performance Characteristics of an Impinging Stream Spray Dryer with Classical Spray Dryer

Parameter	VSD-100[a]	ZT-100[b]
Evaporation capacity, kg H_2O/h	96	80
Air consumption, Nm³/h	3500	3000
Air temperature, °C		
Inlet	148	150
Outlet	74	70
Moisture content, % wb		
Initial	94	98.5
Final	7.0	11.8
Dryer dimensions, m		
Diameter	1.2	3.3
Length	4.0	3.5
Volumetric evaporation capacity, kg/(m³h)	20	2.62
Volumetric heat transfer coefficient, W/(m³ K)	232	33.6

[a]VSD-100, ITMO, Minsk, Belorussia
[b]ZT-100, Germany

Exhaust steam is partially recycled after mixing with fresh superheated steam. Grains treated in this way are suitable for direct use as cattle feed with high digestibility. Table 6 presents the operating characteristics of such ISD installations used for thermal processing of grains.

Figure 18 shows a schematic of a vortex spray dryer (VSD) for suspensions and slurries that operates essentially as a countercurrent, counterrotating impinging stream dryer. The dryer is made as a horizontal cylinder 1.2 m in diameter and 4 m in length with two fluid nozzles located at each end of the dryer. The primary streams of hot air are introduced axially with the nozzles. Secondary air streams are fed tangentially to the primary ones but in opposite directions, which results in a counterrotation of the primary airstreams. Thus, the drying material in the form of liquid or suspension is not only atomized but also brought into a swirling motion by the rotating airstreams. To prevent falling of the sprayed droplets onto the dryer wall (which can result in material overheating), supplementary airstreams are introduced tangentially (in the same direction as the secondary air) through the dryer wall. This part of the dryer wall at both ends is made as a slotted grid formed of suitably directed vanes. After impingement in the middle zone of the dryer, the airstream carrying dry, powdery material is directed to the separation system. Due to intensive evaporation in the impingement zone the inlet air temperature of about 150°C falls rapidly to about 70°C, which allows drying of thermosensitive materials like antibiotics or microorganisms. At this inlet air temperature the evaporative capacity is in the order of 28-kg H_2O/(m³h), which gives about 100 kg of evaporated water per hour. As compared to the classical spray dryers operating under similar conditions, the impinging stream spray dryer has seven times greater evaporative capacity (Table 7), mostly because of the high turbulence in the impingement zone and higher loading of the air.

7. CONCLUSION

ISDs are a new class of dryers for pastes, sludges, and even suspensions. Much research and development remains to be done to understand the fundamental hydrodynamics of the diverse types that are feasible configurations. Scaleup criteria need to be developed and tested. However, there is already enough evidence from pilot and industrial data that demonstrates that ISDs will become increasingly popular in the next decade.

NOMENCLATURE

d_p	particle diameter, m
D	duct diameter, m
G	scale factor
g	acceleration due to gravity, m/s^2
h	heat transfer coefficient, $W/(m^2 K)$
h_v	volumetric heat transfer coefficient, $W/(m^3 K)$
H	distance between accelerating ducts, m
L	dimensionless distance
m_h	mass (holdup), kg
n	number of cycles
ΔP	pressure drop, Pa
r	latent heat of vaporization, kJ/kg
t	time (elapsed), s
t	residence time, s
T	temperature, K (°C)
u	velocity, m/s
V_r	volume of a dryer, m^3
W	mass flow rate, kg/s
x	coordinate, m
X	moisture content (dry basis), kg/kg

Greek Letters

β	volumetric concentration, m^3/m^3
λ	thermal conductivity, $W/(m K)$
μ	dynamic viscosity, $kg/(m s)$
ν	kinematic viscosity, m^2/s
ρ	mass density, kg/m^3
ζ	friction factor

Subscripts

cr	critical
0	initial/flow duct
p	particle (droplet)
s	solid (suspension)
t	terminal
v	volumetric
x	local
max	maximum
1	inlet
2	outlet

Superscripts

s small scale
l large scale
$*$ dimensionless parameter

Dimensionless Numbers

$\text{Eu} = \Delta P/(u^2 \rho)$ Euler number
$\text{Fr} = u_0/(gD)$ Froude number
$\text{Gu} = (T - T_p)/T$ Gukhman number
$\text{Nu} = hd_p/\lambda$ Nusselt number
$\text{Pr} = c\mu/\lambda$ Prandtl number
$\text{Re} = u_0 D/\nu$ Reynolds number
$\text{Re}_p = u_0 d_p/\nu$ Particle Reynolds number
$\text{Re}_r = (u \pm u_p) d_p/\nu$ Reynolds number (based on relative velocity)
$\text{Re}_t = u_t d_p/\nu$ Reynolds number (based on terminal velocity)

REFERENCES

1. Mujumdar, A. S., Impingement Drying. In: *Handbook of Industrial Drying*, ed. A. S. Mujumdar, Marcel Dekker, New York, 1987, pp. 461–474.
2. Kudra, T., and Mujumdar, A. S., Impingement Stream Dryers for Particles and Pastes, *Drying Technology*, 7(2):219–266, 1989.
3. Tamir, A., Impingement Streams (ISD) and their Application to Drying. *Drying '92*, ed. A. S. Mujumdar, Elsevier Science Publishers, Amsterdam, 1992.
4. Tamir, A., and Kitron, A., Application of Impinging Streams in Chemical Engineering Processes—A Review, *Chem. Eng. Communications*, 50:241–330, 1987.
5. Tamir, A., Impinging Stream Contactors; Fundamentals and Application. In: *Advances in Transport Phenomena*, Elsevier Science Publishers, Amsterdam, 1992.
6. Elperin, I. T., and Meltser, V. L., Apparatus for Thermal Treatment of Dispersed Materials, USSR Patent No. 596792, 1978.
7. Kitron, A., Buchman, R., Luzatto, K., and Tamir, A., Drying and Mixing of Solids and Particles' RDT in Four-Impinging-Streams and Multistage Two-Impinging-Streams Reactors, *Ind. Eng. Chem. Res.*, 26:2654–2641, 1987.
8. Enyakin, Yu. P., Depth of Penetration of Solid- or Liquid-Phase Particles in Opposing Gas-Suspension Jets, *J. Eng. Phys.*, 14(6):512–514, 1986.
9. Elperin, I. T., Meltser, V. L., Pavlovskij, L. L., and Enyakin, Yu. P., Transfer Phenomena. In: *Impinging Streams of Gaseous Suspensions*, Nauka i Tekhnika, Minsk, 1972 (in Russian).
10. Elperin, I. T., Meltser, V. L., Levental, L. I., Phateev, G. A., and Enyakin, Yu. P., Computer Calculation of Particle Motion in Impinging Streams of Dispersed Materials, *Izvestia AN BSSR*, (2):95–100, 1972 (in Russian).
11. Levental, L. I., Meltser, V. L., Baida, M. M., and Pisarik, N. K., Investigation of Solid Particle Motion in Reverse Gas-Solid Suspension. In: *Investigation of Transport Processes in Dispersed Systems*, ITMO AN BSSR, Minsk, 1981, pp. 69–76 (in Russian).
12. Soloviev, M. I., Suspending and Transportation of Granular Materials in Horizontal Ducts, *Inzhenerno Phyzicheski Zhurnal*, 7(10):62–67, 1964.
13. Kitron, Y., and Tamir, A., Performance of a Coaxial Gas-Solid Two-Impinging-Streams (TIS) Reactor: Hydrodynamics, Residence Time Distribution and Drying Heat Transfer, *Ind. Eng. Chem. Res.*, 27:1760–1767, 1988.

14. Taubman, V. A., Gornev, B. L., Meltser, V. L., Pastushenko, B. L., and Savinkin, V. I., *Contact Heat Exchangers*, Khimia, Moscow, 1987, p. 252 (in Russian).
15. Elperin, I. T., Enyakin, Yu. P., and Meltser, V. L., Experimental Investigation of Hydrodynamics of Impinging Gas-Solid Particles Streams. In: *Heat and Mass Transfer*, eds. A. V. Luikov and B. M. Smolsky, Minsk, 1968, pp. 454–468 (in Russian).
16. Dzyadzio, A. M., and Kemmer, A. S., *Pneumatic Transport in Grain-Processing Manufactures*, Kolos, Moscow, 1967 (in Russian).
17. Tamir, A., Vertical Impinging Streams and Spouted Bed Dryers: Comparison and Performance, *Drying Technology*, 7(2):183–204, 1989.
18. Kitron, Y., and Tamir, A., Characterization and Scale-up of Co-axial Impinging Stream Gas-Solid Contactors, *Drying Technology*, 8(4):781–810, 1990.
19. Luzatto, K., Tamir, A., and Elperin, I. T., A New Two-Impinging-Streams Heterogeneous Reactor, *AIChE J.*, 30(4):600–608, 1984.
20. Meltser, V. L., and Pisarik, N. K., Interphase Heat Transfer in Impinging Single and Two-Phase Streams, *Proc. XI All Russian Conference on Heat and Mass Transfer*, 6(1):132–135, 1980 (in Russian).
21. Meltser, V. L., *Scientific Principles and Application of Enhanced Processes in Impinging Streams*, ITMO, Minsk, Bielorussia (in press).
22. Strumillo, Cz., and Kudra, T., *Drying; Principles, Applications and Design*, Gordon and Breach Sci. Publ., New York, 1987, p. 345.
23. Tamir, A., *Impinging-Stream Reactors*, Elsevier Science Publishers, Amsterdam, 1994.
24. Meltser, V. L., Gurevich, G. L., Starovoitenko, E. I., and Deryugin, A. I., The Calculation Method of Thermal Processing of Wet Particles in Reversive Gas-Solid Streams. In: *Heat and Mass Transfer Investigations in Drying and Thermal Processing of Capillary-Porous Materials*, ITMO AN BSSR, Minsk, 1985, pp. 131–138 (in Russian).
25. Meltser, V. L., and Tutova, E. G., Drying of Crystalline Lysine in Reversible Impinging Streams, *Biotekhnologia*, 2:70–74, 1986 (in Russian).
26. Tamir, A., and Shalmon, B., Scale-up of Two-Impinging Streams (TIS) Reactors, *Ind. Eng. Chem. Res.*, 27:238–242, 1988.
27. Meltser, V. L., Kudra, T., and Mujumdar, A. S., Classification and Design Considerations for Impinging Stream Dryers, *Proc. Int. Forum on Heat and Mass Transfer—Int. Drying Symposium*, 1:181–184, 1992.

18
Infrared Drying

Cristina Ratti
Planta Piloto de Ingeniería Química
Bahía Blanca, Argentina

Arun S. Mujumdar
McGill University
Montreal, Quebec, Canada

1. INTRODUCTION

Drying is the most common and most energy-consuming industrial operation. With literally hundreds of variants actually used in drying of particulate solids, pastes, continuous sheets, slurries, or solutions, it provides the most diversity among chemical engineering unit operations.

One of the increasingly popular, but not yet common, methods of supplying heat to the product for drying is infrared radiation (IR). Although this type of heat transmission was used incidentally in the past accompanying other types of heat transfer during dehydration, infrared dryers are now designed to utilize radiant heat as the primary source (1).

The most common current applications of IR drying are in dehydration of coated films and webs and to correct moisture profiles in drying of paper and board. Theoretical work and laboratory-scale experimental results on IR drying of paints, coatings, adhesives, ink, paper, board, textiles, and so on can be found in the literature (e.g., Refs. 2–5).

On the other hand, reports on IR drying applied to other products like foodstuffs, wood, or sand are not very common as yet. Most published data on IR drying of foods comes from the former Soviet Union, the United States, and the East European countries (6). Ginzburg (7) described IR drying of grains, flour, vegetables, pasta, meat, fish, and so on and showed that IR drying can be successfully applied to foodstuffs. There are many current industrial applications of drying agricultural produce by IR. Sandu (8) pointed out as advantages of IR drying in foods the versatility of IR heating, simplicity of the required equipment, easy accommodation of the infrared heating with convective, conductive, and microwave heating, fast transient response, and also significant energy savings. Experimental and theoretical works on IR drying of opaque and semitransparent materials (silica sand, brick, brown coal, graphite suspensions, and slurry of surplus activated sludge) have been performed by Hasatani et al. (9,10).

The purpose of this chapter is to give a general review of IR drying with special reference to industrial applications. A detailed description of this process as applied to paper drying can be found in Chapter 31 of this handbook. Note that many direct as well as indirect dryers can be modified to accommodate IR heaters. Indeed, combined convective and IR dryers have been shown to be very attractive. Also, IR heating can be coupled effectively with vacuum operation to permit removal of evaporated moisture. IR heating may be applied continuously or intermittently (in space or time) to save energy and often to improve product quality.

2. BASIC PRINCIPLES

2.1. Theory

Transmission of electromagnetic radiation does not need a medium for its propagation. The wavelength spectrum of the radiation depends on the nature and temperature of the heat source. Every body emits radiation due to its temperature level. It is called *thermal radiation* because it generates heat that is located in the wavelength range of 0.1–100 μm within the spectrum. Infrared radiation falls in this category and is conventionally classified as near infrared (0.75–3.00 μm), medium infrared (3.00–25.00 μm), and far infrared (25–100 μm) (8).

Thermal radiation incident upon a body may be *absorbed* and its energy converted into heat, *reflected* from the surface, or *transmitted* through the material following the balance.

$$\rho + \alpha + \tau = 1 \tag{1}$$

where ρ is the reflectivity, α the absorptivity, and τ the transmissivity. For monochromatic incident radiation, these properties are called *spectral* and when that radiation is polychromatic they are defined as *total* (8). Based on the transmissivity, materials may be classified depending mainly on the physical state of the body where the radiation impinged. A body that does not allow the radiation to be transmitted through it is called *opaque* and is characterized by $\tau = 0$. Examples of these are most solids. On the other hand, liquids and some solids like rock salt or glass have a defined transmissivity so they are "transparent" to radiation.

The reflection may be "regular" (also termed *specular*) or "diffuse" depending on the surface finish of the material. In the regular case, the angle of incidence of the radiation is equal to the angle of reflection due to a highly polished surface or a smooth surface. When the surface has roughnesses larger than the wavelength, radiation is reflected diffusely in all directions.

Generally, solid bodies absorb all of the radiation in a very narrow layer near the surface. This is a very important consideration in modeling the heat transfer process, since mathematically this concept transforms a term within the energy balance into a boundary condition. An ideal body that absorbs all of the incident energy without reflecting or transmitting is called a *black body*, for which $\alpha = 1$.

The total amount of radiation emitted by a body per unit area and time is called *total emissive power* E (11) and depends on the temperature and the surface characteristics of the body. This energy is emitted from a surface in all directions and at all wavelengths. A black body is also defined as one that emits the maximum radiation per unit area and its emissive power E_b depends only on its temperature. The emissivity of a body ε is then defined as the ratio of its total emissive power to that of a black body at the same temperature, $\varepsilon = E/E_b$.

As pointed out above, the total emissive power has energy from all the wavelengths in the spectrum of the radiation. On the other hand, the monochromatic emissive power E_λ is the radiant energy contained between wavelengths λ and $\lambda + d\lambda$ (12). For a black body, this power is expressed by (Planck's law of radiation)

$$E_{b,\lambda} = \frac{2\pi c^2 h \lambda^{-5}}{\exp\left(\dfrac{ch}{\kappa\lambda T}\right) - 1} \tag{2}$$

so the monochromatic emissivity of a body is defined as $\varepsilon_\lambda = E_\lambda/E_{b\lambda}$. Kirchhoff's law states that under thermodynamic equilibrium (which requires all surfaces to be at the same temperature), the monochromatic absorptivity and emissivity of a body are equal.

Equation [2] has a maximum that is related to the temperature by the following expression (Wein's displacement law):

$$\lambda_{\max} T = 2897.6 \ \mu K \tag{3}$$

Equation [2] may be integrated over all wavelengths to obtain the total emissive power for a black body (Stefan–Boltzmann law):

$$E_b = \int_0^\infty E_{b,\lambda} d\lambda = \sigma T^4 \tag{4}$$

where σ is the Stefan–Boltzmann constant.

A *grey body* is defined as one that has the same emissivity over the entire wavelength spectrum. Thus, Kirchhoff's law may be applied to grey bodies independently of their temperature.

Heat exchange by radiation between two black bodies at different temperatures may be obtained using the Stefan–Boltzmann law and is expressed by:

$$Q_r = A_i F_{ij} \ \sigma \ (T_i^4 - T_j^4) \tag{5}$$

where F_{ij} is the shape or view factor between surfaces i and j. By definition this geometrical factor takes into account the part of the total radiation emitted by the surface i that is intercepted by surface j. To calculate theoretically this factor is rather complicated but for most common geometries there are charts and formulas available in the literature (11,12). A useful equation, known as the *reciprocity theorem*, relates the shape factors and the areas of both surfaces through the following equation:

$$A_i F_{ij} = A_j F_{ji} \tag{6}$$

If the change of energy by radiation is between N bodies, the shape factors must follow the relation

$$\sum_{j=1}^{N} F_{ij} = 1 \tag{7}$$

In fact, very few bodies behave as black bodies, so a more realistic assumption is to simulate it as a grey body. The net radiation between two grey bodies is then given by the following equation:

$$Q_r = \frac{\sigma(T_i^4 - T_j^4)}{\left(\dfrac{\rho_i}{\varepsilon_i A_i} + \dfrac{1}{A_i F_{ij}} + \dfrac{\rho_j}{\varepsilon_j A_j}\right)} \tag{8}$$

Also, it must be noted that sometimes the electromagnetic radiation that impinge a body may be attenuated inside the body by scattering along with absorption. Scattering takes into account the electromagnetic radiation that may undergo a change in direction and a partial loss or gain of energy (13, p. 420). Suppose that I_λ represents a spectral radiation impinging normally a layer of material where it is absorbed and scattered, so the intensity of monochromatic radiation is attenuated following the relationship (called Bouguer's law, see Ref. 13, p. 413):

$$I_\lambda(z) = I_\lambda(0) \exp\left[-\int_0^z K_\lambda(z^*)dz^* \right] \qquad [9]$$

where K_λ is the extinction coefficient, z the distance, and $I_\lambda(0)$ is the radiation at the surface of the body. The extinction coefficient depends on temperature, pressure, composition, and the wavelength of the incident radiation. It may be pointed out that Eq. [9] is also termed *Lambert's law* or the *Bougher–Lambert law* and *Beer's law* when the extinction coefficient is put in mass terms, but it must not be confused with *Lambert's cosine law*.

2.2. Radiation Properties of Materials

The design and modeling of any process always requires knowledge of the materials involved. Specifically for infrared drying, this fact may be the clue to accomplish a safe and efficient process because radiation properties of both the radiator and the material to be dried must be matched in order to obtain the most efficient results.

Emissivity, absorptivity, reflectivity, and transmissivity are the key radiation properties. The relative magnitudes of α, ρ, and τ depend not only on the material, its thickness, and its surface finish, but also on the wavelength of the radiation (11). Nevertheless, the emission of electromagnetic waves is only a material property.

Generally, electrical conductors (e.g., metals) show an increase in emissivity with an increase in the wavelength of radiation. On the other hand, nonconductors such as asbestos, cork, wood, concrete, and so on show the opposite trend. Also, the emissivity of many bodies shows directional properties but, as the data of this variation is scarce, a good approximation is to assume an average value for $\varepsilon/\varepsilon_n = 1.2$ for polished metallic surfaces and 0.96 for nonmetallic (11).

For practical purposes, only a mean value of the emissivity or absorptivity over the direction is required. Sieber (14) obtained experimental data on total emissivity of opaque materials dependent on the temperature of the source. Many authors (7,11) have reproduced these results graphically (Fig. 1). The behavior of electrical conductors and nonconductors with temperature of the radiator can be approximately interpreted from the dependency of the monochromatic emissivity on wavelength and the relationship between temperature of the radiator and the wavelength.

At radiation temperatures in the range of 227°C to 620°C, the total reflectivity of polished pure silver is between 0.98 and 0.968, for polished pure gold from 0.982 to 0.965 in the same temperature range, and for polished aluminum the reflectivity varies from 0.961 to 0.943 in a temperature range from 223°C to 577°C (12). This is the reason why reflectors of radiation lamps are made of a thin layer of silver, and polished aluminum is used as a facing material for internal partitions in equipment for infrared radiation (7). It may be noted that for the construction of the equipment for infrared drying and in selecting the reflectors for radiator lamps, opaque materials with high reflectivity are required.

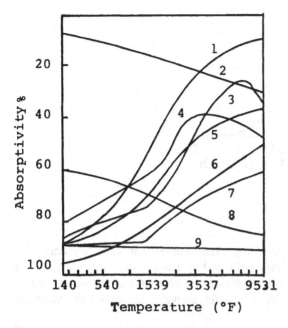

Figure 1 Absorptivity of some materials as a function of temperature (1, fireproof clay, white; 2, aluminum; 3, wood; 4, cork; 5, asbestos; 6, porcelain; 7, concrete; 8, graphite; 9, roofing silvers).

The material to be dried by IR requires a low reflectivity in order to minimize the power required to heat it and, depending on the specific drying process, a high or medium absorptivity. When drying paints or coatings, a high absorptivity of the material is usually better, but in drying thick moist materials such as foodstuffs it is necessary to increase the transmissivity in order to avoid an extremely intensive heating and thermal damage of the surface. It is important to point out that if the absorptivity of a material is low, its transmissivity is high, and vice versa.

Experimental data on properties like absorptivity and transmissivity of moist materials are not frequently encountered in the literature. In addition to the dependency with wavelength and thickness, they also depend on the water content. One of the most extensive reports on experimental data of these properties can be found in Ginzburg (7).

The variation of absorptivity of moist materials with wavelength is difficult to estimate without experimental data. Foodstuffs, as an example, are complex mixtures of different large biochemical molecules and polymers, inorganic salts, and water (8) and the absorption bands of each of these constituents are not the same. Generally, many fully wet materials have their minimum absorptivity at those wavelengths at which water has its maximum transmissivity, pointing out the important role that water plays in radiation absorption. Figure 2 shows the infrared absorption spectrum of liquid water. For many materials, transmissivity is higher at lower wavelengths (7).

As drying proceeds, the material being dried suffers a change in its radiation properties, increasing its reflectivity and consequently lowering its absorptivity at low water contents. It is then possible to change adequately the temperature of the emitter in order to improve the absorption of radiation during drying.

The transmissivity decreases with an increase of layer thickness, while absorptivity is

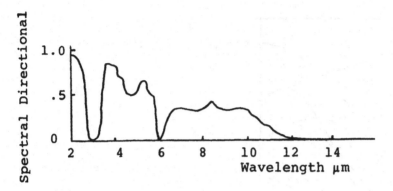

Figure 2   Absorption spectrum of water.

increased. An approximate way of representing experimental transmissivity data as a function of thickness is presented by Ginzburg (7). Tables 1 and 2 show a compilation of experimental data from the literature (7,8,11) about the variation of transmissivity of foodstuffs and other materials commonly dried, respectively, in regard to thickness, water content, and wavelength.

Figure 3 shows the radiant energy peaks for quartz tungsten filament at 2500 K and

Table 1   Transmissivity of Selected Foodstuffs

Product	Spectral peak or λ (μm)	T_r(°C)	Thickness (mm)	W(%)	τ or $τ_λ$(%)
Bread made of			2.00		1.0400
wheat flour	—	400	2.50	—	0.9700
			3.50		0.4600
			5.00		0.0000
Dough made of			1.00		5.7000
wheat flour	—	Mirror	2.75	44.0	1.7600
		lamp	5.00		0.3900
			9.00		0.0268
Tomato paste	1.075		0.50	60.0	95.0
	1.075	—	0.50	70.0	91.4
	1.075		0.50	85.0	55.8
	1.190		0.50	85.0	46.7
	1.350	—	0.50	85.0	38.5
	3.400		0.50	85.0	30.8
Potatoes	1.100	—	2.00	80.5	50.0
	1.100		8.00	80.5	16.0
Potato starch	1.100		1.00	11.8	13.0
	1.100	—	2.00	11.8	5.0
	1.100		8.00	11.8	0.23
Beet	1.100	—	2.00	85.5	40.0
	1.100		6.00	85.5	18.0
Fruit gel	1.100	—	2.00	30.0	18.0
(marmalade)	1.100		7.00	30.0	2.5

Table 2 Transmissivity of Selected Materials

Product	Spectral peak or λ (μm)	Thickness (mm)	τ or τ_λ (%)
Cigarette paper (raw)	1.075		73.0
	3.220	–	48.0
	6.150		20.0
Paper, thick (moist wall paper)	1.190		18.0
	3.750	–	10.0
	6.150		7.0
Wool cloth, dry	1.190		14.3
	3.220	–	12.3
	7.750		2.0
Sand	1.190	3.00	4.8
	6.150	3.00	2.0
Wood	1.190	1.50	4.3
	6.150	1.50	0.7

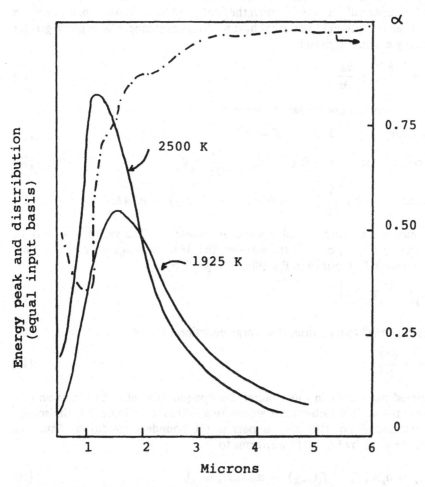

Figure 3 Energy peak and distribution (equal input basis) of quartz tungsten filaments together with the absorption bands of a slice (10 mm) of potato with 74.5% water content (-·-··-).

1925 K together with the spectral absorptivity of potato with 74.5% water content (8) and 10 mm thickness. To avoid overheating of the surface and to allow the radiation to penetrate into the product, it would be better to choose the heat source at 2500 K because its maximum is located in the wavelength at which the absorptivity of the foodstuff is not very high. On the other hand, if the lower heat source temperature is used, the surface may be damaged (scorched) due to intense surface heating.

3. STEADY INFRARED RADIATION

3.1. Modeling the Process

Only the energy balance equation must be modified to include IR in the calculation procedure for drying, while the mass balance equation remains the same as that for purely convective drying. As the driving force for mass transfer depends on temperature, the high temperatures achieved by the drying surface with IR heating enhance the drying rate. In order to simplify the modeling of IR drying, it is convenient to start with IR drying of a flat, partially "transparent" particle in which the heat conduction is one dimensional. In the special case when the internal to external heat transfer resistance ratio is much greater than 0.1, the energy balance is given by

$$\rho_m C_{psh} \frac{\partial T}{\partial t} = k \frac{\partial^2 T}{\partial z^2} + \frac{\partial q_a}{\partial z} \qquad [10]$$

with the following boundary and initial conditions:

Initial condition: $t = 0, \forall z \qquad T = T_0$ [11]

Boundary conditions: $z = 0$ (center) $\dfrac{\partial T}{\partial z} = 0$ [12]

$z = Z$ (surface) $kA \dfrac{\partial T}{\partial z} \Big|_{z=z} = h_g A(T_g - T \rfloor_{z=z}) - n_w A \Delta H_s$ [13]

The above model is commonly called *semitransparent* (9). The variable q_a may be obtained by averaging Eq. [9] over all the wavelengths. If the average net radiation that impinges on a surface is Q_r (defined in Eq. [8]), q_a is given by

$$q_a = Q_r \exp\left[-\int_0^z \overline{K} \, dz^*\right] \qquad [14]$$

If the particle is "opaque" to radiation, the energy equation becomes

$$\rho_m c_{psh} \frac{\partial T}{\partial t} = k \frac{\partial^2 T}{\partial z^2} \qquad [15]$$

As was pointed out earlier in this chapter, an opaque solid absorbs radiation in a narrow zone near the surface rather than attenuating within its volume; so, the second term on the right side of Eq. [10] must appear in the boundary condition. Thus, the boundary condition given by Eq. [13] transforms to

$$kA \frac{\partial T}{\partial z} \Big|_{z=z} = h_g A(T_g - T \rfloor_{z=z}) - n_w A \Delta H_s + Q_r \qquad [16]$$

while the other boundary condition and the initial condition remain the same. This model is called *opaque* (9).

In the case that only an average particle temperature is required or that the internal-to-external heat transfer resistance ratio is smaller than 0.1 (which means that the temperature profiles inside the particle can be considered to be almost uniform), the following simplification of Eq. [15] is permissible:

$$\rho_m C_{psh} \frac{\partial T}{\partial t} = k \frac{1}{\Delta z} \left(\frac{\partial T}{\partial z} \right) = \frac{1}{\Delta z} \left(k \frac{\partial T}{\partial z} \right) \Big]_{z=z} \tag{17}$$

Inserting Eq. [16] into Eq. [17], the energy equation becomes

$$m C_{psh} \frac{dT}{dt} = h_g A (T_g - T) - n_w A \Delta H_s + Q_r \tag{18}$$

The model equations to represent the behavior of a dryer are an extension of the above explanation for IR drying of a single particle. In paper, paint, or textile drying, continuous dryers are commonly used. The equations describing continuous IR dryers can be found in the literature (3,5). Nevertheless, a batch dryer may also be used for deep bed drying of foodstuffs because such materials have a solid structure from the beginning of the process (15). The modeling equations for a batch dryer with circulation of air through the bed can be developed as follows.

Food particles experience shrinkage during drying, but since shrinkage is not appreciably affected by air or particle temperature (16) it may be assumed that IR drying does not affect shrinkage differently from purely convective drying. The key assumptions employed in this model are

1. One-dimensional transport of heat and mass.
2. Uniform velocity distribution in the dryer (plug flow of drying air).
3. Adiabatic system (well insulated).
4. Conduction heat transfer between particles in the bed and contact diffusion are negligible.
5. Shrinking particles.
6. The air is completely transparent to radiation. Although in the drying process the air contains water vapor that absorbs IR radiation, this may be neglected because the amount of water vapor is small relative to the air. As an example, at a dry bulb temperature of 60°C and 40% relative humidity, the absolute humidity is 0.053-kg water/kg dry air.

To avoid the problem of shrinkage, a moving coordinate system that follows the movement of the shrinkage particles may be used, taking as the basis a differential control volume that always contains the same amount of dry mass as the initial amount. This coordinate system is expressed as (16)

$$d(z/L_0) = \frac{\rho_{s,0}(1 - \varepsilon_{l0})}{\rho_s(1 - \varepsilon_l)} d\Lambda \tag{19}$$

Then, the resulting equations that represent batch fixed bed IR drying are

Mass balance in the gas phase:

$$\left[\frac{\partial Y}{\partial t} \right]_\Lambda = \frac{n_w a_v (1 - \varepsilon_l)}{\rho_a \varepsilon_l} - \frac{1}{SL_0} \frac{G_s}{\rho_a \varepsilon_l} \frac{\rho_s(1 - \varepsilon_l)}{\rho_{s,0}(1 - \varepsilon_{l0})} \frac{\partial Y}{\partial \Lambda} \tag{20}$$

Mass balance in the solid:

$$\left[\frac{\partial X}{\partial t}\right]_{\Lambda} = -\frac{n_w a_v}{\rho_s} \tag{21}$$

Energy balance in the solid:

$$\left[\frac{\partial T_s}{\partial t}\right]_{\Lambda} = \frac{a_v}{\rho_s(1 + X)Cp_{sh}} [h_g(T_g - T_s) - n_w\Delta H_s + Q'_r] \tag{22}$$

Energy balance in the gas phase:

$$\left[\frac{\partial T_g}{\partial t}\right]_{\Lambda} = -\frac{h_g a_v(1 - \varepsilon_l)}{\rho_a \varepsilon_l Cp_{ah}} (T_g - T_s) - \frac{1}{SL_0} \frac{G_s}{\rho_a \varepsilon_l} \frac{\rho_s(1 - \varepsilon_l)}{\rho_{s,0}(1 - \varepsilon_{l0})} \frac{\partial T_g}{\partial \Lambda} \tag{23}$$

where Q'_r is defined from Eq. [8] as

$$Q'_r = \frac{\sigma(T_e^4 - T_s^4)}{\dfrac{(1 - \varepsilon_s)}{\varepsilon_s} + \dfrac{1}{F_{se}} + \dfrac{(1 - \varepsilon_e)}{\varepsilon_e(A_e/A_s)}} \tag{24}$$

The initial profile of the four variables in the fixed bed is stated as

$$\Lambda = 0 \begin{array}{l} X = X_0 \\ T_s = T_{s0} \\ Y = Y_{g0} \\ T_g = T_{g0} \end{array} \qquad \Lambda \neq 0 \begin{array}{l} X = X_0 \\ T_s = T_{s0} \\ Y = Y_{sat}(T_{s0}) \\ T_g = T_{s0} \end{array} \tag{25}$$

and the values of T_g and Y_g at $\Lambda = 0$ are set in T_{g0} and Y_{g0} for constant inlet conditions.

Some advances in the understanding of the radiation process have been made in recent years. As an example, Parrouffe (17) has demonstrated on the basis of extensive experimental data that one may, within engineering accuracy, use the analogy between heat and mass transfer to estimate the convective heat or mass transfer coefficients even in the presence of intense radiative heat flux on the evaporating surface. Appropriate corrections must be employed, however, for the high evaporative mass flux at the surface.

3.2. Advantages and Limitations

Many authors have pointed out the advantages and disadvantages of using IR drying (6,18-20). In fact IR drying has many positive attributes, the main one being the reduction in drying time. Figure 4 shows the effect of infrared drying compared to conventional convective drying of acoustic tiles (20). Also, infrared drying offers a solution to problems that seemed to be unsolvable in the past, such as those associated with the carrying of volatile organic compounds from solvent-based paints by the exhaust hot air in conventional convective dryers. A summary of the advantages of IR drying follows:

1. High efficiency to convert electrical energy into heat for electrical IR.
2. Radiation penetrates directly into the product without heating the surroundings.
3. Uniform heating of the product.
4. Easy to program and manipulate the heating cycle for different products and to adapt to changing conditions.
5. Leveling of the moisture profiles in the product and low product deterioration.
6. Ease of control.
7. IR sources are inexpensive compared to dielectric and microwave sources; they have a long service life and low maintenance.

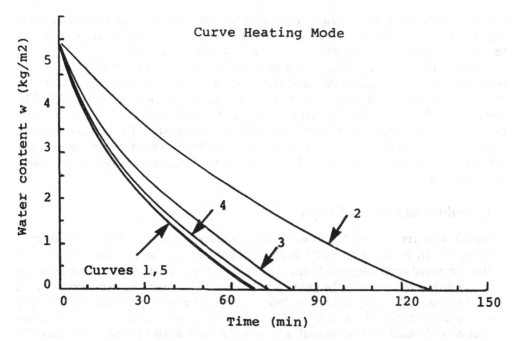

Figure 4 Drying times according to the heating process employed (1: Intermittent Infrared, $q = 10 \text{ kW/m}^2$; 2, 3, 4, 5: Convective drying with $h_g = 15, 50, 100,$ and $200 \text{ W/m}^2\text{k}$, respectively).

8. Directional characteristics that allow drying of selected parts of large objects.
9. Occupies little space and may easily be adapted to previously installed conventional dryers.
10. Low-cost technology.

On the other hand, the disadvantages are

1. Scaling up of the heaters is not always straightforward.
2. Essentially surface dryers. Nevertheless, a great effort is being done to improve this technology in order to adapt it to drying thick materials with successful results.
3. The testing of the equipment must be carried out in the plant to assure a successful design.
4. Potential fire hazards must be considered in design and operation.

4. INDUSTRIAL INFRARED DRYING APPLICATIONS

4.1. Applications of Infrared in Industry

IR heating is widely used in industry for surface drying or dehydration of thin sheets such as textiles, paper, films, paints, stove enamels, and so on. Specifically in the automotive industry, the infrared baking for paint-on-metal application is the most successful. Another sector in which IR plays an important role is in the pulp and paper industry. As an example, in Sweden this sector imposes higher demands in energy consumption than any other, and a relative new method that not only improves paper quality but also achieves energy-efficient drying is to use electrically operated IR radiation heating (21).

The disposal of hazardous waste is a less general IR application but, for instance, the

IR drying of metal hydroxide sludge minimizes the cost impact of increasingly stringent disposal restrictions. When the metal content of sludge is of interest to recycling, it may be one way of changing a costly by-product into a profitable one (22).

Although IR drying of thick, porous materials has not yet been fully developed, some researchers showed, mainly by means of experimental results, that one of the applications in which long-wave IR heating is most efficient is in dehydration of foods (7,23). The Japanese food industry uses this type of heating for drying of seaweed, curry sauce, carrots, and pumpkins, among other things (23). Other important IR applications in the food area (not specifically in drying) are cooking soybeans, cereal grains, cocoa beans and nuts, precooking rice, bacon, and barley grains for "ready-to-eat" products, braising meat, and frying.

4.2. Industrial Infrared Dryers

Although IR dryers may be of the batch or continuous type, the latter is the most common arrangement. IR ovens for dryers are usually designed and constructed from standard IR sections arranged and integrated to the conventional dryers in such a way that IR radiation is directly intercepted by the product to be dried. These sections are selected on the basis of the particular application. It is desirable to test the product in a lab-scale IR oven under simulated conditions and to design the large-scale unit on the basis of the experimental data obtained. To accomplish a reliable design it is also necessary to know the efficiency of conversion from electric to infrared energy of the radiators used in the plant (unless gas-fired IR heaters are used). The main data required are the intensity of radiation and the residence time (19, p. 289) but, although oven style and cross-section are easily determined, on the other hand the selection of the heat source, time/temperature cycle, and the power density requires oven design experience (24).

Airflow is required in IR ovens for two primary purposes:

1. Air movement to cool and protect oven walls and terminals
2. Oven exhaust to remove smoke, moisture, solvents, hazardous vapors, and so on.

The decision of using natural or forced convection and the amount of airflow rate to meet the appropriate cooling effect must be adjusted to the specific application.

Infrared Radiation Sections

Infrared sections basically consist of a heat source (called a *radiator* or *emitter*), a reflector, source sockets, electrical connections, and a shell in which the parts of each section are built together (Figure 5). The main component is the radiator, which, depending on the mode of heating, may be classified as electrically heated or gas fired.

Electrically Heated Radiators. For electrically heated radiators the infrared radiation is obtained by passing an electric current through a resistance that raises its temperature (6). The most common are metal sheath radiant rods, quartz tubes, and quartz lamps. A typical cross-section of a tube emitter is sketched in Figure 6-a.

One of the most important characteristics of such emitters is the *radiant efficiency*, which may be defined as the percentage of radiant output from a heat source referred to the energy input. There is a positive relationship between this efficiency and the temperature of the radiator. Also, as was pointed out previously, there exists an inverse relationship between this temperature and its peak energy wavelength. The peak wavelength can be controlled by changing the temperature of the source so that different types of emitters

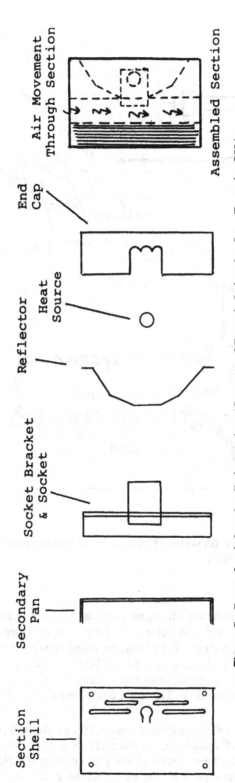

Figure 5 Parts of an infrared radiation section. (Courtesy of Fostoria Industries, Inc., Fostoria, OH.)

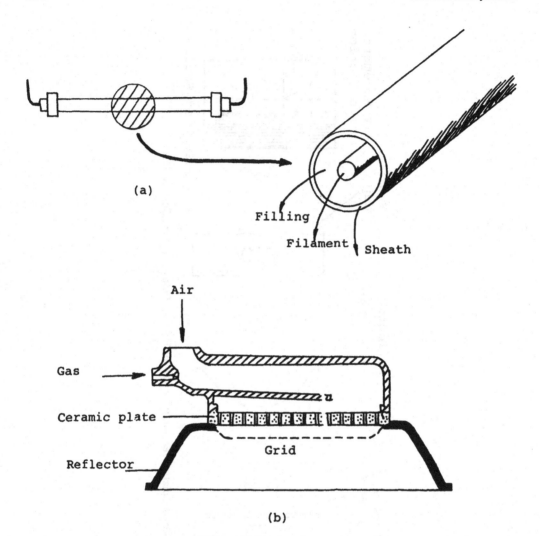

Figure 6 Types of radiators: (a) sketch of an electric infrared radiation source; (b) sketch of a gas-fired infrared radiation source.

operating at the same temperature will all have the same peak wavelength as well as other characteristics like penetration and color sensitivity. Figure 7 presents the relationship between voltage and temperature of the radiator together with the efficiency curve and Figure 8 shows graphically the heatup and cool-down rate of response of the more common emitters, which can be an important criterion in the selection of the proper source for a particular application. The main characteristics of the electric emitters are shown in Table 3.

Gas-Fired Radiators.   Gas-fired radiators consist of a perforated plate (metal or refractory) that is heated by gas flames in one of the surfaces so the plate raises its temperature and emits radiant energy. The porosity of the plate determines the temperature of the other surface to ensure a safe process. Figure 6-b shows a sketch of this type of radiator (18, p. 250). The temperature of such a radiator is generally between 1500°C to 1700°C

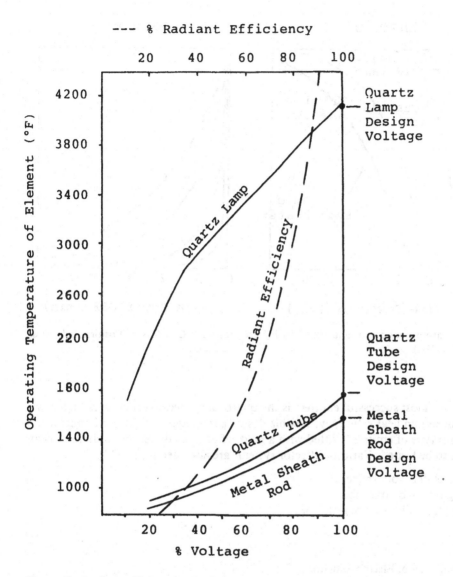

Figure 7 Radiant efficiency and relationship between voltage and temperature of various radiators. (Courtesy of Fostoria Industries, Inc., Fostoria, OH.)

with wavelengths from 2.7 to 2.3 μm (18, p. 249). The radiant efficiency of such radiators is typically about 60%.

For practical purposes, choosing an emitter involves consideration of the following factors (25):

1. Absorption characteristics of the material being heated
2. Power density of the radiating area "seen" by the product
3. Ratio of convected heat to radiant heat
4. Nature of the installation
5. Type of control required

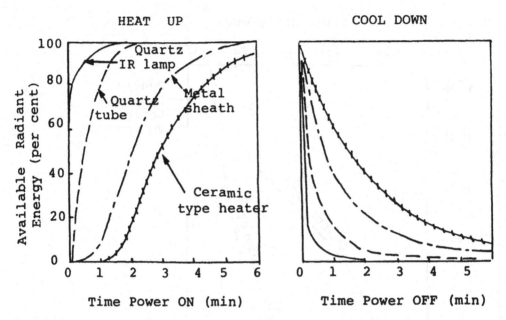

Figure 8   Heat-up/cool-down time cycles for infrared sources. (Courtesy of Fostoria Industries, Inc., Fostoria, OH.)

One of the most successful emitters is the quartz lamp because it ensures high power densities, maximum heat efficiency, flexible design parameters, and ease of controllability. Also, this type of emitter is fitted with a gold reflector to direct the radiation toward the product to be heated. Various reflector systems are also used (6, p. 217):

Individual metallic/gold reflectors
Individual gilt twin quartz tube
Flat metallic/ceramic cassette reflectors

Table 3   Properties of Electric Radiators

	Type of emitter		
	Metal sheath	Quartz tube	Quartz lamp
Sheath material	Stainless steel	Translucent quartz	Clear quartz
Filament	Nickel-chrome	Nickel-chrome	Tungsten
Sheath diameter	3/8 in	3/8 in, 5/8 in, 7/8 in	3/8 in
Filling material	Insulating powder	Air	Inert gas
Maximum filament temperature	1800°F	1800°F	4000°F
Peak wavelength	—	2.3 μ	1.15 μ
Voltage	240 volts	Up to 600 volts	~600 volts
Wattage	60 W per linear inch	30, 60, or 90 W pli	100 W pli
Radiant efficiency	50%	60%	86%

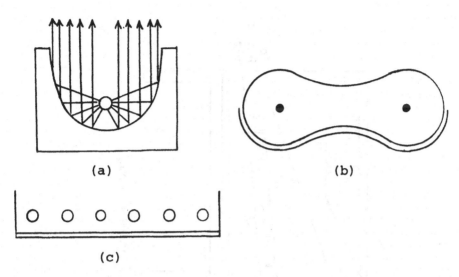

Figure 9   Different types of reflectors: (a) individual reflector; (b) individual gilt twin quartz tube; (c) flat metallic/ceramic cassette reflector.

Figure 9 shows a sketch of such reflectors. The material and the shape of the reflector determines its efficiency. Reflector materials must have high reflectivity, resist corrosion, heat, and moisture, and be easily cleaned. They must also maintain the high reflectivity over a long period of time.

Recent Developments in Design of Infrared Radiation Dryers

Although an IR dryer can be built using an existing convective dryer by putting in the appropriate number of radiators over the product to be dried so as to direct the radiation on it, this technology is being improved and new combinations of dryers have appeared in the market. As examples of these new trends and applications, Figures 10 and 11 show two industrial IR dryers. Figure 10 is a high-velocity hot air impingement oven with infrared electric heaters mounted between adjacent nozzles specially for drying adhesives and inks on papers, foam, composite web substances, and so on (26). Figure 11 is a gas-heated IR dryer for metal hydroxide sludge volume reduction (27). Another promising technology is the combination of intermittent infrared with continuous convection heating (20) for drying thick porous materials such as panels made of wood and acoustic tiles.

4.3. Costs

The capital costs per kilowatt installed depending on the different heating modes are presented in Table 4 (28). Specifically for IR drying the radiators are generally the main cost of a dryer. Table 4 also presents an approximate relationship among the costs of different types of emitters. A lamp radiator has a life of 2000 to 3000 hours while a sheathed element's life is from 5000 to 10,000 hours (19, p. 290). The replacement of the radiator elements is the main maintenance item. The figures in this table should be taken as guidelines rather than as precise costs. With changes in technology, these figures are likely to change with time. The costs are 1992 figures for Quebec, Canada.

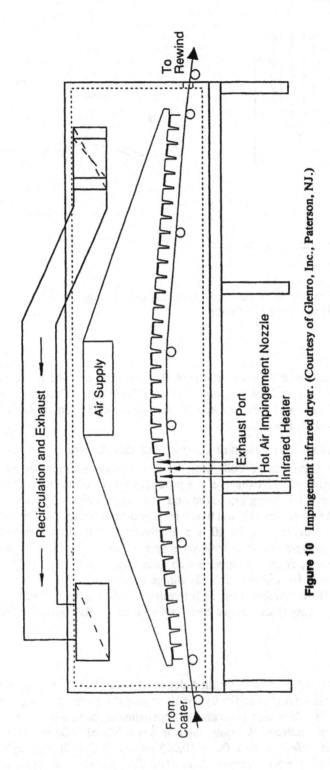

Figure 10 Impingement infrared dryer. (Courtesy of Glenro, Inc., Paterson, NJ.)

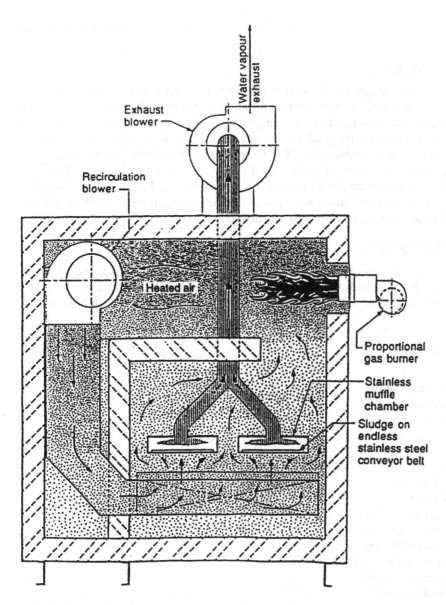

Figure 11 Infrared dryer for treatment of sludge. (Courtesy of JWI, Inc., Holland, MI.)

Table 4 Capital Costs for Different Modes of Heating and Types of Emitters (Quebec, Canada, 1992)

Mode of heating	Capital cost	Capital cost (equal basis) $/kW referred to the highest
Convection	$300/kW	--
Radiofrequency	$2000/kW	–
Infrared	$500/kW	–
Electric lamp	–	0.49
Electric sheathed filament	–	1.00
Gas-fired emitter	–	0.93

5. CONCLUSION

In view of several advantages it is likely that IR drying in combination with convection or vacuum will become increasingly popular. Intermittent (spatial or time-wise) supply of IR heating has the potential merit of saving energy, reducing air consumption, and enhancing the quality of heat-sensitive products. Dryers for continuous sheets or large surfaces (e.g., planks) utilizing a combination of impinging jets and radiant heaters spaced between jets have already proven to be industrially viable. Also, a combination of radiant heating under vacuum operation is technically a sound process for drying certain products. Much fundamental and industrial research and development needs to be carried out to exploit fully the potential of IR drying technologies.

ACKNOWLEDGMENTS

The authors gratefully acknowledge the information provided by Glenro, Incorporated (Paterson, NJ), Fostoria Industries, Incorporated (Fostoria, OH), and JWI, Incorporated (Holland, MI), and for their permission to reproduce relevant figures in this chapter.

NOMENCLATURE

A area
a_v surface area/volume
C_{pah} humid air specific heat
C_{psh} humid solids specific heat
c speed of the light
E total emissive power
F view factor
G_s dry air flow rate
h Planck constant
h_g heat transfer coefficient
I intensity of radiation
K extinction coefficient
$\overline{K}$ average extinction coefficient
k thermal conductivity
L_0 initial bed height
m mass
n_w water mass flux
Q_r heat exchange between bodies
q_a heat absorbed
S dryer transversal section
T temperature
t time
W water content (wet basis)
X solid water content (dry basis)
Y air absolute humidity
z distance

Subscripts

b black body
e emitter

g	gas phase
i	surface
j	surface
n	in a direction normal to the surface
0	initial
r	radiator
s	solid
sat	at saturation
λ	monochromatic

Greek

α	absorptivity
ΔH_s	heat of sorption
ε	emissivity
ε_l	bed porosity
κ	Boltzmann constant
Λ	moving coordinate
λ	wavelength
λ_{max}	wavelength where E is maximum
ρ	reflectivity
ρ_a	air density
ρ_m	solid density
ρ_s	solid mass concentration
σ	Stefan–Boltzmann constant
τ	transmissivity

REFERENCES

1. Williams-Gardner, A., *Industrial Drying*, Chemical and Process Engineering Series, I. L. Hepner, Series Ed., Leonard Hill, London, 1971.
2. Navarri, P., Gevaudan, A., and Andrieu, J., Preliminary Study of Drying of Coated Film Heated by Infrared Radiation. In *Drying '92*, A. S. Mujumdar (Ed.), Elsevier, Amsterdam, 1992, p. 722.
3. Kuang, H., Chen, R., Thibault, J., and Grandjean, B. P. A., Theoretical and Experimental Investigation of Paper Drying Using Gas-Fired IR Dryer. In *Drying '92*, A. S. Mujumdar (Ed.), Elsevier, Amsterdam, 1992, p. 941.
4. Therien, N., Cote, B., and Broadbent, A. D., Statistical Analysis of a Continuous Infrared Dryer, *Textile Research Journal*, 61:193–202 (April 1991).
5. Cote, B., Broadbent, A. D., and Therien, N., Modelisation et Simulation du Séchage en Continu des Couches Minces par Rayonnement Infrarouge, *Can. J. Chem. Eng.*, 68:786–794 (October 1990).
6. Hallström, B., Skjöldebrand, C., and Trägårdh, C., *Heat Transfer and Food Products*, Elsevier Applied Science, London, 1988.
7. Ginzburg, A. S., 1969. *Application of Infra-red Radiation in Food Processing*, Chemical and Process Engineering Series, Leonard Hill, London, 1969.
8. Sandu, C., Infrared Radiative Drying in Food Engineering: A Process Analysis, *Biotechnology Progress*, 2(3):109–119 (1986).
9. Hasatani, M., Harai, N., Itaya, Y., and Onoda, N., *Drying Technology*, 1(2):193–214 (1983).
10. Hasatani, M., Itaya, Y., and Miura, K., *Drying Technology*, 6(1):43–68 (1988).
11. Kreith, F., *Principles of Heat Transfer*, International Textbook Company, Scranton, PA, 1965.

12. Welty, J. R., Wicks, C. E., and Wilson, R. E., *Fundamentals of Momentum, Heat and Mass Transfer* (3rd Ed.), John Wiley and Sons, New York, 1984.
13. Siegel, R., and Howell, J. R., *Thermal Radiation Heat Transfer*, McGraw-Hill, New York, 1972.
14. Sieber, W., Zusammensetzung der von Werk-und Baustoffen Zurückgeworfenen Wärmestrahlung, *Z. Tech. Physik*, 22:130–135 (1941).
15. Heldman, D. R., and Singh, R. P., *Food Process Engineering*, Avi Publishing, Westport, CT, 1981.
16. Ratti, C., Design of Dryers for Vegetable and Fruit Products, Ph.D. thesis (in Spanish), Universidad Nacional del Sur, Bahía Blanca, Argentina, 1991.
17. Parrouffe, J. M., Combined Radiative and Convective Drying of a Capillary Porous Solid, Ph.D. thesis, McGill University, Montreal, Quebec, Canada, 1992.
18. van't Land, C. M., *Industrial Drying Equipment. Selection and Application*, Marcel Dekker, New York, 1991.
19. Nonhebel, M. A., and Moss, A. A. H., *Drying of Solids in the Chemistry Industry*, Butterworths, London, 1971.
20. Dostie, M., Séguin, J.-N., Maure, D., Ton-That, Q.-A., and Châtigny, R., Preliminary Measurements on the Drying of Thick Porous Materials by Combination of Intermittent Infrared and Continuous Convection Heating. In *Drying '89*, A. S. Mujumdar and M. A. Roques (Eds.), Hemisphere, New York, 1989.
21. Hannervall, L., Mets, V., and Gustafson, I., Electric Heating in Drying of Paper and Cardboard with Infrared Technique, Electrotech '92 Meeting Proceedings, Montreal, June 14–18, 1992.
22. Davis, M., and D. Wachter, "Tuning" Dryers for Peak Efficiency, *Process Industry Journal* (January 1992).
23. Kimura, Y., Nagayasu, T., and Kutsuzawa, S., Maximizing the Effect of Long Wave Infrared Heating and Applying It to the Food Industry, Electrotech '92 Meeting Proceedings, Montreal, June 14–18, 1992.
24. Fostoria Industries, Inc., *Technical Product Catalog*, Fostoria Industries, Inc., Fostoria, OH, 1992.
25. Bischof, H., The Answer Is Electrical Infrared, *Journal of Microwave and Electromagnetic Energy*, 25(1):47–52 (1990).
26. Glenro, Inc., *Technical Product Catalog*, Glenro, Inc., Paterson, NJ, 1992.
27. JWI, Inc., *Technical Product Catalog*, JWI, Inc., Holland, MI, 1992.
28. Dostie, M., Optimization of a Drying Process Using Infrared Radio Frequency and Convection Heating. In *Drying '92*, A. S. Mujumdar (Ed.), Elsevier, Amsterdam, 1992, p. 679.

19
Drying of Foodstuffs

Shahab Sokhansanj
University of Saskatchewan, Saskatoon, Saskatchewan, Canada

Digvir S. Jayas
University of Manitoba, Winnipeg, Manitoba, Canada

1. INTRODUCTION

The removal of moisture from solids is an integral part of food processing. Almost every food product is dried at least once at one point of its preparation. The main objectives of dehydration are summarized as follows.

1.1. Extended Storage Life

A dry food product is less susceptible to spoilage caused by the growth of bacteria, molds, and insects. The activity of many microorganisms and insects is inhibited in an environment in which the equilibrium relative humidity is below 70%. Likewise, the risk of unfavorable oxidative and enzymatic reactions that shorten the shelf life of food is reduced.

1.2. Quality Enhancement

Many favorable qualities and nutritional values of food or feed products may be enhanced by drying. Palatability is improved, and likewise digestibility and metabolic conversions are increased. Drying also changes color, flavor, and often the appearance of a food item. The acceptance to that change varies by the end user.

1.3. Ease of Handling

Packaging, handling, and transportation of a dry product are easier and cheaper because the weight and the volume of a product are less in its dried form. A dry product flows easier than a wet product; thus gravity forces can be utilized for loading and unloading and short-distance hauling.

1.4. Further Processing

Food products are dried for improved milling, mixing, or segregation. A dry product takes far less energy than a wet product to be milled. A dry product mixes with other materials more uniformly and is less sticky compared with a wet product.

1.5. Sanitation

Drying has also been used as a means of food sanitation. Insects and other microorganisms are destroyed during the application of heat and moisture diffusion. The sanitation aspect of drying is a time-temperature phenomena (25). The temperature should be at least 60°C for a short duration of 3 to 5 minutes. Lower temperatures, to a minimum of 48°C, can be used for disinfestation but the treatment duration should last at least for 24 hours or longer. Disinfestation in high temperature rotary and tunnel dryers has been studied and results have been published by Sokhansanj et al. (26) and Sokhansanj and Wood (27).

2. MOISTURE IN FOODS

The volatile part of a food item can be termed *moisture*. Moisture in the form of water molecules is bonded to various parts of the product in varying ways as follows: (a) ionic groups, such as carboxyl and amino acids, and (b) hydrogen groups, such as hydroxyl and amides. In high-moisture foods in which moisture contents are more than 50% wet basis, unbound free water exists in the interstitial pores and in intercellular spaces. The descending order of difficulty to remove water from the product also follows the above order.

A food product is in equilibrium with its surroundings when its internal vapor pressure becomes in equilibrium with the outside vapor pressure. The moisture content of the product at this stage is called the *equilibrium moisture content* (EMC). The corresponding vapor pressure in surrounding air at the same temperature is called the *equilibrium vapor pressure*. The ratio of the equilibrium vapor pressure to the saturation vapor pressure is known as the *equilibrium relative humidity* (ERH), or water activity.

A plot of EMC versus ERH usually has a sigmoid (S) shape. The reason is that the affinity of the solid to moisture and the ease of moisture adsorption or desorption depend upon the way moisture bonds to the solid (19). Starting from a bone-dry solid exposed to a humid environment, in a range of 5–10% moisture content, a single layer of water molecules is formed. The next stage is formation of a multilayer of water molecules, during which the slope of the isotherm curve is gradual. The last stage is filling of the capillary pores by condensed water.

The equilibrium moisture content varies with temperature. Several plots of EMC versus ERH are drawn for various temperatures. Each plot is often called an *isotherm*. Samples of isotherms of various food items are given in Figure 1.

Food demonstrate a hysteresis phenomenon during adsorption and desorption processes. Irreversible physical and chemical changes are responsible for this behavior. Often the EMC of a food product in a given condition is varied by two percentage points in EMC depending on whether EMC is obtained by removing or adding moisture.

Equilibrium moisture contents of foods generally are found experimentally. The following equation has been found to fit the experimental data (20):

$$\frac{P_v}{M(P_s - P_v)} = \frac{1}{M_e C} + \frac{C-1}{M_e C} \frac{P_v}{P_s} \qquad [1]$$

Drying of Foodstuffs

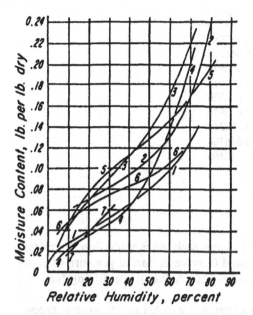

Figure 1 Selected water sorption isotherms (1, egg solids, 10°C; 2, beef, 10°C; 3, codfish, 30°C; 4, coffee, 10°C; 5, starch gel, 25°C; 6, potato, 28°C; 7, orange juice). (From Reference 1.)

where P_v and P_s are water vapor pressure and saturated vapor pressure, respectively. C is called the energy constant. M and M_e are the moisture and the equilibrium moisture content, respectively. These values can be found by performing drying experiments on a thin layer of sample of the product and plotting $P_v/M(P_s - P_v)$ versus P_v/P_s. The order of magnitude of C and M_e are 13.0 and 0.10, respectively.

An equation that has been widely used in the dehydration of cereal grains is of the form

$$M_e = E - F \ln [-R(T + C) \ln (RH)] \qquad [2]$$

Values of the constants E, F, and C are given in Table 1. R is the universal gas constant, and its value is 1.987 in SI units.

The role of moisture in food drying and storage is expressed in terms of water activity. Water activity in a moist food is defined in a similar manner as the relative humidity is defined in moist air, that is, the ratio of vapor pressure to the saturated vapor pressure at the same temperature. The water activity of nonfat dry milk has been found as a function of moisture content and temperature as follows:

$$a_w = 1 - \exp(-2.482T^{-0.735}M^{5.392 \times 10^{-5}T^{1.771}})$$

Water activity relates to the chemical activity of moisture in the food during drying and storage. Oxidation activity is only possible at water activities higher than 0.4 and the rate of inactivation of other organisms requires a water activity of 0.7 or higher. Some enzymatic activities may continue at low levels of water activities of 0.1–0.3 but their reaction rate decreases at low water activities.

Table 1   Coefficients for Use in EMCE RH Equation [2]

Product	C	E	F
Barley	91.323	0.368149	0.402787
Beans	120.098	0.480920	0.066826
Corn	30.205	0.379212	0.058970
Rice	35.703	0.325535	0.046015
Sorghum	102.849	0.391444	0.050970
Soybean	24.576	0.375314	0.066816
Wheat, durum	112.350	0.415593	0.055318
Wheat, hard	50.998	0.395155	0.056788
Wheat, soft	35.662	0.308163	0.042360

Source: From Reference 21

3.   AIR-VAPOR RELATIONSHIP

Dry air and water vapor exert a certain pressure upon each other when they are mixed. These pressures are called partial pressures. The difference in partial pressure of water vapor in the air and the pressure of the moisture in the product is the driving force for drying.

The interrelationships between air and water vapor are called *psychrometric properties* (3). Changes in these properties are shown by a psychrometric chart. A sample of the chart is given in Figure 2. The psychrometric terms given in the chart are defined as follows.

3.1.   Dry Bulb Temperature

Dry bulb temperature is the air temperature indicated by an ordinary thermometer. It is given on the horizontal axis of the psychrometric chart.

3.2.   Humidity Ratio

Humidity ratio or absolute humidity is the ratio of the weight of water vapor to the weight of dry air. Humidity ratios are on the vertical axis of the psychrometric chart.

3.3.   Relative Humidity

The ratio of vapor pressure to saturation vapor pressure is called *relative humidity*. Relative humidity is an indication of the maximum moisture that the moist air can hold at a given temperature. The 100% relative humidity line constitutes the extreme left-side boundary of the psychrometric chart.

3.4.   Dew Point

Dew point temperature is the temperature at which condensation occurs when the air is cooled at a constant humidity ratio and at constant pressure. Dew point is read on the dry bulb axis corresponding to the saturation relative humidity curve.

3.5.   Wet Bulb Temperature

Wet bulb temperature is the temperature shown by a thermometer with a liquid container that is wrapped by a dampened cloth and exposed to the moving air. Air must blow at a

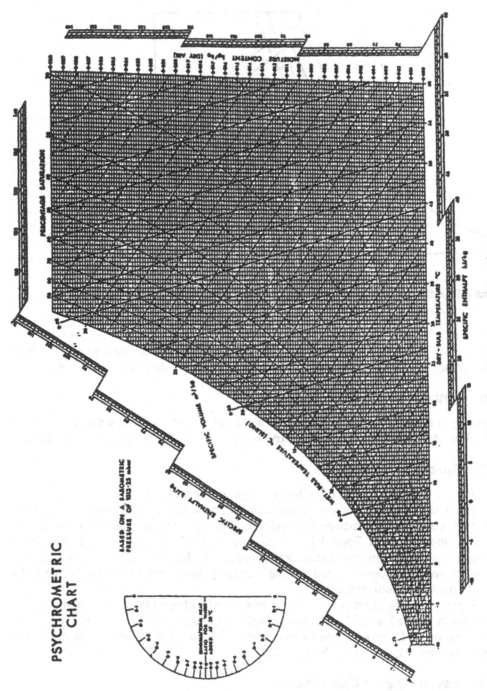

Figure 2   Sample of the psychrometric chart.

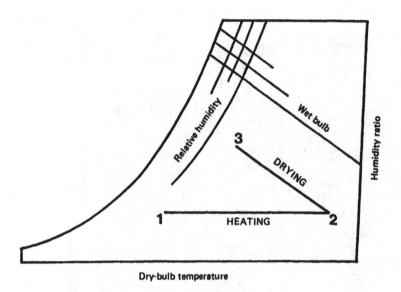

Dry-bulb temperature

Figure 3   Adiabatic drying on a psychrometric chart.

speed of 5 m/s (18 km/h) over the moistened cloth in order to obtain a correct wet bulb temperature. Wet bulb temperatures are the slanted lines on the psychrometric chart.

3.6.   Enthalpy

Enthalpy is the heat content of the moist air. Heat content is based upon a convenient 0°C. Enthalpy lines are almost parallel to the wet bulb lines on the psychrometric chart.

3.7.   Specific Volume

Specific volume is the volume of each unit weight of the moist air.

The utility of a psychrometric chart in an adiabatic dehydration process is shown in Figure 3. Point 1 is the ambient air condition. Point 2 is the air after being heated to a certain temperature T_2. Point 3 is the exit condition of drying air after it has passed through the dryer. Note that in the drying process air has been cooled while taking up moisture from the product. Often the line between points 2 and 3 is a straight line and is parallel to the wet bulb temperature.

A psychrometric chart is quite useful when used within its limitations. For instance, the chart in Figure 2 is valid only for atmospheric pressure. It also does not handle heats of formation and crystallization or temperature of the food. When vapors other than water are involved, a separate chart must be used.

3.8.   Psychrometric Calculations

Equations used to compute properties of moist air are as follows:

Saturation vapor pressure P_s as a function of absolute temperature T

$$\ln(P_s) = 31.9602 - \frac{6270.3605}{T} - 0.46057 \ln(T) \qquad 255.83 \le T \le 273.16$$

$$\ln(P_s/R) = \frac{A + BT + CT^2 + DT^3 + ET^4}{FT - GT^2} \qquad 273.16 \le T \le 533.16$$

$R = 22105649.25, A = -27405.526, B = 97.5413, C = -0.146244,$
$D = 0.12558 \times 10^{-3}, E = -0.48502 \times 10^{-7}, F = 4.34903, G = 0.39381 \times 10^{-2}.$

Humidity ratio, H_i,

$$H = \frac{0.6219 P_v}{P_{atm} - P_v} \qquad \begin{array}{l} 255.38 \le T \le 533.16 \\ P_{atm} = 101330\,\text{Pa} \end{array}$$

Specific volume

$$V_{sa} = \frac{287T}{P_{atm} - P_v} \qquad \begin{array}{l} 255.38 \le T \le 533.16 \\ P_{atm} = 101330\,\text{Pa} \end{array}$$

Relative humidity, *rh*

$$rh = \frac{P_v}{P_s}$$

The units of T are Kelvin and pressures are in Pa.

One usually has the initial air conditions specified in terms of the relative humidity and temperature. Drying calculations are carried out in terms of the humidity ratio. Atypical use of psychrometric equations may follow the following sequences: Use T to calculate P_s, use *rh* and P_s to calculate P_v, and use P_v and P_{atm} to calculate the humidity ratio H. For more equations relating other properties of moist air, see American Society of Agricultural Engineers (ASAE) standards (28).

4. DRYING MODELING AND CALCULATIONS

4.1. General

A mathematical model with which the dryer and process parameters can be studied is of extreme utility to the industry. A reliable model often will prevent or minimize costly mistakes in prototype development. The model also can be utilized for the control of process, specifically in adaptive and feed-forward control strategies. With the proliferation of low cost and powerful computers, the simulation models are useful. In this chapter the emphasis is on the presentation of formulas and equations rather than the conventional charts and graphs in drying calculations.

The following is a list of significant parameters that influence the performance of a dryer (2):

Process air variables
    Airflow rate
    Drying air temperature
    Drying air humidity ratio
Product variables
    Product throughput
    Initial and final moisture contents
    Material sizes and size distribution
Dimensional variables
    Width, height, or diameter of the dryer

Length of the dryer and the number of passes
Dryer configuration

These variables can be controlled or adjusted. Many variables, such as aerodynamic properties of the drying materials, exposed surface area of the product, drying rate characteristics, and the equilibrium moisture content, are specific and cannot be controlled.

A comprehensive drying model must include all the aforementioned variables. It must include the psychrometric and thermal properties of air, vapor, and liquid water.

The development of a drying simulation model is described in the following. Simplifying assumptions and limitations inherent in the model are as follows:

1. A uniform feed rate, with a uniform moisture content
2. A uniform size product
3. No heat loss or heat gain through the dryer walls
4. No external heat and mass fluxes other than those between the air and the material in the dryer

4.2. Heat Balance Equation

Heat balance over a layer of drying particles over the time interval dt, with 0°C as the reference temperature for enthalpy, is written as

Energy lost by air = Energy gained by material

$$AG \, dt \, (i - i') = A \, d \, dz \, [(C_p + MC_w) \, (T_g' - T_g) + dM \, C_w T_g'] \tag{3}$$

Refer to the "Nomenclature" section for definitions.

The enthalpy i can be expressed as

$$i = 1005 T_a + X_a (1820 \, T_a + 2{,}501{,}000) \tag{4}$$

Let

$$F = -\frac{d \, dz}{G \, dt} \tag{5}$$

then,

$$\begin{aligned}
T_a' (1005 + 1820 X_a') = {} & T_a (1005 + 1820 \, X_a) + T_g' [F(C_p + M' C_w)] \\
& + T_g \{ -F[C_p + (M' - dM) C_w] \} \\
& + 2{,}501{,}000 (X_a - X_a')
\end{aligned} \tag{6}$$

4.3. Heat Transfer Equation

The heat transfer equation describes what happens to the heat that is transferred from drying air to material. This heat raises the temperature of solids and evaporates moisture from the product:

$$\begin{aligned}
H \, dt \, \frac{(T_a + T_a') - (T_g + T_g')}{2} \\
= d[(C_p + M' C_w) \, (T_g' - T_g) + (-dM)(i_v' - C_w T_g)]
\end{aligned} \tag{7}$$

where the enthalpy of the water vapor (i_v') at T_a' is given by

$$i_v' = 1820 T_a' + 2{,}501{,}000 \tag{8}$$

4.4. Moisture Balance

Taking a moisture balance over a time interval dt for the layer dz yields

$$A \, dz \, d(M - M') = AG \, dt \, (X'_a - X_a) \tag{9}$$

Equations [6] to [9] must be solved simultaneously for a particular configuration of dryer. The configuration may involve stationary batch, concurrent flow, or counterflow, or it may be a crossflow dryer. In each instance the values of dz and dt are chosen such that the model predicts the physical setup. Equations [6], [7], and [9] describe the air temperature T_a, air humidity X_a, grain moisture content M, and grain temperature T'_g. An additional relationship is needed to describe the drying rate of material. The drying rate will be described in the following section.

4.5. Drying Rates

Moisture is transferred from inside a moist material to the outside surface, where it evaporates. High-moisture foods with a moisture content of more than 50% wet basis may demonstrate two distinct drying rates: (a) a constant drying rate, and (b) a falling drying rate (23). The curve in Figure 4 depicts these two periods. The methods of estimating drying rates for each period are different. The moisture content at which the food material demonstrates a change from constant rate drying to falling rate drying is called the critical moisture content. There may be several critical moisture contents for a particular food item.

4.6. Constant Rate Drying Period

During constant rate drying, moisture is always available at the surface such that resistance to moisture removal is only the rate at which the moisture can evaporate. The constant rate drying period is described by the relationship

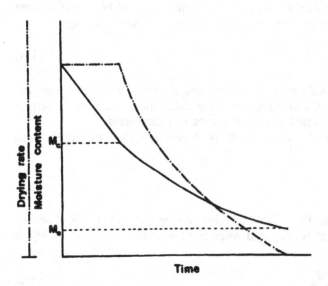

Figure 4   Drying curve for high-moisture foodstuff.

$$\frac{dM}{dt} = \frac{HA(T_a - T_w)}{L} = K_m A (X_w - X_a) \tag{10}$$

H and K_m are coefficients that describe heat transfer and moisture transfer conditions at the surface. Heat transfer and moisture transfer are physical phenomena that are similar in their mathematical representations. Therefore, the relationships similar to those estimating H values can be used to find the K_m values. Coefficients H and K_m are related to each other by Lewis number Le as

$$Le = \frac{H}{K_m C_p} \tag{11}$$

Under conditions in which pressures are in the range of atmospheric (about 100 kPa) and temperatures are under 100°C, Le is nearly equal to 1. Hence, K_m can be found from H values as

$$K_m = \frac{H}{C_p} \tag{12}$$

Several empirical formulas for drying rate during a constant rate period have been developed with experimental data. For example, the following experimental relationships are given in high-temperature alfalfa drying. When $T_a > 200°C$,

$$\frac{dM}{dt} = -K_0 = 0.03066 + 0.0004113 T_a \tag{13}$$

Equation [13] signifies the fact that at temperatures beyond 200°C, the rate of drying is constant as long as the drying temperature remains constant.

4.7. Falling Rate Drying Period

After all the water at the surface of the material has been exhausted, the moisture is diffused from the internal parts of the product to the surface. The amount of water at the surface becomes progressively scarce. As a result the drying rate will be slower as time progresses. The following relationship can be used to describe the rate of drying:

$$\frac{dM}{dt} = -KA \frac{dP}{dx} \tag{14}$$

where K is a moisture transfer (diffusion) coefficient, and dp/dx is the driving force for the moisture movement in terms of water vapor pressure within the product. The moisture can be in the form of vapor or of liquid or a combination of these forms.

It is difficult to solve Eq. [14] for many products. A modified form of this equation is presented as

$$\frac{dM}{dt} = \frac{\pi^2 D}{4L_x^2} (M - M_e) \tag{15}$$

Diffusion coefficient D is constant for the entire falling rate period. L_x is the thickness of the solid. A variation of Eq. [15] is proposed when the ratio $(M - M_e)/(M_i - M_e)$ is greater than 0.6, as

$$\frac{dM}{dt} = \frac{H(T_a - T_w)}{2dL_x L} \frac{M - M_e}{M_i - M_e} \tag{16}$$

The applicability of Eq. [16] is limited to slab or geometrically well-defined products. A much simpler expression for the average moisture content of food material is written as

$$\frac{dM}{dt} = -k(M - M_e)$$ [17]

In writing Eq. [17], two assumptions have been made: (a) the distribution of moisture within the product is uniform, and (b) the drying rate is much dependent on the drying constant k and the equilibrium moisture M_e. The value of k must be found experimentally. Usually a thin layer of product is fully exposed to a highly controlled hot air, and then Eq. [17] is fitted to the experimental drying data. For example, a relationship given for alfalfa is of the form

$$dc = 0.00034e^{0.02028T_a}$$

where $200°C > T_a > 80°C$. Similar relationships for many foods and feeds are available in the literature. See Reference 3 for cereal grains.

When the assumption of "uniformity of moisture in the product" is no longer valid because of the product size, then the following procedure must be used to calculate drying rate and drying times. As it was mentioned earlier, moisture transfer is analogous to heat transfer. Therefore, the well-established methods of heat transfer calculations can be utilized for drying (moisture transfer) calculations. For example, the moisture gradients can be estimated using moisture transfer graphs similar to those of Heisler charts for heat transfer. The charts for finding moisture content at the center of an infinite slab, cylinder, and sphere are given in Figure 5. The use of these charts has been demonstrated by numerical examples in Reference 5.

As an alternative to the charts, empirical equations have been developed. The data for the development of these equations are essentially those of the charts. The equations for use are as follows. For a slab-shaped food product,

$$M_c = R_p \exp(-S_p F_0)$$ [18]

For a cylindrical food product,

$$M_c = R_c \exp(-S_c F_0)$$ [19]

For a spherical food product,

$$M_c = R_s \exp(-S_s F_0)$$ [20]

where M_c is the moisture content dry basis at the center of the slab, cylinder, or sphere. Other constants are defined as

$$R_p = 0.2884A(M) + 0.007927A(M^2) - 0.05287A^2(M^{0.5})$$
$$+ 0.01696A^2(M/3) - 0.09581[A(M^2)]^{0.5} + 1.0018$$ [21]

$$S_p = 0.4608A(M) - 0.08319A(M^2) + 0.03752A(M^3) + 0.8153A(M/3)$$
$$+ 0.2439A^2(M^{0.5}) + 0.001745$$ [22]

$$R_c = 0.7398A(M/3) - 0.01723A(M^3) - 0.2151A^2(M/3) + 1.004$$ [23]

$$S_c = 0.6517A(M) + 2.3186A(M/3) + 0.5124A^2(M^{0.5}) + 0.004211$$ [24]

$$R_s = 1.0847A(M/3) - 0.2204A^2(M^{0.5}) - 0.1222A^2(M/3)$$
$$+ 0.1041[A(M)]^{0.5} + 0.9806$$ [25]

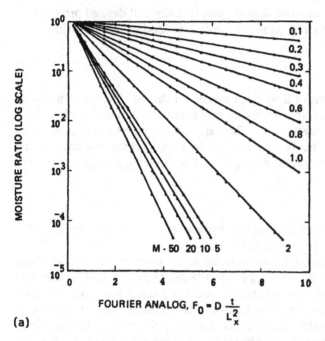

Figure 5 Moisture content distribution during dehydration, at the center of (a) an infinite slab; (b) a cylinder; and (c) a sphere. (From Reference 4.)

$$S_s = 0.1970A(M^2) + 51772A(M/3) + 0.6585A^2(M^{0.5}) \qquad [26]$$
$$+ 0.3960A(M)A(1/M) + 0.003257)$$

where A is arctangent function, F_0 is $Dt/(L_x)^2$, and M is mass transfer resistance ratio defined as $(D\,d/K_m L_x)$ and it is similar to $(K/H\,L_x)$ in heat transfer. Equations [18] through [26] are valid for $0.1 < M_c < 40$. For a full description, see Reference 4.

4.8. Heat Transfer Coefficient

The heat transfer coefficient describes the rate of heat transfer between a food product and the flowing gas. It is mainly dependent on the gas flow and solid configuration. For a laminar flow over a flat food product, Nusselt number Nu can be estimated by

$$Nu = 0.664 \, (Re)^{0.5} \, (Pr)^{0.33} \qquad [27]$$

where H can be found from

$$H = \frac{K_f Nu}{L_x} \qquad [28]$$

For natural or free convection, in which the temperature differential is the main cause of heat transfer from the surface of a solid to air, the following relationships can be used:

$$Nu = k(Gr\ Pr)^a \qquad [29]$$

The constants for vertical flat surfaces and vertical cylinders are as follows: for $10^4 < Gr\ Pr < 10^9$, $k = 0.59$, and $a = 0.25$, and for $10^9 > Gr\ Pr$, $k = 0.021$, and $a = 0.4$; for

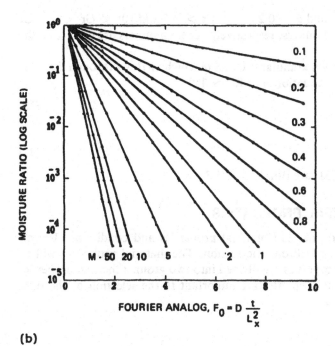

(b)

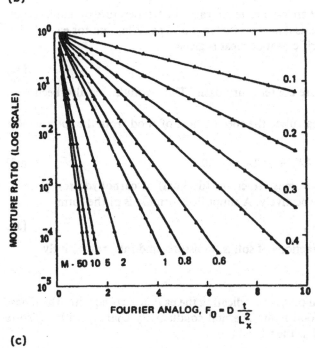

(c)

a horizontal cylinder, $k = 0.525$ and $a = 0.25$ when $Pr > 0.5$ and $10^3 < Gr < 10^9$. Gr and Pr are Grashof and Prandtl numbers, respectively. These numbers are defined in the "Nomenclature" section.

The heat transfer equation for a granular product in which airflows through the mass can be estimated by the following equation. For $Re > 350$,

$$H = 1.06 \, Re^{-0.41} \tag{30}$$

and for $Re \leq 350$,

$$H = 1.95 \, Re^{-0.51} \tag{31}$$

The units of H in Eqs. [30] and [31] are Btu/h $=$ ft^2°F.

5. PHYSICAL AND THERMAL PROPERTIES

Drying calculations are based on air and material properties and conditions. The air properties were discussed in the psychrometric section. The material properties will be reviewed in this section. These properties are divided into two groups: thermal properties and physical properties. Physical properties are important in the selection of the right drying system for the food material.

5.1. Specific Heat

Specific heat is the amount of heat energy needed to raise the temperature of a unit weight of material by one degree; in SI units it is expressed in joules per kilogram for each degree Celsius temperature rise. The specific heat of meat is given by

$$C_p = 0.4 + 0.006M \tag{32}$$

where M is the percentage moisture content, dry basis. This formula has also been used for juices.

A more general equation to calculate the specific heat of food from the specific heat of its constituents is given by

$$C_p = 0.34X_c + 0.37X_p + 0.4X_f + 0.2X_a + 1.0X_m \tag{33}$$

where X_c, X_p, X_f, X_a, and X_m are mass fractions (decimal) of carbohydrates, protein, fat, ash, and moisture content, respectively. A simplified formula is of the form

$$C_p = 0.5X_f + 0.3X_s + 1.0X_m \tag{34}$$

where X_s, X_m, and X_f are mass fractions of solids, moisture, and fats, respectively.

5.2. Thermal Conductivity

Thermal conductivity is a material property indicating the ease and speed with which heat can be transferred through the food item. Thermal conductivity changes with moisture content. The following equation has the widest use:

$$K = K_1 \left[\frac{1 - (1 - aK_s/K_l)^b}{1 + (a - 1)b} \right] \tag{35}$$

where

$$a = \frac{3K_l}{2K_l + K_s} \tag{36}$$

$$b = \frac{X_s}{X_s + X_l} \tag{37}$$

K_l = thermal conductivity of the liquid component of the product
K_s = thermal conductivity of the solids
X_l = mass fraction of the liquid
X_s = mass fraction of the solids

A more specific equation, which has been applied to fruit juices and sugar solutions, is

$$K = (307 + 0.645T - 0.001T^2)[0.46 + 0.054(\%M)]10^{-3} \tag{38}$$

where T is temperature in °F, and K is in Btu/(h·ft·°F).

Thermal conductivity of a bulk of material filled with air can be estimated by the equation

$$K = K_g v_f + K_s(1 - v_f) \tag{39}$$

where
v_f = void fraction
K_g = thermal conductivity of air
K_s = thermal conductivity of solids

Other physical properties required in dryer calculations are dry density dd, wet density wd, and density at any moisture content:

$$d = dd + (wd)M_d \tag{40}$$

where M_d is the dry basis moisture content.

5.3. Particle Size

The size of food particles and its variation play a very important role in the design of a drying system. Particles can be defined as fine or coarse. The size of a particle influences the movement and residence time in the dryer. The size also controls the rate and amount of moisture removed or held by the material. A powder with an average size of 50 microns holds more moisture than a larger granular product with an average size of 5 mm.

Particle size and its variations are measured by passing the material through standard sieves. The mass fraction of material kept on each sieve is plotted against the sieve size. This plot usually takes the form of a normal plot from which a mean and standard deviation for the particles can be calculated (see Standard S319.2 in Ref. 28). Since particle size usually has a wide range, a logarithmic scale is used to express the size of particles.

Another important size factor in drying is the specific surface area that is defined as the ratio of surface area of particles in a unit volume or a unit of mass. One may estimate the surface area of a particle from πD^2 where D is the diameter of sphere of the same volume as that of the particle. The surface area per a unit volume can be estimated from (29)

$$A = \frac{6\lambda w}{\rho_p D_p}$$

where λ is a shape factor that is about 1.75 for most irregular shaped particles, w is the mass of particulate material (kg), ρ_p (kg/m³) is the particle density, and D_p (m) is the particle size determined by sieving technique.

5.4. Density

Density of food material is expressed in two ways, solid density and bulk density. Solid density is the mass of the solid over the volume of the solid excluding the air voids.

$$\rho_s = \frac{m}{V_s}$$

Bulk density is the mass of solids over the bulk volume of the solids including interparticle and intraparticle air voids:

$$\rho_b = \frac{m}{V}$$

Solid density is measured by a method by which air voids can be excluded from measurements. Air comparison pycnometry uses differential air pressure for volume measurements where air pressures up to 1 kPa are used.

Bulk density is measured by filling a container of known volume with particles. The mass of the particles in the container is then measured. Care must be paid to a uniform packing method to fill the container.

The bulk density and solid density values are used in the following equation to estimate the bulk porosity of the granular food materials:

$$\varepsilon = 1 - \frac{\rho_b}{\rho_s}$$

Bulk density of moist food material ranges from 0.3 to 0.5.

6. METHODS OF DRYING

Drying methods have been evolved around every product's specific requirement. The process takes many forms and uses many different kinds of equipment. In general, drying is performed by two basic methods: (a) adiabatic processes and (b) nonadiabatic processes. In adiabatic processes the heat of vaporization is supplied by the sensible heat of air in contact with the material to be dried. In nonadiabatic processes, the heat of evaporation is supplied by radiant heat or by heat transferred through walls in contact with the material to be dried. Dehydration may also be accomplished by mechanical dewatering. However, in this chapter dehydration due only to adiabatic or nonadiabatic as defined above will be described.

In all drying methods the product must be brought in contact with a medium, which is often air, in order to remove the moisture from the product surface and its surroundings.

7. DRYERS

7.1. Introduction

The diversity of food products has introduced many types of dryers to the food industry. Some of the products to be dehydrated and possible dryer types are listed in Table 2.

Methods of supplying heat and transporting the moisture and the drying product are the basic variations among different types of dryers. In general, the types of dryers and their operational design characteristics are reviewed.

Table 2 Food and Feed Products and the Most Suitable Dryer Types

Product	Dryer type
Vegetables, confectionery, fruits	Compartment and tunnel
Grass, grain, vegetables, fruits, nuts, breakfast cereals	Conveyor band
Grass, grain, apple, lactose, poultry manure, peat, starch	Rotary
Coffee, milk, tea, fruit purees	Spray
Milk, starch, predigested infant foods, soups, brewery and distillery by-products	Film drum
Cereal grains	Moving or stationary packed beds
Starch, fruit pulp, distillery waste products, crops	Pneumatic
Coffee, essences, meat extracts, fruits, vegetables	Freeze and vacuum
Vegetables	Fluidized bed
Juices	Foam mat
Apples and some vegetables	Kiln

7.2. Sun Drying

Throughout the world, sun drying is a popular drying method. The sun's radiant heat evaporates moisture. The traditional drying method consists of spreading a thin layer of the product on a smooth pad. The product is stirred and turned over at intervals. Drying proceeds well in warm, dry weather. The temperature of the product during sun drying ranges from 5°C to 15°C above ambient temperature, and drying time may extend up to 3–4 weeks, as for raisins and apricots. Color, shape, and the initial and desired final moisture conditions of the product influence product temperatures and drying time.

Sun drying is widely practiced in grain drying. Figure 6 shows the relative absorptivity of various grains compared with the absorptivity of an ideal black surface, which is 1 on the scale in Figure 6.

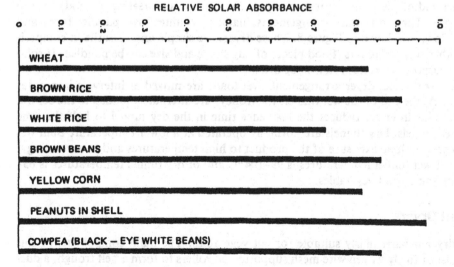

Figure 6 Relative solar absorbance of several grains. (From Reference 6.)

Solar energy is also used to dry fruits and grains indirectly. In this method a solar collector captures solar energy and raises the air temperature. The warmed air is directed to a conventional batch or continuous dryer. In a flat-plate air-type solar collector, the temperature rise is usually about 5°C over a 24-h period for airflows used in the near-ambient temperature drying of granular products. The air may be heated further by conventional fuels if high-temperature drying is desired. Overdrying and susceptibility to spoilage are major problems in solar drying of deep beds of cereal and oil seeds.

7.3. Cabinet Dryers

A cabinet dryer can be a small batch tray dryer. Heat from the drying medium (hot air) to the food product is transferred by convection. The convection current passes over the product, not through the product. It is suitable for dehydration of fruits, vegetables, and meat and its products.

It is relatively easy to set and control the optimum drying conditions in cabinet dryers. For this reason, various heat-sensitive food materials can be dried in small batches. The heat source is usually steam batteries or steam coils. The air from a centrifugal fan is passed through the coils and then baffled across the trays loaded with the product. The trays may either be loaded onto trolleys in stacks of 10–12 or may be stacked individually into the slots of the cabinet. The movement of trays may be manual or mechanically assisted depending on the size and capacity of the dryer. The hot air in almost all cabinet dryers is introduced at the top, and provision is made to recycle the air so that its total drying potential is utilized by the time it discharges to the atmosphere. A schematic illustration of a cabinet dryer and its principle of operation is shown in Figure 7.

7.4. Tunnel Dryers

Tunnel dryers are of many different configurations in general having rectangular drying chambers. The number of tunnels in a dryer is quite variable and can be as high as 100. Truck loads of the wet material are moved at intervals into one end of the tunnel. The whole string of trucks is periodically advanced through the tunnel until these are removed at the other end of the tunnel. Air movement, circulation, and heating methods vary in tunnel dryers. Three different arrangements, namely, counterflow, parallel flow, and combined flow, are shown in Figure 8. These dryers are simple and versatile in comparison with other types of dryers. Food pieces of any shape and size can be handled. If solid trays are incorporated, fluids can also be dried.

In the three-tunnel dryer arrangement, wet loads are moved at intervals into outer wet tunnels. At the end of the wet tunnels the trolleys are alternately loaded to the central dry tunnel. This in effect reduces the residence time in the dry tunnel to half the time spent in wet tunnels. Dry tunnels are normally operated at a lower temperature than wet tunnels to prevent long exposure of the product to high temperatures and to save energy. The range of wet tunnel temperatures is 99–104°C and of dry tunnel temperatures is 65–71°C for drying of root vegetables.

7.5. Belt Dryers

The belt dryer is particularly suitable for cut vegetables. The main components of this dryer consist of finely woven wire mesh supported on rollers to form a belt trough, a duct supplying hot air at the bottom of the trough, and a rotary cleaning brush. The arrange-

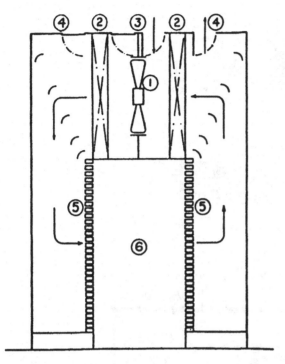

Figure 7 Schematic illustration of a cabinet dryer (1, circulating fan, fully reversible; 2, heater batteries; 3, vented air inlet ports; 4, vented air exhaust ports; 5, adjustable louver walls; 6, truck space). (From Reference 5.)

ment of these components is shown in Figure 9. The belt trough is inclined sideways at an angle of 15–20° to facilitate the gravity unloading of dried product. The airflow rate provides an air cushion but not so high as to fluidize the product. Soft materials can be dried without mashing or rounding their cut edges. Since the product is partly supported by the hot air, it requires relatively uniform-size wet material. If material is not of uniform size, large particles may not be supported well and small particles may be blown out of the trough.

The drying air passes through material at the rate of 0.6–1.4 m/s. From 60% to 90% of air is recirculated in each section. Temperatures are usually limited by the heat sensitivity of the material but seldom exceed 300°C. A higher temperature may damage the lubrication of the conveyor's moving parts. The fan power is about 1 kW/m of the belt width.

The main problem in these dryers is that the sticky materials will lodge and stick to the chains and links of the band. The band must be washed frequently to prevent air blockage.

7.6. Conveyor Band Dryers

A conveyor band dryer consists of single or several perforated wire mesh aprons or conveyor bands as the main component. The wet material is fed evenly at the feed end and is conveyed along the length of the dryer. As in a belt trough dryer, hot air is forced through the bed of moving material. However, unlike the belt trough dryer air is not

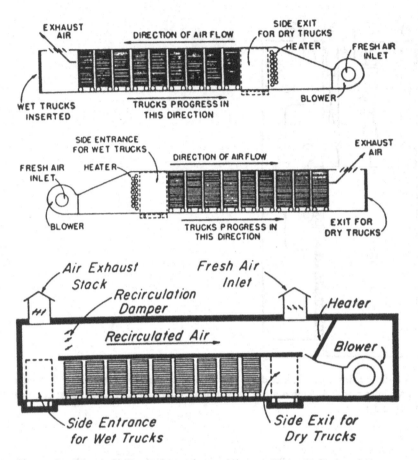

Figure 8   Schematic illustrations of tunnel dryers. (From Reference 5.)

forced at high rates to support wet material. The dried product is continuously discharged at the end of the dryer. The air characteristics, such as temperature and relative humidity, may be adjusted throughout the passage to satisfy the drying characteristics of the product. The direction of movement of air through the permeable bed of the product may be either upward or downward. The thickness of the bed of the product is kept at 25–250 mm. This thin product layer and higher airflow rates result in the uniform drying of the product. Two configurations of band dryer (two stage and multideck) are shown in Figure 10.

7.7. Spray Dryers

Spray dryers are used for dehydrating fluids. The fluid is introduced in the heated airstream in spray form. Dried product is separated from the airstream and is collected for further processing. The design of spray dryers ranges from very simple to very complex, depending upon the fluid. The main differences in the designs are the variations in atomizing devices, in airflow patterns, in air heating systems, and in separating and collecting systems.

The main components of a spray dryer are a drying chamber, an atomization and

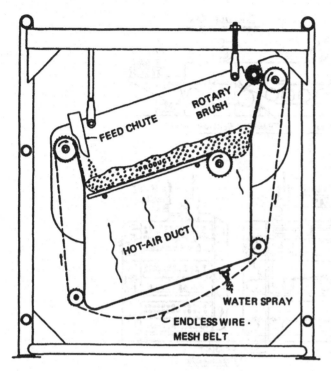

Figure 9 Schematic cross-section of belt trough dryer. (From Reference 1.)

dispersion device to introduce fluid in small droplets into the drying chamber, an air heating and blowing system, and a device to separate and collect the dried product from the air. Air heating may be done by direct firing of gas or liquid fuels in the airstream or by indirect heating through a heat exchanger. Direct firing may contaminate the product; the indirect system results in lower efficiency. Two configurations of spray dryers are shown in Figure 11.

Pressure spray heads, two-fluid nozzles, and centrifugal atomizers are used for dispersion of the fluid particles. Dried product is separated by different methods depending on the make of the dryer. Some dryer designs use cyclone separators; others employ settling chambers. In a cyclone separator, air is introduced tangentially at the top, swirls downward, and reverses the direction of flow to exhaust from the top. In the settling chamber, direction of airflow is changed to facilitate the separation of dried product under force of gravity.

Design Equations for Spray Dryers

Energy Requirements. Energy requirement in horsepower (hp) for three different types of the atomizers is given as follows.

1. Pressure nozzles:

$$E = 7QP \times 10^{-4} \tag{41}$$

where

Q = liquid flow rate, gal/min (U.S. gallon)
P = pressure drop over the nozzle, psi

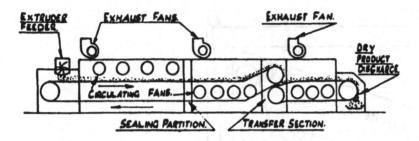

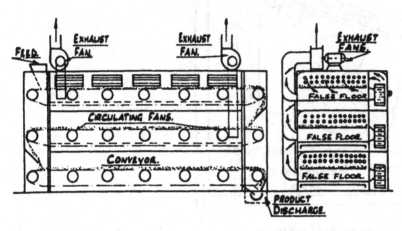

Figure 10 Schematic views of a two-stage (top) and multideck (bottom) conveyor band dryer. (From Reference 7.)

2. Air blast atomizer:

$$E = 0.136GT_a[0.5M_c^2 + 2.5(1 - P^{0.286})] \qquad [42]$$

where

G = air mass flow rate, lb m/s
T_a = temperature, R
M_c = Mach number, as defined in the "Nomenclature" section

3. Rotary atomizers:

$$E = 42.5Q\,(DN')^2 \times 10^{-12} \qquad [43]$$

where

Q = liquid flow rate, lb m/h
N' = rotation speed of the atomizer, rpm
D = diameter of the atomizer, ft

Droplet Size. The droplet size obtained after the atomization process is given by

$$MD = \frac{B}{C}\,\frac{FNyd}{\sin(T)P}\,(d_a)^{-1/6} \qquad [44]$$

where

MD = mean droplet diameter, μm
B = constant, varies from 26.3 to 43 and is found experimentally

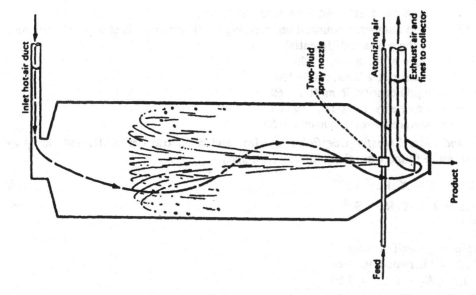

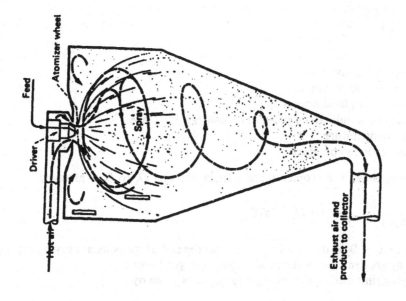

Figure 11 Two configurations of spray dryers.

C = discharge coefficient for a nozzle size, ~0.4

FN = fluid number, a constant representing the flow rate divided by the square root of the pressure differential, ~0.8

T = angle of the spray, ~120

P = pressure gradient, psi, ~100

d = liquid density, lb m/ft^3, ~60

d_a = air density, lb m/ft^3, ~0.065

y = surface tension, dyn/cm, ~50

Heat and Mass Transfer Coefficients. Heat and mass transfer coefficients in spray drying can be found from the following formulas:

$$Nu = 2 + K_1 \, Re^{1/2} \, Pr^{1/3} \tag{45}$$

$$Sh = 2 + K_2 \, Re^{1/2} \, Sc^{1/3} \tag{46}$$

where

Nu = Nusselt number

Sh = Sherwood number

$K_1 = K_2 \sim 0.52 = 0.60$

Drying Time. The drying time required for a droplet to vanish completely during drying is given by

$$t_c = \frac{d_l L D_0^2}{8 K_f (T_a - T_w)} \tag{47}$$

where

d_l = liquid density

L = heat of vaporization

D_0 = initial droplet diameter

K_f = thermal conductivity of air at droplet/air interface

T_a = air temperature

T_w = wet bulb temperature

The drying time for a droplet to reach from diameter D_1 to D_2 is given by

$$t_c = \frac{L}{8 K_f (T_a - T_w)} \, (d_1 D_1^2 - d_2 D_2^2) \tag{48}$$

where the diameter D_2 is the droplet diameter at the end of the constant drying rate period and d_1 and d_2 are densities of the product at the stages 1 and 2.

The drying time during the falling rate period is given by

$$t_f = \frac{d_p D_c L (M_c - M_e)}{6H \, dT} \tag{49}$$

where

d_p = particle density

D_c = diameter at the critical moisture content M_c

M_c = critical moisture content (db)

M_e = equilibrium moisture content (db)

dT = time-averaged temperature difference between particle and air during the falling rate period

7.8. Freeze Dryers

Upon heating of substances in the frozen state liquefication generally precedes vaporization. However, in a process called *sublimation*, the liquid state is bypassed and frozen particles are changed directly to the vapor state. Freeze dryers work on the principle of direct evaporation of the ice in the product to be dried. For direct sublimation, the temperature and the vapor partial pressure is the ice point below the triple point of water. The triple point of water is at approximately 0°C and 4.59-mm Hg absolute pressure.

The process of freeze drying is completed in two stages: (a) freezing and (b) high-vacuum drying. Both these stages are highly expensive; therefore, freeze drying is feasible only for highly valued foods such as fish, meat, chicken, and coffee. This method may also be a necessity for highly heat-sensitive food products. The drying is done at low temperatures; therefore, the heat degradation of the nutrient is minimal and the product obtained is of high quality. The dehydration of vegetables is possible with this method.

Design Equations in Freeze Drying

The drying time required to freeze dry the product can roughly be estimated by either of the following equations.

1. Formula based on heat transfer:

$$t = \frac{LL_x^2}{4K(T_a - T_I)} \tag{50}$$

where

L_x = thickness of the product, ft
L = latent heat of sublimation, Btu/lbm
K = thermal conductivity of product, Btu/°F·ft·h
T_a = air temperature, °F
T_I = ice temperature, °F

2. Formula based on mass transfer:

$$t = \frac{BL_x^2}{16K} \tag{51}$$

where B is the mass of water per unit volume of the product (lb/ft^3) and K is the thermal conductivity of the material.

7.9. Drum Dryers

In drum dryers the surface of a pair of rotating hot drums is coated with the liquid or semiliquid form of the product. Drums are heated by steam or by direct firing inside the drum. The drums rotate slowly and over the course of about 300° of rotations the product is dried. The dried product is scraped off with doctor blades in flakes or sheets.

Drum dryers are classified on the basis of the number of drums as single-drum dryers and double-drum dryers. A typical feed arrangement for a drum dryer is given in Figure 12. The product is collected in the trough and is conveyed toward one end of the drum. By completely enclosing the system, these dryers may be operated under vacuum for heat-sensitive materials that may otherwise spoil at atmospheric pressure.

Foods containing high amounts of sugars may still be in the molten stage when they approach the doctor blades. High-velocity air flowing countercurrent to the drum rotation, a blast of chilled air directed on the product just preceding the doctor blades, and a

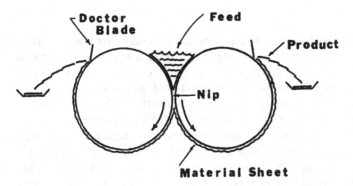

Figure 12 A typical feed arrangement of a drum dryer.

supply of low-humidity air near the scraping and collecting area are special features to deal with viscous materials.

7.10. Foam Mat Dryers

Foam mat dryers can only be used to dehydrate liquid foods that are capable of forming stabilized foams. This requirement imposes a restriction on the wide applicability of the method in the food industry because only a few foods, such as whole milk, have a foam-stabilizing capability. Fruit juices have been dried successfully by adding foam stabilizers. Vegetable gum and soluble protein have successfully been used in the food industry as foam-stabilizing agents.

The liquid food is first foamed and then spread in mats on a perforated or solid support to be dried in hot airstreams. Foaming is accomplished by agitating the fluid product with the airstream. The foam structure is required to last during drying so that the dried product is easily ground to powder. The thickness of the foam layer is maintained at about 0.1–0.5 mm. The layer is dried in a few minutes at temperatures of about 65–70°C. The foamed product is spread on perforated floor craters as the airstream is forced through the bed. The cratering of foam exposes more surface to the airstream when air passes through the holes of the screen and through cratered product around the screen hole.

7.11. Vacuum Dryers

In vacuum drying the boiling point of water is lowered below 100°C by reducing the pressure. If the atmospheric pressure is reduced 100 times, then the boiling point will be around 0°C. The degree of vacuum and the temperature for drying depend on the sensitivity of the material to drying rate and temperatures.

Vacuum drying is one of the most expensive methods of drying. Its costs are comparable to freeze drying but are higher than other methods. Because they are expensive, vacuum dryers often serve as a secondary dryer. The moisture content of high moisture food is reduced to 20–25% by a conventional method, such as hot air drying, and then vacuum is applied to bring the moisture down to 1–3%.

Because of the reduced pressures, transfer of heat depends on methods other than convection. Radiation and/or conduction are other modes; however, conduction may not be efficient because the drying materials shrink, thus reducing the contact area. This method of drying is not very common in the food industry because of high costs. The method has been applied for dehydration of citrus juices, apple flakes, and various heat-sensitive products in which the ascorbic acid retention is important.

The product is dried under vacuum at low temperatures. Temperatures range between 35°C and 60°C. Vacuum-dried products are quite hygroscopic. Special care is needed during packing of the materials to protect against absorption of moisture. This method could be applied to batch as well as to continuous systems. In batch systems, product in trays is dried in a compartment maintained at pressures reduced below atmospheric.

In continuous systems, the product is spread over a stainless steel belt and passed through a vacuum vessel with the help of rollers. The dried product is collected at the other end. This method is also used in conjunction with foam mat drying, described previously.

The vacuum drying technique is also used for the extraction and concentration of essences and flavors.

7.12. Fluidized Bed Dryers

The drying air is introduced at high velocities through the particulate material. The velocity is raised to the point that the particulate material is fully suspended in the hot airstream. All particles are completely exposed to drying air, resulting in high rates of heat transfer. The high rate of heat transfer results in instantaneous evaporation of moisture at the entry point. The dryer mainly consists of a chamber with a perforated bottom through which the air is forced at high velocities. If a glass window is made in the chamber, the start of fluidization can be observed visually. If this is not possible, then the flattening of the pressure drop curve with increasing air velocity indicates the onset of fluidization (Fig. 13). This system is more suitable for batch drying but could also be applied to continuous drying of particulate materials.

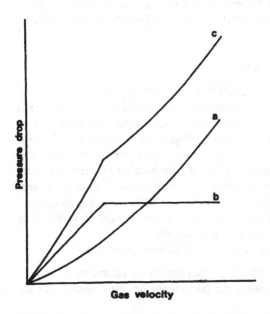

Figure 13 Pressure drop through fluidized beds (a, pressure drop through perforated base plate; b, pressure drop through granular bed; c, combined pressure drop [a and b]). (From Reference 7.)

Food materials that are suitable for fluidized bed drying have the following characteristics:

1. The average particle size should be between 10 μm and 20 mm. Very fine particles
   tend to lump together.
2. The particle size distribution should be reasonably narrow. A wide range of particle
   sizes makes the selection of a gas velocity nearly impossible. A vibrated fluid bed
   sometimes overcomes this problem when it is impossible to have a uniform-size
   material.
3. The particle shape must be regular.
4. A lump of particles must disintegrate easily upon fluidization.

7.13. Microwave Dryers

In microwave drying the product is exposed to very high frequency electromagnetic
waves. The transfer of these waves to the product is similar to the transfer of radiant
heat. As a result of high-frequency waves, water molecules are polarized and tend to
change orientation. In the process of orientation, sufficient heat to expel moisture from
the product is generated.

Microwave generators designed for drying operate under two frequencies, 100 and
900–2500 MHz. As the frequency is increased, the heat generated by the system is decreased. Microwave applications have tended to be in tempering, blanching, and cooking
of food. However, there are some new installations for drying purposes. The drying of
pasta is perhaps the best known of these, in which microwave is used because normal
drying must be very slow to avoid surface cracking of this diffusion-limited material.
Since high-frequency heating evaporates water inside the material and causes convecting
cooling on the surface, the outside surface remains wet until the end of the drying cycle;
hence, cracking does not occur.

Another advantage of microwave is an increase in the rate of kill of bacteria due to
the speedy temperature rise. Microwave drying under vacuum is becoming an established
method of drying citrus fruit concentrates.

7.14. Superheated Steam Drying of Foods

Superheated steam drying has long been recognized as a drying method that leads to
nonpolluting and safe drying at low energy consumption. The principle of the process is
based on using superheated steam for drying coupled with a vapor recompression cycle to
recover heat. The schematic of the system is shown in Figure 14. It consists of a heat
treatment chamber, a compressor, a heat exchanger, and a number of blowers. The
drying medium is superheated steam that runs in a closed loop picking up moisture from
product in the heat treatment chamber and condensing the evaporated water in a heat
exchanger. Part of the saturated steam is bled off from the main recycling line, compressed, and its heat is transferred to the recycled steam to the treatment chamber. The
warm condensate can be used for heating purposes or discarded.

Superheated steam drying has been credited with the following advantages:

1. Improved drying efficiency, as much as 50% less than conventional systems.
2. Pollution free, because it is a closed system no particulate or obnoxious gases are
   emitted to the environment.
3. No product oxidation, because there is no direct contact of hot oxygen with the
   product.

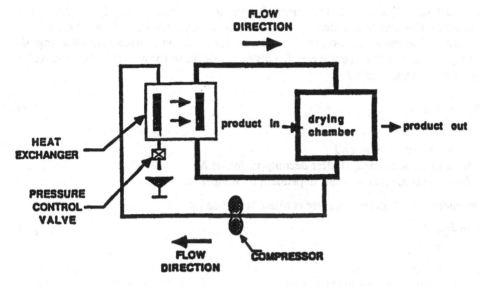

Figure 14   Schematic of superheated steam dryer.

4. Phytosanitary; steam heat is superior to dry air in destroying all stages of insects, molds, and other microorganisms
5. Better degree of control of the dryer operation; by adjusting the quantity of steam bled into the compressor, the degree of dryness of the final product can be adjusted.
6. Superior nutritional quality of feed; steam treatment and hydrolysis of product improve digestibility of feeds by animals.

Several plants based on the vapor recompression cycle have been constructed and tested in Europe. One system uses a horizontal conveyor band in a tunnel drying setup (Bertin, France). Another manufacturer has developed a fluidized bed dryer using high pressure steam (NIRO, Denmark). The results of both systems have been encouraging.

8.   NUTRIENT LOSSES DURING DEHYDRATION OF FOOD PRODUCTS

Many food products of nutritional importance, such as fruit juices, milk, eggs, and vegetables, are dehydrated in the food industry. During dehydration the product is generally above room temperature but well below sterilization temperature. The added heat and exposure time of the product at elevated temperatures affect the nutrient quality of

Table 3   Factors that Influence Food Quality during Drying

Chemical	Physical	Nutritional
Browning reaction	Rehydration	Vitamin loss
Lipid oxidation	Solubility	Protein loss
Color loss	Texture	Microbial survival
	Aroma loss	

the food products. The types of food degradation during drying are listed in Table 3 (30). The influence of drying parameters on the food degradation will be discussed below.

In terms of chemical kinetics, the loss of nutrient can be visualized as the decomposition of a particular chemical compound. This decomposition for a single monomolecular reaction may be expressed by

$$N \xrightarrow{\ k\ } DN \tag{52}$$

where

N = nutritive compound
DN = compound formed after decomposition of N
k = reaction rate constant, dependent on temperature

The dependence of k on temperature is often described by

$$k = k_0 e^{E/RT} \tag{53}$$

where

k_0 = constant
E = activation energy of the reaction
R = gas constant
T = absolute temperature

Since Eq. [52] represents the loss of nutrient N, the rate of loss of the nutrient can be described as

$$\frac{d[N]}{dt} = -k[N] \tag{54}$$

where

$[N]$ = concentration of compound N
$\dfrac{d[N]}{dt}$ = rate of loss of nutrient N

Now if it is assumed that the constant k is a true constant and the concentration of N changes only because of the reaction, then Eq. [54] may be integrated with appropriate boundary conditions to give Eq. [55]:

$$-\ln \frac{[N]}{[N]_0} = kt \tag{55}$$

where $[N]_0$ is the initial concentration of nutrient N.

As the temperature of the product increases, the reaction rate constant is increased (Eq. [53]). The dependence of reaction constant on temperature implies that low-temperature dehydration would produce products with less nutrient disintegration. The vacuum and freeze dehydration processes thus would result in less nutrient degradation compared with other drying processes. A longer constant rate drying period also increases nutrient retention because, owing to evaporative cooling, product is at rather lower temperatures.

The moisture in the product can be held as bound or free water. Water acts as solvent for the chemicals of nutritional importance present in the product. As water is removed, the concentration of the chemicals increases. The loss of nutrient described in Eqs. [52] through [55] is concentration dependent and thus would increase as dehydration progresses. On the other hand, some of the water-soluble compounds may act as catalysts to the decomposition process. These catalytic effects are greatly reduced as the moisture is removed. The increased viscosity of the solution hinders the mobility of the catalyst.

Some nutrients, such as riboflavin, are not very soluble in water. During the dehydration process these compounds form saturated solutions and may be precipitated. Precipitation minimizes the loss of nutrients because of the reduced concentration.

The scarcity of the kinetic data precludes predicting exactly the interaction of water and chemical degradation of the nutrients. A brief review of the qualitative aspects and nutrient retention during dehydration is presented next.

8.1. Proteins

Not much information is available to quantitatively describe the losses of proteins during the dehydration process. The effects of heat treatment on the composition and nutritive value of herring meal are presented in Table 4. Because of scattered conditions used in these experiments no set pattern could be established, but it could be seen that the available lysine and pepsin digestibility is substantially decreased as the moisture content of the meal during heating was increased or the heating time was prolonged. The data also indicate that the high-temperature–high-moisture contents during drying do not affect the nutritional value of the proteins significantly. Studies based on rat NPU (net protein utilization) have shown no significant difference between freeze drying and rapid drying cod meal at 110–115°C.

8.2. Milk Lysine Loss

The loss of milk lysine in spray drying ranges from 3% to 10%, whereas in drum drying the loss ranges from 5% to 40%. During spray drying product particles are small and are dried rapidly. Even though the air temperatures may be high, the milk particles are not heated to high temperatures. In drum drying the milk particles at medium to low moistures are in contact with the hot drum, and therefore the loss of lysine is greater.

Table 4 Effect of Heat and Moisture on Fish Meal

			\multicolumn{4}{c}{Percentage of freeze-dried control}
\multicolumn{3}{c}{Conditions of treatment}			\multicolumn{2}{c}{NPU}
(°C)	Moisture (%)	Time (min)	Available lysine (%)	Pepsin digestibility (%)	Rats (%)	Chicks (%)
96	7.7	30	94	88	—	98.6
	8.8	60	96	84	—	102.0
	10.8	120	87	76	—	98.1
	36.0	60	87	71	97.7	98.6
116	6.4	120	94	78.1	95.3	96.6
	7.5	60	100	78.2	97.0	98.8
	8.4	30	96.0	80.0	97.4	99.7
132	2.5	120	97	58.4	91.8	97.1

Source: From Reference 18.

8.3. Water-Soluble Vitamins

Unstable water-soluble vitamins, such as ascorbic acid, are sensitive to drying. The loss of ascorbic acid is dependent on the presence and type of heavy metals, such as copper and iron, light, water activity level in the product, dissolved oxygen, and the temperature of drying. The losses of ascorbic acid during drying have been between 10% and 50%.

A significant difference has been found in the loss of ascorbic acid when food product is dried on metal trays versus wood trays. The difference was attributed mainly to the faster heat transfer rate and thus faster drying through the metal tray. However, the losses were less with the metal grid compared with the wood grid under similar drying conditions.

Many studies of the losses of thiamine have been reported. Depending on the moisture content of the product and the exposure temperatures and time, losses up to 89% have been reported. At 63°C, 20 h are required for a 50% loss of thiamine in dried pork. It has also been shown that for samples held at 49°C at moisture contents of 0%, 2%, 4%, 6%, and 9%, the losses of thiamine were 9, 40, 80, 90, and 89%, respectively. Thiamine losses during drying from different sources have been summarized in Table 5.

The loss of other water-soluble vitamins are also reported to vary widely from one study to another. Results from some of these studies are summarized in Table 6.

In general, it can be said that losses of water-soluble vitamins during conventional drying are less than 20%. For vegetable dehydration, the losses are less than 5% for all vitamins except ascorbic acid.

Table 5 Thiamine Losses in Drying

Product	Conditions	Loss (%)	Reference nos.
Freeze-dried pork	?	30	9
Freeze-dried chicken	?	5–6	10
Freeze-dried pork	−40°C	5	11
Freeze-dried chicken	1000 μm Hg	5	11
Freeze-dried beef	?	5	11
Air-dried pork	?	50–70	12
Beans[a]	Air dried	5	13
Cabbage	Air dried	9	13
Corn	Air dried	5	13
Peas	Air dried	3	13
Snap beans	Air dried	5	14
Beets	Air dried	5	14
Corn	Air dried	5	14
Peas	Air dried	5	14
Rutabagas	Air dried	5	14
Carrots	Air dried	29	14
Potato	Air dried	25	14
Bean powders	Double-drum dried at 93°C for 30 s	20	15

[a]The loss reported for vegetables does not include blanching loss.

Table 6 Losses of Vitamins during Drying

Vitamin	Product	Loss (%)	Reference nos.
Vitamin B$_6$	Freeze-dried fish	0–30	16
Panthothenic acid	Freeze-dried fish	20–30	16
Riboflavin, niacin, and panthothenic acid	Vegetables	<10	14
Riboflavin	Freeze-dried chicken	4–8	10
Pyridoxine, niacin, and folic acid	Double-drum dried at 93°C for 30 s; bean powder	20	15

8.4. Fat-Soluble Vitamins

Fat-soluble vitamins are expected to degrade by the oxidation mechanism. The degradation is superimposed by direct thermal reactions during drying of the food products. This is evidenced by comparing air drying and freeze drying of carrots (Table 7).

For freeze-dried orange juice, losses of β-carotene of the order of 4% have been reported. Losses of other fat-soluble vitamins, such as A, D, and E, during drying have been negligible. Losses of vitamin up to 5% have been reported in oil seed drying. It could be concluded that the vitamin losses during drying are mainly related to water-soluble vitamins.

In cereal drying, not much work has been done on the quality aspects of drying. Most of the literature is based on germination loss due to heat treatment. The recommendations for safe drying temperatures are varied over a wide range among different countries. In England, for seed or malting grains, it is recommended that the inlet hot air temperature should not exceed 40°C and 43°C for moisture contents of less than and greater than 24% wet basis, respectively.

8.5. Nonenzymatic Browning

A model describing the rate of nonenzymatic browning in skim milk for the range of temperature from 35 to 130°C and 3 to 5% d.b. moisture content. The equation takes the following form:

$$\frac{dB}{dt} = k_0 \exp\left(-\frac{E}{RT}\right)$$

with $k_0 = \exp\left(38.53 + \frac{15.83}{m}\right)$, $\quad \frac{E}{R} = 13157.19 + \frac{90816.51}{m^3}$

Table 7 Losses of Carotene in Drying Carrots

Drying process	Average loss (%)		Reference nos.
	Total β-carotene	trans-β-carotene	
Tray air drying	26	40	17
Freeze drying	10	20	18

where

$\dfrac{dB}{dt}$ = the rate of nonenzymatic browning

k_0 = the arhenuis constant

E = activation energy (kcal/mol)

R = the gas constant

T = absolute temperature (K)

m = moisture content (% d.b.)

NOMENCLATURE

a	constant
a	local velocity of sound, m/s
A	cross-sectional area of the dryer, m^2
A	arctangent
B	mass of water per unit volume of product, kg/m^3
C	energy constant
C	constant
C	discharge coefficient of nozzle
C_w	specific heat of water, J/kg°C
C_p	specific heat of dry product, J/kg°C
d	density of dry product, kg/m^3
d_a	density of air, kg/m^3
d_l	density of liquid, kg/m^3
dd	dry density, kg/m^3
dM	moisture removed, kg/kg
dP/dx	vapor pressure gradient, Pa/m
dt	time step, s
dT	temperature increment, °C
dz	small segment along dryer, m
D	diffusion coefficient, m^2/s
D	diameter of atomizer orifice, m
D_0	initial droplet diameter, m
DN	compound formed after decomposition of nutritive compound N
d_1	density of product at stage 1, kg/m^3
d_2	density of product at stage 2, kg/m^3
D_1	diameter of droplet at stage 1, m
D_2	diameter of droplet at stage 2, m
E	constant
E	activation energy of reaction, J
E	energy requirement, hp
F	constant
FN	fluid number
g	acceleration due to gravity, m/s^2
G	air mass flow rate, $kg/m^2 \cdot s$
H	volumetric heat transfer coefficient, W/m^3°C
H_w	absolute humidity at wet bulb temperature (saturation temperature), kg/kg
i	enthalpy, J/kg
i_v	enthalpy of water vapor, J/kg

k	drying constant, l/s
k	reaction rate constant
k_0	drying constant, l/s
K_m	mass transfer coefficient
K	moisture transfer (diffusion) coefficient
K	thermal conductivity of food material, $J/m \cdot h \cdot K$
K_f	thermal conductivity of air at droplet/air interface, $J/m \cdot h \cdot K$
K_g	thermal conductivity of air
K_l	thermal conductivity of liquid portion, $J/m \cdot h \cdot K$
K_s	thermal conductivity of solid portion, $J/m \cdot h \cdot K$
K_1	constant
K_2	constant
L	latent heat of vaporization, J/kg
L_x	thickness of solid, m
M	moisture content, dry basis, kg/kg
M	mass resistance ratio
M_c	moisture content at the center of slab-shaped, cylindrical, or spherical food particles, kg/kg
M_d	dry basis moisture content, kg/kg
MD	mean droplet diameter, μm
M_e	equilibrium moisture content
N	nutritive compound before decomposition
$[N]$	concentration of nutrient N, kg/m^3
N'	rotation speed of atomizer, rpm
(π)	constant, 3.14159
P	pressure drop over nozzle, Pa
P_a	atmospheric pressure, Pa
P_v	water vapor pressure, Pa
P_s	saturation vapor pressure, Pa
Q	liquid flow rate, gal/min
R	universal gas constant, 1.987
RH	relative humidity, decimal
R_c	constant for cylinder
R_p	constant for infinite plate
R_s	constant for sphere
S_c	constant for cylinder
S_p	constant for infinite plate
S_s	constant for sphere
t	time required to freeze dry the product, min
t_c	drying time during constant rate period, min
t_f	drying time during falling rate period, min
T	temperature, K
T	angle of spray, rad
T_a	air temperature, °C
T_g	product temperature, °C
T_I	ice temperature, °C
T_s	surface temperature, °C
T_w	wet-bulb temperature, °C
T_{inf}	temperature at free stream conditions, °C

u	velocity, m/s
v_f	void fraction
w_d	density of moist product, kg/m^3
X_a	humidity ratio in the air, kg/kg
X_a	mass fraction of ash, decimal
X_c	mass fraction of carbohydrates, decimal
X_f	mass fraction of fat, decimal
X_l	mass fraction of liquids, decimal
X_m	mass fraction of moisture, decimal
X_p	mass fraction of protein, decimal
X_s	mass fraction of solids, decimal
X_w	mass fraction of water, decimal
y	surface tension, dyn/cm
Gr	Grashof number, $g(T_s - T_{inf})L_x^3/2$
Le	Lewis number, $H/K_m C_p$
M_c	Mach number, u/a
Nu	Nusselt number, HL_x/K
Pr	Prandtl number, $C_p\mu/K$
Re	Reynolds number, duL_x/μ
Sh	Sherwood number, $K_m L_x/D$
α	Volume coefficient of expansion, 1°C
μ	Dynamic viscosity, kg/m·s
ν	Kinematic viscosity, m^2/s

Prime (′) means one step further into the dryer

REFERENCES

1. Van Arsdel, W. B., and M. J. Copley. *Food Dehydration*, Vol. 1. AVI Publ. Co., Inc., Westport, Connecticut, 1973.
2. Bakker-Arkema, F. W., R. C. Brook, and L. E. Lerew. Cereal grain drying. In *Advances in Cereal Science and Technology*, Vol. III. American Association of Cereal Chemists, Minneapolis, 1977, pp. 1–90.
3. Brooker, D. B., F. W. Bakker-Arkema, and C. W. Hall. *Drying Cereal Grains*. AVI Publ. Co., Inc., Westport, Connecticut, 1974.
4. Ramaswamy, H. S., and K. V. Lo. *Simplified Relationships for Moisture Distribution during Drying of Regular Solids. ASAE Paper 81-103*. Am. Soc. Agric. Engineers, St. Joseph, Michigan, 1981.
5. Heldman, D. R. Food Process Engineering. AVI Publ. Co., Inc., Westport, Connecticut, 1975.
6. Arinze, E. A., G. Schoenau, and F. W. Bigsby. *Solar Energy Absorption Properties of Some Agricultural Products. ASAE Paper No. 79-3071*. Am. Soc. Agric. Engineers, St. Joseph, Michigan, 1979.
7. Gardner, A. W. *Industrial Drying*. Leonard Hill, London, 1971.
8. Myklestad, O., J. Bjornstad, and L. Njaa. Effects of heat treatment on composition and nutritive value of herring meal. *Fiskerdinetoratets Skrifter Ser. Technol. Undersok* 5(10):1–15, 1972.
9. Karmas, E., J. E. Thompson, and D. B. Peryam. Thiamin retention in freeze-dehydrated irradiated pork. *Food Tech.* 16:107–108, 1962.
10. Rowe, D. M., G. J. Mountrey, and I. Prudent. Effect of freeze drying on the thiamin, riboflavin and niacin content of chicken muscle. *Food Tech.* 17:1449–1450, 1963.

11. Thomas, M., and D. H. Calloway. Nutritional value of dehydrated food. *J. Am. Diet Assoc.* 39:105–116, 1961.

12. Calloway, D. H. Dehydrated foods. *Nutr. Rev.* 20:257–260, 1962.

13. Harris, R. S., and H. von Loesecke. *Nutritional Evaluation of Food Processing.* John Wiley and Sons, New York, 1960.

14. Hein, R. E., and I. J. Hutchings. Influence of processing on vitamin, mineral content and biological availability in processed foods. Council Foods Nutrition, AMA and AMA-Food Liaison Comm, New Orleans, 1971.

15. Miller, C. F., D. Quadagni, and S. Kow, Vitamin retention in beam powders: Cooked, canned and instant. *J. Food Sci.* 38:493–495, 1973.

16. Schroeder, H. Losses of vitamins and trace minerals resulting from processing and preservation of foods. *Am. J. Clin. Nutr.* 24:562–572, 1971.

17. Della Monica, E. S., and P. E. McDowell. Comparison of beta-carotene content of dried carrots prepared by three dehydration processes. *Food Tech.* 19:141–143, 1965.

18. Sweeney, J. P., and A. C. Marsh. Effect of processing on provitamin A in vegetables. *J. Am. Diet. Assoc.* 59:238–243, 1971.

19. Van Arsdel, W. B., and M. J. Copley. *Food Dehydration,* Vol. 2. AVI Publ. Co., Inc., Westport, Connecticut, 1973.

20. Charm, S. E. *The Fundamentals of Food Engineering.* AVI Publ. Co., Inc., Westport, Connecticut, 1963.

21. ASAE. *Standards 1984.* Am. Soc. Agric. Eng., St. Joseph, Michigan, 1984.

22. O'Callaghan, J. R., D. J. Menzies, and P. H. Bailey, Digital Simulation of agricultural drier performance. *J. Agric. Eng. Res.* 16(3):223–244, 1971.

23. Hall, C. W. *Drying and Storage of Agricultural Crops.* AVI Publ. Co., Inc., Westport, Connecticut, 1980.

24. Nellist, M. E. *Safe Temperatures for Drying Grain. Report No. 29.* National Institute of Agricultural Engineering, Silsoe, England, 1978.

25. Sokhansanj, S., V. S. Venkatesan, H. C. Wood, J. F. Doane, and D. T. Spurr. Thermal kill of wheat midge and Hessian fly (*Diptera: Cecidomyiidae*). *Postharvest Biology and Technology* 2:65–71 (1992).

26. Sokhansanj, S., H. C. Wood, and V. S. Venkatesan. Simulation of thermal disinfestation of hay in rotary drum dryers. *Transactions of the ASAE* 33(5):1647–1651 (1990).

27. Sokhansanj, S., and H. C. Wood. Simulation of thermal and disinfestation characteristics of a bale dryer. *Drying Technology—An International Journal* 9(3):643–656 (1991).

28. ASAE. *ASAE Standards.* 40th Edition. American Society of Agricultural Engineers, St. Joseph, Michigan, 1993.

29. Earle, R. L. *Unit Operations in Food Processing.* Pergamon Press, Toronto, 1983.

30. Okos, M. R. *Design and Control of Energy Efficient Food Drying Processes with Specific Reference to Quality.* U.S. Department of Energy Report DOE/ID/12608-4, Washington, DC, 1989.

20
Drying of Agricultural Products

Vijaya G. S. Raghavan
Macdonald Campus of McGill University
Ste.-Anne-de-Bellevue, Quebec, Canada

1. INTRODUCTION

Grain has been an important agricultural commodity and primary food source in many regions for centuries. The present distribution of the world's population has made strong demands on grain handling technology. Irrespective of whether it is international trade or distribution within a country, grain needs low moisture levels for safe storage. The most common method of preserving grain has always been by drying. In the days of premechanized agriculture, enough grain was usually stored by hanging ears of corn in barn lofts and attics to meet the needs of a community. As mechanization of agriculture spread to meet the needs of a population that was rapidly growing and urbanizing, mechanical methods for drying large quantities of grain were needed. Grain now travels thousands of miles either in large grain-carrying ships or in different types of carriers on wheels and must reach its destination in a high-quality state. Proper drying of these huge quantities of grain is a prerequisite to safe storage and delivery.

In 1980–1981 an estimated 727 million tons of coarse grains were produced in the world (1). Estimating harvested grain moisture and storage moisture to be in the range of 20% to 30% and 10% to 13%, respectively, 92 million tons of water had to be removed from this crop. This water removal process is very energy consumptive. Therefore, it becomes evident that the efficiency of grain drying, with respect to both energy and time, has important economic consequences for both grain producers and consumers.

Further, there are several advantages for making the grain dryer a standard part of the harvesting system in the Western agricultural sector. First, when a grain dryer is used, extra hours of harvesting each year are possible, potentially reducing the farmer's overall machinery investment. Second, earlier harvesting is possible with a grain dryer, allowing a crop to be harvested nearer to its ideal moisture content for minimizing field loss. This permits farmers to do a better job of weed control through timely chemical application and tillage practices for the following year's crop. Third, weather damage and losses due to wildlife may be reduced by harvesting at the tough or damp stages and then drying the grain. Last, proper drying and aeration of tough or damp grain reduces or eliminates spoilage problems in storage due to hot spots and insect infestations.

Not all grain dryers are suited to a given geographical area and farm. The choice of a system depends upon the annual volume produced, the marketing pattern, the type of farm, and the kind and capacity of existing facilities. This chapter is intended to provide an introduction of the various types of grain dryers presently available on the market so that the reader may understand how a given dryer is selected for a given farming operation. An attempt has also been made to indicate the importance of solar drying, nonconventional methods of drying, and some aspects of hay drying.

2. CROP CONDITIONING

Although the purpose of this chapter is primarily to discuss "heated air" dryers, some mention will be made about the four crop moisture reduction methods, usually called *crop conditioning*, since they are sometimes used in place of or in combination with heated air drying.

2.1. Aeration

Aeration consists essentially of moving small amounts of unheated air through a pile of grain in order to equalize the grain temperature and prevent moisture migration in bins exposed to drastic changes in ambient temperature. It may also be used to cool grain after drying, to keep damp grain cool until it can be dried, to remove storage odors, or to distribute fumigants in the grain mass.

Aeration is usually carried out in a storage bin that is equipped with a fan, duct system, and perforated floor along with exhaust vents to provide escape for moist air. Whether or not the ventilating air is blown upward or sucked down through the grain is largely a matter of choice. Upward ventilation is more commonly used, although there are advantages and disadvantages to each of these methods. An important advantage of using upward ventilation is that it allows storage temperatures to be measured easily because the warmest grain is always at the top of the pile. The recommended airflow rate for normal aeration of shelled corn, soybeans, and small grains at 125 Pa (0.5 in of water) is 5 m^3/h per m^3 of grain (0.1 cfm/bu) (2). However, for aerating damp grain at 500 to 750 Pa (2 to 3 in of water), flow rates of approximately 50 m^3/h per m^3 of grain (1 cfm/bu) are needed (3).

It is important to note that the aeration fan should not be run when the relative humidity of the ambient air is too high. For example, during fall and winter the operator should select days when the average relative humidity is less than or equal to 70% (4) and the air temperature is more than -1.1°C (30°F). It should be further noted that bins of 40 m^3 (1000 bu) or less generally do not require aeration if the grain is put in dry.

2.2. Natural Air Drying

Natural air drying employs a similar setup but higher airflow rates than those used for aeration. Typical rates for a storage depth of 1.2 to 1.8 m (4 to 6 ft) of small grains, peas, and beans, and shelled and ear corn are, respectively, 150 to 250 m^3/h per m^3 (3 to 5 cfm/bu) and 250 to 500 m^3/h per m^3 (5 to 10 cfm/bu) (2).

2.3. In-Storage Drying with Supplemental Heat

In-storage drying with supplemental heat involves drying of a relatively large batch of grain in situ (i.e., in the storage bin). It is carried out in bins of varying capacity up to 100 tons (5). Ventilation is accomplished by blowing slightly heated air, 4°C to 12°C (7–22°F) above ambient temperature through a duct system or through one centrally placed cylinder, as is the case for batch drying. Drying by this method usually requires continuous operation of the ventilation system for about one to three weeks.

In-storage drying may also be carried out on a bar floor provided with a powerful fan and a satisfactory system of floor and lateral ducts. Airflow rates range from 80 to 165 m^3/h per ton of grain. The advantages of this method are its low cost and simplicity.

2.4. Multistage Drying

The term *multistage drying* refers to any process that uses high-temperature drying in combination with aeration or natural air drying. An outline of two such processes, dryeration and combination drying, follows.

Dryeration

Dryeration is the term referring to the two-stage process by which grain is dried in a heated air dryer to within about 2% of its "dry" moisture content and then moved to an aerating bin (3). Here, it is allowed to steep for about 10 hours. This allows time for moisture within the kernels to move to the outside for easier removal. Aeration at airflows in the order of 25–50 m^3/h per m^3 of grain (0.5–1 cfm/bu) is then maintained for about 12 hours. The advantages of this system are (a) the ability to use higher drying temperatures since the grain does not remain in the high-temperature dryer until it is completely dry; (b) capacity increases of up to 60% of the grain drying system are possible since no cooling time in a high-temperature dryer is required; (c) the last few moisture percentage points, which are especially difficult to remove, are removed in the bin using the heat already contained within the grain, resulting in fuel savings of 20% or more; and (d) the grain quality is improved by slower cooling, which results in fewer heat stress cracks.

If the air is blown up through the grain, there is often a considerable amount of condensation on the roof and walls of the bin. The grain must therefore be moved to another bin for storage. The amount of condensation on the roof can be reduced by pulling the air down through the grain or by cooling the grain immediately after it comes out of the dryer.

Combination Drying

Combination drying is an extension of the dryeration process and is used primarily when grain with very high moisture (>25%) has to be dried (6). A high-temperature dryer is used to reduce the grain moisture content to about 19% to 23%. The grain is then moved to a bin dryer in which drying is completed using natural air or supplemental heat. With this method the output of the high-temperature dryer is increased to two or three times that obtained when it is used for complete drying. In addition, energy requirements may be reduced by as much as 50%. Airflows for the bin drying portion of the process are between 45 and 90 m^3/h/m^3 of grain (0.9 and 1.8 cfm/bu).

Selection of dryeration or combination drying will depend on the amount of grain to be dried, its initial moisture content, and the cost of energy and capital investment involved. If small amounts of grain at relatively low moisture contents are to be dried, the purchase of equipment for combination drying would not be warranted. Combination drying is more suited to high moisture contents and large volumes of grain. In all cases, for bins 100 m³ or larger, aeration ducts large enough for airflows of at least 36 m³/h per m³ of grain (0.7 cfm/bu) should be provided. Since fully perforated bin floors allow the greatest variety of options, their installation should be seriously considered on all large, new storage bins.

3. ARTIFICIAL HEATED AIR DRYING

An important consideration when dealing with any heated air process is drying temperature. Suggested drying temperatures vary depending on what the grain's use is to be. Table 1 lists a few recommendations for natural and heated air drying and the maximum drying temperatures to be used on grain for seed, commercial use, and animal feed. Drying time and airflow rate are also important. However, these vary according to the drying temperature and type of dryer used.

The two major types of heated air grain dryers are bin dryers and portable dryers. Bin dryers are available in batch, recirculating, and continuous categories, whereas portable dryers are available commonly in nonrecirculating and recirculating types.

3.1. Bin Dryers

Bin dryers are manufactured in many sizes and capacities and are used to obtain various drying rates. They are usually operated with lower airflow rates than other types, and hence are more energy efficient. They do, however, have a slower drying rate than most others. The general philosophy of bin dryer size selection is to be able to dry as much grain in 24 hours as will be harvested in a normal day.

Batch Dryers

The least expensive setup for drying is the one using the "batch-in-bin" process. The main components of this system are a bin with a perforated floor, a grain spreader, a fan and heater unit, a sweep auger, and an underfloor unloading auger (Fig. 1). The heater fan starts when the first load of grain is put in and continues to operate as long as is required to lower the average grain moisture content to the desired level.

Drying rate depends on several variables, such as drying time, grain depth, temperature of the heated air, and airflow rate. Final depth is selected by noting the pressure drop in a manometer. Airflow rates are determined from charts supplied with the fan unit. Usually a rate of 450 m³/h per m³ of grain (9 cfm/bu) is recommended for efficient drying. For a given grain depth, raising the air temperature speeds up drying but increases the chance of overdrying near the floor. Hence a safe air temperature is chosen for the crop being dried considering its initial moisture content (Table 1).

Before being stored, newly dried grain must be cooled. This is done by shutting off the heat and using the dryer fan to blow cool air over the grain, or by transferring the warm grain to an aerated storage bin and letting it cool there.

One variation of the batch-in-bin process is to use alternate heating and cooling cycles. This reduces the moisture differential between the drier grain near the perforated floor and the damper grain near the top of the grain column.

Some bin dryers have overhead, perforated, cone-shaped drying floors supported about one meter below the roof (see Fig. 2). A heater fan unit is installed below the perforated floor and blows warm air up through the grain.

When one batch of dry grain is dropped to a perforated floor at the bottom of the bin where it is cooled by an aeration fan, the next batch is loaded and dried on the dryer

Table 1 Recommendations for Drying Grain with Natural Air and Heated Air

	Ear corn	Shelled corn	Wheat	Oats	Barley	Sorghum	Soybeans	Rice	Peanuts
Maximum moisture content of crop at harvesting for satisfactory drying:									
With natural air, %[a]	30	25	20	20	20	20	20	25	45–50
With heated air, %[a]	35	35	25	25	25	25	25	25	45–50
Maximum moisture content[b] of crop for safe storage in a tight structure, %[a]	13	13	13 (Seed wheat, 12%)	13 (Seed oats, 12%)	13	12	11	12	13
Maximum relative humidity of air entering crop that will dry crop down to safe storage level when natural air is used for drying, %	60	60	60	60	60	60	65	60	75
Maximum safe temperature of heated air entering crop for drying when crop is to be used for									
Seed, °C	43	43	43	43	41	43	43	43	32
Sold for commercial use, °C[a]	54	54	60	60	41	60	49	43	32
Animal feed, °C[a]	82	82	82	82	82	82			

Source: From Reference 19.

[a]Moisture contents on wet basis: (a) higher temperatures than those listed may be used when the corn is dried under carefully controlled conditions so that the maximum temperature of the kernels does not exceed 54°C at any time; (b) if there is any possibility that the crop may be sold, use the lower temperature as listed for commercial use.

[b]If the products are to be stored for long periods, the moisture content should be 1% to 2% lower than shown in this tabulation.

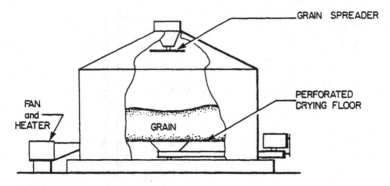

Figure 1 A typical batch dryer bin.

floor above. Cool, dry grain is transferred to another storage bin via an underfloor auger. The advantage of this system is that drying can continue while the grain is being cooled and transferred.

Vertical stirring augers may be added to bin dryers, which not only promote more uniform drying, but also permit a higher airflow rate, thus increasing the drying rate for a given crop. Although stirring augers may result in slightly lower fuel efficiencies, the increased drying rate, reduction in overdrying at the bottom, and the larger batch size outweigh this disadvantage.

Recirculating Dryers

In the recirculating type of dryer, grain is constantly mixed while drying. One example of a recirculating dryer is shown in Figure 3. A slanted floor causes the grain to move toward a vertical auger situated in the center of the dryer. The auger picks up the grain and delivers it to the top of the grain bin. The result is a more uniformly dried crop than that obtained using nonrecirculating types.

The dryer shown in Figure 4 is used as a recirculating batch or continuous flow dryer. When being used as a recirculator, an "under grain" sweep auger moves grain to the center of the perforated bin floor where it is picked up by a vertical auger and delivered to a grain spreader. When the dryer is operated as a continuous flow dryer, the grain traveling up the vertical auger is transferred to an aeration bin via an inclined auger.

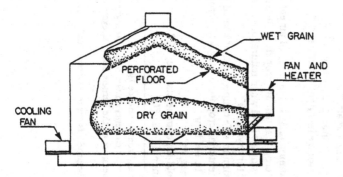

Figure 2 A bin dryer with an overhead drying floor.

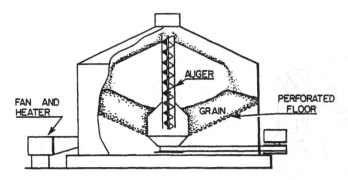

Figure 3   A recirculating batch bin dryer.

Continuous Flow Dryers

Although there are many types of continuous flow dryers, one of the more common types uses two to four vertical grain columns through which hot air is forced perpendicularly to grain flow (Fig. 5). The grain is loaded at the top and passed down both sides of the hot and cold plenums before entering the unloading augers. Grain flow rate is either manually controlled or controlled by a thermostat near the outside of the grain column. As fan capacity is decreased or column width increased, more efficient use of heat results; however, the grain moisture differential between the inside and outside layer increases.

Some continuous flow dryers use three fans and three plenums, each with individual temperature controls. These may be run with two heating sections and one cooling section or else all three with heat, in which case the grain must be cooled in an aerated bin (Figs. 6-a and 6-b).

Farm Fans, Indianapolis, Indiana,* has a series of dryers of this type that they term continuous multistage dryers, ranging in capacity from about 5 to 27 tons/h (265 bu/h to 1220 bu/h) based on drying and cooling corn from 25% to 15% moisture.

A number of companies recycle drying and/or cooling air. Two common techniques of accomplishing this are shown in Figure 6. Some manufacturers use the system shown

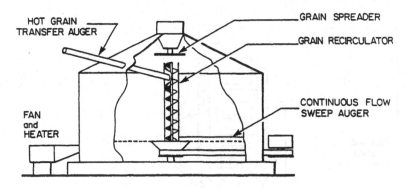

Figure 4   A recirculating bin dryer.

---

*Mention of proprietary products in this article does not imply any recommendation or endorsement of their particular brands.

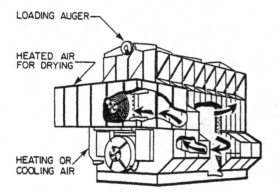

Figure 5 A typical stationary continuous flow dryer with an air recirculating system.

in Figure 6-a. Here, ambient temperature air is drawn through the grain in the cooling section and then passed through the fan heater unit of the midsection. This system results in more energy saving than the system shown in Figure 6-b due to the fact that air from the first heating section is recycled. Its disadvantage is that chaff and fine material may be drawn into the midsection hot air plenum, necessitating frequent cleaning.

Most continuous flow dryers are of the stationary type, although some of the smaller-size dryers are portable. For example, Gilmore and Tatge Manufacturing Company, Incorporated, Clay Center, Kansas, make a concentric cylinder type portable dryer that handles 7.8 ton/h (350 bu/h) based on moisture removal from 20.5% to 15.5%. Grain column width on many of these dryers is 0.30 m as compared to the 0.45 m found on the GT-Tox-o-Wik recirculating batch dryers. It should be noted that the moisture differential across the grain column is lowered as its width is decreased. A thinner column therefore means that, for a given average moisture content, the inner layer is overdried less. Thus, using a continuous flow dryer might be of some benefit when drying heat-sensitive small grains such as wheat, oats, and barley.

Another type of continuous flow dryer is the parallel flow dryer in which the grain

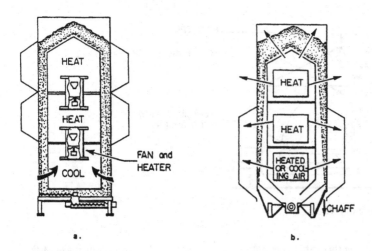

Figure 6 Heat recovery systems: (a) reverse cooling; (b) one-way airflow.

moves in the same direction as the hot airflow. This results in more uniform drying and reduces the danger of heat damage. Furthermore, since no screens are used in parallel flow dryers, small seed crops can be dried without leakage.

Continuous flow dryers are not well suited for the drying of small quantities of different types of grain because startup and emptying of these dryers is inefficient. Accurate moisture control is difficult to achieve until a uniform flow is established. Continuous flow dryers are best in situations in which large quantities of grain must be dried without frequent changes from one type to another.

3.2. Portable Dryers

Portable dryers generally appeal to the farmer who has grain bins scattered in various locations, or who does custom drying off his or her farm. Portable dryers without a proper grain-handling system may be used to fill an immediate need in an emergency situation; however, they are normally not used where drying is beneficial but not necessary due to the inconvenience of setup and dismantling of the system. The two types of portable batch dryers are nonrecirculating and recirculating.

Nonrecirculating Dryers

Most nonrecirculating types of dryers are of the fully enclosed concentric cylinder type. These are loaded at the top and drying is accomplished by blowing hot air radially through a column of grain. Similar to the batch-in-bin drying system, the inside grain layer (the layer near the hot air plenum) becomes overdried while the outside layer remains underdried. Nevertheless, as the grain is removed from the dryer, the damp and dry grain are mixed so that a satisfactory product results for further use.

Other types of portable nonrecirculating batch dryers exist, such as wagon or truck box dryers. These use a heater fan unit similar to that used for bin drying, and which is connected to smaller air ducts suspended at mid-height of the box or located on its floor. If suspended air ducts are utilized, exhaust ducts on the floor are a necessity.

Automatic dryers of the above-mentioned types are equipped with thermostats or timers to control heating and unloading cycles. No manual supervision is required if a completely mechanized grain-handling system is used.

Recirculating Dryers

Portable recirculating batch dryers are essentially the same as nonrecirculating models except for a central auger that picks up grain near the bottom of the column and deposits it at the top (see Fig. 7). A complete recirculation of grain occurs roughly every 15 minutes.

Most common dryers of this type come in sizes ranging from 10 to 18.5 m^3 (300 bu to 525 bu) bin capacity. These dryers are often used by medium-size farms in eastern North America that cannot afford a more expensive continuous flow model. The dryers may be used for virtually any crop, provided that the maximum safe drying temperature is not exceeded. However, their disadvantage is that constant augering can cause damage to certain seeds such as beans, peas, and malting barley, especially when they are nearly dry.

4. DRYER SELECTION

Selection of a continuous flow, batch, or batch-in-bin dryer depends largely on the amount of grain to be dried and the facilities a farmer already has available when the dryer is purchased. For example, a farmer who already has a good size storage bin and

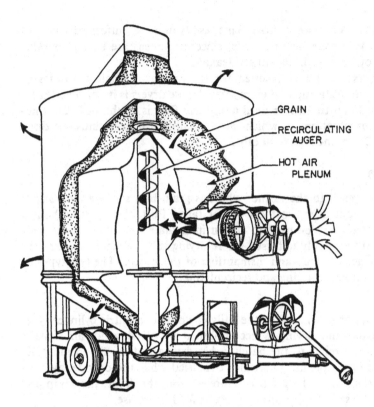

GRAIN

RECIRCULATING
AUGER

HOT AIR
PLENUM

Figure 7   A typical portable batch dryer.

only has a small volume of grain to store would probably use an in-storage dryer instead of buying a portable dryer and "wet grain" holding bin. This system, however, would not be suitable for farms larger than 160 ha.

The recommendations concerning the type of system to be used can be made based on the annual production of a farm as illustrated in Table 2. Although the capacity range presented here is for corn, it can also be extended to other grains and cereals. It must be further noted that these recommendations were made for farmers in the central United States area. For eastern Canada, where temperatures are cooler and humidity is higher, the figures for natural air drying presented in Table 2 are slightly higher. In Quebec, for

Table 2   Recommended Drying System Based on the Annual Farm Production at Harvest

Annual production[a]	Type of drying system[a]
22 to 60 tons (100 to 2700 bu)	Natural air drying
60 to 445 tons (2700 to 20,000 bu)	Natural air drying with supplemental heat
445 to 1556 tons (20,000 to 70,000 bu)	Batch-in-bin dryers
Above 1556 tons (>70,000 bu)	Portable and continuous flow dryers

[a]Based on D. I. Chang, D. S. Chung, and T. O. Hodges, *Grain Dryer Selection Model*, *Am. Soc. Agr. Eng. Paper No. 79-3519*, St. Joseph, Michigan, 1979.

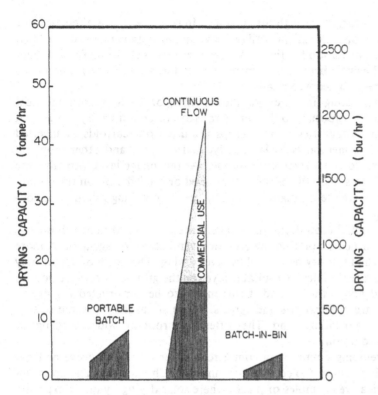

Figure 8 Drying rates based on corn dried from 25% to 15% in various types of commercially available dryers.

example, corn at harvest may have 35% moisture content with an average around 30%, whereas in western Canada and the midwestern United States, it is usually harvested at moisture contents in the low 20% range. This factor should therefore be considered when selecting a suitable drying system.

The crops most often dried in Canada and the United States by artificial means are corn (maize) and beans. Wheat, oats, and barley are harvested in the dry season and usually come off the field at a moisture content suitable for safe storage. If need be, the grain may be dried with natural air on sunny, warm days.

Most of the dryers in Canada are found in Quebec and Ontario (7), many of these being portable batch types. Larger, continuous flow models may be found at co-ops across the country or on the larger farms in southwestern Ontario where farmers are growing 320–360 ha of their own crop. Drying rates possible in different commercially available dryers are summarized in Figure 8.

5. SOLAR ENERGY IN DRYING

An alternative that is being encouraged in hot, dry countries of Asia and Africa is solar drying. Solar heat is trapped with a solar collector constructed from an aluminum sheet painted black. The collector may be fixed to the drying bin in such a way that an air space exists between it and the bin wall. Energy absorbed by the collector heats the ventilating air by a few degrees as it is forced through the air space. In North America, these types of dryers have been known to operate satisfactorily with grain moisture contents up to 25%, even on cloudy days (5). The reason for this is that solar energy is about half visible light and half infrared rays, the latter being able to penetrate clouds. On rainy days and nights supplemental heat may be supplied electrically.

In countries where harvesting time occurs at the beginning of the dry season, the most popular method of drying is exposure to the sun. Crops are often left to dry in the field before harvesting. In some countries various crops are dried on scaffolds or inverted latticework cones. Another method is to lay paddy, maize, cobs, and other crops on heaps of stubble and then to cover them with stubble. At the village level, probably the most common practice is to spread the harvested threshed or shelled crop on the ground or on a specially prepared area (e.g., matting, sacking, mud/cow dung mixture, or concrete) exposed to the sun.

In humid countries, initial crop drying may take place as outlined above; however, further drying is accomplished by placing the crop in a ventilating storage area. A more effective type of drying than sun drying is shallow layer drying. This form of drying may be achieved by spreading the produce in a suitable layer on the ground or on wire bottom trays that are supported above the ground. Cribs may also be constructed for drying maize on the cob or unthreshed legumes and cereals. These are usually oriented so that the long axis is facing the prevailing wind. They often have roofs or wide overhangs to protect the drying crop from rain.

In most warm, developing countries, a commercial dryer is too expensive and not essential enough for a single farmer to consider its purchase. China, India, and countries on the continent of Africa are examples of places where solar drying by direct exposure or by a cheaply constructed collector is employed.

6.   ARTIFICIAL DRYING IN DEVELOPING COUNTRIES

Where humidity is too high to allow grain to be adequately dried by natural means, it is necessary to supply heat to the drying crop. The most popular forms of artificial drying may be categorized according to the depth of grain being dried. These are (a) deep-layer drying, (b) in-sack drying, and (c) shallow-layer drying.

Deep-layer dryers consist of silo bins (rectangular warehouses) fitted with ducting or false floors through which air is forced. Depths of up to 3.5 m of grain may be dried at one time (8).

An in-sack dryer is made of a platform that contains holes just large enough to hold jute sacks full of grain. Heated air is blown up through the holes (and grain) via a heater/fan unit. The platform may be constructed from locally available material. A typical oil-fired unit that deals with two to five tons of grain is equipped with a fan that delivers 9700 m^3/h of air heated to 14°C above ambient temperature and consumes about 4.5 liters of oil per hour (8). For two-ton loading, the moisture removal rate is about 1% per hour.

Shallow-layer dryers are those consisting of trays, cascades, or columns in which a thin layer of grain is exposed to hot air. In these dryers, the hot airstream is at the highest safe temperature and the amount of drying is determined by the length of time the grain is allowed to remain in the dryer, either as a stationary bach or as a slow-moving stream. Due to the fact that the layer of grain being dried is thin (less than about 0.20 m), no significant moisture gradient develops through the grain (8). This means that the drying temperature is limited only by the possibility of heat damage to the grain.

Another simple but effective type of artificial dryer utilizes a locally built platform dryer in which the products of combustion of local fuel are not allowed to pass through the grain. The heated air passes through the produce by means of natural air movement or convection currents. One such dryer built at Mokwa, Nigeria, uses a pit (which became the hot air plenum) covered by the drying floor, the firebox being located outside the plenum chamber.

Yet another type of dryer is the horizontal dryer, which contains a number of chambers, each being divided by horizontal, equidistant, screen-bottomed trays placed on horizontal pivots. Damp grain is placed on the top tray in a layer 0.16–0.18-m deep and is tipped to the next set of trays after an initial drying period. Since this type of dryer is normally operated as a batch dryer, it is an advantage to have two cooling chambers per unit so that one batch may be loaded into the dryer while the other is being removed from the machine. A typical setup of this type would include a double drying chamber, a cleaning unit, and augers or elevating units for tilling the dryer and elevating the grain to storage.

7. NONCONVENTIONAL METHODS

Recent increases in the cost of fossil fuels have prompted researchers to investigate and develop more energy efficient dryers (9–12). One attempt at reducing fuel cost was to pass unheated air through large beds of absorbent material such as silica gel before passing it through the grain. The problem with this was that the gel itself had to be dried at high temperatures, making the operation expensive.

Other methods related to enhancement of heat transfer techniques are also being studied. Particle-particle heat transfer is one such technique that has led to the design of many experimental dryers. Richard and Raghavan have dealt with this topic extensively (13). They discuss the theoretical aspects, experimental data, and demonstrate the potential of this method. The main advantage of this type of dryer is its rapidity. Following this concept, a continuous flow conduction grain processor/dryer was developed in the late 1980s at McGill University, Quebec, Canada; it is shown in Figure 9 and fully described in a paper and two patents by Pannu and Raghavan (14–16). It is based on particle-particle heat transfer and was designed to control mixing and heating time and provide ease of separation of the grain and the particulates.

The dryer consists of three concentric conical drums rotating about a common axis. The inner cone is fitted with a propane burner and buckets to carry the heat transfer medium toward the hottest part of the flame. The particulate medium then flows into the second drum, where it is mixed with moist grain and carried in the opposite direction by the helical walls of the second drum. As the mixture proceeds toward the opposite end, it is separated by screen mesh, the grain being carried on to the second cone outlet while the particulate medium drops into the outer cone and is recirculated to the heating chamber. The outer cone is insulated with R-10 glass wool to reduce heat losses. Grain residence time, and therefore the heating and moisture removal rates, are controlled by adjusting the rotating speed of the unit.

The dryer performance can be adjusted by varying different parameters such as the heating medium, medium/grain mass ratio, grain moisture, medium temperature, and angular velocity. The possibility of improving drying performance by using zeolites (molecular sieves) rather than sand was investigated (17). It was found that the difference in moisture removed between the molecular sieves and sand was a function of residence time. When using sand, the relative humidity in the dryer reached saturation within 2

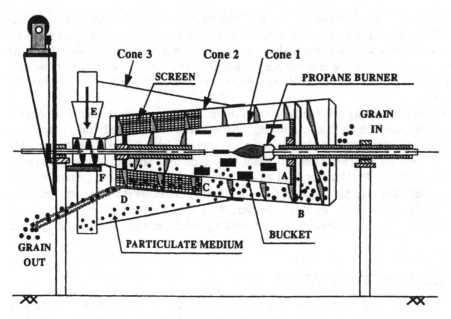

Figure 9 Schematic diagram of conduction grain processor.

minutes (>90%), whereas it dropped to a steady value of 10–20% when zeolites were used. Thus the differences in moisture removal increased with time.

The high heat transfer efficiency makes the unit, using sand as a particulate, suitable for heat treatment applications such as pasteurization, precooking, insect eradication, and other applications in which moisture removal is not a priority. With zeolites as the transfer medium, moisture removal rates double, thus bringing drying efficiency up to standards.

The enhanced drying performance using molecular sieves is encouraging. However, questions can be raised as to possible hazards associated with synthetic zeolite residues and as to nutritional quality of corn dried at higher temperatures.

The moisture removal increase was 50% to 130% higher for zeolites than sand, depending on the operating conditions. Upon nutritional analysis, no differences were found in digestibility or acceptability, as indicated by daily feed intake (18). Unavailable proteins tended to increase with medium temperature; however, the increase was not significant.

8. HAY DRYING

Hay is also an important crop, like grain and cereals. It is estimated that in any given year in Canada and the United States, 30% of hay is lost during harvesting and storage. Proper drying and handling techniques might reduce these losses. The hay is usually field dried to approximately 40% moisture and then dried to 20% with barn hay dryers. By employing suitable management techniques for barn drying systems, harvest losses can be reduced, produce can be harvested at its optimum stage of growth, and storage losses can be minimized. Although these advantages are acceptable, the number of barn drying systems has not increased in recent years because of the difficulty of providing a handling system compatible with the harvesting method. Forced air drying and heated air drying systems are generally used for hay drying. Further information on hay drying is given in Reference 19.

ACKNOWLEDGMENTS

The author greatly appreciates the contribution of K. Anderson for obtaining the information required for the preparation and compilation of this chapter. The author is indebted to the following individuals for their contributions during the preparation of this manuscript: R. Langlois for his drafting; P. Alvo, S. Gameda, V. Orsat, and F. Taylor for proofreading; and R. Haraldsson and Y. Gariepy for typing the manuscript. Finally, the author thanks the following companies for providing information on different types of dryers: Beard Industries, Frankfort, Indiana; Caldwell Manufacturing Company, Kearney, Nebraska; Farm Fans, Indianapolis, Indiana; Gilmore and Tatge Manufacturing Company, Incorporated, Clay Center, Kansas; Long Manufacturing N.C. Incorporated, Tarboro, North Carolina; Martin Steel Corporation, Mansfield, Ohio; and Mathews Company, Crystal Lake, Illinois.

REFERENCES

1. Food and Agriculture Organization of the United Nations. 1980. *FAO Production Yearbook*, Vol. 34, FAO Publications, Rome, pp. 85–109.
2. Agriculture Canada. 1962. Drying and conditioning. In *Agricultural Materials Handling Manual*, Part 3, The Queen's Printer, Ottawa, pp. 1–31.
3. Foster, G. H. 1984. Drying cereal grains. In *Storage of Cereal Grains and Their Products*, Amer. Society of Cereal Chem., Inc., St. Paul, Minnesota, pp. 79–116.
4. Brooker, D. B., Bakker-Arkema, F. W., and Hall, C. W. 1974. Grain drying systems. In *Drying Cereal Grains*, AVI Publishing Company, Inc., Westport, Connecticut, pp. 145–184.
5. Nash, M. J. 1978. Cereal grains, legume grains and oil seeds. In *Crop Conservation and Storage*, Pergamon Press, New York, pp. 27–79.
6. Friesen, O. H. 1981. *Heated-Air Grain Dryers*, Information Services Agriculture Canada Publication 1700, Ottawa, pp. 3–25.
7. Otten, L., Brown, R., and Anderson, K. 1980. A study of a commercial crossflow grain dryer. *Can. Agr. Eng.* 22(2):163–170.
8. Hall, D. W. 1970. Handling and storage of food grains in tropical and subtropical areas. Food and Agriculture Organization of the United Nations, pp. 1–198.
9. Meiring, A. G., Daynard, T. B., Brown, R., and Otten, L. 1977. Dryer performance and energy use in corn drying. *Can. Agr. Eng.* 19(1):49–54.
10. Mittal, G. S., and Otten, L. 1981. Evaluation of various fan and heater management schemes for low temperature corn drying. *Can. Agr. Eng.* 23(2):97–100.
11. Mujumdar, A. S., and Raghavan, G. S. V. 1984. Canadian research and development in drying—A survey. In *Drying '84*, Hemisphere/McGraw-Hill, New York.
12. Sturton, S. L., Bilanski, W. K., and Menzies, D. R. 1981. Drying of cereal grains with the dessicant Bentonite. *Can. Agr. Eng.* 23(2):101–104.
13. Richard, P., and Raghavan, G. S. V. 1984. Drying and processing by immersion in a heated particulate medium. In *Advances in Drying*, Vol. 3, Hemisphere, New York, pp. 39–70.
14. Pannu, K., and Raghavan, G. S. V. 1987. A continuous flow particulate medium grain processor. *Can. Agr. Eng.* 29(1):39–43.
15. Raghavan, G. S. V., and Pannu, K. S. Méthode et appareil de séchage et de traitement à la chaleur d'un matériau à l'état granulaire. *Canada Patent No. 1 254 381*, 1989.

16. Raghavan, G. S. V., and Pannu, K. S. Method and apparatus for drying granular material. *U.S. Patent No. 4 597 737*, July 1, 1986.
17. Raghavan, G. S. V., Alikhani, Z., Fanous, M., and Block, E. 1988. Enhanced grain drying by conduction heating using molecular sieves. *Trans. of the ASAE* 31(4):1289–1294.
18. Alikhani, Z., Raghavan, G. S. V., and Block, E. 1990. Effect of particulate medium drying on nutritive quality of corn. *Can. Agr. Eng.* 33:79–84.
19. Hall, C. W. 1980. *Drying and Storage of Agricultural Crops*, AVI Publishing Company, Inc., Westport, Connecticut.

21
Drying of Fruits and Vegetables

K. S. Jayaraman and D. K. Das Gupta
Defence Food Research Laboratory, Siddarthanagar, Mysore, India

1. INTRODUCTION

From the point of view of consumption, *fruits* are plant products with aromatic flavor that are naturally sweet or normally sweetened before eating (1). Apart from providing flavor and variety to human diet, they serve as important and indispensable sources of vitamins and minerals although they are not good or economic sources of protein, fat, and energy. The same is true of vegetables, which also play an important role in human nutrition in supplying certain constituents in which other food materials are deficient and in adding flavor, color, and variety to the diet (2).

After moisture, carbohydrates form the next most abundant nutrient constituent in fruits and vegetables, and are present as low molecular weight sugars or high molecular weight polymers like starch, and so on. The celluloses, hemicelluloses, pectic substances, and lignin characteristic of plant products together form dietary fiber, the value of which in human diet is increasingly being realized in recent years, especially for the affluent society of the Western countries. Virtually all human's dietary vitamin C, an important constituent of human diet, is obtained from fruits and vegetables, some of which are rich in pro-vitamin A (beta-carotene) (e.g., mango, carrot, etc.). They are important suppliers of calcium, phosphorus, and iron.

Fruits and vegetables have gained commercial importance and their growth on a commercial scale has become an important sector of the agricultural industry. Recent developments in agricultural technology have substantially increased the world production of fruits and vegetables. Consequently a larger proportion of several important commodities is being handled, transported, and marketed all over the world than before with concomitant losses calling for suitable postharvest techniques for storage and processing to ensure improved shelf life. Production and consumption of processed fruits and vegetables are also increasing.

2. POSTHARVEST TECHNOLOGY OF FRUITS AND VEGETABLES

2.1. World Production

The present world production of fruits (excluding melons) according to Food and Agriculture Organization (FAO) estimates was about 341.9 million metric tons (mMT) in 1990 (3). Brazil, with a production of about 30.1 mMT (8.8%), is a leading producer of fruits in the world. India, with 27.8 mMT (8.1%), occupies second position, followed by the United States (7%) and China (6.3%). The average consumption in the United States and Western Europe is high, around 10 kg/head/year for citrus and banana and below 1 kg for other fruits. The daily per capita consumption of fruits in India is 40 g against 120 g recommended, the big gap attributed to postharvest losses.

World production of vegetables (including melons) is about 441.8 mMT. The major vegetable-producing countries in 1990 were China, India, the Soviet Union, the United States, Turkey, Japan, and Italy. China is the largest producer, accounting for about 117.1 mMT (26.5%), while India is second, contributing about 51 mMT (11.5%). The per capita availability in India is around 120–130 g/day against 300 g recommended.

2.2. Losses

Most fruits and vegetables contain more than 80% water and are therefore highly perishable. Water loss and decay account for most of their losses, which are estimated to be more than 30–40% in the developing countries in the tropics and subtropics (1) due to inadequate handling, transportation, and storage facilities. For example, the total losses in India are estimated at 20% to 25%, worth more than U.S. $1.5 billion annually. Apart from physical and economic losses, serious losses do occur in the availability of essential nutrients, notably vitamins and minerals.

The need to reduce postharvest losses of perishable horticultural commodities is of paramount importance for developing countries to increase their availability, especially in the present context when the constraints on food production (land, water, and energy) are continually increasing. It is being increasingly realized that the production of more and better food alone is not enough and should go hand in hand with suitable postharvest conservation techniques to minimize losses, thereby increasing supplies and availability of nutrients besides giving the economic incentive to produce more (1).

2.3. Role of Preservation

One of the prime goals of food processing or preservation is to convert perishable foods such as fruits and vegetables into stabilized products that can be stored for extended periods of time to reduce their postharvest losses. Processing extends the availability of seasonal commodities, retaining their nutritive and esthetic values, and adds variety to the otherwise monotonous diet. It adds convenience to the products. In particular it has expanded the markets of fruit and vegetable products and ready-to-serve convenience foods all over the world, the per capita consumption of which has rapidly increased during the past two to three decades.

Several process technologies have been employed on an industrial scale to preserve fruits and vegetables, the major ones being canning, freezing, and dehydration. Among these, dehydration is especially suited for developing countries with poorly established low-temperature and thermal processing facilities. It offers a highly effective and practical means of preservation to reduce postharvest losses and offset the shortages in supply.

2.4. Preservation by Drying

The technique of dehydration is probably the oldest method of food preservation practiced by humankind. The removal of moisture prevents the growth and reproduction of microorganisms causing decay and minimizes many of the moisture-mediated deteriorative reactions. It brings about substantial reduction in weight and volume, minimizing packing, storage, and transportation costs and enables storability of the product under ambient temperatures. These features are especially important for developing countries and in military feeding and space food formulations.

A sharp rise in energy costs has promoted a dramatic upsurge in interest in drying worldwide over the last decade. Advances in techniques and development of novel drying methods have made available a wide range of dehydrated products, especially instantly reconstitutable ingredients, from fruits and vegetables with properties that could not have been foreseen some years ago. The growth of fast foods has fueled the need for such ingredients. Due to changing life styles, especially in the developed world, there is now a great demand for a wide variety of dried products with emphasis on high quality and freshness besides convenience.

This chapter is intended to provide a comprehensive account of the various drying techniques and appliances developed and applied over the years specifically for the dehydration of fruits, vegetables, and their products. Theoretical and practical aspects of drying as applied to foodstuffs in general have been covered by Sokhansanj and Jayas in the earlier edition of the *Handbook of Industrial Drying* (4). Therefore, discussion will be restricted to fruit and vegetable drying besides quality changes during drying and storage as specifically applied to these commodities.

3. GENERAL MORPHOLOGICAL FEATURES

Fruits and vegetables have certain morphological features quite distinct from other natural materials used as food that greatly influence their behavior during processing and preservation, especially by drying. A brief enumeration of their important features is therefore useful to the understanding of the various pretreatments and drying techniques and accompanying quality changes to be discussed subsequently.

3.1. Diversity of Form

The diversity of form shown by fruit and vegetable structures is extremely wide (5). Among the vegetables we have representatives of all the recognizable morphological divisions of the plant body—whole shoots, roots, stems, leaves, petioles, inflorescence, fruits, and so on. Fruits, though falling naturally into a single large group, may also be classified on simple structural grounds into a number of distinct types—from the single-seeded drupes to many-seeded berries, pomes, hespiridiums, and pepos, besides multiple types.

3.2. Types of Tissue Forming the Structural Framework

The various chemical constituents are distributed through a highly complex structural framework built up from a number of different kinds of tissue. Most of the normal metabolic activity of the plant is carried out in relatively unspecialized tissue called *parenchyma*, which generally makes up the bulk of the volume of all soft, edible plant structures. The outermost cell layer, the *epidermis*, protects the surface, while mechanical support is provided by highly specialized cells called *collenchyma* and *sclerenchyma*. Water, minerals, and organic products of metabolism are transported from one part of the plant to another through the so-called vascular tissues, *xylem* and *phloem*, which also show a high degree of specialization and characteristic anatomical differences.

3.3. Chemical and Physical Properties of Tissues Influencing Quality

The different tissues described above differ from one another not only in structure but in chemical composition and, most important of all, in physical properties. They vary in their resistance to the effects of cooking and processing treatments. The textural quality of fruits and vegetables is therefore determined in no small part by the fine structure of individual tissues and by the relative proportions in which they are present.

The ground tissue or parenchyma of plants is composed largely of undifferentiated cells, each cell with a prominent cell wall and its internal volume occupied by a single, large, sap-filled vacuole. The protein and lipid materials are largely contained in a thin layer of living cytoplasm lying between the cell wall and the vacuole. The cytoplasm has the consistency of a viscous fluid and in the intact living state shows properties of semipermeability that are responsible for the osmotic behavior of the cells.

The polysaccharide materials other than starch are largely confined to the cell wall and include cellulose, hemicelluloses, and pectic materials. Between the walls of adjacent cells is a very thin layer of material called *middle lamella*, which is largely pectic in nature and acts as an intercellular cement. Its composition largely determines the firmness with which adjacent cells adhere to each other, with important bearing on the textural quality. As much as 50–60% of the cellulose in the plant cell wall is in crystalline condition, interconnected through noncrystalline regions and organized into fine, unbranched threads called *microfibrils*, which form the structural skeleton of the cell wall and largely determine its physical properties, conferring considerable tensile strength combined with moderate elasticity. Hemicelluloses and pectic materials together make up about 30% of the total volume of the cell wall and their removal leads to a marked shrinkage of the wall and a change in physical properties. The pectic substances also show characteristic changes with the stage of maturity and ripening.

The protective tissue, the *epidermis*, contains an extracellular, water-impermeable surface layer called the *cuticle* in which cell walls are thickened and impregnated with lipidlike materials (wax and cutin). In the supporting tissue, the *sclerenchyma*, the cell walls are lignified, which imparts greater rigidity and hardness, and the fiber cells persist unchanged after cooking, giving rise to fibrousness or stringyness of texture in products like asparagus and green bean.

3.4. Other Organized Structures Influencing Quality

Besides the main structural features enumerated above, there are certain other kinds of organized structures, the occurrence of which can have an important bearing on the quality of fruit and vegetable products. Small bodies of variable shape called *plastids* occur in the cytoplasm of most living plant cells; these act as centers for accumulation of various products of cellular metabolism. Some, for example, contain pigments; these are referred to as *chromoplasts*, the commonest type of which is the *chloroplast*, in which the chlorophylls and associated pigments are contained. These chloroplasts are present in all green tissues. The only other pigments that occur in special chromoplasts are the carotenoids. Colorless plastids called *leucoplasts* act as centers for accumulation of storage materials like starch, which occurs in the form of grains of different shapes. The starch in grains is partly crystalline, which is normally destroyed during blanching or cooking and is gelatinized going into solution in the cell sap. This is one of the main effects of heat processing of starch-containing products.

4. PRETREATMENTS FOR DRYING

Fruits and vegetables are subjected to certain pretreatments with a view to improve drying characteristics and minimize adverse changes during drying and subsequent storage of the products. These include alkaline dips for fruits and sulfiting and blanching for fruits and vegetables (6).

4.1. Alkaline Dip

The *alkaline dip* involves immersion of the product in an alkaline solution prior to drying and is used primarily for fruits that are dried whole, especially prunes and grapes. A sodium carbonate or lye solution (0.5% or less) is usually used at a temperature ranging from 93.3°C to 100°C (1). It facilitates drying by forming fine cracks in the skin. Oleate esters constitute the active ingredients of commercial dip solutions used for grapes. They accelerate moisture loss by causing the wax platelets on the grape skin to dissociate, thus facilitating water diffusion.

4.2. Sulfiting

Sulphur dioxide treatments are widely used in fruit and vegetable drying as sulphur dioxide is by far the most effective additive to avoid nonenzymatic browning (7). It also inhibits various enzyme-catalyzed reactions, notably enzymic browning, and acts as an antioxidant in preventing loss of ascorbic acid and protecting lipids, essential oils, and carotenoids against oxidative deterioration during processing and storage. It also helps in inhibition and control of microorganisms, especially microbial fermentation of sugars in fruits such as sun-dried apricots as encountered during prolonged drying. It has the advantage of allowing higher temperatures, hence shorter drying times, to be used. It is intended to maintain color, prevent spoilage, and preserve certain nutritive attributes until marketed.

Fruits for dehydration are often treated with gaseous SO_2 from burning sulphur as used in the manufacture of dried apricots, peaches, bananas, raisins, and sultanas. Alternatively, apple slices are generally dipped in solutions of the additive (prepared by dissolving sodium bisulfite or SO_2 in water) and may receive an extra treatment with gaseous SO_2 during drying.

Treatment of vegetables with SO_2 gas is impractical. Sulfite solutions are preferred as the most practical method of controlling absorption. As vegetables are blanched prior to drying, generally the additive is incorporated at the blanching stage either in the blanch liquor if the vegetable is to be dipped or as a spray in the case of steam blanching.

Sufficient SO_2 must be absorbed by the prepared material to allow for losses that occur during drying and subsequent storage. The various methods of application of SO_2 result in varying levels of uptake, which is a function of SO_2 concentration, length of treatment and time allowed for draining, size and geometry of the food, and the pH of the blanch liquor or spray. Drying times in excess of 12 h for fruits and vegetables and of several days as in sun drying of fruits, necessitate use of large amounts of SO_2. It has been shown that only 35–45% of the additive initially incorporated is measurable after

Table 1 Suggested Sulphur Dioxide Levels in
Dried Vegetables

Vegetable	SO$_2$ (ppm)
Beans	500
Cabbages	1500–2500
Carrots	500–1000
Celery	500–1000
Peas	300–500
Potato granules	250
Potato slices	200–500
Sweet potatoes (diced)	200–500
Beets	Not necessary
Corn	2000
Peppers[a]	1000–2500
Horseradish	Destroys flavor

Source: From Reference 7
[a]0.2% antioxidant BHA gives better color retention

drying. The subsequent loss of SO$_2$ from dried products occurring during storage determines the practical shelf life with respect to spoilage through nonenzymatic browning.

Tables 1 and 2 show suggested levels for SO$_2$ in vegetables and fruits, respectively, after the completion of drying (7).

4.3. Blanching

Blanching consists of a partial cooking, usually in steam or hot water, prior to dehydration. It is intended to denature enzymes responsible for bringing about undesirable reactions that adversely affect product quality such as enzymic browning and oxidation during processing and storage. The effectiveness of the treatment is judged by the degree of enzyme inactivation. Thus, activity of polyphenoloxidase is followed in fruits, that of catalase in cabbage and of peroxidase in other vegetables. The other beneficial effects produced by blanching include (6) reduced drying time, removal of intercellular air from the tissues, softening of texture, and retention of carotene and ascorbic acid during storage. Commercially both continuous and batch-type blanchers are employed, involv-

Table 2 Suggested Sulphur Dioxide Levels in
Dried Fruits

Fruit	SO$_2$ (ppm)
Apples	1000–2000
Apricots	2000–4000
Peaches	2000–4000
Pears	1000–2000
Raisins	1000–1500

Source: From Reference 7

ing 2- to 10-min exposure to live steam. Series blanching in hot water is also used, in which the solids content of the water is maintained at an equilibrium level to minimize leaching losses.

A modification to steam blanching is the individual quick blanch (IQB) process (8) comprised of heating and holding steps, producing less effluent than conventional steam blanching. Heating is done in a condensing steam unit with food particles one layer deep and enough heat is added to raise the mass average temperature of the product high enough to inactivate enzymes (above 87.7°C). The second step is adiabatic holding in which pieces leaving the heating section achieve a uniform temperature and sufficient time is provided to inactivate enzymes and to yield the desired texture.

In addition to water and steam blanching, use of microwave energy was demonstrated to be a convenient and effective method of blanching (9) and superior in retention of ascorbic acid. The texture of rehydrated, microwave-blanched freeze-dried spinach was firm, chewy, and highly acceptable.

The prevalence of water blanchers in the industry necessitates the comparison of different types of blanching for their energy utilization. On the basis of a theoretical requirement of 134 kg of steam per 10^3 kg of raw vegetables, energy efficiency of a steam blancher was estimated at 5%, a hydrostatic steam blancher at 27%, an IQB unit at 85%, and a water blancher at 60% (10).

5. DRYING TECHNIQUES AND EQUIPMENT

5.1. Dehydration

Dehydration involves the application of heat to vaporize moisture and some means of removing water vapor after its separation from the fruit/vegetable tissue. Hence it is a combined/simultaneous heat and mass transfer operation for which energy must be supplied.

Several types of dryers and drying methods, each better suited for a particular situation, are commercially used to remove moisture from a wide variety of fruits and vegetables (11). While sun drying of fruit crops is still practiced for certain fruits such as prunes, grapes, and dates, atmospheric dehydration processes are used for apples, prunes, and several vegetables. Continuous processes, such as tunnel, belt-trough, and fluidized bed, are mainly used for vegetables. Spray drying is suitable for fruit juice concentrates and vacuum dehydration processes are useful for low-moisture, high-sugar fruits.

Factors on which the selection of a particular dryer/drying method depends include form of raw material and its properties, desired physical form and characteristics of the product, necessary operating conditions, and operation costs.

Three basic types of drying processes may be recognized as applied to fruits and vegetables: sun drying and solar drying; atmospheric drying including batch (kiln, tower, and cabinet dryers) and continuous (tunnel, belt, belt-trough, fluidized bed, explosion puff, foam-mat, spray, drum, and microwave heated) processes; and subatmospheric dehydration (vacuum shelf belt/drum and freeze dryers). Recently the scope has been expanded to include use of low-temperature/low-energy processes like osmotic dehydration.

In the following sections only a few types of dryers and drying techniques of importance to fruit and vegetable drying are briefly discussed. Detailed information on their design, operation, and economics may be obtained from references quoted in the relevant sections.

5.2. Solar Drying

One of the oldest uses of solar energy since the dawn of civilization has been the drying and preservation of agricultural surpluses. It was also the cheapest means of preservation by which water activity was brought to a low level so that spoilage would not take place. It has been used mainly for drying of fruits such as grapes, prunes, dates, and figs.

There is no accurate estimate of the vast amount of material dried using this traditional technique. Since the method was simple and was originated and utilized in most of the developing countries, its acceptance created no problem. But there were many technical problems associated with the traditional way of drying in the direct sun. These problems include rain and/or cloudiness; contamination from dust and by insects, birds, and animals; lack of control over drying conditions; and possibility of chemical, enzymic, and microbiological spoilage due to long drying times. The recent increase in the cost of fossil fuels associated with depletion of the reserve and scarcity has led to renewed interest in solar drying.

Methods of solar drying having varying levels of technological sophistication have been surveyed and the thermodynamics of conventional dehydration compared with those of solar drying by Szulmayer (12). Bolin and Salunkhe (13) have exhaustively reviewed the drying methods using solar energy alone and with an auxiliary energy source, besides discussing the quality (nutrient) retention and economic aspects. They suggested that to produce high quality products with economic feasibility, the drying should be fast. Drying time can be shortened by two main procedures: by raising the product temperature so that moisture can be readily vaporized, while at the same time the humid air is constantly being removed, and by treating the product to be dried so that moisture barriers such as dense hydrophobic skin layers or long water migration paths will be minimized. Developments in solar drying of fruits and vegetables up to 1990 have been reviewed by Jayaraman and Das Gupta (14).

To design a solar dryer for drying fruits and vegetables, two important stages are to be considered: to heat the air by the radiant energy from sun and to bring this heated air in contact with the material inside a chamber to evaporate moisture.

Solar dryers are generally classified (15) according to their heating modes or the manner in which the heat derived from solar radiation is utilized. These classes include sun or natural dryers, direct solar dryers, indirect solar dryers, hybrid systems, and mixed systems.

Sun or Natural Dryers

Solar or natural dryers make use of the action of solar radiation, ambient air temperature, and relative humidity and wind speed to achieve the drying process.

Solar Dryers—Direct

In direct solar dryers the material to be dried is placed in an enclosure with a transparent cover or side panels. Heat is generated by absorption of solar radiation on the product itself as well as on the internal surfaces of the drying chamber. This heat evaporates the moisture from the drying product. In addition it serves to heat and expand the air, causing the removal of the moisture by the circulation of air.

Solar Dryers—Indirect

In indirect solar dryers, solar radiation is not directly incident on the material to be dried. Air is heated in a solar collector and then ducted to the drying chamber to dehydrate the product. Generally flat-plate solar collectors are used for heating the air for low and moderate temperature use. Efficiency of these collectors depends on the design and operating conditions. The main factors that affect collector efficiency are heater configuration, airflow rate, spectral properties of the absorber, air barriers, heat transfer coefficient between absorber and air, insulation, and insolation. By optimizing these factors a high efficiency can be obtained. More sophisticated designs of flat-plate collectors are now available. Imre described such collectors and their efficiency (16).

Hybrid Systems

Hybrid systems are dryers in which another form of energy, such as fuel or electricity, is used to supplement solar energy for heating and/or ventilation.

Mixed Systems

Mixed systems include dryers in which both direct and indirect modes of heating have been utilized (Fig. 1). Several experimental methods were evaluated for the solar dehydration of fruits (apricots): (a) wooden trays; (b) solar troughs of various materials designed to reflect radiant energy onto drying trays; (c) natural convection, solar-heated cabinet dryers with slanted plate heat collectors; (d) dryers incorporating inflated polyethylene (PE) tubes as solar collectors; and (e) PE semicylinders either incorporating a fan blower to be used in inflated hemispheres or incorporating a similar dome used as a solar collector, the air from which is blown over fruit in a cabinet dryer (17). Method (d) was found to be cheap, 38% faster than sun drying, and could be used as a supplementary heat source for conventional dehydrators.

Solar drying incorporating a desiccant bed for heat storage has been used to dry

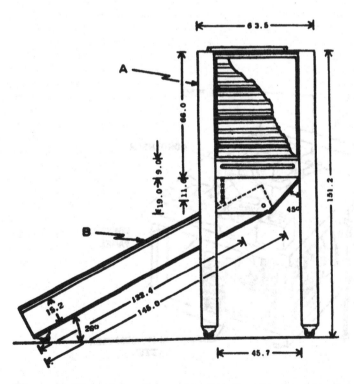

Figure 1 Dimensions of a combined-mode solar dryer (A, dryer; B, solar collector). (From Reference 13.)

fruits and vegetables (18). Hot air up to 27°C above ambient was obtained in a single glass-covered collector with an airflow of about 140 kg/h and raised to 52°C for airflow of 25 kg/h. In the absorbent circuit which used a double glass-covered collector, temperature differences were 10% higher. Other forms of heat storage involving use of natural materials such as water, pebbles or rocks, and the like, and salt solutions or absorbents have also been used.

Design and construction of a dryer was described (19) to utilize solar energy in the two-step osmovac dehydration of papaya consisting of a 56-by-25-by-25-cm plexiglass (3.8 cm thick) and a portable condenser vacuum unit (Fig. 2). Solar osmotic drying had higher drying rates and sucrose uptake than in the nonsolar runs. Similarly, drying rates from solar vacuum drying were about twice those of nonsolar vacuum drying.

For commercial success a solar dryer should be economically feasible. But, in general, solar energy systems are capital intensive. In these dryers, although operating costs are low, large investments have to be made on equipment. The prime economic problem is to balance annual cost of extra investment against fuel savings. Therefore solar drying could be economical only if the equipment cost is decreased or in the event of fuel cost escalation.

5.3. Hot Air Drying

Presently most of the dehydrated fruits and vegetables are produced by the technique of hot air drying, which is the simplest and most economical among the various methods. Different types of dryers have been designed, made, and commercially used based on this technique.

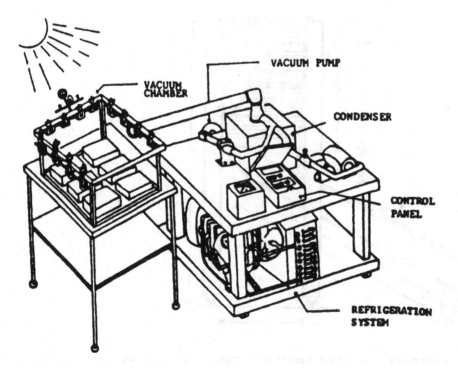

Figure 2   Components built for solar vacuum drying. (From Reference 19.)

In this method, heated air is brought into contact with the wet material to be dried to facilitate heat and mass transfer; convection is mainly involved. Two important aspects of mass transfer are the transfer of water to the surface of the material being dried and the removal of water vapor from the surface. The basic concepts, various methods of drying, and different types of hot air dryers are discussed by various authors in review articles and books (1,2,6,20–22).

To achieve dehydrated products of high quality at a reasonable cost, dehydration must occur fairly rapidly. Four main factors affect the rate and total time of drying (21): physical properties of the foodstuff, especially particle size and geometry; its geometrical arrangement in relation to air (crossflow, through-flow, tray load, etc.); physical properties of air (temperature, humidity, velocity); and design characteristics of the drying equipment (crossflow, through-flow, cocurrent, countercurrent, agitated bed, pneumatic, etc.). Choice of the drying method for a food product is determined by desired quality attributes, raw material, and economy.

The dryers generally used for the drying of piece-form fruits and vegetables are cabinet, kiln, tunnel, belt-trough, bin, pneumatic, and conveyor dryers. Among these, the cabinet, kiln, and bin dryers are batch operated, the belt-trough dryer is continuous, and the tunnel dryer is semicontinuous.

Cabinet Dryers

Cabinet dryers are small-scale dryers used in the laboratory and pilot plants for the experimental drying of fruits and vegetables. They consist of an insulated chamber with trays located one above the other on which the material is loaded and a fan that forces air through heaters and then through the material by crossflow or through flow.

Tunnel Dryers

Tunnel dryers are basically a group of truck and tray dryers widely used due to their flexibility for the large-scale commercial drying of various types of fruits and vegetables. In these dryers trays of wet material, stacked on trolleys, are introduced at one end of a tunnel (a long cabinet) and when dry are discharged from the other end. The drying characteristic of these dryers depends on the movement of airflow relative to the movement of trucks, which may move parallel to each other either concurrently or countercurrently, each resulting in its own drying pattern and product properties.

Belt-Trough Dryers

Belt-trough dryers are agitated bed, through-flow dryers used for the drying of cut vegetables of small dimensions. They consist of metal (wire) mesh belts supported on two horizontal rolls; a blast of hot air is forced through the bed of material on the mesh. The belts are arranged in such a way to form an inclined trough so that the product travels in a spiral path and partial fluidization is caused by an upward blast of air.

Pneumatic Conveyor Dryers

Pneumatic conveyor dryers are generally used for the finish drying of powders or granulated materials and are extensively used in the making of potato granules. The feed material is introduced into a fast-moving stream of heated air and conveyed through ducting (horizontal or vertical) of sufficient length to bring about desired drying. The dried product is separated from the exhaust air by a cyclone and/or filter. Jayaraman et al. described a pneumatic dryer in which an initial high temperature (160–180°C for 8 min) drying of piece-form vegetables was done up to 50% moisture, resulting in expansion and porosity in the products that hastened finish drying in a conventional cabinet dryer besides significantly reducing rehydration times and increasing rehydration coefficients of the products (Table 3) (23).

5.4. Fluidized Bed Drying

The fluidized bed (FB) type of dryer was originally used for the finish drying of potato granules. In FB drying, hot air is forced through a bed of food particles at a sufficiently high velocity to overcome the gravitational forces on the product and maintain the particles in a suspended (fluidized) state (20). Fluidizing is a very effective way of maximizing the surface area of drying within a small total space. Air velocities required for this will vary with the product and more specifically with the particle size and density. A major limitation is the limited range of particle size (diameter usually 20 μm–10 mm) that can be effectively fluidized. The bed remains uniform and behaves as a fluid when the so-called Froude number is below unity.

The theory and food applications of fluidized bed drying have been discussed in many textbooks and articles (6,20–22,24,25). Apart from the commercial drying of peas, beans, and diced vegetables, it is also used for drying potato granules, onion flakes, and fruit juice powders. It is often used as a secondary dryer to finish the drying process initiated in other types of dryers. It can be carried out as a batch or continuous process with a variety of modifications.

Advantages of fluidized bed drying are high drying intensity; uniform and closely controllable temperature throughout; high thermal efficiency; time duration of the material in the dryer may be chosen arbitrarily; elapsed drying time is usually less than other types of dryers; equipment operation and maintenance is relatively simple; the process can be automated without difficulty; and, compact and small, several processes can be combined in an FB dryer (6).

Heat transfer in FB drying could be improved by increasing gas velocity. But, at higher velocities, the particles are transported out of bed and voidage in the bed increases, reducing the volumetric effectiveness of the equipment. From the viewpoint of good gas-to-solid contact, this is undesirable because most of the gas passes around the layers of particles without effective contact.

Another drawback of conventional fluidized bed drying is that the maximum gas

Table 3 Process Conditions for High-Temperature, Short-Time Pneumatic Drying of Vegetables and Rehydration Characteristics of Products

| Material | Moisture content (%) | | | | Optimum HTST drying | | Rehydration time | Rehydration Coefficient |
	Raw	Cooked/ blanched	HTST treated	Final dried	Temp (°C)	Time (min)		
Potatoes	82.2	83.3	59.3	4.1	170	8	5	0.94
Green peas	71.1	72.5	38.3	3.4	160	8	5	1.06
Carrots	89.3	91.0	52.9	4.2	170	8	5	0.50
Yams	76.6	78.3	50.2	3.9	180	8	6	1.01
Sweet potatoes	73.6	78.6	53.8	5.3	170	8	2	1.06
Colocasia	80.2	83.3	54.2	4.9	170	8	2	0.98
Plantains, raw	80.8	83.3	58.8	4.6	170	8	4	0.97

Source: From Reference 23

velocity is closely related to the physical characteristics of the food particles such as shape, surface roughness, bulk density, and firmness. The maximum gas velocity controls the amount of heat delivered to the bed, since for foods there is usually a critical maximum gas temperature for processing.

The centrifugal fluidized bed (CFB) was designed (26,27) to overcome the limitations of piece size and heat requirements encountered in a conventional FB dryer by subjecting the food particles during fluidization to a centrifugal force greater than the gravitational force. This had the effect of increasing the apparent density of the particles and allowing smooth, homogenous fluidization. Smooth fluidization could be achieved at any desired gas velocity by varying the centrifugal force. The other advantages provided by CFB include increasing the gas velocity can provide improved heat transfer at moderate gas temperature without the problem of heat damage, and large pressure drops across the grid supporting the bed are not needed to obtain smooth fluidization. It was demonstrated to be effective for extremely high rate drying of high-moisture, low-density, sticky, piece-form foods.

A modified centrifugal fluidized bed dryer (CFBD) developed consisted of a cylinder with perforated walls, rotating horizontally about its axis in a high velocity, heated crossflow airstream (Fig. 3) (27). Piece-form product to be dried was fed into one end of the rotating cylinder, moved along the cylinder in almost plug-flow manner through the hot air blast, passing crossflow through the perforated walls, and discharged from the other end of the cylinder. On the downstream side (relative to the airflow) within the cylinder, the pieces were held as a fixed bed against the wall by the additive forces of frictional air drag and centrifugation. At high rpm and/or low air velocity, the centrifugal

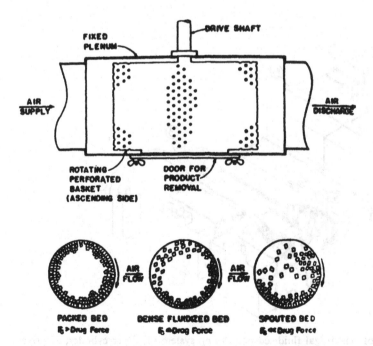

Figure 3 Modified design of centrifugal fluidized bed dryer allows for lower pressure drops and better heat economy. As the air velocity is increased, the degree of fluidization changes from packed to spouted. (From Reference 27.)

force on any particle was greater than the drag force of the entering airstream and each particle remained fixed in place. If the air velocity was increased or the rpm decreased, dense phase fluidization was obtained on the upstream side of the bed because the drag force on the pieces was equal to or slightly greater than the opposing centrifugal force. If the air velocity was further increased, transport of the particles across the cylinder occurred as in a spouted bed. Centrifugal force obtained through cylinder rotational speed to give 3 to 15 Gs allowed the use of air velocities up to 15 m/s or higher, many times greater than can be employed in conventional FBs.

Carrots, potatoes, apples, and green beans dried in this modified CFB at an air velocity of 2400 ft/min and 240°F showed that a weight reduction of 50% could be achieved in less than 6 min for all items. In comparison with a tunnel dryer with a crossflow air velocity of 780 ft/min, 160°F temperature, and 2-lb/ft^2 tray loading, it was shown that average drying rate in a modified CFB (air velocity 2400 ft/min) was 5.3 times the crossflow value. This increase in drying rate (3 times the theoretical value) was due to high efficiency of the air-to-particle contact achieved in the CFB.

A continuous CFBD was further designed (Fig. 4) with a dryer surface of approximately 21 ft^2 in the form of a rotating perforated stainless steel cylinder (10-in diameter and 100-in long) with an open area of 45% and teflon-coated inside (28). The cylinder could be rotated at speeds up to 350 rpm ($F_e = 17.4 \times G$) through a belt drive and tilted between 0° to 6° from the horizontal to help control the residence time of material being dried. Centrifugal fans with steam heaters enabled air temperatures up to 140°C.

Table 4 gives the performance data from trials for drying bell peppers, beets, carrots, cabbages, onions, and mushrooms using a CFBD (28,29). Good continuous operation

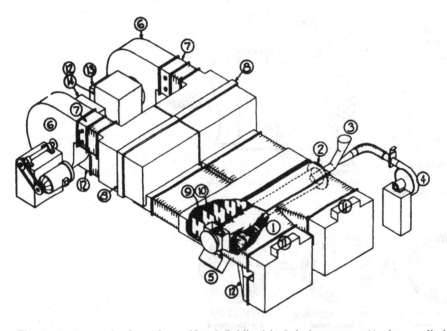

Figure 4 Isometric view of centrifugal fluidized bed drying system (1, dryer cylinder; 2, drive pulley; 3, aspiration feeder; 4, feeder blower; 5, discharge chute; 6, air blower; 7, air discharge damper; 8, steam coil heater; 9, plenum; 10, air vent; 11, vent port; 12, recirculating duct; 13, make-up air; 14, blower intake). (From Reference 28.)

Table 4 Operating Conditions for Drying Some Vegetables in Continuous Centrifugal Fluidized Bed Dryer

Commodity	Feed rate (kg/h)	Discharge rate (kg/h)	Moisture (%)		Weight reduction (%)	Temp (°C)	Air velocity (m/s)
			Feed	Discharge			
Bell pepper, diced	142	71	93.4	86.1	53	71	15.3
Beet, diced	133	74	84.6	74.5	40	99	15.3
Carrot							
Flaked	109	79	88.9	84.6	28	93	15.3
Diced	130–150	–	89.5	82.0	46	100–140	15.3
Cabbage, shredded	90–200	–	93.3	88.0	44	100–140	15.3
Onion, sliced	150–160	–	87.7	82.5	35	100–140	15.3
Mushroom, diced	230	–	95.3	91.3	48	100–140	15.3

Source: From References 28 and 29

was achieved for a 1-h period. Feed rates and evaporation (kg/h) are given for a range of dryer sizes in Table 5 for cabbages, carrots, onions, and mushrooms (29).

While the CFBD can take the product to any degree of dryness, it is considered best suited for the rapid removal of moisture (30–50% weight reduction in about 5-min exposure) during the early stages of drying of piece-form vegetables as a predryer, to be followed by a conventional tray or band dryer for later stages of evaporation in which the rate of moisture removal is governed by diffusion and high velocity is no longer advantageous. Incorporated upstream of the existing dehydration line, it increases overall output with a saving in floor space.

By using a whirling fluidized bed containing inert particles like glass beads, it was found feasible to dry coarse-size and sticky materials like diced potatoes and carrots (30). A novel type of FB dryer, known as a toroidal fluidized bed, reported to be manufactured in the United Kingdom (22), could be used for a number of processes such as cooking, expanding, roasting, and drying. A high-velocity stream of heated air entering the base of the process chamber through blades or louvers that imparted a rotary motion to the air created a compact, rotating bed of particles that varied in depth from a few mm to in excess of 50 mm. High rates of heat and mass transfer could be attained, resulting in rapid drying. This dryer could be utilized for a wide range of particle sizes and shapes of materials like peas, beans, diced potatoes, and carrots and operated on a continuous or batch basis.

5.5. Explosion Puffing

The technique of explosion puffing was initially developed to fulfill the objective of dehydrating relatively large pieces of fruits and vegetables that would reconstitute rapidly; the system would be operable at a cost comparable to conventional hot air drying. The method, adequately described and extensively reviewed in several articles (21,31), consisted of initially partially dehydrating the fruit and vegetable pieces, then imparting a porous structure by explosion puffing, and subsequently drying to a low moisture content. Initial drying was required to reduce the moisture content to a level so that disintegration did not occur during explosion puffing. Since uniformity was essential for optimum results, an equilibration step was desirable after the partial drying. As an opera-

Table 5   Feed Rates and Evaporation (kg/h) for a Range of Continuous Centrifugal Fluidized Bed Dryer Sizes

Dryer size (m)	Cabbages		Carrots		Onions		Mushrooms	
	Feed	Evaporation	Feed	Evaporation	Feed	Evaporation	Feed	Evaporation
0.305 dia × 2.13	133	58	130	54	156	44	231	106
0.50 dia × 5.0	658	285	643	266	771	215	1143	425
0.65 dia × 6.5	1263	546	1232	511	1478	412	2190	1003
0.80 dia × 8.0	2130	921	2077	861	2491	696	3693	1694
1.00 dia × 10.0	3719	1607	3628	1504	4352	1216	6451	2958

Source: From Reference 29

tional step integrated in hot air dehydration at moisture contents of 15–35%, explosion puffing created porosity in food pieces and speeded up hot air drying, modifying or eliminating diffusion controlled drying as the rate-controlling step. The case hardening problem was minimized so that processors could dry large pieces economically in shorter times, lessening browning potential. Also increased overall volume recovery on rehydration was reported compared with hot air drying. Batch models with output of 180 kg/h of 1-cm diced potatoes or carrots were designed and tested.

The gun used in batch model explosion puffing was essentially a rotating cylindrical pressure chamber that was fitted with a quick-release lid, and was heated externally. The rotational speed of the gun was fixed to give an optimal tumbling action of the charge. This speed (33 rpm) was about 40% of the critical speed, that is, the speed at which the centrifugal force and gravitational force are equal and no tumbling takes place. In the gun, the pieces exposed to 10–70 psig steam so that they were quickly heated and their remaining water was superheated relative to atmospheric pressure. When the pieces were suddenly discharged to the atmosphere, the rapid pressure drop caused some of the water within the pieces to flash into steam. The escaping steam caused channels and fissures, thus imparting a porous structure to the pieces. Commodities that were successfully dehydrated by this method include potatoes, carrots, beets, cabbages, sweet potatoes, apples, and blueberries.

A continuous explosive puffing system (CEPS) with 680-kg/h capacity was designed by separating the heating and puffing functions and successfully tested (32). The three subassemblies that were unique to the system were the feed chamber, the heating chamber, and the discharge chamber (Fig. 5). The use of CEPS resulted in better process control, improved product quality, and reduced labor costs. Once the system feed rate, feed moisture content, internal pressure, internal temperature, and discharge rate reached

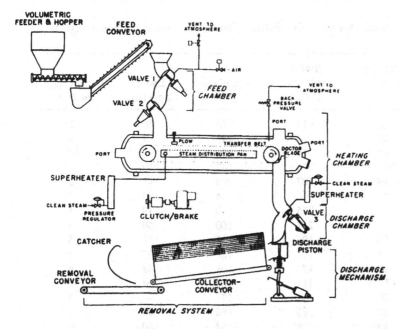

Figure 5 Schematic diagram identifying major components of continuous explosion puffing systems. (From Reference 32.)

steady state, it operated with minimal care and needed only occasional operational adjustment (31).

Energy evaluation based on steam consumption showed a 44% reduction in steam consumption when a CEPS was used to dehydrate apple pieces as compared with conventional dehydration; this is attributed to the time saved for drying from 20% to less than 3% moisture. Process cost for EPS is reported to be similar to the cost of conventional hot air drying. Table 6 gives processing conditions (batch vs. continuous) for a number of fruits and vegetables (33).

5.6. Foam Drying

The foam drying process is limited to specific products, such as fruit powders, for preparation of instant drinks. Techniques like vacuum puff drying, foam mat drying, microflake dehydration, and foam spray drying have been described and reviewed earlier (6,21). Among these, the foam mat drying process has received considerable attention.

Foam mat drying, originally developed by Morgan, involves drying thin layers of stabilized foam from liquid food concentrates in heated air at atmospheric pressure. Foam is prepared by the addition of a stabilizer and a gas to the liquid food in a continuous mixer. It can be dried in a continuous belt tray dryer. Good quality powders capable of instant rehydration were made experimentally from tomatoes, oranges, grapes, apples, and pineapples (20).

Foam formation is the primary requirement of this process. Two characteristics required for foam stability are consistency and film-forming ability. Film-forming components used in the drying of fruits and vegetables are glyceryl monostearate, solubilized soya protein, and propylene glycol monostearate. Drying time and temperatures depend on the product being dried; most fruit juices required about 15 min at 160°F to dry to

Table 6 Process Conditions for Explosive Puffing (Batch/Continuous) of Some Vegetables and Fruits

Commodity	Puffing moisture (%)	Steam pressure (kPa)	Temperature (°C)	Dwell time (s)	Rehydration time (min)
Potatoes	25	414	176	60	5
Carrots	25	275	149	49	5
Yams	25	241	160	75	10
Beets	20–26	276	163	120	5
Peppers	19	207	149	45	2
Onions	15	414	154	30	5
Celery	25	275	149	39	5
Rutabagas	25	241	160	60	6
Mushrooms	20	193	121	39	5
Apples	15	117	121	35	5
Blueberries	18	138	204	39	4
Cranberries	17–26	138	163	64	3
Strawberries	25	90	177	—	3
Pineapples	18	83	166	60	1
Pears	18	228	154	60	5

Source: From Reference 33

about 2% moisture. Air velocity and humidity had no appreciable effect on the time required.

Foam mat drying has two definite advantages (34). First, the use of foam greatly speeds up moisture removal and permits drying at atmospheric conditions in a stream of hot air in a short time. Second, though the product may be sticky at drying temperatures, it can be transferred to a cooling zone and crisped before it is scraped off the surface.

5.7. Microwave Drying

In microwave drying, heat is generated inside the food materials by the interaction of chemical constituents of foods and radiofrequency energy (915 MHz and 2450 MHz). Use of this type of energy found its application in the finish drying of potato chips. Much of the work on the drying of fruits and vegetables utilizing microwave energy was described by Decareau (35).

Advantages of using microwave energy are penetrating quality, which effects a uniform heating of materials upon which radiation impinges; selective absorption of the radiation by liquid water; and capacity for easy control so that heating may be rapid if desired. It can reduce drying time, particularly when the size of the piece is such that a conventional drying method is not feasible. However, the high cost per unit of energy compared with the conventional energy and the high initial cost of equipment limits its use for drying.

Microwave vacuum drying of concentrates of oranges, lemons, grapefruits, pineapples, strawberries, and others has been described (35). One full-scale plant was in operation for the vacuum drying of orange and grapefruit juices utilizing a 48-kW, 2450-MHz unit that dried 63° Brix orange juice concentrate to 2% moisture in 40 min.

5.8. Spray Drying

The spray drying method is most important for drying liquid food products and has received much experimental study. *Spray drying* by definition is the transformation of a feed from a fluid state into a dried form by spraying into a hot, dry medium (36). In general it involves atomization of the liquid into a spray (by a nozzle) and contact between the spray and the drying medium (hot air), followed by separation of dried powder from the drying medium (by a cyclone separator). Applicable to a wide range of products, there is no single, standardized design for the spray dryer common to all. Each product is treated individually and the dryer is designed to suit the product specifications. Principles and applications of this technique are well described in the literature (20–22,24,36).

Applications of spray drying to fruit and vegetable products are very limited. Fruit juices, pulps, and pastes can be spray dried with additives. Special care must be taken to design the drying chamber as well because during postdrying, handling, and packing operations the products are both hygroscopic and thermoplastic. Fruits that have been spray dried include tomatoes, bananas, and, to a limited extent, citrus fruit, peaches, and apricots.

Tomato pulp is a typical example of a product that is very difficult to dry as the powder is sticky and poses a caking problem. A spray drying plant capable of producing a free-flowing product that on reconstitution compares favorably with tomato paste has been designed featuring a co-current drying chamber having a jacketed wall for air cooling and a conical base. Cooling air intake is controlled to enable close maintenance of wall temperature in the range of 38–50°C. The paste is sprayed into the drying air entering the chamber at a temperature of 138–150°C.

A wide range of vegetables can be spray dried following homogenization and the powders can be readily used in dry soup mixes. As yet, there is limited interest in spray drying of vegetables though the drying process is not different from fruits and standard equipment can be used. Jayaraman and Das Gupta spray dried a number of fruit juices in admixture with whole milk or yogurt (37).

5.9. Drum Drying

Drum drying is an important and inexpensive drying technique suitable for a wide range of products—liquid, slurry, and puree. The material to be dried is applied as a thin layer to the outer surface of a slowly revolving hollow drum (made of iron or stainless steel) heated internally by steam (24). The principle and types of drum dryers have been discussed by a number of authors (20,22,24,25).

The success of drum drying depends on the application of a uniform film of maximum thickness. The high rate of heat transfer is obtained by direct contact with the hot surface and the equipment may be used under atmospheric or vacuum condition (21). It is mainly used for the manufacture of potato flakes. Its usefulness for dehydration of fruits (particularly fruits high in sugar and low in fiber content) is limited by the high temperature required. Thin sheets of very dry fruit are usually so hygroscopic that it has been necessary to overdry under severe heating conditions to compensate for the later pick up. The fruit was therefore usually heat damaged.

Modification of the drum dryer and operating technique effected to overcome the problem encountered in conventional drum drying of fruits like tomato flakes involved the use of a partial shroud and low humidity collection zones (Fig. 6) (38). The shroud,

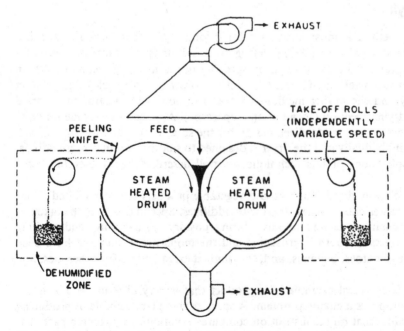

Figure 6 Modified drum dryer showing the two primary modifications: the partial shroud and the low humidity collection zones. (From Reference 38.)

concentric with the bottom of the drums, has a 1-in clearance through which ambient air flows at 1000 fpm in a direction countercurrent to the drum rotation. The product collection zones are maintained at about 15–20% RH. The dryer is operated best at 3 to 3-1/3 rpm at 280°F to produce a product with 5–7% moisture with a commonly used drum separation gap of 0.008–0.01 in when hot to be subsequently finish dried in bins at a lower temperature to 2.5–3.0% moisture content.

A pilot-plant, double-drum dryer modified to produce low-moisture flakes from a wide range of fruit purees has been described (39). Products with a relatively high fiber content such as apples, guavas, apricots, bananas, papayas, and cranberries could be dried successfully without additives. Purees with a low fiber content such as raspberries, strawberries, and blueberries required the addition of fiber (low methoxyl pectin, up to 1%) to aid in the sheet formation at the doctor blade. The modification consisted of incorporation of variable-speed take-off rolls, cool airflow directed at the doctor blade area, and a ventilation system to remove saturated air from the area beneath the drums.

A process for manufacture of instant, drum-dried flakes from tropical sweet potato puree was evaluated using a Buflovac laboratory model atmospheric double drum dryer internally steam heated at 35 psig (40). The drums revolved at 1.73 rpm with a clearance between drums of 0.305 mm. It was found that pretreatment with alpha-amylase improved the drying characteristics of the puree.

A mathematical model was predicted for drum drying of mashed potatoes on the basis of primary process parameters such as drum speed, steam pressure, number of spreader rolls, wet and dry bulb temperatures, mash moistures, and drum dimension (41).

5.10. Freeze Drying

Freeze drying, which involves a two-stage process of first freezing of water of the food materials followed by application of heat to the product so that ice can be directly sublimed to vapor, is already a commercially established process. Sublimation from ice to water vapor can only be accomplished below the triple point of water, that is, at 4.58 torr at a temperature of approximately 32°F. Since the moisture removal does not pass through a liquid phase, the structure of the product remains in a more acceptable state. In addition, drying takes place without exposing the product to excessively high temperatures.

The advantages of freeze drying are shrinkage is minimized; movement of soluble solids within the food material is minimized; the porous structure of the product facilitates rapid rehydration; and retention of volatile flavor compounds is high. It has therefore proved to be the superior method of dehydration for many fruits. The major limitation to its commercial application is its very high capital and processing costs and the need for special packaging to avoid oxidation and moisture pick up. Industrial application includes some exotic fruits and vegetables, soup ingredients, mushrooms, and orange juice. Much of the recent work is directed toward freeze-dried fruit juices and vegetables like spinach and carrots.

Essential components of a freeze dryer include the vacuum chamber, condenser, and vacuum pump. As in other forms of drying, freeze drying represents coupled heat and mass transfer. For the analysis of this operation, Karel considered three cases that represent three basic types of possibilities in vacuum freeze drying: (a) heat transfer and mass transfer pass through the same path (dry layer) but in opposite directions; (b) heat transfer occurs through the frozen layer and mass transfer through the dry layer; and (c) heat generation occurs within ice (by microwaves) and mass transfer through the dry layer (24). Principles and applications of freeze drying are described in detail in many books and articles (20–22,25,42–44).

Another aspect that determines the structure of food materials, particularly fruit juices, during freeze drying is the phenomenon of collapse. Freezing of food materials causes aqueous solution to be separated into two phases: ice crystals and concentrated aqueous solution. The properties of this concentrated aqueous solution depend on composition, concentration, and temperature. If during drying the temperature is very low, the mobility in the extremely viscous concentrated phase is so low that no structural changes occur during drying. But, if the temperature is above a critical level (known as the *collapse temperature*), mobility of the concentrated solution phase may be so high that flow and loss of original structure occurs. This is known as the phenomenon of collapse and was investigated in detail by several workers.

Initially, commercial freeze dryers were batch operated but later development aimed toward more economical and labor-saving continuous processes. There are two basic types of continuous dryers (24): (a) tray dryers in which the product is stationary on a tray and the trays are continuously moved through the dryer and (b) dynamic or trayless dryers in which the particles of the product are moved through dryers (i.e., belt freeze dryers, circular plate dryers, vibratory dryers, fluidized bed dryers, and spray freeze dryers).

Atmospheric freeze drying is one possible technique for reducing the cost of the process that has attracted widespread attention in recent times. The advantages of atmospheric freeze drying have been listed as (a) it is true freeze drying – the product is held below its freezing temperature for the entire process; (b) it requires simple equipment (conventional fluidized bed can be used); drying capacity can be increased as with conventional drying equipment; (c) volatile flavors are better retained in the absence of vacuum and elevated temperatures; and (d) as with vacuum freeze drying, inert gas can be used in the process, thus removing air from contact with food (45).

For studying the feasibility of atmospheric freeze drying of a number of foods, including a few vegetables and fruits, an equipment system consisting of two desiccator chambers, a heat reservoir, two blowers, a refrigeration unit, a heater, conveyor system, and suitable ducting was used (46). There were two major phases in the operation of a carrier-sublimation process: drying and desiccator regeneration (Fig. 7). Only foods of low sugar content could be freeze dried by this method. Temperature was the most important operating variable, as found by the time required to remove 90% water from french cut beans, which ranged from 3.3 h at a mean gas temperature of 22.5°F to 6.9 h at 14.7°F.

From an analytical model (42), atmospheric freeze drying was shown to be mass transfer controlled and to have a slow drying rate when compared with conventional freeze drying. However, the increase in drying time experimentally observed was in part due to the decrease in internal resistance to heat and mass transfer because of product shrinkage and the presence of liquid vapor transport (47). The process could be designed into a continuous system and should produce a higher quality product by eliminating the harmful influence of excessive temperature in the semidry portion of the food product (48).

Atmospheric freeze drying of several foods, including mushrooms and carrots, was investigated in a fluidized bed of finely divided adsorbent that combined adsorption and fluidization, achieving improved heat and mass transfer and shorter drying time than vacuum drying (49). Products could be dried economically using very simple equipment.

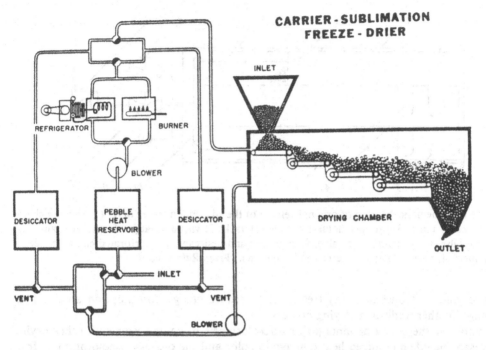

Figure 7 Diagram indicating how nonvacuum freeze drying would be done commercially. Gas circulates through a progressively deeper product bed to increase contact time as drying progresses; desiccant takes water from air and then is regenerated. (From Reference 46.)

Bell and Mellor developed an adsorption freeze-drying process that depended upon the removal of water vapor by a desiccant rather than by refrigeration coils (50). The desiccant created a high and well-sustained vapor pressure driving force, particularly at moderately low temperature, since the equilibrium water vapor pressure of the desiccant decreased as the temperature was lowered. The water vapor in adsorption freeze drying did not condense as ice, but was adsorbed into the porous structure of the adsorbent. The process consisted of a chamber in which the air pressure was reduced, a product rack to hold the samples, and a perforated container of desiccant that required regeneration (Fig. 8). Defrosting, drying the chamber, and vacuum pretesting were not required because the inside of the chamber remained dry.

5.11. Osmotic Dehydration

Osmotic dehydration is a water removal process that consists of placing foods, such as pieces of fruits or vegetables, in a hypertonic solution. Since this solution has higher osmotic pressure and hence lower water activity, a driving force for water removal arises between solution and food, while the natural cell wall acts as a semipermeable membrane. As the membrane is only partially selective, there is always some diffusion of solute from the solution into the food and vice versa. Direct osmotic dehydration is therefore a simultaneous water and solute diffusion process (51). Up to a 50% reduction in the fresh weight of the food can be achieved by osmosis. Its application to fruits and, to a lesser extent, to vegetables has received considerable attention in recent years as a technique for production of intermediate moisture foods (IMF) and shelf-stable products (SSP) or as

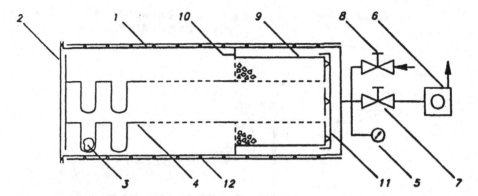

Figure 8 Schematic diagram of original version of the adsorption freeze dryer (1, vacuum chamber; 2, cover plate; 3, product holder; 4, product rack; 5, vacuum gauge; 6, vacuum pump; 7, shutoff valve; 8, air admittance valve; 9, container and desiccant; 10, perforated end of desiccant container; 11, baffle plate; 12, heater and insulation). (From Reference 50.)

a predrying (preconcentration) treatment to reduce energy consumption and/or heat damage in other traditional drying processes.

Some of the stated advantages of direct osmosis in comparison with other drying processes include minimized heat damage to color and flavor, less discoloration of fruit by enzymatic oxidative browning, better retention of flavor compounds, and less energy consumption since water can be removed without change of phase. However, products cannot be dried to completion solely by this method and some means of stabilizing them is required to extend their shelf lives.

Many workers have studied the different aspects of osmotic dehydration: the solutes to be employed, the influence of process variables on drying behavior, the opportunity to combine osmosis with other stabilizing techniques, and the quality of the final products. The osmotic agents used must be nontoxic and have a good taste and high solubility besides low a_w. Sugar in different concentrations is widely used. Common salt is an excellent osmotic agent for vegetables.

The quantity and the rate of water removal depend on several variables and processing parameters. In general it has been shown that the weight loss in osmosed fruit is increased by increasing solute concentration of the osmotic solution, immersion time, temperature, solution/food ratio, specific surface area of the food, and by using vacuum, stirring, and continuous reconcentration. Also, to obtain the same a_w reduction, time tended to decrease exponentially as the temperature increased.

An empirical equation suggested correlating weight loss of golden delicious apple (half rings) with time was given as (52)

$$\theta = [(90 - B)/100] \, (e^{F/25} - 1) e^{163/T - 32}$$

where

θ = time, hours
B = sucrose syrup concentration, °Brix
T = sucrose syrup temperature, °F
F = weight reduction, percent of original weight

Evaluation of the kinetics of osmotic drying of grapes and calculation of the drying constant using a simplified diffusion equation showed the activation energy of dehydration process to be about 5.2 kcal/mole, which was very close to that of the air drying process, confirming the similar nature of the rate-limiting phenomena (53). The drying rate in equimolar solution was different according to the different solutes used with a scale inversely proportional to the molecular weight. However, due to the infinite solubility of glycerol, higher concentration of this solute allowed achievement of the highest drying rates.

Several models were proposed to show the effect of concentration of osmotic solution and temperature on the rate of water loss and gain of osmotic agent. Thus, a model developed (54) for calculation of osmotic mass transport data for potato and water activity to equilibrium in sucrose solutions for the concentration range 10–70% and solution/solids range 1–10 showed that, at equlibrium, there was an equality of water activity and soluble solids concentration in the potato and in the osmosis solution. A linear relationship existed between normalized solids content (NSC) and $\log (1 - a_w)$ and was given by

$$\text{NSC} = 6.1056 + 2.4990 \log (1 - a_w).$$

Another model developed (55) for solute diffusion in osmotic dehydration of apple based on solids gain divided by water content M as a function of rate constant K, time (t), and a constant A was given as $M = Kt + A$. A relationship was established in the form of $K = T^{1.40} C^{1.13}$, where rate parameter K is related to temperature T at different sucrose concentrations C. The average activation energy of the process was 28.2 kJ mole^{-1}.

The effects of solution concentration, osmosis time, and the osmosis temperature were studied in the osmotic dehydration of pineapple in sucrose solution (56). The solute diffusion was analyzed by Magee's model. The effect of sucrose concentration C on rate parameter K was given by power law regression equation as $K = 4.15 \times 10^{-4} C^{1.51}$ at 20°C.

An empirical equation derived based on osmotic dehydration of apple slices could predict rate of osmosis F, that is, percentage of dehydration of any given fruit slices of specific size with time T, given the concentration of sugar (% B) and the temperature as follows (57):

$$F = 31.8 - 0.307B - (0.56 - 0.016B)t - 2.10^{-9.26/B} - 1(T - 0.3)^{0.54}$$
$$- 0.00425t$$

where F = decrease in mass %, was valid for $B = 60$–75%, $t = 40$–80°C, and $T = 0.5$–4.5 h.

Direct osmosis of different fruits at 70° Brix sugar at atmospheric and low pressure (about 70 mm Hg) revealed higher drying rates with the latter. Addition of a small amount of NaCl to different osmotic solutions increased the driving force of the drying process.

Apple cubes submitted to high-temperature, short-time (HTST) osmosis in sugar at 60–80°C for 1 to 20 min showed osmosis to be greatly accelerated by high temperature, since the water loss in apples after 1–3 min HTST osmosis was the same as that given by 2-h treatment at ambient temperature and HTST osmosis completely inactivated the enzymes.

Partial dehydration of fruits and vegetables by osmosis using various osmotic agents has been employed prior to drying by other conventional methods, namely, hot air convection drying, high-temperature fluidized bed drying, vacuum drying, freeze drying, and dehydro-freezing as a means of reducing processing time and limiting energy consumption besides improving sensory characteristics.

Osmotic dehydration has been utilized for developing intermediate moisture fruits stabilized solely by a_w control with added antimycotic preservative, as well as shelf-stable products with higher a_w stabilized by a combination preservation technique involving a_w and pH control plus heat pasteurization, due to simplicity of the operations involved, economy, and low energy inputs.

Intermediate moisture food (IMF) is characterized by a water activity low enough to prevent the growth of bacteria and by conditions minimizing the potential for growth of other microorganisms (24). Some traditional foods such as sun-dried fruits from apricots, prunes, dates, and figs belong to this category. Water activity range and moisture content vary generally between 0.7 to 0.9 and 20–40%, respectively, in such types of foods. Microorganisms that can grow in that range of water activity are prevented by addition of preservatives. One advantage of IMF is that it does not require refrigeration for storage and can be eaten as such without rehydration.

Modern IMF production techniques may be classified as follows (58): (a) moist infusion, in which normal moisture solid food pieces are soaked and/or cooked in an appropriate solution to result in a final product having the desired a_w level; (b) dry infusion, in which solid food pieces are first dehydrated following which they are infused by soaking in a solution containing the desired osmotic agents; and (c) blending, in which the components are weighed, blended, and cooked and extruded or otherwise combined to result in a finished product of the desired a_w.

Various product optimization processing steps are usually included to produce variance in these procedures. These include means for a_w adjustment in the finished product by either dehydration or evaporation, microbiological stabilization by heating or use of chemical additives, prevention of enzymatic deterioration by blanching and of physical/chemical deterioration by addition of antioxidants, chelators, emulsifiers or stabilizers, and nutritive adjustments by inclusion of appropriate nutrients.

Preparation and shelf stability of several intermediate moisture tropical fruits have been reported by Jayaraman (59) using moist infusion (desorption) with solutions containing glycerol and sugar or sugar alone together with antimycotics or by using sugar coupled with partial hot air drying. In the case of vegetables (e.g., carrots, cauliflower) infusion with a combination of salt, sugar, and glycerol coupled with partial drying to 50% moisture was used.

Development of shelf-stable fruits has been reported in an effort to satisfy some current needs such as packaging material different from the metal can (e.g., flexible films), low or no use of chemicals, keeping of natural fresh taste, and low energy consumption. The term *shelf-stable product* (SSP) was introduced by Leistner for high-moisture food products ($a_w > 0.90$) that are storable without refrigeration although receiving a mild heat treatment. The method consists basically of hindering or reducing the action of the various possible modifying factors (nonenzymatic browning, enzyme catalyzed reactions, and bacteria, yeast, and mold growth) with various limiting factors (a_w, pH, temperature, thermal treatments, pressure, and E_h) that act as a series of "hurdles" against the modifying agents until their cumulative effect is sufficient to achieve shelf stability.

Shelf-stable dehydrated peaches and apricots were obtained (60) by blanching and immersion in 70° Brix sugar syrup to achieve 22–26° Brix and an a_w of 0.95–0.96 followed by vacuum packing in flexible pouches and pasteurization at 75°C for 40 min. Similarly, a process for stabilizing pineapple slices for ambient storage was developed (61) by immersing the steam-blanched slices in glucose syrup (39–76% w/w; syrup : fruit = 1 : 1.37) with the addition of potassium sorbate and sodium bisulphite to give 1000 and 150 ppm, respectively, and an equilibrium a_w of 0.97 and pH of 3.1 in the product.

6.   QUALITY CHANGES DURING DRYING AND STORAGE

6.1.   Loss of Vitamins (Vitamins A and C)

Fruits and vegetables are the major sources of vitamin C (ascorbic acid) and provitamin A (beta-carotene) besides minerals. It is, therefore, quite understable that to determine the efficacy of dehydration techniques scientists have primarily investigated and compared the effect such techniques have on these nutrients.

Comparison of beta-carotene content of carrots dried by three different methods (explosive puffing, air drying, and vacuum freeze drying) (62) showed total beta-carotene content of explosive puffed product to compare favorably with vacuum freeze-dried product, while air-dried product had a lower content (Table 7). Of the transisomers of beta-carotene (provitamin A) available in the blanched carrots, 60% was retained by conventionally dried and explosive-dried products while about 80% was retained by the freeze-dried products.

Effect of process variables such as freezing, blanching, sulfiting, and drying methods (bin drying and cabinet drying) on the stability of beta-carotene and xanthophylls of dehydrated parsley was studied (63). Before dehydration, blanched parsley contained significantly more pigment than the corresponding unblanched samples. However, after bin drying, the unblanched parsley had significantly more beta-carotene than the blanched material. Greater retention of beta-carotene in various plant materials after blanching has been attributed generally to inactivation of enzymes. Loss of natural antioxidants could be the possible explanation for the lower values in the blanched material after drying. Xanthophyll changes in unblanched parsley followed the same pattern.

A systematic study (64) carried out to determine the effect of different processing methods and storage conditions on nutrient retention (provitamin A and vitamin C) of apricots showed (Table 8) that loss of vitamin A was 13–14% in sun-dried halves of apricots while there was no loss in dehydrated carrots. Sulfuring had no effect on carotene retention. Loss of vitamin A during drum drying of carrots was not significant (9–10%). Loss of beta-carotene was found to be maximum (30%) during sun drying of sulfured apricot leather compared with dehydrated and shade-dried products. Loss of vitamin C during drying was substantial (74% to 91%) in all modes of drying studied.

The effect of predrying treatments, dehydration, storage, and rehydration was studied (65) on the retention of carotene in green peppers and peaches during home dehydration. Carotene was completely retained in the case of green peppers. In peaches, 72.7% of the carotene was retained after predrying treatment, which decreased to 37.3% after dehydration. Retention of ascorbic acid during predrying treatment and dehydration depended on the nature of food. Thus, in the case of green peppers, most losses occurred during storage whereas dehydration was responsible for most of the loss in the dipped peaches.

In general, rapid drying retained a greater amount of ascorbic acid than slow drying. Thus vitamin C content of vegetable tissue is greatly reduced during a slow sun-drying process, while dehydration, especially by spray drying and freeze drying, reduced these losses. The effect of sun drying on the ascorbic acid content of 10 Nigerian vegetables showed that there was 21–58% loss depending on the nature of the vegetables (66).

Table 7 Total and All *Trans* Beta-Carotene of Carrots

Sample	Before cooking				After cooking			
	Range of concentration[a]	Average percent	Percent present as all *trans* beta-carotene	Percent of original amount of all *trans* beta-carotene	Range of concentration[a]	Average percent[c]	Percent present as all *trans* beta-carotene	Percent of original amount of all *trans* beta-carotene
Fresh	622–1000	100	100	100	980–1860[d]	100	72	100
Blanched	775–1050	108	95	100	–	–	–	100
Explosive puff dried	500–900	81[b]	72	60	805–1060	60	77	64
Vacuum freeze dried	526–890	85[b]	89	80	870–1125	65	84	76
Conventional air dried	600–830	74[b]	80	60	636–987	49	77	52

Source: From Reference 62

[a] Micrograms beta-carotene per g solids (moisture free basis, MFB) (range for 5 lots)
[b] Percent based on beta-carotene of blanched fresh carrots
[c] Percent based on beta-carotene of cooked fresh carrots
[d] Represents an increase in beta-carotene of 57–85% above original concentration

Table 8 Vitamin A and C Content of Processed Apricots

Treatment	Vitamin A IU/100 g (MFB)	Vitamin C mg/100 g (MFB)
Fresh apricots	20,900	67.2
Halves		
Sulfured, sun dried	18,200 (13)[a]	17.4 (74)[a]
Unsulfured, sun dried	18,000 (14)	3.0 (96)
Sulfured, dehydrated	20,900 (0)	15.9 (76)
Sulfured, pureed, drum dried	16,400 (22)	12.2 (82)
Heated, pureed		
Concentrated, drum dried	18,800 (10)	7.5 (89)
Sulfured, drum dried	19,000 (9)	6.3 (91)
Canned	12,200 (42)	—

Source: From Reference 64
[a]Percent loss

Oxidative changes would be expected to be minimum in freeze-dried samples since freeze drying is a low-temperature process operating under vacuum. A study (67) of the changes in quality of compressed carrots prepared in combinations of freeze drying and hot air drying showed that values of ascorbic acid ranged from 15–97 mg/100 g for the totally air-dried samples to 33.3 mg/100g for the totally freeze-dried samples (Table 9). In the case of carotenes also the totally hot air drying treatment had the lowest value (34.16 mg/100 g) and totally freeze-dried samples had the highest value (70.37 mg/100 g) (Table 10).

In general it is difficult to compare the losses in vitamins during dehydration because retention of vitamins depends on the nature of foods, predrying treatments given (sulfuring, blanching methods), and the conditions of drying (techniques, time, temperature).

Table 9 Effect of Drying Treatment on Ascorbic Acid and Alpha tocopherol of Dehydrated Carrots (Mg/100 g Dry Weight Basis)[a]

Treatment (% moisture)	Ascorbic acid		Alpha tocopherol	
Fresh	85.28		3.41	
Totally freeze dried	33.39	a	3.45	a
Totally air dried	15.97	d	0.04	f
Freeze dried (3%), mist plasticized (10%), air dried	32.76	a	2.98	b
Freeze dried (10%), air dried	27.71	b	1.42	c
Freeze dried (20%), air dried	16.78	cd	1.13	d
Freeze dried (30%), air dried	16.38	cd	1.10	d
Freeze dried (40%), air dried	20.38	c	0.96	d
Freeze dried (50%), air dried	17.49	cd	0.55	e

Source: From Reference 67
[a]Means within columns followed by the same letter are not significantly different at the 5% level according to Duncan's multiple range test

Table 10   Effect of Drying Treatment on Carotene Content of Dehydrated Carrots (Mg/100 g Dry Weight Basis)[a]

Treatment (% moisture)	Alpha-carotene		Beta-carotene		Total carotene	
Fresh	14.4		52.06		66.20	
Totally freeze dried	15.66	a	54.71	a	70.37	a
Totally air dried	6.67	e	27.50	f	34.16	f
Freeze-dried (3%), mist						
Plasticized (10%), air dried	10.61	d	40.47	e	51.08	e
Freeze dried (10%), air dried	12.81	b	49.40	b	62.21	b
Freeze dried (20%), air dried	11.73	c	44.49	d	56.22	d
Freeze dried (30%), air dried	11.42	cd	47.22	c	58.68	c
Freeze dried (40%), air dried	11.02	cd	44.89	d	55.91	d
Freeze dried (50%), air dried	10.52	d	40.23	c	50.81	e

Source: From Reference 67
[a]Means within columns followed by the same letter are not significantly different at the 5% level according to Duncan's multiple range test

6.2. Loss of Natural Pigments (Carotenoids and Chlorophylls)

Color is an important quality attribute in a food to most consumers. It is an index of the inherent good qualities of a food and association of color with acceptability of food is universal. Among the natural color compounds, carotenoids and chlorophylls are widely distributed in fruits and vegetables. Preservation of these pigments during dehydration is important to make the fruit and vegetable product attractive and acceptable. Both the pigments are fat soluble although they are widely distributed in aqueous food systems.

Carotenoids are susceptible to oxidative changes during dehydration due to the high degree of unsaturation in their chemical structure. The major carotenoids occurring in food are carotenes and oxycarotenoids (xanthophylls).

Leaching of soluble solids during blanching had considerable effect on the stability of carotenoids of carrots during drying and subsequent storage (68). Carotenoid destruction increased with increased leaching of soluble solids (Table 11). Investigation of the effects

Table 11   Effect of Leaching of Soluble Solids on Destruction of Carotenoids during Dehydration of Leached Carrots

Treatment	Loss in soluble solids (g/100 g, dry weight basis)	Carotenoid content (μg/g, dry weight basis)		Carotenoid loss (%)
		Before dehydration	After dehydration	
Blanched	2.7	1062	1007	5.2
Water dipped	6.9	1227	1125	8.3
Detergent dipped	7.5	1259	1143	9.0
Water washed	11.9	1442	1285	10.9
Detergent washed	14.5	1537	1339	12.9

Source: From Reference 68

of water activity, salt, sodium metabisulfite, and Embanox-6 on the stability of carotenoids in dehydrated carrots shows that carotenoid pigments were most stable at 0.43 a_w and addition of salt, metabisulfite, and Embanox-6 helped in stabilizing carotenoids in dehydrated carrots (Table 12) (69).

Sulphur dioxide was found to have a pronounced protective effect on carotenoids of unblanched carrots during dehydration (70). Dehydrated, sulfited, unblanched carrots contained about 2.9 times more carotenoids than dehydrated unblanched carrots that had not been sulfited (Table 13). Treatment with SO_2 gave additional protection to carotenoids of blanched carrots during dehydration and effectiveness of SO_2 increased with an increase in SO_2 content.

The importance of chlorophyll in food processing is related to the green color of vegetables. Many studies have been made on the changes of chlorophyll during processing and storage but little is known about the pigment behavior in low-moisture systems such as dehydrated vegetables. Generally, it was found that chlorophyll was quite stable in low-moisture systems. Degradation of chlorophyll depended on temperature, pH, time, enzyme action, oxygen, and light. The most common mechanism of chlorophyll degradation is its conversion to pheophytin in the presence of acid. Although the pathways of this degradative reaction are well known, a method for its stabilization is not well established.

Water activity has been shown to have a definite influence on the rate of degradation of chlorophyll in freeze-dried, blanched spinach puree (71). At 37°C and an a_w higher than 0.32, the most important mechanism of chlorophyll degradation was conversion to pheophytin. The complex kinetics of chlorophyll-a degradation at 37°C are shown in Figure 9 for the entire range of water activities. At a_w lower than 0.32 the rate of pheophytin formation in spinach was low. The rate of chlorophyll-a transformation was 2.5 times faster than chlorophyll-b. Study of the degradation of chlorophyll as a function of a_w, pH, and temperature in a spinach system during storage showed that even in the dry state the elimination of a magnesium atom and transformation of chlorophyll into pheophytin was very sensitive to pH changes (72). Effect of temperature on the rate of chlorophyll-a degradation at water activity 0.32 and pH 5.9 is shown in Figure 10.

6.3. Browning and Role of Sulphur Dioxide

One obstacle always encountered by the food technologists in the dehydration and long-term storage of dehydrated fruits and vegetables is the discoloration due to browning. Browning in foods is of two types: enzymatic and nonenzymatic. In the former, the enzyme polyphenol oxidase catalyzes the oxidation of mono and ortho diphenols to form quinones that cyclize, undergo further oxidation, and condense to form brown pigments (melanins). In the dehydration of fruits and vegetables, blanching destroys the causative enzymes and prevents subsequent enzymatic browning. Sulphur dioxide and sulfites act as inhibitors of enzyme action during preblanching stages. Presence of SO_2 retards browning of dehydrated fruits and vegetables, especially when the enzymes have not been heat inactivated (e.g., freeze-dried products).

Nonenzymatic browning (NEB), also known as Maillard reaction, describes a group of diverse reactions between amino groups and active carbonyl groups leading eventually to the formation of insoluble, brown, polymeric pigments, collectively known as *melanoidin pigments*. The basic reactions that lead to the browning are well documented in literature. These reactions are sometimes desirable but in many instances are considered to be deleterious not only due to formation of unwanted color and flavor but also due to the loss of nutritive value through the reactions involving the alpha-amino group of lysine

Table 12 Effect of NaCl, Na$_2$S$_2$O$_3$, and Embanox-6 on Total Carotenoids, TBA Value, and Nonenzymic Browning in Air-Dried Carrots

Storage period (months)	Control			Salt treated			Salt + metabisulphite treated			Salt + metabisulphite + Embanox-6 treated		
	Carotenoids (µg/g)	TBA value	NEB	Carotenoids (µg/g)	TBA value	NEB	Carotenoids (µg/g)	TBA value	NEB	Carotenoids (µg/g)	TBA value	NEB
0	1120	0.12	0.08	1137	0.12	0.06	1114	0.10	0.05	1135	0.09	0.05
3	505	0.92	0.14	669	0.83	0.10	691	0.64	0.08	827	0.28	0.09
6	316	1.38	0.21	416	0.92	0.15	449	0.78	0.18	620	0.46	0.14
9	222	1.50	0.28	308	1.05	0.24	353	0.92	0.22	408	0.58	0.18

Source: From Reference 69

TBA value, mg of malonaldehyde per kg substance; NEB, nonenzymic browning reported as optical density at 420 nm

Table 13 Effect of Concentration of SO_2 on Carotenoid Content of Dehydrated Carrot of 5% Moisture Content during Storage at 37°C

Blanching time (min)	Initial SO_2 content ($\mu g \cdot g^{-1}$)	Carotenoid content after dehydration ($\mu g \cdot g^{-1}$)	Carotenoids remaining (%) Storage time (days)				
			60	120	180	300	440
0	0	464	68.0	51.1	43.0	36.2	33.1
0	1723	1296	87.5	76.5	69.4	62.6	55.5
1	2325	1360	92.5	85.0	79.4	69.0	62.0
2	2330	1350	88.7	79.4	71.1	61.7	55.0
5	0	1202	77.5	62.5	56.1	50.2	48.2
5	1584	1298	80.5	67.4	60.5	54.0	50.2
5	2357	1308	87.0	76.1	68.6	58.5	52.0
5	9621	1380	89.9	80.0	73.1	62.8	54.0

Source: From Reference 70

moieties and other groupings in proteins. It is a major deteriorative mechanism in dry foods and is sensitive to water content. It is influenced by the types of reactant sugars and amines, pH, temperature, and a_w.

The addition of sulfites during the predrying step is the only effective means available at present of controlling NEB in the dried fruit and vegetable product. Sulfite is considered to be a safe additive to incorporate into fruit and vegetable products up to certain permissible limits. However, recently there are reports on the hypersensitivity of a few individuals to the ingested sulfite. Numerous attempts are therefore being made to find alternative means to prevent browning reactions.

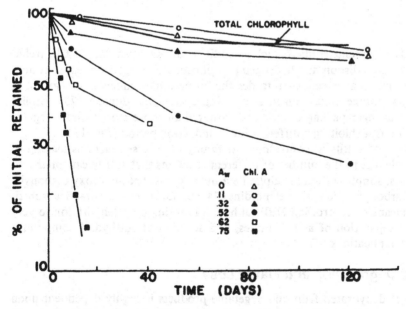

Figure 9 Degradation of chlorophyll in spinach (37°C, air atmosphere). Total chlorophyll curve is average for all storage conditions. (From Reference 71.)

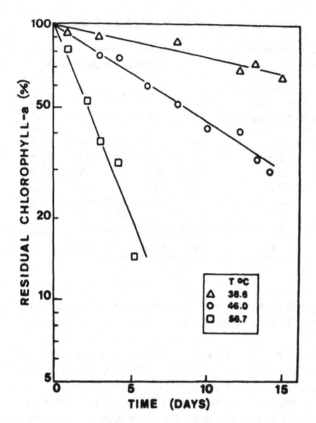

Figure 10 Degradation of chlorophyll-a in spinach as a function of temperature ($a_w = 0.32$; pH $= 5.9$). (From Reference 72.)

Among various treatments studied, such as addition of SO_2, cysteine, $CaCl_2$, trehalose, manganese chloride, disodium dihydrogen pryophosphate, oxygen scavenger pouch, and so on, the only ones that effectively retarded the formation of undesirable pigment in dried apples during storage were oxygen scavenging and sulphur dioxide (73). Apples stored in oxygen scavenger packages darkened slower than those stored under regular atmospheric conditions, exhibiting a different initial induction period (Fig. 11).

The effectiveness of sulfite in controlling the family of diverse reactions leading to browning is probably due to the number of different reactions that sulfite can enter into with reducing sugars, simple carbonyls, alpha beta-dicarbonyls, beta-hydroxy carbonyls, beta-unsaturated carbonyls, and with melanoidins (74). So far there is no practical substitute for SO_2 as a means of controlling NEB, although lowering pH, dehydration to very low water activity, separation of active species, and addition of sulfhydryl compounds might have limited applications (7).

6.4. Oxidative Degradation and Flavor Loss

The acceptability of dehydrated fruit and vegetable products is highly dependent upon their flavor attributes. Loss of desirable flavor is the limiting characteristic for most dehydrated products. The natural flavor constituents are subjected to much variation and

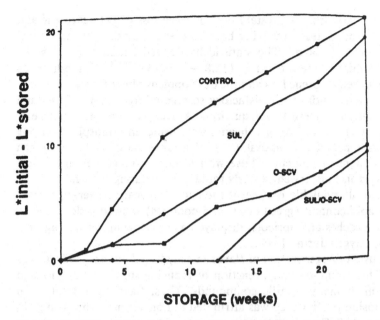

Figure 11   Effect of in-package oxygen scavenger on dried apple darkening during storage at 30°C (ΔL^* of 8 = observable change). (From Reference 73.)

loss during predrying operations, drying, and storage. Conditions generally responsible for the destruction of natural flavors include rough handling, delay in processing, exposure to light, high temperature, and certain chemicals. Flavor retention is especially important in products in which the principal flavor constituents are volatile oils, as in onions.

Based on comparison of loss of flavor, air-dried and freeze-dried vegetables were divided into three main groups (75): (a) vegetables with nonvolatile flavors (e.g., peas, beans, and cauliflower); (b) vegetables with volatile flavors (e.g., onions); and (c) vegetables with both volatile and nonvolatile flavors (most root vegetables like carrots, turnips, etc.). Flavor defects in dehydrated products were, however, not solely due to volatile losses. Chemical reactions, especially oxidation and nonenzymatic browning, greatly contributed to flavor deterioration.

In general, freeze-dried products had more preferable flavors than air-dried ones except in the case of onions, for which an air-dried product had a stronger flavor due to entrapment of volatile oils by shrinkage. Leeks and celery showed similar behavior.

Staling and off flavors developed during storage of both air-dried and freeze-dried vegetables. The degree of change was mainly related to temperature of storage and moisture content of the dried vegetables. Air-dried peas (6–7% moisture) developed off flavor at 15°C after 15–18 months. At about 20°C, shelf life was reduced to 9–12 months and at 37°C the period was 2–3 months. Comparatively, freeze-dried vegetables were much more sensitive to storage conditions because the highly porous texture allowed easy entry of air and stale flavor developed rapidly. For example, freeze-dried carrots developed off flavor after 1 month in air at 20°C. At 30°C the oxygen level had to be reduced to 0.1% to give a storage life of 6 months.

The off flavor that developed when dehydrated carrots were stored in the presence of oxygen was described as similar to violets and was suggested to be due to beta-ionone formation (76). A direct relationship was established between loss of beta-carotene and off flavors in a series of storage tests involving both carotene analysis and taste panel assessment for natural and off flavor. The statistically derived equation relating off flavor (t = taint) and residual carotene C is $t = (12.8 - 0.0013\ C^2)^{1/2}$. This indicates a nonlinear relationship between carotene loss and the development of off flavor.

Absence of oxygen was essential for satisfactory storage of freeze-dried fruits and vegetables. Excellent retention of fresh flavor quality was achieved in a series of freeze-dried foods of plant origin in "zero" oxygen head space, using an atmosphere of 5% hydrogen in nitrogen with palladium catalyst (77). Vegetable items took up oxygen chiefly as a function of pigment content. Those with a high carotene content (sweet potatoes, spinach, and carrots) underwent a fairly rapid uptake during the first 15 to 40 weeks and had consumed all available oxygen at the end of 1 year. Lesser-pigmented vegetables with a lower lipid content (green beans and potatoes) showed a slow, steady uptake. Two fruit items, peaches and apricots, displayed a very slow uptake, using only 30% to 50% of available oxygen during 1 year.

One of the major causes of degeneration of flavor in dehydrated potato products was the Maillard reaction. This amino-carbonyl reaction of reducing sugar and amino acid resulted in the formation of many volatile compounds. Thus, flavor deterioration in potatoes during the explosion puffing step was attributed to nonenzymatic browning. In the puffing gun, potatoes at 30% moisture were subjected to a temperature condition conducive to NEB, which resulted in the formation of volatile aldehydes. On the other hand, dominant, rancid off flavor that developed during the storage of dried potato products was due to autoxidation of potato lipids (78), giving hexanal as a major volatile product. Use of BHA alone or with BHT effectively retarded the autoxidation of explosion-puffed potatoes, keeping oxidative off flavors below threshold levels for up to 12 months in storage as compared to 3 months for air-packed samples without antioxidant. The incorporation of a scavenger pouch packaging system (H_2-palladium catalyst), although very effective in antioxidative effect, was severely limited because of pinhole leaks.

Prepared foods present the advantage that their flavor characteristics can be modified by formulation. Spices, for example, contribute not only to enhance flavor, but retard oxidation in some cases (79).

6.5. Texture and Reconstitution Behavior

The problem of hot air drying, which is still the most economical and widely used method for dehydrating piece-form vegetables and fruits, is the irreversible damage to the texture, leading to shrinkage, slow cooking, and incomplete rehydration. Many commercially dehydrated vegetables exhibit a dense structure with most capillaries collapsed or greatly shrunk, which affects the textural quality of the final product.

The possible causative factors suggested by different workers are loss of differential permeability in the protoplasmic membrane, loss of turgor pressure in the cell, protein denaturation, starch crystallinity, and hydrogen bonding of macromolecules. Texture of air-dried vegetables deteriorates during storage if the product is exposed to high temperature or if inadequately dehydrated. Even the freeze-drying technique has failed to produce an acceptable dehydrated product from celery. Damage generally occurred during freezing, drying, storage, and reconstitution.

Water removal affects many aspects of cell structure; histological studies were generally carried out to assess the membrane integrity. Pedlington and Ward, in studies on air-dried carrots, parsnips, and turnips, observed several changes, including a loss in the selective permeability of cytoplasmic membranes of cells responsible for maintaining turgidity and crisp texture of vegetables (80). They found loss of water to result in rigidity of cell walls and to their slow collapse by the stresses set up by shrinkage of neighboring cells.

Jayaraman et al. studied the effect of sugar and salt on the texture of dehydrated cauliflower (81). They found that in treated, dehydrated florets there were 80% intact cells as compared with 0% in the untreated, dehydrated florets due to tissue collapse resulting in disruption of cell walls and loss of cell integrity. Khedkar and Roy found a higher reconstitution ratio in cabinet-dried raw mango slices as compared with sun-dried slices (82); this was due to less rupture of cells during cabinet drying (36.4%) than sun drying (67.3%).

Different dehydration techniques were tried to improve the rehydration behavior of dehydrated piece-form fruits and vegetables. Generally, it was observed that the greater the degree of drying, the slower and less complete was the degree of rehydration. Dehydration techniques used to improve the rehydration qualities of dehydrated fruits and vegetables include those aimed at reducing the drying time or involving use of additives like salt and polyhydroxy compounds such as sugar and glycerol as a predrying treatment.

Dehydrated carrots puffed and dried in a centrifugal fluidized bed unit absorbed 2 1/2 parts by weight of water and appeared completely rehydrated in 5 min while the unpuffed controls absorbed 1 1/2 parts and still had hard centers (27). Jayaraman et al. found rehydration ratio, coefficient of rehydration, and reconstitution time of HTST pneumatic-dried vegetables to be much superior to those of directly cabinet-dried samples (23).

Effect of additives on the rehydration qualities of dehydrated vegetables was studied by Neumann (83) and Jayaraman et al. (81). A combined predrying treatment of sodium carbonate and sucrose (60%) produced the best rehydrated celery, with a rehydration percentage of 71% and the dices were well filled out with texture remaining tender to firm (83). Similarly, a presoaking treatment in a combined solution of salt and sugar at 4°C for 16 h prior to cabinet drying markedly increased the rehydration percentage of cauliflower and reduced the shrinkage as compared with control without treatment (81).

Study of the rehydration ratios of forced air dried compressed carrots after partially freeze drying to different moisture levels showed the drying treatment significantly affected rehydration ratios in all cases (67). The sample that was freeze dried to 50% moisture, compressed, and then air dried had the highest ratio and was the quickest to rehydrate. In comparison, the totally freeze-dried and hot air dried compressed carrots showed much lower values of rehydration ratios. These observations were supported by scanning electron microscopy, which showed collapse of cellular structure and tissue coagulation to act as a barrier for rehydration.

Levi et al. observed that pectin, one of the major cell wall and intercellular tissue components, played a significant role in the rehydration capacity of dehydrated fruits (84).

6.6. Influence of Water Activity

During the last three decades water activity a_w has played a major role in many aspects of food preservation and processing. It is defined as the ratio of the vapor pressure of water P in the food to the vapor pressure of pure water P_o at the same temperature ($a_w = P/P_o$). Next to temperature, it is now considered as probably the most important parameter having a strong effect on deteriorative reactions. The effect of water activity was studied not only to define the microbial stability of the product but also on the biochemical reactions in the food system and its relation to its stability. It has become a very useful tool in dealing with water relations of foods during processing.

It is now well known that microorganisms cannot grow in the dehydrated food system when the water activity range is less than or equal to 0.6–0.7, but other reactions, enzymatic and nonenzymatic (e.g., lipid oxidation, nonenzymatic browning, etc.), that cause change in color, flavor, and stability continue during processing and storage. Water activity has become the most useful parameter that can be used as a reliable guide to predicting food spoilage or to determine the drying end point required for a shelf-stable product.

The relationship between equilibrium moisture content and water activity, known as the *sorption isotherm*, is an important characteristic that influences many aspects of dehydration and storage. It can be constructed graphically or derived mathematically. The shape of the isotherm generally determines the storage stability of the dehydrated product. This concept is used to establish product specifications for the effective drying, packaging, and storage of foods.

Adsorption isotherms of potatoes were of sigmoid shape and were affected by drying method, temperature, and addition of sugar (85). The freeze-dried product absorbed more water vapor than the vacuum-dried materials (Fig. 12). The sorption isotherm prepared from fresh and freeze-dried Thompson seedless grapes indicated a hysteresis loop at both the upper and lower moisture level (86). The isotherm of sun-dried grapes was slightly lower than that of vacuum-dried grapes.

Both lipid oxidation and NEB are greatly influenced by a_w (87). Autoxidation of lipids occurs rapidly at low a_w levels, decreasing in rate as a_w is increased until in the 0.3–0.5 range and increasing thereafter beyond 0.5 a_w. Most rapid browning can be expected to occur at intermediate a_w levels in the 0.4–0.6 range. Whether or not it is minimized at

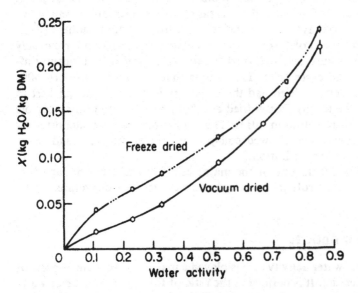

Figure 12 Adsorption isotherms at 25°C of freeze-dried and vacuum-dried Norchip potatoes. (From Reference 85.)

the lower or upper portion of this range depends significantly on the specific solutes used to poise a_w, the nature of the food (especially amino compounds and simple sugars that might be present), as well as the pH and a_w of the product. Interestingly, at a_w levels that minimize browning, autoxidation of lipids is maximized.

The kinetics of chlorophyll-a transformation was studied as a function of time at different water activities at 38.6°C (Fig. 13) (72). For $a_w > 0.32$ the most important mechanism of chlorophyll degradation was the transformation into pheophytin; this had a first-order dependence on pH, water activity, and pigment concentration.

Carotenoids in freeze-dried carrots were relatively more stable in the range of 0.32 to 0.57 a_w (69). The maximum stability was near 0.43 a_w (corresponding to an equilibrium moisture content of 8.8–10%). Increase in the rate of carotenoid destruction was greater at lower a_w than at higher a_w.

The kinetics of quality deterioration in dried onion flakes (nonenzymatic browning and thiosulfinate loss) and dried green beans (chlorophyll-a loss) were studied as a function of water activity and temperature and empirical equations/mathematical models developed that successfully predicted the shelf life of the dried products as a function of temperature and a_w (Tables 14 and 15) (88). Above the monolayer (a_w 0.32 to 0.43) for onion, increasing moisture contents resulted in greater reaction rates for browning and

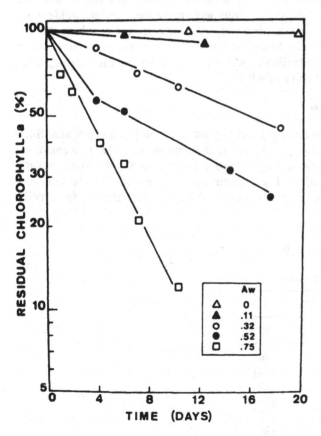

Figure 13 Degradation rate of chlorophyll-a in spinach as a function of time at different water activities (pH = 5.9; temperature 38.6°C). (From Reference 72.)

Table 14 Actual (and Predicted) Shelf Life (Days) of Dried
Onion Flakes Based on Browning and Thiosulfinate Loss at
Different Temperatures

a_w	Browning			Thiosulfinate loss		
	20°C	30°C	40°C	20°C	30°C	40°C
0.32	>631	474	59	>631	369	66
	(4778)	(472)	(63)	(1619)	(306)	(55)
0.43	593	83	22	631	136	40
	(600)	(69)	(21)	(585)	(139)	(38)
0.56	183	31	17	298	84	27
	(190)	(33)	(17)	(288)	(82)	(29)

Source: From Reference 88

thiosulfinate loss. Very little browning was observed over a storage period of 631 days at 20°C and $a_w = 0.33$, while all other samples stored at 30°C and 40°C and $a_w = 0.43$ and 0.59 deteriorated to unacceptable levels within this time period. Similarly, in the case of green beans, the destruction of chlorophyll-a (pheophytinization) was found to be the principal factor responsible. The dried green beans were considered unacceptable when more than 30% loss of chlorophyll-a was observed, the concentration at which the dull olive-green color began to predominate. Since conversion of chlorophyll-a to pheophytin is an acid catalyzed reaction, the availability of water was essential and therefore a_w could be expected to influence the rate of chlorophyll loss.

6.7. Microbiological Aspects

Drying is the oldest method of preserving food against microbiological spoilage. Since presence of water is essential for enzymic reactions, the removal of water prevents these reactions and the activities of contaminating microorganisms present. Removal of water increases the solute concentration of the food system and thus reduces the availability of water for microorganisms to grow. There is a lower limit of water activity for specific

Table 15 Actual (and Predicted) Shelf Life
(Days) of Dried Green Beans Based on
Chlorophyll-a Loss at Different Temperatures

a_w	Temperature (°C)		
	20	30	40
0.32	>637	273	86
	(962)	(282)	(84)
0.43	478	143	45
	(452)	(146)	(38)
0.56	150	61	25
	(148)	(56)	(26)

Source: From Reference 88

microorganisms to grow; for complete microbiological stability, water activity of the system should be below 0.6.

Drying, however, is also an effective means of preserving microorganisms in a viable state, even though their numbers may be reduced and a proportion sublethally damaged (89). Survival during and after drying will depend upon the physicochemical conditions experienced by microorganisms, such as temperature, a_w, pH, preservatives, oxygen, and so on. The survival of food spoilage organisms may give rise to problems in a reconstituted food item, but survival of foodborne pathogens must be viewed much more seriously.

With a view to minimize organoleptic changes in foods during drying, time and temperatures are kept as short and as low, respectively, as feasible. The process of drying, whether by freeze drying, hot air drying, solar drying, or by high temperature (e.g., spray or drum drying) is not per se lethal to all microorganisms and many may survive. The more heat-resistant organisms are the more likely survivors (e.g., bacterial spores, yeasts, molds, and thermoduric bacteria). Thus there is a strong possibility for microbial growth, including pathogens, before the a_w of the product falls below the critical level for each organism.

Vegetables, because of their greater proximity to soil and lower acidity and sugar content as compared with fruits, predominately have more bacterial populations. A majority of the species have been found to be common soil- and water-borne bacteria of the genera *Bacillus* and *Pseudomonas*. Some workers have found other types of bacteria such as coliforms and bacteria of the genera *Achromobacter*, *Clostridium*, *Micrococcus*, and *Streptococcus* from different dehydrated vegetables.

Factors that influence markedly the microbial population of dehydrated vegetables include the microbial quality of fresh produce; the method of pretreatment of the vegetables (peeling, blanching, etc.); the time elapsed between preparation of the vegetables and start of the dehydration process; the time involved in the dehydration of the vegetables; the temperature of dehydration; the moisture content of the finished product; and the general level of sanitation in the dehydration plant (90). Blanching, if sufficient to inactivate enzymes, would reduce the contamination of the fresh produce to an insignificant figure. Reduction in total count during blanching was found to be greater than 99.9%.

Coliforms and enterococci are commonly used as indicators of unsanitary conditions in food processing. Clarke et al. isolated enterococci from 18 out of 35 dehydrated vegetable samples (91). They found coliforms in 18 and enterococci and coliforms in 15 samples (Table 16). Statistical analysis showed a positive correlation between number of enterococci and coliforms. The predominant species recovered from enterococci was *Streptococcus faicium* (60%) and from coliforms was *Aerobacter* (56%).

Fanelli et al. surveyed a number of commercially available vegetable soups and found that the maximum total number of bacteria was less than 50,000 per g and the mean and median total bacterial numbers were very low (92). The numbers of coliforms, yeasts, molds, and aerobic species were also low (Table 17).

Dehydrated onion, which is an important commercial flavoring ingredient, is not blanched before dehydration. Its microbiology was therefore extensively investigated. Total plate count (TPC) was less than 100,000 per g in 76% of the slices from the belt dryer and only 52% of the sample in the tunnel-dried product (93). In both cases, average bacterial spore count was 12,000 per g. Many workers have variously reported the presence of *Bacillus*, *Pseudomonas*, *Aerobacter*, *Lactobacillus*, and *Leuconostoc* species (94). Exposure to ethylene oxide gas was found effective in reducing the relatively high TPC, but future application of this gas is in doubt because of its toxic hazards.

Table 16 Estimates of the Bacterial Populations in Dehydrated Vegetable Materials

Sample no.	Vegetable	Plate count[a] Total/g[b]	Coliform/g[c]	Enterococcus/g[d]
1	Carrot, diced	63,000	400	27,000
2	Celery, stalk, diced	21,000,000	2,400,000	270,000
3	Celery, stalk, diced	36,000	15	5
4	Coriander, ground	1,300,000	19,000	30
5	Ginger, ground	24,000,000	50	35
6	Onion, green, minced	38,000	50	25
7	Onion, powdered	280,000	250	120
8	Onion, powdered	870,000	10,000	90
9	Onion, powdered	5,700,000	1,900	50
10	Pepper, black, ground	12,000,000	50	5
11	Peppers, green, bell	230,000	34,000	11,000
12	Peppers, sweet, flakes	15,000	420	2,100
13	Vegetables, mixed, flakes	7,000	800	830
14	Vegetables, mixed, flakes	12,000	20	150
15	Vegetables, mixed, flakes	140,000	3,100	1,200
16	Carrot, diced	550	<10	15
17	Onion, chopped	31,000	<10	5
18	Pepper, sweet, bell	7,600	<10	40
19	Garlic, powdered	2,800,000	450	<10
20	Onion, minced	1,700	20	<10
21	Onion, powdered	38,000	15	<10
22	Cabbage, diced	1,700	<10	<10
23	Celery, flakes	14,000	<10	<10
24	Celery, flakes	30,000	<10	<10
25	Onion, chopped	41,000	<10	<10
26	Onion, flakes	1,200	<10	<10
27	Onion, green, minced	870	<10	<10
28	Onion, powdered	32,000	<10	<10
29	Onion, toasted, minced	50,000	<10	<10
30	Onion, toasted, minced	8,200	<10	<10
31	Parsley, flakes	460	<10	<10
32	Parsley, flakes	1,200	<10	<10
33	Parsley, flakes	2,100	<10	<10
34	Peppers, red, bell	3,600	<10	<10
35	Peppers, red, cracked	10	<10	<10

Source: From Reference 91
[a]Numbers reported as averages of duplicate platings
[b]Standard methods agar; plates incubated 48 h ± 3 h at 32°C
[c]Violet-red bile agar; plates incubated 18–24 h at 32°C
[d]Citrate azide agar; plates incubated 72 h at 37°C

Table 17 Microbiological Examination of Vegetable, Vegetable Noodle, Tomato, Tomato Vegetable, Tomato Vegetable Noodle, and Cream Vegetable Dry Soup Mixes

Brand, type	No. samples	Number of microorganisms per gram of dry sample											
		Total			Coliforms			Yeasts and molds			Aerobic spores		
		Range	Mean	Median	Range	Mean	Median	Range	Mean	Median	Range	Mean	Median
A, vegetable	8	2900–8300	5500	4700	<10–230	54	28	<10–300	59	30	240–11,000	2100	1500
B, vegetable	4	6900–18,000	11,000	12,000	<10–85	38	28	<10–15	11	10	240–460	350	350
C, vegetable	4	2000–24,000	12,000	12,000	<10	<10	<10	20–55	39	28	93–1100	440	280
B, vegetable	4	6700–31,000	14,000	10,000	10–50	21	12	35–320	170	150	240–930	520	460
D, Italian vegetable	1		2600			<10			<10			240	
D, French vegetable	2	27,000–4600	37,000		<10	<10		<10–60	35		>24,000	>24,000	
A, cream vegetable	10	89,000–530,000	300,000	260,000	<10–790	230	150	<10–90	37	33	4600–>24,000	12,000	11,000
D, cream vegetable	4	2700–8000	5400	5400	<10	<10	<10	<10–10	10	10	460–4600	2100	1800
E, tomato	8	5000–23,000	12,000	9300	<10–2000	53	33	<10–35	18	10	240–>24,000	7600	4600
E, tomato	8	450–11,000	3900	2400	<10–10	<10	<10	<10–210	63	25	43–2400	780	350
F, tomato	6	4100–40,000	13,000	8200	<10	<10	<10	<10–45	18	10	2400–11,000	5600	3500

Source: From Reference 92

Alternatively, the application of gamma radiation at a level 0.2 to 0.4 Mrad was suggested to sterilize onion powder without any detrimental effect.

6.8. Factors Affecting Storage Stability

The shelf-life of dehydrated fruits and vegetables depends on many deleterious reactions, which in turn depend on the specific nature of the food materials, storage conditions, and nature of packaging. The undesirable changes that occur are due to off flavors, browning, and loss of pigments and nutrients as enumerated above. Knowledge of the causes of these reactions is highly necessary to improve the shelf life of the dehydrated products.

Villota et al., in their review on the storage stability of dehydrated foods, discussed the factors mainly responsible for deterioration, that is, moisture, storage temperature and period, oxygen, and light (95). They compiled the literature data on storage stability of several dehydrated products, which included dehydrated fruits, vegetables, and fruit and vegetable powders, based on method of drying, additional treatment, storage conditions, time required for appearance of earliest defects, and the state of other factors at times of unacceptability.

Moisture content is a very important parameter influencing the stability of dehydrated foods. It has been suggested that the optimal amount of water for long-term storage corresponds in most dehydrated foods to the Brunauer-Emmett-Teller (BET) monolayer value. On the other hand, items such as freeze-dried spinach, cabbage, and orange juice were reported to be more stable at a zero moisture content, while items like potatoes and corn had maximum stability at the monomolecular moisture content. It appeared that optimal moisture content could not be predicted with precision on the basis of theoretical considerations.

Another important factor affecting storage stability of dehydrated foods is temperature and period of storage. Generally, the storage stability bears an inverse relationship to storage temperature, which affects not only the rate of deteriorative reaction (enzyme hydrolysis, lipid oxidation, nonenzymatic browning, protein denaturation), but also the kind of spoilage mechanism.

It is well established that elimination of oxygen by packing in an inert atmosphere such as nitrogen contributes to extending the storage stability of many dehydrated products. However, in certain products like spray-dried powders, in which a large surface area is exposed to air during processing, some entrapment of oxygen occurs in the final product and packing under inert atmosphere results in very little improvement. Storing in "zero" oxygen headspace, using an atmosphere of 5% hydrogen in nitrogen with a palladium catalyst, is reported to result in superior quality retention. Further, since oxidation of lipids and vitamins like ascorbic acid, riboflavin, thiamine, and vitamin A and loss of pigments such as carotenoids and chlorophyll are initiated or accelerated by light, adequate packaging needs to be provided to protect such dehydrated foods from light.

REFERENCES

1. D. K. Salunkhe and B. B. Desai, *Postharvest Biotechnology of Fruits*, CRC Press, Cleveland, Ohio (1984).
2. D. K. Salunkhe and B. B. Desai, *Postharvest Biotechnology of Vegetables*, CRC Press, Cleveland, Ohio (1985).
3. *FAO Year Book Production*, Food and Agricultural Organization of the United Nations, Rome, *44*:125 (1991).
4. S. Sokhansanj, and D. S. Jayas, Drying of foodstuffs. In *Handbook of Industrial Drying*, 1st Ed. (A. S. Mujumdar, ed.), Marcel Dekker, New York, p. 517 (1987).

5. R. B. Duckworth, *Fruit and Vegetables*, Pergamon Press, Oxford p. 38 (1966).

6. D. K. Salunkhe, and H. R. Bolin, Developments in technology and nutritive value of dehydrated fruits, vegetables and their products. In *Storage, Processing and Nutritional Quality of Fruits and Vegetables* (D. K. Salunkhe, ed.), CRC Press, Cleveland, Ohio, p. 39 (1974).

7. J. Dunbar, Use of sulphur dioxide in commercial drying of fruits and vegetables, *Food Tech. New Zealand*, *21*(2):11 (1986).

8. J. L. Bomben, W. C. Dietrich, D. F. Farkas, J. S. Hudson, E. S. DeMarchena, and D. W. Sanshuck, Pilot plant evaluation of individual quick blanching (IQB) for vegetables, *J. Food Sci.*, *38*:590 (1973).

9. N. M. Quentzer and E. E. Burns, Effect of microwave steam and water blanching in freeze-dried spinach, *J. Food Sci.*, *46*(2):410 (1981).

10. J. L. Bomben, Effluent generation, energy use and cost of blanching, *J. Food Process Eng.*, *1*(4):329 (1977).

11. L. P. Somogyi and B. S. Luh, Dehydration of fruits. In *Commercial Fruit Processing*, 2nd Ed. (J. G. Woodroof and B. S. Luh, eds.), AVI Publishing, Westport, Connecticut, p. 353 (1986).

12. W. Szulmayer, From sundrying to solar dehydration. I. Methods and equipment, *Food Tech. Australia*, *23*:440 (1971).

13. H. R. Bolin and D. K. Salunkhe, Food dehydration by solar energy, *CRC Crit. Rev. Food Sci. Tech.*, *16*:327 (1982).

14. K. S. Jayaraman and D. K. Das Gupta, Dehydration of fruits and vegetables—Recent developments in principles and techniques, *Drying Tech.*, *10*:1 (1992).

15. T. A. Lawland, Agricultural and other low temperature applications of solar energy. In *Solar Energy Handbook* (J. F. Kreider and F. Kreith, eds.), McGraw-Hill, New York, p. 18.1 (1981).

16. L. L. Imre, Solar drying. In *Handbook of Industrial Drying*, 1st Ed. (A. S. Mujumdar, ed.), Marcel Dekker, New York, p. 357 (1987).

17. H. R. Bolin, C. C. Huxoll, and D. K. Salunkhe, Fruit drying by solar energy, *Confructa*, *25*(3/4):147 (1980).

18. J. V. Carbonell, F. Pinega, and J. L. Pena, Solar drying of food products. III. Description of a pilot dryer and evaluation of a flat plate collector, *Revista de Agroquimica y Tecnologia de Alimentos*, *23*(1):107 (1983).

19. J. H. Moy and M. J. L. Kuo, Solar osmovac dehydration of papaya, *J. Food Process Eng.*, *8*(1):23 (1985).

20. W. B. Van Arsdel and M. J. Copley, *Food Dehydration*, Vol. 2, AVI Publishing, Westport, Connecticut (1973).

21. S. D. Holdsworth, Dehydration of foodstuffs—A review, *J. Food Tech.*, *6*:331 (1971).

22. J. G. Brennen, Dehydration of foodstuffs. In *Water and Food Quality* (T. M. Hardman, ed.), Elsevier Applied Science Publishers, London, p. 33 (1989).

23. K. S. Jayaraman, V. K. Gopinathan, P. Pitchamuthu, and P. K. Vijayaraghavan, Preparation of quick cooking dehydrated vegetables by high temperature short time drying, *J. Food Tech.*, *17*(6):669 (1982).

24. M. Karel, Dehydration of foods. In *Principles of Food Science Part II. Physical Principles of Food Preservation* (M. Karel, D. R. Fennema, and D. R. Lund, eds.), Marcel Dekker, New York, p. 309 (1974).

25. D. R. Heldman and R. P. Singh, Food dehydration. In *Food Process Engineering*, 2nd Ed., AVI Publishing, Westport, Connecticut, p. 261 (1981).

26. M. E. Lazar and D. F. Farkas, Centrifugal fluidized bed drying—A review of its application and potential in food processing. In *Drying '80*, Vol. 1 (A. S. Mujumdar, ed.), Hemisphere, New York, p. 242 (1980).

27. G. E. Brown, D. F. Farkas, and E. S. De Marchena, Centrifugal fluidized bed blanches, dries and puffs piece-form foods, *Food Tech.*, *26*(12):23 (1972).

28. P. F. Hanni, D. F. Farkas, and G. E. Brown, Design and operating parameters for a continuous centrifugal fluidized bed dryer (CFB), *J. Food Sci.*, *41*(5):1172 (1976).

29. M. W. Cannon, New dryer for vegetables, *Food Tech. New Zealand, 13*(9):28 (1978).
30. J. L. Baxeires, Y. S. Yow, and H. Gilbert, Study of the fluidized bed drying of various food products, *Lebensm.-Wiss. u.-Technol., 16*:27 (1983).
31. J. F. Sullivan and F. C. Craig, Jr., The development of explosion puffing, *Food Tech., 38*(2): 52 (1984).
32. W. K. Heiland, J. F. Sullivan, R. P. Konstance, J. C. Craig, Jr., J. Cording, Jr., and N. C. Aceto, A continuous explosion puffing system, *Food Tech., 31*(11):32 (1977).
33. M. F. Kozempel, J. F. Sullivan, J. C. Craig, Jr., and R. P. Konstance, Explosion puffing of fruits and vegetables, *J. Food Sci., 54*(3):772 (1989).
34. R. E. Berry, C. J. Wagner, O. W. Brisset, and M. K. Veldhuis, Preparation of instant orange juice by foam mat drying, *J. Food Sci., 37*:803 (1972).
35. R. V. Decareau, *Microwaves in the Food Processing Industry*, Academic Press, Orlando, p. 79 (1985).
36. K. Masters, *Spray Drying Handbook*, 4th Ed., George Godwin, London (1985).
37. K. S. Jayaraman and D. K. Das Gupta, Preparation and storage stability of some instant fruit flavoured milk and lassi beverage powders, *Beverage and Food World (India), 16*(4):15 (1989).
38. M. E. Lazar and J. C. Miers, Improved drum-dried tomato flakes are produced by a modified drum drier, *Food Tech., 25*:830 (1971).
39. J. A. Kitson and D. R. MacGregor, Drying fruit purees on an improved pilot plant drum dryer, *J. Food Tech., 17*(2):285 (1982).
40. M. Manlan, R. F. Mathews, R. P. Bates, and S. K. O'Hair, Drum drying of tropical sweet potatoes, *J. Food Sci., 50*(3):764 (1985).
41. M. F. Kuzempel, J. F. Sullivan, J. C. Craig, Jr., and W. W. Heiland, Drum drying of potato flakes—A predictive model, *Lebensm.-Wiss. u.-Technol., 19*(3):193 (1986).
42. C. J. King, *Freeze Drying of Foods*, CRC Press, Cleveland, Ohio (1971).
43. S. A. Goldblith, L. Rey, and W. W. Rothmayr, *Freeze Drying and Advanced Food Technology*, Academic Press, London (1975).
44. J. Lorentzen, Freeze drying: The process, equipment and products. In *Developments in Food Preservation*, Vol. 1 (S. Thorne, ed.), Applied Science Publishers, London, p. 153 (1981).
45. G. J. Malecki, P. Shinde, A. I. Morgan, Jr., and D. F. Farkas, Atmospheric fluidized bed freeze drying, *Food Tech., 24*(5):93 (1970).
46. T. H. Woodward, Freeze drying without vacuum, *Food Eng. (New York), 35*(6):96 (1963).
47. F. W. Schmidt, Y. S. Chen, M. Kirby-Smith, and J. H. MacNeil, Low temperature air drying of carrot cubes, *J. Food Sci., 42*(5): 1294 (1977).
48. D. R. Heldman and G. A. Honer, An analysis of atmospheric freeze drying, *J. Food Sci., 39*: 147 (1974).
49. K. E. Yassin and H. Gibert, Atmospheric freeze-drying of foods in a fluidized bed of finely divided adsorbent, *Proc. 6th Int. Congr. Food Sci. Tech., 1*:208–209 (1983).
50. G. A. Bell and J. D. Mellor, Further developments in adsorption freeze drying, *CSIRO Food Res. Qrtly., 50*(2):48 (1990).
51. C. R. Lerici, D. Mastrocola, A. Sensidoni, and M. Dalla Rosa, Osmotic concentration in food processing. In *Preconcentration and Drying of Food Materials* (S. Bruin, ed.), Elsevier Science Publishers B.V., Amsterdam, p. 123 (1988).
52. D. F. Farkas and M. E. Lazar, Osmotic dehydration of apple slices: Effect of temperature and syrup concentration on rates, *Food Tech., 23*:688 (1969).
53. M. Riva and C. Peri, Osmotic dehydration of grapes, *Proc. 6th Int. Congr. Food Sci. Tech., 1*:179–180 (1983).
54. A. Lenart and J. M. Flink, Osmotic concentration of potato. I. Criteria for the end point of the osmosis process, *J. Food Tech., 21*(2):307 (1984).
55. T. R. A. Magee, A. A. Hassaballah, and W. R. Murphy, Internal mass transfer during osmotic dehydration of apple slices in sugar solution, *Irish J. Food Sci. Tech., 7*(2):147 (1983).

56. M. S. Rahman and J. Lamb, Osmotic dehydration of pineapple, *J. Food Sci. Tech.* (*India*), 27(3):150 (1990).

57. K. Videv, S. Tanchev, R. C. Sharma, and V. K. Joshi, Effect of sugar syrup concentration and temperature on the rate of osmotic dehydration of applies, *J. Food Sci. Tech.* (*India*), 27(5):307 (1990).

58. N. D. Heidelbaugh and M. Karel, Intermediate moisture food technology. In *Freeze Drying and Advanced Food Technology* (S. A. Goldblith, L. Rey, and W. W. Rothmayr, eds.), Academic Press, London, p. 619 (1975).

59. K. S. Jayaraman, Development of intermediate moisture tropical fruit and vegetable products—Technological problems and prospects. In *Food Preservation by Moisture Control* (C. C. Seow, ed.), Elsevier Applied Science Publishers, London, p. 175 (1989).

60. E. Maltini, D. Torreggiani, G. Bertolo, and M. Stecchini, Recent developments in the production of shelf stable fruit by osmosis, *Proc. 6th Int. Congr. Food Sci. Tech.*, *1*:177-178 (1983).

61. S. M. Alzamora, L. N. Gerschenson, P. Cerrutti, and A. M. Rojas, Shelf-stable pineapple for longterm non-refrigerated storage, *Lebensm.-Wiss. u.-Technol.*, 22(5):233 (1989).

62. E. S. Della Monica and P. E. McDowell, Comparison of beta-carotene content of dried carrots prepared by three dehydration processes, *Food Tech.*, *19*:1597 (1965).

63. M. D. Nutting, H. J. Neumann, and J. R. Wagner, Effect of processing variables on the stability of beta-carotene and xanthophylls of dehydrated parsley, *J. Sci. Food Agr.*, *21*:197 (1970).

64. H. R. Bolin and A. E. Stafford, Effect of processing and storage on provitamin A and vitamin C in apricots, *J. Food Sci.*, *39*:1034 (1974).

65. T. Desrosiers, T. G. Smyrl, and G. Paquette, Retention of carotene in green peppers and peaches after a home dehydration process, *Can. Inst. Food Sci. Tech.*, *18*(2):144 (1985).

66. A. A. Adenike, Ascorbic acid retention of stored dehydrated Nigerian vegetables, *Nutrition Reports Int.*, *24*(4):769 (1981).

67. E. R. Shadle, E. E. Burns, and L. J. Talley, Forced air drying of partially freeze dried compressed carrot bars, *J. Food Sci.*, *48*(1):193 (1983).

68. A. K. Baloch, K. A. Buckle, and R. A. Edwards, Effect of processing variables on the quality of dehydrated carrot. II. Leaching losses and stability of carrots during dehydration and storage, *J. Food Tech.*, *12*:295 (1977).

69. S. S. Arya, V. Natesan, D. B. Parihar, and P. K. Vijayaraghavan, Stability of carotenoids in dehydrated carrots, *J. Food Tech.*, *14*:579 (1979).

70. A. K. Baloch, K. A. Buckle, and R. A. Edwards, Effect of sulphur dioxide and blanching on the stability of carotenoids of dehydrated carrots, *J. Sci. Food Agr.*, *40*:179 (1987).

71. F. Lajolo, S. R. Tannenbaum, and T. P. Labuza, Reaction at limited water concentration. 2. Chlorophyll degradation, *J. Food Sci.*, *36*:850 (1971).

72. F. M. Lajolo and U. M. L. Marquez, Chlorophyll degradation in a spinach system at low and intermediate water activities, *J. Food Sci.*, *47*:1995 (1982).

73. H. R. Bolin and R. J. Steele, Nonenzymatic browning in dried apples during storage, *J. Food Sci.*, *52*(6):1654 (1987).

74. D. J. McWeeny, M. E. Knowles, and J. F. Hearne, The chemistry of nonenzymatic browning in foods and its control by sulphites, *J. Sci. Food Agr.*, *25*:735 (1974).

75. D. H. Palmer, A. W. Taylor, and M. K. Withers, Flavor, texture and color of air dried and freeze dried vegetables, *Proc. 1st Int. Congr. Food Sci. Tech.*, 4:37 (1965).

76. M. E. Falconer, M. J. Fishwick, D. G. Land, and E. R. Sayer, Carotene oxidation and off flavour development in dehydrated carrot, *J. Sci. Food Agr.*, *15*:897 (1964).

77. S. J. Bishov, A. S. Henick, J. W. Giffee, I. T. Nid, P. A. Prell, and M. Wolf, Quality and stability of some freeze dried foods in "zero" oxygen head space, *J. Food Sci.*, *36*:532 (1971).

78. R. P. Konstance, J. F. Sullivan, F. B. Talley, M. J. Calhoun, and J. Craig, Jr., Flavor and storage stability of explosion puffed potatoes: Autoxidation, *J. Food Sci.*, *43*:411 (1978).

79. J. M. Tuomy and W. Fitzmaurice, Effect of ingredients on the oxygen uptake of cooked, freeze-dried combination foods, *J. Agr. Food Chem.*, *19*(3):500 (1971).

80. S. Pendlington and J. P. Ward, Histological examination of some air dried and freeze dried vegetables, *Proc. 1st Int. Congr. Food Sci. Tech.*, 4:55 (1965).

81. K. S. Jayaraman, D. K. Das Gupta, and N. Babu Rao, Effect of pretreatment with salt and sucrose on the quality and stability of dehydrated cauliflower, *Int. J. Food Sci. Tech.*, 25:47 (1990).

82. D. M. Khedkar and S. K. Roy, Histological evidence for the reconstitutional property of dried/dehydrated raw mango slices, *J. Food Sci. Tech.* (*India*), *17*(6):276 (1980).

83. H. J. Neumann, Dehydrated celery; Effects of predrying treatments and rehydration procedures on reconstitution, *J. Food Sci.*, *37*:437 (1972).

84. A. Levi, N. Ben-Shalom, D. Plat, and D. S. Reid, Effect of blanching and drying on pectin constituents and related characteristics of dehydrated peaches, *J. Food Sci.*, *53*(4):1187 (1988).

85. G. Mazza, Moisture sorption isotherm of potato slices, *J. Food Tech.*, *17*:47 (1982).

86. H. R. Bolin, Relation of moisture to water activity in prunes and raisins, *J. Food Sci.*, *45*: 1190 (1980).

87. J. A. Troller, Water activity and food quality. In *Water and Food Quality* (T. M. Hardman, ed.), Elsevier Applied Science Publishers, London, p. 1 (1989).

88. C. M. Samaniego-Esguerra, I. F. Boag, and G. L. Robertson, Kinetics of quality deterioration in dried onions and green beans as a function of temperature and water activity, *Lebensmi-Wiss. u.-Tech.*, *24*(1):53 (1991).

89. P. A. Cribbs, Microbiology of dried foods. In *Concentration and Drying of Foods* (D. MacCarthy, ed.), Elsevier Applied Science Publishers, London, p. 89 (1986).

90. R. H. Vaughn, The microbiology of dehydrated vegetables, *Food Res.*, *16*:429 (1951).

91. W. S. Clarke, Jr., G. W. Reinbold, and R. S. Rambo, Enterococci and coliforms in dehydrated vegetables, *Food Tech.*, *20*(10):113 (1966).

92. M. J. Fanelli, A. C. Peterson, and M. F. Gunderson, Microbiology of dehydrated soups. I. A survey, *Food Tech.*, *19*(1):83 (1965).

93. J. M. Sheneman, Survey of aerobic mesophilic bacteria in dehydrated onion products, *J. Food Sci.*, *38*:206 (1973).

94. H. Heath, The microbiology of onion products, *Food, Flavg. Ingr. Proc. Packg.*, *5*(1):22 (1983).

95. R. Villota, I. Saguy, and M. Karel, Storage stability of dehydrated food: Evaluation of literature data, *J. Food Qual.*, *3*:123 (1980).

22
Osmotic Dehydration of Fruits and Vegetables

Piotr P. Lewicki and Andrzej Lenart
Warsaw Agricultural University, Warsaw, Poland

1. INTRODUCTION

Among the food products present on the market there is a group of foodstuffs that resembles dried products by its microbial stability, but it cannot be classified as dry food. Its water content is quite high, thus the food is plastic and chewy. Dried plums, figs, and raisins, candied fruits, cottage ham, and dry sausage are good examples of this type of food.

This food contains 20% to 50% water and is called *intermediate moisture food* (IMF). It is microbiologically stable, but quite susceptible to chemical changes. The IMF undergoes Maillard nonenzymatic browning at a higher rate than the dry product. It is less pliant to oxidation. If processing parameters applied do not denature proteins, some enzymatic activity can be still present in the IMF.

The microbial stability of the IMF is due to sufficiently low water activity in the material. The lowering of water activity can be achieved in two ways, either by addition of humectants or by removal of solvent (i.e., water). The first way is mostly unacceptable by consumers since it needs large amounts of sodium chloride, sugars, or polyols to be added to food. Moreover, this way is limited by nutritional and toxicological restraints. The other way is energy intensive, hence the final product is expensive.

The use of osmosis allows both ways of decreasing water activity in food to be applied simultaneously. The permeability of plant tissue is low to sugars and high molecular weight compounds, hence the material is impregnated with the osmoactive substance in the surface layers only. Water, on the other hand, is removed by osmosis and the cell sap is concentrated without a phase transition of the solvent. This makes the process favorable from the energetic point of view. The flux of water is much larger than the countercurrent flux of osmoactive substance. For this reason the process is called *osmotic dehydration* or *osmotic dewatering*.

The food produced by this method has many advantageous features:

It is ready to eat and rehydration is not needed.

The amount of osmoactive substance penetrating the tissue can be adjusted to individual requirements.

The chemical composition of the food can be regulated according to needs.

The mass of the raw material is reduced, usually by half.

The osmotic dehydration does not reduce water activity sufficiently to hinder the proliferation of microorganisms. The process extends, to some degree, the shelf life of the material, but it does not preserve it. Hence, the application of other preservation methods, such as freezing, pasteurization, or drying is necessary. However, processing of osmotically dehydrated semiproducts is much less expensive and preserves most of the characteristics acquired during the osmosis.

2.   THE NATURE OF OSMOTIC PRESSURE

Each substance is characterized, from the thermodynamic point of view, by internal energy, which is a sum of energy of all molecules and the energy of their reciprocal interactions. The change of internal energy of the substance in the system caused by the change of its concentration by one mole is called the *chemical potential*, expressed by the equation

$$\mu_i = \frac{\partial U}{\partial n_i}\bigg)_{S,V,n_j} \tag{1}$$

where

μ_i = chemical potential of the species i
U = internal energy
n_i = concentration of species i
S = entropy
V = volume
n_j = concentration of species j

The interaction of a system with its surroundings or with another system is manifested by the energy exchange. This exchange proceeds until the equilibrium state is achieved, which is the state in which chemical potentials of a system and its surroundings or of two systems are the same.

Chemical potential is a function of concentration, temperature, and pressure. Under isothermal conditions it is solely determined by concentration and pressure. The increase of solute concentration decreases the chemical potential of a solvent, which can be brought to the initial value by the increase of pressure. Hence, a pure solvent can be in equilibrium with a solution if the solution is under appropriate pressure.

Excess pressure needed to reach the state of equilibrium between pure solvent and a solution is called *osmotic pressure*. In the equilibrium state,

$$\mu_{solvent}^{solvent}\bigg)_{P_1} = \mu_{solvent}^{solution}\bigg)_{P_2} \tag{2}$$

and the osmotic pressure is expressed by the formula:

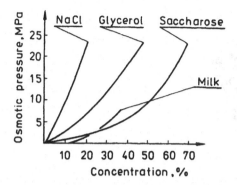

Figure 1   Relationship between concentration and osmotic pressure of a solution.

$$\Pi = -\frac{RT}{V} \ln a \tag{3}$$

where

  R = gas constant
  T = absolute temperature
  V = molar volume
  a = activity

Since in foods water is the solvent, Eq. [3] can be simplified to

$$\Pi = -4.6063 \cdot 10^5 T \ln a_w \tag{4}$$

where a_w is water activity.

Osmotic pressure is related to molar mass of the solute; the smaller the mass is, the higher the pressure at the same concentration will be. Electrolytes show higher osmotic pressure than nonelectrolytes since each ion affects the chemical potential of a solvent.

Relationship between concentration and osmotic pressure is shown on Figure 1. Osmotic pressure has an inhibitory effect on microorganisms. Most bacteria do not proliferate at $\Pi > 12.7$ MPa, yeasts at $\Pi > 17.3$ MPa, and molds at $\Pi > 30.1$ MPa. Hence, the shelf life of foods can be regulated by the osmotic pressure of the solution in the material.

3.   THE STRUCTURE OF PLANT TISSUE

Fruits and vegetables undergoing processing come from different parts of a plant. They are roots (carrots, parsley, beet roots), stems (kohlrabi, potatoes), shoots (asparagus, onions), leaves (cabbages, spinach), flowers (cauliflower, broccoli), fruits (tomatoes, cucumbers, pumpkins, apples, pears, plums, green bean), and seeds (green peas, beans). All these parts of a plant consist of cells that are highly specialized. A group of cells that is designated to play a special role in a plant is called *tissue*.

In general, three types of tissue are recognized. *Epidermal tissue* forms the outermost layer of cells that are thick walled and covered, in many cases, with cuticle containing a waxy substance known as *cutin*. *Parenchymatous tissue* forms the essential part of organs and serves to produce and store nutritional substances. The cells are predominantly large, thin walled, and highly vacuolated. *Vascular tissue* is composed of the ducts that carry a

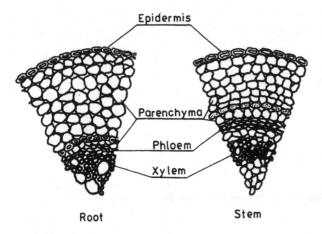

Figure 2 Arrangements of tissues in plant organs.

solution of minerals and nutritious substances in a plant. The arrangement of these tissues in different parts of a plant is shown on Figure 2.

A plant cell can be simply pictured as a unit consisting of two main components: the cell wall and the protoplast (Fig. 3). Cellulose is the main component of the cell wall. Its content is between 60–90% and depends strongly on the stage of maturity. Other components include pectins, hemicellulose, and mineral compounds. The cell wall is permeable to water and low molecular weight compounds and is not a barrier in solute transport from/to the cell. The cell wall is perforated and the channels are filled with thin strands of protoplasm, assuring the contact between protoplasts of neighboring cells. These strands of protoplasm are called *plasmodesmata*. The diameter of the strands is 20–70 nm and the average contact area can be estimated as 0.2 m^2/m^2 of the cell wall (1).

The protoplast is composed of protoplasm enclosed in a membrane called *plasmalemma*, vacuoles, and other structural elements such as the nucleus, plastids, and so on. The plasmalemma is a protein-lipid layer that regulates the contact between the protoplast and the environment. It is 7.5–10 nm thick (2), permeable to water, and selectively

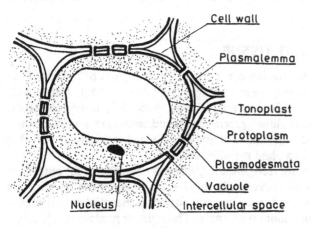

Figure 3 A plant cell (simplified).

permeable to other substances. Protoplasm is a colloidal solution of proteins and lipoproteins in water. The vacuole is suspended in protoplasm and is enclosed in a membrane called the *tonoplast*. It contains a solution of minerals, sugars, and other organic compounds in water.

Most cells have dimensions between 10 and 100 μm. Depending on their function they are loosely or closely packed in a tissue. Usually, parenchyma cells are loosely arranged in the tissue and intercellular spaces are formed. It is estimated that the volume occupied by cell walls and intercellular spaces accounts for 7–10% of the tissue volume (3). Intercellular spaces form a continuous system of channels that is filled with air.

A particular type of tissue is the vascular one. It contains *xylem* and *phloem*, which form bundles. Xylem consists in elongated cells with perforated end walls that no longer contain viable protoplasm. Xylem, in other words, forms open dead vessels that provide a way of transportation for minerals and water from roots to other parts of a plant. Phloem consists in elongated viable cells that have sieve end plates. Phloem translocates a solution of sugars, amino acids, and other nutritious substances.

4. PLASMOLYSIS

A solution in a vacuole has an osmotic pressure that pushes protoplasm and plasmalemma toward the cell wall. The protoplast is tightly pressed to the cell wall and the cell is in a turgor state. The difference between the osmotic pressure in the cell and in its surroundings is called the *turgor pressure*.

If the cell and the surroundings have the same osmotic pressure then turgor pressure is zero and the system is in thermodynamic equilibrium. Osmotic pressure of the surroundings lower than that of the cell causes transfer of water into the cell. The cell swells, but the rigid cell wall limits the extent of swelling. A cell placed in a *hypertonic solution* (osmotic pressure higher than that of the cell) will lose water. The dehydration of a protoplast causes decrease of its volume and, in consequence, detachment of plasmalemma from the cell wall. This process is called *plasmolysis* (Fig. 4). Since the cell wall is permeable the volume between the cell wall and plasmalemma fills with the hypertonic solution.

5. TRANSPORT OF WATER IN PLANT TISSUE

Osmotic dehydration occurs on a piece of material and not on a single cell. Hence, it should be assumed that the piece exists in all kinds of plant tissue. As a rule a skin is removed from the raw material; therefore, epidermal cells and cuticle are absent in most cases. A piece of fruit or vegetable thus will contain parenchymatous and vascular tissue and intercellular spaces, as well.

From the process point of view a plant material can be considered as a capillary-

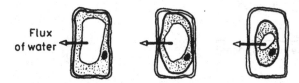

Flux of water

Figure 4 Plasmolysis.

porous body that is divided internally in numerous repeating units (Fig. 5). Some capillaries and pores are filled with a solution, while others are empty (i.e., contain air). Most capillaries and pores are open. Repeating units can exchange water between each other.

The internal structure of a body is not a homogeneous one as far as transport of water is considered. Cell walls are built from microfibrils, and intermicrofibrillar spaces are some 10 nm in cross-section (4). These spaces are large enough to allow water, ions, and small molecules to pass through them. Since cell walls are interconnected in the tissue, a continuous matrix capable of transporting water and small molecules is formed. This continuum is called the *apoplast*.

The structure of the matrix is modified by pectins and glycoproteins that give rise to fixed charges in the wall. The Donnan free space is formed in the apoplast, and it can bind ions preferentially or it can hinder their passage through the wall. Under certain conditions the apoplast can behave like an ion exchanger (4). This property can be especially important when osmotic dehydration is done in electrolytes.

Plasmalemma is the next barrier to the mass transfer in the tissue. In a majority of cells, protoplasm of neighboring cells is interconnected through plasmodesmata and another continuous network is formed. The system of protoplasts and connecting plasmodesmata is widely known as *symplast*. Since plasmodesmata permit the passage of solutes (1), they undoubtedly permit the passage of water also.

Each vacuole is enclosed in tonoplast and they are not interconnected. Hence, vacuoles form a discontinuity in the system.

Classic theory of water transport in plants assumed the movement of water from the vacuole of one cell to the vacuole of the neighboring cell; the driving force was the difference in water chemical potential. Recently two ways water is transported in a plant have been recognized: apoplasmic and symplasmic (Fig. 6). The hydraulic conductivity of cell walls is of the order of 10^{-2} m/(s·Pa), while for plasmalemma it is 1–6 10^{-4} m/(s·Pa) (4). The difference is two orders of magnitude. The diffusive permeability coefficient for ions is 10^{-3}–10^{-4} m/s for cell walls and 10^{-9}–10^{-10} m/s for plasmalemma (4). It is generally agreed that the cell walls provide the major pathway of water movement in plant material. The ratio of volume flows in the apoplasmic and symplasmic (vacuole-to-vacuole) pathways is of the order of 50 : 1 in leaf tissue (1). For the root cortex, the ratio is lower.

The capillary and porous system of the body exists in vascular tissue and intercellular spaces. Xylem forms an open conduit of relatively low hydraulic resistance that is filled with diluted mineral solution. Phloem exists in cells with a width ranging from 10 to 70

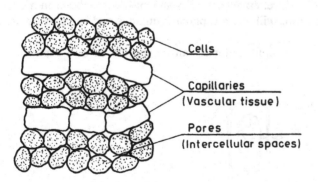

Figure 5   A plant material as a capillary-porous body.

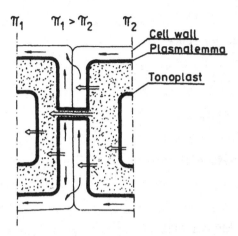

Figure 6 Apoplasmic and symplasmic transport of water.

μm and a length from 100 to 500 μm in dicotyledons (5). Their turgor is around 2 MPa (beet root is 1.83 MPa) with a pressure gradient of 0.02–0.03 MPa/m (6,7). Since phloem transports substances of very different molecular weight, shape, charge, and surface activity along with water, it is presumed that the mechanism is an osmotically driven solution flow (7).

The ability of phloem to transport solutions is large. In the following viable plants, a 1 cm^2 of cross-section of phloem cells is able to transport (8)

Solanum: 2.1–4.5 g d.m./h
Pyrus, *Prunus*: 0.6 g d.m./h
Cucurbita: 3.3–4.8 d.m./h

On the basis of these data it has been calculated that translocation was about four orders of magnitude faster than diffusion. Hence, the osmotic pressure flow seems to be the most probable mechanism of solution transportation in plants.

The intercellular system of channels has the volume dependent on the kind of tissue. In potatoes, it occupies 1% to 3%, while in beet root 25% of volume is attributed to cell walls and intercellular spaces (9).

There is no doubt that all these structures of the transport system in the plant tissue will participate in the process of osmotic dehydration.

Contacting plant tissue with the hypertonic solution, a sequence of mass transfer processes can be envisaged (Fig. 7) as follows:

In intercellular spaces a capillary suction will occur. The channels will fill in with the hypertonic solution and the gas phase will be compressed until the equilibrium state will be achieved.

Xylem and phloem containing solutions of lower osmotic pressure than the hypertonic solution will be penetrated by the osmoactive substance by diffusion. Osmotic pressure flow can also take place.

Cell walls in contact with hypertonic solution will lose water due to diffusion and osmotic flow. Osmoactive substance will penetrate cell walls by diffusion.

Change of osmotic pressure in xylem and phloem and the dewatering of the cell walls will initiate the symplasmic movement of water in the material. The dehydration of the cells will take place and plasmolysis will be induced.

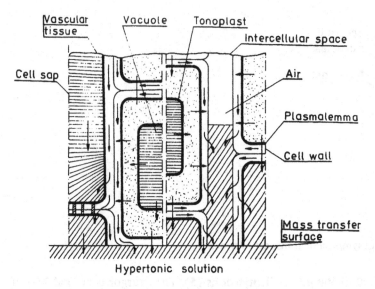

Figure 7   Possible ways of transport of water in a plant material during osmotic dehydration.

The sequence presented above also suggests the kinetics of the osmotic dehydration process. As long as the mass transfer processes is not strongly dependent on the symplasmic pathway, the water flux will predominate over the osmoactive substance flux. This is due to osmotic pressure flow, which will reduce the countercurrent diffusion of osmoactive substance but it will not strongly affect the diffusive flux of water since the self-diffusion of water in a solution is of the same order of magnitude as that for solute. When plasmolysis occurs and the hypertonic solution fills in the volume between cell walls and plasmalemma, the process of dewatering will be substituted by the impregnation of the tissue. The flux of osmoactive substance will be equal to or will surpass the flux of water.

The above picture of the mechanism of osmotic dehydration suggests that the plasmalemma resistance to mass transfer affects the process to only a small extent. The process will be rather dependent on the internal resistance to osmotic flow and apoplast dewatering and, to some extent, on external resistance to mass transfer.

6.   MODELING THE OSMOTIC DEHYDRATION PROCESS

From the previous description of the structure of the plant material and processes that can be involved in the mass transfer between plant tissue and the osmotic solution, it is evident the modeling of osmotic dehydration is not simple.

The first major effort to describe an osmotic mass transfer in plant tissue based on its cellular properties was done by Philip (10–12). He assumed a linear aggregation of cells and a nondiffusible, nonpermeating solute. Molz and Hornberger extended this model to include a diffusible permeating solute (13). In both models the transport across biological membranes was applied and that assumption made the models fail in describing osmotic phenomena in a whole piece of tissue.

Molz and Ikenberry built a model in which mass fluxes in the cell wall and intercellular space were allowed (14). The model was completed by Molz by accounting for symplastic and apoplastic transport (15). Further development of Molz's model included the discrete cellular geometry and anatomy of the tissue, hence the behavior of each individual cell could be simulated (16). Philip's and Molz's models have been developed to describe the behavior of the plant tissue under normal growing conditions, that is, when the osmotic pressure in the tissue is higher than that in the surroundings and the plant cell is in the turgor state.

Little work has been done on modeling the osmotic dehydration process. Mostly the theory of molecular diffusion in the solid has been used to predict the water loss during the process. An unsteady unidirectional diffusion described by the second Fick's equation was used to quantify the process by the effective diffusivity. The resulting diffusivity is generally correlated with the concentration and the temperature of the hypertonic solution (17,18).

The models based on the second Fick's equation do not necessarily simulate the osmotic dehydration process. In this process countercurrent fluxes of water and the osmoactive substance occur. Moreover, a flux of soluble solids from the tissue accompanies a flux of water (Fig. 8). Hence, there is a simultaneous mass transfer and probable interactions between flows cannot be taken into account. The estimated effective diffusivities are affected by the countercurrent flows and they cannot be used to predict the contribution of each flux to the process. Moreover, in these models the resistance at the surface of the solid is assumed to be negligible, thus the whole resistance to mass transfer is in the solid. Last, the models do not take into account the possible effect of the living cell on the mass transfer process.

Besides all these limits, the models based on the second Fick's equation proved to be quite successful. Hawkes and Flink plotted the normalized solids content of apple versus the square root of time and obtained a straight line, the slope of which was called the *mass transfer coefficient* (17). This approach was used in numerous works (18–26).

Recently Toupin et al. (27,28), using the irreversible process thermodynamics, developed a model in which the cell membrane characteristics, the cell volume changes, tissue shrinkage, and internal volumetric rearrangements and diffusion of nonpermeating and permeating species are taken into account. The set of equations solved numerically showed the model as satisfactorily representing the behavior of parenchymatous storage tissue undergoing osmotic dehydration. Moreover, the simulations have shown that the cell membrane represents the major resistance to mass transfer in such systems. The model needed simultaneous adjustment of four constants to obtain a good fit, hence its practical usefulness is rather questionable.

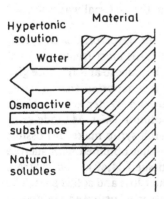

Figure 8   Mass transfer in osmotic dehydration.

Marcotte et al. improved Toupin's model by giving a closer thermodynamic description of forces involved in the osmotic dehydration process (29). The transmembrane transport is modeled on the basis of irreversible thermodynamics while transport in the intercellular space is modeled by relations derived from the second Fick's equation.

The models based on the irreversible process thermodynamics show that the cell membrane (plasmalemma) represents the major resistance to mass transfer. This is contradicted by findings of Raoult et al. (30), who showed that membranes are not necessary for osmotic dehydration and merely diffusive properties of the material are responsible for high water flux with only marginal sugar penetration. The following mechanism is suggested by these authors.

At the beginning of the process the removal of water concentrates the superficial layer of solute in the surface of the material. This layer is detrimental to further solute incorporation but is favorable to water removal since it creates a pronounced concentration gradient (31-33). The compartmental model was developed that provided good fit for the different situations tested. The solution of the set of differential equations was done by numerical methods.

It is also possible to describe the process by empirical correlations. However, these correlations are limited to the investigated product and treatment conditions, and they cannot be used to model the process in general.

The lack of full understanding of mechanisms involved in the osmotic dehydration process makes it difficult to control the main variables of the process. Hence, current technologies are somewhat empirical.

7. OSMOACTIVE SUBSTANCES

Osmoactive substances used in food must comply with special requirements. They have to be edible with accepted taste and flavor, nontoxic, inert to food components, if possible, and highly osmotically active. Sucrose, lactose, glucose, fructose, maltodextrins, and starch or corn syrups are commonly used in osmotic dehydration of fruits and vegetables. Honey, glycerol, plant hydrocolloids, and sodium chloride were also tested (34-36).

Quality of the final product was the main aim of most experiments testing the suitability of different osmoactive substances. Their technological applicability is estimated on water loss rate and final water content in the material. Usually, saturated solutions, or the solutions at the same concentrations, are compared. Flink used two criteria to rate different osmoactive substances: water loss and the amount of the substance penetrating the material being osmosed (37). Lowering of water activity in the material was also used as an indicator of suitability of an osmoactive substance (38).

Solutions of sugars are mostly used to dehydrate fruits and glycerol, starch syrup, and sodium chloride are used for vegetables (39-42). Sucrose is the most frequently used substance (43-45). It can be substituted in part by lactose (17). Glucose and fructose give a similar dehydration effect (46,47). In other publications it is reported that fructose increases the dry matter content by 50% as compared with sucrose. Water activity of the final product was also lower with fructose as a hypertonic solution (48). Maltodextrins are less effective than sucrose at the same concentration. Starch syrup makes it possible to have similar final water content in dehydrated material as that obtained with sucrose but at a much lower influx of osmoactive substance into tissue (44,49). The dextrose equivalent of the syrup affected strongly the ratio between water loss and solids gain. The effect of the kind of osmoactive substance on the water content of osmosed material is presented in Figure 9 (50).

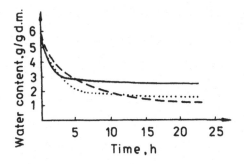

Figure 9 The effect of osmoactive substance on the course of osmotic dehydration of apples at 30°C (—— glucose; ------ saccharose; · · · · starch syrup). (From Reference 50.)

Mixtures of osmoactive substances are also used. Maltini et al. used sucrose and starch syrup in a ratio of 1 : 1 (51). Lerici et al. dehydrated apples in a solution containing 42% fructose, 52% sucrose, 3% maltose, 3% polysaccharides, and 0.5% sodium chloride in dry matter (52). Mastrocola et al. used solutions containing sucrose and fructose in varying proportions (53). Water loss was similar for all solutions tested but the penetration of the osmoactive substance was different. Peaches dehydrated in solutions of glucose and fructose were especially suitable to pasteurization (54).

Sodium chloride was used to dehydrate vegetables. Speck et al. used 10% solutions to dehydrate carrots (55). Lewicki et al. used 15% NaCl to dehydrate carrots and potatoes (56). Adambounou et al. dehydrated paprika, tomatoes, and eggplant in saturated salt solutions, getting water activity as low as 0.8 (57). Use of sodium and potassium chlorides made it possible to regulate sodium and potassium content in dehydrated corn and green peas. The use of sodium chloride is questionable from the nutritional point of view. Hence, only low concentrations of sodium chloride in syrups are recommended.

It has been found that the addition of low molecular weight substances such as sodium chloride, malic acid, lactic acid, and hydrochloric acid in concentrations of 1% to 5% to sugars or starch syrups improves the process of osmotic dehydration. In general they promote removal of water from the material. Calcium chloride and malic acid were added to sucrose to improve the texture of osmosed apples (58).

8. PROCESSING PROCEDURES FOR OSMOTIC DEHYDRATION

Osmotic dehydration can be done basically in two ways: by a static or a dynamic process. In a static process material is mixed with an osmoactive substance, which can be used as crystals or solution, and the mixture is left motionless until the desired water loss is achieved. It has been shown that the mass transfer resistance in this method is higher than that observed in a dynamic process (59,60).

In a dynamic process, the mixture is mixed; different methods of mixing can be used. Movement of food particles in a stationary solution (Fig. 10), mixing of the whole suspension (Fig. 11), and the flow of the osmoactive substance through the stationary layer of food pieces (Fig. 12) are the commonly used designs of the dynamic process. If crystals of the osmoactive substance are used, the fluidized bed is the solution for the dynamic process. It has been shown that the rate of motion has little effect on the rate of osmotic dehydration (61,62). It is just sufficient to induce motion of particles or solution in the system to have increased mass transfer rates.

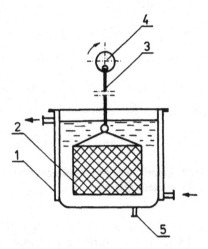

Figure 10 Osmotic dehydrator with a vibrating basket (1, jacketed vessel; 2, basket; 3, shaft; 4, eccentric; 5, spout).

The rate and efficiency of the process are dependent on such parameters as the kind and concentration of the osmoactive substance; the weight ratio of the solution to food; the kind of osmosed material, its size, and shape; temperature and pressure; and the pretreatment of the material prior to osmosis.

The rate of osmosis increases with increase of concentration of the osmoactive substance (Fig. 13) (63). The weight loss of mango and papaya is linearly dependent on sucrose concentration up to 60% (64). At higher concentrations a lower rate is observed and the impregnation of fruits with saccharose is high. Ponting used 65–70% sucrose

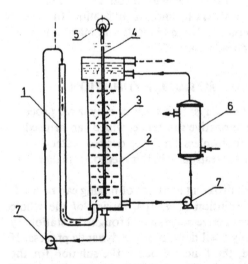

Figure 11 Osmotic dehydrator with a vibrating plate mixer (1, feed leg; 2, vessel; 3, vibrating mixer; 4, shaft; 5, eccentric; 6, heat exchanger; 7, pump). (Adapted from Reference 62.)

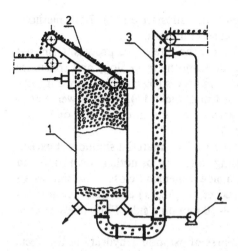

Figure 12 Osmotic dehydrator—a packed bed unit (1, vessel; 2, redler conveyor; 3, feed leg; 4, pump). (Adapted from Reference 62.)

solutions to dehydrate apples (45). The rate of dehydration of apples in starch syrup was not affected by its concentration within 42% and 50% (65) and with its repeated use. Pinnavaia et al. recommend 70% starch syrup for osmodehydration of apples (44). A high rate of dehydration of carrot and potato was obtained with 15% sodium chloride solution (66,67).

Crystalline osmoactive substance is used at a weight ratio of 1 : 1 to fruits (45). For solutions, investigations were done at weight ratios 1 : 1 to 1 : 6 (65). Osmotic dehydration of fruits and vegetables is recommended to be done at a weight ratio 1 : 4 to 1 : 5 of food to osmoactive solution (68).

Water withdrawn from the material dilutes the hypertonic solution. Hence, it is important to keep its concentration constant, either by a continuous evaporation of excess

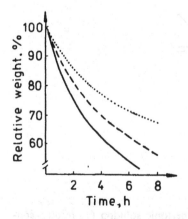

Figure 13 The effect of sugar syrup concentration on the course of osmotic dehydration of apples at 50°C (——, 70°Bx; -----, 60°Bx; · · · · · , 50°Bx). (Adapted from Reference 63.)

water (62,65) (Fig. 14) or by dissolution of osmoactive substance (65). Both methods make it possible to use the same hypertonic solution several times.

Shape and size of the material affect pronouncedly the rate of the process. Osmosed fruits and vegetables have different forms that come from the technology and consumer requirements (40). Plums were dehydrated whole or in halves (69); apples were cut in 12 segments (45,63) or sliced in slices 3 mm (70) or 3–4 mm thick (17). Peaches were cut in 6 or 8 segments and pears in 8 segments (45). Carrots were cut in cubes of 5 or 10 mm (55,71). Potatoes were sliced 50 and 10 mm thick (72).

Lenart and Lewicki have shown that the thickness of the material should not exceed 10 mm (61). Taking into account further processing following osmotic dehydration and use of the product, they considered a cube with a side dimension close to 10 mm as an optimal size and shape for most materials. Lewicki et al. (42,73) and Lerici et al. (74) dehydrated apples, carrots, and potatoes as cubes 8–10 mm on a side. Flink likewise used this shape in most of his studies (75).

Temperature has a substantial effect on the course of osmotic dehydration. It affects not only the rate of the process but also influences the chemical composition and properties of the product. Increased temperature increases the rate of chemical reactions and mass transfer processes as well. Viscosity of hypertonic solution is lowered and the diffusion coefficient of water increases with the increase of temperature.

Andreotti et al. recognize a temperature of 43°C as the optimal for osmotic dehydration of cherries and pears in glucose or glucose-fructose syrup (19). They recommended to osmose apricots at 20°C. Garcia et al. osmotically dehydrated bananas and papaya at 60°C (76). Temperatures in the range of 40°C to 60°C were used in osmotic dehydration of bananas (57). Apples were dehydrated at temperature 35–55°C (65) and 30–90°C (50) (Fig. 15). It has been shown that the increase of temperature in the range of 40°C to 80°C substantially shortens the time of dehydration (62). However, increased temperature promotes penetration of osmoactive substance into the tissue.

A high-temperature, short-time (HTST) process was proposed for osmotic dehydra-

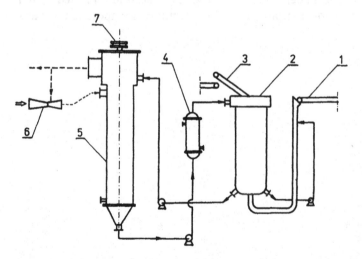

Figure 14 Osmotic dehydration with reconcentration of hypertonic solution (1, feeding conveyor; 2, osmotic dehydrator; 3, redler conveyor; 4, heat exchanger; 5, scraped surface evaporator; 6, thermocompressor; 7, driven wheel; —— flow of hypertonic solution; ------, vapor; ⇒, high pressure steam; _._._., heating steam).

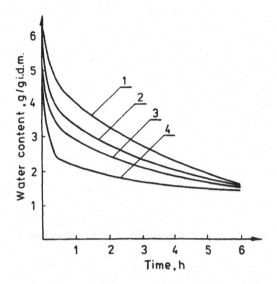

Figure 15 The effect of temperature on the course of osmotic dehydration of apples in saccharose solution (1, 30°; 2, 50°C; 3, 70°C; 4, 90°C). (From Reference 50.)

tion by Mastrocola et al. (53), Lerici et al. (74), and Levi et al. (77). The process is conducted at 65–90°C at a time of 1 to 20 min. The degree of dehydration is equivalent to that at 20°C lasting for 2 hours. The HTST process also gives the effect of blanching, which inactivates enzymes and removes part of the air from the intercellular space.

To obtain a high ratio between water loss and solids gain (Fig. 16), a temperature

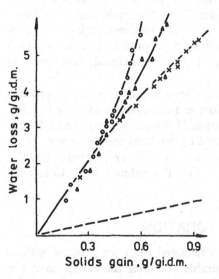

Figure 16 Relationship between water loss and solids gain in apples during osmotic dehydration in saccharose solution (O – O, 20°C; △–△, 30°C; + – +, 40°C; ------, diagonal). (From Reference 78.)

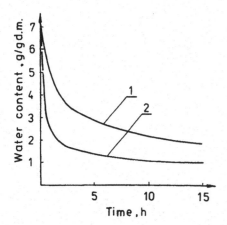

Figure 17   The effect of the kind of material on the course of osmotic dehydration in saccharose solution at 20°C (1, apple; 2, carrot).

between 20°C and 40°C is recommended (52,78,79). The degree of dehydration is regulated by the time of osmosis.

The course of mass loss versus time is curvilinear. The highest rates occur at the beginning of the process. According to Lenart optimal time of apple dehydration is 5–6 h at 20°C, 3–4 h at 30°C, and 1–1.5 h at 40°C (80). The reduction of material mass is some 35% and sugaring of the tissue is small. A reduction of mass by 50% can be achieved after 2.5 to 3 h of osmosis at 50°C.

Reduction of pressure during osmotic dehydration increases the rate of the process (17). It has also been observed that low pressure facilitates penetration of the osmoactive substance into the tissue (81).

It is well recognized that the characteristics of the material undergoing dehydration by osmosis affects strongly the course of the process. Some materials lose water faster than others under identical conditions (Fig. 17). The penetration of the osmoactive substance differs markedly (Fig. 18); hence, properties of the product and its consumer acceptance are strongly affected by the initial properties of the material, supposedly by its tissue structure.

Pretreatment of the material prior to osmosis affects the course of the process. Blanching of carrots and potatoes reduces water loss and increases solids gain (42,56). Hence, its effect is detrimental to osmotic dehydration of these materials. The influx of osmoactive substance into the tissue can be hindered by artificial semipermeable membranes covering the material. Substances such as pectins or starch can be used (41,82). It has been reported that immersion of some materials in $CaCl_2$ solution prior to osmotic dehydration affects the properties of the product.

9.   ENERGY ASPECTS OF OSMOTIC DEHYDRATION

Osmotic dehydration is distinctive in that water is removed from the product without undergoing the phase change. It offers a considerable potential for energy saving in comparison with convection drying. This potential has hardly been ever evaluated.

Most publications consider energy consumption in a drying process that is preceded by osmotic dehydration (83) or analyze the process under laboratory conditions (66,84).

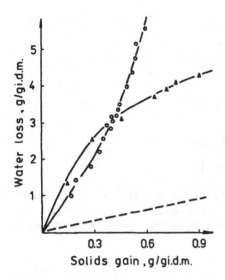

Figure 18 The effect of the kind of material on the relationship between water loss and solids gain in osmotic dehydration in saccharose solution at 20°C (O – O, apple; △ – △, carrot; ------, diagonal).

Energy consumption in the osmotic dehydration process arises from the following (85):

Heating of the material and osmoactive solution to the required temperature and making up a heat loss.

Solution mixing or pumping and recirculation and, depending on the variant applied,

Dissolution of hypertonic substance in a diluted solution.

Evaporation of water in an appropriate evaporator.

The problem of reconstitution of diluted hypertonic solution is the important one. The diluted solution can be completed with osmoactive substance to the desired concentration and used several times. Nevertheless, the surplus solution that is equal to the amount of water removed from the material must be managed. The amount of surplus solution depends on the degree of dewatering of the material.

It is estimated that dissolution of osmoactive substance in a hypertonic solution needs some 1 kJ/kg of water removed from the material. Hence this process affects energy consumption in osmotic dehydration negligibly.

The amount of water removed during osmotic dehydration is not large. Processing of 1 ton of fruits or vegetables per hour will give, at the most, 450 kg of surplus solution (i.e., some 5.5 tons of water evaporated per day). Hence, a single-effect evaporator with vapor recompression will meet the needs.

Energy use in the evaporator consists of electric energy for syrup circulation and heat for water evaporation. Electric energy use is estimated to be equal to 10 kJ/kg of evaporated water, and heat consumption is some 1.8 MJ/kg of evaporated water.

The increase of temperature shortens the osmotic process. The use of energy for syrup mixing or circulation is estimated as 17.2, 10.0, and 4.3 kJ/kg of water removed at temperatures 20°C, 30°C, and 40°C, respectively. To keep the process running at a desired temperature, a supply of heat is necessary. Depending on the amount of water

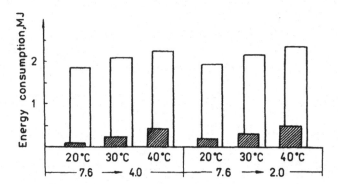

Figure 19 Energy use in osmotic dehydration expressed per kilogram of removed water. Temperature of the process and the degree of dewatering are the parameters (▨, diluted hypertonic solution completed with osmoactive substance; ☐, diluted hypertonic solution concentrated in evaporator).

removed from the material, the heat supply amounts to 180–240 kJ/kg at 30°C and 380–500 kJ/kg of water removed at 40°C.

Energy consumption in osmotic dehydration of fruits and vegetables under industrial conditions is estimated to be between 100 and 2400 kJ/kg of water removed (Fig. 19), depending on the temperature of the process and the way the surplus solution is managed. It is worthwhile to notice that convection drying needs some 5 MJ/kg of evaporated water, which is at least twice as much as is needed in osmotic dehydration.

10. PRODUCT CHARACTERISTICS

Osmotic dehydration is a complex process of countercurrent mass transfer between the plant tissue and hypertonic solution. This leads to dehydration of the material and changes in its chemical composition as well. Hence, it must be expected that the properties of the material dehydrated by osmosis will differ substantially from those dried by convection.

The flux of osmoactive substance penetrating the osmosed tissue changes its chemical composition. It has been shown that the content of sucrose increases in cell sap during osmotic dehydration (17,70), and the sucrose flux is increased by the presence of sodium chloride (72). On the other hand, use of starch syrup gives only a small influx of sugars to the material (65). Sodium chloride penetrates tissue very effectively, hence contacting of the material with this substance leads to salting rather than to dewatering of the tissue (42,56,72). There is also a flux of native substances leaving the tissue. Concentration of organic acids is lowered by 29–40% (70) and native sugars are replaced by sucrose (35,52).

Penetration of an osmoactive substance, except sodium chloride, is a surface process. Sugars penetrate to the depth of 2–3 mm while changes in water content are observed up to the depth of 5 mm (Fig. 20) (48,86,87). Lee and Salunkhe using saccharose labeled with C^{14} showed that the penetration depth is some 1 mm (88). When sodium chloride is used, it penetrates carrot tissue to a depth exceeding 12 mm.

Concentration of the cell sap and influx of osmoactive substance lower the water activity in the tissue (Table 1). The water-binding capacity of the tissue is also affected by the osmotic process, although changes are observed in surface layers only (90). Osmotic

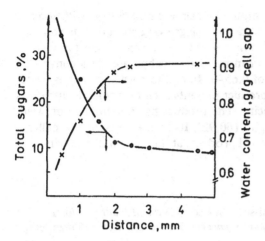

Figure 20 The depth of osmotic substance penetration and cell sap concentration in apple osmosed in 68.5% saccharose solution at 40°C for 4 hours. (From Reference 86.)

dehydration done for 0.5 h led to a sixfold decrease of water-binding forces at the surface of apple in comparison to the raw material (91). Water removal from the tissue by osmosis shows a much stronger effect on water-binding forces than the convection drying done to the same final water content.

As it has been stated previously osmotic dehydration cannot be treated as a food preservation process per se. It is a pretreatment that removes a certain amount of water from the material; to achieve shelf stability a further processing of the product is needed. Hence, the interaction of osmotic dehydration with further processing is important for quality assurance.

Use of osmotic dehydration practically eliminates the need to use preservatives such as sulfur dioxide in fruits (34). The process removes a substantial amount of air from the tissue, thus blanching prior to osmotic dehydration also can be omitted.

It has been shown that apples dried by osmosis and then frozen compared favorably with the conventional dehydrofrozen fruits (65). Osmotic dehydration preceding freeze drying shortens the time of the process and yields fruits superior to those not treated by osmosis (17,37,75). Osmotic dehydration followed by vacuum drying gives products that are very stable upon storage (45).

Table 1 Water Activity of Food Dehydrated by Osmosis

Product	Water activity
Apple	0.909
Apricot	0.946
Carrot	0.912
Cherry	0.931
Kaki	0.950
Peach	0.949
Strawberry	0.954

Source: From Reference 89

Most research has been directed toward combining osmotic dehydration with convection drying. The approach is of special interest due to the growing consumer demand for commodities in the freshlike state. The IMF comply well with consumer expectations.

Fruits and vegetables treated by osmosis can be further dehydrated in a convection dryer to lower the water activity to the level of 0.65–0.90. At those water activities, water content in the material is still high and the product presents such organoleptic attributes as chewiness, softness, elasticity, and plasticity. The product has a natural color, well-preserved flavor, and high retention of vitamins (60,92). Its shrinkage is much smaller if compared with that observed in convection-dried products at the same water activity.

REFERENCES

1. R. M. Spanswick, Symplasmic transport in tissues. In *Encyclopedia of Plant Physiology. Vol. 2. Transport in Plants II, Part B. Tissues and Organs* (U. Luttge and M. G. Pitman, eds.), Springer Verlag, Berlin, p. 35 (1976).
2. A. Nason and R. L. Dehaan, *The Biological World*, J. Wiley, New York (1973).
3. F. B. Salisbury and C. Ross, *Plant Physiology*, Wadsworth Publ. Co., Belmont, CA (1969).
4. A. Lauchli, Apoplasmic transport in tissues. In *Encyclopedia of Plant Physiology. Vol. 2. Transport in Plants II, Part B. Tissues and Organs* (U. Luttge and M. G. Pitman, eds.), Springer Verlag, Berlin, p. 3 (1976).
5. N. V. Parthasarathy, Sieve-element structure. In *Encyclopedia of Plant Physiology, Vol. 2. Transport in Plants I. Phloem Transport* (M. H. Zimmermann and J. A. Milburn, eds.), Springer Verlag, Berlin, p. 3 (1975).
6. H. Ziegler, Nature of transported substances. In *Encyclopedia of Plant Physiology, Vol. 2. Transport in Plants I. Phloem Transport* (M. H. Zimmermann and J. A. Milburn, eds.), Springer Verlag, Berlin, p. 59 (1975).
7. J. A. Milburn, Pressure flow. In *Encyclopedia of Plant Physiology, Vol. 2, Transport in Plants I. Phloem Transport* (M. H. Zimmermann and J. A. Milburn, eds.), Springer Verlag, Berlin, p. 328 (1975).
8. M. J. Canny, Mass transfer. In *Encyclopedia of Plant Physiology, Vol. 2, Transport in Plants I. Phloem Transport* (M. H. Zimmermann and J. A. Milburn, eds.), Springer Verlag, Berlin, p. 139 (1975).
9. R. J. Poole, Transport in cells of storage tissue. In *Encyclopedia of Plant Physiology, Vol. 2, Transport in Plants II. Part B. Tissues and Organs* (U. Luttge and M. G. Pitman, eds.), Springer Verlag, Berlin, p. 227 (1976).
10. J. R. Philip, The osmotic cell, solute diffusibility and the plant water economy, *Plant Physiol.*, *33*:264 (1958).
11. J. R. Philip, Propagation of turgor and other properties through cell aggregations, *Plant Physiol.*, *33*:271 (1958).
12. J. R. Philip, Osmosis and diffusion in tissue: Half-times and internal gradients, *Plant Physiol.*, *33*:275 (1958).
13. F. J. Molz and G. M. Hornberger, Water transport through plant tissues in the presence of a diffusible solute, *Soil Sci. Soc. Am. Proc.*, *37*:883 (1973).
14. F. J. Molz and E. Ikenberry, Water transport through plant cell and cell walls: Theoretical development, *Soil Sci. Soc. Am. Proc.*, *38*:699 (1974).
15. F. J. Molz, Water transport through plant tissue: The apoplasm and symplasm pathways, *J. Theor. Biol.*, *59*:277 (1976).
16. F. J. Molz, D. V. Kerns, C. M. Peterson, and J. H. Dane, A circuit analog model for studying qualitative water relations of plant tissue, *Plant Physiol.*, *64*:712 (1979).
17. J. Hawkes and J. M. Flink, Osmotic concentration of fruit slices prior to freeze dehydration, *J. Food Proc. Preserv.*, *2*:265 (1978).
18. J. Conway, F. Castaigne, and X. V. Pickard, Mass transfer considerations in the osmotic dehydration of apples, *Can. Inst. Food Sci. Technol.*, *16*:1 (1983).

19. R. Andreotti, M. Tomasicchino, and L. Machiavell, La disidratazione parziale della frutta per osmosi, *Ind. Conserve*, *58*:88 (1983).

20. M. Tomasicchino, R. Andreotti, and A. De Giorgi, Desidratazione parziale della frutta per osmosi. II. Ananas, fragole e susine, *Ind. Conserve*, *61*:108 (1986).

21. D. Torreggiani, R. Giangiacomo, G. Bertolo, and E. Abbo, Ricerche sulla disidratazione osmotica della frutta. I. Idoneita varietale delle ciliege, *Ind. Conserve*, *61*:101 (1986).

22. M. Dalla Rosa, G. Pinnavaia, and C. R. Lerici, La disidratazione della frutta mediante osmosi diretta. II. Esperienze di laboratorio su alcumi generi di frutta, *Ind. Conserve*, *57*:3 (1982).

23. R. Giangiacomo, D. Torreggiani, and E. Abbo, Osmotic dehydration of fruit. Part I. Sugars exchange between fruit and extracting syrups, *J. Food Proc. Preserv.*, *11*:183 (1987).

24. D. Torreggiani, R. Giangiacomo, G. Bartolo, and E. Abbo, Ricerche sulla disidratazione osmotica della frutta. II. Idoneita varietale della albicocche, *Ind. Conserve*, *61*:226 (1986).

25. T. R. A. Magee, A. A. Hassaballah, and W. R. Murphy, Internal mass transfer during osmotic dehydration of apple slices in sugar solutions, *Ir. J. Food Sci. Technol.*, *7*:147 (1983).

26. C. I. Beristain, E. Azuara, R. Cortes, and H. S. Garcia, Mass transfer during osmotic dehydration of pineapple rings, *Int. J. Food Sci. Technol.*, *25*:576 (1990).

27. C. J. Toupin, M. Marcotte, and M. Le Maguer, Osmotically-induced mass transfer in plant storage tissue. A mathematical model. Part I, *J. Food Engng.*, *10*:13 (1989).

28. C. J. Toupin and M. Le Maguer, Osmotically-induced mass transfer in plant storage tissue. A mathematical model. Part II, *J. Food Engng.*, *10*:97 (1989).

29. M. Marcotte, C. J. Toupin, and M. Le Maguer, Mass transfer in cellular tissues. Part I. The mathematical model, *J. Food Engng.*, *13*:199 (1991).

30. A.-L. Raoult, F. Lafont, G. Rios, and S. Guilbert, Osmotic dehydration: Study of mass transfer in terms of engineering properties. In *IDS '89* (A. S. Mujumdar and M. A. Roques, eds.), Hemisphere Publ. Co., Washington, DC, p. 487 (1990).

31. A.-L. Raoult-Wack, S. Guilbert, M. Le Maguer, and G. Rios, Simultaneous water and solute transport in shrinking media Part 1. Application to dewatering and impregnation soaking process analysis (osmotic dehydration), *Drying Technol.* 9:589 (1991).

32. A.-L. Raoult-Wack, F. Petitdemange, F. Giroux, S. Guilbert, G. Rios, and A. Lebert, Simultaneous water and solute transport in shrinking media. Part 2. A compartmental model for the control of dewatering and impregnation soaking process, *Drying Technol.* 9:631 (1991).

33. A.-L. Raoult-Wack, O. Botz, S. Guilbert, and G. Rios, Simultaneous water and solute transport in shrinking media Part 3. A tentative analysis of the spatial distribution of the impregnating solute in the model gel , *Drying Technol.* 9:631 (1991).

34. J. D. Ponting, G. G. Watters, R. R. Forrey, R. Jackson, and W. L. Stanley, Osmotic dehydration of fruits, *Food Technol.*, *20*:125 (1966).

35. C. R. Lerici, D. M. Dalla Rosa, and G. Pinnavaia, Direct osmosis as pretreatment to fruit drying. In *Proceedings of 2nd Europ. Conf. on Food Chem.*, pp. 287–296 (1983).

36. D. Torreggiani and E. Maltini, Ricerche sulla disidratazione osmotica della frutta. I. Idoneita varietable della ciliege, *Ind. Conserve*, *61*:101 (1986).

37. J. M. Flink, Dehydrated carrot slices: Influence of osmotic concentration on drying behaviour and product quality, *Food Proc. Engng.*, *1*:412 (1980).

38. C. R. Lerici and D. Mastrocola, Nuove tecnologie nella transformazione della frutta. 1. La disidratazione osmotica, *Notizario Tecnico*, *18*:2 (1985).

39. A. Lenart, Osmotyczne odwadnianie produktów spożywczych, *Przem. Spoż.*, *30*:86 (1976).

40. A. Lenart, Aspekty inżynieryjne osmotycznego odwadniania jabłek, *Przem. Spoż.*, *30*:216, (1976).

41. P. P. Lewicki, A. Lenart, and W. Pakuła, Influence of artificial semi-permeable membranes on the process of osmotic dehydration of apples, *Ann. Warsaw Agric. Univ.*, *Food Technol. and Nutr.*, *16*:17 (1984).

42. P. P. Lewicki, A. Lenart, and D. Turska, Diffusive mass transfer in potato tissue during osmotic dehydration, *Ann. Warsaw Agric. Univ.*, *Food Technol. and Nutr.*, *16*:25 (1984).

43. T. L. Adambounou and F. Castaigne, Deshydration partielle par osmose des bananes et determination de courbes de sorption isotherme, *Lebensm.-Wiss. u. Technol.*, *16*:230 (1983).

44. G. Pinnavaia, M. Dalla Rosa, and C. R. Lerici, La disidratazione mediante osmosi diretta per

la valorizzazione di produtti vegetali. In *Atti 2° Convegno Nazionale Nutrizione Ambiente Lavoro* (M. Coachi and T. G. Modena, eds.), p. 313 (1983).

45. J. D. Ponting, Osmotic dehydration of fruits—Recent modifications and applications, *Process. Biochem.*, *8*:18 (1973).

46. E. Maltini and D. Torreggiani, La disidratazione osmotica. II. Possibilita d'impiego della concentrazione osmotica per la preparazione di conserve di frutta. In *Monografia No. 4. Progressi della Tecniche di Disidratazione di Frutta e Ortaggi*, CNR-IPRA, Rome, p. 149 (1985).

47. H. Sarosi and A. Polak, Possibilities of application of osmotic drying in the food industry, *Tudomanyos Kozlemenyck. Elelmiszeripari Foiskola*, *6*:63 (1976).

48. H. R. Bolin, C. C. Huxsoll, and R. Jackson, Effect of osmotic agents and concentration on fruit quality, *J. Food Sci.*, *48*:202 (1983).

49. A. Lenart and E. Gródecka, Wpływ rodzaju substancji osmotycznej na kinetykę suszenia owiewowego jabłek i marchwi, *Ann. Warsaw Agric. Univ.*, *Food Technol. and Nutr. 18*:27 (1989).

50. A. Lenart and P. P. Lewicki, Osmotic dehydration of apples at high temperature. In *IDS '89* (A. S. Mujumdar and M. Roques, eds.), Hemisphere Publ. Co., New York, p. 501 (1990).

51. E. Maltini, D. Torreggiani, G. Bertolo, and M. Stecchini, Recent developments in the production of shelf-stable fruit by osmosis. *Proceedings of 6th Int. Congress Food Sci. and Technol.*, pp. 177–178 (1983).

52. C. R. Lerici, N. Pepe, and G. Pinnavaia, La disidratazione della frutta mediante osmosi diretta. I. Resultati di esperienze effettuate in laboratorio, *Ind. Conserve*, *52*:1 (1977).

53. D. Mastrocola, C. Severini, and C. R. Lerici, Prove di disidratazione per via osmotica della carota, *Ind. Conserve*, *62*:33 (1987).

54. R. Andreotti, Conservazione di pesche parzialmente disidrate per osmosi diretta, *Ind. Conserve*, *60*:96 (1985).

55. P. Speck, F. Escher, and J. Solms, Effect of salt pretreatment on quality and storage stability of air-dried carrots, *Lebensm.-Wiss. u. Technol.*, *10*:308 (1977).

56. P. P. Lewicki, A. Lenart, and M. Małkowska, Diffusive movement of substance in the carrot dehydrated osmotically in the sodium chloride solution, *Ann. Warsaw Agric. Univ.*, *Food Technol. and Nutr.*, *17*:45 (1987).

57. T. L. Adambounou, F. Castaigne, and I. C. Dillon, Abaissement de l'activite de l'eau de legumes tropicaux par deshydration osmotique partielle, *Sci. Aliments*, *3*:551 (1983).

58. M. W. Hoover and N. C. Miller, Factors influencing impregnation of apple slices and development of a continuous process, *J. Food Sci.*, *40*:698 (1975).

59. G. W. Hope and D. G. Vitale, *Osmotic Dehydration*, Int. Devel. Res. Center Monogr., IDRC-004e, Ottawa, Canada, p. 1 (1972).

60. D. R. Bongirwar and A. Sreenivasan, Studies on osmotic dehydration of banana, *J. Food Sci. and Technol.*, *14*:104 (1977).

61. A. Lenart and P. P. Lewicki, Osmotyczne odwadnianie jabłek w cyrkulującym syropie cukrowym, *Zesz. Probl. Post. Nauk Roln.*, *243*:223 (1980).

62. V. Pavasović, M. Stefanović, and R. Stefanović, Osmotic dehydration of fruit. In *Drying '86* (A. S. Mujumdar, ed.), Hemisphere Publ. Co., New York, Vol. 2, p. 761 (1986).

63. D. F. Farkas and M. E. Lazar, Osmotic dehydration of apple pieces: Effect of temperature and syrup concentration on rates, *Food Technol.*, *23*:688 (1969).

64. J. H. Moy, N. B. Lau, and A. M. Dollar, Effects of sucrose and acids on osmovac-dehydration of tropical fruits, *J. Food Proc. and Preserv.*, *2*:131 (1978).

65. J. E. Contreras and T. G. Smyrl, An evaluation of osmotic concentration of apple rings using corn syrup solids solutions, *Can. Inst. Food Sci. Technol. J.*, *14*:310 (1981).

66. P. P. Lewicki, A. Lenart, and G. Młynarczyk, Porównanie energochłonności usuwania wody z jabłek metodąsuszenia konwekcyjnego i odwadniania osmotycznego, *Przem. Ferm. i Owoc.-Warz.*, *24*(8/9):24 (1980).

67. A. Lenart and J. M. Flink, An improved proximity equilibration cell method for measuring water activity of food, *Lebensm.-Wiss. u. Technol.*, *16*:84 (1983).

68. A. Lenart and J. M. Flink, Osmotic concentration of potato, I. Criteria for the end-point of the osmosis process, *J. Food Technol.*, *19*:45 (1984).

69. W. M. Camirand, R. R. Forrey, K. Popper, F. P. Boyle, and W. C. Stanley, Dehydration of membrane coated foods by osmosis, *J. Food Agric. Chem.*, *19*:472. (1968).

70. G. M. Dixon, J. J. Jen, and V. A. Paynter, Tasty apple slices result from combined osmotic-dehydration and vacuum-drying process, *Food Proc. Develop.*, *10*:634 (1976).

71. G. Mazza, Dehydration of carrots. Effect of pre-drying treatments on moisture transport and product quality, *J. Food Technol.*, *18*:113 (1983).

72. M. N. Islam and J. M. Flink, Dehydration of potato. II. Osmotic concentration and its effect on air drying behaviour, *J. Food Technol.*, *17*:387 (1982).

73. P. P. Lewicki, A. Lenart, and G. Młynarczyk, Suszenie osmotyczno-owiewowe jabłek. I. Analiza kinetyki odwadniania osmotycznego jabłek, *Przem. Ferm. i Rol.*, *21*(10):21 (1977).

74. C. R. Lerici, D. Mastrocola, and G. Pinnavaia, Esperienze di osmosi diretta ad alta temperatura per tempi brevi, *Ind. Conserve*, *61*:223 (1986).

75. J. M. Flink, Methods for improving freeze drying economy and their influence on product quality, Report to Danish Agric. Veterin. Res. Counc., Copenhagen, p. 1 (1981).

76. R. Garcia, J. F. Menchu, and C. Rolz, Tropical Fruit Drying, A Comparative Study. In *Proceedings of 4th Int. Cong. Food Sci. Technol.*, pp. 13–14 (1974).

77. A. Levi, S. Gagel, and B. Juven, Intermediate moisture tropical fruits products for developing countries. I. Technological data on papaya, *J. Food Technol.*, *18*:667 (1983).

78. A. Lenart and P. P. Lewicki, Kinetics of osmotic dehydration of the plant tissue. In *Drying '87* (A. S. Mujumdar, ed.), Hemisphere Publ. Co., New York, Vol. 2, p. 239 (1987).

79. A. Lenart and P. P. Lewicki, Mechnizm odwadniania osmotycznego tkanki roślinnej, Materiały XVII Sesji Naukowej Komit Techn. i Chem. Żywn. PAN, Łódź, pp. 558–561 (1986).

80. A. Lenart, *Sacharoza Jako Czynnik Modyfikujący Osmotyczno-Owiewowe Utrwalanie Jabłek*, Wydawnictwo SGGW, Warszawa (1988).

81. M. Dalla Rosa, G. Pinnavaia, and C. R. Lerici, La disidratazione della frutta mediante osmosi diretta. II. Esperienze di laboratorio su alcuni generi di frutta, *Ind. Conserve*, *57*:3 (1982).

82. R. D. Scott, Dehydration and packaging of foodstuffs by dialysis, U.S. Patent 3 758 313 (1983).

83. J. F. Girod, A. Collignan, A. Themelin, and A.-L. Raoult-Wack, Energy study of food processing by osmotic dehydration and air drying, *Procédés de Déshydration-Imprégnation par Immersion (DII) dans des Solutions Concentrées*, Montpellier (1990).

84. A. Lenart and P. P. Lewicki, Energy consumption during osmotic and convective drying of plant tissue, *Acta Alim. Polonica*, *14*:65 (1988).

85. P. P. Lewicki and A. Lenart, Energy consumption during osmo-convection drying of fruits and vegetables. In *Drying of Solids* (A. S. Mujumdar, ed.), Inter. Sci. Publ., New York, and Oxford & IBH Publ. Co., New Dehli, pp. 354–367 (1992).

86. A. Lenart and P. P. Lewicki, Dyfuzja wody w tkance jabłka podczas osmotycznego odwadniania, *Zesz. Nauk. SGGW-AR, Technol. Roln.-Spoż.*, *14*:33 (1981).

87. A. Lenart, Wpływ właściwości tkanki roślinnej na dyfuzjęsubstancji osmoaktywnych, *Zesz. Probl. Post. Nauk Roln.*, *297*:351 (1986).

88. C. Y. Lee and D. K. Salunkhe, Sucrose penetration in osmo-freeze dehydrated apple slices, *Curr. Sci.*, *37*:297 (1968).

89. M. Le Maguer, Osmotic dehydration: Review and future directions. *Progress in Food Preservation Processes*, CERIA, Brussels, Vol. 1, p. 283 (1988).

90. A. Lenart, P. P. Lewicki, and Z. Pałacha, Water binding in the apple tissue during its diffusive processing. In *Drying '86* (A. S. Mujumdar, ed.), Hemisphere Publ. Co., New York, p. 516 (1986).

91. Z. Pałacha, A. Lenart, and P. P. Lewicki, Desorpcja wody z jabłka po obróbce dyfuzyjnej w zakresie aktywności wody 0.9–1.0, *Zesz. Nauk. Politech. Łódz. Inż. Chem.*, *14*:56 (1987).

92. C. R. Lerici, G. Pinnavaia, M. Dalla Rosa, and D. Mastrocola, Applicazione dell'osmosi diretta nella disidratazione della frutta, *Ind. Aliment.*, *22*:184 (1983).

23
Evaporation and Spray Drying in the Dairy Industry

Jan Písecký
Niro A/S
Soeborg, Denmark

1. INTRODUCTION

The task of the dairy industry is to ensure the regular supply of milk products to the population. The original purpose of milk powder manufacture was to assist this service. Dried milk products, in comparison with their liquid equivalents, have far longer keeping qualities, occupy less volume, have much lower weight, and have no special cooling requirements during transport and storage. It is therefore possible to compensate for the seasonal and local variations of both production and consumption of milk by transferring the milk from peak to low production season and from highly productive countries to highly populated, low-production areas.

Today the milk powder industry is a specialized branch of the dairy industry. The main products are still conventional dairy products in dry form as skim and whole milk powder. In addition, a great number of new products have been developed for special applications in households, food processing industries, and agriculture, for instance, baby-food products, fat-filled milk powders, both for human consumption and cattle feeding, whey and modified whey powders (demineralized, fat filled, and others), dried casein and caseinates, protein concentrate powders, permeate powders, coffee whiteners, cocoa beverage powders, and ice cream powder. Special technologies have been developed to improve functional properties of the products, that is, to produce them in a form that is more convenient for final use, such as instant (easy to reconstitute) products, milk powders with high water binding or whipping properties (for bakeries), and powders with high bulk density (saving shipping volume).

Transforming liquid milk into powder requires removal of almost all the water, the amount of which many times exceeds the weight of the final product. During this process the milk changes its physical structure and appearance considerably. Thus a single water-removing process cannot ensure optimum performance throughout the whole range of water removal.

The basic water removal processes in the milk powder industry are vacuum evaporation, which removes the first part of water and transforms a thin liquid milk into a concentrate of high viscosity, and spray drying, transforming the concentrate into droplets and evaporating water from them to obtain dry powder.

A third water removal process was introduced about two decades ago, fluid bed drying, in which, by removing the last part of water, the wet powder from the spray dryer is transformed into the final dry powder. The fluid bed soon became an integrated component of modern milk drying installations.

2. EVAPORATION

Evaporation in the milk powder industry means a process in which part of the water is removed from the liquid milk by boiling to such an extent that the resulting concentrate still is easily pumpable. The amount of water removed by evaporation represents 86% (whole milk) to 95% (whey) of total water removed. There are two main reasons for using evaporation prior to spray drying.

It is a far more economical process, so removal of as much water as possible by evaporation improves the total heat economy.

It has a positive influence on many qualitative properties of final powders.

2.1. Principle

The most simple evaporator is a kettle placed on a hot plate. The milk is heated to the boiling point, and evaporation takes place from the surface. However, when boiling milk under such conditions a skin forms on the surface, burnt deposits are created on the bottom, and milk has a tendency to boil over. Such an operation will take a long time because of small evaporation surface, and an undesirable serious deterioration will take place due to the excessive exposure to heat.

To avoid these problems it is obvious that

1. The boiling temperature must be reduced to below the temperature causing harmful changes to the milk.
2. The evaporative surface must be increased to achieve a fast evaporation, that is, to reduce the heat exposure time.
3. The heating medium should not have a much higher temperature than that of the milk boiling point to avoid burning.
4. Therefore, the heating surface must be enlarged.

These considerations resulted in the development of vacuum evaporators heated by saturated steam. A large number of designs have been developed.

Single-Stage Evaporation

The first evaporator in the dairy industry was a batch evaporator, described in Section 2.3 (Fig. 1). This evaporator works as a single stage. In such a system, provided that the product is at the boiling temperature and that there is no heat loss, the amount of created vapor is equal to the amount of used heating steam. The specific steam consumption is 1 kg steam per kg evaporated water. This is usually expressed as specific steam consumption of 100%.

Multistage Evaporation

The vapor from the water evaporated from milk contains almost all the energy that was applied in the form of steam and can be utilized to evaporate more water. This is done by

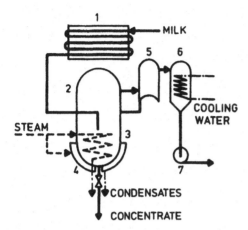

Figure 1 The principle of single-stage batch evaporation: (1, pasteurizer; 2, evaporator body; 3, steam coils; 4, steam jacket; 5, separator; 6, condenser; 7, vacuum pump).

adding a further evaporative stage in which the vapor from the first stage is introduced into the second-stage heating space to evaporate more water from the already preconcentrated milk from the first stage. Altogether, 2 kg water is evaporated using 1 kg steam; that is, specific steam consumption is 50% (Fig. 2).

Using the same system, further evaporative stages can be added. The theoretical specific steam consumption of an n-stage evaporator is $100/n\%$. In practice, the steam consumption is always somewhat higher owing to the heat losses, increase of boiling point, pasteurization, and other factors. The steam-saving effect obtained by adding an additional stage is declining; each additional stage brings less saving than the previous one.

Thermal Vapor Recompression

Thermal vapor recompression (TVR) is another way of reutilizing the vapor from the milk. The thermocompressor compresses the vapor from low pressure to a higher pressure, increasing the temperature level and at the same time transferring it from the boiling space to the heating space. The principle of a thermocompressor is shown in Figure 3. It consists of an ejector through which live steam passes at high velocity, sucking the vapor from the separator. The margin in steam economy obtained by a thermocompressor depends on the compression ratio (the amount of vapor to the amount of live steam). In multistage evaporators, it is also dependent on the number of stages across which it is working.

Mechanical Vapor Recompression

The vapors can also be reutilized by mechanical compression. This is done by an axial or radial flow compressor, possibly multistage or rotary piston. It may be driven by electric motors, fuel engines, or steam turbines. Mechanical vapor recompression (MVR) evaporators are relatively expensive in both investment and maintenance. On the other hand, they have very good heat economy comparable with up to 15-stage evaporators. Generally, the economy of the MVR evaporator depends very much on the kind of energy used for driving the compressor and, of course, on the costs of this energy.

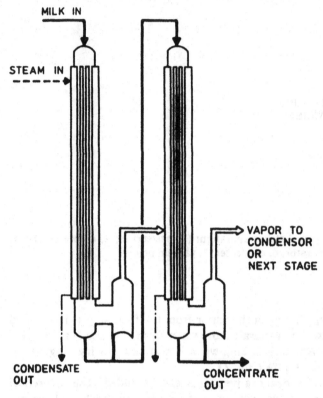

Figure 2   The principle of two-stage evaporation.

2.2. Components of an Evaporator

The essential parts of an evaporator are as follows:

1. A heat exchanger in which the heat transfer from the heating medium to the pro-
    cessed liquid and the boiling and evaporation take place. As a heat exchanger, a
    simple vessel with steam jacket and coils can be used. For continuous operation,
    tubular, plate, or swept-surface heat exchangers are used.
2. A separator to separate the concentrate from the vapor. Milk evaporation plants use
    centrifugal separators.
3. A condenser to condense the vapor. Water is used as the cooling medium, either

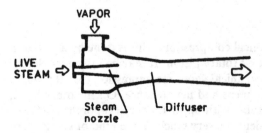

Figure 3   The principle of the thermocompressor.

directly (mixing condenser) or indirectly using a tubular heat exchanger (surface condenser).

4. A vacuum pump to create and maintain the vacuum, such as a water-ring pump or a steam jet vacuum pump (similar construction to the thermocompressor).
5. A pasteurizer (straight or spiral tube) is also an important part of the milk evaporator to ensure the microbiologic quality.

The assembling of these components in an evaporator appears in Figure 1.

2.3. Evaporator Types

Batch Evaporator

The batch evaporator was also called a vacuum pan (Fig. 1). The main parts are the pasteurizer, evaporator body, heating system, consisting of a steam jacket and steam coils, separator, condenser, and vacuum pump. It was originally used for condensed products. When used for preconcentration prior to spray drying, the maximum concentration obtainable is about 40%. The residence time in such an evaporator is up to several hours, so higher concentrations therefore result in an excessive viscosity increase (age thickening).

Recirculation Vertical Tube Evaporator

The heat exchanger of the recirculation vertical tube evaporator (Fig. 4) is a tube calandria, surrounded by the jacket. Heating steam is introduced into the jacket and heats the outer surface of the tubes. The milk rises upward in the tubes, driven by the vapor created by boiling. The mixture of vapors and concentrated milk is separated in a centrifugal separator, the vapor is condensed in the condenser, and the concentrate is fed back to the bottom of the heat exchanger. This evaporator works continuously. The concentrate is drawn off from the return pipe, and liquid milk is introduced to the bottom of the heating body.

Recirculative tube evaporators were the most important type before the falling-film evaporators, and they were constructed with up to three stages. The residence time was still long, and the maximum obtainable concentration suitable for spray drying was therefore about 45%.

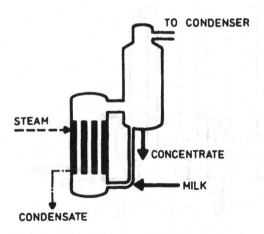

Figure 4 Recirculation vertical tube evaporator.

Evaporator with Mechanical Film Creation

The evaporator with mechanical film creation appeared in the dairy industry after World War II, and it brought a considerable improvement of the concentrate quality together with an increased concentration. The retention time was very short (less than 10 s), and it was therefore possible to concentrate milk to over 50% solids and still maintain a low viscosity. This type works exclusively as a single stage and has therefore a high specific steam consumption (110–120%). Nowadays, it is applied only for special duties, for instance as a finisher to achieve a high concentration or for highly viscous products.

Plate Evaporator

The advantage of a plate evaporator, which uses a plate-heat exchanger as the heating system, is low height requirement. Another advantage is that the evaporative capacity can be easily altered by adding plates. On the other hand, it is not very suitable for highly viscous liquids, and the practically obtainable concentration with milk is about 46–48%.

Falling-Film Evaporator

During the last 30 years, the falling-film evaporator has become the dominating type in the dairy industry. The heating surface is formed by a great number of tubes through which the milk passes downward, forming a thin film. The heating steam enters the jacket surrounding the tubes at the top, condenses on the outer surface of the tubes, and leaves at the bottom as condensate. The milk in the tubes is boiling, and the vapors created maintain and accelerate the downward movement of the film. Modern multistage falling-film evaporators are built with up to 15 m (49 ft) long tube and with five to eight stages. The flow sheet of a seven-stage evaporator with TVR is shown in Figure 5. It evaporates altogether 1300 kg water using 100 kg steam; thus the specific steam consumption is 7.7%. Falling-film evaporators achieve a concentration well above 50%.

2.4. Instrumentation and Automation

The modern multistage evaporator requires a high level of instrumentation and automatic control to ensure a feed supply of constant rate and concentration to the spray dryer. The

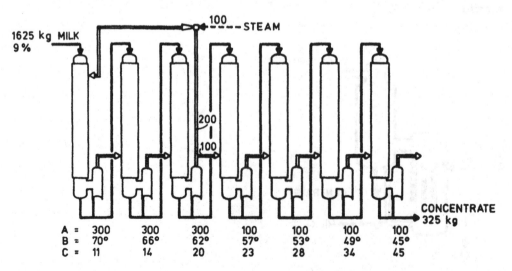

A =	300	300	300	100	100	100	100
B =	70°	66°	62°	57°	53°	49°	45°
C =	11	14	20	23	28	34	45

Figure 5 An example of a seven-stage evaporator with thermal vapor recompression (NIRO A/S): A, vapor in kg; B, boiling temperature (°C); C, concentration (%).

most important parameters, which have to be automatically controlled, are raw milk flow, main steam pressure, pasteurization temperature, vacuum, and solids content of the concentrate.

3. SPRAY DRYING

3.1. Principle

Spray drying is characterized by the transformation of a liquid feed, containing dissolved and/or dispersed solids, into a great number of small droplets, which are exposed to a fast current of hot air. Because of the very large surface area of the droplets, water evaporation takes place almost instantaneously. The droplets are transformed into dry powder particles, which are then separated from the utilized air.

3.2. Drying Gas: Terms

The drying gas used in the milk industry is atmospheric air, heated to provide the necessary heat for evaporation. The evaporation of water proceeds under adiabatic conditions. In such a system all sensible heat given up by the drying air is utilized for evaporation of water, which becomes a part of the drying air. The enthalpy of the air remains constant. This process can be analyzed using the h/x diagram by means of which the following terms can be explained (see Fig. 6):

Saturation line, defined by $\phi = 1$, dividing the chart into a zone of unsaturated air and zone of fog.
Dry bulb temperature T_G, the temperature of air above the saturation line as measured by thermometer.
Absolute humidity x, the amount of moisture, kg/1 kg dry air.

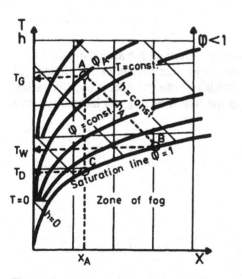

Figure 6 The principle of the h/x diagram. The air characterized by point A has the dry bulb temperature T_G, enthalpy h_A, absolute humidity x_A, relative humidity ϕ_A, wet bulb temperature T_W, and dew point temperature T_D.

Relative humidity ϕ, the ratio of the partial pressure of water vapor to the water vapor pressure at the saturation point at the same temperature.

Wet bulb temperature T_W is a characteristic of the air of dry bulb temperature T_G expressing the temperature of the saturated air ($\phi = 1$) that has the same enthalpy ($h =$ constant).

Dew point temperature T_D is another characteristic of the air of dry bulb temperature T_G, expressing the temperature of saturated air ($\phi = 1$) that has the same absolute humidity ($x =$ constant).

For practical spray dryer operation, the following terms are usually used:

Inlet temperature: Hot drying air temperature in °C (°F) when entering the drying system.

Outlet temperature: exhaust air temperature in °C (°F) when leaving the drying system.

Ambient temperature: The temperature in °C (°F) of supply air before entering the system (prior to air filter).

Feed concentration: The feed is the concentrate supplied to the dryer. The concentration is expressed as a percentage of the total solids content.

Moisture content of the product: Usually expressed as a percentage on a wet basis, kg moisture per 100 kg product.

Water activity a_w: The relative humidity of the air in equilibrium with a powder of given moisture content. It is temperature dependent.

2.3. Spray Dryer Components and Types

The main components of a spray dryer are

1. Drying chamber
2. Air disperser with hot air supply system
3. Atomizing device with feed supply system
4. Powder recovery system
5. Powder aftertreatment system
6. Instrumentation and automation

Drying Chamber

The shape of the drying chamber together with the positions of the atomizing device, air disperser, exhaust air outlet duct, and powder discharge duct determine the airflow pattern and product flow. Taking just main characteristics into consideration, the various chamber types used in the milk industry differ in the following (figures in brackets refer to types shown on Figure 7).

Chamber shape
    Vertical cylindrical [1,2,3,4]
      Low profile [1,3,4]
      Tall form [2]
    Horizontal box type [5]
Product discharge
    Together with exhaust air [3,4]
    Partially separated from exhaust air [1,2,5]
Product transport to the discharge point
    By gravity, conical bottom [1,2]
    Mechanically, flat bottom [3,4,5]

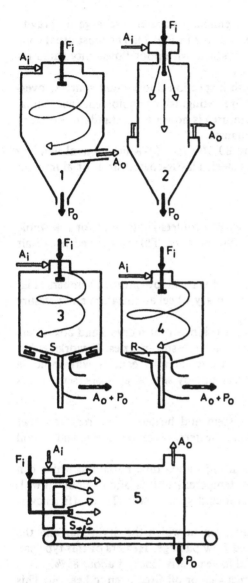

Figure 7 Drying chamber types (A, drying air; F, feed; P, product; i, inlet; o, outlet; S, scraper remover; R, rotating suction duct).

Airflow
 Rotary downward [1,3,4]
 Streamline downward [2]
 Streamline horizontal [5]
Air-spray mixing
 Co-current [1,2,3,4,5]
 Countercurrent

The development of the two-stage drying system in the early 1970s (referred to later) was very quickly adopted by most spray dryer manufacturers. This resulted in the fact that dryers nowadays mostly use the chamber types, enabling powder discharge by gravity partially separated from the drying air (see types 1 and 2 in Fig. 7). The most usual cone angle is 60° or sometimes 50° or even 40° for products with poor flowability (high fat products).

The drying chamber is further equipped with inspection doors, light sources, over-pressure vents, and air-sweep doors. The safety fire-extinguishing equipment (water noz-zles activated by an increase in the outlet temperature) is nowadays a standard accessory often required by authorities and insurance companies.

The drying chamber is usually insulated by 80–100 mm (3–4 in) mineral wool to reduce the radiation loss and clad with stainless steel, plastic-coated steel, or aluminum plates.

Air Disperser and Hot Air System

Good mixing of the hot drying air with the spray of droplets is essential for the whole process and has a decisive influence on the product quality. This is ensured by an air disperser, and there are in principle two types:

1. Air disperser with rotary airflow, which is used in vertical low-profile chambers (Fig. 7, types 1, 3, and 4). It can operate with both rotary wheel atomization and pressure nozzle atomization.
2. Air disperser with streamline airflow, which is used in tall form dryers and box dryers (Fig. 7, types 2 and 5) operating exclusively with pressure nozzles. It works either with a single nozzle placed in the center of a circular airstream or with multiple nozzles distributed symmetrically. Some dryers employ several air dispersers of this type in one chamber.

The supply air is filtered before entering the system and heated to the required inlet temperature. The heating is either indirect by steam heaters, oil or gas ovens, and hot oil heat exchangers or direct by gas or electricity.

The most usual way of heating is by steam using finned tube radiators, which can heat the air about 10°C (18°F) below the steam temperature. It is often combined with preheaters using condensates from the evaporator and spray dryer. Their efficiency is about 98–99%.

Indirect heating by gas or oil is done in an oven in which the drying air and the combustion gases from the burner have separated flow passage. Heaters of this type can achieve up to 250°C (482°F) air temperature and have an efficiency of about 85%.

The hot oil liquid-phase heater uses an indirect gas- or oil-fired oven to heat oil. This is then introduced into a heat exchanger placed in the drying airflow. It is mostly used for boosting when a higher temperature than obtainable by available steam is required.

Direct gas heating is accomplished by a gas burner placed in the drying airflow. The resulting hot drying air thus contains the products of combustion. It is a cheap and efficient system, allowing very high temperatures. The drying air contains, however, some nitrogen oxides formed catalytically on the hot metal surface of the burner. There is some risk of product contamination by nitrites and nitrosamines. It is therefore not allowed in many countries for production of products for human consumption. New types of burn-ers, so called LO-NOX, are diminishing this contamination considerably. Another disad-vantage is the moisture of combustion, which increases the humidity of the drying air, to be compensated for by increasing the outlet temperature by about 1°C for each 80°C direct heating.

Electrical heating is used mostly for small pilot plants or for boosting.

Atomizing Device with Feed Supply System

The purpose of an atomizing device is to transform the feed into a large number of droplets of well-defined size distribution. There are three systems of atomization.

Rotary Wheel. The rotary wheel is a partially closed bowl rotating around the axis of symmetry. The feed enters close to the center and is accelerated by the centrifugal force through the vanes to the periphery. Rotary atomizers work with a peripheral speed of 100–200 m/s (330–660 ft/s). The vanes are usually radial of cylindrical or rectangular shape (Fig. 8).

A special wheel developed for milk products is shown on Figure 9. It has curved vanes accomplishing the removal of air bubbles from the feed by centrifugal force. The milk powder produced with this wheel has a lower content of occluded air and a higher bulk density than the powder from the radial-vaned wheels.

Another possibility for achieving high bulk density and almost air-free powder is by the steam-swept wheel. Low-pressure steam is introduced into the wheel and replaces the air/feed interface by a steam/feed interface.

The industrial rotary wheel atomizers are driven by electrical motors using a belt drive or a gearbox drive. They have a diameter of 150–400 mm (6–16 in) and a speed of rotation of 7000–15,000 rpm.

Pressure Nozzle. In the pressure nozzle the pressure energy supplied by high-pressure pumps is converted into kinetic energy. There are two types of pressure nozzles (Fig. 10), one with a swirl chamber and another with a grooved insert core. Both types are commonly used in milk dryers.

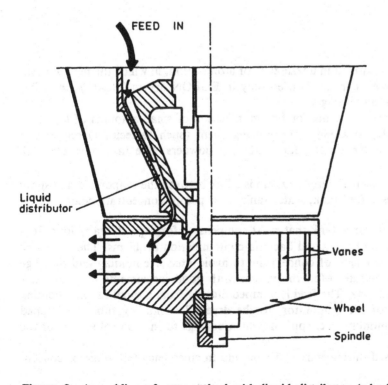

Figure 8 Assembling of rotary wheel with liquid distributor (wheel with radial rectangular vanes).

Figure 9 Rotary wheel with curved vanes (top plate removed). (Courtesy of NIRO A/S, Copenhagen, Denmark.)

Two-Fluid Nozzle. The two-fluid nozzle type of atomization, in which the feed is atomized by a high-velocity air current, is used only in EGRON dryers (Nestlé type). This atomizing air is also the hot drying air.

The two-fluid nozzle is suitable for fine atomization of small amounts of liquid. It is applied, for example, to atomize the wetting agent (such as lecithin) on powder when producing cold water-soluble, fat-containing powders, like instant whole milk powder.

Feed Supply System. The feed supply system is a link between the evaporator and spray dryer, which comprises a feed tank, water tank, feed pump, concentrate heater, filter, and feed line.

The feed tank is a balance tank that must compensate for capacity variations. It is advisable to install two tanks to avoid bacteria contamination problems. The feed tank should operate with a low level of concentrate to avoid excessive holding and thus age thickening of the concentrate. Modern systems with good instrumentation can work without any special feed tank. The feed is pumped directly from a small vacuum holding tube, which is a part of the evaporator, to the dryer. The holding tube is equipped with maximum and minimum level control, which belongs to the control system of the evaporator.

A water tank is used during start and stop and in emergency (shortage of concentrate).

The feed pump for rotary atomizers can be a centrifugal pump with control valve, a

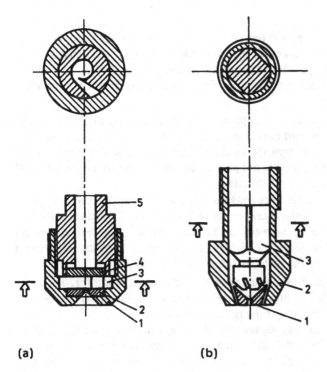

Figure 10 Essential parts of pressure nozzles: (a) With swirl chamber (1, nozzle body; 2, orifice insert; 3, swirl chamber; 4, end plate; 5, screw pin); (b) With grooved insert core (1, orifice insert; 2, nozzle body, 3, grooved core insert).

monopump, or possibly a gear pump with speed control. For pressure atomization the high-pressure, multiposition pump, preferably with speed control or possibly with bypass control, is used.

If the feed is homogenized it is advantageous to use the speed-controlled homogenizer as the feed pump. In combination with the pressure nozzle atomization, the homogenizer must be able to supply the total pressure high enough to cover both duties, 50–150 bar (730–2200 psi) for homogenization and 150–250 bar (220–3670 psi) for atomization.

Preheating the feed to 60–80°C (140–176°F) is highly recommended. It reduces the viscosity and makes it possible to increase the feed concentration. Also, it increases the capacity, improves the economy, and has a positive influence on the powder quality. It should be applied just prior to atomization in order to avoid the increase of viscosity due to heat-accelerated age thickening.

The recommended heater types are spiral tube heat exchangers or scraped surface heaters. The former can also be made in high-pressure execution and placed in the feed line prior to the pressure nozzles.

The cheap and effective way of heating is direct steam injection (DSI). Food-grade steam has to be used. It is usually installed directly in the feed line.

The feed line is a stainless steel pipe dimensioned for a feed velocity of 1.0–1.5 m/s (3–5 ft/s) or up to 3 m/s (10 ft/s) if a high-pressure pump is used.

Powder Recovery System

The exhaust air from some spray dryer types contains the total powder production. In modern systems a partial separation takes place in the chamber, and the exhaust air then contains 5–50% of the produced powder (depending on the product). This part of the powder is often referred to as *fines*, consisting of the smallest size fraction. The most used primary separator is a cyclone.

Cyclone. A cyclone is a centrifugal dust separator in which the dust-containing airstream is brought into spiral downward movement (vortex motion) inside the cylindrical or conical body and finally reversed upward to the discharge tube at the center of the cyclone top (Fig. 11). The powder particles are pressed to the cyclone wall by centrifugal force leaving through the airlock placed at the cone bottom.

The efficiency of cyclone separation depends on many factors, such as type of product, type of process, and size and design parameters of the cyclone. The usual stack loss of spray dryers with cyclones varies between 0.1% (fat powders) to several percent (caseinates). An airlock rotary valve with a slightly conical shape, allowing adjustment of the gap between the housing and the rotor, is shown in Figure 12.

Bag Filter. A bag filter can be used as a primary or secondary dust separator. The latter is more common. A bag filter is not an advantage for hygiene reasons, as the powder collected in the filter is exposed for a long time to the exhaust air.

The use of bag filters as secondary separators after the cyclones has become more and more common, especially for installations placed in populated areas, where strict environmental pollution control is required. Bag filters have a considerably better efficiency than cyclones.

Bag filters consist of a great number of filter bags through which the dirty air is forced. The fines powder adheres to the bags. Periodically the filter bags are cleaned by a

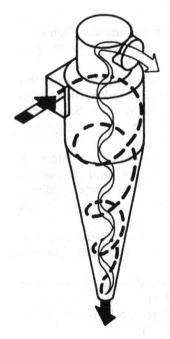

Figure 11   The principle of a cyclone.

Figure 12 Rotary valve with conical body. (Courtesy of NIRO A/S, Copenhagen, Denmark.)

reversed flow of air or shaking action, and the powder is collected in a hopper under the bag sections.

Wet Scrubber. A wet scrubber is used exclusively as a secondary dust separator. The washing liquid is sprayed into the exhaust air, leaving the cyclones in a venturi-type tube. The washing liquid absorbs the dust and is separated from the cleaned air in a centrifugal separator. As washing liquid, either water or the processed liquid product before entering the evaporator can be used. This is advantageous from the economy point of view, as the loss is converted into product. On the other hand, such an operation requires careful control and intermediate cleanings to avoid bacteria contamination problems. Process liquids used for washing are skim milk and whey only.

Powder Aftertreatment

The method of powder aftertreatment determines the structure and quality of the product that can be manufactured in a given installation.

Pneumatic Conveying System. The dried product leaving the spray dryer is usually discharged at several points, at the chamber cone outlet and the cyclones. A pneumatic conveying system is the most common device to collect the powder and to cool it. The powder is separated from the transport air in an end-cyclone. A spray dryer with pneumatic conveying system is shown in Figure 13.

Fluid Bed System. Fluid bed aftertreatment can be used either for cooling or for after-drying with subsequent cooling.

In a fluid bed the air passes upward through many closely placed small holes in a horizontal perforated plate. By this great number of small airstreams the powder is brought to fluidization (movement similar to boiling) and behaves as a fluid.

In order to achieve the required afterdrying or cooling effect, relatively large amounts of air must be used. The air passes through the holes at a velocity of about 20 m/s (66

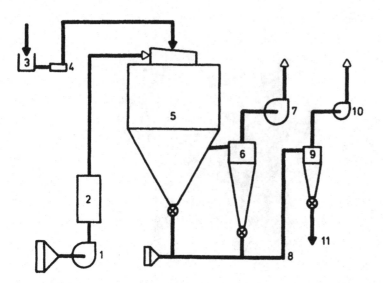

Figure 13 Flow sheet of a spray dryer with pneumatic conveying system (1, air supply fan; 2, air heater; 3, feed tank; 4, feed pump; 5, drying chamber; 6, main cyclone; 7, exhaust air fan; 8, conveying duct; 9, transport cyclone; 10, conveying air fan; 11, product outlet).

ft/s). The perforated area (i.e., total hole area expressed as a percentage of total plate area) must be chosen to achieve the fluidizing velocity adjusted to the type of product. Normally it is 0.5–2.0%.

This requires a highly perforated plate area, and fluid beds therefore have the shape of a long horizontal tube, which is separated by the perforated plate into lower clean airspace and upper fluidizing space. The whole unit is vibrated to ensure good fluidization. Without vibration the milk powders tend to exhibit the channeling effect.

A modern type of dairy fluid bed is shown in Figure 14. The perforated plate in this fluid bed is fully welded to the housing. It is corrugated to obtain the necessary strength. The holes in the perforated plate are oriented in the direction of the powder flow to ensure a good emptying at the end of the run.

The flow sheet of a spray dryer with cooling bed is shown in Figure 15. The cooling bed is supplied with ambient air in the first section and conditioned air in the second section.

The flow sheet of a spray dryer with fluid bed afterdrying and cooling system is shown in Figure 16. The inlet air temperature for the drying fluid bed is usually 80–100°C (176–212°F).

Fines Return System.   The fines from all cyclones are collected by the pneumatic conveying system and separated in an end-cyclone. From there they fall into the blow-through valve, which is a part of the pressure air transport system. There is usually a flow diversion valve, which makes it possible to introduce them back to the fluid bed for nonagglomerated products or into the chamber for agglomeration. These alternatives are shown in Figures 15 and 16.

For agglomeration it is necessary to return the fines to the wet zone of the spray dryer, close to the atomizing device. The agglomeration takes place by collisions of the fines with droplets and wet particles. In order to get a good agglomeration it is essential

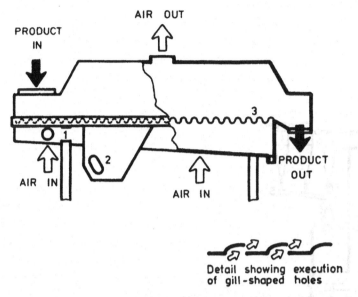

Figure 14 Modern design of sanitary vibrating fluid bed (1, springs; 2, vibrator; 3, corrugated perforated plate).

to introduce the fines to the right position, by proper velocity and direction. Many systems have been developed; some of them are shown in Figure 17.

Lecithination System. Whole milk and similar fat-containing powders are difficult to reconstitute in cold water, even if they are agglomerated. Such powders are covered by a thin layer of free fat, which makes them hydrophobic. Coating by lecithin wetting agents

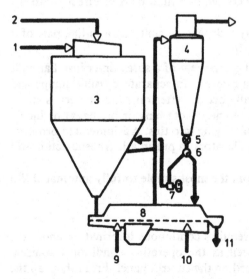

Figure 15 Flow sheet of a spray dryer with cooling bed system (1, hot air supply; 2, feed supply; 3, drying chamber; 4, main cyclone; 5, rotary valve; 6, flow diversion valve; 7, blow-through valve with blower; 8, fluid bed; 9, ambient air supply; 10, conditioned air supply; 11, product outlet).

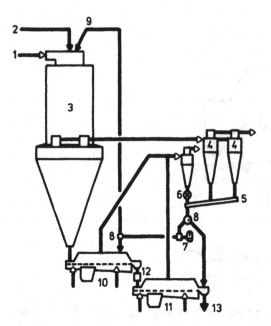

Figure 16 Flow sheet of a spray dryer with afterdrying and cooling fluid bed system: (1, hot air supply; 2, feed supply; 3, drying chamber; 4, main cyclones; 5, vibrating conveyor; 6, rotary valve; 7, blow-through valve with blower; 8, flow diversion valve; 9, fines return duct; 10, afterdrying fluid bed; 11, cooling bed; 12, lecithination equipment; 13, product outlet.

makes them wettable and cold-water soluble. Lecithin is usually applied in the amount of 0.2% (of powder) in 30% solution in butter-oil. It is sprayed by means of a two-fluid nozzle onto the concentrated powder stream between two fluid beds or directly into the fluidized layer.

The lecithination unit can be either a fully independent unit or an in-line part of a spray dryer-fluid bed system (as shown in Fig. 16).

Sieving, Storage, and Bagging Off.   The final component of a spray dryer installation is usually a sifter (shaking, vibrating, or rotating mesh) to separate possible lumps and hard, oversize particles, which may occasionally occur. Afterward the powder is either directly bagged off or filled into transport containers (tote bins or big bags) or blown into storage silos before final bagging off or packaging into tins. Agglomerated products are transported in big containers or by bucket elevators as pneumatic transportation will break down the agglomerates.

There are many systems on the market from the most simple to fully automated silo storage and bagging off systems.

Instrumentation and Automation

The operator running a spray dryer complex requires continuous information about the operation of all individual components as well as the operating conditions. A usual accessory is a mimic diagram of the dryer placed on the control panel. From there all the motors can be started or stopped, and the mimic diagram indicates their operation. The dials and scales on the control panel show the most important running conditions, such

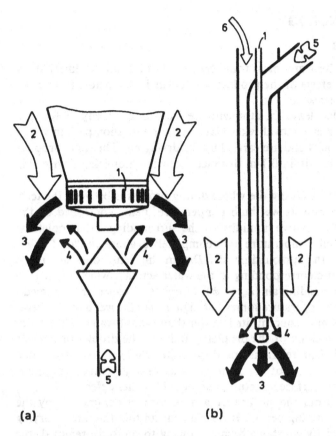

(a)   (b)

Figure 17 Introduction of fines to the wet zone for agglomeration: (a) rotary wheel; (b) pressure nozzle (1, feed supply; 2, drying air; 3, atomizing cloud; 4, fines cloud; 5, fines air transport; 6, cooling air).

as temperatures, pressures, motor amperages, and airflows. Some of this information is also recorded to analyze the operation at a later stage.

Most spray dryers nowadays have at least some of the parameters automatically controlled. The inlet temperature of the drying air is controlled by regulating the steam pressure or the oil or gas flow. The outlet temperature is controlled by the speed of the feed pump. As the spray dryer complex also consists of an evaporator, it is useful to control automatically the spray dryer feed concentration.

The degree of instrumentation and automation varies from the most simple, manually operated systems to the most sophisticated, fully automated systems in which all the components of the production line are controlled by microprocessors and computers. Such systems can be started by a one-push-button operation and can control both the sequence of starting individual motors and components as well as the adjustment of the required running conditions.

The modern high-capacity evaporator-spray dryer-fluid bed installations require a high degree of instrumentation and automation to ensure trouble-free operation and standard product quality. Computer control is therefore becoming more common.

3.   MILK POWDER TECHNOLOGY

3.1.   Drying Milk Droplets

When spray drying milk, large heat and mass transfers take place in an extremely short period of time. Severe quality defects of the product may occur if the factors promoting degradation are not known or respected.

The milk concentrate droplets leave the atomizing device with a velocity of 100–200 m/s. The greatest part of the water removal takes place in the retardation path in which the droplets lose their initial velocity and are carried by the drying air. The small droplets evaporate 90% of their moisture within 0.1-m distance from the atomizing device; big droplets need about a 1-m path.

The temperature of the droplet during the whole drying process is between the temperature of the surrounding air and its wet bulb temperature, being controlled by the droplet water activity and by the relative humidity of the surrounding air. Droplets of water ($a_w = 1$) retain the wet bulb temperature until completely evaporated; "dry" product will get the temperature of the surrounding air. *Dry* in this sense does not mean moisture free, but with a moisture corresponding to the water activity, which is equal to the relative humidity of the surrounding air ($a_w = \phi$). Droplets of concentrate immediately after atomization will get a temperature somewhat higher than the wet bulb temperature, as the water activity of the milk concentrate is lower than that of water. With water removal proceeding, the water activity of the droplet is falling. The droplet or particle moisture and temperature and their relationships during the whole process (so-called droplet temperature-moisture history) is of utmost importance as to potential heat degradation, particle structure, particle surface structure, and possibly other defects.

The evaporation of water from the surface of a milk droplet commences by the so-called constant (or first) rate drying period. It does not mean that the rate of drying and droplet temperature are strictly constant, because, owing to moisture removal, the water activity is decreasing. The droplet at this stage, however, is still a fluid in which the moisture can migrate easily from the droplet interior to the surface and keep it moist.

In a later stage the moisture content achieves a critical value and the droplet loses the character of a fluid and becomes a wet solid. The critical moisture content of milk products is believed to be between 30% and 15% moisture. It is characterized by a sudden occurrence of a moisture gradient across the droplet diameter. The controlling factor of the rate of drying begins to be the rate of diffusion of moisture through the particle (rather than droplet). The rate of drying begins to decline. This period is called the falling (or second) rate drying period. The rate of heat transfer exceeds that of mass transfer, and the particle begins to heat up faster. The moisture and temperature distribution become uneven, and a hard crust forms on the surface.

The transformation of a milk droplet into a final dried particle is accompanied by a weight loss of about 50%, which under ideal conditions will be followed by about 60% loss of volume and 25% reduction of droplet diameter. During the early drying stage, the droplet follows closely the ideal weight-volume-diameter reduction relationships and retains a globular shape. When hard-shell formation on the surface occurs, the size of the particle is more or less defined.

The presence of air in atomized droplets also has a big influence on the final shape and structure. There is always some air in the droplet, depending on the aeration of the feed between the evaporator and spray dryer, the method of atomization (a rotary wheel gives more air than pressure nozzles), and the kind and conditions of the feed (mainly the foaming ability, influenced by the concentration and state of whey proteins). High-heat product and high-concentration feed with high temperature has a much lower foaming ability than low-heat product that is not highly concentrated.

Thus, by controlling the technological conditions prior to spray drying and by the method of atomization (special wheels or pressure nozzle), the amount of air incorporated in the droplets can be reduced considerably.

The air in the droplets has a big influence on the final shape of the particle. Depending on the droplet size, volume of air bubbles, and droplet temperature history, the droplet will expand, shrink, collapse, form balloons, or break down into pieces. The air that remains in the droplet forms the so-called vacuoles in particles and is referred to as occluded air (ml per 100 g solids). It is undesirable because it decreases the bulk density, affects the reconstitution properties, and makes inert gas packaging more difficult.

The more desirable way of drying to avoid heat degradation and expansion of incorporated air is to keep the constant rate of drying period as long as possible to achieve low surrounding air temperature at the critical point. The efforts to achieve such conditions resulted in the development of the two-stage drying method.

3.2. Single-Stage Drying

In single-stage drying the total removal of water is achieved solely by spray drying. The removal of the last portion of moisture with this process proceeds slowly and is costly. For instance, the drying of skim milk concentrate of 50% total solids using an inlet air temperature of 200°C (392°F) to a powder with 3.6% moisture requires 33% more air and more energy than drying to only 7.0% moisture. The excess of moisture corresponding to the difference between 7.0% and 3.6% makes 4.1% of the total evaporation and needs 33% more energy.

The last phase of drying may also be harmful to the quality. Thus, single-stage dryers have to operate with conditions that will keep the droplet temperature reasonably low. This means the use of a relatively low inlet temperature and feed concentration, especially when drying sensitive and high-quality products. This affects the economy.

3.3. Two-Stage Drying

Two-stage drying is spray drying to a moisture content about 2–5% higher than the required final moisture content followed by a fluid bed drying to remove the excess moisture. The outlet air temperature from the spray dryer is 15–25°C (27–45°F) lower than in the single-stage process. Consequently, the surrounding air temperature at the critical moisture content and particle temperature is correspondingly lower. This process therefore allows an increase of the inlet temperature and feed concentration above values that are prohibitive for single-stage drying. This contributes further to improving the economy.

The excess of moisture is removed by fluid bed drying. With this method, the drying air and thus the heat for evaporation is supplied gradually in accordance with the rate of moisture diffusion. The temperature of the powder from the drying chamber falls in the fluid bed and continues to decrease during further drying. It begins to rise again when the moisture content approaches the final value. However, no heat damage can take place under these conditions.

As to the heat economy, second-stage fluid bed drying of course requires some energy. This is, however, only 30–50% of the energy saved in the first stage. Thus, in comparison with single-stage drying, when all other parameters are the same, it saves at least 10% energy. It may save much more when utilizing the other advantages of this method, namely, the possibility of increasing the air inlet temperature and feed concentration.

Two-stage drying also has some limitations. It can be applied, for instance, to skim milk, whole milk, precrystallized whey concentrates, caseinates, and similar powders. The level of moisture of the powder leaving the chamber has some limits, given by the stickiness of the powder. With rising moisture content the sticking temperature falls. The *sticking temperature* is defined as the temperature at which the powder begins to stick to the dryer surface and form lumps. It is dependent on powder composition. An increase in amorphous lactose content, lactic acid content, and moisture decreases the sticking temperature.

For skim milk and whole milk the maximum moisture content from the spray dryer is about 7–8% to ensure that the product can be collected from the dryer by gravity and discharged to the fluid bed without lumps. Any mechanical treatment of such powders is undesirable, and flat bottom dryers with scrapers therefore cannot be applied.

For whey drying, two-stage drying is possible only if the whey concentrate is precrystallized. The content of amorphous lactose is thereby reduced. For other products it must be decided on a case-to-case basis whether at all or possibly to which extent the two-stage drying method may be applied.

3.4. Fluid Bed Technology

The first application of the fluid bed in milk dryers was as a cooling bed for the production of high-fat powders (30–75% fat). It was found that such powders with fat contents higher than 30% could not be processed successfully on a dryer with a pneumatic conveying system as this frequently became blocked.

After introducing a fluid bed it was recognized that there was a distinct difference in the structure of a fluid bed-treated powder compared with the powder from the pneumatic conveying system. The fluid-bed-processed powder was distinctly more coarse and more freely flowing as it was partly agglomerated. These agglomerates are created in the chamber by the so-called primary agglomeration and are broken down to individual particles when using a pneumatic conveying system. The fluid bed treatment is gentle and preserves the agglomerates.

Moreover, a fluid bed exhibits the classification effect by blowing off the nonagglomerated particles from big particles and agglomerates. This classification effect, which enables a desired size fraction to be blown off and recycled for agglomeration back into the chamber, is controlled by the fluidizing velocity. It can vary between 0.1 and 1.0 m/s and has the following rough guidelines: caseinates and protein powder, 0.1 m/s; skim milk powder, 0.2–0.3 m/s; whole milk powder, 0.3–0.4 m/s; high-fat milk powder, 0.4–0.6 m/s; and special applications, up to 1.0 m/s. The inlet air temperature for drying in a fluid bed is usually 80–100°C (176–212°F).

Cooling beds work mostly with two sections. The first section uses ambient air and the second section conditioned air by cooling down to 5–8°C (41–46°F) followed by a slight reheating.

There is danger that the powder will pick up some humidity from the air if cooled to too low a temperature. The water activity of final milk powders is between 0.15 and 0.20. There is a temperature equilibrium between the air and powder. It means that the powder will begin to pick up the moisture from the air if cooled down below 40–34°C (104–93°F) when using air of T_D 8°C (46°F) and 35–28°C (95–82°F) when using air of T_D 5°C (41°F). The moisture increase is not too high if the final powder temperature is not less than about 5°C (9°F) below these temperatures.

3.5. Application Examples

There can seldom be found two fully identical spray drying plants. The design is often tailored to the need of the powder manufacture as to capacity, range of products, required product quality, available energy sources, environmental requirements, climatic conditions, and possibly other specific circumstances.

The flow sheets for basic types of spray drying plants shown in Figures 13, 15, and 16 are referred to when presenting the performance examples on some basic products. However, the use of other components than those shown on the flow sheets is also possible without substantial change in the performance. For example, rotary wheel atomization can be replaced by pressure nozzles, or a low-profile dryer can be replaced by a tall form dryer with pressure nozzle atomization.

The capacities of industrial installations vary from 500 to over 10,000 kg/h (1100–18,350 lb/h) water evaporation. A distinct trend of capacity increase can be seen on spray dryers built during the last two decades. Although the average capacity of spray dryers built in the 1960s was about 500–1000 kg/h (1100–2200 lb/h) water evaporation, it is today about four times greater.

Product characteristics in the examples in Tables 1, 2, and 3 refer to the following analytic methods (4):

Final moisture: The loss of weight expressed in the percentage determined by oven drying at 105°C for 3 h

Free moisture: As above, using the temperature 87°C for 6 h

Table 1 Some Examples of the Performance of Various Spray Drying Systems in the Manufacture of Skim Milk Powder

Process	Pneumatic conveying system	Cooling bed system	Two-stage process[a]	Straight-through process[b]
Powder type	Ordinary	Dustless	Ordinary	Instant
Inlet temperature				
°C	180–230	180–200	200–230	180–200
°F	356–446	356–392	392–446	356–392
Outlet temperature				
°C	94–100	96–100	84–89	84–88
°F	169–212	173–212	183–192	183–190
Feed concentration (%)	45–50	48–50	48–52	48–50
Moisture from chamber (%)	–	–	6–7	6–7
Final moisture (%)	3.6–4.0	3.6–4.0	3.6–4.0	3.6–4.0
Bulk density[b] (g/ml)	0.60–0.70	0.50–0.55	0.68–0.80	0.45–0.50
Wettability (s)	–	30–120	–	10–30
Dispersibility (%)	–	70–90	–	85–95
Mean particle size (μm)	30–50	120–160	50–80	150–200

[a]The spray dryer with an afterdrying-cooling fluid bed system can be operated by two alternative methods for recycling of fines. When producing nonagglomerated products, the fines are recycled to the fluid bed. This process is currently called a two-stage drying process. For agglomerated products the fines are returned to the spray dryer. This process is known as the straight-through drying process. Both alternatives are shown in Figure 16.
[b]The conveying of the agglomerated products by any kind of air transport is undesirable as it will break down the agglomerates. Thus the product characteristics refer to products collected directly from the installations into tote bins. For ordinary products, the characteristics refer to the quality after air transport to storage silos.

Table 2 Some Examples of the Performance of Various Spray Drying Systems in the Manufacture of Whole Milk Powder

Process	Pneumatic conveying system	Cooling bed system	Two-stage process[a]	Straight-through process[b]
Quality	Ordinary	Partially agglomerated	Ordinary	Agglomerated instant[c]
Inlet temperature				
°C	180	180	180–200	180
°F	356	356	356–392	356
Outlet temperature				
°C	95–97	95–97	76–80	78–82
°F	203–207	203–207	169–176	172–180
Feed concentration (%)	48–50	48–50	48–52	48–52
Moisture from chamber (%)	–	–	5–6	5–6
Final moisture (%)	2.5–3.0	2.5–3.0	2.5–3.0	2.5–3.0
Bulk density[b] (g/ml)	0.58–0.63	0.50–0.55	0.62–0.68	0.45–0.50
Wettability (s)	–	–	–	10–30[c]
Dispersibility (%)	–	–	–	90–100[c]
Mean particle size (μm)	30–50	120–160	50–80	150–200

[a]The spray dryer with an afterdrying-cooling fluid bed system can be operated by two alternative methods for recycling of fines. When producing nonagglomerated products, the fines are recycled to the fluid bed. This process is currently called a two-stage drying process. For agglomerated products the fines are returned to the spray dryer. This process is known as the straight-through drying process. Both alternatives are shown in Figure 16.

[b]The conveying of the agglomerated products by any kind of air transport is undesirable, as it will break down the agglomerates. Thus the product characteristics refer to products collected directly from the installations into tote bins. For ordinary products, the characteristics refer to the quality after air transport to storage silos.

[c]The instant characteristics of whole milk powder refer to the product treated by 0.2% lecithin. The equipment for this operation is also shown in Figure 16.

Total moisture: As determined by the Carl Fischer method, which also determines the water of crystallization

Bulk density: The reciprocal value of the volume of 100 g milk powder measured in a 250 ml glass cylinder after tapping to minimum volume, multiplied by 100

Wettability: The wetting time in seconds of 10 g of skim milk powder (13 g of whole milk powder) when dropped on the level of 100 ml water of 20°C

Dispersibility: The percentage of dissolved powder after stirring 10 or 13 g (skim or whole milk powder) in 100 ml water at 25°C for 20 s and pouring the reconstituted milk through a 150 μm sieve (5)

The survey of running conditions and product characteristics for the production of skim milk powder is shown in Table 1, for whole milk powder in Table 2, and for whey powder in Table 3.

4. NONCONVENTIONAL SPRAY DRYERS

There are three types of spray dryers on the market, which differ considerably from the conventional installations described. The background idea of these types is more extensive utilization of the two-stage drying principle; the moisture content of the powder leaving

Table 3   Some Examples of the Performance of Various Spray Drying Systems in the Manufacture of Sweet Whey Powder

Feed	Pneumatic conveying system		Straight-through process[a] precrystallized[c]	Modified straight-through process[a,b] precrystallized[c]
	Nonprecrystallized	Precrystallized[c]		
Quality	Ordinary	partially caking	Noncaking	Noncaking
Inlet temperature				
°C	180	200	185	150–160
°F	356	392	365	302–320
Outlet temperature				
°C	90	92–95	80	50–55
°F	194	198–203	176	122–131
Feed concentration (%)	42	50–55	50–60	50–54
Moisture from chamber (%)	—	—	5	10–54
Final moisture				
Free (%)	3.0	2.5	2.0	2.5
Total (%)	—	4.0–5.0	5.0	5.0
Bulk density (g/ml)	0.6–0.7	0.6–0.7	0.55–0.65	0.4–(0.7)[d] 1000–3000[d]
Mean particle size (μm)	30–40	40–50	100–150	(40–80)

[a]The spray dryer with an afterdrying-cooling fluid bed system can be operated by two alternative methods for recycling of fines. When producing nonagglomerated products, the fines are recycled to the fluid bed. This process is currently called a two-stage drying process. For agglomerated products the fines are returned to the spray dryer. This process is known as the straight-through drying process. Both alternatives are shown in Figure 16.

[b]This process requires a transport belt between the chamber outlet and fluid bed inlet, which gives 3–6-min holding time to achieve the aftercrystallization of lactose in the wet powder. The belt is not shown in Figure 16.

[c]This process requires precrystallization of the concentrate to transform the amorphous lactose into α-lactose monohydrate. It is done in tanks with good agitation by holding at 30°C (86°F) for several hours and subsequent cooling to 15–10°C (59–50°F). The whole crystallization process requires 6–16 h and allows 60–80% of the total lactose content to be crystallized.

[d]The product leaving the plant has low bulk density and consists of big agglomerates. It is usually ground by hammer mill. Figures in brackets refer to milled powder.

the primary drying stage in these plants is considerably higher than that possible in conventional dryers.

4.1. Integrated Belt Dryer

The integrated belt dryer is a box-type multinozzle spray dryer with a downward multi-streamline airflow (Fig. 18) marketed by NIRO. The bottom of the chamber is formed by a transport belt, which is a mesh made of polyester material. The wet powder from the spray drying stage settles on the belt and is slowly transported to further sections for afterdrying and cooling. The drying air passes through the powder layer and the belt to the cyclones. This type is claimed to be suitable for thermoplastic powders and to have low stack loss and good economy.

4.2. Streamline Airflow Multistage Dryer

The streamline airflow multistage dryer was developed by NIRO A/S, Copenhagen, Denmark. It is a multinozzle dryer with downward streamline airflow (Fig. 19). The chamber base is formed by a stationary nonvibrating fluid bed, operating with high fluidizing

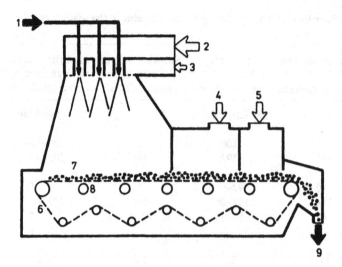

Figure 18 Integrated belt dryer (1, feed to nozzles; 2, primary drying air; 3, ceiling cooling air; 4, secondary drying air; 5, cooling air; 6, textile screen mesh belt conveyor; 7, powder layer; 8, exhaust air to cyclones; 9, product outlet).

velocity. The wet particles with high moisture content from the primary stage fall directly into the fluidized powder layer, where secondary drying takes place. The final drying and cooling are accomplished in a conventional vibrating fluid bed. This dryer operates with very high inlet air temperatures and low outlet air temperatures. It is suitable for all types of dried milk products. The advantages of this dryer are good product quality, high economy, and low space requirement.

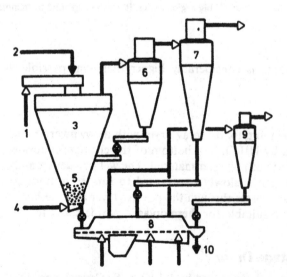

Figure 19 Streamline airflow multistage dryer (NIRO A/S) (1, primary drying air; 2, feed supply; 3, drying chamber; 4, secondary drying air; 5, stationary fluid bed with powder; 6, primary cyclone; 7, secondary cyclone; 8, vibrating fluid bed; 9, fluid bed cyclone; 10, product outlet).

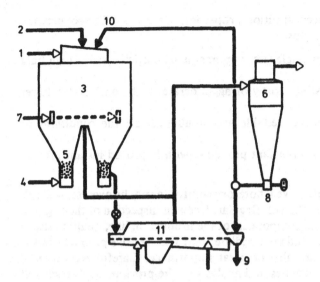

Figure 20 Rotary airflow multistage dryer (NIRO A/S) (1, primary drying air; 2, feed supply; 3, drying chamber; 4, secondary drying air; 5, stationary fluid bed with powder; 6, cyclone; 7, wall-sweep air; 8, blow-through valve with blower; 9, product outlet; 10, fines return; 11, vibrating fluid bed).

4.3. Rotary Airflow Multistage Dryer

As the previous dryer, the rotary airflow multistage dryer (Fig. 20) was also developed by NIRO A/S, Copenhagen, Denmark, and utilizes a stationary fluid bed, forming the base of the drying chamber for a second drying stage. It differs from the previous dryer as it applies the downward rotary airflow and can work with either rotary wheel or pressure nozzles. The airflow rotation is further supported by tangential inlets of the wall sweep air. The holes in the perforated plate of the ring-shaped stationary fluid bed are oriented tangentially as well, so that the fluidized powder layer is kept in rotation. The third drying and cooling stage is performed in a conventional vibrating fluid bed. For some applications, however, this dryer can work just with a pneumatic conveying system instead of a vibrating fluid bed as a "compact two-stage dryer." It can process all types of powdered milk products. The application of the rotary wheel atomization makes this dryer especially suitable for precrystallized whey products.

5. QUALITY CONTROL

The final products must meet the requirements prescribed by quality standard specifications (such as national or factory standards, and agreements between manufacturer and buyer). These contain the specific requirements as to the individual bacteriological, sensory, chemical, and physical properties.

   As well as the hygienic standards the most important general properties are moisture content, solubility index, and bulk density. The suitability of a product to a given purpose is defined by a great number of functional properties such as instant properties (wettability and dispersibility), flowability, whey protein nitrogen index, friability, thermostability, hygroscopicity, and degree of caking.

As to physical structure and reconstitution properties, the dried milk products are classified into the following main groups:

1. Ordinary products consisting of single (nonagglomerated) particles with a high bulk density
2. Agglomerated products, consisting mostly of agglomerates with considerably lower bulk density
3. Instant products, which are easily wettable and soluble (under the conditions of specific use)
4. Dustless products, also called semi-instant products, with improved instant properties, but not fully instant

The importance of well-organized effective laboratory control cannot be overemphasized. Testing final product quality is not sufficient. Regular, frequent inspection of the hygienic condition of the whole factory and all components of the production line, quality control of raw milk and other raw materials, and intermediate control of decisive product properties on critical production points are also of great importance. Careful recording of production condition and product qualities and analysis of the previous day's results at daily meetings between production and laboratory personnel are absolutely essential to avoid quality problems.

REFERENCES

1. K. Masters, *Spray Drying Handbook*, 3rd Edition, George Godwin, Ltd., London, John Wiley and Sons, New York, 1979.
2. H. G. Kessler, *Food Engineering and Dairy Technology*, Verlag A. Kressler, Freising, West Germany, 1981.
3. V. Westergaard, *Milk Powder Technology, Evaporation and Spray Drying*, NIRO A/S, Copenhagen, 1980.
4. I. Haugaard-Soerensen, J. Krag, J. Pisecky, and V. Westergaard, *Analytical Methods for Dry Milk Products*, 4th Edition, A/S NIRO ATOMIZER, Copenhagen, 1978.
5. International Dairy Federation, Determination of the dispersibility and wettability of instant dried milk, International IDF Standard 87:1979, 1979.

Printed in the United States
by Baker & Taylor Publisher Services